The ARRL
Operating
Manual

Sixth Edition

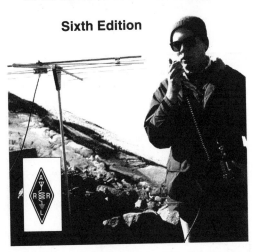

Edited by: Paul Danzer, N1II

Composition and Proofreading:

Shelly Bloom, WB1ENT
Paul Lappen
Joe Shea
Steffie Nelson, KA1IFB

Cover Design: Sue Fagan

ABOUT THE COVER

Clockwise, from left

Adjusting the satellite discone at W5NN, Field Day 1996

Roger, KA7EXM, enjoying an ARRL VHF Contest from the north face of Oregon's Mount Hood (*photo courtesy W7ZOI*)

KB3BHV, the Delaware Valley OMIK Field Day station 1996

W6YL Field Day site 1994

Background: The K1ZM antenna farm in New York State features an 80-meter 4-Square and three 160-meter quarter-wave verticals

Back cover: WD8MLN, the Macedonia ARC's first annual Field Day operation

Published by: The American Radio Relay League Newington, CT 06111 USA

Contents

Foreword

Introduction

1   **Shortwave Listening**
Curt Phillips, W4CP

2   **The Amateur Radio Spectrum**
Paul L. Rinaldo, W4RI

3   **Basic Operating**
Bill Jennings, K1WJ, Paul Danzer, N1II, and the ARRL Staff

4   **Antenna Orientation**
Chuck Hutchinson, K8CH

5   **DXing**
Bill Kennamer, K5FUV

6   **The Internet**
Stan Horzepa, WA1LOU

7   **Contests**
Clarke Greene, K1JX, and the ARRL Staff

8   **Operating Awards**
Steve Ford, WB8IMY

9   **HF Digital Communications**
Steve Ford, WB8IMY

10   **Packet Radio**
Stan Horzepa, WA1LOU

11   **FM and Repeaters**
Brian Battles, WS1O

12   **VHF/UHF Operating**
Michael R. Owen, W9IP

13   **Satellites**
Jon Bloom, KE3Z, and Steve Ford, WB8IMY

14   **Emergency Communications**
Richard Regent, K9GDF, and the ARRL Staff

15   **Traffic Handling**
Maria L. Evans, KT5Y

16   **Image Communications**
Bruce Brown, WA9GVK, and Ralph Taggart, WB8DQT

17   **References**

Index

Foreword

Many hams will remember *The Operating Manual* of years ago—thin and covered in orange, it contained everything the active ham needed to know about operating a station. CW, AM and perhaps SSB and a noisy RTTY machine were your choices—your only choices.

How technology has changed the way we do ham radio! It's no exaggeration to say that the References chapter of this new 6th Edition has more information than the entire book did in the early days of *The Operating Manual*!

Some modes evolve, while others arrive on the scene or change rapidly. CW operators still use the same Q signals they used decades ago, while PACTOR and CLOVER weren't even invented until relatively recently. Whether you're active on one or more of the newer digital modes, you enjoy CW so much that you still don't own a microphone—or you're someplace in between—you'll find what you need in this book. The authors we've assembled for this 6th Edition have written and compiled information that's complete, clear and concise. The goal is to help you, the reader, find and use the information you need when you need it—whether it's the Friday before a big contest weekend, or the *Mir* space station will be within range in an hour.

Aside from the many updates and revisions to tables, lists and the text, this edition includes one brand new chapter, on the Internet, and two that have been thoroughly revised and updated, "Shortwave Listening" and "DXing." In addition, this edition includes explanations of the use of antenna orientation and propagation prediction software, invaluable tools for all active hams.

If you should have a favorite operating aid, Web site, piece of software or need for information that you don't find in this book, I hope you'll let us know. The feedback form at the back is designed for that purpose. We can already predict that the next edition will cover new modes and activities that aren't available today. With the benefit of your comments and suggestions, we can make that next edition even more useful.

David Sumner, K1ZZ
Executive Vice President
June 1997

Introduction

Welcome to the Sixth Edition

The Internet—Those fascinating digital modes—PCs in the ham shack—The vagaries of the sunspot cycle.

All of these affect what we do as hams. We've produced this new edition of *The ARRL Operating Manual* to reflect the way hams operate in the late 1990s. Some chapters have been completely rewritten and the others modified and updated. In addition, all the tables and lists have been brought up to date.

A few of the more significant changes to this edition:

Curt Phillips, W4CP, has updated his **Chapter 1—Shortwave Listening**. It has new information that applies both to the low sunspot numbers we are seeing as this edition is published, and for the upward swing we expect shortly.

Chapter 4—Antenna Orientation, now includes a discussion of computer tools and Web sites. The latitude, longitude and antenna bearing table has been updated for the new DX prefixes.

ARRL DXCC Manager Bill Kennamer, K5FUV, has written a new **Chapter 5—DXing**. It has the tips and techniques you need whether you're on the DXCC Honor Roll or whether you're content with the occasional DX you can manage with your marginal geographic location and antenna system.

Amateur Radio's newest resource, **the Internet**, is the subject of a brand new **Chapter 6**. ARRL Contributing Editor Stan Horzepa, WA1LOU, took time from his packet activities to write this fresh look at wired ham radio.

Chapter 16—Image Communications has been brought up-to-date with more information on today's PC-based slow-scan equipment and operating techniques.

We've also added something else—a **Ham Desktop Reference**. This handy 24-page booklet, bundled with the book, brings the most popular tables and operating aids to your fingertips. Many hams will keep it on top of the operating table, while the book itself remains nearby in a bookcase or desk drawer. We hope you find it useful.

Many of the revisions made for this edition are based on comments and suggestions from the active hams on the ARRL HQ staff. If they had difficulty finding something or felt a different approach to a subject would be more effective, they didn't hesitate to let the Editor know about it. If you'll do the same, via the feedback form at the back or publications e-mail address (**pubsfdbk@arrl.org**), we can continue to improve *The ARRL Operating Manual*.—N1II

Acknowledgments

The Editor wishes to thank all the authors who contributed to this and previous editions of this book. In addition, ARRL Technical Advisor Dennis Bodson, W4PWF, provided valuable input, as did several members of the ARRL HQ staff, particularly Senior Lab Engineer Zack Lau, W1VT, Regulatory Information Specialist John Hennessey, N1KB, Assistant Field Services Manager Steve Ewald, WV1X, and Awards Manager Eileen Sapko.

The ARRL At Your Service

ARRL Headquarters is open from 8 AM to 5 PM Eastern Time, Monday through Friday, except holidays. Our address is: 225 Main St, Newington, CT 06111-1494. You can call us at 860-594-0200, or fax us at 860-594-0259.

If you have a question, try one of these Headquarters departments . . .

	Telephone	Electronic Mail
QST Delivery	860-594-0338	circulation@arrl.org
Publication Orders	888-277-5289	pubsales@arrl.org
Only, 8 AM to 9 PM Eastern Time (toll free)		
Regulatory Info	860-594-0323	reginfo@arrl.org
Exams	860-594-0300	vec@arrl.org
Educational Materials	860-594-0301	ead@arrl.org
Contests	860-594-0252	contest@arrl.org
Technical Questions	860-594-0214	tis@arrl.org
Awards	860-594-0288	awards@arrl.org
DXCC/VUCC	860-594-0234	dxcc@arrl.org
Advertising	860-594-0207	ads@arrl.org
Media Relations	860-594-0328	newsmedia@arrl.org
QSL Service	860-594-0274	buro@arrl.org
Scholarships	860-594-0230	foundation@arrl.org
Emergency Communication	860-594-0261	k1ce@arrl.org

You can send e-mail to any ARRL Headquarters employee if you know their name or call sign. The second half of every Headquarters e-mail address is **@arrl.org**. To create the first half, simply use the person's call sign. If you don't know their call sign, use the first letter of their first name, followed by their complete last name. For example, to send a message to John Hennessee, N1KB, Regulatory Information Specialist, you could address it to **jhennessee@arrl.org** or **N1KB@arrl.org**.

If all else fails, send e-mail to **hq@arrl.org** and it will be routed to the right people or departments.

ARRL on the On-Line Services

We maintain accounts on these major on-line services and check for mail several times daily.

Service	ARRL Address
CompuServe	70007,3373
America Online	HQARRL1

Downloadable files for the new ham, upgrader, instructor or disabled ham are featured in the various on-line ham radio file libraries.

ARRL BBS

The ARRL Hiram Bulletin Board System is as close as your telephone. Hiram offers more than a thousand software files for your enjoyment. You can also use Hiram to send messages to anyone at Headquarters—or to other hams who frequent the BBS. Hiram accepts up to four simultaneous connections at rates from 1200 to 28,800 baud. Fire up your modem and call 860-594-0306.

Technical Information Server

If you have Internet e-mail capability, you can tap into the ARRL Technical Information Server, otherwise known as the *Info Server*. To have user instructions and a handy index sent to you automatically, simply address an e-mail message to:

info@arrl.org

Subject: **Info Request**

In the body of your message enter:

    HELP
    SEND INDEX
    QUIT

ARRL on the World Wide Web

You'll also find the ARRL on the World Wide Web at:

http://www.arrl.org/

At the ARRL Web page you'll find the latest W1AW bulletins, a hamfest calendar, exam schedules, an on-line ARRL Publications Catalog and much more. We're always adding new features to our Web page, so check it often!

CHAPTER 1

Shortwave Listening

Shortwave listening (also known as "world band" listening) has experienced a resurgence of interest in the last several years. This is due both to the increased availability of good performing, moderately priced shortwave receivers and the widening realization that, in an information society, the unique information available on the shortwave bands cannot be ignored.

CURT PHILLIPS, W4CP
6200 VICKY DR
RALEIGH, NC 27603

There is no official definition of the shortwave bands, but they are most often considered to comprise the frequencies between 3 and 30 MHz. For hobbyist purposes, the monitoring of frequencies between 100 kHz and 30 MHz can be considered *shortwave listening*, notwithstanding that the AM broadcast band is in the *medium wave* spectrum and frequencies below that are technically considered *long wave*.

The ham bands within the shortwave bands are also referred to as the high frequency or "HF" bands, as opposed to the very high frequency (VHF) and ultra high frequency (UHF) bands that are above 30 MHz.

There's a good chance that you're a licensed ham, or you're considering becoming one. But shortwave listening is just... LISTENING. As an Amateur Radio operator, you can transmit RF, not just simply receive it. So, what does shortwave listening have to do with ham radio? Isn't shortwave listening a step backward?

The short answer is no. The diverse, worldwide communications that you have access to on the shortwave bands provides a wealth of possibilities that *supplement* the choices provided by Amateur Radio. What shortwave listening can offer you depends in part when you joined our hobby—and how. Let me explain.

In the "old" days, new recruits entered the hobby through direct exposure to the shortwave bands. Before they took their exams, many new hams spent hours listening to shortwave as they honed their CW skills. The code practice sessions provided by the ARRL flagship station W1AW and the other CW activity on the shortwave bands were the prime source of CW practice fodder. Bear in mind that this was before the days of cassette tape machines and personal computers.

After they were on the air, their shortwave experience was limited. For a period of time from about the mid-1950s until the late 1970s, ham transceivers tuned the Amateur Radio portions of the shortwave spectrum only. Why? The technology of radio equipment during that period was such that optimum performance was most cost-effectively obtained by rigs that covered a small portion of the radio spectrum. So hams that may have started with a general coverage receiver quickly "downgraded" to a ham-band only receiver.

Today the most popular method of entry into Amateur Radio is the Technician license. This license doesn't require learning the Morse code or offer operating privileges below 30 MHz. For these hams, shortwave listening offers a chance to become acquainted with operations on the HF ham bands.

So, Technicians have a new world to explore and veterans can reacquaint themselves with the fun they may have missed. But for hams of any vintage, there are an amazing number of intriguing signals that can be heard on the shortwave bands. Indeed, since no license is required to listen, everyone in the family can enjoy the diverse offerings of the shortwave bands.

WHAT'S OUT THERE?

Traditionally, the HF bands have been the target of short wave listeners, and this chapter has the traditional name. Today, technology has sent traditional radio signals in several directions. The HF bands are still very crowded with signals, but during periods of low sunspot activity, propagation—especially at night—reduces the probability of catching that elusive weak station. Under these conditions, most activity occurs under 10 MHz. This makes the lower HF short-wave bands sound like 20 meters during a contest!

Many stations have migrated, such as many public service stations (fire, police and other local municipal agencies) from the old 30-40 MHz bands to VHF, UHF and trunked 800 MHz systems. Later in this chapter scanners—and addition to the customary HF receiving capability—and the spectrum from 30 to 1000 MHz is discussed. Other stations that used to distribute audio and RTTY information on HF, have moved to satellites, where they can still be monitored.

Finally, an old mode—dating back to the days before HF was technically feasible—has received increased attention. The low frequencies, below the broadcast band, provides many people with interesting listening. Although many countries have abandoned the ship-to-shore communication, previously

Table 1-1
Shortwave Frequency Guide

All frequencies shown in kHz

Frequencies	Service	Comments
535—1705	AM Broadcast	Standard North America AM
1705—1800	Fixed Service	Land/Mobile/Marine
1800—2000	Amateur 160 Meters	
2000—2107	Maritime Mobile	
2107—2170	Fixed Service	Land/Mobile/Marine
2170—2194	Land Mobile Service	
2194—2300	Fixed Service	
2300—2495	Shortwave Broadcast	120 Meters
2495—2505	Time Standard	
2505—2850	Fixed Service	Land/Mobile/Marine
2850—3155	Aeronautical Mobile	Transoceanic Flights
3155—3200	Fixed Service	
3200—3400	Shortwave Broadcast	90 Meters
3400—3500	Aeronautical Mobile	Transoceanic Flights
3500—4000	Amateur 80/75 Meters	
3900—4000	Shortwave Broadcast	75 Meters, Not in Region 2
4000—4000	Time Standard	
4000—4063	Fixed Service	
4063—4438	Maritime Mobile	Ship/Shore
4438—4650	Fixed Service	
4650—4750	Aeronautical Mobile	Transoceanic Flights
4750—5060	Shortwave Broadcast	60 Meters
5005—5450	Fixed Service	
5450—5730	Aeronautical Mobile	Transoceanic Flights
5730—5950	Fixed Service	
5950—6200	Shortwave Broadcast	49 Meters
6200—6525	Maritime Mobile	Ship/Shore
6525—6765	Aeronautical Mobile	Transoceanic Flights
6765—7000	Fixed Service	
7000—7300	Amateur 40 Meters	
7100—7300	Shortwave Broadcast	41 Meters, Not in Region 2
7300—8195	Fixed Service	
8195—8815	Maritime Mobile	Ship/Shore
8815—9040	Aeronautical Mobile	Transoceanic Flights
9040—9500	Fixed Service	
9500—9900	Shortwave Broadcast	31 Meters
9775—9995	Fixed Service	
10005—10100	Aeronautical Mobile	Transoceanic Flights
10100—10150	Amateur 30 Meters	CW/Data Only
10100—11175	Fixed Service	
11175—11400	Aeronautical Mobile	Transoceanic Flights
11400—11650	Fixed Service	
11650—12050	Shortwave Broadcast	25 Meters
12050—12330	Fixed Service	
12330—13200	Maritime Mobile	Ship/Shore
13200—13360	Aeronautical Mobile	Transoceanic Flights
13360—13600	Fixed Service	
13600—13800	Shortwave Broadcast	New WARC Allocation
13800—14000	Fixed Service	
14000—14350	Amateur 20 Meters	
14350—14995	Fixed Service	
15010—15100	Aeronautical Mobile	Transoceanic Flights
15100—15600	Shortwave Broadcast	19 Meters
15600—16460	Fixed Service	
16460—17360	Maritime Mobile	Ship/Shore
17360—17550	Fixed Service	
17550—17900	Shortwave Broadcast	16 Meters
17900—18030	Aeronautical Mobile	Transoceanic Flights
18030—18780	Fixed Service	
18780—18900	Maritime Mobile	Ship/Shore
18900—19680	Fixed Service	
19680—19800	Maritime Mobile	Ship/Shore
19800—21000	Fixed Service	
21000—21450	Amateur 15 Meters	
21450—21850	Shortwave Broadcast	13 Meters
21850—22000	Aeronautical Mobile	
22000—22720	Maritime Mobile	Ship/Shore
22720—23200	Fixed Service	
23200—23350	Aeronautical Mobile	
23350—24990	Fixed Service	
24890—24990	Amateur 12 Meters	Shared with Fixed Service
25010—25330	Petroleum Industry	
25330—25600	Government Frequency	
25600—26100	Shortwave Broadcast	11 Meters
26100—26480	Land Mobile Service	
26480—26950	Government	
26950—26960	International Fixed Service	
26960—27410	Citizen's Band	Channels start at 26965 kHz
27410—27540	Land Mobile Service	
27540—28000	Government	
28000—29700	Amateur 10 Meters	
29700—29800	Forestry Service	
29800—29890	Fixed Service	
29890—29910	Government	
29910—30000	Fixed Service	

found in the 400-500 kHz region, beacons, arctic and sub-arctic area and European broadcasting, weather and some data services still use these bands. Contact The Longwave Club of America, 45 Wildflower Road, Levittown, PA 19057 for information on this specialized SWL activity.

UNIQUE PERSPECTIVE ON THE NEWS

For the past several years, the United States has been sending troops on peace keeping and humanitarian missions around the world. The United States Army and Marines have served on extended missions in Bosnia, Haiti and Somalia. Air Force and naval assets are shuttled back and forth from the area around Iraq, as the threat there ebbs and flows. Boatloads of Cuban expatriates flood into Florida and are relocated to such places as Guantanamo Bay.

The shortwave listener has a ringside seat to events such as these. The military frequencies are full of traffic as troops and equipment are moved. The Voice of America broadcasts the American views of the world situation. Opposing viewpoints are heard from Radio Iraq International and Radio Habana

ITU

Radio waves can not be kept within a country's borders and radio is a means of international communication, so it was evident very early in the development of radio that international treaties and agreements must be established to prevent chaos on the radio waves.

The organization that evolved into the radio coordinating council for the world is the International Telecommunications Union. They allocate frequency ranges for various uses, and the assignment of specific operating frequencies is left to the governments of the individual nations.

As a part of their authority, the ITU sets the frequency ranges where the various services operate, but many countries are not bashful about moving their national broadcasting services outside the specified bands. That's why you need to scan outside the listed frequency ranges. You'll find many fascinating signals lurking there.

The various frequency ranges are traditionally referred to using wavelength in meters of one of the component frequencies. For example, 49 meters is the wavelength of 6122 kHz. However, the 49 meter "band" extends from 5950-6200 kHz.

Cuba. The shortwave stations of other nations weigh in with their analysis. Despite all the high tech, satellite equipment available, transmissions on the "Mystic Star" frequencies from government VIP aircraft can be heard as they shuttle about on their diplomatic missions.

Even for events occurring within the United States, shortwave radio can provide information substantially before it is released by the traditional media. When President Ronald Reagan was shot, listeners monitoring the radio traffic between the White House and Vice President George Bush heard reports on the President's condition long before it was broadcast on the major networks.

Listeners to the Israeli broadcast station Kol Yisroel heard the warning for the gas attack from Iraq as the first missiles were detected in flight—and the warning was broadcast to the Israeli public on all medium and shortwave transmitters. Many parts of the US news media were unaware of the attack until an hour or two later.

In our turbulent world there seem to always be hot spots where shortwave radio provides a direct access to the information. The broadcast news of the major networks tells you what *they* think is important, filtered through their perspective and edited to fit into a brief newscast. The shortwave bands bring you news direct from the source, and in as much detail as they care to provide.

During the Russian coup in August 1991 and the armed uprising of old-line Soviets in the legislature during September 1993, those monitoring Radio Moscow got quite a different perspective on the internal situation in Russia than those confined to the popular media. In both cases, it was evident that the pro-democratic forces maintained control of Russia's international broadcasting apparatus, portending a favorable outcome for them while the American broadcasters were very negative on their chances.

After the end of the Cold War, leaders in several Eastern Bloc countries told of how the Voice of America, Radio Free Europe, Radio Liberty and other Western shortwave broadcasts had helped them overcome the new blackouts and censorship that had been common in their countries.

The ability to listen directly to broadcasts from countries around the world can encourage students to have greater interest in history, geography and foreign languages. Therefore, shortwave listening is an excellent way to both encourage the younger generation in their educational pursuits *and* get them interested in radio as a hobby.

But it's not just wars and military actions where the shortwave bands offer special news access. In December 1986, shortwave listeners were treated to the exciting communications supporting the flight of the *Voyager*, the first non-stop, non-refueled airplane flight around the world.

RADIO DRAMA

One art form that might have died without the support of the international shortwave broadcasters is the radio drama. On these stations, radio dramas are not only surviving but thriving.

The BBC's two weekly offerings, "Play of the Week" and "Thirty-Minute Drama" are excellent, the latter having been consistently chosen as one of the Ten Best Shows on shortwave by a major shortwave publication.

The religious show "Unshackled" (broadcast on HCJB and several other stations) is the longest-running shortwave radio drama, and sounds about the same today as it did when first heard almost 30 years ago. "Unshackled" retains an "old-timely" aura (swelling organ music, a terse announcer) about

it, so listening to it is to be time-warped back to the heyday of radio dramas in the 1930s. "Unshackled" is produced by Pacific Garden Mission, located in Chicago.

MUSIC

On shortwave you can hear African rhythms, the sitar of India and other exotic music direct from the source. The national shortwave broadcasts present a range of regional music that is not available via any other media, along with commentaries and information on its composition and history. Often, unique music from a country is used as a station's signature just prior to its identification, or broadcast for a few minutes just before the start of scheduled programming.

LANGUAGE LESSONS

With the prevalence of the English language in the world, almost every international broadcaster offers English language broadcasts. But, if you are fluent in another tongue, the shortwave bands will give you ample opportunity to exercise your skill. For the would-be fluent, many stations offer introductory "courses" on their native language.

DXing

The activities we have discussed so far focus on the content of the shortwave signals. But there is a sector of the shortwave listening hobby that concentrates on the technical challenge of receiving hard to hear, distant (DX) stations. Shortwave Dxers pride themselves on obtaining official verification of their reception (QSL cards) and competing for various awards.

THE IONOSPHERE

Someone accustomed to the line-of-sight nature of the UHF/VHF bands (and who may have forgotten the HF theory they studied for their exam!) might wonder how the shortwave signals make it around the world. To answer this question we need to briefly examine the nature of the ionosphere.

The ionosphere is a layer of the Earth's atmosphere approximately 80 to 300 miles above Earth's surface where free ions and electrons can cause some radio waves to be reflected or refracted. In brief, at the VHF region and above (above 30 MHz) the radio waves will most often pass through the ionosphere and out into space. Within the shortwave spectrum, radio waves are often reflected or refracted (bent) back to the Earth's surface. As you drop below the shortwave frequencies, the ionosphere tends to absorb and attenuate the signal so as to essentially eliminate any refraction. It is the "bouncing" of radio signals off the ionosphere (sometimes via several bounces) that allows shortwave signals to travel (propagate) around the world.

The ability of radio waves to be refracted over long distances by the ionosphere is affected by the time of day, time of year and the number of sunspots. Speaking very generally, higher frequencies (closer to 30 MHz) work best during the daylight hours and during the summer. During late evenings and in the winter, the lower frequencies (close to 3 MHz and even below) begin to peak in efficiency. This phenomenon can be observed on the medium wave AM broadcast band (0.5 MHz-1.7 MHz). During the day you hear mainly local stations. In the evening, distant stations can be clearly heard. During the summer, however, even evening reception of distant AM broadcast stations, even at night, is often difficult and complicated by high noise levels created by thunderstorms and other weather phenomena.

These daylight/darkness propagation patterns are amplified

by the sunspot cycle. During the peak of the 11-year sunspot cycle, the high frequencies are at their best for long distance communication. At sunspot minimums, the capabilities of the lower frequencies become more prominent. In 1997, the latest sunspot cycle seemed to hit rock bottom, and most HF SWL activity was concentrated below 10 MHz, and occasionally below 7 MHz! The higher shortwave frequencies are projected to peak again near the end of the century.

WWV: Propagation and Time Information

If you are interested in detailed propagation information and forecasts, or simply want to know the *exact* time, station WWV is the station for you. Located in Fort Collins, Colorado, it is operated by the Time and Frequency Division of the National Institute of Standards and Technology (formerly the National Bureau of Standards). Their 24-hour-a-day broadcast provides a precise and accurate time signal around the clock and technical information on propagation conditions.

Additional services include standard time intervals, standard frequencies, geophysical alerts, marine storm warnings, Global Positioning System (GPS) information, UT1 time corrections and BCD Time Codes.

Another station, WWVH is located in Hawaii. Both stations broadcast on 2.5, 5, 10, 15 and 20 MHz.

Voice announcements are made from WWV and WWVH once every minute. Since both stations can be heard in some locations, a man's voice is used on WWV, and a woman's voice is used on WWVH to reduce confusion. The WWVH announcement occurs first, at about 15 seconds before the minute. The WWV announcement follows at about 7.5 seconds before the minute. Though the announcements occur at different times,

Economics

Many countries around the world maintain standard time and frequency stations. For over a year, I have attempted to compile a list of those operating on HF. However, with the world-wide phenomenons of decreased government budgets and political changes, the list seemed to undergo revisions almost weekly. Major changes were either carried out or promised for 1997 and 1998. Areas of the world that have undergone major political changes, such as the CIS (previously USSR) also have made major changes in their time and frequency stations, with more forecast for the near future.

Many non-US stations broadcast on multiples of 5 MHz, or close to these frequencies, and are covered up by the US stations. Therefore, this chapter has concentrated on the major US and Canadian stations, which seem to stay fairly stable in their broadcast schedule and immediate future plans.

To identify one of these non-US time and frequency stations, a general purpose frequency guide such as *Ferrell's Confidential Frequency List*, published by Listening In, P.O. Box 123, Park Ridge, NJ 07656, may help.— *N1II*

the tone markers are transmitted at the exact same time from both stations. However, they may not be received at exactly the same instant due to differences in the propagation delays from the two station sites. The announced time is "Coordinated Universal Time" (UTC).

The international agreement that established UTC in 1972 also specified that occasional adjustments of exactly 1 second

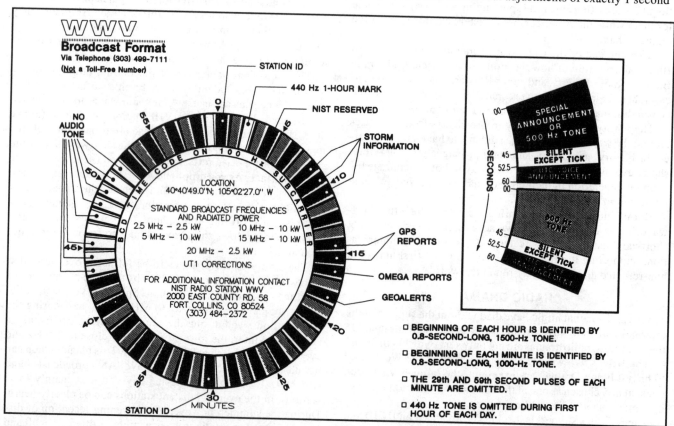

WWV broadcasts continuously. They have a Web site at **http://www.boulder.nist.gov** that includes a virtual reality tour of WWV and WWVH. *(chart courtesy of the National Institute of Standards and Technology)*

will be made to UTC so that UTC should never differ from a particular astronomical time scale, UT1, by more than 0.9 second. This was done as a convenience for some time-broadcast users, such as boaters, using celestial navigation, who need to know time that is based on the rotation of the Earth. These occasional 1 second adjustments are known as "leap seconds." When deemed necessary by the international Earth Rotation Service in Paris, France, the leap seconds are inserted into UTC, usually at the end of June or at the end of December, making that month 1 second longer than usual. Typically, a leap second has been inserted at intervals of 1 to 2 years.

The broadcasts from WWV and WWVH include a time code. The time code signal is 100 Hz away from the main carrier and is called a subcarrier. The code pulses are sent out once per second. With a good signal from a fairly high-quality receiver, you can hear the time code as a low rumble in the audio. HF receivers that receive and decode this signal can automatically display the time of day.

WWV/WWVH also broadcasts propagation information at 18 minutes past each hour. Unfortunately, the information they provide isn't quite as simple as "Propagation will be good today" or even "Propagation will be fair under 15 MHz today." Instead, they give you *indices*. With a little effort, you can turn the indices into a meaningful prediction of radio conditions.

Understanding the indices does take a bit of study. Both the *ARRL Handbook* and the *ARRL Antenna Book* offer some details on how to interpret the propagation data provided by WWV.

Since WWV and other time and frequency stations broadcast with substantial power around the clock on a variety of frequencies, just using them as a beacon can provide you with information on propagation. A quick listen to how strong the signal from Boulder, Colorado, is on the East Coast on 5, 10 and 15 MHz gives an immediate, if imprecise, reading on how the ionosphere is doing. Another good source for time signals and "beacon" use is Canada's CHU. Located in Ontario, it transmits on 3.330, 7.335 and 14.670 MHz.

Without doing any calculations or checking sunspot cycles, there is one propagation characteristic on which you can depend; propagation tends to get substantially worse just before the top of the hour. As you listen to an unidentified station and anxiously await the station identification at the top of the hour, a law of the universe known as "Murphy's Law" comes into play. Murphy's Law, well known to hams and engineers, predicts that the more rare the station is, the worse propagation will become as the station ID nears. For predicting propagation at other times however, the WWV information can be very helpful.

Propagation also can be humbling at other times. When propagation is at its worst, Radio Fredonia can put a megawatt into a 20 dB gain curtain array antenna and you can crank up your $20,000 ex-CIA receiver connected to the best receiving antenna and you'll still hear NOTHING! When the signal's not there, it's just not there. All the radio horsepower in the world can't change that. When this occurs, it's time to either switch to another frequency (which may help) or switch to another hobby for a while, like fishing or commiserating with the locals on your favorite 2 meter repeater.

GADGETS GALORE

Ham radio operators tend to be inveterate gadgeteers, and the shortwave listening hobby brings a whole new set of interesting radios for you to consider.

The past decade has seen the introduction of ever smaller

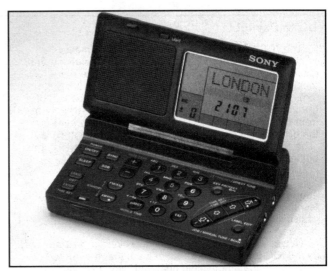

The Sony ICF-SW100 would fit in your pocket, but is packed with features.

and more conveniently sized shortwave radios, culminating in the "Walkman" sized Sony ICF-SW100 (pictured). The SW100 is a full coverage (0.15-30 MHz) shortwave receiver featuring 50 memories and the much desired synchronous detection (more on this technology later). And the new generation of "handy-talkie" sized UHF/VHF scanners include the shortwave spectrum (see sidebar, The DC-to-Daylight Radios Are Here!). In addition to their gadget appeal, the new generation of portable radios have sufficient sensitivity and selectively (and the sideband reception capability) to allow you to monitor the activities on the HF ham bands when you're away from home (and your transmitter).

The range and variety of portable shortwave receivers available today offer a whole new world of electronic goodies for the hobbyist to chose from. If you've got the money, there's definitely plenty of interesting places to put it.

CHOOSING A RECEIVER

Many current HF amateur transceivers feature a *general coverage receiver* (usually 0.5-30 MHz). If you own one you're just a spin of the dial away from the joys of shortwave listening. If you're planning to purchase an amateur transceiver in the near future, make general coverage receive capability a

The SW-8, made by Drake, is designed to be used on a tabletop, with comfortable-sized controls. (*photo courtesy of The R.L. Drake Co.*)

DC to Daylight

The "DC-to-Daylight" Receiver Is Here! Almost.

"If the broadcast receiver were a very special one that could continue to tune higher in frequency (there are technical reasons why this is impractical without switching), you would find many different groups, or bands, of frequencies, used by many different services." This quote from the ARRL's *How to Become a Radio Amateur* from the distant past has been used to begin a section called "The Super-Special Wide-Range Receiver Is Here" in past editions of the *Operating Manual*.

These editions have discussed the evolution of receiver technology up to the point of continuous coverage receivers spanning from 3 to 30 MHz, and indeed this technology is now so prevalent that virtually all ham transceivers incorporate a full coverage shortwave receiver within them.

However the ultimate in receiver coverage was always called "DC-to-daylight." The receivers that are out today may not hit that impossible goal, but they *do* cover a frequency range so wide as to boggle the mind. As amazing as it may seem, several models come in small hand-held size packages!

At the low frequency end, these wide-range receivers bottom out at 30 to 500 kHz, which is about typical for a general coverage shortwave receiver. But at the high frequency end, these receivers soar past the 30 MHz shortwave boundary deep into the VHF/UHF region. All of them go well past 1 GHz (1000 MHz) and the most advanced of them *exceed* 2 GHz.

Actually, these receivers have sprung from an explosion in UHF/VHF scanner technology. Monitors of the multitudinous services in this spectrum have sought portable radios that had an expansive number of memories and covered all the popular voice modes. For years scanners have encroached upon shortwave territory, going as low as 25 MHz. By the mid-1990s, scanners began to expand into full shortwave spectrum coverage.

One of the first entries into this category was the AOR AR-1000, covering 500 kHz to 1300 MHz in 5 kHz steps with 1000 memories. It is small in size (6.7"×2.6"×1.4") and provides AM, narrow-band FM (NBFM) and wide-band FM (WBFM). The AR-1500 was essentially the same radio, but with a beat frequency oscillator (BFO) for reception of single side-band (SSB) and continuous wave (CW) signals.

Soon thereafter, ICOM introduced the R1, an even smaller radio (4"×1.9"×1.4"), which covered 100 kHz to 1300 MHz in steps as small as 0.5 kHz. However, it only demodulated AM, FM and WBFM and had 100 memories.

As early entries in this super high-tech arena, each of these radios had some deficiencies like poor sensitivity over part of the specified range, intermodulation problems and complicated programming. Although each of these have their ardent defenders, their well publicized liabilities limited their appeal.

In wasn't long afterward before American readers of the British publication *Shortwave Magazine* noticed some ads for a small (6.1"×2.6" ×1.5") radio from a little known manufacturer, the Yupiteru MVT-7100. This radio covered 100 kHz to 1650 MHz in 50 Hz steps. Because it used true carrier injection to provide SSB reception, the full range of modes (AM/FM/WBFM/LSB/USB) were programmable in its 1000 memories. A conversation about the radio commenced on one of the national on-line services, and soon several enthusiasts had ordered 7100s from an accommodating English dealer.

In the wake of the rave reviews that followed on the on-line services, more of the Yupiteru 7100s were ordered from England and a phenomenon had started. Finally here was a radio that was reasonably simple to use (considering all it would do) and that provided good sensitivity throughout its frequency range. A review on the shortwave show "Spectrum" actually said it had as good sensitivity

on shortwave as the Sony 2010, but that may be a bit of an overstatement. However, it does provide shortwave reception approximately commensurate with the respected Sangean ATS-803 (which, like it, doesn't have synchronous detection) in a much smaller package. And it's also a full featured VHF/UHF scanner!

The Yupi's position alone at the top of the heap was relatively short-lived. About a year after the advent of the 7100, AOR introduced the AR-8000, which covers 100 kHz to 1.9 GHz in 50 Hz steps, with 1000 channels and true single sideband, as with the MVT-7100 and in almost exactly the same size package. However, it also features a four line alpha-numeric display and computer interface capability.

Recent entrants into this race are the Yupiteru MVT-9000 and the ICOM IC-R10. The MVT-9000 covers 530 kHz to 2039 MHz and has twin VFOs. The R-10 is a little smaller than the MVT-9000 or AR-8000, and covers from 500 kHz to 1300 MHz. The R-10 also features a computer interface, while the 9000 does not. Both of these rigs have 1000 memories and an alpha-numeric display with a real-time bandscope.

Hand held, this ICOM IC-R10 covers 500 kHz to 1300 MHz, with AM. FM, WBFM, USB, SSB and CW receive capabilities.

As covered in the section on "Scanning," the FCC requires that new radios with coverage in the 800 MHz range have the cellular phone frequencies exorcised from them. Yupiteru has not chosen to comply with this yet, so only the ICOM and AOR units are readily available in the United States, although at this writing some Yupiteru radios are being shipped to the U.S. from other countries.

Increasingly, table-top radios are being introduced with wide-ranging frequency coverage. ICOM has recently introduced the IC-R8500, while AOR's new entry is the AR-5000. With the hand-held radios, although their shortwave capability was acceptable, it was well known that their performance had been optimized for the VHF/UHF frequencies. But these two table-top radios defy description. From all reports, they are excellent performers throughout their frequency coverage. As with their smaller brothers, cellular frequency coverage is deleted.

So hobbyists now have numerous choices in the "DC-to-daylight" category. The hand-helds sell in the $500 to $600 range, and the ICOM and AOR desktop units are both about $1900.

Admittedly, receivers with a "DC-to-daylight" type frequency range have been available to the government and military for years, but in desk crushing sizes and weights and at a cost approaching that of a Rolls-Royce. Even those ostensibly marketed to hobbyists carried price tags in the $5000 range.

For those without such a deep bank account, the advent of these affordable, full-frequency receivers gives a ham the capability to monitor everything from 2 meter repeaters, local police and fire activity, the BBC and 80 meter sideband nets from one radio small enough to easily hang from a belt. The radio hobbyist never had it so good!

requirement. You won't regret it.

However, if you're just interested in "browsing" the short-wave bands, one of the new portables on the market may be for you.

Widely considered the best of the low cost portables are the mid-sized (approximately 7.5"×4.75"×1.5", 1.5 pounds) Sony ICF-SW7600G with synchronous detection (note the "G") and the Grundig Yacht Boy 400. Both of these are available at a street price of under $200 each.

More money buys more features and performance, such as the DXers delight Sony ICF-2010, renown for its weak signal capability, and the aurally pleasing Grundig Satellite 700, both of which have synchronous detection. Both of these radios are relatively large portables, approximately 12"×6.5"×2.5" and weighing approximately 4 pounds each. The *street price* (early 1997) of the Sony and Grundig is approximately $350 and $400, respectively.

The Drake SW-8A and the Lowe HF-150 are in a category know as "porta-tops." Both of these offer performance approaching that of a desk-top receiver in a portable (albeit a bit bulky) package. These two receivers are in the $700 price range. The Lowe HF-150 is now famous as the choice of late night television host David Letterman.

The previously mentioned Sony ICF-SW100 (4.5"×0.7"×2.9"; 8 ounces) offers performance reputed by its owners to rival the renowned Sony 2010 in a remarkably small package. It carries a street price of about $350.

Be aware that many of the least expensive shortwave radios can't demodulate single sideband, the mode of choice for HF ham radio and for utility/military transmissions. If these transmissions are of interest to you, be sure the radio you choose has SSB capability. All of the radios listed here can demodulate SSB.

SYNCHRONOUS DETECTION

So what is this much-hyped synchronous detection? It's a radio technology that can stabilize signals with fluctuating signal strengths, and provide a clearly readable signal in the presence of severe adjacent channel interference, with a minimum degradation in audio quality. Since it is only used for AM reception, even a well informed ham who doesn't listen to short-wave broadcasting probably won't know about it (perhaps that's an oxymoronic phrase... "a well informed ham who doesn't listen to shortwave broadcasting.") Even hams who once transmitted AM may not know about it, since in the "AM days," synchronous detection didn't exist.

Amplitude modulation has three constituent parts; two identical sidebands and a carrier. If the shortwave receiver can't get an adequate reception of both sidebands and the carrier, the signal quality can quickly deteriorate and become unreadable. The synchronous detection circuit automatically locks onto the transmitted carrier, mutes it and substitutes a carrier generated internally in the radio. Since this carrier is strong and stable, the effects of fluctuations on the received signal are reduced. Also as a part of this process it is now possible to receive a clear signal using only one of the sidebands. Therefore, the user can tune to the sideband opposite to any station whose close-by transmission is splattering over on the station you want to hear (adjacent channel interference), which improves readability in a crowded band. Also, as synchronous detection substitutes its own carrier, any signal whose carrier is not precisely on the same frequency is substantially attenuated, without the loss in fidelity that a narrow filter would entail.

In actual use, the results can be dramatic. Signals that are unintelligible because of adjacent channel interference can become easily and clearly readable with synchronous detection. Weak stations can be made much more readable when their fading carrier is replaced internally. There can be a difference of opinion about any technology, but few hobbyists who have heard a radio with a well designed synchronous detection system deny its substantial benefit for the reception of AM signals.

FROM "DEATH BEFORE DIGITAL" to "DIGITAL FOREVER"

Many long time hams once believed that "real hams don't use digital readouts." However, with one exposure to a good digital readout, they quickly changed their minds. Particularly for someone new to shortwave listening, a digital readout is a tremendous aid. The ability to accurately determine the frequencies of unidentified signals makes looking them up in frequency guides a breeze. Also, digital readout assists in "repeatability"—the ability to be sure that you can return to precisely the same frequency you were listening to last week.

With the vagaries of propagation, the signal might not be strong enough to listen to, but at least you will be sure you are on the correct frequency. With the old, less precise analog frequency readouts, when you couldn't hear the signal you were seeking, it was difficult to determine whether it was bad propagation or you were just off frequency.

Getting to the frequency you want is often easier with digital receivers. The calculator or telephone-type keypad on many of them allows you to directly punch in the frequency you want and instantly be on frequency. This makes jumping from frequency to frequency to check propagation or different programs.

But it can get easier still. Typical among the modern digital shortwave receivers are *memories*, where frequencies and modes can be stored for quick access. Thus your favorite BBC frequency and other often used frequencies can be accessed by pressing a couple of buttons. This makes manual scanning of multiple frequencies (like the low traffic utility frequencies) a breeze.

Another advantage of radios with digital readouts is their use of frequency synthesizers, which most often all but eliminate drift. "Drift" is the tendency of a radio to change the frequency as its internal components warm up. While a digital readout does not ensure that the radio uses frequency synthesis, the two are found together so commonly as to be virtually synonymous.

OLDIES BUT GOODIES?

Some well meaning hams may advise you to get an older shortwave receiver at a hamfest to get your start in shortwave listening. They may speak with nostalgia of spinning the dials on a Hammarlund, Hallicrafters, National, etc. back in their early days in radio.

Although it sounds good at first, there are a couple of problems with this advice. Those older receivers don't have digital readouts or synchronous detection. They also don't have the stability of frequency synthesis; it was a common practice in the old days to let receivers warm up for an hour or so before doing any serious listening, to get the drift stabilized.

Another situation negatively impacting the attractiveness of older receivers is that in recent years they have become the target of collectors. What this means is that the good ones aren't cheap and the cheap ones aren't good, and weren't very good even when they were new. Given the appetite of collectors, increasingly even the bad ones aren't cheap.

Certainly, these old receivers have a special appeal and that special glow that comes from a hernia-inducing big box full of tubes. They can also provide much of the heat your shack requires in the winter. A late 1950s Hammarlund HQ-180 and a 90 pound military surplus Collins/EAC R-390A may have a place of honor in many shacks. But while these older receivers have excellent sensitivity and selectivity, you can usually achieve better performance and ease of use (not to mention lower cost) with modern receivers.

If there is any exception to this "rule," it is in medium wave and long wave spectrum. Many of the newer shortwave receivers tend to suffer from performance drop off at the AM broadcast band and below. The best of the oldies are great performers on these frequencies. If you intend to be especially active in medium or longwave monitoring, or if you must have that "real radio" look, a well selected oldie can be a goodie. Just don't expect these radios to be inexpensive, and the cost of replacement tubes might make you gulp twice!

It "only" covers 100 kHz to 2000 MHz, so there is not much you will miss with this ICOM IC-R8500. Scanning and 1000 memories included, of course!

ANTENNAS

Put as plainly as possible, if you're using an HF transceiver, an 80 meter or 40 meter dipole will work well for shortwave listening. If you use a portable shortwave receiver, adequate reception of the international broadcasters is often possible using the attached whip antenna.

If you want maximum performance, put up a wire outside as high as possible and as long as possible. Practically, a wire between 25 feet and 75 feet long and 20 feet (perhaps between two trees) will work well. If your portable receiver doesn't have an antenna jack, just use an alligator clip to connect the external antenna to the whip. Connecting an outside antenna to the whip will sometimes overload the receiver, so you may have to experiment to determine the optimum combination for the signals you want to hear.

While you may achieve acceptable results with just the whip attached to a portable radio, don't underestimate the value of a good antenna system. Many people like to spend most of their money on the radio. It's certainly the element of the station that will draw "oohs and ahs" from the uninitiated. However, the more experienced operator knows that, while the antennas may not be as photogenic as radios, it is easier to get high quality results with a good radio and a good antenna than with the best radio and a lousy antenna!

Living in a place where it is difficult to erect an outside antenna shouldn't prevent you from enjoying shortwave listening. An "invisible" outside antenna (constructed using a wire so thin as to be nearly invisible) can bring useful results. Invisible antennas are covered in many ham radio and shortwave radio antenna texts.

Reading and experimenting will help you achieve the optimum performance for your location and equipment. Two books that contain many ideas for invisible—or at least inconspicuous—antennas are *Your Ham Antenna Companion*, by Paul Danzer, N1II and *Low Profile Amateur Radio*, by Jim Kearman, KR1S. Both are published by the ARRL.

THE INTERNATIONAL SHORTWAVE BROADCASTERS

The international broadcasters are the best known and easiest to find stations on the shortwave bands. They are often sponsored by governments, who are willing to spend substantial sums to spread their viewpoint around the world. Their booming signals, not uncommonly in the half megawatt range, provide music, news (sometimes tainted with propaganda), educational and other entertainment programming. Not only are their signals usually strong, their transmissions tend to be confined to identifiable frequency ranges.

Another reason that the international broadcasters may be easy to hear, is that the transmitter may not be located where you think it is. Though you may be listening to Radio Netherlands, Deutsche Welle or Radio Taiwan, the transmitter might be located off the north coast of South America (the Netherlands Antilles is a popular location) or even from within North America, via a rented transmitter site! Those listening for program content won't care where the transmitter is located, but DXers often specifically seek those transmissions that originate far from their locations.

Interval Signals

If you listen to an international shortwave broadcaster at the beginning of their transmission or at the top of the hour, you may hear a repeating series of tones. These tones are known as the "interval signals." They're designed to help you identify the station when they are transmitting in a language foreign to you, or when the signal is weak. The familiar tones that the National Broadcasting Company (NBC) still uses are a domestic example of an interval signal. In addition to interval signals, there is another type of musical identification; the musical signature.

Interval signals are short and often played repeatedly (sometimes to the point of irritation) for 2-5 minutes just before the top of each hour. The instrumentation of an interval signal is simple, usually just a piano, celesta, carillon or electronic keyboard. The musical signature is typically more fully orchestrated and is played just once per period. Most often interval signals are played hourly and musical signatures only at sign-on and sign-off, but there are numerous exceptions.

The tones or music are often related to the culture and history of the originating nation. The Voice of America uses

"Yankee Doodle Dandy" and Radio Australia uses "Waltzing Maltilda," while the BBC uses a very regal sounding tune called "Lillibulero." Many of the European nations use classical music for their musical signatures. Not surprisingly, the religious stations often use hymns.

Hearing and identifying interval signals is another sub-set of the shortwave listening hobby. The level of interest is indicated by the fact that one "pirate" radio station uses the bogus call *WLIS*—We Love Interval Signals.

WORLD TIME

If your operating has been confined to the VHF bands, you may not have much experience with what is known as *World Time*. Since shortwave signals reach around the world, the listing of the time of the transmissions can be a problem. Whose time do you use?

To eliminate this confusion, the time at the Royal Observatory in England at the zero degree meridian (formerly at Greenwich) is used. It was called Greenwich Mean Time (GMT) for years, but now it is known as Coordinated Universal Time (UTC), World Time, Universal Time or Zulu (from the phonetic for the letter Z, which the military uses to indicate UTC).

UTC was established by international agreement in 1972, and is governed by the International Bureau of Weigh and Measures (BIPM) in Paris, France. It differs from local time by a specific number of hours. The number of hours depends on the number of time zones between your location and the location of the zero meridian. For time zones in the United States, UTC is 5 hours ahead of EST, 6 hours ahead of CST, 7 hours ahead of MST and 8 hours ahead of PST. That is, at 5 AM EST, the UTC time is 10 AM. When local time changes from daylight saving to standard time, or vice versa, UTC does not change. However, the difference between UTC and local time does change—by 1 hour.

UTC is expressed using the 24-hour clock system, sometimes called "military time." The hours are numbered beginning with 00 hours at midnight through 12 hours at noon to 23 hours and 59 minutes just before the next midnight. So 7 PM is 1900 hours.

Another attribute that you must remember when using UTC is that (for the United States at least), since UTC reaches midnight before the American time zones, *the date also changes earlier*. Thus, since 7:00 PM EST (1900 EST) is midnight UTC, at 7:01 PM (1901) EST on December 31st, it is 0001 UTC *on January 1st*. Many people keep 24 hour clocks set to UTC at their listening post, which makes logging the correct time a breeze. But even experienced hobbyists sometimes forget to advance the date after 0000 UTC. Additional information on UTC is available in Chapter 3.

THE MAJOR PLAYERS

Before we delve into some of the major international broadcasters, a word about names is in order. Since we are dealing with foreign countries and foreign languages, these stations can be called by either the native name or the English variant of it. There are no definite rules, but some standards have arisen. The "Voice of Germany" is almost always referred to as Deutsche Welle (and as with the classical composer Wagner, the "w" is pronounced as a "v"). The national stations of Cuba and Holland are referred to almost equally by their native and English names, Radio Habana Cuba/Radio Havana Cuba and Radio Nederland/Radio Netherlands respectively. The French are very sensitive about maintaining the influence and purity of their language; therefore, their national broadcaster is always

called "Radio France International" and never "Radio Francaise" or some other variant. (A note to our French readers... although true, that was a joke.) HCJB's nickname "The Voice of the Andes" is occasionally called "La Voz De Los Andes." So if you see a name unfamiliar to you, be sure it's not just a native language rendition of a station whose name you already know.

BBC

If you can't hear the BBC on your shortwave receiver, the radio isn't working. The British Broadcasting Corporation's worldwide network of shortwave transmitters was established to serve the British Empire back in the days when the sun never set on it (it was known then as the Empire Service), and they continue to put out powerhouse signals all across the shortwave spectrum. The BBC remains the standard by which all shortwave broadcasting operations are measured.

Long before the days of cable TV and CNN, the around the 24-hour-a-day news reports of the BBC from around the globe provided listeners with virtually constant access to late breaking information from important events. Even in today's news rich environment, those without cable TV or wanting an alternative viewpoint depend on the World Service (its name today) of the BBC. The program content does, however, often represent the British government point of view, particularly that of the conservative Foreign Office.

Reliance on the BBC extends to more than news. If you want to know the proper pronunciation of a word, listen for it on the BBC; reportedly, they have several people whose full time job it is to determine the correct English (perhaps as opposed to "American") pronunciation of any new word that comes up on their broadcasts.

Voice of Russia (Radio Moscow)

If you ever listened to Radio Moscow prior to 1987, you *must* listen to a few broadcasts from Radio Moscow *today*, even if you aren't interested in shortwave listening. Borrow a portable shortwave receiver or go to a friend's shack if you don't have shortwave capability. The change in the content of their newscasts and programming will flabbergast you. If fact, they've even changed their name. They are now called the Voice of Russia.

In the old days, the only times they weren't slamming the United States and the western world in general was when they were broadcasting fascinating reports on "The Wheat Harvest in the Ukraine," or some other such thrilling subject. They also were reported to be very tough on announcers whose tongues were loose. One announcer who slipped and called the Soviet action in Afghanistan "the invasion" was never heard from again. (Evidently, the official line was that they were in Afghanistan for humanitarian reasons, using Soviet battle tanks to plow the fields and the heavily armed Hind helicopters for crop spraying.)

But as changes took place in Russia from 1987 to 1991, so did Radio Moscow change. As previously mentioned, monitoring Radio Moscow during the August 1991 coup and the insurrection of the legislature in September 1993 provided more accurate information than virtually any of the outside news sources.

Now the Voice of Russia sounds very much like many of the western broadcasters. Most shocking to those who listened years ago, is hearing the frank discussions of problems within Russia. Not to make light of their problems (every country has its share), but to go from such a controlled purveyor of the party

The VOA (Voice of America) site near Greenville, NC remains a tourist attraction. Wouldn't it be nice to rent these antennas for a contest? *(photo courtesy of VOA)*

line to a relatively independent broadcaster has been an amazing transformation.

A downside of the changes is that Russia, and the Commonwealth of Independent States (CIS) that replaced the Soviet Union, are reducing the amount of programming on the Voice of Russia. Some of the transmitter sites are being used by the newly independent states for their independent broadcasts. Other transmitter sites have been rented to anyone who will pay their price. At present, the Voice of Russia is still a major presence on the shortwave bands, but substantially less so than in its heyday. What the future holds is very uncertain, but whatever happens, monitoring the evolution of the Voice of Russia is a fascinating experience.

Voice of America

The Voice of America (VOA) was established immediately after the end of World War II. After spending the war years listening to the German propaganda broadcasts orchestrated by Joseph Goebbels, the US government was especially sensitive to broadcast content. As a result, the Voice of America was forbidden to create broadcasts for consumption within the United States. The idea was to remove the temptation for the government to propagandize the general population.

However, in recent years this view has moderated and now VOA provides an excellent amount of programming information, even providing a computer bulletin board system (BBS) and a site on the Internet. Though the VOA definitely broadcasts from the American point of view, it is generally respected as an authoritative and respected source of information. VOA broadcasts were a critical lifeline to the outside world for those behind the "Iron Curtain" during the Cold War.

The end of the cold war also has been difficult for the VOA. The programming priority has shifted from Eastern Europe to emerging democracies of Africa and Latin America, but there is less urgency (and less money) supporting these broadcasts. The large VOA transmitter site in Bethany, Ohio, was decommissioned in late 1994, and a transmitter and receiver site were recently decommissioned near Greenville, North Carolina. The remaining site near Greenville is the VOA's sole remaining voice on the East Coast.

Should you be in the North Carolina area and have the time to drive to Greenville, a trip to the remaining transmitter site will surprise and astound you. If you call them (919-752-7115) they can set up a tour, but just driving by one of the sites will allow you to see an antenna field that no radio enthusiast will forget. Cut into the North Carolina pines is an open field approximately 3 miles long and 3 miles wide. The huge transmitter building sits about $1/2$ mile off the narrow secondary road, and off in the distance are groups of four 400 foot towers set in a semi-circle with about a 1 mile radius spanning over 180

The Voice of America facility in Greenville supports worldwide broadcast of VOA programming material. *(photo courtesy of VOA)*

On-line Computer Resources for SWLs

The books, magazines and shortwave shows listed in the main body of this chapter provide the shortwave listener with valuable information, but shortwave's ties to fast breaking events and the vast quantities of information make the type of communication provided by interconnected computers the ideal method of information exchange.

How else can you exchange detailed, exhaustive information with other hobbyists around the world, 24 hours a day, in a format that allows you to quickly turn it into convenient printed "hard copy"? Where else can you be in instant communication with the programmers at major international broadcasters overseas? All this and more can be found among the electronic information resources available today via computers and modems. Although well-equipped radio hobbyists almost always have a microcomputer at their disposal, only now are many of them becoming aware of how much additional usefulness connecting it to a telephone line can add.

Long before all the hype about the "information superhighway" has been networks of computer information resources that can substantially increase the enjoyment of hobby radio. The integration of telecommunications capability and microcomputers has provided an important new way to transmit and exchange information.

Although independent computer bulletin board systems (BBSs) and commercial services continue to provide a valuable source of radio information, the explosion of activity on the Internet make it the primary source of information today. A chapter on using the Internet is provided later in this book.

Internet allows for access to information in several ways. The most common methods of information dissemination are newsgroups, electronic mail, the World Wide Web and files.

Newsgroups

No matter how knowledgeable you are about your special interests in radio or how long you have practiced your hobby, on occasion you will be confronted with a seemingly unsolvable question or need information on some topic unfamiliar to you. Electronic information services can provide you with a forum to get input on your questions from hobbyists from all over the country. A corollary to the Internet called USENET provides public message areas called "newsgroups."

On BBSs and the commercial services they may be called conference areas or SIGs (Special Interest Groups). The user chooses the newsgroup with the most appropriate title and types in ("posts") the question, usually addressed to "ALL." Anyone accessing the newsgroup (nationwide or even internationally on USENET) after that will see the message. Within days or hours, the original question will usually receive responses, the responses may produce more questions and responses and will often yield an in-depth exploration of the topic by a number of knowledgeable (and some not so knowledgeable) responders.

Of course, the responders are self-selecting, so there are no guarantees as to their accuracy and expertise. But this same caveat applies to many seminars, books, magazine articles and the like. In an electronics analogy, this is known as the "signal-to-noise" ratio, or "noise factor." The number of unknowledgeable or belligerent responders (and there are a few belligerent ones) cause the noise factor, and the higher their number compared to knowledgeable and reasonable respondents, the higher the "noise factor."

Over time reading the messages in a conference area, the discerning user will be able to determine which participants' information is trustworthy. Newsgroups of interest to shortwave listeners include:

    rec.radio.shortwave
    rec.radio.scanner
    rec.radio.swap
    alt.radio.pirate
    alt.radio.scanner
    rec.radio.amateur.antenna
    rec.radio.amateur.boatanchors
    rec.radio.amateur.digital
    rec.radio.amateur.equipment
    rec.radio.amateur.misc

Electronic Mail

The newgroups are generally public posting areas (where everyone can read them), but if you have a private message you can send it somewhat confidentially to one person or several through e-mail (Electronic mail). The "somewhat" modifier is used because, in addition to the intended recipient(s), the administrators of the service, some people at Internet servers (and anyone they designate) can see the message.

Files

People being introduced to computer communications are usually surprised by the amount of free programs available. In many computer or electronics magazines, there are ads for diskettes of software available for from $3 to $9 per disk. Most of this software can be downloaded (transferred to your computer via the phone line) from computer bulletin boards (BBSs) for free. The range of this "shareware" is as broad as the range of regularly distributed commercial software, and the quality and level of sophistication is often as good or better. Spreadsheet, communications, database and word processing programs are all available, as well as a seemingly endless supply of games.

The quality of this software varies widely, but you can't complain about the cost of acquiring it. Although the number of programs dealing with radio and electronics is not as large as those on other topics, they are growing in number. Some of them can be very useful and some are worth exactly what you paid for them, but they all provide some insight in the interface of computers and radios.

Given the volume of shortwave information, sometimes it is too voluminous for the conference areas and often in these cases it is compiled and placed in a file area. In many cases the author of the software gives the rights of usage freely or only requests donations from satisfied users. Other authors consider shareware an alternative method of distributing and selling commercial software (allowing you to try the software instead of advertising), and require a modest payment if you use the program regularly. Of course, given the circumstances, you're on the honor system.

The World Wide Web

There are many Web pages on shortwave and other radio topics, and they come and go almost daily. To get their addresses, use a search engine (see Chapter 6).

GETTING STARTED IN TELECOMPUTING

The basic tools needed for telecomputing are a computer with a serial interface, communications (or terminal) software, a modem and a telephone line. A typical voice-type telephone line is all that is required from the phone company. Once you get on-line, you'll never again lack for reading material on your favorite hobby. In fact, you'll soon have to begin dealing with an excess of information, "information overload." That's a nice problem to have. For additional information on the internet, see Chapter 6 of this book.

degrees of arc. Each of these sets of 4 towers holds a massive "curtain array" antenna. To change the direction of their signal beam, they don't rotate an antenna; they just switch to an antenna pointing in the needed direction. Visitors never fail to be impressed.

The Voice of America headquarters in Washington, D.C. also has an excellent tour, albeit one that is less impressive from a radio hardware point of view.

Radio Netherlands

The "Happy Station," home of the famous "Happy Station" program for well over half a century, is one of the most popular shortwave broadcasters in the world. Radio Netherlands International dates back to 1927 when Philips Radio Laboratories began broadcasting to the Dutch West Indies. In 1994, Radio Netherlands celebrated the 75th anniversary of Dutch radio, contending (in a friendly way) that the Dutch radio pioneer Steringa Idzerda invented radio. . . or radio broadcasting. . . or at least regularly scheduled radio broadcasts.

Even though not many encyclopedias recall Idzerda's work back in 1918, Radio Netherlands had a fun celebration of his work and the radio equipment and techniques of the early days. Their goal was to prove that "75 years on, there's a lot of life left in international radio." That's a sentiment sure to increase their popularity.

HCJB, Quito, Ecuador

Broadcasting from the equator at an elevation of over 10,000 feet, the evangelical Christian "Voice of the Andes" dates back to the early 1930s. Compared to the sometimes strident type of Christian broadcasting common within the United States, the style of HCJB is relatively understated. HCJB also has many programs popular with those who do not necessarily agree with the beliefs of their sponsors, World Radio Missionary Fellowship of Opa Locka, Florida.

To combat some atmospheric problems caused by high power broadcasting at such a high altitude, engineers at HCJB invented the cubical quad antenna in the late 1930s, an antenna that has become very popular with ham radio operators. HCJB sends a booming signal into the United States, and is typically an early catch for a new shortwave listener

Radio Habana Cuba

The Cold War lives on at RHC, so don't tune in here to hear any praise of the western world in general and United States in particular! Because of its proximity to the United States, it generally puts in a solid signal. Over the past few years, there have been times when the signal suffered from distortions and weaknesses that have been attributed to power brownouts and lack of spare parts. Russia no longer pours money and resources into the island nation, which has led to fuel and resource crises. The former Soviet Union is not infrequently criticized on the air for abandoning the Communist philosophy (and with it, their substantial support for Cuba).

This will be an interesting station to monitor to observe the changes that are sure to come in Cuba. And given its location, the implications for its northern neighbor (us!) give listeners in the United States substantial motivation to follow the trends.

Deutsche Welle

The "Voice of Germany" was for many years the voice of "West" Germany. In the many years since World War II it has emerged from the shadows of the propaganda broadcasts of Radio Berlin. It is a consistent source of news and analysis of

events on the continent. Though its world-wide resources aren't as vast as those of the BBC, it is an emerging leader in international reporting and provides the political perspective of this country in the post-Cold War world.

Radio France Internationale

Anchored by its amazing ALLISS rotating monster antenna and recently increased transmitter output, RFI is on its way to establishing itself as a power in international broadcasting.

It retains strong African coverage from its colonial days and its clever programming treads a fine line between spontaneity and professionalism. For French language students, they offer an excellent source of practice direct from an unimpeachable source. Selected frequencies in kHz are 6175, 9805, 11670, 11705, 15365, 17620 and 17795.

SHORTWAVE SHOWS FOR THE HOBBYIST

To keep up with the fast changing world of shortwave radio, the serious listener requires access to information resources that can quickly disseminate updated data. One excellent resource for this is Internet (see the sidebar "On-line Computer Resources for SWLs"), but these require the use of a computer and modem.

However, the shortwave broadcasters themselves offer shows on a regular basis (usually weekly) that help the SWL stay on top of what's happening. The only equipment the listener needs to access this information is their shortwave receiver.

The breadth of topics varies with each show, but among those covered are the latest happenings in the shortwave world, tips on improving the proficiency and enjoyment of shortwave listening, human interest stories relating to SWLing and other related communications and electronics information. Information on satellite broadcasting and reception has been added to many of these shows.

In addition, some of the "mailbag shows" include SWL re-

Table 1-2
Mystic Star Frequencies

Radio frequencies used by high-ranking military and government officials when traveling on official business.

All frequencies shown in kHz

3032	6817	9023	11413	13585	20053
3046	6803	9026	11441	13710	20154
3076	6918	9043	11460	13823	20313
3071	6927	9120	11466	13960	22723
3116	6993	9158	11484	14715	23265
3144	7316	9180	11488	14902	25578
4721	7690	9270	11498	14913	26471
4731	7735	9320	11545	15015	
4742	7765	9958	11596	15036	
4760	7813	9991	11615	15048	
5688	7858	10112	11627	15091	
5700	7997	10427	12324	15687	
5710	8040	10530	12317	16080	
5760	8060	10583	13201	16117	
5800	8162	10881	13204	16320	
5820	8170	11035	13214	16407	
6683	8967	11055	13215	17385	
6715	8992	11118	13241	17480	
6738	8993	11176	13247	17972	
6756	9007	11180	13412	17993	
6757	9014	11210	13440	18027	
6760	9017	11226	13455	18218	
6790	9018	11249	13457	19047	
6812	9020	11407	13485	20016	

ports that have been mailed to the station. They also give listening tips, frequency changes and other information of interest to shortwave listeners.

Here's a selected list of the most popular shows:

MediaScan, on Radio Sweden (formerly called *Sweden Calling Dxers*) is the world's oldest radio program about international broadcasting. Radio Sweden has presented this roundup of radio news, features, and interviews on Tuesdays since 1948. It is currently broadcast on the first and third Tuesdays of the month.

Media Network on Radio Netherlands is a staple of shortwave enthusiasts. If it's on the air in this solar system, producer Jonathan Marks and the "Media Network" team are listening! Each Thursday (repeated Fridays) this award-winning survey of communication developments draws from its network of more than 157 regular contributors around the globe, including Jim Cutler, Mike Bird, Arthur Cushen, Diana Janssen, Victor Goonetilleke and their "secret weapon," Andy Sennitt. Andy Sennitt just happens to be editor of the venerable (and Amsterdam based) *World Radio-TV Handbook*, the creation of which keeps him plugged into high quality information sources around the world. Media Network also covers the latest in the world of technology.

DXers Unlimited is a program broadcast by Radio Habana Cuba. Host Arnie Coro, CO2KK, devotes a considerable amount of his time on topics of interest to those new to shortwave radio. "DXers Unlimited" is totally nonpolitical, and probably the only program on RHC about which that can be said. "DXers Unlimited" is broadcast twice a week, on Sundays and Wednesdays.

DX PartyLine from HCJB (Saturdays) is for "dial-twiddlers" and those who "like to listen to faraway countries just for the thrill of hearing a distant signal." Ken MacHarg hosts this popular show, which also has regular features for shortwave newcomers.

Ham Radio Today is also offered by HCJB. As its name suggests, "Ham Radio Today" focuses on issues of interest to hams, although in acknowledging the wide interests of hams, it does stray in to non-ham but related subjects including the history of radio. It is hosted by John Beck every Wednesday.

Spectrum is a weekly show covering the "spectrum" of electronics, ham radio, shortwave radio and related subjects. It is broadcast on independent shortwave station *WWCR* (Nashville, Tennessee) and a network of medium wave (AM broadcast) stations early Sunday mornings (Saturday evenings, in United States time zones).

World of Radio can be heard on various shortwave stations throughout the week, including WWCR and Radio of Peace International (RFPI in Costa Rica). Produced and hosted by Glenn Hauser, this show focuses rather narrowly but in depth on radio monitoring in shortwave and mediumwave. The program is a mix of updates on changes in programming and frequencies of the international broadcasters, with an occasional special program on a particular aspect of the shortwave hobby.

DXing with Cumbre is a relative newcomer to the scene, but it has already attracted a loyal audience. Host Marie Lamb provides the latest shortwave radio information for "hard-core" DXers. It is broadcast weekends on independent shortwave stations WHRI, WRMI and a few medium-wave outlets.

LISTENING AND THE LAW

The first rule of radio monitoring has always been that no license is required to listen. In general, there are virtually no rules regulating listening, so there was little to know, from a professional and legal standpoint.

The guiding rule of shortwave listening has always been the Communications Act of 1934 (Section 705), whose rule was simple: you could listen to anything you wanted to, but unless the transmission was by a broadcaster, a hobbyist (amateur, Citizens Band, etc.) or a ship in distress you were not to divulge the contents of the transmission nor use it for personal gain (don't get excited... the average hobbyist can listen for eons and not hear anything that can be readily used for personal gain). In fact, the spirit of the law was observed such that no one was prosecuted for telling their listening friends about an interesting signal they heard last night, as long as there was no negative effect to the parties of the transmission.

In 1986, due to pressure from lobbyists from the cellular telephone industry, Congress passed a law called the Electronic Communications Privacy Act of 1986 (ECPA). This law made it illegal to listen to cellular phone calls and several other types of communication that normally take place in the VHF/UHF spectrum and not generally of interest to SWLs. However, it also made it illegal to listen to remote broadcast transmissions and studio to transmitter links, communications that sometimes take place in the 26-30 MHz range.

There have been numerous reports of the reception of studio to transmitter communications, often hundreds of miles from the studio site. In one publicized example during the peak in the sunspot cycle, a Florida radio station was receiving (and soliciting!) reception reports from across the United States from its 100 watt shortwave studio to transmitter link. Interesting and innocent though this occurrence may be, it is illegal now and prudence would indicate that the legally minded hobbyist avoid it—or at least avoid a confession that such illegal listening took place (which is what a reception report would be). To date there are no known instances of the casual reception of such stations being prosecuted, but just because such a silly law has been ignored so far does not ensure that some techno-noramus (technological ignoramus) lawyer won't seek to make his reputation on such a case in the future. Don't you be the guinea pig for such legal action.

UTILITIES AND MILITARY

Despite how the name sounds, shortwave utilities monitoring is not listening in to crews from your local electric and gas company. In the broad sense of the term, utility monitoring is listening to any of the vast array of non-Amateur Radio transmissions not intended for public consumption. The respected *Klingenfuss Guide to Utility Stations* calculates that 77% of the shortwave bands are dedicated to the utility services.

Within the realm of utility monitoring falls maritime communications, aeronautical communications, commercial fixed stations and military transmissions on the air, land and sea.

This activity requires a little more sophisticated level of equipment than shortwave broadcast monitoring. Because these transmissions are only intended for a select audience, the power level is lower than most broadcasters. Thus reception requires a better antenna, a more sensitive receiver—or both. In addition, most utility's voice transmissions are in upper sideband (USB), a mode not included in many low-end shortwave receivers. Radioteletype (RTTY) and the digital modes require outboard decoding equipment (usually including a computer) and a receiver with good selectivity and stability.

Military monitoring is a subset of utilities monitoring. It encompasses the full spectrum of the utilities world: upper sideband, RTTY and digital modes as well as unique and secret transmission modes.

K4ZAD'S Shortwave Voice Utility Sampler

(All frequencies are in kHz)

2182	Marine Emergency Calling Channel
2598	Canadian CG Marine Information Broadcasts
2670	USCG Marine Information Broadcasts
3413	Aero Weather—Shannon Ireland
3485	Aero WX—New York, NY and Gander, Newfoundland
4065	Inland River Towboats—WCM—Cincinnati
4125	Marine Ship Calling
4149	Marine Simplex Utility Channel 4B
4372	US Navy
4381	Great Lakes Ore Boats—WLC—Rogers City, MI (Ships on 4089)
4582	Civil Air Patrol—Emergency Channel
4722	RAF Aero Weather—Continuous
4725	US Air Force—Global High Frequency System
4742	RAF—Architect
5015	US Army Corps of Engineers—Net at 8:00 AM ET, M-F
5211	Federal Emergency Management Agency —Primary Night Channel
5505	Aero Weather—Shannon Ireland—Continuous
5598	Air Traffic Control—North Atlantic—NY, Gander, Shanwick
5616	Air Traffic Control—North Atlantic—Gander, Shanwick
5680	Search & Rescue Channel—Worldwide
5692	US Coast Guard—Chopper Ops.
5696	US Coast Guard—Air Ops.
5841	US Anti-Drug Agents
6215	Marine Ship Calling Ch./Utility Ch. (Ch. 606) —Shore on 6516
6230	Marine Simplex Utility Channel 6C
6510	River Towboats—WCM—Cincinnati (Skeds at 14:30-15:00 ET)
6577	Air Traffic Control—Caribbean—NY
6604	Aero WX—New York, NY and Gander, Newfoundland
6628	Air Traffic Control—North Atlantic—NY, Santa Maria
6676	Aero WX—Sydney, Singapore, Bangkok, Bombay
6679	Aero WX—Honolulu, Tokyo, Auckland, Hong Kong
6697	US Navy
6720	US Navy
6738	US Air Force—Global High Frequency System
6753	Canadian Military WX—Edmonton, Trenton, St. Johns
6812	USAF—A prime frequency for Air Force One
7527	US Anti-Drug Agents
7535	US Navy
7635	CAP—Nationwide Frequency (Command Net Weekdays at 1600 UTC)
8125	FAA—Eastern Net (Wednesdays at 10:45 AM ET)
8176	Sydney, Australia. Marine Radio —VIS—(early mornings)
8213	River Towboats—WCM—Cincinnati (Skeds at 13:00-14:00 ET)
8255	Marine Ship Calling (Ch. 821) —Shore Stations on 8779
8297	Marine Simplex Utility Channel 8B
8794	Great Lakes Ore Boats—WLC— Rogers City, MI—Ships on 8270
8825	Air Traffic Control—North Atlantic—NY, Gander, Shanwick
8828	Aero WX—Honolulu, Tokyo, Auckland, Hong Kong
8846	Air Traffic Control—Caribbean—NY
8864	Air Traffic Control—North Atlantic—Gander, Shanwick
8867	Air Traffic Control—Pacific—Honolulu, Auckland, Sydney, Nandi
8903	Air Traffic Control—Pacific & Africa
8906	Air Traffic Control—North Atlantic—NY, Santa Maria
8912	US Anti-Drug Agents
8957	Aero Weather—Shannon Ireland—Continuous
8967	US Air Force—Global High Frequency System
8980	US Coast Guard—Chopper Ops.
8984	US Coast Guard—Air Ops.
8993	US Air Force—Global High Frequency System
9023	Canadian Military & USAF NORAD
9032	RAF—Architect
10051	Aero WX—New York, NY and Gander, Newfoundland
10493	Federal Emergency Management Agency— Primary Day Channel
10780	USAF—NASA Support—Cape Radio— Primary Day Channel
11176	US Air Force—Global High Frequency System
11191	US Navy—Air Operations—Hershey at Key West, FL
11195	US Coast Guard—Air Ops.
11198	US Coast Guard—Chopper Ops.
11200	RAF Aero Weather—Continuous
11205	US Navy
11233	Canadian Military
11234	RAF—Architect
11255	US Navy
11267	US Navy
11279	Air Traffic Control—North Atlantic—NY, Gander, Shanwick
11282	Air Traffic Control—Pacific—San Francisco, Honolulu
11309	Air Traffic Control—North Atlantic—NY, Santa Maria
11384	Air Traffic Control—Pacific—Honolulu, Tokyo, Hong Kong
11387	Aero WX—Sydney, Singapore, Bangkok, Bombay
11396	Air Traffic Control—Caribbean—NY
11494	US Anti-Drug Agents
12290	Marine Ship Calling Ch. (Ch. 1221)—Shore Stations on 13137
12359	Marine Simplex Utility Channel 12C
13201	US Air Force—Global High Frequency System
13257	Canadian Military
13261	Air Traffic Control—Pacific—Honolulu, Auckland, Sydney, Nandi
13264	Aero Weather—Shannon Ireland—Continuous
13270	Aero WX—New York, NY and Gander, Newfoundland
13282	Aero WX—Honolulu, Tokyo, Auckland, Hong Kong
13288	Air Traffic Control—Pacific—San Francisco, Honolulu
13297	Air Traffic Control—Caribbean—NY
13300	Air Traffic Control—Pacific—Honolulu, Tokyo, Hong Kong
13306	Air Traffic Control—North Atlantic—NY, Gander, Shanwick
13312	Anti-Drug Agents and FAA Flight Tests and Commercial Flight Tests
13330	Air—Long Distance Operational Control —NY, Houston
13354	Air Traffic Control—Pacific—San Francisco, Honolulu
13457	FAA—Western Net (Wednesdays at 10:30 AM MT)
15015	US Air Force—Global High Frequency System
15867	US Anti-Drug Agents
16420	Marine Ship Calling Channel (Ch. 1621) —Shore Stations on 17302
16534	Marine Simplex Utility Channel 16C
17904	Air Traffic Control—Pacific—Honolulu, Tokyo, Hong Kong
17946	Air Traffic Control—North Atlantic—NY, Gander, Shanwick
17975	US Air Force—Global High Frequency System
18009	US Navy
22060	Marine Ship Calling Channel (Ch 2221) —Shore Stations on 22756
22171	Marine Simplex Utility Channel 22E
23287	US Navy

Frequencies courtesy of Tom McKee, K4ZAD, from his book, *The Other Shortwave* (Key Research, PO Box 846G, Cary, NC 27512).

Typical of today's new ham transceivers, this Kenwood TS-570D includes a 3-kHz to 30 MHz receiver. It is a full featured ham transceiver, and you can make use of these features when listening on the shortwave bands.

To those willing to go that extra step to acquire the equipment necessary for utilities monitoring, the rewards are many. The Mystic Star frequencies can yield very interesting listening as members of the Executive Branch and VIPs travel about the world. Although the planes that serve as Air Force One are equipped with sophisticated satellite communications gear, the shortwave equipment is still on-board and occasionally used for routine traffic or to achieve a certain diplomatic effect. Occasional, un-encoded phone patches can be heard from a variety of aircraft, including Air Force 1.

Monitoring the Land and the Sea

Despite all the high-tech equipment available to the U.S. Air Force, the Global High Frequency System remains a backbone of communications. Long ago, during the height of the Cold War, when many first heard the transmission, "Skyking, Skyking, this is McDill, do not answer...", they were ready to head for the bomb shelter. They knew these transmissions concerned the (now defunct) Strategic Air Command and the American global nuclear forces, but did not realize this type of transmission occurred frequently.

Even with the major reorganization of the military air commands that took place in June 1992, the "Skyking..." transmissions can still be heard. However, frequency lists compiled prior to June 1992 can not be considered accurate. Those listed on the accompanying "Voice Utility Sampler" has been compiled since the reorganization. Though many of these transmissions are cryptic, those transmissions that are "in the clear" can be fascinating.

The civilian air communications also provide interesting listening, and they are a little easier to understand. When flying over land, civilian aircraft use VHF frequencies, but on transoceanic and transpolar flights the shortwave frequencies come into use. Since English is the language of international aviation, translation is seldom a problem for the American listener.

The frequencies listed as "Air Traffic Control" on the "Utility Sampler" are used to transmit in-flight information such as weather queries, fuel consumption, updated estimated times of arrival and other related data. These frequencies are not in constant use; if you want to hear some activity on these bands choose a busy time (a time that you have reason to believe that air traffic will be heavy on the route you have chosen) and "park" on the associated frequency or do a little scanning.

For instance, on the frequency list, 5598, 5616, 11279, 13306 and 17946 (all kHz) are listed for the North Atlantic.

Remember, these transmissions are on upper sideband. Prime listening times (when the bulk of the jets are in the air) is generally 7 PM-12 AM EST (0000-0500 UTC) and 7 AM-12 PM EST (1200-1700 UTC).

During this period, just sit on one of these frequencies (below 10 MHz evenings, above 10 MHz mornings), or use the memory function of your receiver to occasionally switch between them. Since there is no continuous "chatter" on these types of frequencies it may take a while, but sooner or later you will come across some transmissions. Obviously, patience is required for this aspect of the hobby.

Some of the easiest catches in civilian shortwave aviation transmissions are the VOLMET or aviation weather transmissions. Listed as "Aero WX" (for Aero Weather) on the accompanying "Utility Sampler," they also list the times of broadcasts. For example, New York Radio transmits on its assigned frequencies (3485 kHz, 6604 kHz, 10051 kHz, etc.) from the top of each hour to 20 minutes past each hour and from 30 minutes past each hour to 50 minutes past each hour.

As with aircraft, both VHF and shortwave frequencies are used for maritime communications. While most aircraft transmissions use "back-and-forth" simplex type transmission typical on Amateur Radio, quite a bit of the maritime communications take place in "full duplex," where two frequencies are used simultaneously, and only one side of the communication can be heard on any given frequency.

However, there are some simplex frequencies and the "Utility Sampler" contains some of both. The U.S. Coast Guard Information frequency of 2670 kHz is a good one to check when conditions are favorable for a frequency that low.

RTTY and Digital Modes

Many of the signals heard on the utility bands are radioteletype and other digital modes. If you are active on VHF packet radio, you may already have some of the equipment you need to monitor these signals.

Many popular packet TNCs such as the AEA PK-232MBX, PK-900 and the Kantronics KAMPlus can decode a variety of digital signal types, including radioteletype. These, along with your computer, will allow you to monitor transmission from news agencies, diplomatic messages, marine traffic, meteorological data and other information that is best transferred by means suitable for hard copy.

Although monitoring radio teletype and the digital modes can be complicated, the ham bands are the best place to start and this book has detailed information on monitoring RTTY and packet. See Chapters 9 and 10 for a full description on how to decipher the digital modes.

Numbers Stations On Shortwave

If you tune around in the utility shortwave frequencies, sooner or later you will encounter strange stations sending groups of numbers or letters in Spanish, English, Russian or German. These stations are known as *numbers stations*.

Ever since these stations were first noticed in the early 1960s, people have speculated as to their purpose. Does the transmission contain encoded weather forecasts or shipping information? The possibilities are endless. Some people even thought the numbers transmissions were part of a secret project communicating with UFOs!

Over time a consensus developed that these transmissions were messages from various intelligence agencies to their agents in the field. As radio direction finding techniques located transmission sites in both (the former) East and West Germany, Nicaragua and Cuba, as well as on some military bases in the United States, this theory seemed largely confirmed. More recently, former "spies" have related how these transmissions were used to convey instructions to them. However, it continues to puzzle listeners that the level of numbers station activity has not appreciably decreased with the end of the Cold War, and that the stations on both sides of the former German border continue to operate.

Numbers stations are found as low as 2 MHz to as high as 26 MHz, using (CW) and voice transmissions in both AM and SSB. Though they can be heard at any time, activity tends to peak from about 0000 UTC to 0800 UTC.

PIRATES AND CLANDESTINE BROADCASTERS

Two unique types of shortwave stations are pirate and clandestine broadcasters. Though both generally operate illegally, they are quite different in their purposes.

Clandestine broadcasters are specifically political in nature. When a political group is out of power in a country, they will sometimes set up a shortwave station to broadcast to those sympathetic to their cause. These stations may be located within the target country, often illegally, or may be located in a nearby sympathetic country. In this case they may be operating legally, even though the host country may not officially acknowledge their existence (to avoid official diplomatic hostility with the target country).

Central America has traditionally been a hotbed of clandestine broadcasts, with stations appearing and disappearing as the fortunes of their sponsoring groups change within the target countries. Africa and the Middle East also has had a sizable number of clandestine stations over the years, but since North America is far from their target audiences, they can be very hard to hear.

These stations generally broadcast in the native language of the target country, so for Central American stations Spanish is the norm. However, some clandestine stations want to obtain money and support from within the United States. Such stations may use English, even when English is rarely spoken within the target country.

Pirate broadcasters are sometimes politically oriented, but typically they are people broadcasting for the joy of broadcasting. Their enthusiasm notwithstanding, the quality of pirate broadcasts range from that rivaling the professionals to sounding like a drunk turned loose in a radio studio—and indeed, they sometimes *are* inebriated!

The most common programming "format" of pirate stations are satire and comedy (sometimes unintentional). When political views are expressed, they tend to be extreme (at either end of the spectrum). Pirate stations are most active evenings on weekends and holidays, They tend to keep erratic schedules to help them avoid discovery by the FCC. To prevent FCC monitors from getting an accurate fix on their location, their transmissions tend to be no longer than 30 minutes to 1 hour in length. The most common frequencies for pirate operations have been 7415 kHz and 7485 kHz, but recently legitimate broadcast stations have begun transmissions on these frequencies causing the pirates to move.

Remember, pirates are generally using old Amateur Radio transmitters from the 1950s running 100-200 watts into temporary antennas. The major international broadcasters may run from 100,000 to 500,000 watts into sophisticated antenna arrays. When the two clash on a frequency, there is little doubt whose signal will triumph.

Now, most pirate stations are operating between 6925 and 6970 kHz (6955 is the favorite) so that's the best place to start tuning. Other fruitful places to tune are around 13900 and 15050 kHz on weekend afternoons and 1620 kHz evenings. Most signals are AM, but occasionally upper sideband broadcasts will be heard. Most weekends there will be between 2 and 20 North American pirates to be found (the most on holiday weekends), but since they don't have powerful signals, a good antenna and receiver are needed to hear them. Since pirates are already operating illegally, they aren't bound by FCC obscenity rules, so by forewarned!

While pirates would contend that they operate in the spirit of "good fun," the FCC does NOT take a casual attitude towards pirate broadcasters. Fines in excess of $10,000 are not uncommon and the revocation of any FCC licenses held (including Amateur Radio licenses) is virtually automatic. Whatever the appeal of running your own shortwave broadcast station may be, it's not worth your Amateur Radio career.

KEEPING TRACK—SHORTWAVE LOGS

One of the traditional rules of scientific experimentation has been, "If you don't record it, it didn't happen." The tradition of keeping a list of the stations heard, the date and time, the frequency, conditions and transmission content in a *logbook* has been continued since the days when radio transmission and reception *were* scientific experiments.

For many years, the keeping of a logbook by licensed ham radio operators was the law. The Federal Communications Commission used to mandate that licensed ham radio operators keep a meticulous log of their transmissions and communications, but those regulations have long since been relaxed.

Despite the less restrictive regulations, the logbook still has many useful purposes in both ham radio and shortwave listening. The log is a permanent record of the reception achievements of the operator and the station. It is a reference resource for active frequencies, propagation and the comparison of different radios and antennas..

QSLs

So you've heard a rare station. How do you prove it? One way is to send a report of the reception to the station, including the date, time, frequency and information on the reception quality and programming content. This is essentially the same information you record in a logbook.

If they find your information to be accurate, often they will send you a confirmation. Sometimes this will be in the form of a letter (especially with a utility type station), but the international broadcasters traditionally send a picture postcard in verification. These cards have become known as "QSL cards" (from the ham radio "Q" signal meaning, "acknowledging receipt.) Collecting QSL cards is another sub-set of shortwave listening, along with collecting the sometimes exotic stamps that come with them.

The QSL card is a courtesy of the transmitting station, and to increase the chances of receiving one and to defray the mailing costs of the broadcaster, many SWLs include an IRC (International Reply Coupon) to help defray the postage cost.

An IRC can be exchanged for first class postage in any country and is available from most larger post offices.

The SINPO Code

The SINPO code is a way of quantifying reception conditions in a five-digit code, especially for use in reception reportsto broadcasters. The ham radio equivalent is the RST system for Readability, Strength and Tone. The SINPO components cover Signal strength, Interference (from other stations), Noise (from atmospheric conditions), Propagation disturbance (or Fading due to propagation, which sometimes leads to it being called the SINFO code), and Overall merit. The code is as follows:

In recent years, many broadcasters have tried to steer listeners away from the SINPO code and toward the simpler SIO code. SIO deletes the extremes (1 and 5) and the noise and propagation categories, which were confusing to too many people to be useful. In sending reports to stations other than large international broadcasters, who are likely tounderstand the codes, it is better to simply describe reception conditions in words.

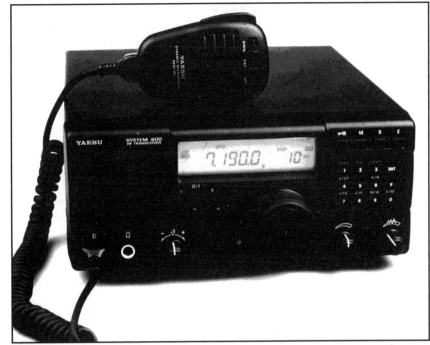

It looks deceptively simple, but there is a lot of capability packed into this Yaesu FT-600 HF transceiver, including 50 kHz to 30 MHz receive coverage and a computer interface.

Value	(S)ignal	(I)nterference	(N)oise	(P)ropagation	(O)verall
5	excellent	none	none	none	excellent
4	good	slight	slight	slight	good
3	fair	moderate	moderate	moderate	fair
2	poor	severe	severe	severe	poor
1	barely audible	extreme audible	extreme	extreme	unusable

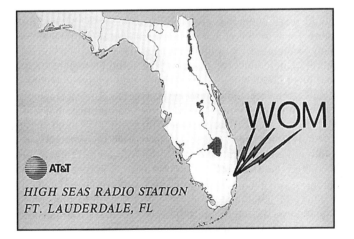

WOM handles high-seas radiotelephone traffic from a computerized control station in Florida. Log one of their many transmissions or those of their sister stations (KMI in California and WOO in New Jersey) and they will QSL.

SHORTWAVE VS SATELLITES

Some people have predicted the end of shortwave radio as satellite technology becomes more prevalent. Certainly many countries have cut back on their shortwave operations, both to conserve funds and because they've switched their programming to satellites. Military and diplomatic operations have switched much of their communications to satellites. So is shortwave listening on its deathbed?

No. One major difference between shortwave transmissions and satellite transmissions is *infrastructure*. For satellite operations to work, a lot of complex equipment has to perform in concert. Satellite dishes have to be aimed precisely at both ends of the signal path. To make matters worse, the satellites have to function in one of the harshest environments imaginable. As the old saying goes, "the more complicated they make the plumbing, the easier it is to clog up the pipes." That's why even the military, with their desire for secure communications, can still be heard frequently on the shortwave bands.

The need for infrastructure also requires approval from those who control the infrastructure. When Operation Desert Storm commenced, the Iraqi government immediately took control of the satellite uplinks. From that moment onward, they had exclusive control of all the information coming out of the country.

In the past few years, both Iraq and Saudi Arabia have outlawed the private ownership of satellite dishes, to better control the information being received by their citizens.

Shortwave communication, on the other hand, is not so easy to control. Relatively simple equipment is all you need: a transmitter, a receiver and the appropriate antennas (and a little help from the ionosphere). While it would be hard for someone in a controlled country to hide a satellite dish, a portable shortwave receiver and antenna can be very easily concealed.

So for all the convenience and quality that satellite operations provide, there will always be a need for shortwave operations. Your shortwave receivers won't lack for signals for a long time.

UHF/VHF SCANNING

The hobby of scanning, the monitoring of the VHF-UHF "public service" frequencies with multi-channel scanning receivers, has exploded from the minor sub-set of shortwave listening that it was two decades ago into the primary activity of many hobbyists. Many Amateur Radio operators have been enticed to sample these by the increasingly common ability of 2 meter and 440 MHz transceivers to receive frequencies substantially above and below the confines of these two ham bands.

THE LAW

It is unfortunate, but appropriate given recent events, that an overview of the legal aspects of scanning should be discussed prior to delving into other aspects of the hobby.

For years scanning was relatively unregulated, governed primarily by the same law as shortwave listening, namely the Communications Act of 1934—it was illegal for the listener to divulge or profit by monitored communications not intended for the general public.

But it also is now illegal to listen to cellular phone transmissions. It is not illegal to own radios that cover the cellular frequencies, but you're not supposed to listen to them. This law, the Electronic Communications Privacy Act of 1986 (ECPA '86), also makes it illegal to "*intercept* (listen to) ...all forms of common carrier communication, except cordless phones and tone-only paging communications, and of any non-common carrier or private radio communications when they are encrypted [or] scrambled..." This law was primarily passed based on lobbying pressure from the cellular phone manufacturers and service providers, to give their users a (false) sense that their unscrambled conversations were private and secure. In 1994, a law was passed that made it illegal to listen to cordless phone activity, but not baby monitor transmissions that share many of the same frequencies.

As of April 1994, it is illegal to import or manufacture scanners for resale that can receive or can *be easily modified* to

The Radio Shack PRO-2042 covers bands from 25 MHz to 1300 MHz, with the cellular frequencies locked out. It has a variety of scanning modes and 1000 memories. *(photo courtesy of Radio Shack)*

receive the cellular phone frequencies. What constitutes "easily modified" has been the topic of much discussion, but in general it has to be harder than just clipping a diode (as was required on the old Radio Shack PRO-2004/5/6 series). It is still not illegal to own or resell a used scanner that can cover these frequencies, possibly because many old television sets can receive some of the cellular phone frequencies with their UHF tuner. Since the ban, some people have successfully ordered uncensored scanners from sources in Canada or England, but this is a chancy proposition; such a scanner might possibly be confiscated or returned by Customs agents at the border.

The monitoring of transmissions from state, federal and local law enforcement agencies, public safety services, aviation, business, the military and most other sources that are "readily accessible to the public" is not illegal. It is up to you to determine why our law makers consider a police transmission on the 800 MHz band "readily accessible to the public," and the cellular phone transmission a few megahertz up the band not "readily accessible."

The practical effect of this law has been minimal, since the pertinent law enforcement agencies have made it clear that they have more important things to do than to become the radio Gestapo. Generally it would be difficult to prove that a person had listened to a prohibited transmission unless they confessed, or recorded their eavesdropping session but this is the law and as with all laws you ignore it at your own risk.

Of course, this is exactly what happened in early 1997. A couple in Florida cruising in their automobile monitored a cell phone transmission of a conference call of Republican politicos and happened to have a tape recorder available that they used to record the activity. They then forwarded the tape to their Democratic congressman, who gave it to the press and embarrassed the Republicans.

It was a Democratically controlled House of Representatives that passed the laws of 1986 and 1994, so when the embarrassed Republicans now in charge called for hearings on cellular phone eavesdropping, the hobby had few (no?) friends among the attending members of Congress. No further laws on this have been passed yet, but anyone interested in scanning should pay close attention to legal developments. Interestingly though, there have been no reports of the Florida couple or anyone involved in this episode being prosecuted under the existing 1986 law.

Another area of scanning law that the prospective hobbyist should investigate is the legality of scanning while mobile. Using a scanner in an automobile is subject to restrictions in many locales (including being prohibited in Florida). If you are interested in scanning on the move, you definitely should check your local laws.

In some places where mobile scanners are prohibited, licensed Amateur Radio operators are exempted from the law. And by mandate and exemption of the FCC, any ham radio transceiver can be legally used in mobile operation by an Amateur Radio operator, even if it also receives frequencies outside the ham bands. This has caused some scanner enthusiasts to obtain their ham license, just so they can enjoy their scanner hobby with less restrictions. Unfortunately, some overzealous local police have seized ham VHF/UHF transceivers because they were capable of scanning, and recovery by the hams involved proved difficult and tedious—even in one well publicized case where the scan capability was used to help find an individual who persistently jammed the local police radio system!

Print Resources for the Shortwave Listener

With the vast frequencies in the shortwave spectrum, the ever-changing international implications of shortwave radio and the constant changes in technology, many information sources can be helpful. The previously mentioned shortwave shows for the hobbyist can be helpful and on-line computer services are available for those on the much touted "information highway" (see "On-line Resources for SWLs), but what if you're a beginner with no radio or modem? For these people and many others, the multitudinous print resources for the shortwave listener come to the rescue.

The two print mainstays of shortwave broadcast listening are the *WRTH* or *World Radio TV Handbook* (BPI Communications, 1515 Broadway, New York, NY 10036) and *Passport to World Band Radio* (available from the ARRL), both published yearly. *Passport* is considered easier to navigate for the beginner, while the *WRTH* is considered the "bible," the definitive source for shortwave information.

Passport offers a listing of broadcasts by frequency, and a 24 hour tour of selected transmission to North America. *WRTH* provides a listing of transmissions by language and details of everything you could possibly want to know about each individual broadcaster. If you're new to shortwave, you should get a copy of each. Don't let your $300-$500 receiver go under-utilized for lack of the information in a couple of $20 books. If you are like many people you may find that they are both so useful in their own special ways that they each warrant a yearly purchase.

Introductory books on the subject include *The Shortwave Listening Guidebook* by Harry Helms, AA6FW (HighText Publications, Inc., P.O. Box 1489, Solana Beach, CA 92075) and *The Complete Shortwave Listener's Handbook* by Hank Bennett, Andrew Yoder, et al (TAB Books, Division of McGraw-Hill, Inc., 860 Taylor Station Rd., Blacklick, OH 43004). Helms' book (first published in 1990) is an excellent survey of all aspects of shortwave listening, and is highly recommended.

Those of us with a copy of Len Buckwalter's *The Fun of Short-Wave Radio Listening* (Copyright 1965, now out of print) might question the attempt to promote *The SWL's Handbook* (first published in 1974) as the first complete text on shortwave listening, but its position as a founding informational resource on shortwave listening is undisputed. However, the current edition suffers from the patchwork editing of four different authors. This can cause reader schizophrenia, especially if you read Hank

Bennett's stories of shortwave listening in World War II and subsequent, and mistakenly attribute them to young pirate radio expert Andrew Yoder (who was barely born before the Vietnam "conflict" was over). At least those of us with the original 1974 Hank Bennett solo edition can tell the difference. Still, in any edition *The Complete Shortwave Listener's Handbook* is a valuable reference.

Monthly publications can help the SWL stay on top of the ever-changing shortwave world, and the two standard monthlies are *Popular Communications* (76 North Broadway, Hicksville, NY 11801) and *Monitoring Times* (P.O. Box 98, Brasstown, NC 28902). *PopComm* is a slickly produced shortwave newsmagazine that can be found on many newsstands, while *Monitoring Times* use of newsprint allows for slightly more current information in each issue. Both magazines cover the full gamut of shortwave listening interests, from international broadcasting to military to utilities to digital modes.

For those interested in specific information on utilities and the digital modes, two frequency guides stand out. *Ferrell's Confidential Frequency List* (Listening In, P.O. Box 123, Park Ridge, NJ 07656) carries a full listing by frequency of the usage, mode, call, location and type signal from 1602 kHz to 25545 kHz. It also has a reverse listing sorted by call. *(If you work for a defense contractor, don't try to take this book into the office. I did, and had real problems with a member of the security force. They take a dim view of carrying around anything marked "Confidential"! —N1II)*

The *Klingenfuss Guide to Utility Stations* (Klingenfuss Publications, Hagenloher Str. 14, D-72070 Tuebingen, Germany) has its own frequency list, in addition to detailed listings of RTTY and FAX services and schedules and other miscellaneous but useful information.

For those interested in the content of the shows on the international broadcasting stations, the *Guide To Shortwave Programs* by Kannon Shanmugan (Grove Enterprises, P.O. Box 98, Brasstown, NC 28902) is an invaluable resource. It lists the frequencies and programs that are on the air at any time of the day or year. Though many have tried to lay claim to the title, this is perhaps the closest approximation of a *TV Guide* for the shortwave bands. It also has a special section on shortwave shows for the hobbyist.

Should you decide to delve deeply into propagation theory and the use of the indices, *The New Shortwave Propagation Handbook* by George Jacobs, W3ASK, Theodore J. Cohen, N4XX and Robert Rose, K6GKU (CQ Publishing, Inc., 76 North Broadway, Hicksville, NY 11801) is an excellent text on the subject.

WHAT CAN YOU HEAR?

Listening to police calls are one of the first things people think of when they hear the word "scanner," so much so that frequently they are called "police scanners." And certainly, police and law enforcement operations are a very interesting part of scanner listening.

You can listen to car chases, the hunt for missing children, hostage negotiations, the breathless voice of an undercover cop chasing a suspect on foot and shoot-outs, sometimes hearing the gunfire in the background. For some of their most secretive activities, law enforcement agencies use a virtually unbreakable digital voice scrambling system, but most activities are "in the clear." Of course, in addition to the adrenaline inducing tension situations, you can hear such routine activities as the change of shift roll-call and the apprehension of speeders and drunk drivers.

But listening to law enforcement is far from the only use for scanners. Today the fire department is not just fighting fires,

although listening to those operations can be as thrilling as the law enforcement radio traffic. The fire department is also the first responder for many diverse types of incidents including hazardous waste spills, people trapped both high and low and UFO incidents (yes, even those).

Listening to the fire department sometimes can give you information crucial to your health and well-being. Messages on Internet told of how scanner monitors in the San Francisco Bay area were able to monitor the progress, direction and danger areas of the wide spread fires that swept through that area a few years ago, giving them vital information long before it was broadcast on the local news.

Fire departments are particularly interesting, since often they use a low-power repeater located on a fire truck. Thus, personnel equipped with low-power hand-held radios on a fire scene work though the fire-truck repeater, parked at the scene. The local repeater connects both the local personnel and those at the fire dispatch station. You may never hear one of these low-power fire repeaters unless you are located within a few blocks of a fire or fire investigation.

Listening to rescue squad activities and the "Lifeflight" helicopters transporting the critically injured to the life-saving facilities of a hospital puts you right in the middle of real life-and-death situations. . . who needs the doctor shows on television?

Yet there is much more.

You can listen to the "Blue Angels" as they coordinate their intricate air ballet at an airshow, and to race car drivers as they discuss strategies with their pit crews at the races. At major airports, the scanner's frequencies are full of radio traffic from planes arriving and departing.

During inclement weather, monitoring can give you information on road conditions from the Highway Patrol, radio traffic reporters (before it is broadcast) and state and city road crews. If the electricity is out for any reason, a battery powered scanner can help you track the utility company workers as the try to restore power.

Since so many businesses use radio in their activities, you can hear construction crews, security forces, taxi companies, and other business in your area. You can even intercept the important communications between the order-taker and the customer at fast food restaurants, and learn if the car ahead of you is ordering a medium or large order of fries!

FREQUENCIES

The specific frequencies used by agencies vary with the locality, and there are books published with this information. One of the best and best known is *POLICE CALL* (now called *POLICE CALL PLUS*); it is available at Radio Shack. Often a local Radio Shack or other electronics store will have a photocopied list of frequencies active in the area. If they have one, they will give you a copy of the list if you buy a scanner from them (and sometimes even if you don't). Try to find a scanner buff at a local Amateur Radio club meeting: this can be a good way to learn the hot frequencies. For those "on-line," Internet and computer bulletin boards are a source for many frequency lists.

For general information on what signals are where, here's a sampling of the frequencies ranges with the most activity between the ham bands:

30-50 MHz—This is called the VHF low-band. Most users are gradually leaving this band because it requires a relatively large antenna and as with Citizens Band (CB is at 26.9-27.4 MHz, is just below it) long range propagation via the ionosphere can be a problem. Still, some highway patrol (including North Carolina's) and sheriff's offices continue operations in this frequency range on FM, so it's good that most scanners include it. Quite often, police officers don't realize skip can bring in signals from other parts of the country, and it can be amusing to hear them argue about "who owns this frequency."

Baby monitors (with which people essentially "bug" their own homes) and cordless telephones also transmit within this band. Given their very low power, if you hear them they are almost certainly close by, but if it's a cordless phone you should immediately tune past it since listening to it would be against the law.

108-136 MHz—This is the aviation band and almost all operations are in AM. Primarily used by civilian aircraft, this band is omitted from the many scanners that only have FM demodulators. This can be a very interesting band, even if you don't live near an airport. A commercial airliner at 30,000 feet has quite a range (line of sight). Several of the selectable mode

AM/FM scanners expand the upper end of this tuning range to 144 MHz, to receive the FM military mobile operations often found between 136-144 MHz, and some restrict the lower end of this tuning range to 118 MHz, since there is very little voice activity from 108-118 MHz. If you have a choice, get a scanner that covers this range since many airports broadcast continuous weather (*ATIS* —automated terminal information system) on some of these frequencies.

150-174 MHz—This is called the VHF High band and is the most active and diverse of the scanner bands. Within this range FBI, Secret Service, police, fire, and rescue squad stations can be found, as well as less exciting business, taxi, paging, railroad and other general purpose radio services. In addition, the marine VHF frequencies fall within this range. Transmissions are primarily FM.

225-400 MHz—Military operations in AM, primarily aviation, are found in this range that is omitted on almost all but most expensive and continuous coverage scanners.

406-420 MHz—Operations by the federal government can be found in this frequency range, including transmissions from Air Force One, the Bureau of Alcohol, Tobacco and Firearms and the FBI. This FM band is common on most recent vintage scanners.

450-470 MHz—Law enforcement agencies predominate on FM in this range, although some general purpose radio services can also be found here. There is some migration from this band to 800 MHz trunked systems (discussed below), but an enormous amount activity remains on here.

470-512 MHz—This is called the T-band, because the primary allocation in this range is for UHF TV. Rarely used for general radio service, scanner coverage here is most often used to receive the FM audio for TV channels 14-20. Occasionally a wireless microphone's transmissions can be heard on these frequencies.

806-960 MHz—This is probably the fastest growing area of the scanning spectrum, but some of the less expensive scanners don't cover it. Coverage of this range is quite rare in scanners more than 10 years old.

Cellular phones occupy parts of this range, along the high end of the UHF TV channels. As was previously mentioned, it is illegal to listen to cellular phone transmissions, which is why so many people are now so interested in it. "Informed sources" claim that most cellular phone conversations are boring, although occasionally details of a drug deal or an illicit romantic rendezvous can be overheard.

The Personal Communication Systems (PCS) that are currently being marketed in metropolitan areas also operate in this frequency range. However, their signals are digital in nature and not decipherable with equipment that is readily available.

The latest (and most expensive) cordless phones also operate in this range. Some of these units use a digital or spread-spectrum transmission method, which makes them almost impossible to monitor. Regardless of the mode, however, it is illegal to listen to them.

TRUNKED RADIO

Many municipalities are installing "trunked radio systems" that operate in the 800 MHz frequency range. For reasons beyond the scope of this article, a trunked system uses radio spectrum more efficiently than conventional systems and can provide for better communication between participating agencies. In a trunked system, many agencies may share a group of frequencies, but through a sophisticated data transmission system (much like packet radio), each agency hears only the radio traffic considered appropriate for their mission.

Trunked systems can be difficult to monitor, because frequencies are often switched. For example, in a trunked system, a given conversation between a police officer and dispatcher may switch frequencies every time the mike button is released. If there is only one conversation taking place, a fast scanner may find the next active frequency rapidly enough for the conversation to be monitored essentially continuously, but if other conversations are underway the scanner may stop on a different conversation. If the listener quickly presses the "scan" button and can recognize the voices, the monitoring can be continued reasonably well.

The trunking system by GE/Ericsson transmits a series of tones after each transmission specifically to make it more difficult for scanners to follow the activity. Again, if the listener remains actively involved in operating the scanner, the conversation can be followed, but it isn't easy.

At least one company (Uniden) is offering a scanner that can decode the data signals and follow the conversations of the trunked systems manufactured by Motorola. Another company sells add-on circuitry for scanners that will help defeat the GE/Ericsson trunking tones, but nothing yet that will actively follow frequency changes. Stay tuned for more changes and more capability.

EQUIPMENT

Multi-band, programmable scanners are so prevalent and reasonably priced now that crystal controlled scanners almost have to be given away at hamfests. Even if someone would give you a 16 channel crystal controlled scanner, by the time you bought the crystals for it (if none of the existing crystals could be used) you would have spent almost enough to purchase a good low-end programmable rig.

The one exception to this is the avid scanner listener who wants to constantly monitor just one frequency. For instance, the hobbyist may want to hear every transmission on the police or fire dispatch channel, and yet continue to listen to other activity. In this case, a crystal controlled scanner with just one crystal can be used to cost effectively monitor the critical frequency continuously, and programmable scanners can be used for all other listening.

Programmable scanners are generally priced from $100 to $500, but well-heeled enthusiasts are generally opting for a rig in the burgeoning "DC-to-daylight" category (see sidebar). The favorites among the traditional desktop scanners (with no shortwave capability) are the Radio Shack PRO-2042 and the Uniden Bearcat BC-9000XLT.

Within the moderate price range, more expensive units are characterized by broader frequency coverage and more channels. At the top of this range, common features include a priority channel and frequency search capability, which allows for general band scans to seek out new active frequencies. Most of the popular scanners have reasonable sensitivity and selectivity, so the choice of which to buy is reduced to frequency coverage, number of channels and the user convenience features. A number of amazingly high performance scanners are now available as hand-held (walkie-talkie type) units, and definitely merit your consideration if portability would be useful. A few to consider are the Bearcat "Scancat" 150, and the Radio Shack PRO-60, both of which can be found for under $200.

Some good sources for scanners, frequency lists and scanner accessories are Radio Shack, Communications Electronics (PO Box 1045, Ann Arbor, MI 48101), Electronic Equipment Bank (137 Church St. NW, Vienna, VA 22180 and Scanner World (10 New Scotland Ave., Albany, NY 12208). Advertisements from these and other scanner dealers can be found in *Popular Commu-*

nications and *Monitoring Times*. The addresses for these two magazines can be found in the section on "Print Resources for the Shortwave Listener."

ANTENNAS

Radio hobbyists show a pronounced tendency to spend lots of time and money on their equipment, and a relatively small amount of time and money on their antennas. Most scanners are sold with a whip antenna that attaches to the radio and many scanner listeners never get beyond using it. This type of antenna would be only marginally effective in the best of locations, but because they are attached to the radio they can't be placed to best effect. At least they usually are oriented vertically, which corresponds to the vertical polarization of most VHF transmissions.

An outdoor antenna can enable you to listen to transmissions that are totally impossible to receive on an indoor whip. An outdoor antenna at 30 feet may provide a range of 50 to 60 miles, compared to an indoor whip's range of 5 to 10 miles.

If you are active on the UHF/VHF ham bands, you may already know that selecting the proper coax is increasingly important as you go to the higher frequencies. Certain types of coax, such as the popular RG-58, literally "soaks up" (attenuates) signals in the UHF/VHF range, with signal loss getting worse the higher in frequency you go. Coax that is marginally useful on the 2 meter band can be terrible on 440 MHz and virtually worthless if you are monitoring in the 800 MHz range. If you are going to need more that 30 feet or so of coax, consult a ham knowledgeable in UHF operation or check the *ARRL Handbook* or *ARRL Antenna Book* for information on what types of coax are best for the UHF ranges.

Despite their usefulness, there are times when you just can't use an outdoor antenna. Apartment dwellers and residents of areas with restrictive covenants have to contend with this situation constantly, but even for those with outdoor access there are times when using an outdoor antenna isn't prudent.

Thunderstorms present a particular dilemma. Leaving the scanner connected to an outside antenna during a thunderstorm is an open invitation for lightning damage, but the police, fire and rescue frequencies are particularly active then. The solution is an effective indoor antenna. In this case, the antenna supplied with the radio can allow you to monitor the activity while providing some protection for your valuable radio.

THE NEWS WHILE IT HAPPENS

Monitoring the VHF/UHF scanner frequencies can keep you in touch with breaking events, inform you about local happenings that might not make the news and open up a whole new world of radio activity. Today, no radio shack can be considered well equipped unless it includes *at least* one scanner.

THIS IS JUST THE BEGINNING

If the foregoing has sparked your interest in shortwave listening, then this is just the beginning for you. Amateur Radio in its broadest sense, the enjoyment of all facets of radio, is a tremendously rewarding and educational hobby, and the shortwave listening portion of the hobby is a vital part of it.

Shortwave listening has the appeal to attract people of all types, from nationally renowned figures such as Jackie Gleason and David Letterman to the kid down the block. Many hams were once the "kid down the block," and shortwave listening was their entrance to Amateur Radio—and remains a constant and immensely satisfying part of it.

Experience all that SWLing has to offer. Find a radio and give the shortwave spectrum a try. The world awaits you.

CHAPTER 2

The Amateur Radio Spectrum

PAUL RINALDO, W4RI
ARRL WASHINGTON OFFICE

The electromagnetic spectrum is a limited resource. Every kilohertz of the radio spectrum represents precious turf that is blood sport to those who lay claim to it. Fortunately, the spectrum is a non-depletable resource—one that if misused can be restored to normal as soon as the misuse stops. Every day, we have a fresh chance to use the spectrum intelligently.

Simply because the radio spectrum has been used in a certain way, changes are not impossible. Needs of the various radio services evolve with technological innovation and growth. There can be changes both in the frequency-band allocations made to the Amateur Service and how we use them. Also, to our benefit, signals transmitted from one spot on Earth propagate only to certain areas, so specific frequencies may be reused numerous times throughout the globe. Amateur Radio is richly endowed with a wide range of bands starting at 1.8 MHz and extending to above 300 GHz. Thus, we enjoy a veritable smorgasbord of bands with propagational "delicacies" of every type—direct wave, ground wave, sky wave, tropospheric scatter, meteor scatter—and we can supplement this with moon reflections and active artificial satellites.

Why do you need to know any of this? Well, you don't. But if you know the history of our frequency allocations, who the players are and what the process is, you may be able to help influence the process in the future.

INTERNATIONAL REGULATION OF THE SPECTRUM

Amateur Radio frequency band allocations don't just happen. Band allocation proposals must first crawl through a maze of national agencies and the International Telecommunication Union (ITU) with more adroitness than a computer-controlled mouse. Simultaneously, the proposals have to run the gauntlet of the competing interests of other spectrum users.

Treaties and Agreements

To bring some order to international relationships of all sorts, nations sign treaties and agreements. Otherwise (with respect to international communications) chaos, anarchy and bedlam would vie for supremacy over the radio spectrum. Pessimists think we already have some of that, but they haven't any idea how bad it could be *without* international treaties and agreements. The primary ones that affect the Amateur Service are:

• The International Telecommunication Convention, signed at Nairobi on November 6, 1982.

• The Radio Regulations annexed to the International Telecommunication Convention were signed at Geneva on December 6, 1979, and entered into force on January 1, 1982.

• Partial revisions of the Radio Regulations, relating to space and radio astronomy, were signed at Geneva on November 8, 1963, and July 17, 1971. They became effective on January 1, 1965, and January 1, 1973, respectively.

• The United States-Canada Agreement on Coordination and Use of Radio Frequencies above 30 MHz was agreed to in an exchange of notes at Ottawa on October 24, 1962. A revision to the Technical Annex to the Agreement, made in October 1964 at Washington, was put into effect in a note signed by the United States on June 16, 1965. Canada signed the note on June 24, 1965. It became effective on June 24, 1965. Another revision to this Agreement to add Arrangement E (between the Department of Communications of Canada and the National Telecommunications and Information Administration and the Federal Communications Commission of the United States on the use of the 406.1 to 430 MHz band in Canada-US Border Areas) was made by an exchange of notes signed by the United States on February 26, 1982, and Canada on April 7, 1982.

The International Telecommunication Union

The origins of the International Telecommunication Union (ITU) trace back to the invention of the telegraph in the 19th century. To establish an international telegraph network, it was necessary to reach agreement on uniform message handling and technical compatibility. Bordering European countries worked out some bilateral agreements. This eventually led to creation of the ITU at Paris in 1865 by the first International Telegraph Convention, which yielded agreement on basic telegraph regulations.

Starting with a conference in Berlin in 1906, attention focused on how to divide the radio spectrum by specific uses to

Glossary of Spectrum Management

Parenthetical abbreviations following definitions indicate the following sources:
- (RR) ITU *Radio Regulations*
- (NTIA) National Telecommunications and Information Administration *Manual of Regulations and Procedures for Federal Radio Frequency Management*

Access—Freedom to use, or protocol for using, the spectrum. (See Code, Frequency and Time Division Multiple Access.)

Adjacent Channel—The channel immediately above or below the reference channel.

AIRS—ARRL Interference Reporting System—collection of information regarding out-of-band transmissions in the amateur bands (formerly called Intruder Watch).

Alligator—An unbalanced repeater with a "big mouth" (high-power transmitter) and "small ears" (insensitive receiver).

Allocation (of a frequency band)—Entry in the Table of Frequency Allocations of a given frequency band for the purpose of its use by one or more (terrestrial or space) radiocommunication services or the radio astronomy service under specified conditions. This term shall also be applied to the frequency band concerned. (RR)
Note: Allocations are distributions of frequency bands to specific radio services. On the highest level, allocations are made by competent (legally qualified) World Administrative Radio Conferences. Certain specific uses for the spectrum are grouped together so that the probability of interference between major types of radio services is minimal. Domestic (US) allocations are made by the FCC and NTIA in coordination within the limits of the *Radio Regulations* and using broad principles... (Feller, *Planning an Electromagnetic Environment Model for Spectrum Management*, 1981, FCC)

Allotment (of a radio frequency or radio frequency channel)—Entry of a designated frequency channel in an agreed plan, adopted by a competent Conference, for use by one or more administrations for a (terrestrial or space) radiocommunication service in one or more identified countries or geographical areas and under specified conditions. (RR)
Note: Allotments are distributions of frequencies to areas or to countries. They provide for the orderly use of certain frequency bands. (Feller, ibid)

Amateur-Satellite Earth Station—An earth station in the Amateur-Satellite Service.

Amateur-Satellite Service—A radiocommunication service using space stations on earth satellites for the same purposes as those of the Amateur Service. (RR)

Amateur-Satellite Space Station—A space station in the Amateur-Satellite service.

Amateur Service—A radiocommunication service for the purpose of self-training, intercommunication and technical investigation carried out by amateurs, that is, by duly authorized persons interested in radio technique solely with a personal aim and without pecuniary interest. (RR)

Amateur Station—A station in the Amateur Service. (RR)

AMSL—Height above mean sea level.

Amplitude Compandored Single Sideband (ACSSB)—A single-sideband transmission system which uses a pilot to tell the receiver how much expansion is required at each instant, in order to recover the dynamic range of the original signal.

Assigned Frequency—The center of the Frequency Band assigned to a station.

Assigned Frequency Band—The frequency band within which the emission of a station is authorized; the width of the band equals the necessary bandwidth plus twice the absolute value of the frequency tolerance. Where space stations are concerned, the assigned frequency band includes twice the maximum Doppler shift that may occur in relation to any point of the Earth's surface. (RR)

Assignment (of a radio frequency or radio frequency channel)—Authorization given by an administration for a radio station to use a radio frequency or radio frequency channel under specified conditions. (RR)
Note: Assignments are frequency distributions to specific stations. When making assignments, administrations authorize a station to use one or more specific frequencies under specified conditions. (Feller, ibid) With rare exceptions, the FCC does not assign specific frequencies within the Amateur Service but permits amateur stations to operate on any frequency within allocated frequency bands.

Attenuation Ratio—The magnitude of the path loss.

Authorized Bandwidth—Authorized bandwidth is . . . the necessary bandwidth (bandwidth required for transmission and reception of intelligence) and does not include allowance for transmitter drift or Doppler shift. (NTIA)

Automatic Control—The use of devices and procedures for control of a station when it is transmitting so that compliance with the FCC Rules is achieved without the control operator being present at a control point. (FCC Rules, Section 97.3)

Automatic Power Control—Dynamic control of transmitter power level in order to limit interference to other stations and allow frequency reuse.

Automatic Transmitter Identification System (ATIS)—A device built into a transmitter that sends a unique identifier without action by the operator.

Availability—The characteristic denoting whether the resource is ready for immediate use, or the probability related to a period of time.

Balanced System—A system, usually a repeater, whose transmitting and receiving subsystems are designed for the same quality signal at the same range. (See *Alligator* and *Rabbit*.)

Band—(1) A range of frequencies. (2) One of the ranges of frequencies allocated to the Amateur Service.

Bandwidth—The width of a frequency band outside of which the mean power of the total emission is attenuated at least 26 dB below the mean power of the total emission, including allowances for transmitter drift or Doppler shift. (FCC Rules, Section 97.3)

Beacon, Engineering—A transmission or frequency used to convey technical information about the host system. In a satellite, it provides data on the health and welfare of the spacecraft.

Beacon (packet radio)—Periodic transmission of a frame giving station identification information—an early practice now deprecated as it adds to channel congestion.

Beacon, Propagation—A transmitting station that continuously or periodically emits a signal to permit listeners to determine whether a path is open and its quality.

Beam (antenna)—The major lobe of an antenna radiation pattern.

Beamwidth (antenna)—The width in degrees azimuth between the half-power points in the major lobe of the antenna radiation pattern.

Broadband—(See *Wideband*.)

Broadcasting—Transmissions intended for reception by the general public, either direct or relayed. (FCC Rules, Section 97.3)

Calling Frequency—A frequency used only for establishing contact, after which the stations change to another frequency to carry out their communication. Some band plans identify national and DX calling frequencies.

Carrier—A continuous radio frequency capable of being modulated by a baseband signal or subcarrier.

Carrier Power (of a radio transmitter)—The average power supplied to the antenna transmission line by a transmitter during one radio frequency cycle taken under the condition of no modulation. (RR)

Carrier-to-Noise Ratio—The ratio of the amplitude of the carrier vs the amplitude of noise, usually expressed in dB.

CCIR—International Radio Consultative Committee, a former agency of the International Telecommunication Union, replaced by ITU-R.

CCITT—International Telegraph and Telephone Consultative Committee, a former agency of the International Telecommunication Union, replaced by ITU-T.

Cell—A single service zone within a geographical grid. The network often has the structure of a honeycomb. This is the basis of the cellular telephone system which has base stations in the center of each cell to serve mobiles throughout the cell. Individual cells, if regular shapes, may be hexagonal, circular or elliptical. Cell shapes also may be modified according to terrain.

Cellular—A UHF radio system wherein groups of frequencies in a cell are reused in others in a controlled pattern and wherein a central computer is used to hand off control of a given mobile unit from cell to cell by means of landline trunks.

Centimetric Waves—Radio frequencies between 3 and 30 GHz. (SHF)

Channel—(1) A one- or two-way transmission path divided from any others in frequency or time. (2) A band of frequencies designated for a specific use.

Channel Loading—The total occupancy of a channel by all users affected by each other's operation.

Channel Spacing—The frequency interval between successive (usually adjacent) channels, measured from center to center or between other characteristic frequencies.

Channel, Voice (Speech) Grade—A frequency channel that can faithfully pass frequencies in the 300-3000 Hz audio range.

Characteristic Frequency—A frequency which can be easily identified and measured in a given emission. A carrier frequency may, for example, be designated as the characteristic frequency. (RR) (See also *Reference Frequency*.)

CISPR—International Special Committee on Radio Interference, an organ of the International Electrotechnical Commission (IEC).

Class of Emission—The set of characteristics of an emission, designated by standard symbols, e.g., type of modulation, modulating signal, type of information to be transmitted, and also if appropriate, any additional signal characteristics. (RR)

Closed Repeater—A repeater where use by nonmembers is discouraged.

Cochannel Interference—Interference on the same channel.

Code Division Multiple Access (CDMA)—A technique to permit multiple stations to access a given band by using different spread spectrum codes.

Coded Squelch—A system wherein radio receivers are equipped with devices which allow audio signals to appear at the receiver output only when a carrier modulated with a specific signal is received. (NTIA)

Coordinated Station Operation—The repeater or auxiliary operation of an amateur station for which the transmitting and receiving frequencies have been implemented by the licensee in accordance with the recommendation of a frequency coordinator. (FCC Rules, Section 97.3)

Coordinated Universal Time (UTC)—Time scale, based on the second (SI), as defined and recommended by the CCIR and maintained by the International Time Bureau (BIH). For most practical purposes associated with the Radio Regulations, UTC is equivalent to mean solar time at the prime meridian (0° longitude), formerly expressed in GMT. (RR)

Coordination—The process of ascertaining from other users whether a proposed use of a radio frequency can occur without causing harmful interference.

Coordination Area—The area associated with an earth station outside of which a terrestrial station sharing the same frequency band neither causes nor is subject to interfering emissions greater than a permissible level. (RR)

Coordination Contour—The line enclosing the coordination area. (RR)

Coordination Distance—Distance on a given azimuth from an earth station beyond which a terrestrial station sharing the same frequency band neither causes nor is subject to interfering emissions greater than a permissible level. (RR)

Critical Frequency—The frequency below which vertically directed radio waves are reflected back to earth.

Crossband—Two-way communication in which the frequencies lie in two distinctly separate frequency bands. (Compare In-band.)

Decametric—Radio frequencies between 3 and 30 MHz. (HF)

Decimetric Waves—Radio frequencies between 300 MHz and 3 GHz. (UHF)

Deviation (frequency)—The difference in frequency from the center of the channel to the upper and lower excursions of the carrier frequency.

Disaster Communication—Emergency communication relating to an unforeseen calamity.

Doppler shift—The change in apparent frequency resulting from relative motion between the source and observer.

Drift (frequency)—An undesired change in frequency, usually as a result of temperature change in frequency-determining components.

Duplex Operation—Operating method in which transmission is possible simultaneously in both directions of a telecommunication channel. (RR)

Duty Cycle—The ratio of the *on* time to the total time of an intermittent operation. (FM is said to have a 100% duty cycle, as its transmitter is at full power, when transmitting, regardless of the amount of modulation.)

Dynamic Frequency Sharing (or real-time frequency management)—Operation on a secondary basis where there is no possibility of a claim for interference-free communication made possible with frequency-agile transmitting and receiving equipment. (CCIR Report WARC-92, Section 5.4)

Dynamic Range—The ratio to the maximum signal level (with a specified amount of distortion) to the noise level (or minimum signal level), usually expressed in dB.

Earth Station—A station located either on the Earth's surface or within the major portion of the Earth's atmosphere and intended for communication:
—with one or more space stations; or
—with one or more stations of the same kind by means of one or more reflecting satellites or other objects in space. (RR) An amateur station located on, or within 50 km of, the Earth's surface intended for communications with space stations or with other Earth stations by means of one or more objects in space. (FCC Rules, Section 97.3)

Effective Isotropic Radiated Power (EIRP)—The product of the power supplied to the antenna and its gain relative to an isotropic antenna.

Equivalent Isotropically Radiated Power (EIRP or e.i.r.p.)—The product of the power supplied to the antenna and the antenna gain in a given direction relative to an isotropic antenna (absolute or isotropic gain). (RR)

Electromagnetic Compatibility (EMC)—The ability of electronic equipment to function in its environment without introducing additional disturbances into that environment detrimental to other electronic equipment.

Emergency Communication—Any Amateur Radio communication directly relating to the immediate safety or life of individuals or the immediate protection of property. (FCC Rules, Section 97.3)

Emission—Radiation produced, or the production of radiation, by a radio transmitting station. (RR)

European Council of Postal and Telecommunications Administrations (CEPT)—The European regional organization of post and telecommunications organizations.

Extremely High Frequencies (EHF)—Radio frequencies between 30 and 300 GHz. (Millimetric)

Facsimile—A form of telegraphy for the transmission of fixed images, with or without half-tones, with a view to their reproduction in a permanent form. In this definition the term telegraphy has the same general meaning as defined in the Convention. (RR)

Fading—Changes in signal strength generally caused by variations in the propagation medium.

Fading, Flat—Fading in which all frequencies within a channel fade together.

Fading, Selective—Fading in which frequencies within a channel fade somewhat independently.

Frequency Agility—The ability to shift the frequency of a transmitter and/or receiver rapidly to achieve a desired effect.

Frequency Coordinator—An entity, recognized in a local or regional area by amateur operators whose stations are eligible to be auxiliary or repeater stations, that recommends transmit/receive channels and associated operating and technical parameters for such stations in order to avoid or minimize potential interference. (FCC Rules, Section 97.3)

(Glossary continued on next page)

Frequency Diversity—A form of diversity reception using simultaneous transmission on two or more frequencies to take advantage of the tendency that the different frequencies fade independently.

Frequency Division Multiple Access (FDMA)—A technique to permit multiple stations to access a given band by breaking the band into channels.

Frequency, Input—(repeater usage) The center frequency to which the repeater receiver is tuned.

Frequency Offset—The difference (in hertz) between two frequencies, such as: (a) between one used for transmitting and one used for receiving, (b) between a beat-frequency and the signal of interest in a receiver, or (c) between a specific transmission frequency and the nominal channel frequency.

Frequence Optimum de Travail (FOT)—A radio frequency 15% below the maximum usable frequency that will provide communications 90% of the days. (Also called *Optimum Working Frequency, OWF*.)

Frequency, Output—(repeater usage) The center frequency to which the repeater transmitter is tuned.

Frequency Registration—The act of making a written record of frequency assignment or usage in a central registry or in a published form for distribution.

Frequency Reuse—The ability to use a frequency in a different geographical location without harmful interference.

Frequency Sharing—The common use of the same portion of the radio frequency spectrum by two or more users where a probability of interference exists. (NTIA)

Frequency-Shift Telegraphy—Telegraphy by frequency modulation in which the telegraph signal shifts the frequency of the carrier between predetermined values. (RR)

Frequency Tolerance—The maximum permissible departure by the center frequency of the frequency band occupied by an emission from the assigned frequency or, by the characteristic frequency of an emission from the reference frequency. The frequency tolerance is expressed in parts in 10^6 or in hertz. (RR)

Full Carrier Single-Sideband Emission—A single-sideband emission without suppression of the carrier. (RR)

Geostationary Satellite—A geosynchronous satellite whose circular and direct orbit lies in the plane of the Earth's equator and which thus remains fixed relative to the Earth; by extension, a satellite which remains approximately fixed relative to the Earth. (RR)

Geostationary-satellite Orbit (GSO)—The orbit in which a satellite must be placed to be a geostationary satellite. (RR)

Geosynchronous Satellite—An earth satellite whose period of revolution is equal to the period of rotation of the Earth about its axis. (RR)

Guard Band—A narrow band of frequencies surrounding a (usually weak-signal) channel (or band) that are not to be used in order to prevent interference.

Guard Channel (or Frequency)—A (calling) frequency designated to be monitored continuously or within scheduled times.

Harmful Interference—Interference which endangers the functioning of a radionavigation service or of other safety services or seriously degrades, obstructs, or repeatedly interrupts a radiocommunication service operating in accordance with these Regulations.

Hectometric Waves—Radio frequencies between 300 kHz and 3 MHz. (MF)

High Band—VHF FM jargon for the 2-meter Amateur Radio band.

Highest Possible Frequency (HPF)—A radio frequency 15% above the maximum usable frequency, considered to be usable 10% of the days.

High Frequency (HF)—Radio frequencies between 3 and 30 MHz. (Decametric)

Image Frequency—(superheterodyne receiver) An undesired signal frequency that differs from the desired frequency by 2 times the intermediate frequency (IF).

In-band—Two or more frequencies within the same audio- or radio-frequency band.

Independent Sideband—Modulation of a carrier in which each sideband carries different intelligence.

Interference—Unwanted signals. (See *Harmful Interference*)

International Amateur Radio Union (IARU)—The federation of national Amateur Radio societies worldwide founded in Paris in 1925. It has three regional organizations: Region 1) Africa, Europe, the Commonwealth of Independent States, Middle East (excluding Iran) and Mongolia; Region 2) North and South America, including Hawaii and the Johnston. Its International Secretariat is located in Newington, Connecticut.

Inter-American Telecommunication (CITEL)—The regional telecommunications organization of the Americas, with headquarters in Washington, DC.

International Frequency Registration Board (IFRB)—Formerly, an ITU agency that recorded frequency assignments made by different countries, replaced by RRB.

Intersymbol Interference (ISI)—Extraneous energy from the signal in one keying interval that interferes with another keying interval. (Data transmission)

Inverse-Square Law—(radio) The power varies with distance inversely as the square of that distance (a basic law of free-space propagation which says that as the distance is doubled, the signal will be one fourth or 6 dB weaker).

Jamming—Intentional radiation of a signal for the purpose of imparing the usefulness of radio reception.

Key Clicks—Undesired switching transients beyond the necessary bandwidth of a Morse code radio transmission caused by improperly shaped modulation envelopes.

Kilometric Waves—Radio frequencies between 30 and 300 kHz. (LF)

Left-Hand (or Anti-Clockwise) Polarized Wave—An elliptically or circularly polarized wave, in which the electric field vector, observed in the fixed plane, normal to the direction of propagation, whilst looking in the direction of propagation, rotates with time in a left-hand or anti-clockwise direction. (RR)

Low Band—VHF FM jargon for the 6-meter Amateur Radio band. HF jargon for the 160, 80 or 40-meter bands.

Low Earth Orbit (LEO)—A nongeostationary satellite orbit close to the Earth.

Lowest Usable (High) Frequency (LUF)—The minimum radio frequency that will provide satisfactory ionospheric communication between two points 90% of the days.

Low Frequencies (LF)—Radio frequencies between 30 and 300 kHz. (Kilometric)

Machine—VHF FM jargon for a repeater.

Maximum Usable Frequency (MUF)—The highest radio frequency that will provide satisfactory ionospheric communication between two points at a given time 50% of the days.

Mean Power (of a radio transmitter)—The average power supplied to the antenna transmission line by a transmitter during an interval of time sufficiently long compared with the lowest frequency encountered in the modulation taken under normal operating conditions. (RR)

Medium Frequencies (MF)—Radio frequencies between 300 kHz and 3 MHz. (Hectometric)

Meteor Burst Communications—Communications by the propagation of radio signals reflected by ionized meteor trails. (NTIA)

Meteor Scatter—Propagation via ionized meteor trails.

Metric Waves—Radio frequencies between 30 and 300 MHz. (VHF)

Microcell—A small cell covering a few city blocks.

Microwaves—Radio frequencies above 1 GHz.

Millimetric Waves—Radio frequencies between 30 and 300 GHz. (EHF)

Modulation—Variation of one signal with another, usually to convey intelligence.

Multipath—Propagation of a radio signal via two or more distinct paths, such that signals (at certain frequencies and times) combine out of phase and produce a distorted output in the receiver.

Myriametric Waves—Radio frequencies between 3 and 30 kHz. (VLF)

Narrow-band—(1) A band of frequencies which are smaller than a reference bandwidth. (2) A band of frequencies which lie within the passband of a receiver or measuring device.

National Radio Quiet Zone—The area in Maryland, Virginia and West Virginia bounded by 39° 15'N on the north, 78° 30'W on the east, 37° 30'N on the south and 80° 30'W on the west. (FCC Rules, Section 97.3)

Necessary Bandwidth—For a given class of emission, the width of the frequency band which is just sufficient to ensure the transmission of information at the rate and with the quality required under specified conditions. (RR)

Net Control Station—A radio station responsible for real-time direction of a radio net, including maintenance of circuit discipline necessary to expeditious handling of traffic.

Net Manager—An individual responsible for the mission and resource management of a radio net.

NTIA—National Telecommunications & Information Administration of the US Department of Commerce.

Occupied Bandwidth—The width of a frequency band such that, below the lower and above the upper frequency limits, the mean powers emitted are each equal to a specified percentage B/2 of the total mean power of a given emission. Unless otherwise specified by the CCIR for the appropriate class of emission, the value of B/2 should be taken as 0.5%. (RR)

Open Repeater—A repeater where transient operators are welcome.

Optimum Working Frequency (OWF)—(Same as *Frequence Optimum de Travail, FOT*)

Out-of-band Emission—Emission on a frequency or frequencies immediately outside the necessary bandwidth which results from the modulation process, but excluding spurious emission. (RR)

Paging—One-way signaling to portable receivers for alerting or transmission of brief messages.

Pair (frequency)—Associated input and output frequencies.

Peak Envelope Power (of a radio transmitter)—The average power supplied to the antenna transmission line by a transmitter during one radio frequency cycle at the crest of the modulation envelope taken under normal operating conditions. (RR)

Period (of a satellite)—The time elapsing between two consecutive passages of a satellite through a characteristic point on its orbit. (RR)

Pilot—A signal to control the characteristics of a transmission. (For example, a 3.1-kHz pilot is used for automatic frequency control reference and to indicate the amount of expansion needed upon reception of an amplitude compandored single-sideband signal.)

Polarization—The alignment of the electric lines of force of a wave with respect to earth. If the lines are perpendicular to earth, the polarization is said to be vertical; if parallel to earth, horizontal. Polarization may also be elliptical or circular (right-hand or left-hand).

Polarization Discrimination—Reduction in interference between two systems using antenna polarizations that are orthogonal to each other. Examples: one antenna vertical, the other horizontal; one antenna right-hand circular, the other left-hand circular.

Polarization Diversity—A form of diversity reception using separate vertically and horizontally polarized antennas to take advantage of the tendency of vertical and horizontal components to rotate and thus fade independently.

Power Flux-Density (p.f.d.)—Power density per unit cross-sectional area per unit of frequency. Usually expressed in $dB(W/m^2/Hz)$, although bandwidths other than 1 Hz, such as 4 kHz or 1 MHz, may be seen.

Propagation—The motion of a radio wave.

Protection Ratio—The minimum value of the wanted-to-unwanted signal ratio, usually expressed in decibels, at the receiver input determined under specified conditions such that a specified reception quality of the wanted signal is achieved at the receiver output. (RR)

Rabbit—An unbalanced repeater with "big ears" (sensitive receiver) and "small mouth" (low-power transmitter).

Radiation—The outward flow of energy from any source in the form of radio waves. (RR)

Radiocommunication—Telecommunication by means of radio waves. (RR)

Radiocommunication Service—A service . . . involving the transmission, emission and/or reception of radio waves for specific telecommunication purposes. In these regulations, unless otherwise stated, any radiocommunication service relates to terrestrial radiocommunication. (RR)

Radio Regulations Board (RRB)—A component of the ITU that records frequency assignments made by different countries.

Raster—The pattern of spacing of channels in a specific group or band.

Reciprocity—(1) (radio propagation) The ability of radio signals to traverse the same path in both directions with the same results. (2) (mobile radio) Transmitting and receiving subsystems that have equal capability to work over the same path loss. (See *Balanced System*.)

Reduced Carrier Single-Sideband Emission—A single-sideband emission in which the degree of carrier suppression enables the carrier to be reconstituted and to be used for demodulation. (RR)

Reference Frequency—A frequency having a fixed and specific position with respect to the assigned frequency. The displacement of this frequency with respect to the assigned frequency has the same absolute value and sign that the displacement of the characteristic frequency has with respect to the center of the frequency band occupied by the emission. (RR) (See also *Characteristic Frequency*.)

Roaming—Moving between cells. (cellular radio)

Roving—Moving between grid squares. (amateur VHF-microwave)

Scanner—A radio receiver that can be programmed to sample different frequencies in a systematic sequence.

Selectivity—The ability of a radio receiver to discriminate against frequencies outside the desired passband.

Service Area—The area served by a radio station (typically a broadcast transmitter, base station or repeater).

Shift (frequency)—The difference between the frequency of the mark and space frequencies in frequency-shift-keyed modulation.

Sideband—The group of frequencies above or below the main carrier frequency which contain the product of modulation.

Signal-to-Interference (S/I) Ratio—The ratio of the amplitude of the desired vs interfering signals.

Signal-to-Noise Ratio—The ratio of the amplitude of the signal vs noise, usually expressed in dB.

Silent Period—A designated time during which all stations must cease transmission and listen for distress traffic.

Simulcast—Simultaneous transmission on two or more frequencies or modulation modes. (Also known as quasi-synchronous transmission.)

Single-Sideband Emission—An amplitude modulated emission with one sideband only. (RR)

Site Noise—Noise, external to the receiver, that limits reception at a location—a composite of co-channel interference, power-line noise, machine emissions, and spurious emissions from transmitters.

Space Station—A station located on an object which is beyond, intended to go beyond, or has been beyond, the major portion of the Earth's atmosphere. (RR) An amateur station located more than 50 km above the Earth's surface. (FCC Rules, Section 97.3)

Space Telecommand—The use of radiocommunication for the transmission of signals to a space station to initiate, modify or terminate functions of equipment on an associated space object, including the space station. (RR)

Space Telemetry—The use of telemetry for the transmission from a space station of results of measurements made in a spacecraft, including those relating to the functioning of the spacecraft. (RR)

Spectrum Efficiency—The ratio of the information delivered to the spectrum resources (bandwidth, time and physical space) used.

Spectrum Engineering—The science of effective use of the radio spectrum to achieve the desired results economically.

(Glossary continued on next page)

Spectrum Management—Efficient and judicious use of the radio spectrum through application of techniques such as spectrum engineering, frequency coordination, monitoring and measurement, computer data base, and analysis.

Spectrum Metric—A unit of measure of spectrum-resource use.

Spectrum Occupancy—A frequency is said to be occupied when the signal strength exceeds a given threshold. Spectrum Occupancy is normally expressed in terms of probability or as a percent.

Splatter—A form of adjacent-channel interference caused by overmodulation on peaks.

Spread Spectrum—A signal structuring technique that employs direct sequence, frequency hopping or a hybrid of these, which can be used for multiple access and/or multiple functions. This technique decreases the potential interference to other receivers while achieving privacy and increasing the immunity of spread spectrum receivers to noise and interference. Spread spectrum generally makes use of a sequential noise-like signal structure to spread the normally narrowband information signal over a relatively wide band of frequencies. The receiver correlates the signals to retrieve the original information signal. (NTIA)

Spurious Emission—Emission on a frequency or frequencies which are outside the necessary bandwidth and the level of which may be reduced without affecting the corresponding transmission of information. Spurious emissions include harmonic emissions, parasitic emissions, intermodulation products and frequency conversion products, but exclude out-of-band emissions. (RR)

Station—One or more transmitters or receivers or a combination of transmitters and receivers, including the accessory equipment, necessary at one location for carrying on a radiocommunication service, or the radio astronomy service. Each station shall be classified by the service in which it operates permanently or temporarily. (RR)

Subband—A part of a (frequency) band, usually having the same purpose or operator privileges.

Super High Frequencies (SHF)—Radio frequencies between 3 and 30 GHz. (Centimetric)

Suppressed Carrier Single-Sideband Emission—A single-sideband emission in which the carrier is virtually suppressed and not intended to be used for demodulation. (RR)

Telecommand—The use of telecommunication for the transmission of signals to initiate, modify or terminate functions of equipment at a distance. (RR)

Telecommunication—Any transmission, emission or reception of signs, signals, writing, images and sounds or intelligence of any nature by wire, radio, optical or other electromagnetic systems. (RR)

Telegraphy—A form of telecommunication which is concerned in any process providing transmission and reproduction at a distance of documentary matter, such as written or printed matter or fixed images, or the reproduction at a distance of any kind of information in such a form. For the purposes of the Radio Regulations, unless otherwise specified therein, telegraphy shall mean a form of telecommunication for the transmission of written matter by the use of a signal code. (RR)

Telemetry—The use of telecommunication for automatically indicating or recording measurements at a distance from the measuring instrument. (RR)

Telephony—A form of telecommunication set up for the transmission of speech or, in some cases, other sounds. (RR)

Television—A form of telecommunication for the transmission of transient images of fixed or moving objects. (RR)

Time-Bandwidth Product—The duration of a transmission multiplied by the bandwidth of the signal—a measure of how much bandwidth is used for how long.

Time Division Multiple Access (TDMA)—A technique to permit multiple stations to access a given band by nonoverlapping time sequenced bursts.

Transmitter Power—The peak envelope power (output) present at the antenna terminals (where the antenna feed line, or if no feed line is used, the antenna, would be connected) of the transmitter. The term "transmitter" includes any external radio frequency power amplifier which may be used. Peak envelope power is defined as the average power during one radio frequency cycle at the crest of the modulation envelope, taken under normal operating conditions. (FCC Rules, Section 97.3)

Trunking—A channel-utilization scheme whereby two stations automatically can select an unused channel out of a group.

Tuning, Tune-Up—Emission of a signal for the purpose of adjusting a radio transmitter.

Ultra High Frequencies (UHF)—Radio frequencies between 300 MHz and 3 GHz. (Decimetric)

Unwanted Emissions—Consist of spurious emissions and out-of-band emissions. (RR)

Very High Frequencies (VHF)—Radio frequencies between 30 and 300 MHz. (Metric)

Very Low Frequencies (VLF)—Radio frequencies between 3 and 30 kHz. (Myriametric)

Vestigial Sideband (VSB)—The sideband of an AM transmission that has been substantially suppressed and which rolls off gradually with distance from the carrier frequency.

Wideband—(1) A band of frequencies which are larger than a reference bandwidth. (2) A band of frequencies which exceed the passband of a receiver or measuring device.

World Administrative Radio Conference (WARC)—Former term for ITU radio conferences, of which WARC-92 was the last. (See WRC)

World Radiocommunication Conference (WRC)—New term for ITU radio conferences starting in 1993.

Working Frequency—A frequency designated for exchange of communications after contact is made on a calling frequency or guard channel.

minimize interference. In 1927, the ITU Washington Radio Conference of 27 maritime states was the beginning of international regulation of Amateur Radio. At that conference, the Table of Frequency Allocations was first devised, and six harmonically related bands from 1.715 to 60 MHz were allocated to amateurs. Some of the amateur bands were revised in this and subsequent conferences held in Cairo (1938), Atlantic City (1947) and Geneva (1959), in consideration of the needs of other radio services.

In 1932, a Telegraph and Telephone Conference and a Radiotelegraph Conference in Madrid merged into a single International Telecommunication Convention. The Radio Regulations resulting from this conference explicitly prohibited amateurs from handling international third-party messages unless specifically permitted by the two countries involved. US amateurs may exchange third-party messages with the countries listed in Chapter 5.

In Atlantic City in 1947, radical changes were made in the organization of the Union. The ITU became a specialized agency of the United Nations—the oldest such agency—and ITU headquarters was transferred from Berne to Geneva.

At Nice in 1989, the ITU established a High Level Committee (HLC) to review the structure and functioning of the Union. They produced a report in 1991 entitled "Tomorrow's ITU: The Challenges of Change," that said the ITU must reorganize itself and adopt new working methods to keep pace with technological change. A new ITU Constitution and new Convention were approved at the Addition Plenipotentiary Conference, Geneva, 1992. The new organization and procedures were fully implemented by 1994. The new structure of the ITU is shown in Fig 2-1.

Plenipotentiary Conferences

The ITU has Plenipotentiary Conferences every four years. "Plenipotentiary" is a conference that is fully empowered to do business. The conferences determine general policies, review the work of the Union, revise the Convention if necessary, elect the Members of the Union to serve on the ITU Council, and elect the Secretary-General, Deputy Secretary-General, the Directors of the Bureaus and members of the Radio Regulations Board.

The Three ITU Sectors

The Telecommunication Standardization Sector (ITU-T) studies technical, operating and tariff questions, and issues recommendations relating to the telephone and data networks. World Telecommunications Standardization Conferences (WTSCs) are held every four years. Familiar recommendations issued by ITU-T include the V-series on modems and X-series on data protocols. The work of the sector is done through its Study Groups and the Telecommunication Standardization Bureau (TSB).

The Telecommunication Development Sector (ITU-D) primarily assists developing countries in modernizing their telecommunications infrastructure. Its World Telecommunication Development Conferences (WTDCs) meet every four years. It has two Study Groups and a Telecommunication Development Bureau (BDT).

The Radiocommunication Sector (ITU-R) revises the international Radio Regulations, studies technical and operational questions, and issues recommendations relating to radio communication. The organization of the Radiocommunication Sector is shown in Fig 2-2. Radiocommunication Bureau (BR) Director Richard C. Kirby, WØLCT, was succeeded by Robert W. Jones, VE3CTM, in January 1995. The League and the International Amateur Radio Union monitor the work of the Radiocommunication Sector primarily through participation in Study Group 8, the Radiocommunication Advisory Group (RAG), Radiocommunication Assembly (RA), Conference Preparation Meetings (CPMs) and World Radiocommunication Conferences (WRCs)—

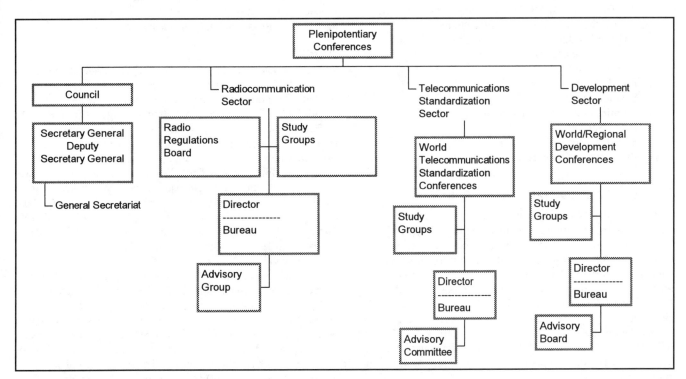

Fig 2-1—Organization chart of the International Telecommunication Union (ITU).

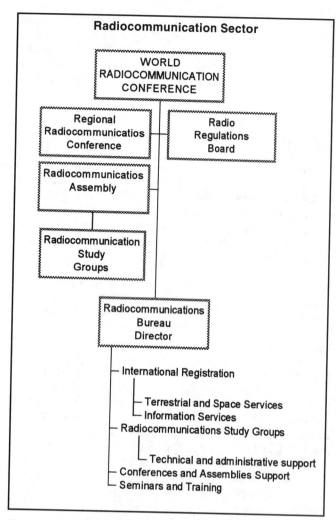

Radiocommunication Sector

WORLD RADIOCOMMUNICATION CONFERENCE

Regional Radiocommunicatios Conference

Radio Regulations Board

Radiocommunicatios Assembly

Radiocommunication Study Groups

Radiocommunications Bureau Director

— International Registration

— Terrestrial and Space Services
— Information Services
— Radiocommunications Study Groups

— Technical and administrative support
— Conferences and Assemblies Support
— Seminars and Training

Fig 2-2—Organization chart of the Radiocommunication Sector of the ITU.

now held every two years. ARRL contributes to the United States preparation for these ITU meetings and conferences.

Study Groups of the Radiocommunication Sector are:
1. Spectrum management
3. Radio-wave propagation
4. Fixed-satellite service
7. Science services
8. Mobile, radiodetermination and amateur services
9. Fixed Service
10. Broadcasting (sound)
11. Broadcasting (television)

World Administrative Radio Conferences

It is the Federal Communication Commission's responsibility to determine the spectrum requirements of the non-federal government users in the United States. In preparation for WARCs and WRC-95, the Commission solicited advice through a series of Notices of Inquiry (NOIs) and temporary industry Advisory Committees. The FCC then issued a Notice of Proposed Rule Making (NPRM) and a Report and Order (R&O) with its proposals for the conference. In the process, the Commission's proposals were coordinated with other federal agencies and the proposals to be sent to the ITU. A US delegation was formed to develop and implement strategies for acceptance of the US proposals at the conference.

WARC-79

Hundreds of individuals and clubs submitted comments at several stages of the proceeding. The League's comments alone totaled nearly 200 pages. The Commission's proposals were documented in a 560-page Report and Order.

Formal adoption of WARC-79 objectives for Amateur Radio came at a series of meetings of the IARU Regions in 1975-1976. At a meeting in 1976, IARU's then-President Eaton established an IARU worldwide working group which met several times. It produced a model position paper in English, French and Spanish--the three working languages of the ITU. The approach used in preparing for the WARC-79 proved to be sound. At the conference, Amateur Radio emerged with not only its existing bands but with important new allocations at 10, 18, and 24 MHz as well. There was relaxation of restrictions on some existing amateur bands, and additional segments were earmarked for amateur satellites.

WARC-92

There were a number of WARCs between 1972 and 1992 that dealt with the needs of certain radio services, but not affecting the amateur services. At WARC-92, held in Malaga-Torremolinos, Spain, the agenda included an item for possible expansion of bands used for HF broadcasting.

The League urged the FCC to propose realignment of the amateur and broadcasting bands around 7 MHz. This resulted in a US proposal to give amateurs in all three Regions the band 6900-7200 kHz and broadcasters a band above 7200 kHz. The proposal to give amateurs 300 kHz of spectrum worldwide was not on the agenda, as such, but thought to be possible as a consequence of satisfying the broadcasting agenda item. The conference declined to modify the allocations around 7 MHz but adopted a Recommendation that a future conference should consider a realignment.

World Radiocommunication Conferences

From 1993 forward, ITU World Radiocommunication Conferences (WRCs) were to be held every two years with agendas agreed at the previous WRC and confirmed by the ITU Council. The only function of the first WRC held in 1993 in Geneva was to agree on the agenda for WRC-95 and develop a tentative agenda for WRC-97.

WRC-95

WRC-95 had several purposes: to rewrite the Radio Regulations as proposed by the Voluntary Group of Experts (VGE), to make limited allocations to the mobile-satellite service (MSS), to set the agenda for WRC-97 and to prepare a tentative agenda for WRC-99.

For the amateur services, WRC-95 was mainly defensive— that is, protection of our allocations. As most of the proposals for new MSS allocations were of US origin, the defense of amateur and amateur-satellite bands against MSS industry proposals to share the 70-cm, 23-cm and 13-cm bands was accomplished mainly within the FCC's Advisory Committee Informal Working Groups through the Department of Defense, ARRL and AMSAT-NA. There was preliminary consideration of the 420-423 MHz band for allocation to MSS by countries in the Americas but effectively opposed by the US Government.

Although not on the WRC-95 agenda, the elimination of the Morse code requirement in the Radio Regulations for amateurs to be licensed to operate on frequencies below 30 MHz was introduced by New Zealand. Several country delegations supported the principle but insisted that it could not be treated because it wasn't on the agenda. The IARU delegation argued that it would need three years to have the matter considered by each of its three Regional organizations. Discussions by the delegations of New

Zealand, Germany, the US and IARU resulted in agreement that Article S25 (formerly Article 32, which contains the Morse code requirement) be placed on the tentative agenda for WRC-99.

WRC-97

As the FCC began its preparations for WRC-97 in early 1996, there was widespread agreement among those in government and industry that WRCs every two years was cause to change how the United States gets ready for radio conferences. It was recognized as an almost continuous process. The FCC Advisory Committee (AC) was made permanent, although its internal structure for Informal Working Groups (IWGs) can adapt to the agendas of coming WRCs. The age-old procedure of Notices of Inquiry was abandoned as too slow to meet the needs of the rapid pace of change in telecommunications. NOIs were replaced by IWG meetings open to the public, meeting announcements and minutes of past meetings on the FCC's World Wide Web pages, and a series of public notices announcing draft proposals that survived the IWG, AC and FCC staff review process.

The WRC-97 agenda calls for more MSS allocations, space services and HF broadcasting service allocations and setting the agenda for WRC-99. By May 1996, a spokesperson for MSS low-earth-orbit (LEO) satellites below 1 GHz (also known as "little LEO's") presented a list of candidate bands. This included the amateur 2-meter and 70-cm bands, and the ARRL strongly objected. The up-to-date status of this threat is carried in *QST* and on the League's World Wide Web page (**http://www.arrl.org**)

WRC-99

As this is written, WRC-99 is two years away and the agenda won't be known until the last days of WRC-97. Nevertheless, it is shaping up to be an important conference for the amateur services. First, there is likely to be a review of Article S25 and not just limited to the Morse code requirement. Secondly, this could be the conference that realigns the amateur and broadcasting bands around 7 MHz, the objective being to have a 300-kHz-wide band in all three Regions.

Inter-American Telecommunication Commission (CITEL)

CITEL is the regional telecommunications organization for the Americas. It is a permanent commission under the Organization of American States (OAS), with a secretariat in Washington, DC. The CITEL Assembly meets every four years, while its committees meet once or more times yearly. ARRL and IARU Region 2 are active participants in Permanent Consultative Committee no. 3 (PCC.III - Radiocommunications). Others are COM/CITEL (executive committee), PCC.I (public telecommunication networks) and PCC.II (broadcasting).

TELECOMMUNICATIONS REGULATION IN THE UNITED STATES

The Communications Act of 1934, as amended, provides for the regulation of interstate and foreign commerce in communication by wire or radio. This Act is printed in Title 47 of the US Code, beginning with Section 151.

Federal Communications Commission

The FCC is responsible, under the Communications Act, for regulating all telecommunications except that of the Federal government. This includes the Amateur Radio Service and the Amateur-Satellite Service, which are regulated under Part 97

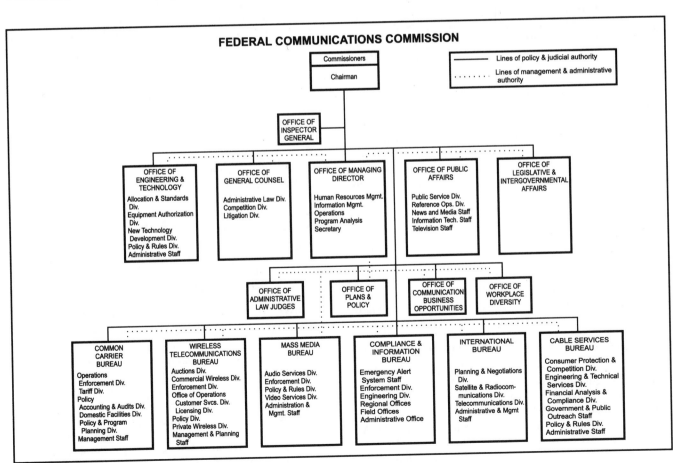

Fig 2-3—Organization chart of the Federal Communications Commission (FCC).

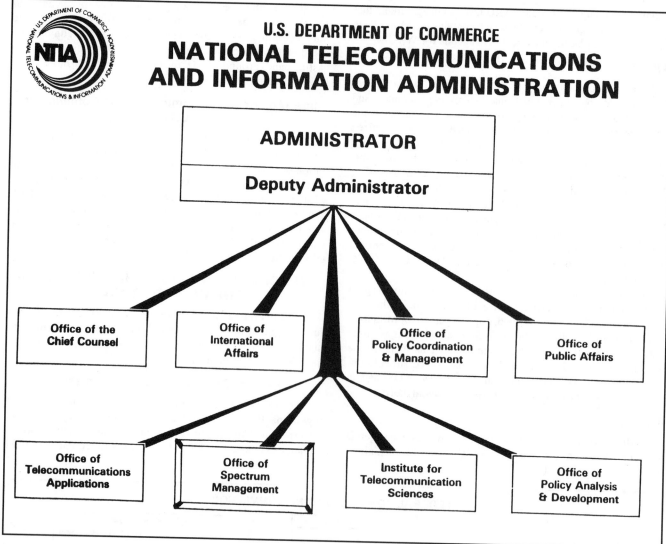

U.S. DEPARTMENT OF COMMERCE
NATIONAL TELECOMMUNICATIONS AND INFORMATION ADMINISTRATION

ADMINISTRATOR

Deputy Administrator

Office of the Chief Counsel	Office of International Affairs	Office of Policy Coordination & Management	Office of Public Affairs
Office of Telecommunications Applications	Office of Spectrum Management	Institute for Telecommunication Sciences	Office of Policy Analysis & Development

Fig 2-4—Organization chart of the National Telecommunications and Information Administration (NTIA).

of the Commission's rules. (Part 97 is available in *The FCC Rule Book,* published by the ARRL.)

The new internal structure of the FCC in place in late 1994 is shown in Fig 2-3. The FCC components of most concern to Amateur Radio are:

- The five Commissioners
- Wireless Telecommunications Bureau
- Special Services Division
- Licensing Division, Gettysburg, PA
- International Bureau
- Office of Engineering and Technology
- Compliance and Information Bureau, Regional Offices, Field Offices and Monitoring System

Federal Government Telecommunications

The functions relating to assignment of frequencies to radio stations belonging to, and operated by, the United States Government were assigned to the Assistant Secretary of Commerce for Communications and Information (Administrator, National Telecommunications and Information Administration-NTIA) by Department of Commerce Organization Order 10-10 of May 9, 1978. See Figs 2-4 and 2-5 for NTIA organizational charts. Among other things, NTIA:

- coordinates telecommunications activities of the Executive Branch
- develops plans, policies and programs relating to international telecommunications issues
- coordinates preparations for US participation in international telecommunications conferences and negotiations
- develops, in cooperation with the FCC, a long-range plan for improved management of all electromagnetic spectrum resources, including jointly determining the National Table of Frequency Allocations.
- conducts telecommunications research and development.

Interdepartment Radio Advisory Committee

The Interdepartment Radio Advisory Committee (IRAC) traces back to a mutual agreement of Government departments on June 1, 1922. Its status is as defined on December 10, 1964, and continued per Executive Order 12046 of March 27, 1978. The NTIA appoints the officers of the IRAC and chairpersons of its subcommittees. It is composed of representatives appointed by the following member departments and agencies: Agriculture, Air Force, Army, Coast Guard, Commerce, Energy, Federal Aviation Administration, Federal Emergency Management Agency, General Services Administration, Health and Human Services, Interior, Justice, National Aero-

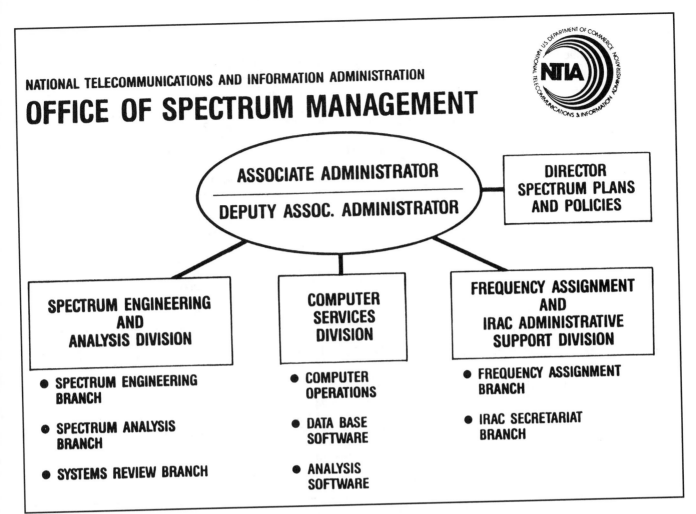

NATIONAL TELECOMMUNICATIONS AND INFORMATION ADMINISTRATION

OFFICE OF SPECTRUM MANAGEMENT

ASSOCIATE ADMINISTRATOR
_____
DEPUTY ASSOC. ADMINISTRATOR

DIRECTOR
SPECTRUM PLANS
AND POLICIES

SPECTRUM ENGINEERING AND ANALYSIS DIVISION

- SPECTRUM ENGINEERING BRANCH
- SPECTRUM ANALYSIS BRANCH
- SYSTEMS REVIEW BRANCH

COMPUTER SERVICES DIVISION

- COMPUTER OPERATIONS
- DATA BASE SOFTWARE
- ANALYSIS SOFTWARE

FREQUENCY ASSIGNMENT AND IRAC ADMINISTRATIVE SUPPORT DIVISION

- FREQUENCY ASSIGNMENT BRANCH
- IRAC SECRETARIAT BRANCH

Fig 2-5—Organization chart of the NTIA Office of Spectrum Management.

nautics and Space Administration, National Science Foundation, Navy, State, Treasury, US Information Agency, US Postal Service and Veterans Administration.

The IRAC assists the Assistant Secretary of Commerce in assigning frequencies to US government radio stations and in developing and executing plans, procedures, and technical criteria. The FCC must coordinate with IRAC any actions that might affect the Federal government's use of the radio spectrum. It is reasonable to expect that FCC dockets involving such sharing will take a bit longer to consider than those solely under the FCC's wing.

Bibliography

Baldwin and Sumner, "The Geneva Story," *QST,* Feb 80, pp 52-61.

Feller, *Planning an Electromagnetic Environment Model for Spectrum Management* (Washington, DC: FCC, June 26, 1981).

ITU, *Radio Regulations* (Geneva, 1990), as amended by the Final Acts of WARC-92.

Kleinschmidt and Rinaldo, "WARC-92: What it Means to You," *QST,* Jun 91, p 16.

NTIA, *Manual of Regulations and Procedures for Federal Radio Frequency Management* (Washington, DC: Department of Commerce, May 1989).

Palm, *The FCC Rule Book,* ARRL, Newington, CT.

Rinaldo, "League WARC-92 Preparations Continue," *QST,* May 90, p 55.

Rinaldo, "WARC-92: Inside the United States Delegation," *QST,* May 92, p 28.

Rinaldo, "The 1995 World Radio Conference Geneva", *QST,* Feb 96, p 56.

Sumner, "FCC WARC Proposals Finalized," *QST,* Feb 79, p 55.

Sumner, "ITU Lays Technical Foundation for WARC-79," *QST,* Mar 79, p 56.

Sumner, "WARC-92 Draft Proposals Released," *QST,* Nov 90, p 42.

Sumner, "WARC Agenda Released," *QST,* Aug 90, p 9.

Sumner, "WARC-92 Finds Room for New Radio Services," *QST,* May 92, p 25.

AMATEUR FREQUENCY ALLOCATIONS AND BAND PLANS

There are bands of frequencies allocated to the Amateur Service extending from 1800 kHz (73 kHz in the UK) to over 300 GHz. Table 2-1 will give you an overview of the radio spectrum, some nomenclature and the Amateur Service bands allocated in the ITU *Radio Regulations.*

Table 2-4 is a more-detailed, band-by-band picture of the radio spectrum. The part labeled "Allocations" is an extract of the ITU *Radio Regulations* Table of Frequency Allocations. All footnotes that pertain to the Amateur Service and other services sharing the same frequency bands are included. The part labeled "Amateur Radio Band Plans" includes the band plans formally adopted by the IARU Regions and the

Table 2-1

The Electromagnetic Spectrum with Amateur Service Frequency Bands by ITU Region

Wave-length	Frequency		Nomen-clature	Metric Band	Amateur Radio Bands by ITU Region		
					Region 1	Region 2	Region 3
1 mm	300 GHz						
			EHF	1 mm	241-250	241-250	241-250
			Milli-	2 mm	142-149	142-149	142-149
		M	metric	2.5 mm	119.98-120.02	119.98-120.02	119.98-120.02
		i		4 mm	75.5-81	75.5-81	75.5-81
		c		6 mm	47-47.2	47-47.2	47-47.2
1 cm	30 GHz	r					
		o	SHF	1.2 cm	24-24.25	24-24.25	24-24.25
		w	Centi-	3 cm	10-10.5	10-10.5	10-10.5
		a	metric	5 cm	5.65-5.85	5.65-5.925	5.65-5.85
		v		9 cm		3.3-3.5	3.3-3.5
10 cm	3 GHz	e					
		s	UHF	13 cm	2.3-2.45	2.3-2.45	2.3-2.45
			Deci-	23 cm	1240-1300	1240-1300	1240-1300
			metric	33 cm		902-928	
				70 cm	430-440	430-440	430-440
1	300 MHz						
			VHF	1.25 m		222-225	
			Metric	2 m	144-146	144-148	144-148
				6 m		50-54	50-54
10	30 MHz						
			HF	10 m	28-29.7	28-29.7	28-29.7
			Deca-	12 m	24.89-24.99	24.89-24.99	24.89-24.99
			metric	15 m	21-21.45	21-21.45	21-21.45
				17 m	18.068-18.168	18.068-18.168	18.068-18.168
				20 m	14-14.350	14-14.350	14-14.350
				30 m	10.1-10.150	10.1-10.150	10.1-10.150
				40 m	7-7.1	7-7.3	7-7.1
				80 m	3.5-3.8	3.5-4	3.5-3.9
100	3 MHz						
			MF	160 m	1.81-1.85	1.8-2	1.8-2
			Hecto-				
			metric				
1000	300 kHz						
			LF				
			Kilo-				
			metric				
10,000	30 kHz						
			VLF				
			Myria-				
			metric				
100,000	3 kHz						

Note: This table should be used only for a general overview of where ITU Amateur Service and Amateur-Satellite Service frequencies fall within the radio spectrum. They do not necessarily agree with FCC allocations; for example, the 70-cm band is 420-450 MHz in the United States. Refer to subsequent tables for more specific information.

ARRL. In most of the high-frequency bands there are no ARRL band plans that were adopted by the ARRL Board of Directors.

There is a tendency for band plans to lag reality. This is due in part to the time needed to research, invite and digest comment from amateurs, arrive at a mix that will serve the diverse needs of the amateur community, and adopt a formal band plan. This is a process that can take a year or more on the national level and a similar period in the IARU. Nevertheless, new communications modes or popularity of existing ones can make a year-or-two-old band plan look obsolete. Such revolutionary change has taken place recently with the popularity of data communications, particularly in the 20 and 2-meter bands. Changes of this magnitude cause the new users to scramble for frequencies and some of the existing mode users to draw their wagons in a circle. The national societies, their staffs and committees, and the IARU have the job of sorting out the contention for various frequencies and preparing new band plans. Fortunately, we have not exhausted all possible ways of improving our management of the spectrum so that all Amateur Radio interests can be accommodated.

The 160-Meter Band

Those who have a fascination for 160-meter DX should take a close look at Chapter 5. It contains a country-by-country listing of the frequency allocations in this band. The basic problem with allocations in this band has been competition with the Radiolocation Service. New pressures are possible as a result of planned expansion of the medium-frequency broadcast band in the 1605-1705 kHz range.

The 80-Meter Band

While US amateurs enjoy the use of 3500-4000 kHz, not all countries allocate such a wide range of frequencies to the 80-meter band. There are fixed, mobile and broadcast operations, particularly in the upper part of the band.

The 40-Meter Band

The 40-meter band has a big problem: International broadcasting occupies the 7100-7300 kHz (41-meter) band in many parts of the world. During the daytime particularly when sunspots are high, broadcasting does not cause much interference to US amateurs. However, at night, especially when sunspot activity is low, the broadcast interference is heavy. Some countries allocate only the 7000-7100 kHz band to amateurs. Others, particularly in Region 2, allocate 7100-7300 kHz as well, which at times is subject to interference. The result is that there is a great demand for frequencies in the 7000-7100 kHz segment. The effect is that there are two band plans overlaid on each other: ours is spread out over 7000-7300 kHz and a foreign one that compresses everything into 7000-7100 kHz.

The 30-Meter Band

The 30-meter band is an excellent band for CW and digital modes. The only problem is that US amateurs must not cause harmful interference to the fixed operations outside the US. This restricts transmitter power output to 200 watts and is one reason for not having contests on this band.

The 20-Meter Band

The workhorse of DX is undoubtedly the 20-meter band. It offers excellent propagation to all parts of the world throughout the sunspot cycle and is virtually clean of interference from other services.

The 17-Meter Band

The 18068-18168 kHz band was awarded to amateurs on an exclusive basis, worldwide, at WARC-79. It was made available for amateur use in the US in January 1989.

The 15 and 12-Meter Bands

The 21000-21450 and 24890-24990 kHz bands are excellent for DX during the high part of the sunspot cycle. They also offer some openings throughout rest of the sunspot cycle.

The 10-Meter Band

This is an exclusive amateur band worldwide. Its popularity has risen sharply as a result of FCC PR Docket 86-161, better known as Novice Enhancement.

The VHF and Higher Bands

The 6-meter band is not universal, but the trend seems to be toward allocating it to amateurs as TV broadcasting vacates the 50-54 MHz band. Recent new 6-meter privileges for certain European amateurs adds to the worldwide interest in this band. It is also excellent for amateur exploitation of meteor-scatter communications using various modes including packet radio.

Two meters is heavily used throughout the world for CW, EME, SSB, FM and packet radio. Satellites occupy the 145.8-146 MHz segment.

In 1991, the FCC reallocated the 220-222 MHz band to the land mobile service. In partial compensation for that loss of amateur spectrum, acting on petition of the League, the Commission allocated the 219-220 MHz to the Amateur Radio Service on a secondary basis. There are special provisions to protect domestic waterways telephone systems using that band. US amateurs have a primary allocation at 222-225 MHz, which is largely used for repeaters.

The 70-cm band is prime UHF spectrum. The 430-440 MHz band is virtually worldwide, whereas the 420-430 MHz and 440-450 MHz bands are not. Frequencies around 432 are used for weak-signal work, including EME, and the 435-438 MHz band is for amateur satellites. The 70-cm band is the lowest-frequency band that can be used for fast-scan television and spread spectrum emissions.

As a result of Congressional action, the 2305-2310 MHz band is subject to auction for other radio services, thus giving rise to concern about Amateur Radio retention and useful access to this band. The League is pursuing primary status for the Amateur Service in the 2300-2305 MHz band and continued use of the 2305-2310 MHz band.

The 33-cm band (902-928 MHz) is widely shared with other services, including Location and Monitoring Service (LMS), which is primary, and ISM (industrial, scientific and medical) equipment applications. A number of low-power devices including spread spectrum local area networks operate in this band under Part 15 of the FCC's Rules.

The 1240-1300 MHz band is used by amateurs for essentially all modes, including FM and packet. By regulation, the 1260-1270 MHz segment may be used only in the Earth-to-space direction when communicating with amateur satellites.

While the Amateur Service has a secondary allocation in the 2300-2450 MHz band in the international tables, in the United States, the allocation is 2390-2400 MHz and 2402-2417 MHz primary, and 2300-2310, 2400-2402 and 2417-2450 secondary. Most of the weak-signal work in the US takes place around 2304 MHz, while much of the satellite activity is in the 2400-2402 MHz segment.

The remaining microwave and millimeter bands are the

Fig 2-6—This map shows channel spacing in kHz. Please check with your Regional Frequency Coordinator for more information.

territory of amateur experimenters. It is important that the Amateur Service and the Amateur-Satellite Service use these bands, and contribute to the state-of-the-art in order to retain them. There is growing interest on the part of the telecommunications industry and the space science community to fully exploit the 20-95 GHz spectrum.

FREQUENCY COORDINATION

From the earliest days of Amateur Radio, hams simply listened on a frequency to see if it was in use, and if not, went ahead and transmitted on that frequency. Amateurs operating on HF still use this simple listen-before-transmit procedure of randomly *assigning* themselves a frequency for use over a short period of time. This simple procedure has served us well over the years, but is no longer the only "rule of the road."

HF CW and phone nets historically have had problems finding frequencies on which to conduct their regular operations. Some have come up with strong arguments supporting the need for nets to have priority over a two-station contact on the same frequency. The rationale is that nets have many stations that would be difficult to move to another frequency and that the needs of the many translate to more public good than just a few ragchewers. The other side of the argument is that the two stations have a right to conduct their communication. The principle of priority becomes a bit clearer when one set of users has emergency traffic and the other does not. Nevertheless, no one (including the FCC and the ARRL) has expressed willingness to say who gets a specific HF channel and who does not. This is rooted in the knowledge that stations and their needs come and go hour by hour, that ionospheric propagation changes hour by hour, and that most hams can bend a little in frequency and time to let both sets of users carry on.

Over time it became necessary to introduce a mild spectrum-management technique of frequency registration to let hams know where and when nets normally operate. This information is published annually in the ARRL *Net Directory*. This appearance of a listing in the *Net Directory* does not indicate an *assignment* by any authority. It is particularly useful when researching new frequencies for net or contest operations. It is of only limited use to individuals wishing to conduct random contacts, as the listen-before-transmit technique is more practical.

The Days Before Repeater Coordination

Two-meter FM repeaters became the rage in the 1960s. Just put your repeater on 34/94 (146.34 MHz input, 146.94 MHz output) and there was no problem. . . until someone else wanted to do the same thing nearby. "We got here first, so we are in the right; they are in the wrong. Why don't *they* find another frequency pair? What pair? That's *their* problem. No, that's not the way.

The early repeater contention problems were solved by having the latecomers find new frequencies. New transceivers typically were supplied with crystals for (146) 34, 52, 76 and 94. There were repeaters with inputs on 34 and outputs on 76 to avoid local simplex operation on 94. But, 76 became paired with a 16 input, and the 600-kHz offset was well on its way to becoming standard. Much of the early spadework in repeater band planning was done by the Texas VHF FM Society.

By early 1976, the repeater count in the US was more than 2000. The ARRL VHF Repeater Advisory Committee (VRAC) agreed on band plans that named repeater pairs in the 10-meter, 6-meter, 2-meter, 1.25-meter and 70-cm bands.

Many early tube-type amateur FM transceivers had seen service in the back of a police car or taxi. The land mobile services had been using 15-kHz deviation and had spaced their channels in a 30-kHz raster (30 kHz between channel center frequencies) but were in the process of converting to 5-kHz deviation and 15-kHz channel spacing. Amateurs followed suit by first setting up 30-kHz channel spacing with so-called *wideband* FM (15-kHz) then changing to 15-kHz channel spacing, which in that context was known as *narrow-band* FM. The new channels created in between the original 30-kHz spaced channels were called *splinter channels*. The change was accomplished in some older rigs by backing off on the transmitter deviation and somehow tightening up the receiver bandwidth. The problem was solved by new solid-state rigs from Japan, first with a few crystals and room for more, later with synthesizers.

Synthesized transceivers were able to tune to any frequency within the 2-meter band in 5-kHz steps. This opened up another possibility: It was now feasible to try some different channel spacings without causing people to buy several hundred dollars in new crystals. The result today is parts of the US use different spacings. A map showing 2-meter channel spacing throughout the US and Canada is shown in Fig 2-6.

This spreading out throughout the 2-meter and other bands to accommodate more repeaters had its limits, of course. It was clear that there would come a day when all the channels would be full, particularly in the metropolitan areas. What would happen then? Would latecomers just fire up new repeaters on top of established ones? Chaos! That was to be avoided at all costs, so amateurs actually began talking to each other. The prime movers were the repeater operators; they banned together and appointed *repeater frequency coordinators* to keep out the "upstarts." This was fine unless you were one of the upstarts.

Amateur Repeater Coordination Councils

When it looked as if there would be contention for the finite number of repeater pairs, repeater operators established *repeater coordination councils*. Typically, the members of the council were the trustees of the area repeaters and perhaps others (if membership was open to them) who were planning repeaters or just concerned. The democratic processes began to work. The established repeater operators were able to keep their frequencies, and the newcomers somehow could be accommodated until all the pairs were full. Most areas had a *frequency coordinator*, who had the administrative job of keeping records on all existing repeaters. The frequency coordinator kept sufficient information, and theoretically had the skills to determine whether a proposed repeater would interfere with existing repeaters. If, after polling established repeater operators, there appeared to be no unacceptable interference, the frequency coordinator would *coordinate* the new repeater. This added a new term to the Amateur Radio lexicon: *coordinated repeater*.

What if a "noncoordinated" repeater owner is not satisfied with the decision made administratively by the coordinator? There are ways to solve this problem: The owner or trustee could become a member of the council—perhaps even volunteer for president. This way he or she could try to influence the frequency coordinator into favorable action. Never fire a frequency coordinator who is doing the job; where could you find another competent person to put in long hours for no pay?

But wait a minute! Is Repeater Coordination Council X legitimate? Don't they just represent the existing repeater owners, not prospective ones, not the repeater users, and not the other hams who might want to use the same frequencies for other purposes? Some have been challenged. Some have come and gone. A few frustrated individuals have started rival coordination entities. For a discussion of what happens when a frequency coordinator "doesn't," see the FM/RPT column in September 1986 *QST*.

Okay, "Let's ask the FCC," someone says. The FCC is not enthusiastic about this (and it is usually undesirable and not recommended from the amateur point of view). How would you like to be called in to settle a fight? You know what the FCC's bottom line will be: "Why don't you guys get together and work this thing out. *OUT.*"

FCC PR Docket 8S-22

Coordinating one's choice of frequencies for a repeater has never been a licensing requirement. By early 1985 demand for frequencies had grown to the point where even an FCC predisposed to prefer "unregulation" decided something needed to be done. In February 1985, after many formal comments and replies the FCC issued PR Docket 85-22, which stated in part: ". . . . Most of the reported cases of amateur repeater-to-repeater interference appear to involve one or more non-coordinated repeaters. In the past two years we have had to resolve amateur repeater-to-repeater interference disputes in which at least one non-coordinated repeater was involved and in which the parties to the dispute could reach no amicable solution. We attribute the growing number of instances of amateur repeater-to-repeater interference and the need for increased intervention in these matters to the mounting pressure, which develops as the desirable repeater frequencies become fully assigned.

". . . . The commission should not recognize a single entity, such as a national frequency coordinator for amateur repeater operation. Such coordination activities should be performed by local, or regional, volunteer frequency coordinators with appropriate support to these coordinators to be provided by the League . . . . "

This FCC docket became the rationale behind the current Part 97 rules that affect amateur repeater coordination.

The League's Role in Coordination

While the ARRL successfully argued before the FCC against a national frequency coordinator, there remains a significant role for the League in the coordination process. For one thing, the ARRL has for years published the *Repeater Directory*. By 1994 it had over 20,000 entries including repeaters, packet radio and beacons.

The data base used in preparation of the *Directory* is accessible to frequency coordinators via a toll-free phone line. The design of the National Repeater Data Base (NRDB) was proposed by The Mid-Atlantic Repeater Council (TMARC), and was placed in service as designed in April of 1989—just in time for its "trial by fire"—the 1990 edition of the *Directory*!

After a year of use, it was determined that some changes and streamlining were necessary for the system. A new data base was designed by the Mid-America Coordination Council (MACC) in June of 1991. This "user-friendly" system is running smoothly and allows coordinators to actively use the data for coordination of repeaters near bordering states, thus lessening the chance for interference.

The fields used by the coordinators are as follows:

Band	29-52-144-222-440-ATV-902-1240-PKT-BEA
Output	146.94, etc
Input	146.34, etc
State	2-letter designators
Location	City/town
Call	Amateur call sign of repeater
Sponsor	Club or sponsoring organization
Last Update	Current date
Source	Coordinator or Council
Geographic Area	County
Notes	Open-Linked-Autopatch-etc

With the fields set up in this manner the coordinators have access to all pertinent coordination data, and the *Repeater Directory* can make use of the "general" information for publication annually.

Frequency coordination in the US continues to evolve. Over the course of 1995 and 1996 the National Frequency Coordinator's Council (NFCC) came into being and is beginning to take root and grow—with support from ARRL.

The Board of Directors of the National Frequency Coordinator's Council (NFCC) met on September 13, 1996 at the ARRL National Convention in Peoria, IL, with members of the ARRL's Ad-Hoc Repeater Committee, for the purpose of negotiating the terms of a MOU (Memorandum of Understanding) that could be recommended for approval to both organizations.

The participants were pleased to announce a consensus on the text of a proposed MOU which provides for the recognition, development, and enhancement of amateur radio frequency coordination. The NFCC Board of Directors unanimously approved and authorized the Council President to sign the proposed MOU. The ARRL Committee unanimously agreed to recommend approval of the proposed MOU to the full ARRL Board.

The MOU provides a workable mechanism for the improvement of amateur frequency coordination, the development of standards, policies and procedures, and the expeditious initiation of necessary regulatory and legislative reform. The MOU was signed by both the ARRL president, Rod Stafford, KB6ZV, and NFCC President Owen Wormser, K6LEW, in January 1997. What began as a national meeting of regional frequency coordinators, the ARRL and a representative from the FCC in St. Charles, Missouri way back in October of 1995 had finally come to be.

PROPAGATION BEACONS

Over the last several years the number of propagation beacons has increased, and with the increase, came almost daily changes. In Region 1 the latest band plan provides a separate beacon allocation in most HF bands and in almost every VHF and higher bands. Printed lists are obsolete almost as fast as they can be compiled.

Beacons provide a valuable resource to those who are interested in propagation. DXers and just about every one who wants to know what is going on. Even though a particular frequency is not designated for beacons in your region or area, it is very good practice to keep off the beacon allocations so as not to interfere with those who are using the frequency.

NCDXF/IARU International Beacon Network

The Northern California DX Foundation, in cooperation with the IARU, has established a widespread, multi-band beacon network. At present this network operates on the 14, 18, 21, 24 and 28-MHz bands. A sequence of power levels from 0.1 to 10 watts on a time sequence is transmitted. NCDXF can be contacted at P.O. Box 2368, Stanford, CA 94309-2368. A full description with the most up-to-date status is on their web page at **http://www.ncdxf.org/beacon.htm**

Table 2-3
Amateur Frequency Allocations and Band Plans

Unlike FCC rules, which must be obeyed if the operator is to avoid official sanctions (possibly including loss of license), band plans are voluntary guidelines developed through representative mechanisms. Regional band plans were first developed, and are periodically reviewed and updated, by the regional organizations of the International Amateur Radio Union (IARU). The IARU Administrative Council serves as a coordinator of regional policies, and seeks worldwide agreement whenever possible. However, because each region is autonomous, subtle differences in the regional band plans exist which cannot be easily explained in a simple chart. The charts of regional band plans shown here are intended to summarize the band plans of the three IARU regions, and should not be assumed to be definitive.

Emissions are to be confined to the band segments specified for them. See end of table for Emission Designators.

The regions differ in their treatment of data modes. In general, the term "DATA" used here includes Baudot, AMTOR, ASCII, G-TOR, PACTOR and similar modes. Where a segment of a band is specifically designated for packet radio, the intent generally is to indicate a preference that packet radio in that band should be limited to that segment.

SSTV generally includes FAX.

Footnote designations beginning with the letter "S" were adopted by the ITU at WRC-95.

Region 3 has adopted the following policy with regard to its band plans:

(1) In all cases of conflict between a band plan and the national regulations of a country, the latter shall prevail.

(2) Nothing in these band plans shall be construed as prohibiting different national arrangements, provided that harmful interference is not caused to stations in the countries operating in accordance with the regional band plans.

(3) Notwithstanding item 2 above, member societies of Region 3 are strongly urged to use these regional band plans as a basis for their national band plans.

Allocations							
INTERNATIONAL			UNITED STATES				
Region 1 kHz	Region 2 kHz	Region 3 kHz	Band kHz	National Provisions	Government Allocation	Non-Government Allocation	Remarks
1800-1810 RADIOLOCATION S5.93 1810-1850 AMATEUR S5.98 S5.99 S5.100 S5.101	1800-1850 AMATEUR	1800-2000 AMATEUR FIXED MOBILE except aeronautical mobile RADIONAVIGATION Radiolocation	1800-1900			AMATEUR	
1850-2000 FIXED MOBILE except aeronautical mobile S5.92 S5.96 S5.103	1850-2000 AMATEUR FIXED MOBILE except aeronautical mobile RADIOLOCATION RADIONAVIGATION S5.102	S5.97	1900-2000	US290	RADIOLOCATION	RADIOLOCATION Amateur	

S5.92—Some countries of Region 1 use radiodetermination systems in the bands 1606.5-1625 kHz, 1635-1800 kHz, 1850-2160 kHz, 2194-2300 kHz, 2502-2850 kHz and 3500-3800 kHz, subject to agreement obtained under Article 14/No. S9.21. The radiated mean power of these stations shall not exceed 50 W.

S5.93—Additional allocation: in Angola, Armenia, Azerbaijan, Belarus, Bulgaria, Georgia, Hungary, Kazakhstan, Latvia, Lithuania, Moldova, Mongolia, Nigeria, Uzbekistan, Poland, Kyrgyzstan, Slovakia, the Czech Republic, Russia, Tajikistan, Chad, Turkmenistan and Ukraine, the bands 1625-1635 kHz, 1800-1810 kHz and 2160-2170 kHz are also allocated to the fixed and land mobile services on a primary basis, subject to agreement obtained under Article 14/No. S9.21

S5.96—In Germany, Armenia, Azerbaijan, Belarus, Denmark, Estonia, Finland, Georgia, Hungary, Ireland, Israel, Jordan, Kazakhstan, Latvia, Lithuania, Malta, Moldova, Norway, Uzbekistan, Poland, Kyrgyzstan, Slovakia, the Czech Republic, the United Kingdom, Russia, Sweden, Tajikistan, Turkmenistan and Ukraine, administrations may allocate up to 200 kHz to their amateur service in the bands 1715-1800 kHz and 1850-2000 kHz. However, when allocating the bands within this range to their amateur service, administrations shall, after prior consultation with administrations of neighbouring countries, take such steps as may be necessary to prevent harmful interference from their amateur service to the fixed and mobile services of other countries. The mean power of any amateur station shall not exceed 10 W.

S5.97—In Region 3, the Loran system operates either on 1850 kHz or 1950 kHz, the bands occupied being 1825-1875 kHz and 1925-1975 kHz respectively. Other services to which the band 1800-2000 kHz is allocated may use any frequency therein on condition that no harmful interference is caused to the Loran system operating on 1850 kHz or 1950 kHz.

S5.98—Alternative allocation. in Angola, Armenia, Austria, Azerbaijan, Belarus, Belgium, Bulgaria, Cameroon, the Congo, Denmark, Egypt, Eritrea, Spain, Ethiopia, France, Georgia, Greece, Italy, Kazakhstan, Lebanon, Lithuania, Luxembourg, Malawi, Moldova, Uzbekistan, the Netherlands, Syria, Kyrgyzstan, Russia, Somalia, Tajikistan, Tanzania, Tunisia, Turkmenistan, Turkey and Ukraine, the band 1810-1830 kHz is allocated to the fixed and mobile, except aeronautical mobile, services on a primary basis.

S5.99—Additional allocation: in Saudi Arabia, Bosnia and Herzegovina, Iraq, The Former Yugoslav Republic of Macedonia, Libya, Slovakia, the Czech Republic, Romania, Slovenia, Chad, Togo and Yugoslavia, the band 1810-1830 kHz is also allocated to the fixed and mobile, except aeronautical mobile, services on a primary basis.

S5.100—In Region 1, the authorization to use the band 1810-1830 kHz by the amateur service in countries situated totally or partially north of 40° N shall be given only after consultation with the countries mentioned in Nos. S5.98 and S5.99 to define the necessary steps to be taken to prevent harmful interference between amateur stations and stations of other services operating in accordance with Nos. S5.98 and S5.99.

S5.101—Alternative allocation: in Burundi and Lesotho, the band 1810-1850 kHz is allocated to the fixed and mobile, except aeronautical mobile, services on a primary basis.

S5.102—Alternative allocation: in Argentina, Bolivia, Chile, Mexico, Paraguay, Peru, Uruguay and Venezuela, the band 1850-2000 kHz is allocated to the fixed, mobile, except aeronautical mobile, radiolocation and radionavigation services on a primary basis.

US290—In the band 1900-2000 kHz, amateur stations may continue to operate on a secondary basis to the Radiolocation Service, pending a decision as to their disposition through a future rule making proceeding in conjunction with implementation of the Standard Broadcasting Service in the 1625-1705 kHz band.

FCC Part 97 Privileges

License Class	Terrestrial location of the amateur radio station			Limitations*
	Region 1 kHz	Region 2 kHz	Region 3 kHz	
General Advanced Extra	1810-1850	1800-2000	1800-2000	

* See §§97.301-307 of the FCC rules for further details on authorized privileges for each license class, frequency sharing requirements and authorized modes.

Amateur Radio Band Plans

Region 1 kHz	Region 2 kHz	Region 3 kHz	ARRL* kHz
1810-1838 CW	1800-1830 CW/DATA	1800-1830 CW	1800-1840 CW/RTTY/NB
	1830-1840 CW/DATA DX CW Window	1830-1834 CW/DX/RTTY	
		1834-1840 CW	
1838-1840 CW/DATA except packet			1830-1850 Intercontinental QSOs only
1840-1842 DATA/Phone/CW except packet	1840-1850 Phone/CW DX Phone WIndow	1840-2000 Phone/CW	1840-2000 Phone SSTV WB CW
1842-2000 Phone/CW	1850-2000 Phone/CW		

*Adopted by the ARRL Board of Directors January 1986.

Allocations

INTERNATIONAL			UNITED STATES				
Region 1 kHz	Region 2 kHz	Region 3 kHz	Band kHz	National Provisions	Government Allocation	Non-Government Allocation	Remarks
3500-3800 AMATEUR S5.120 FIXED MOBILE except aeronautical mobile S5.92 3800-3900 FIXED AERONAUTICAL MOBILE (OR) LAND MOBILE 3900-3950 AERONAUTICAL MOBILE (OR) S5.123 3950-4000 FIXED BROAD-CASTING	3500-3750 AMATEUR S5.120 S5.119 3750-4000 AMATEUR S5.120 FIXED MOBILE except aeronautical mobile (R) S5.122 S5.124 S5.125	3500-3900 AMATEUR S5.120 FIXED MOBILE 3900-3950 AERONAUTICAL MOBILE BROADCASTING 3950-4000 FIXED BROADCASTING S5.126	3500-4000	S5.120		AMATEUR	

S5.92—Some countries of Region 1 use radiodetermination systems in the bands 1606.5-1625 kHz, 1635-1800 kHz, 1850-2160 kHz, 2194-2300 kHz, 2502-2850 kHz and 3500-3800 kHz, subject to agreement obtained under Article 14/No. 9.21. The radiated mean power of these stations shall not exceed 50 W.

S5.119—Additional allocation: In Honduras, Mexico, Peru and Venezuela, the band 3500-3750 kHz is also allocated to the fixed and mobile services on a primary basis.

S5.120—For the use of the bands allocated to the amateur service at 3.5 MHz, 7.0 MHz, 10.1 MHz, 14.0 MHz, 18.068 MHz, 21.0 MHz, 24.89 MHz and 144 MHz in the event of natural disasters, see Resolution 640.

S5.122—Alternative allocation: in Argentina, Bolivia, Chile, Ecuador, Paraguay, Peru and Uruguay, the band 3750-4000 kHz is allocated to the fixed and mobile, except aeronautical mobile, services on a primary basis.

S5.123—Additional allocation: in Botswana, Lesotho, Malawi, Mozambique, Namibia, South Africa, Swaziland, Zambia and Zimbabwe, the band 3900-3950 kHz is also allocated to the broadcasting service on a primary basis, subject to agreement obtained under Article 14/No. 9.21.

S5.124—Additional allocation: in Canada, the band 3950-4000 kHz is also allocated to the broadcasting service on a primary basis. The power of broadcasting stations operating in this band shall not exceed that necessary for a national service within the frontier of this country and shall not cause harmful interference to other services operating in accordance with the Table.

S5.125—Additional allocation: in Greenland, the band 3950-4000 kHz is also allocated to the broadcasting service on a primary basis. The power of the broadcasting stations operating in this band shall not exceed that necessary for a national service and shall in no case exceed 5 kW.

S5.126—In Region 3, the stations of those services to which the band 3995-4005 kHz is allocated may transmit standard frequency and time signals.

FCC Part 97 Privileges

License Class	Terrestrial location of the amateur radio station			Limitations*
	Region 1 kHz	Region 2 kHz	Region 3 kHz	
Novice Technician Plus	3675-3725**	3675-3725**	3675-3725**	
General	3525-3750 —	3525-3750 3850-4000	3525-3750 3850-3900	
Advanced	3525-3750 3775-3800	3525-3750 3775-4000	3525-3750 3775-3900	
Extra	3500-3800	3500-4000	3500-3900	

*See §§97.301-307 of the FCC rules for further details on authorized privileges for each license class, frequency sharing requirements and authorized modes.

**The Novice band was moved from 3700-3750 to 3675-3725 kHz March 16, 1991, per FCC Docket 90-100.

Amateur Radio Band Plans

Region 1 kHz	Region 2 kHz	Region 3 kHz	ARRL kHz
3500-3510 DX CW Contests	3500-3510 DX CW	3500-3510 DX CW	3500-3580 CW
3510-3560 CW Contests	3510-3525 CW	3510-3535 CW	
	3525-3580 CW (Phone permitted)	3535-3775† Phone/CW	
3560-3580 CW			
3580-3590 CW/DATA	3580-3620 DATA/CW (Phone permitted)		3580-3620 DATA/CW
3590-3600 CW/DATA/Packet			3590 DX/RTTY
3600-3620 DATA/Phone/CW Contests			
3620-3650 Phone/CW Contests	3620-3635 Packet/DATA/CW (Phone permitted)		3620-3635 Packet/DATA/CW
	3635-3775 Phone (CW permitted)		3635-3750 CW
3650-3700 Phone/CW			
3700-3730 Phone/CW Contests			
3730-3740 SSTV/Fax/Phone/CW Contests			
3740-3775 Phone/CW Contests			3750-3790 Phone
3775-3800 DX Phone/CW	3775-3800 DX Phone (CW permitted)	3775-3800 DX/Phone/CW	3790-3800 Phone DX Window
	3800-3840 Phone (CW permitted)	3800-3900 Phone/CW	3800-4000 Phone
	3840-3850 SSTV/Phone (CW permitted)		3845 SSTV
	3850-4000 Phone (CW permitted)		

† In Region 3 countries where the total band available nationally is 100 kHz or less, phone operation may commence at 3525 kHz.

INTERNATIONAL			UNITED STATES				
Region 1 kHz	Region 2 kHz	Region 3 kHz	Band kHz	National Provisions	Government Allocation	Non-Government Allocation	Remarks
7000-7100	AMATEUR S5.120 AMATEUR-SATELLITE S5.140 S5.141		7000-7100	S5.120		AMATEUR S5.120 AMATEUR-SATELLITE	
7100-7300 BROADCASTING	7100-7300 AMATEUR S5.120 S5.142	7100-7300 BROADCASTING	7100-7300	S5.120 S5.142		AMATEUR	

S5.120—For the use of the bands allocated to the amateur service at 3.5 MHz, 7.0 MHz, 10.1 MHz, 14.0 MHz, 18.068 MHz, 21.0 MHz, 24.89 MHz and 144 MHz in the event of natural disasters, see Resolution 640.

S5.140—Additional allocation. in Angola, Iraq, Rwanda, Somalia and Togo, the band 7000-7050 kHz is also allocated to the fixed service on a primary basis.

S5.141—Alternative allocation: in Egypt, Eritrea, Ethiopia, Guinea, Libya, Madagascar and Malawi, the band 7000-7050 kHz is allocated to the fixed service on a primary basis.

S5.142—The use of the band 7100-7300 kHz in Region 2 by the amateur service shall not impose constraints on the broadcasting service intended for use within Region 1 and Region 3.

FCC Part 97 Privileges				
	Terrestrial location of the amateur radio station			
License Class	Region 1 kHz	Region 2 kHz	Region 3 kHz	Limitations*
Novice Technician Plus	7050-7075	7100-7150	7050-7075	
General	7025-7100	7025-7150 7225-7300	7025-7100	
Advanced	7025-7100	7025-7300	7025-7100	
Extra	7000-7100	7000-7300	7000-7100	

*See §§97.301-307 of the FCC rules for further details on authorized privileges for each license class, frequency sharing requirements and authorized modes.

Amateur Radio Band Plans			
Region 1 kHz	Region 2 kHz	Region 3 kHz	ARRL kHz
7000-7035 CW	7000-7035 CW	7000-7025 CW	7000-7080 CW
		7025-7030 DATA/CW	
		7030-7040 DATA/Phone/CW	
7035-7040 DATA except packet/SSTV/Fax/CW	7035-7040 CW/DATA with other regions		
7040-7045 DATA except packet/SSTV/Fax/Phone/CW	7040-7050 CW/packet with other regions	7040-7100 Phone/CW	7040 DX RTTY
7045-7100 Phone/CW	7050-7080 Phone (CW permitted)		
	7080-7100 DX Phone (CW permitted)		7080-7100 DATA/CW
	7100-7120 DATA/Phone (CW permitted)	7100-7300 Phone/CW (Secondary operation where permitted by the administration)	7100-7105 Packet/CW
	7120-7165 Phone (CW permitted)		7105-7150 CW
	7165-7175 SSTV/Phone (CW permitted)		7150-7300 Phone
	7175-7300 Phone (CW permitted)		7171 SSTV

INTERNATIONAL			UNITED STATES				
Region 1 kHz	Region 2 kHz	Region 3 kHz	Band kHz	National Provisions	Government Allocation	Non-Government Allocation	Remarks
10100-10150	FIXED Amateur S5.120		10100-10150	US247 S5.120		AMATEUR	

S5.120—For the use of the bands allocated to the amateur service at 3.5 MHz, 7.0 MHz, 10.1 MHz, 14.0 MHz, 18.068 MHz, 21.0 MHz, 24.89 MHz and 144 MHz in the event of natural disasters, see Resolution 640.

US247—The band 10100-10150 kHz is allocated to the fixed service on a primary basis outside the United States and Possessions. Transmissions of stations in the amateur service shall not cause harmful interference to this fixed service use and stations in the amateur service shall make all necessary adjustments (including termination of transmission) if harmful interference is caused.

FCC Part 97 Privileges				
	Terrestrial location of the amateur radio station			
License Class	Region 1 kHz	Region 2 kHz	Region 3 kHz	Limitations*
General Advanced Extra	10100-10150	10100-10150	10100-10150	

*See §§97.301-307 of the FCC rules for further details on authorized privileges for each license class, frequency sharing requirements and authorized modes.

Amateur Radio Band Plans			
Region 1 kHz†	Region 2 kHz	Region 3 kHz	ARRL kHz
10100-10140 CW	10100-10130 CW	10100-10140 CW	10100-10130 CW
10140-10150 CW/DATA except Packet	10130-10140 DATA/CW	10140-10150 DATA/CW	10130-10140 DATA/CW
	10140-10150 Packet/DATA/CW		10140-10150 Packet/DATA/CW

†In Region 1 only:
SSB may be used in the 10 MHz band during emergencies involving the immediate safety of life and property and only by stations actually involved in the handling of emergency traffic.
The band segment 10120-10140 kHz may be used for SSB transmissions in the area of Africa south of the equator during local daylight hours.
It is recommended that unmanned stations using digital modes shall avoid the use of the 10 MHz band.
News bulletins on any mode should not be transmitted on the 10 MHz band.

Allocations

INTERNATIONAL			UNITED STATES				
Region 1 kHz	Region 2 kHz	Region 3 kHz	Band kHz	National Provisions	Government Allocation	Non-Government Allocation	Remarks
14000-14250	AMATEUR S5.120 AMATEUR-SATELLITE		14000-14250	S5.120		AMATEUR AMATEUR-SATELLITE	
14250-14350	AMATEUR S5.120 S5.152		14250-14350	S5.120		AMATEUR	

S5.120—For the use of the bands allocated to the amateur service at 3.5 MHz, 7.0 MHz, 10.1 MHz, 14.0 MHz, 18.068 MHz, 21.0 MHz, 24.89 MHz and 144 MHz in the event of natural disasters, see Resolution 640.

S5.152—Additional allocation: in Armenia, Azerbaijan, Belarus, China, Cote d'Ivoire, Georgia, the Islamic Republic of Iran, Kazakhstan, Moldova, Uzbekistan, Kyrgyzstan, Russia, Tajikistan, Turkmenistan and Ukraine, the band 14250-14350 kHz is also allocated to the fixed service on a primary basis. Stations of the fixed service shall not use a radiated power exceeding 24 dBW.

FCC Part 97 Privileges

License Class	Terrestrial location of the amateur radio station			Limitations*
	Region 1 kHz	Region 2 kHz	Region 3 kHz	
General	14025-14150 14225-14350	14025-14150 14225-14350	14025-14150 14225-14350	
Advanced	14025-14150 14175-14350	14025-14150 14175-14350	14025-14150 14175-14350	
Extra	14000-14350	14000-14350	14000-14350	

*See §§97.301-307 of the FCC rules for further details on authorized privileges for each license class, frequency sharing requirements and authorized modes.

Amateur Radio Band Plans

Region 1 kHz	Region 2 kHz	Region 3 kHz	ARRL kHz
14000-14060 CW Contests	14000-14070 CW	14000-14070 CW	14000-14070 CW
14060-14070 CW			
14070-14089 DATA/CW	14070-14095 DATA/CW	14070-14099.5 DATA/CW	14070-14095 DATA/CW
14089-14099 Packet/DATA/CW	14095-14099.5 Packet/DATA/CW		14095-14099.5 Packet/DATA/CW
14099-14101 Beacons	14099.5-14100.5 Beacons	14099.5-14100.5 Beacons	14099.5-14100.5 Beacons
14101-14112 Packet/Phone/CW	14100.5-14112 Packet/DATA/ Phone/CW permitted	14100.5-14112 DATA/CW/Phone	14100.5-14112 Packet/DATA/CW
14112-14125 Phone/CW	14112-14150 Phone/CW permitted International traffic	14112-14225 Phone/CW	14112-14150 CW
14125-14300 Phone/CW Contests	14150-14225 Phone/CW permitted		14150-14350 Phone
14230 SSTV/Fax calling frequency	14225-14235 SSTV/Phone/CW permitted	14225-14235 SSTV/Phone/CW	14230 SSTV
	14235-14250 Phone/CW permitted	14235-14350 Phone/CW	
14300-14350 Phone/CW	14250-14340 Phone/CW permitted International traffic		
	14340-14350 Phone Emergency (CW permitted) International traffic		

Allocations

INTERNATIONAL			UNITED STATES				
Region 1 kHz	Region 2 kHz	Region 3 kHz	Band kHz	National Provisions	Government Allocation	Non-Government Allocation	Remarks
18068-18168	AMATEUR S5.120 AMATEUR-SATELLITE S5.154		18068-18168	S5.120		AMATEUR AMATEUR-SATELLITE	

S5.120—For the use of the bands allocated to the amateur service at 3.5 MHz, 7.0 MHz, 10.1 MHz, 14.0 MHz, 18.068 MHz, 21.0 MHz, 24.89 MHz and 144 MHz in the event of natural disasters, see Resolution 640.

S5.154—Additional allocation: in Armenia, Azerbaijan, Belarus, Georgia, Kazakhstan, Moldova, Uzbekistan, Kyrgyzstan, Russia, Tajikistan, Turkmenistan and Ukraine, the band 18068-18168 kHz is also allocated to the fixed service on a primary basis for use within their boundaries, with a peak envelope power not exceeding 1 kW.

FCC Part 97 Privileges

License Class	Terrestrial location of the amateur radio station			
	Region 1 kHz	Region 2 kHz	Region 3 kHz	Limitations*
General Advanced Extra	18068-18168	18068-18168	18068-18168	

*See §§97.301-307 of the FCC rules for further details on authorized privileges for each license class, frequency sharing requirements and authorized modes.

Amateur Radio Band Plans

Region 1 kHz	Region 2 kHz	Region 3 kHz	ARRL kHz
18068-18100 CW	18068-18100 CW	18068-18100 CW	18068-18100 CW
18100-18109 DATA/CW	18100-18105 DATA/CW	18100-18110 DATA/CW	18100-18105 DATA/CW
18109-18111 Beacons	18105-18110 Packet/DATA/CW		18105-18110 Packet/DATA/CW
18111-18168 Phone/CW	18110-18168 Phone/CW	18110-18168 Phone/CW	18110-18168 Phone/CW

Allocations

INTERNATIONAL			UNITED STATES				
Region 1 kHz	Region 2 kHz	Region 3 kHz	Band kHz	National Provisions	Government Allocation	Non-Government Allocation	Remarks
21000-21450	AMATEUR S5.120 AMATEUR-SATELLITE		21000-21450	S5.120		AMATEUR AMATEUR-SATELLITE	

S5.120—For the use of the bands allocated to the amateur service at 3.5 MHz, 7.0 MHz, 10.1 MHz, 14.0 MHz, 18.068 MHz, 21.0 MHz, 24.89 MHz and 144 MHz in the event of natural disasters, see Resolution 640.

FCC Part 97 Privileges

License Class	Terrestrial location of the amateur radio station			
	Region 1 kHz	Region 2 kHz	Region 3 kHz	Limitations*
Novice Technican Plus	21100-21200	21100-21200	21100-21200	
General	21025-21200 21300-21450	21025-21200 21300-21450	21025-21200 21300-21450	
Advanced	21025-21200 21225-21450	21025-21200 21225-21450	21025-21200 21225-21450	
Extra	21000-21450	21000-21450	21000-21450	

* See §§97.301-307 of the FCC rules for further details on authorized privileges for each license class, frequency sharing requirements and authorized modes.

Amateur Radio Band Plans

Region 1 kHz	Region 2 kHz	Region 3 kHz	ARRL kHz
21000-21080 CW	21000-21070 CW	21000-21070 CW	21000-21070 CW
	21070-21090 DATA/CW	21070-21125 DATA/CW	21070-21090 DATA/CW
21080-21100 DATA/CW	21090-21125 Packet/DATA/CW		21090-21100 Packet/DATA/CW
21100-21120 Packet/DATA/CW			21100-21200 CW
21120-21149 CW	21125-21149.5 CW	21125-21149.5 CW	
21149-21151 Beacons	21149.5-21150.5 Beacons	21149.5-21150.5 Beacons	
21151-21450 Phone/CW	21150.5-21200 Phone/CW permitted International traffic	21150.5-21335 Phone/CW	
21340 SSTV/Fax calling frequency	21200-21300 Phone/CW permitted		21200-21450 Phone
	21300-21335 Phone/CW permitted International traffic		
	21335-21345 SSTV/Phone (CW permitted)	21335-21345 SSTV/Phone/CW	21340 SSTV
	21345-21440 Phone (CW permitted)	21345-21450 Phone/CW	
	21440-21450 Phone Emergency (CW permitted)		

Allocations

INTERNATIONAL			UNITED STATES				
Region 1 kHz	Region 2 kHz	Region 3 kHz	Band kHz	National Provisions	Government Allocation	Non-Government Allocation	Remarks
24890-24990	AMATEUR S5.120 AMATEUR-SATELLITE		24890-24990	S5.120		AMATEUR AMATEUR-SATELLITE	

S5.120—For the use of the bands allocated to the amateur service at 3.5 MHz, 7.0 MHz, 10.1 MHz, 14.0 MHz, 18.068 MHz, 21.0 MHz, 24.89 MHz and 144 MHz in the event of natural disasters, see Resolution 640.

FCC Part 97 Privileges

License Class	Terrestrial location of the amateur radio station			
	Region 1 kHz	Region 2 kHz	Region 3 kHz	Limitations*
General Advanced Extra	24890-24990	24890-24990	24890-24990	

*See §§97.301-307 of the FCC rules for further details on authorized privileges for each license class, frequency sharing requirements and authorized modes.

Amateur Radio Band Plans

Region 1 kHz	Region 2 kHz	Region 3 kHz	ARRL kHz
24890-24920 CW	24890-24920 CW	24890-24920 CW	24890-24920 CW
24920-24929 DATA/CW	24920-24925 DATA/CW	24920-24930 DATA/CW	24920-24925 DATA/CW
24929-24931 Beacons	24925-24930 Packet/DATA/CW		24925-24930 Packet/DATA/CW
24931-24990 Phone/CW	24930-24990 Phone/CW	24930-24990 Phone/CW	24930-24990 Phone/CW

Allocations

INTERNATIONAL			UNITED STATES				
Region 1 kHz	Region 2 kHz	Region 3 kHz	Band kHz	National Provisions	Government Allocation	Non-Government Allocation	Remarks
28000-29700 AMATEUR AMATEUR-SATELLITE			28000-29700			AMATEUR AMATEUR-SATELLITE	

FCC Part 97 Privileges

License Class	Terrestrial location of the amateur radio station			
	Region 1 kHz	Region 2 kHz	Region 3 kHz	Limitations*
Novice Technician Plus	28100-28500	28100-28500	28100-28500	
General Advanced Extra	28000-29700	28000-29700	28000-29700	

* See §§97.301-307 of the FCC rules for further details on authorized privileges for each license class, frequency sharing requirements and authorized modes.

Amateur Radio Band Plans

Region 1 kHz	Region 2 kHz	Region 3 kHz	ARRL kHz
28000-28050 CW	28000-28070 CW	28000-28050 CW	28000-28070 CW
28050-28120 DATA/CW	28070-28120 DATA/CW	28050-28150 DATA/CW	28070-28120 DATA/CW
28120-28150 Packet/DATA/CW	28120-28189 Packet/DATA/CW		28120-28189 Packet/DATA/CW
28150-28190 CW		28150-28190 CW	
	28189-28200 CW New Beacon Band	28190-28200 CW New Beacon Band	
28190-28199 Regional Time-Shared Beacons			
28199-28201 Worldwide Time-Shared Beacons	28200-28300 CW Old Beacon Band	28200-28300 CW Old Beacon Band	28190-28300 CW Beacon Band
28201-28225 Continuous Duty Beacons	28300-28670 Phone CW permitted	28300-28675 Phone/CW	28300-29300 Phone
28225-29200 Phone/CW 28680 SSTV/Fax calling frequency	28670-28690 SSTV/Phone CW permitted	28675-28685 SSTV/Phone/CW	28680 SSTV
		28685-29300 Phone/CW	
	28690-29300 Phone CW permitted		
29200-29300 DATA/NBFM packet/Phone/CW			
29300-29510 Satellite Downlink	29300-29510 Satellite	29300-29510 Satellite	29300-29510 Satellite
29510-29700 Phone/CW	29510-29700 FM Phone and repeaters	29510-29700 WB/FM/CW	29510-29590 Repeater inputs
			29600 FM simplex calling frequency
			29610-29690 Repeater outputs

Allocations							
INTERNATIONAL			UNITED STATES				
Region 1 MHz	Region 2 MHz	Region 3 MHz	Band MHz	National Provisions	Government Allocation	Non-Government Allocation	Remarks
47-68 BROADCASTING S5.163 S5.164 S5.165 S5.169 S5.171	50-54 AMATEUR S5.166. S5.167 S5.168 S5.170		50-54			AMATEUR	

S5.163—Additional allocation: in Armenia, Azerbaijan, Belarus, Estonia, Georgia, Hungary, Kazakhstan, Latvia, Lithuania, Moldova, Mongolia, Uzbekistan, Kyrgyzstan, Slovakia, the Czech Republic, Russia, Tajikistan, Turkmenistan and Ukraine, the bands 47-48.5 MHz and 56.5-58 MHz are also allocated to the fixed and land mobile services on a secondary basis.

S5.164—Additional allocation: in Albania, Germany, Austria, Belgium, Bosnia and Herzegovina, Bulgaria, Côte d'Ivoire, Denmark, Spain, Finland, France, Gabon, Greece, Ireland, Israel, Italy, Jordan, Lebanon, Libya, Liechtenstein, Luxembourg, Madagascar, Mali, Malta, Morocco, Mauritania, Monaco, Nigeria, Norway, the Netherlands, Poland, the United Kingdom, Senegal, Slovenia, Sweden, Switzerland, Swaziland, Syria, Togo, Tunisia, Turkey and Yugoslavia, the band 47-68 MHz and in Romania, the band 47-58 MHz, are also allocated to the land mobile service on a primary basis. However, stations of the land mobile service in the countries mentioned in connection with each band referred to in this footnote shall not cause harmful interference to, or claim protection from, existing or planned broadcasting stations of countries other than those mentioned in connection with the band.

S5.165—Additional allocation: in Angola, Cameroon, the Congo, Madagascar, Mozambique, Somalia, Sudan, Tanzania and Chad, the band 47-68 MHz is also allocated to the fixed and mobile, except aeronautical mobile, services on a primary basis.

S5.166—Alternative allocation: in New Zealand, the band 50-51 MHz is allocated to the fixed, mobile and broadcasting services on a primary basis; the band 53-54 MHz is allocated to the fixed and mobile services on a primary basis.

S5.167—Alternative allocation: in Bangladesh, Brunei Darussalam, India, Indonesia, the Islamic Republic of Iran, Malaysia, Pakistan, Singapore and Thailand, the band 50-54 MHz is allocated to the fixed, mobile and broadcasting services on a primary basis.

S5.168—Additional allocation: in Australia, China and the Democratic People's Republic of Korea, the band 50-54 MHz is also allocated to the broadcasting service on a primary basis.

S5.169—Alternative allocation: in Botswana, Burundi, Lesotho, Malawi, Namibia, Rwanda, South Africa, Swaziland, Zaire, Zambia and Zimbabwe, the band 50-54 MHz is allocated to the amateur service on a primary basis.

S5.170—Additional allocation: in New Zealand, the band 51-53 MHz is also allocated to the fixed and mobile services on a primary basis.

S5.171—Additional allocation: in Botswana, Burundi, Lesotho, Malawi Mali, Namibia, Rwanda, South Africa, Swaziland, Zaire and Zimbabwe, the band 54-68 MHz is also allocated to the fixed and mobile, except aeronautical mobile, services on a primary basis.

FCC-Part 97 Privileges

License Class	Terrestrial location of the amateur radio station			
	Region 1 MHz	Region 2 MHz	Region 3 MHz	Limitations*
Technician Technician Plus General Advanced Extra	—	50-54	50-54	

* See §§97.301-307 of the FCC rules for further details on authorized privileges for each license class, frequency sharing requirements and authorized modes.

Amateur Radio Band Plans

Region 1 MHz	Region 2 MHz	Region 3 MHz	ARRL MHz
50.00-50.10 CW	50.00-50.10 CW	50.0-50.10 CW/Beacons	50.00-50.10 CW/Beacons
50.02-50.08 Beacons			50.06-50.08 Automatically Controlled beacons
50.09 Center of CW activity	50.10-50.30 SSB/CW	50.10-54 CW/Phone/WB	50.10-50.30 SSB, CW
50.10-50.50 All narrow-band modes			50.10-50.125 DX window
			50.125 SSB Calling
50.10-50.13 SSB/CW/DX	50.30-50.60 All modes		50.30-50.6 All Modes
			50.4 AM calling
50.11 DX calling	50.60-50.80 Non-voice		50.60-50.80 Nonvoice Communications
50.15 SSB center of activity			50.62 Digital calling
	50.80-51.00 Radio control		50.80-51.00 Radio Control (20 kHz channels)
50.185 Center of activity for crossband working	51.00-51.10 Pacific DX Window		51.00-51.10 Pacific DX Window
			51.12-51.48 Repeater inputs (19 channels)
50.20 Meteor scatter center of activity	51.10-53.00 FM		51.12-51.18 Digital repeater outputs
			51.50-51.60 Simplex (6 channels)
50.50-52.00 All modes			51.62-51.98 Repeater outputs (19 channels)
50.51 SSTV (AFSK)			51.62-51.68 Digital repeater outputs
50.55 Fax working frequency			52.0-52.48 Repeater inputs (except as noted; 23 channels)
50.60 RTTY (FSK)			52.02, 52.04 FM simplex
50.62-50.75 Digital			52.2 TEST PAIR (input)
51.21-51.39 FM repeaters input channels, 20 kHz spacing			52.50-52.98 Repeater outputs (except as noted; 23 channels)
			52.525 Primary FM simplex
51.41-51.59 FM			52.54 Secondary FM simplex
51.51 FM calling frequency			52.7 TEST PAIR (output)
51.81-51.99 FM repeaters output channels, 20 kHz spacing	53.00-54.00 FM and radio control		53.0-53.48 Repeater inputs (except as noted; 19 channels)
			53.0 Remote base FM simplex
			53.02 Simplex
			53.1, 53.2, 53.3, 53.4 Radio remote control
			53.50-53.98 Repeater outputs (except as noted; 19 channels)
			53.5, 53.6, 53.7, 53.8 Radio remote control
			53.52, 53.9 Simplex

Allocations							
INTERNATIONAL			**UNITED STATES**				
Region 1 *MHz*	Region 2 *MHz*	Region 3 *MHz*	Band *MHz*	National *Provisions*	Government *Allocation*	Non-Government *Allocation*	*Remarks*
144-146	AMATEUR S5.120 AMATEUR-SATELLITE S5.216		144-146	S5.120		AMATEUR AMATEUR-SATELLITE	
146-148 FIXED MOBILE except aeronautical mobile (R) S5.217	146-148 AMATEUR S5.217	146-148 AMATEUR FIXED MOBILE S5.217	146-148			AMATEUR	

S5.120—For the use of the bands allocated to the amateur service at 3.5 MHz, 7.0 MHz, 10.1 MHz, 14.0 MHz, 18.068 MHz, 21.0 MHz, 24.89 MHz and 144 MHz in the event of natural disasters, see Resolution 640.

S5.216—Additional allocation: in China, the band 144-146 MHz is also allocated to the aeronautical mobile (OR) service on a secondary basis.

S5.217—Alternative allocation: in Afghanistan, Bangladesh, Cuba, Guyana and India, the band 146-148 MHz is allocated to the fixed and mobile services on a primary basis.

FCC Part 97 Privileges				
	Terrestrial location of the amateur radio station			
License *Class*	Region 1 *MHz*	Region 2 *MHz*	Region 3 *MHz*	*Limitations**
Technician Technician Plus General Advanced Extra	144-146	144-148	144-148	

* See §§97.301-307 of the FCC rules for further details on authorized privileges for each license class, frequency sharing requirements and authorized modes.

Amateur Radio Band Plans

Region 1 MHz
- 144.00-144.035 EME SSB and CW
- 144.035-144.15 CW exclusive
- 144.05 CW calling
- 144.15-144.40 SSB
- 144.30 SSB calling
- 144.40-144.49 Beacons
- 144.49 SAREX uplink
- 144.50 SSTV calling
- 144.50-144.80 All modes
- 144.60 RTTY calling
- 144.70 FAX calling
- 144.75 ATV calling
- 144.80-144.99 Digital
- 144.994-145.1935 FM repeater inputs (12.5 kHz spacing)
- 145.194-145.5935 FM simplex channels (12.5 kHz spacing)
- 145.50 Mobile calling
- 145.594-145.7935 FM repeater outputs (12.5 kHz spacing)
- 145.80-146.00 Satellites

Region 2 MHz
- 144.00-144.275 CW/SSB (DX EME)
- 144.200 SSB calling
- 144.275-144.300 Beacons
- 144.300-144.500 Experimental
- 144.500-144.900 Repeaters
- 144.900-145.100 Digital modes
- 145.100-145.500 Repeaters
- 145.500-145.590 SAREX
- 145.590-145.800 Digital modes
- 145.800-146.00 Satellites
- 146.00-146.400 Repeaters
- 146.400-146.600 FM simplex
- 146.520 FM calling
- 146.600-147.400 Repeaters
- 147.400-147.600 All-mode simplex
- 147.600-148.00 Repeaters

Region 3 MHz
- 144.00-144.035 EME
- 145.80-146.00 Satellite

ARRL MHz
- 144.00-144.05 EME (CW)
- 144.05-144.10 General CW and weak signals
- 144.10-144.20 EME and weak-signal SSB
- 144.20 National SSB calling
- 144.20-144.275 General SSB operation
- 144.275-144.30 Beacons
- 144.30-144.50 New OSCAR subband
- 144.50-144.60 Linear translator inputs
- 144.60-144.90 FM repeater inputs
- 144.90-145.10 Weak signal and FM simplex
- 145.10-145.20 Linear translator outputs
- 145.20-145.50 FM repeater outputs
- 145.50-145.80 Miscellaneous and experimental modes
- 145.80-146.00 OSCAR subband
- 146.01-146.37 Repeater inputs
- 146.40-146.58 Simplex*
- 146.52 National simplex calling
- 146.61-147.39 Repeater outputs
- 147.42-147.57 Simplex
- 147.60-147.99 Repeater inputs

ARRL
*The frequency 146.40 MHz is used in some areas as a repeater input.

The following Packet Radio frequency recommendations were adopted by the ARRL Board of Directors in July 1987.
1) Automatic/unattended operations should be conducted on 145.01, 145.03, 145.05, 145.07 and 145.09 MHz.
   a) 145.01 should be reserved for inter-LAN use.
   b) Use of the remaining frequencies (above) should be determined by local user groups.
2) Additional frequencies within the 2-meter band may be designated for packet-radio use by local coordinators.

Packet Footnotes
Specific VHF/UHF channels recommended above may not be available in all areas of the US.
Prior to regular packet-radio use of any VHF/UHF channel, it is advisable to check with the local frequency coordinator.
The decision as to how the available channels are to be used should be based on coordination between local packet-radio users.

Repeater frequency pairs (input/output):

144.51/145.11	144.81/145.41	146.31/146.91
144.53/145.13	144.83/145.43	146.34/146.94
144.55/145.15	144.85/145.45	146.37/146.97
144.57/145.17	144.87/145.47	146.40 or 147.60/147.00*
144.59/145.19	144.89/145.49	146.43 or 147.63/147.03*
144.61/145.21	146.01/146.61	146.46 or 147.66/147.06*
144.63/145.23	146.04/146.64	147.69/147.09
144.65/145.25	146.07/146.67	147.72/147.12
144.67/145.27	146.10/146.70	147.75/147.15
144.69/145.29	146.13/146.73	147.78/147.18
144.71/145.31	146.16/146.76	147.81/147.21
144.73/145.33	146.19/146.79	147.84/147.24
144.75/145.35	146.22/146.82	147.87/147.27
144.77/145.37	146.25/146.85	147.90/147.30
144.79/145.39	146.28/146.88	147.93/147.33
		147.96/147.36
		147.99/147.39

*Local Option

Some areas use 146.40-146.60 and 147.40-147.60 MHz for either simplex or repeater inputs and outputs. Frequency pairs in those areas are:

147.415/146.415	147.46/146.46	147.505/146.505
147.43/146.43	147.475/146.475	147.595/146.595
147.445/146.445	147.49/146.49	

Suggested 15-kHz splinter channels (input/output):

146.025/146.625	146.295/146.895	147.735/147.135
146.055/146.655	146.325/146.925	147.765/147.165
146.085/146.685	146.355/146.955	147.825/147.225
146.115/146.715	146.385/146.985	147.855/147.255
146.145/146.745	147.615/147.015	147.885/147.285
146.175/146.775	147.645/147.045	147.915/147.315
146.205/146.805	147.675/147.075	147.945/147.345
146.235/146.835	147.705/147.105	147.975/147.375
146.265/146.865		

Simplex frequencies:

*146.415	*146.475	146.535	146.595	147.465	147.525
*146.43	*146.49	146.55	147.42	147.48	147.54
*146.445	*146.505	146.565	147.435	147.495	147.555
*146.46	#146.52	146.58	147.45	147.51	147.57
					147.585

#National Simplex Frequency
*May also be a repeater (input/output). See repeater pairs listing.

Several states have chosen to realign the 146-148 MHz band, using 20 kHz spacing between channels. This choice was made to gain additional repeater pairs.

The transition from 30 to 20 kHz spacing is taking place on a case by case basis as the need for additional pairs occurs. Typically the repeater on an odd numbered pair will shift to 10 kHz, up or down, creating a new set on an even numbered channel. For example, the pair of 146.13/.73 would change to 146.12/.72 or 146.14/.74 while the pairs of 146.10/.70 and 146.16/.76 would be left unchanged.

INTERNATIONAL			UNITED STATES				
Region 1 MHz	Region 2 MHz	Region 3 MHz	Band MHz	National Provisions	Government Allocation	Non-Government Allocation	Remarks
223-230 BROADCASTING Fixed Mobile	220-225 AMATEUR FIXED MOBILE RadiolocationS5.241	223-230 FIXED MOBILE BROADCASTING AERONAUTICAL RADIONAVIGATION	216-220	S5.241 G2 NG152	MARITIME MOBILE Aeronautical-Mobile Fixed Land Mobile Radiolocation	MARITIME MOBILE Aeronautical-Mobile Fixed Land Mobile	Amateur 219-220 MHz
S5.243 S5.244 S5.246 S5.247	225-235 FIXED MOBILE	Radiolocation S5.250	222-225	US243 S5.241	Radiolocation	AMATEUR	

S5.241—In Region 2, no new stations in the radiolocation service may be authorized in the band 216-225 MHz. Stations authorized prior to 1 January 1990 may continue to operate on a secondary basis.

S5.243—Additional allocation: in Somalia, the band 216-225 MHz is also allocated to the aeronautical radionavigation service on a primary basis, subject to not causing harmful interference to existing or planned broadcasting services in other countries.

S5.244—Additional allocation: in Oman, the United Kingdom and Turkey, the band 216-235 MHz is also allocated to the radiolocation service on a secondary basis.

S5.246—Additional allocation: in Spain, France, Israel and Monaco, the band 223-230 MHz is allocated to the broadcasting and land mobile services on a primary basis (see No. S5.33) on the basis that, in the preparation of frequency plans, the broadcasting service shall have prior choice of frequencies; and allocated to the fixed and mobile, except land mobile, services on a secondary basis. However, the stations of the land mobile service shall not cause harmful interference to, or claim protection from, existing or planned broadcasting stations in Morocco and Algeria.

S5.247—Additional allocation: in Saudi Arabia, Bahrain, the United Arab Emirates, Jordan, Oman, Qatar and Syria, the band 223-235 MHz is also allocated to the aeronautical radionavigation service on a primary basis.

S5.250—Additional allocation: in China, the band 225-235 MHz is also allocated to the radio astronomy service on a secondary basis.

US243—In the band 220-225 MHz, stations in the radiolocation service have priority until 1 January 1990.

G2—In the bands 216-225, 420-450 (except as provided by US217) 890-902 928-942, 1300-1400, 2310-2390, 2417-2450, 2700-2900, 5650-5925, and 9000-9200 MHz, the Government radiolocation is limited to the military services.

NG152—The band 219-220 MHz is also allocated to the amateur service on a secondary basis for stations participating, as forwarding stations, in point-to-point fixed digital message forwarding systems, including inter-city packet backbone networks.

FCC Part 97 Privileges				
	Terrestrial location of the amateur radio station			
License Class	Region 1 MHz	Region 2 MHz	Region 3 MHz	Limitations*
Novice	—	222-225	—	
Technician TechnicianPlus General Advanced Extra		219-220* 222-225		219-220 MHz: point-to-point digital message forwarding stations only. Stations must register with ARRL 30 days prior to activation.

*See §§97.301-307 of the FCC rules for further details on authorized privileges for each license class, frequency sharing requirements and authorized modes.

NOTE: Effective August 28, 1991 the FCC allocated 220 to 222 MHz on an exclusive basis to the Land Mobile Service and allocated 222 to 225 MHz exclusively to the Amateur Radio Service.

Amateur Radio Band Plans

Region 1 MHz	Region 2 MHz	Region 3 MHz	ARRL MHz
			219-220**** Digital Links
			222.0-222.15 Weak-signal modes
			222.0-222.025 EME
			222.05-222.06 Propagation beacons
			222.10 Calling frequency (SSB & CW)
			222.10-222.15 Weak-signal SSB & CW
			222.15-222.25 Local coordinator's option: Weak signal, repeater inputs, control
			222.25-223.38 Repeater inputs**
			223.40-223.52 Simplex***
			223.52-223.64 Digital, packet
			223.64-223.70 Links, control
			223.71-223.85 Local coordinator's option: FM simplex, packet, repeater outputs
			223.85-224.98 Repeater outputs**

**Repeater frequency pairs (input/output):

222.26/223.86	222.50/224.10	222.74/224.34	222.96/224.56	223.18/224.78
222.28/223.88	222.52/224.12	222.76/224.36	222.98/224.58	223.20/224.80
222.30/223.90	222.54/224.14	222.78/224.38	223.00/224.60	223.22/224.82
222.32/223.92	222.56/224.16	222.80/224.40	223.02/224.62	223.24/224.84
222.34/223.94	222.58/224.18	222.82/224.42	223.04/224.64	223.26/224.86
222.36/223.96	222.60/224.20	222.84/224.44	223.06/224.66	223.28/224.88
222.38/223.98	222.62/224.22	222.86/222.46	223.08/224.68	223.30/224.90
222.40/224.00	222.64/224.24	222.88/224.48	223.10/224.70	223.32/224.92
222.42/224.02	222.66/224.26	222.90/224.50	223.12/224.72	223.34/224.94
222.44/224.04	222.68/224.28	222.92/224.52	223.14/224.74	223.36/224.96
222.46/224.06	222.70/224.30	222.94/224.54	223.16/224.76	223.38/224.98
222.48/224.08	222.72/224.32			

***Simplex frequencies:

223.42	223.52	223.62	223.72	223.82
223.44	223.54	223.64	223.74	223.84
223.46	223.56	223.66	223.76	
223.48	223.58	223.68	223.78	
223.50‡	223.60	223.70	223.80	

‡National simplex frequency

****100-kHz channels for point-to point fixed digital message-forwarding systems with a maximum of 50 watts PEP. Notification of ARRL 30 days before operation is required, and coordination with Automated Maritime Telecommunication System (AMTS) Coast Stations may be required; see Section 97.303(e).

Channel A	219.050	Channel F	219.550
Channel B	219.150	Channel G	219.650
Channel C	219.250	Channel H	219.750
Channel D	219.350	Channel I	219.850
Channel E	219.450	Channel J	219.950

Allocations							
INTERNATIONAL				**UNITED STATES**			
Region 1 MHz	Region 2 MHz	Region 3 MHz	Band MHz	National Provisions	Government Allocation	Non-Government Allocation	Remarks
420-430	FIXED MOBILE Except aeronautical mobile Radiolocation 5.269 S5.270 S5.271		420-450	US7 US87 US217 US228 US230 S5.282 S5.286	RADIOLOCATION	Amateur	
430-440 AMATEUR RADIOLOCATION S5.138 S5.271 S5.272 S5.273 S5.274 S5.275 S5.276 S5.277 S5.280 S5.281 S5.282 S5.283	430-440 RADIOLOCATION Amateur S5.271, S5.276 S5.277 S5.278 S5.279 S5.281 S5.282						
440-450	FIXED MOBILE except aeronautical mobile Radiolocation S5.269 S5.270 S5.271 S5.284 S5.285 S5.286				G2 G8	NG135	

S5.269—Different category of service: in Australia, the United States, India, Japan and the United Kingdom, the allocation of the bands 420-430 MHz and 440-450 MHz to the radiolocation service is on a primary basis (see No. S5.33).

S5.270—Additional allocation: in Australia, the United States, Jamaica and the Philippines, the bands 420-430 MHz and 440-450 MHz are also allocated to the amateur service on a secondary basis.

S5.271—Additional allocation: in Armenia, Azerbaijan, Belarus, China, Estonia, Georgia, India, Kazakhstan, Latvia, Lithuania, Moldova, Uzbekistan, Kyrgyzstan, the United Kingdom, Russia, Tajikistan, Turkmenistan and Ukraine, the band 420-460 MHz is also allocated to the aeronautical radionavigation service (radio altimeters) on a secondary basis.

S5.272—Different category of service: in France, the allocation of the band 430-434 MHz to the amateur service is on a secondary basis (see No. S5.32).

S5.273—Different category of service: in Denmark, Libya and Norway, the allocation of the bands 430-432 MHz and 438-440 MHz to the radiolocation service is on a secondary basis (see No. S5.32).

S5.274—Alternative allocation: in Denmark, Norway and Sweden, the bands 430-432 MHz and 438-440 MHz are allocated to the fixed and mobile, except aeronautical mobile, services on a primary basis.

S5.275—Additional allocation: in Bosnia and Herzegovina, Croatia, Finland, The Former Yugoslav Republic of Macedonia, Libya, Slovenia and Yugoslavia, the bands 430-432 MHz and 438-440 MHz are also allocated to the fixed and mobile, except aeronautical mobile, services on a primary basis.

S5.276—Additional allocation: in Afghanistan, Algeria, Saudi Arabia, Bahrain, Bangladesh, Brunei Darussalam, Burkina Faso, Burundi, Egypt, the United Arab Emirates, Ecuador, Eritrea, Ethiopia, Greece, Guinea, India, Indonesia, the Islamic Republic of Iran, Iraq, Israel, Italy, Jordan, Kenya, Kuwait, Lebanon, Libya, Liechtenstein, Malaysia, Malta, Nigeria, Oman, Pakistan, the Philippines, Qatar, Syria, Singapore, Somalia, Switzerland, Tanzania, Thailand, Togo, Turkey and Yemen, the band 430-440 MHz is also allocated to the fixed service on a primary basis and the bands 430-435 MHz and 438-440 MHz are also allocated to the mobile, except aeronautical mobile, service on a primary basis.

S5.277—Additional allocation: in Angola, Armenia, Azerbaijan, Belarus, Bulgaria, Cameroon, the Congo, Djibouti, Estonia, Gabon, Georgia, Hungary, Kazakhstan, Latvia, Mali, Moldova, Mongolia, Niger, Uzbekistan, Pakistan, Poland, Kyrgyzstan, Democratic People's Republic of Korea, Slovakia, the Czech Republic, Romania, Russia, Rwanda, Tajikistan, Chad, Turkmenistan and Ukraine, the band 430-440 MHz is also allocated to the fixed service on a primary basis.

S5.278—Different category of service: in Argentina, Colombia, Costa Rica, Cuba, Guyana, Honduras, Panama and Venezuela, the allocation of the band 430-440 MHz to the amateur service is on a primary basis (see No. S5.33).

S5.279—Additional allocation: in Mexico, the bands 430-435 MHz and 438-440 MHz are also allocated on a primary basis to the land mobile service, subject to agreement obtained under Article 14/No. S9.21.

S5.280—In Germany, Austria, Bosnia and Herzegovina, Croatia, The Former Yugoslav Republic of Macedonia, Liechtenstein, Portugal, Slovenia, Switzerland and Yugoslavia, the band 433.05-434.79 MHz (centre frequency 433.92 MHz) is designated for industrial, scientific and medical (ISM) applications. Radiocommunication services of these countries operating within this band must accept harmful interference which may be caused by these applications. ISM equipment operating in this band is subject to the provisions of No. S15.13.

S5.281—Additional allocation: in French Overseas Departments in Region 2 and India, the band 433.75-434.25 MHz is also allocated to the space operation service (Earth-to-space) on a primary basis. In France and in Brazil, the band is allocated to the same service on a secondary basis.

S5.282—In the bands 435-438 MHz, 1260-1270 MHz, 2400-2450 MHz, 3400-3410 MHz (in Regions 2 and 3 only) and 5650-5670 MHz, the amateur-satellite service may operate subject to not causing harmful interference to other services operating in accordance with the Table (see No S5.43). Administrations authorizing such use shall ensure that any harmful interference caused by emissions from a station in the amateur-satellite service is immediately eliminated in accordance with the provisions of No. S25.11. The use of the bands 1260-1270 MHz and 5650-5670 MHz by the amateur-satellite service is limited to the Earth-to-space direction.

S5.283—Additional allocation: in Austria, the band 438-440 MHz is also allocated to the fixed and mobile, except aeronautical mobile, services on a primary basis.

S5.284—Additional allocation: in Canada, the band 440-450 MHz is also allocated to the amateur service on a secondary basis.

S5.285—Different category of service: in Canada, the allocation of the band 440-450 MHz to the radiolocation service is on a primary basis (see No. S5.33).

S5.286—The band 449.75-450.25 MHz may be used for the space operation service (Earth-to-space) and the space research service (Earth-to-space), subject to agreement obtained under Article 14/No. S9.21.

US7—In the band 420-450 MHz and within the following areas, the peak envelope power output of a transmitter employed in the amateur service shall not exceed 50 watts, unless expressly authorized by the Commission after mutual agreement, on a case-by-case basis, between the Federal Communications Commission Engineer in Charge at the applicable District Office and the Military Area Frequency Coordinator at the applicable military base:

(a) Those portions of Texas and New Mexico bounded on the south by latitude 31° 45' North, on the east by 104° 00' West, on the north by latitude 34° 30' North, and on the west by longitude 107° 30' West;

(b) The entire State of Florida including the Key West area and the areas enclosed within a 200-mile radius of Patrick Air Force Base, Florida (latitude 28° 21' North, longitude 80° 43' West), and within a 200-mile radius of Eglin Air Force Base, Florida (latitude 30° 30' North, longitude 86° 30' West);

(c) The entire State of Arizona;

(d) Those portions of California and Nevada south of Latitude 37° 10' North, and the areas enclosed within a 200-mile radius of the Pacific Missile Test Center, Point Mugu, California (latitude 34° 09' North, longitude 119° 11' West).

(e) In the State of Massachusetts within a 160-kilometer (100 mile) radius around locations at Otis Air Force Base, Massachusetts (latitude 41° 45' North, longitude 70° 32' West).

(f) In the State of California within a 240-kilometer (150 mile) radius around locations at Beale Air Force Base, California (latitude 39° 08' North, longitude 121° 26' West).

(g) In the State of Alaska within a 160-kilometer (100 mile) radius of Clear, Alaska (latitude 64° 17' North, longitude 149° 10' West). (The Military Area Frequency Coordinator for this area is located at Elmendorf Air Force Base, Alaska.)

(h) In the State of North Dakota within a 160-kilometer (100 mile) radius of Concrete, North Dakota (latitude 48° 43' North, longitude 97° 54' West). (The Military Area Frequency Coordinator for this area can be contacted at: HQ SAC/SXOE, Offutt Air Force Base, Nebraska 68113.)

(i) In the States of Alabama, Florida, Georgia and South Carolina within a 200 kilometer (124 mile) radius of Warner Robins Air Force Base, Georgia (latitude 32° 38' North, longitude 83° 35' West).

(j) In the State of Texas within a 200-kilometer (124 mile) radius of Goodfellow Air Force Base, Texas (latitude 31° 25' North, longitude 100° 24' West).

US87—The frequency 450 MHz, with maximum emission bandwidth of 500 kHz, may be used for Government and non-Government stations for space telecommand at specific locations, subject to such conditions as may be applied on a case-by-case basis.

US217—Pulse-ranging radiolocation systems may be authorized for Government and non-Government use in the 420-450 MHz band along the shorelines of Alaska and the contiguous 48 States. Spread spectrum radiolocation systems may be authorized in the 420-435 MHz portion of the band for operation within the contiguous 48 States and Alaska. Authorizations will be granted on a case-by-case basis; however, operations proposed to be located within the zones set forth in US228 should not expect to be accommodated. All stations operating in accordance with this provision will be secondary to stations operating in accordance with the Table of Frequency Allocations.

US228—Applicants of operation in the band 420 to 450 MHz under the provisions of US217 should not expect to be accommodated if their area of service is within the following geographic areas:

(a) Those portions of Texas and New Mexico bounded on the south by latitude 31° 45' North, on the east by 104° 00' West, on the north by latitude 34° 30' North, and on the west by longitude 107° 30' West;

(b) The entire State of Florida including the Key West area and the areas enclosed within a 200-mile radius of Patrick Air Force Base, Florida (latitude 28° 21' North, longitude 80° 43' West), and within a 200-mile radius of Eglin Air Force Base, Florida (latitude 30° 30' North, longitude 86° 30' West);

(c) The entire State of Arizona;

(d) Those portions of California and Nevada south of Latitude 37° 10' North, and the areas enclosed within a 200-mile radius of the Pacific Missile Test Center, Point Mugu, California (latitude 34° 09' North, longitude 119° 11' West).

(e) In the State of Massachusetts within a 160-kilometer (100 mile) radius around locations at Otis Air Force Base, Massachusetts (latitude 41° 45' North, longitude 70° 32' West).

(f) In the State of California within a 240-kilometer (150 mile) radius around locations at Beale Air Force Base, California (latitude 39° 8' North, longitude 121° 26' West).

(g) In the State of Alaska within a 160-kilometer (100 mile) radius of Clear, Alaska (latitude 64° 17' North, longitude 149° 10' West). (The Military Area Frequency Coordinator for this area is located at Elmendorf Air Force Base, Alaska.)

(h) In the State of North Dakota within a 160-kilometer (100 mile) radius of Concrete, North Dakota (latitude 48° 43' North, longitude 97° 54' West). (The Military Area Frequency Coordinator for this area can be contacted at: HQ SAC/SXOE, Offutt Air Force Base, Nebraska 68113.)

(i) In the States of Alabama, Florida, Georgia and South Carolina within a 200-kilometer (124 mile) radius of Warner Robins Air Force Base, Georgia (latitude 32° 38' North, longitude 83° 35' West).

(j) In the State of Texas within a 200-kilometer (124 mile) radius of Goodfellow Air Force Base, Texas (latitude 31° 25' North, longitude 100° 24' West).

G2—In the bands 216-225, 420-450 (except as provided by US217), 890-902, 928-942, 1300-1400, 2310-2390, 2417-2450, 2700-2900, 5650-5925, and 9000-9200 MHz, the Government radiolocation is limited to the military services.

US230—Non-government land mobile service is allocated on a primary basis in the bands 422.1875-425.4875 and 427.1875-429.9875 MHz within 50 statute miles of Detroit, MI, and Cleveland, OH, and in the bands 423.8125-425.4875 and 428.8125-429.9875 MHz within 50 statute miles of Buffalo, NY.

G8—Low power government radio control operations are permitted in the band 420-450 MHz.

NG135—In the 420-430 MHz band the Amateur service is not allocated north of line A. (def. §2.1).

FCC Part 97 Privileges

License Class	Terrestrial location of the amateur radio station			
	Region 1 MHz	Region 2 MHz	Region 3 MHz	Limitations*
Technician Technician Plus General Advanced Extra	430-440	420-450	420-450	

*See §§97.301-307 of the FCC rules for futher details on authorized privileges for each license class, frequency sharing requirement and authorized modes.

The following Packet Radio frequency recommendations were adopted by the ARRL Board of Directors in January 1988.

1) 100-kHz-bandwidth channels:

430.05	430.35	430.65
430.15	430.45	430.85
430.25	430.55	430.95

2) 25-kHz-bandwidth channels:

431.025	441.000	441.050
440.975	441.025	441.075

Amateur Radio Band Plans

Region 1 MHz	Region 2 MHz	Region 3 MHz	ARRL MHz
430.000-431.981 Sub-regional (national bandplanning)			420.00-426.00 ATV repeater or simplex with 421.25-MHz video carrier, control links and experimental
432.000-432.150 CW only			426.00-432.00 ATV simplex with 427.25-MHz video carrier
432.15-432.50 SSB/CW		430.00-431.90	
432.20 SSB center of activity		431.90-432.24 EME	432.00-432.07 EME
432.35 Microwave talkback center of activity			432.07-432.10 Weak-signal CW
432.50-432.60 Linear transponder input			432.10 Calling frequency
432.60-432.80 Linear transponder output		432.24-435.00	432.10-432.30 Mixed mode and weak signal
432.80-432.99 Beacons			432.30-432.40 Beacons
432.994-443.381 Repeater input Region 1 standard (25 kHz spacing)			432.40-433.00 Mixed mode and weak signals
433.394-433.581 FM simplex channels (25 kHz spacing)			433.00-435.00 Auxiliary/repeater links
433.50 Mobile FM calling		435.00-438.00 Satellite	435.00-438.00 Satellite only (internationally)
433.60-434.00 All modes		438.00-450.00	438.00-444.00 ATV repeater input with 439.25-MHz video carrier and repeater links
434.00-434.594 ATV			442.00-445 Repeater inputs and outputs (local option)
434.594-435.981 ATV/repeater output (Region 1 system, 25 kHz spacing)			445.00-447.00 Shared by auxiliary and control links, repeaters and simplex (local option)
435.981-438.000 ATV/satellite service			446.00 National simplex frequency
438.00-440.00 ATV and sub-regional (national bandplanning)			447.00-450.00 Repeater inputs and outputs (local option)

Allocations							
INTERNATIONAL			**UNITED STATES**				
Region 1 MHz	Region 2 MHz	Region 3 MHz	Band MHz	National Provisions	Government Allocation	Non-Government Allocation	Remarks
890-942 FIXED MOBILE except aeronautical mobile BROADCASTING S5.322 Radiolocation	902-928 FIXED Amateur Mobile except aeronautical mobile Radiolocation	890-942 FIXED MOBILE BROADCASTING Radiolocation	902-928	US215 US218 US267 US275 S5.325	RADIOLOCATION	PRIVATE LAND MOBILE AMATEUR	(ISM 915 ± 13 MHz)
S5.323	S5.150 S5.325 S5.326	S5.327			G11 G59		

S5.150—The following bands: 13553-13567 kHz (centre frequency 13560 kHz), 26957-27283 kHz (centre frequency 27120 kHz), 40.66-40.70 MHz (centre frequency 40.68 MHz), 902-928 MHz in Region 2 (centre frequency 915 MHz), 2400-2500 MHz (centre frequency 2450 MHz), 5725-5875 MHz (centre frequency 5800 MHz), and 24-24.25 GHz (centre frequency 24.125 GHz) are also designated for industrial, scientific and medical (ISM) applications. Radiocommunication services operating within these bands must accept harmful interference which may be caused by these applications. ISM equipment operating in these bands is subject to the provisions of No. S15.13.

S5.322—In Region 1, in the band 862-960 MHz, stations of the broadcasting service shall be operated only in the African Broadcasting Area (see Nos. S5.10 to S5.13) excluding Algeria, Egypt, Spain, Libya and Morocco, subject to agreement obtained under Article 14/No. S9.21.

S5.323—Additional allocation: in Armenia, Azerbaijan, Belarus, Bulgaria, Georgia, Hungary, Kazakstan, Latvia, Lithuania, Moldova, Mongolia, Uzbekistan, Poland, Kyrgyzstan, Slovakia, the Czech Republic, Romania, Russia, Tajikistan, Turkmenistan and Ukraine, the band 862-960 MHz is also allocated to the aeronautical radionavigation service on a primary basis until 1 January 1998. Up to this date, the aeronautical radionavigation service may use the band, subject to agreement obtained under Article 14/No. S9.21. After this date, the aeronautical radionavigation service may continue to operate on a secondary basis.

S5.325—Different category of service: in the United States, the allocation of the band 890-942 MHz to the radiolocation service is on a primary basis, (see No.S5.33), subject to agreement obtained under Article 14/No. S9.21.

S5.326—Different category of service: in Chile, the band 903-905 MHz is allocated to the mobile, except aeronautical mobile, service on a primary basis, subject to agreement obtained under Article 14/No. S9.21.

S5.327—Different category of service: in Australia, the allocation of the band 915-928 MHz to the radiolocation service is on a primary basis (see No. S5.33).

US215—Emissions from microwave ovens manufactured on and after January 1, 1980, for operation on the frequency 915 MHz must be confined within the band 902-928 MHz. Emissions from microwave ovens manufactured prior to January 1, 1980, for operation on the frequency 915 MHz must be confined within the band 902-940 MHz. Radiocommunications services operating within the band 928-940 MHz must accept any harmful interference that may be experienced from the operation of microwave ovens manufactured before January 1, 1980.

US218—The band 902-928 MHz available for Location and Monitoring Service (LMS) systems subject to not causing harmful interference to the operation of all Government stations authorized in these bands. These systems must tolerate interference from the operation of industrial, scientific, and medical (ISM) devices and the operation of Government stations authorized in these bands.

US267—In the band 902-928 MHz, amateur radio stations shall not operate within the States of Colorado and Wyoming, bounded by the area of: latitude 39° N to 42° N and longitude 103° W to 108° W.

US275—The band 902-928 MHz is allocated on a secondary basis to the amateur service subject to not causing harmful interference to the operations of Government stations authorized in this band or to the Location and Monitoring Service (LMS) systems. Stations in the amateur service must tolerate any interference from the operations of industrial, scientific and medical (ISM) devices, LMS systems, and the operations of Government stations authorized in this band. Further, the Amateur Service is prohibited in those portions of Texas and New Mexico bounded on the south by latitude 31° 41' North, on the east by longitude 104° 11' West, on the north by latitude 34° 30' North, and on the west by longitude 107° 30' West; in addition, outside this area but within 240 kilometers (150 miles) of these boundaries of White Sands Missile Range the service is restricted to a maximum transmitter peak envelope power output of 50 watts.

G11—Government fixed and mobile radio services, including low power radio control operations, are permitted in the band 902-928 MHz on a secondary basis.

G59—In the bands 902-928 MHz, 3100-3300 MHz, 3500-3700 MHz, 5250-5350 MHz, 8500-9000 MHz, 9200-9300 MHz, 13.4-14.0 GHz, 15.7-17.7 GHz and 24.05-24.25 GHz, all Government non-military radiolocation shall be secondary to military radiolocation, except in the sub-band 15.7-16.2 GHz, airport surface detection equipment (ASDE) is permitted on a co-equal basis subject to coordination with the military departments.

FCC Part 97 Privileges				
	Terrestrial location of the amateur radio station			
License Class	Region 1 MHz	Region 2 MHz	Region 3 MHz	Limitations*
Technician Technician Plus General Advanced Extra	—	902-928	—	

*See §§97.301-307 of the FCC rules for further details on authorized privileges for each license class, frequency sharing requirements and authorized modes.

Amateur Radio Band Plans

Region 1 MHz	Region 2 MHz	Region 3 MHz	ARRL* MHz
			902.0-903.0 Weak signal communications
			902.1 Calling Frequency
			903.0-906.0 Digital
			903.1 Alternate calling frequency
			906.0-909.0 FM repeater inputs
			909.0-915.0 ATV
			915.0-918.0 Digital
			918.0-921.0 FM repeater outputs
			921.0-927.0 ATV
			927.0-928.0 FM simplex and links

*Adopted by the ARRL Board of Directors meeting in July 1989.

The following Packet Radio frequency recommendations were adopted by the ARRL Board of Directors in January 1988 as interim guidance.

Two 3-MHz-bandwidth channels are recommended for 1.5 Mbit/s links. They are 903-906 MHz and 915-918 MHz with 10.7-MHz spacing.

Packet Footnotes
Specific VHF/UHF channels recommended above may not be available in all areas of the US. These channels may interfere with existing weak-signal communications operating on or about 903.100 MHz.

Prior to regular packet-radio use of any VHF/UHF channel, it is advisable to check with the local frequency coordinator.

The decision as to how the available channels are to be used should be based on coordination between local packet-radio users.

Allocations							
INTERNATIONAL			**UNITED STATES**				
Region 1 MHz	Region 2 MHz	Region 3 MHz	Band MHz	National Provisions	Government Allocation	Non-Government Allocation	Remarks
1240-1260			1240-1300	S5.272 S5.333 S5.334	RADIOLOCATION	Amateur	
	RADIOLOCATION RADIONAVIGATION-SATELLITE (Space-to-Earth) S5.329 Amateur						
	S5.330 S5.331 S5.333 S5.334						
1260-1300	RADIOLOCATION Amateur						
	S5.282 S5.330 S5.331 S5.333 S5.334				G56		

S5.282—In the bands 435- 438 MHz, 1260-1270 MHz, 2400-2450 MHz, 3 400-3 410 MHz (in Regions 2 and 3 only) and 5650-5670 MHz, the amateur-satellite service may operate subject to not causing harmful interference to other services operating in accordance with the Table (see No S5.43). Administrations authorizing such use shall ensure that any harmful interference caused by emissions from a station in the amateur-satellite service is immediately eliminated in accordance with the provisions of No. S25.11. The use of the bands 1260-1270 MHz and 5650-5670 MHz by the amateur-satellite service is limited to the Earth-to-space direction.

S5.329—Use of the radionavigation-satellite service in the band 1 215-1 260 MHz shall be subject to the condition that no harmful interference is caused to the radionavigation service authorized under No. S5.331.

S5.330—Additional allocation: in Angola, Saudi Arabia, Bahrain, Bangladesh, Cameroon, China, the United Arab Emirates, Eritrea, Ethiopia, Guinea, Guyana, India, Indonesia, the Islamic Republic of Iran, Iraq, Israel, Japan, Jordan, Kuwait, Lebanon, Libya, Malawi, Morocco, Mozambique, Nepal, Nigeria, Pakistan, the Philippines, Qatar, Syria, Somalia, Sudan, Sri Lanka, Chad, Thailand, Togo and Yemen, the band 1215-1300 MHz is also allocated to the fixed and mobile services on a primary basis.

S5.331—Additional allocation: in Algeria, Germany, Austria, Bahrain, Belgium, Benin, Bosnia and Herzegovina, Burundi, Cameroon, China, Croatia, Denmark, the United Arab Emirates, France, Greece, India, the Islamic Republic of Iran, Iraq, Kenya, The Former Yugoslav Republic of Macedonia, Liechtenstein, Luxembourg, Mali, Mauritania, Norway, Oman, Pakistan, the Netherlands, Portugal, Qatar, Senegal, Slovenia, Somalia, Sudan, Sri Lanka, Sweden, Switzerland, Turkey and Yugoslavia, the band 1215-1300 MHz is also allocated to the radiolocation service on a primary basis.

S5.333—In the bands 1215-1300 MHz, 3100-3300 MHz, 5250-5350 MHz, 8550-8650 MHz, 9500-9800 MHz and 13.4-14.0 GHz, radiolocation stations installed on spacecraft may also be employed for the earth exploration-satellite and space research services on a secondary basis.

S5.334—Additional allocation: in Canada and the United States, the bands 1240-1300 MHz and 1350-1370 MHz are also allocated to the aeronautical radionavigation service on a primary basis.

G56—Government radiolocation in the bands 1215-1300, 2900-3100, 5350-5650 and 9300-9500 MHz is primarily for the military services; however, limited secondary use is permitted by other Government agencies in support of experimentation and research programs. In addition, limited secondary use is permitted for survey operations in the band 2900-3100 MHz.

FCC Part 97 Privileges				
	Terrestrial location of the amateur radio station			
License Class	Region 1 MHz	Region 2 MHz	Region 3 MHz	Limitations*
Novice	1270-1295	1270-1295	1270-1295	
Technician Technician Plus General Advanced Extra	1240-1300	1240-1300	1240-1300	

*See §§97.301-307 of the FCC rules for further details on authorized privileges for each license class, frequency sharing requirements and authorized modes.

Amateur Radio Band Plans

Region 1 MHz	Region 2 MHz	Region 3 MHz	ARRL* MHz
1240.00-1243.25 All modes			1240-1246 (2,3) ATV #1
1243.25-1260.00 ATV			1246-1248 (2) NB FM point-to-point links and digital duplex with 1258-1260
			1248-1252 (2,8,9) Digital communications
			1252-1258 (2,3) ATV #2
			1258-1260 (2) NB FM point-to point links and digital, duplexed with 1246-1252
1260.00-1270.00 Satellites		1260-1270 Satellite	1260-1270 (5,6) Satellite uplinks, experimental, simplex ATV
1270.00-1272.00 All modes		1270-1296	1270-1276 (2,4,9,10) Repeater inputs, FM and linear, paired with 1282-1288, 239 pairs, every 25 kHz, eg, 1270.025, 050, 075, etc.
1272.00-1290.994 ATV			1271/1283 Uncoordinated test pair
1290.994-1291.481 FM repeater inputs (25 kHz spacing)			1276-1282 (2,3) ATV #3
			1282-1288 (2,4,9,10) Repeater outputs paired with 1270-1276
1291.494-1296.00 All modes			1288-1294 (6) WB experimental, simplex ATV
			1294-1295 (7) FM simplex, 25-kHz channels
			1294.5 National FM simplex calling
			1295-1297 NB weak-signal (no FM)
			1295.0-1295.8 SSTV, FAX, ACSSB experimental
			1295.8-1296.0 Reserved for EME, CW expansion
1296.00-1296.15 CW only		1296-1297 EME	1296.0-1296.05 EME exclusive
1296.000-1296.025 Moonbounce			1296.07-1296.08 CW beacons
1296.15-1296.80 SSB and CW			1296.1 CW/SSB calling
1296.80-1296.9875 Beacons			1296.4-1296.6 Crossband linear translator input
1296.994-1297.481 FM repeater outputs			1296.6-1296.8 Crossband linear translator output
1297.494-1297.981 FM simplex			1296.8-1297.0 Experimental beacons (exclusive)
1298.00-1300.00 All modes		1297-1300	1297-1300 Digital communications

Footnotes:
1) Deleted
2) Coordinated assignments required.
3) ATV assignments should be made according to modulation type (for example), VSB-ATV, SSB-ATV, or combination. Coordination of multiple users of a single channel in a local area can be achieved through isolation by means of cross polarization and directional antennas. DSB ATV may be used, but only when local and regional activity levels permit. The excess bandwidths from such users are secondary to the assigned services.
4) Coordinate assignments with 100-kHz channels, beginning at the lower end of the segment until allocations are filled, then assign 50-kHz channels until allocations are filled before assigning 25-kHz channels.
5) Wide bandwidth experimental users are secondary to the satellite service and may be displaced upon the installation of any new satellites. Users are EIRP-limited to the noise floor of the satellites in service and may suffer interference from satellite uplinks.
6) Simplex services only; permanent users shall not be coordinated in this segment. High-altitude repeaters or other unattended fixed operations are not permitted.
7) Voice and non-voice operations.
8) Consult 47 C.F.R. 97.307 (FCC regulations) for allowable data rates and bandwidths.
9) Provide guard bands at the higher frequency end of segments, as required, to avoid interference to ATV.
10) 1274.0-1274.2 and 1286.0-1286.2 MHz are optionally reserved for contiguous Linear Translators supporting multiple narrow bandwidth users. These may also be duplexed with other non-coordinated band segments.

The following Packet Radio frequency recommendations were adopted by the ARRL Board of Directors in January 1988.
1) 2-MHz-bandwidth channels at:
1249.00   1251.00   1298.00

2) 100-kHz-bandwidth channels:
1299.05   1299.45   1299.75
1299.15   1299.55   1299.85
1299.25   1299.65   1299.95
1299.35

3) 25-kHz-bandwidth channels:
1294.025   1294.125
1294.050   1294.150
1294.075   1294.175
1294.100 National packet simplex calling

*Adopted by the ARRL Board of Directors January 1985.

INTERNATIONAL			UNITED STATES				
Region 1 MHz	Region 2 MHz	Region 3 MHz	Band MHz	National Provisions	Government Allocation	Non-Government Allocation	Remarks
2300-2450 FIXED MOBILE Amateur Radiolocation S5.150 S5.282 S5.395	2300-2450 FIXED MOBILE RADIOLOCATION Amateur S5.150 S5.282 S5.393 S5.394 S5.396		2300-2310	US253	G123	Amateur	
			2310-2360	S5.393 US276 US327 US328	Mobile Radiolocation Fixed G2 G120	BROADCASTING- SATELLITE Mobile	
			2360-2390	US276	MOBILE RADIOLOCATION Fixed G2 G120	MOBILE	
			2390-2400		G122	AMATEUR	Part 15
			2400-2402	S5.150 S5.282	G123	Amateur	Part 15 ISM 2450 ± 50 MHz
			2402-2417	S5.150 S5.282	G122	AMATEUR	Part 15 ISM 2450 ± 50 MHz
			2417-2450	S5.150 S5.282	Radiolocation G2 G124	Amateur	Part 15 ISM 2450 ± 50 MHz

S5.150—The following bands:
13553-13567 kHz	(centre frequency 13560 kHz),
26957-27283 kHz	(centre frequency 27120 kHz),
40.66-40.70 MHz	(centre frequency 40.68 MHz),
902-928 MHz	in Region 2 (centre frequency 915 MHz),
2400-2500 MHz	(centre frequency 2450 MHz),
5725-5875 MHz	(centre frequency 5800 MHz), and
24-24.25 GHz	(centre frequency 24.125 GHz)

are also designated for industrial, scientific and medical (ISM) applications. Radiocommunication services operating within these bands must accept harmful interference which may be caused by these applications. ISM equipment operating in these bands is subject to the provisions of No. S15.13.

S5.282—In the bands 435-438 MHz, 1260-1270 MHz, 2400-2450 MHz, 3400-3410 MHz (in Regions 2 and 3 only) and 5650-5670 MHz, the amateur-satellite service may operate subject to not causing harmful interference to other services operating in accordance with the Table (see No. 435). Administrations authorizing such use shall ensure that any harmful interference caused by emissions from a station in the amateur-satellite service is immediately eliminated in accordance with the provisions of No. 2741/S25.11. The use of the bands 1260-1270 MHz and 5650-5670 MHz by the amateur-satellite service is limited to the Earth-to-space direction.

S5.393—Additional Allocation: In the United States of America and India, the band 2310-2360 MHz is also allocated to the broadcasting-satellite service (sound) and complementary terrestrial broadcasting service on a primary basis. Such use is limited to digital audio broadcasting and is subject to the provisions of Resolution 528 (WARC-92).

S5.394—In the United States, the use of the band 2300-2390 MHz by the aeronautical mobile service for telemetry has priority over the other uses by the mobile services. In Canada, the use of the band 2300-2483.5 MHz by the aeronautical mobile service for telemetry has priority over other uses by the mobile services.

S5.395—In France, the use of the band 2310-2360 MHz by the aeronautical mobile service for telemetry has priority over other uses by the mobile service.

S5.396—Space stations of the broadcasting-satellite service in the band 2310-2360 MHz operating in accordance with No. S5.393 that may affect the services to which this band is allocated in other countries shall be coordinated and notified in accordance with Resolution 33. Complementary terrestrial broadcasting stations shall be subject to bilateral coordination with neighbouring countries prior to their bringing into use.

US253—In the band 2300-2310 MHz, the fixed and mobile services shall not cause harmful interference to the amateur service.

US276—Except as otherwise provided for herin, use of the band 2310-2390 MHz by the mobile service is limited to aeronautical telemetering and associated telecommand operations for flight testing of manned or unmanned aircraft, missiles or major components thereof. The following six frequencies are shared on a co-equal basis by Government and associated telecommand operations of expendable and re-usable launch vehicles whether or not such operations involve flight testing: 2312.5, 2332.5, 2352.5, 2364.5, 2370.5, and 2382.5 MHz. All other mobile telemetering uses shall be secondary to the above uses.

US327—The band 2310-2360 MHz is allocated to the broadcasting-satellite service (sound) and complementary terrestrial broadcasting service on a primary basis. Such use is limited to digital audio broadcasting and is subject to the provisions of Resolution 528.

US328—In the band 2310-2360 MHz, the mobile and radiolocation services are allocated on a primary basis until 1 January 1997 or until broadcasting-satellite (sound) service has been brought into use in such a manner as to affect or be affected by the mobile and radiolocation services in those service areas, whichever is later. The broadcasting-satellite (sound) service during implementation should also take cognizance of the expendable and reusable launch vehicle frequencies 2312.5, 2332.5, and 2352.5 MHz, to minimize the impact on this mobile service use to the extent possible.

G2—In the bands 216-225, 420-450 (except as provided by US217), 890-902, 928-942, 1300-1400, 2310-2390, 2417-2450, 2700-2900, 5650-5925, and 9000-9200 MHz, the Government radiolocation is limited to the military services.

G120—Development of airborne primary radars in the band 2310-2390 MHz with peak transmitter power in excess of 250 watts for use in the United States is not permitted.

G122—The bands 2390-2400, 2402-2417 and 4660-4685 MHz were identified for immediate reallocation, effective August 10, 1994, for exclusive non-Government use under Title VI of the Omnibus Budget Reconciliation Act of 1993. Effective August 10, 1994, any Government operations in these bands are on a non-interference basis to authorized non-Government operations and shall not hinder the implementation of any non-Government operations.

G123—The bands 2300-2310 and 2400-2402 MHz were identified for reallocation, effective August 10, 1995, for exclusive non-Government use under Title VI of the Omnibus Budget Reconciliation Act of 1993. Effective August 10, 1995, any Government operations in these bands are on a non-interference basis to authorized non-Government operations and shall not hinder the implementation of any non-Government operations.

G124—The band 2417-2450 MHz was identified for reallocation, effective August 10, 1995, for mixed Government and non-Government use under Title VI of the Omnibus Budget Reconciliation Act of 1993.

FCC Part 97 Privileges

License Class	Terrestrial location of the amateur radio station			
	Region 1 MHz	Region 2 MHz	Region 3 MHz	Limitations*
Technician	2300-2310	2300-2310	2300-2310	
Technician Plus				
General	2390-2450	2390-2450	2390-2450	
Advanced				
Extra				

*See §§97.301-307 of the FCC rules for further details on authorized privileges for each license class, frequency sharing requirements and authorized modes.

Amateur Radio Band Plans

Region 1 MHz	Region 2 MHz	Region 3 MHz	ARRL* MHz
2300.00-2320.00 Subregional (national band plans)			2300.0-2303.0 High-rate data (transmission rate ≥ 4800 bauds, duplex)
			2303.0-2303.5 Packet (transmission rate ≤ 2400 bauds, channel spacing = 25 kHz)
			2303.5-2303.8 TTY, packet (transmission rate ≤ 2400 bauds, bandwidth ≤ 2.5 kHz)
			2303.8-2303.9 Packet (bandwidth ≤ 2.5 kHz), TTY, CW, EME
			2303.9-2304.1 CW, EME
			2304.1-2034.2 CW, EME, SSB
			2304.2-2304.3 SSB, SSTV, fax, packet (bandwidth ≤ 2.5 kHz), AM, AMTOR
			2304.30-2304.32 Propagation beacon network
			2304.32-2304.40 General propagation beacons
			2304.400 Calling frequency
			2304.4-2304.5 SSB, SSTV, ACSSB, fax, packet (bandwidth ≤ 2.5 kHz), AM, AMTOR, experimental (bandwidth ≤ 2.5 kHz)
			2304.5-2304.7 Crossband linear translator input
			2304.7-2304.9 Crossband linear translator output
			2304.9-2305.0 Experimental beacons
			2305.0-2305.2 FM simplex (channel spacing = 25 kHz)
			2305.200 FM simplex calling frequency
			2305.2-2306.0 FM simplex (channel spacing = 25 kHz)
			2306.0-2309.0 FM repeaters (channel spacing = 25 kHz, input)
			2309.0-2310.0 Control and auxiliary links
			2390.0-2396.0 Fast-scan television
			2396.0-2399.0 High-rate data (transmission rate ≥ 4800 bauds, duplex)
2320.00-2320.15 CW			2399.0-2399.5 Packet (transmission rate ≤ 2400 bauds, channel spacing = 25 kHz)
2320-2320.025 Moonbounce			2399.5-2400.0 Control and auxiliary links
			2400.0-2403.0 Satellite
2320.15-2320.80 CW/SSB			2403.0-2408.0 Satellite, high-rate data (transmission rate ≥ 4800 bauds, duplex)
2320.2 SSB center of activity			2408.0-2410.0 Satellite
2320.80-2321.00 Beacons			2410.0-2413.0 FM repeaters (channel spacing = 25 kHz, output)
2321.00-2322.00 FM simplex and repeaters			2413.0-2418.0 High-rate data (transmission rate ≥ 4800 bauds, duplex)
			2418.0-2430.0 Fast-scan television
2322.00-2400.00 All modes			2430.0-2433.0 Satellite
			2433.0-2438.0 Satellite, high-rate data (transmission rate ≥ 4800 bauds, duplex)
2400.00-2450.00 Satellites			2438.0-2450.0 Wide-band FM, fast-scan television, FM television, spread spectrum, experimental

*This band plan for 2300-2310 and 2390-2450 MHz was adopted by the Board of Directors at its January 1991 meeting. See Minute 56.

INTERNATIONAL			UNITED STATES				
Region 1 MHz	Region 2 MHz	Region 3 MHz	Band MHz	National Provisions	Government Allocation	Non-Government Allocation	Remarks
3300-3400 RADIOLOCATION S5.149 S5.429 S5.430	3300-3400 RADIOLOCATION Amateur Fixed Mobile S5.149 S5.430	3300-3400 RADIOLOCATION Amateur S5.149 S5.429	3300-3500	US108 S5.282	RADIOLOCATION	Amateur Radiolocation	
3400-3600 FIXED FIXED-SATELLITE (Space-to-Earth) Mobile Radiolocation S5.431 S5.434	3400-3500 FIXED FIXED-SATELLITE (Space-to-Earth) Amateur Mobile Radiolocation S5.433 S5.282 S5.432				G31		

S5.282—In the bands 435-438 MHz, 1260-1270 MHz, 2400-2450 MHz, 3400-3410 MHz (in Regions 2 and 3 only) and 5650-5670 MHz, the amateur-satellite service may operate subject to not causing harmful interference to other services operating in accordance with the Table (see No S5.43). Administrations authorizing such use shall ensure that any harmful interference caused by emissions from a station in the amateur-satellite service is immediately eliminated in accordance with the provisions of No. S25.11. The use of the bands 1260-1270 MHz and 5650-5670 MHz by the amateur-satellite service is limited to the Earth-to-space direction.

S5.429— Additional allocation: in Saudi Arabia, Bahrain, Bangladesh, Brunei Darussalam, China, the Congo, the United Arab Emirates, India, Indonesia, the Islamic Republic of Iran, Iraq, Israel, Japan, Jordan, Kuwait, Lebanon, Libya, Malaysia, Oman, Pakistan, Qatar, Syria, Democratic People's Republic of Korea, Singapore and Yemen, the band 3300-3400 MHz is also allocated to the fixed and mobile services on a primary basis. The countries bordering the Mediterranean shall not claim protection for their fixed and mobile services from the radiolocation service.

S5.430—Additional allocation: in Armenia, Azerbaijan, Belarus, Bulgaria, Cuba, Georgia, Kazakhstan, China, the Congo, Moldova, Mongolia, Poland, Kyrgyzstan, Romania, Russia, Tajikistan, Turkmenistan and Ukraine, the band 3300-3400 MHz is also allocated to the radionavigation service on a primary basis.

S5.431—Additional allocation: in Germany, Israel, Nigeria and the United

Kingdom, the band 3400-3475 MHz is also allocated to the amateur service on a secondary basis.

S5.432—Different category of service: in Indonesia, Japan and Pakistan, the allocation of the band 3400-3500 MHz to the mobile, except aeronautical

S5.433—In Regions 2 and 3, in the band 3400-3600 MHz, the radiolocation service is allocated on a primary basis. However, all administrations operating radiolocation systems in this band are urged to cease operations by 1985. Thereafter, administrations shall take all practicable steps to protect the fixed-satellite service and coordination requirements shall not be imposed on the fixed-satellite service.

S5.434—In Denmark, Norway and the United Kingdom, the fixed, radiolocation and fixed-satellite services operate on a basis of equality of rights in the band 3400-3600 MHz. However, these Administrations operating radiolocation systems in this band are urged to cease operations by 1985. After this date, these Administrations shall take all practicable steps to protect the fixed-satellite service and coordination requirements shall not be imposed on the fixed-satellite service.

US108—Within the bands 3300-3500 MHz and 10000-10500 MHz, survey operations, using transmitters with a peak power not to exceed five watts into the antenna, may be authorized for Government and non-Government use on a secondary basis to other Government radiolocations operations.

G31—In the bands 3300-3500 MHz, the Government radiolocation is limited to the military services, except as provided by footnote US108.

FCC Part 97 Privileges				
	Terrestrial location of the amateur radio station			
License Class	Region 1 MHz	Region 2 MHz	Region 3 MHz	Limitations*
Technician Technician Plus General Advanced Extra	--	3300-3500	3300-3500	

Amateur Radio Band Plans			
Region 1 MHz	Region 2 MHz	Region 3 MHz	ARRL** MHz
3400-3402 Narrow band CW/EME/SSB			
3402-3456 All modes			3456.3-3456.4 Propagation beacons
3456-3458			
Narrow band CW/EME/SSB			
3458-3475 All modes			

*See §§97.301-307 of the FCC rules for further details on authorized privileges for each license class, frequency sharing requirements and authorized modes.

**A beacon subband was adopted by the ARRL Board of Directors in July 1988.

INTERNATIONAL			UNITED STATES				
Region 1 MHz	Region 2 MHz	Region 3 MHz	Band MHz	National Provisions	Government Allocation	Non-Government Allocation	Remarks
5650-5725 RADIOLOCATION Amateur Space Research (deep space) S5.282 S5.451 S5.453 S5.454 S5.455			5650-5850	S5.282 S5.150	RADIOLOCATION	Amateur	(ISM 5800 ±75 MHz)
5725-5830 FIXED-SATELLITE (Earth-to-space) RADIOLOCATION Amateur S5.150 S5.451 S5.453 S5.455 S5.456	5725-5830 RADIOLOCATION Amateur S5.150 S5.453 S5.455				G2		
5830-5850 FIXED-SATELLITE (Earth-to-space) RADIOLOCATION Amateur Amateur-Satellite (space-to-Earth) S5.150 S5.451 S5.453 S5.455 S5.456	5830-5850 RADIOLOCATION Amateur Amateur-Satellite (space-to-Earth) S5.150 S5.453 S5.455		5850-5925	US245 S5.150	RADIOLOCATION G2	Amateur FIXED-SATELLITE (Earth-to-space)	
5850-5925 FIXED FIXED-SATELLITE (Earth-to-space) MOBILE S5.150	5850-5925 FIXED FIXED-SATELLITE (Earth-to-space) MOBILE Amateur Radiolocation S5.150	5850-5925 FIXED FIXED- SATELLITE (Earth-to-space) MOBILE Radiolocation S5.150					

S5.150—The following bands: 13533-13567 kHz (centre frequency 13560 kHz), 26957-27283 kHz (centre frequency 27120 kHz), 40.66-40.70 MHz (centre frequency 40.68 MHz), 902-928 MHz in Region 2 (centre frequency 915 MHz), 2400-2500 MHz (centre frequency 2450 MHz), 5725-5875 MHz (centre frequency 5800 MHz), and 24-24.25 GHz (centre frequency 24.125 GHz) are also designated for industrial, scientific and medical (ISM) applications. Radiocommunication services operating within these bands must accept harmful interference which may be caused by these applications. ISM equipment operating in these bands is subject to the provisions of No. S15.13.

S5.282—In the bands 435-438 MHz, 1260-1270 MHz, 2400-2450 MHz, 3400-3410 MHz (in Regions 2 and 3 only) and 5650-5670 MHz, the amateur-satellite service may operate subject to not causing harmful interference to other services operating in accordance with the Table (see No. S5.43). Administrations authorizing such use shall ensure that any harmful interference caused by emissions from a station in the amateur-satellite service is immediately eliminated in accordance with the provisions of No. S25.11. The use of the bands 1260-1270 MHz and 5650-5670 MHz by the amateur-satellite service is limited to the Earth-to-space direction.

S5.451—Additional allocation: in the United Kingdom, the band 5470-5850 MHz is also allocated to the land mobile service on a secondary basis. The power limits specified in Nos. S21.2, S21.3, S21.4 and S21.5 shall apply in the band 5725-5850 MHz.

S5.453—Additional allocation: in Saudi Arabia, Bahrain, Bangladesh, Brunei Darussalam, Cameroon, the Central African Republic, China, the Congo, the Republic of Korea, Egypt, the United Arab Emirates, Gabon, Guinea, India, Indonesia, the Islamic Republic of Iran, Iraq, Israel, Japan, Jordan, Kuwait, Lebanon, Libya, Madagascar, Malaysia, Malawi, Niger, Nigeria, Oman, Pakistan, the Philippines, Qatar, Syria, Democratic People's Republic of Korea, Singapore, Swaziland, Tanzania, Chad, and Yemen, the band 5650-5850 MHz is also allocated to the fixed and mobile services on a primary basis.

S5.454—Different category of service: in Armenia, Azerbaijan, Belarus, Bulgaria, Georgia, Kazakhstan, Moldova, Mongolia, Uzbekistan, Kyrgyzstan, Russia, Tajikistan, Turkmenistan and Ukraine, the allocation of the band 5670-5725 MHz to the space research service is on a primary basis (see No. S5.33).

S5.455—Additional allocation: in Armenia, Azerbaijan, Belarus, Bulgaria, Cuba, Georgia, Hungary, Kazakhstan, Latvia, Moldova, Mongolia, Uzbekistan, Poland, Kyrgyzstan, Slovakia, Russia, Tajikistan, Turkmenistan and Ukraine the band 5670-5850 MHz is also allocated to the fixed service on a primary basis.

S5.456—Additional allocation: Germany and in Cameroon, the band 5755-5850 MHz is also allocated to the fixed service on a primary basis.

US245—The Fixed-Satellite Service is limited to International inter-Continental systems and subject to case-by-case electromagnetic compatibility analysis.

G2—In the bands 216-225, 420-450 (except as provided by US217), 890-902, 928-942, 1300-1400, 2310-2390, 2417-2450, 2700-2900, 5650-5925, and 9000-9200 MHz, the Government radiolocation is limited to the military services.

Amateur Radio Band Plans

Region 1 MHz	Region 2 MHz	Region 3 MHz	ARRL MHz
5650-5668 Satellite uplinks			
5668-5670 Narrow-band modes and satellite uplinks			5760.3-5760.4 Propagation beacons
5670-5700 Digital			
5700-5720 ATV			
5720-5760 All modes			
5760-5762 Narrow-band modes			
5762-5790 All modes			
5790-5850 Satellite downlinks			

FCC Part 97 Privileges

License Class	Terrestrial location of the amateur radio station			
	Region 1 MHz	Region 2 MHz	Region 3 MHz	Limitations*
Technician Technician Plus General Advanced Extra	5650-5850	5650-5925	5650-5850	

*See §§97.301-307 of the FCC rules for further details on authorized privileges for each license class, frequency sharing requirements and authorized modes.

Allocations

INTERNATIONAL			UNITED STATES				
Region 1 GHz	Region 2 GHz	Region 3 GHz	Band GHz	National Provisions	Government Allocation	Non-Government Allocation	Remarks
10-10.45 FIXED MOBILE RADIOLOCATION Amateur	10-10.45 RADIOLOCATION Amateur	10-10.45 FIXED MOBILE RADIOLOCATION Amateur	10-10.45	US58 US108 S5.479	RADIOLOCATION	Amateur Radiolocation	
S5.479	S5.479 S5.480	S5.479			G32	NG42	
10.45-10.5	RADIOLOCATION Amateur Amateur-Satellite		10.45-10.5	US58 US108	RADIOLOCATION	RADIOLOCATION Amateur Amateur-Satellite	
	S5.481				G32	NG42 NG134	

S5.479—The band 9975-10025 MHz is also allocated to the meteorological-satellite service on a secondary basis for use by weather radars.

S5.480—Additional allocation: in Costa Rica, Ecuador, Guatemala and Honduras, the band 10-10.45 GHz is also allocated to the fixed and mobile services on a primary basis.

S5.481— Additional allocation: in Germany, Angola, China, Ecuador, Spain, Japan, Morocco, Nigeria, Oman, Democratic People's Republic of Korea, Sweden, Tanzania and Thailand, the band 10.45-10.5 GHz is also allocated to the fixed and mobile services on a primary basis.

US58—In the band 10000-10500 MHz, pulsed emissions are prohibited, except for weather radars on board meteorological satellites in the band 10000-10025 MHz. The amateur service and the non-Government radiolocation service, which shall not cause harmful interference to the Government radiolocation service, are the only non-Government services permitted in this band. The non-Government radiolocation service is limited to survey operations as specified in footnote US108.

US108—Within the bands 3300-3500 MHz and 10000-10500 MHz, survey operations, using transmitters with a peak power not to exceed five watts into the antenna, may be authorized for Government and non-Government use on a secondary basis to other Government radiolocations operations.

G32—Except for weather radars on meteorological satellites in the band 9975-10025 MHz and for Government survey operations (see footnote US108). Government radiolocation in the band 10000-10500 MHz is limited to the military services.

NG42—Non-Government stations in the radiolocation service shall not cause harmful interference to the amateur service.

NG134—In the band 10.45-10.5 GHz non-Government stations in the radiolocation service shall not cause harmful interference to the amateur and amateur-satellite services.

FCC Part 97 Privileges

License Class	Terrestrial location of the amateur radio station			
	Region 1 GHz	Region 2 GHz	Region 3 GHz	Limitations*
Technician Technician Plus General Advanced Extra	10.0-10.5	10.0-10.5	10.0-10.5	

*See §§97.301-307 of the FCC rules for further details on authorized privileges for each license class, frequency sharing requirements and authorized modes.

Amateur Radio Band Plans

Region 1 GHz	Region 2 GHz	Region 3 GHz	ARRL GHz
10.000-10.150 Digital			10.00-10.5
10.150-10.250 All modes			10.3681* Calling frequency
10.250-10.350 Digital			
10.350-10.368 All modes			10.3683-10.3684** Propagation beacons
10.368-10.370 Narrow-band modes			
10.370-10.450 All modes			
10.450-10.500 All modes and satellites			

*Adopted by ARRL Board of Directors January 1987.
**Adopted by ARRL Board of Directors July 1988.

Allocations

INTERNATIONAL			UNITED STATES				
Region 1 GHz	Region 2 GHz	Region 3 GHz	Band GHz	National Provisions	Government Allocation	Non-Government Allocation	Remarks
24-24.05	AMATEUR AMATEUR-SATELLITE S5.150		24-24.05	US211 S5.150		AMATEUR AMATEUR-SATELLITE	
24.05-24.25	RADIOLOCATION Amateur Earth Exploration-Satellite (Active) S5.150		24.05-24.25	US110 S5.150	RADIOLOCATION Earth Exploration Satellite(Active) G59	Amateur Radiolocation Earth Exploration Satellite (Active)	(ISM 24.125 ± 125 MHz)

S5.150—The following bands: 13533-13567 kHz (centre frequency 13560 kHz), 26957-27283 kHz (centre frequency 27120 kHz), 40.66-40.70 MHz (centre frequency 40.68 MHz), 902-928 MHz in Region 2 (centre frequency 915 MHz), 2400-2500 MHz (centre frequency 2450 MHz), 5725-5875 MHz (centre frequency 5800 MHz), and 24-24.25 GHz (centre frequency 24.125 GHz) are also designated for industrial, scientific and medical (ISM) applications. Radiocommunication services operating within these bands must accept harmful interference which may be caused by these applications. ISM equipment operating in these bands is subject to the provisions of No. S15.13.

US211—In the bands 1670-1690, 5000-5250 MHz, and 10.7-11.7, 15.1365-15.35,15.4-15.7, 22.5-22.55, 24-24.05, 31.0-31.3, 31.8-32, 40.5-42.5, 84-86, 102-105, 116-126, 151-164, 176.5-182, 185-190, 231-235, 252-265 GHz, applicants for airborne or space station assignments are urged to take all practicable steps to protect radio astronomy observations in the adjacent bands from harmful interference; however, US74 applies.

G59—In the bands 902-928 MHz, 3100-3300 MHz, 3500-3700 MHz, 5250-5350 MHz, 8500-9000 MHz, 9200-9300 MHz, 13.4-14.0 GHz, 15.7-17.7 GHz and 24.05-24.25 GHz, all Government non-military radiolocation shall be secondary to military radiolocation, except in the subband 15.7-16.2 GHz airport surface detection equipment (ASDE) is permitted on a co-equal basis subject to coordination with the military departments.

FCC Part 97 Privileges				
	Terrestrial location of the amateur radio station			
License Class	Region 1 GHz	Region 2 GHz	Region 3 GHz	Limitations*
Technician Technician Plus General Advanced Extra	24.00-24.25	24.00-24.25	24.00-24.25	

*See §§97.301-307 of the FCC rules for further details on authorized privileges for each license class, frequency sharing requirements and authorized modes.

Allocations							
INTERNATIONAL			UNITED STATES				
Region 1 GHz	Region 2 GHz	Region 3 GHz	Band GHz	National Provisions	Government Allocation	Non-Government Allocation	Remarks
47-47.2	AMATEUR AMATEUR-SATELLITE		47-47.2			AMATEUR AMATEUR-SATELLITE	

FCC Part 97 Privileges				
	Terrestrial location of the amateur radio station			
License Class	Region 1 GHz	Region 2 GHz	Region 3 GHz	Limitations*
Technician Technician Plus General Advanced Extra	47.0-47.2	47.0-47.2	47.0-47.2	

*See §§97.301-307 of the FCC rules for further details on authorized privileges for each license class, frequency sharing requirements and authorized modes.

Allocations							
INTERNATIONAL			UNITED STATES				
Region 1 GHz	Region 2 GHz	Region 3 GHz	Band GHz	National Provisions	Government Allocation	Non-Government Allocation	Remarks
75.5-76	AMATEUR AMATEUR-SATELLITE Space Research (space-to-Earth)		75.5-76			AMATEUR AMATEUR-SATELLITE	
76-81	RADIOLOCATION Amateur Amateur-Satellite Space Research (space-to-Earth) S5.560		76-81	S5.560	RADIOLOCATION	RADIOLOCATION Amateur Amateur-Satellite	

S5.560—In the band 78-79 GHz radars located on space stations may be operated on a primary basis in the Earth exploration-satellite service and in the space research service.

FCC Part 97 Privileges				
	Terrestrial location of the amateur radio station			
License Class	Region 1 GHz	Region 2 GHz	Region 3 GHz	Limitations*
Technician Technician Plus General Advanced Extra	75.5-81	75.5-81	75.5-81	

*See §§97.301-307 of the FCC rules for further details on authorized privileges for each license class, frequency sharing requirements and authorized modes.

Allocations

INTERNATIONAL			UNITED STATES				
Region 1 GHz	Region 2 GHz	Region 3 GHz	Band GHz	National Provisions	Government Allocation	Non-Government Allocation	Remarks
119.98-120.02	EARTH EXPLORATION SATELLITE (passive) FIXED INTER-SATELLITE MOBILE S5.558 SPACE RESEARCH (passive) Amateur S5.341		119.98-120.02	US211 US263 S5.341 S5.558	FIXED INTER-SATELLITE MOBILE EARTH EXPLORATION SATELLITE (Passive) SPACE RESEARCH (Passive)	FIXED MOBILE INTER-SATELLITE EARTH EXPLORATION-SATELLITE (Passive) SPACE RESEARCH (Passive) Amateur	

S5.341—In the bands 1400-1727 MHz, 101-120 GHz and 197-220 GHz, passive research is being conducted by some countries in a programme for the search for intentional emissions of extraterrestrial origin

S5.558—In the bands 54.25-58.2 GHz, 59-64 GHz, 116-134 GHz, 170-182 GHz and 185-190 GHz, stations in the aeronautical mobile service may be operated subject to not causing harmful interference to the inter-satellite service (see No. S5.43).

US263—In the frequency bands 21.2-21.4, 22.21-22.5, 36-37, 50.2-50.4, 54.25-58.2, 116-126, 150-151, 174.5-176.5, 200-202 and 235-238 GHz, the Space Research and the Earth Exploration-Satellite Services shall not receive protection from the Fixed and Mobile Services operating in accordance with the Table of Frequency Allocations.

FCC Part 97 Privileges

	Terrestrial location of the amateur radio station			
License Class	Region 1 GHz	Region 2 GHz	Region 3 GHz	Limitations*
Technician Technician Plus General Advanced Extra	119.98-120.02	119.98-120.02	119.98-120.02	

*See §§97.301-307 of the FCC rules for further details on authorized privileges for each license class, frequency sharing requirements and authorized modes.

Allocations							
INTERNATIONAL			UNITED STATES				
Region 1 GHz	Region 2 GHz	Region 3 GHz	Band GHz	National Provisions	Government Allocation	Non-Government Allocation	Remarks
142-144	AMATEUR AMATEUR-SATELLITE		142-144			AMATEUR AMATEUR-SATELLITE	
144-149	RADIOLOCATION Amateur Amateur-Satellite S5.149 S5.555		144-149	S5.149	RADIOLOCATION	RADIOLOCATION Amateur Amateur-Satellite	

S5.149—In making assignments to stations of other services to which the bands: 13360-13410 kHz, 25550-25670 kHz, 37.5-38.25 MHz, 73-74.6 MHz in Regions 1 and 3, 79.75-80.25 MHz in Region 3, 150.05-153 MHz in Region 1, 322-328.6 MHz*, 406.1-410 MHz, 608-614 MHz in Regions 1 and 3, 1330-1400 MHz*, 1610.6-1613.8 MHz*, 1660-1670 MHz, 1718.8-1722.2 MHz*, 2655-2690 MHz, 3260-3267 MHz*, 3332-3339 MHz*, 3345.8-3352.5 MHz*, 4825-4835 MHz*, 4950-4990 MHz, 4990-5000 MHz, 6650-6675.2 MHz*, 10.6-10.68 GHz, 14.47-14.5 GHz*, 22.01-22.21 GHz*, 22.21-22.5 GHz, 22.81-22.86 GHz*, 23.07-23.12 GHz*, 31.2-31.3 GHz, 31.5-31.8 GHz in Regions 1 and 3, 36.43-36.5 GHz*, 42.5-43.5 GHz, 42.77-42.87 GHz*, 43.07-43.17 GHz*, 43.37-43.47 GHz*, 48.94-49.04 GHz*, 72.77-72.91 GHz*, 93.07-93.27 GHz*, 97.88-98.08 GHz*, 140.69-140.98 GHz*, 144.68-144.98 GHz*, 145.45-145.75 GHz*, 146.82-147.12 GHz*, 150-151 GHz*, 174.42-175.02 GHz*, 177-177.4 GHz, 178.2-178.6 GHz*, 181-181.46 GHz*, 186.2-186.6 GHz*, 250-251 GHz*, 257.5-258 GHz*, 261-265 GHz, 262.24-262.76 GHz*, 265-275 GHz, 265.64-266.16 GHz*, 267.34-267.86 GHz*, 271.74-272.26 GHz* are allocated (* indicates radio astronomy use for spectral line observations), administrations are urged to take all practicable steps to protect the radio astronomy service from harmful interference. Emissions from spaceborne or airborne stations can be particularly serious sources of interference to the radio astronomy service (see Nos. 343/S4.5 and 344/S4.6 and Article 36/S29).

S5.555—Additional allocation: the bands 48.94-49.04 GHz, 97.88-98.08 GHz, 140.69-140.98 GHz, 144.68-144.98 GHz, 145.45-145.75 GHz, 146.82-147.12 GHz, 250-251 GHz and 262.24-262.76 GHz are also allocated to the radio astronomy service on a primary basis.

FCC Part 97 Privileges				
	Terrestrial location of the amateur radio station			
License Class	Region 1 GHz	Region 2 GHz	Region 3 GHz	Limitations*
Technician Technician Plus General Advanced Extra	142-149	142-149	142-149	

*See §§97.301-307 of the FCC rules for further details on authorized privileges for each license class, frequency sharing requirements and authorized modes.

Allocations

INTERNATIONAL			UNITED STATES				
Region 1 GHz	Region 2 GHz	Region 3 GHz	Band GHz	National Provisions	Government Allocation	Non-Government Allocation	Remarks
241-248	RADIOLOCATION Amateur Amateur-Satellite S5.138		241-248	S5.138	RADIOLOCATION	RADIOLOCATION Amateur Amateur-Satellite	(ISM 245 GHz ± 1 GHz)
248-250	AMATEUR AMATEUR-SATELLITE		248-250			AMATEUR AMATEUR-SATELLITE	

S5.138—The following bands: 6765-6795 kHz (centre frequency 6780 kHz), 433.05-434.79 MHz (centre frequency 433.92 MHz) in Region 1 except in the countries mentioned in No. S5.280, 61-61.5 GHz (centre frequency 61.25 GHz), 122-123 GHz (centre frequency 122.5 GHz), and 244-246 GHz (centre frequency 245 GHz) are designated for industrial, scientific and medical (ISM) applications. The use of these frequency bands for ISM applications shall be subject to special authorization by the administration concerned, in agreement with other administrations whose radiocommunication services might be affected. In applying this provision, administrations shall have due regard to the latest relevant ITU-R Recommendations.

FCC Part 97 Privileges

License Class	Terrestrial location of the amateur radio station			
	Region 1 GHz	Region 2 GHz	Region 3 GHz	Limitations*
Technician Technician Plus General Advanced Extra	241-250	241-250	241-250	

*See §§97.301-307 of the FCC rules for further details on authorized privileges for each license class, frequency sharing requirements and authorized modes.

ITU Radio Regulations S4.8

Where, in adjacent Regions or sub-Regions, a band of frequencies is allocated to different services of the same category. . . the basic principle is the equality of right to operate. Accordingly, the stations of each service in one Region or sub-Region must operate so as not to cause harmful interference to services in the other Regions or sub-Regions.

ITU Resolution No. 640

Relating to the International Use of Radiocommunications, in the Event of Natural Disasters, in Frequency Bands Allocated to the Amateur Service

The World Administrative Radio Conference, Geneva, 1979.

considering

a) that in the event of natural disaster normal communication systems are frequently overloaded, damaged or completed disrupted;

b) that rapid establishment of communication is essential to facilitate worldwide relief actions;

c) that the amateur bands are not bound by international plans or notification procedures, and are therefore well adapted for short-term use in emergency cases;

d) that international disaster communications would be facilitated by temporary use of certain frequency bands allocated to the amateur service;

e) that under those circumstances the stations of the amateur service, because of their widespread distribution and their demonstrated capacity in such cases, can assist in meeting essential communication needs;

f) the existence of national and regional amateur emergency networks using frequencies throughout the bands allocated to the amateur service;

g) that, in the event of a natural disaster, direct communication between amateur stations and other stations might enable vital communications to be carried out until normal communications are restored;

recognizing

that the rights and responsibilities for communications in the event of a natural disaster rest with the administrations involved;

resolves

1. that the bands allocated to the amateur service which are specified in No. 510 [S5.120] may be used by administrations to meet the needs of international disaster communications;

2. that such use of these bands shall be only for communications in relation to relief operations in connection with natural disasters;

3. that the use of specified bands allocated to the amateur service by non-amateur stations for disaster communications shall be limited to the duration of the emergency and to the specific geographical areas as defined by the responsible authority of the affected country;

4. that disaster communications shall take place within the disaster area and between the disaster area and the permanent headquarters of the organization providing relief;

5. that such communications shall be carried out only with the consent of the administration of the country in which the disaster has occurred;

6. that relief communications provided from outside the country in which the disaster has occurred shall not replace existing national or international amateur emergency networks;

7. that close cooperation is desirable between amateur stations and the stations of other radio services which may find it necessary to use amateur frequencies in disaster communications;

8. that such international relief communications shall avoid, as far as practicable, interference to the amateur service networks;

invites administrations

1. to provide for the needs of international disaster communications;

2. to provide for the needs of emergency communications within their national regulations.

Emission Designators

Emissions are classified by the following basic characteristics:

(1) First symbol—type of modulation of the main carrier

(1.1) Emission of an unmodulated carrier — N

(1.2) Emission in which the main carrier is amplitude modulated (including cases where subcarriers are angle modulated)

  (1.2.1) Double sideband — A

  (1.2.2) Single sideband, full carrier — H

  (1.2.3) Single sideband, reduced or variable-level carrier — R

  (1.2.4) Single sideband, suppressed carrier — J

  (1.2.5) Independent sidebands — B

  (1.2.6) Vestigial sideband — C

(1.3) Emission in which the main carrier is angle modulated

  (1.3.1) Frequency modulation — F

  (1.3.2) Phase modulation — G

(1.4) Emission in which the main carrier is amplitude and angle modulated either simultaneously or in a pre-established sequence. — D

(1.5) Emission of pulses[1]

  (1.5.1) Sequence of unmodulated pulses — P

  (1.5.2) A sequence of pulses

    (1.5.2.1) modulated in amplitude — K

    (1.5.2.2) modulated in width/duration — L

    (1.5.2.3) modulated in position/phase — M

    (1.5.2.4) in which the carrier is angle modulated during the period of the pulse — Q

    (1.5.2.5) which is the combination of the foregoing or is produced by other means — V

(1.6) Cases not covered above, in which an emission consists of the main carrier modulated, either simultaneously or in a preestablished sequence, in a combination of two or more of the following modes: amplitude, angle, pulse — W

(1.7) Cases not otherwise covered — X

(2) Second symbol—nature of signal(s) modulating the main carrier

(2.1) No modulating signal — 0

(2.2) A single channel containing quantized or digital information without the use of a modulating subcarrier[2] — 1

(2.3) A single channel containing quantized or digital information with the use of a modulating subcarrier[2] — 2

(2.4) A single channel containing analog information — 3

(2.5) Two or more channels containing quantized or digital information — 7

(2.6) Two or more channels containing analog information — 8

(2.7) Composite system with one or more channels containing quantized or digital information, together with one or more channels containing analog information — 9

(2.8) Cases not otherwise covered — X

(3) Third symbol—type of information to be transmitted[3]

(3.1) No information transmitted — N

(3.2) Telegraphy—for aural reception — A

(3.3) Telegraphy—for automatic reception — B

(3.4) Facsimile — C

(3.5) Data transmission, telemetry, telecommand — D

(3.6) Telephony (including sound broadcasting) — E

(3.7) Television (video) — F

(3.8) Combination of the above — W

(3.9) Cases not otherwise covered — X

The following two optional characteristics may be added for a more complete description of an emission:

(4) Fourth symbol—detail of signal(s)

(4.1) Two-condition code with elements of differing numbers and/or durations — A

(4.2) Two-condition code with elements of the same number and duration without error correction — B

(4.3) Two-condition code with elements of the same number and duration with error correction — C

(4.4) Four-condition code in which each condition represents a signal element (of one or more bits) — D

(4.5) Multi-condition code in which each condition represents a signal element (of one or more bits) — E

(4.6) Multi-condition code in which each condition or combination of conditions represents a character — F

(4.7) Sound of broadcasting quality (monophonic) — G

(4.8) Sound of broadcasting quality (stereophonic or quadraphonic) — H

(4.9) Sound of commercial quality (excluding categories K and L below) — J

(4.10) Sound of commercial quality with the use of frequency inversion or band-splitting — K

(4.11) Sound of commercial quality with separate frequency-modulated signals to control the level of demodulated signal — L

(4.12) Monochrome — M

(4.13) Color — N

(4.14) Combination of the above — W

(4.15) Cases not otherwise covered — X

(5) Fifth symbol—nature of multiplexing

(5.1) None — N

(5.2) Code-division multiplex[4] — C

(5.3) Frequency-division multiplex — F

(5.4) Time-division multiplex — T

(5.5) Combination of frequency-division multiplex and time-division multiplex — W

(5.6) Other types of multiplexing — X

Notes

[1]Emissions where the main carrier is directly modulated by a signal that has been coded into quantized form (for example, pulse code modulation) should be designated under (1.2) or (1.3).

[2]This excludes time-division multiplex.

[3]In this context the word "information" does not include information of a constant, unvarying nature such as is provided by standard-frequency emissions, continuous-wave and pulse radars, and so forth.

[4]This includes bandwidth-expansion techniques.

Emission Types

Emissions may also be referred to by the following emission types:

CW
First Symbol: A, C, H, J or R
Second Symbol: 1
Third Symbol: A or B
Also J2A and J2B.

MCW
First Symbol: A, C, D, F, G or R
Second Symbol: 2
Third Symbol: A or B

Phone
First Symbol: A, C, D, F, G, H, J or R
Second Symbol: 1, 2 or 3
Third Symbol: E
Also first symbol B and second symbol 7, 8 or 9.

Image
First Symbol: A, C, D, F, G, H, J or R
Second Symbol: 1, 2 or 3
Third Symbol: C or F
Also first symbol B; second symbol 7, 8 or 9; and third symbol W.

RTTY/Data
First Symbol: A, C, D, F, G, H, J or R
Second Symbol: 1 (2, 7 or 9 above 51 MHz)
Third Symbol: B (RTTY) or D (Data) (W above 51 MHz)
J2B is permitted anywhere F1B is permitted; J2D is permitted anywhere F1D is permitted.

Pulse
First Symbol: P, K, L, M, Q, V or W
Second Symbol: 0, 1, 2, 3, 7, 8, 9 or X
Third Symbol: A, B, C, D, E, F, N, W or X

SS (Spread Spectrum)
First Symbol: A, C, D, F, G, H, J or R
Second Symbol: X
Third Symbol: X

Test (Emissions containing no information)
First Symbol: Any
Second Symbol: Any
Third Symbol: N

CHAPTER 3

Basic Operating

BILL JENNINGS, K1WJ
PAUL DANZER, N1II
and
THE ARRL STAFF

There are many faces to Amateur Radio. Tire of one and there is another awaiting you. This constant stream of new ideas, new challenges and new skills is one reason why so many people become hams and remain in the hobby for the rest of their lives.

There is one basic idea linking the various opportunities in ham radio—that of communication. Talking, sending pictures and data, building and testing transmitters and receivers all assume sooner or later you are going to operate these pieces of equipment in the various modes.

This manual covers all the most popular modes of ham radio. The basics of operating a ham station are discussed in Chapter 3. If you are a new ham or one just rejoining the hobby after a long absence, this material should help you decide what to operate, where to operate and how to operate.

Think of your participation in Amateur Radio as a never-ending journey—there's always something new to explore, always something new to do.

LET'S GET STARTED

Perhaps you already have your license or you anticipate passing the exam shortly. Now it's time to plan to get on-the-air. This can seem to be an overwhelming task, and your head is probably spinning with all kinds of questions ranging from "What kind of antenna should I use?" to "How do I make my first contact?"

Every newcomer needs help. Even experienced "Old Timers" once had those seemingly embarrassing questions when they began. After all, everyone has to start somewhere.

Who Are Hams?

Because of the diverse activities found within ham radio, almost everyone can find something to pique his or her interests. The 14-year-old high school student can keep in contact with friends across town. A 70-year-old retired engineer can chase those last elusive countries to put him on the DXCC (DX Century Club) Honor Roll. A Midwestern homemaker can show off her skills at contesting by racking up consistent QSO rates of 60 contacts per hour in the ARRL November Sweepstakes. You never know *who* you might run into.

Amateur Radio is a Service!

If you have read the Public Service column in *QST*, you've no doubt heard of the exploits of hams from all walks of life who've selflessly donated their time by providing emergency communication. Many, many more hams provide this service than are given recognition by the media, but that's all part of being a ham—the intrinsic reward is the satisfaction of doing a job well.

You may or may not be called at some time to provide this service to your community. But by being prepared, honing your on-the-air operating skills to their sharpest, maximizing your equipment to obtain the best from it, and being prepared to pitch in should the need arise, you will be ready. In so doing, you will derive untold hours of satisfaction from the exciting hobby we know as Amateur Radio.

Experience? Get On the Air

Actual on-the-air operating experience is the best teacher. In this section, we will try to give you enough of the basics to be able to make it through your first QSO. If nothing else, read the following material for additional reference notations. Learn by doing and don't be afraid to ask questions of someone who

If you plan to operate for extended periods of time, you'll need a comfortable, well-arranged station. When the station includes a computer and several rigs, space is at a premium. Here Mark, ON4WW, operates ON4UN. The stick-on notes contain reminders of temporary control settings.

In Pursuit of . . . DX

"I'll never forget the thrill of my first DX contact. It was on CW in the Novice portion of 15 meters. I heard a G3 (a station from England) calling CQ, and no one answered right away so I decided to give it a try. Success! With a lump in my throat and sweat on my palms, I managed to complete my first DX QSO and become hopelessly hooked. DX is great!"

Most hams can tell a similar story. DX is Amateur Radio shorthand for long distance; furthermore, DX is universally understood by hams to be a station in a foreign country. Chasing DX has steadily become one of the most popular activities in our hobby, and you—if you operate on the HF bands—can get in on the fun!

Within your Novice/Technician Plus frequency allocations, you can work DX almost as easily as an Amateur Extra ham. You don't need a super kilowatt station and a huge antenna installation. The beauty of DXing is that your operating skills can overcome deficiencies in your station equipment. Pick the right frequency band at the right time-of-day (or night) and the DX stations will be there—ready to talk to you.

Keep a few tips in mind. If you hear a lone DX station calling CQ, work it. If there is a pileup (a lot of stations trying to work the same station), don't jump right in. Listen for five or six contacts so you can figure out how the DX station is handling the pileup. Remember to keep your calls short, and don't call the DX station after he has gone back to someone else. Try not to send the DX station's call more than once, if at all. It's better to sign just your call. If the DX station is working "tailenders" (stations who call immediately at the end of the DX station's last QSO), try it. If the station is not working tailenders, don't disrupt the proceedings by trying it.

When the DX station is working only stations in a call area other than your own, perhaps conditions are not yet quite right for you. Listen, and follow the DX station's instructions to the letter. Be courteous and fair to all the stations on frequency; using good operating procedures is every bit as important as logging that DX.

Check the "How's DX" column in *QST*. It contains tips on propagation and news of interest to DXers. And don't forget to listen to the weekly W1AW DX bulletin for the latest in DX operations. Good DX!

might be able to help. After all, we're in this hobby together, and assistance is only as far away as the closest ham. How do you develop good operating habits? This *ARRL Operating Manual* is an excellent place to start. An entire chapter is devoted to each of the major Amateur Radio activities, everything from working DX to space communications—each chapter written by a ham with considerable experience in that area of hamdom. Experience, even through the words of others, is a powerful teacher. Notice the table of contents; you'll be amazed at the diversity and amount of good solid reference material you have at your fingertips.

Amateur Radio Publications

The American Radio Relay League (ARRL), publisher of this book, and the only nonprofit organization to serve the over 600,000 Amateur Radio operators (hams) in the United States, produces an entire line of reference manuals and specialty publications. It also publishes the world-renowned monthly journal, *QST*. Just about anything you may need to know—whether it be from the technical or operating sides of the hobby—can be found in an ARRL publication or obtained directly from the HQ staff. The League's basic beginner's pub-

lication is *Now You're Talking!* You only need contact the ARRL (see page viii), for complete information on all of the League's services.

QST Magazine

The ARRL journal, *QST,* published since 1914, is an excellent source of technical, operating, regulatory and general Amateur Radio information. Since it is published each month, *QST* is a timely source of information all hams can use, and a significant benefit of League membership. Contact the ARRL for current membership rates.

On-the-Air Bulletins

Up-to-the-minute information on everything of immediate interest to amateurs—from Federal Communications Commission policymaking decisions to propagation predictions, to news of DXpeditions—is transmitted in W1AW bulletins. W1AW is the Amateur Radio station maintained at ARRL HQ in Newington, Connecticut. Bulletins are transmitted at regularly scheduled intervals on CW (Morse code), phone (voice) and several digital (teletype) modes on various frequencies. In addition there are Morse-code practice broadcasts. A schedule of W1AW transmissions appears regularly in *QST* and may also be obtained from ARRL HQ.

Local Amateurs/Clubs

Don't overlook the greatest source of information of all— the experience of your fellow hams. There is nothing hams like to do better than share the "vast" wealth of experience they have in the hobby. Being human, most hams like to brag a little about their on-the-air exploits. A general question on a particular area of operation to a ham who has experience in that area is likely to bring you all kinds of data. A dedicated DXer will talk for hours on techniques for working a rare DX station. Similarly, a traffic handler will be only too happy to give you hints on efficient net procedures.

For the new ham (or potential ham), the problem may not be in asking the proper questions, but in finding another amateur of whom to ask the questions. That's where Amateur Radio clubs play an important role.

As a potential ham or a new ham, you might not know of an amateur in your immediate area. A good choice would be to try to find the time and meeting place for a nearby Amateur Radio club. What better source of information for the not-yet-licensed and the newly licensed than a whole club full of experienced hams? The Educational Activities Department at ARRL HQ will refer you to an ARRL-affiliated Amateur Radio club near you. The League is more than happy to supply this information. If you are not yet licensed, your local Amateur Radio club is the place to find Morse code and electronics theory courses that will prepare you for the FCC exams.

The ARRL Educational Activities Department offers a wide range of information for prospective hams. They have a toll-free telephone number: 800-32NEWHAM.

Listen Listen Listen...!

"Experience is the best teacher." The most efficient way to learn to do something is to actually pitch right in and give it a try. Since there are rules and an established way of doing things in the Amateur Radio Service, you'll learn most effectively by just plain, old-fashioned listening. Use a receiver (yours or one borrowed from a fellow amateur) and your ears. Tune the amateur frequency bands, and just listen, listen, listen. Copy as many QSOs as you can. Learn how the operators there conduct

Your First Contact

Are your palms sweaty, hands shaking and is there a queasy herd of butterflies (wearing spikes, perhaps) performing maneuvers in your stomach? Chances are you're facing your first Amateur Radio QSO. If so, take heart! Although few may admit it, the vast majority of hams felt the same way before firing up the rig for their first contact.

Although nervousness is natural, there are some preparations that can make things go a bit more smoothly. Practicing QSOs with a friend, or perhaps with several members of the local Amateur Radio club is a good way to ease the jitters.

Another thing some people find helpful is to write down information that you will use during the QSO in advance. One nervous Novice made notes of her own QSO data on index cards. One card contained her QTH and another her name (no kidding!); but after a half dozen contacts, she gradually forgot to use the cards. So will you.

A good security blanket to have is an experienced operator in your radio shack during your first few QSOs. You'll find that after your initial nervousness wears off you are able to do just fine by yourself. Even so, you'll have honored a ham friend (perhaps your teacher or your "Elmer") by allowing him the privilege of sharing your initial QSO. Indeed you may also want your Elmer to help you write up an equipment tune-up checklist so you can put a properly adjusted signal on the air every time, even when Elmer isn't around.

themselves, see what works for other operators and what doesn't, and incorporate those good points in your own operating habits. In other words, "copy the best and forget the rest."

Especially when going onto a new band or trying a new mode, take the time to listen to how those already established operators conduct themselves. Try to recognize good operating habits and incorporate them into your own style. You'll be surprised how much operator savvy you can pick up just by listening. A little common sense goes a long way in helping decide what is and what isn't good operating practice.

On-the-Air Experience for Newcomers

When it is time for you to take the plunge, fire up the rig and make your first QSO, go for it! Nothing can compare with actual on-the-air experience. Remember, everyone on the air today had to make their first QSO at some time or other. They had the same tentativeness as you might have now, but they

Jo Ann Simpson, KA3WPD, was a new operator when this picture was taken. All hams are a little nervous during their first few QSOs.

made it through fine. So will you.

It's the nature of the ham to be friendly, especially towards other members of this wonderful fraternity we all joined when we passed that examination. So trust the operator at the other end of your first QSO to understand your feelings and be as helpful as possible. After all, that operator was in your shoes once, too!

LET'S BUILD A STATION

The task of selecting and buying a ham station can seem overwhelming. You may be new to ham radio. Or perhaps you're planning to get back on the air after an absence of a few years. There are so many different possible station configurations it is hard to choose the very "best" station for you. In addition all hams face constraints such as having a limited budget to spend for ham gear or having restrictions on the size and location of antennas.

The first step in selecting your new station is to make up a list. The list simply answers a few questions. What sort of operation do you want—HF or VHF? How much room—how big can your shack or operating position be? Do you have room for long antennas, such as HF dipoles or high antennas such as HF verticals and VHF/UHF arrays? How much do you plan to spend, including rig, furniture, coax and wire?

Some of these questions may not apply to your situation, and you may need to list the answers to other questions not discussed here. The result of making this list is to give you an idea of what you want as well as your limitations.

The next step is to do some research. You have many sources of information on ham gear:

1) *Radio club members.* Ask members of your local radio club about their personal preferences in gear and antennas. Be prepared for a great volume of input. Every amateur has an opinion on the "best" equipment and antennas. Years of experimentation usually go into finding just the right station equipment to meet a particular amateur's needs. Listen and take note of each ham's choices and reasons for selecting a particular kind of gear; there's a lot of experience, time, effort and money behind each of those choices.

2) *Hands-on experience.* Try to use as many different pieces of gear as possible before you decide. Ask a nearby ham friend or one or more of your fellow radio club members if you can use their station. Try your club station. Note what you like and what features you don't care for in each of the stations you tried.

3) *Advertisements. QST is* chock-full of ads for all the newest up-to-date equipment as well as some premium used gear. Read the ads, and don't be afraid to contact the manufacturers of the gear for further information. Compare specifications and prices to get the best deal. If there is a dealer close to you pay him a visit.

Buying Your Station Equipment

After you've made your final choice (it's not really final, as most hams will trade station equipment several times during their ham careers), consider the sources where you might get the best deal, whether it be a new or used transceiver. Remember to include shipping and handling in the cost, and inquire about warrantee service. Several possible sources of equipment are:

1) *Local amateurs.* Many hams will have spare used gear and may be willing to part with this gear at a reasonable price. Be sure you know what a particular rig is going for on the open market before settling on a final price. If you are new to the

hobby and you buy a rig from a local club member, you may be able to talk him or her into "Elmering" you (helping you) with the rig's installation and operation.

2) *Hamfests/flea markets.* Many radio clubs run conventions called hamfests. Usually one of the big attractions of these events is the flea market or equipment sales. Much used gear, usually in passable shape, can be found at reasonable rates. Local distributors and manufacturers of new ham gear sometimes show up at these events to sell equipment at fair prices.

3) *Local electronics dealers.* If you are lucky enough to live near an electronics distributor who handles a line or two of ham gear, so much the better. The dealer can usually answer any questions you may have, will usually have a demonstration unit and will be only too happy to assist you to buy your new gear.

4) *Mail order.* There are many mail order ham equipment distributors to choose from. Some deal in new equipment, some deal in used equipment and some deal in both. Check the ads in the ham publications, such as those found in *QST,* for the equipment you want. The ads are also an excellent place to find out the going price of a particular piece of gear. Since prices sometimes change very fast many dealers list an 800 number to call for a price. Don't be shy—call their number.

A detailed treatment of used equipment and getting on the air quickly and effectively is in the ARRL's *Now You're Talking!.* The ARRL *Radio Buyer's Sourcebook* contains dozens of in-depth Product Reviews from *QST* magazine. Other *Sourcebook* articles will help you understand how the tests were conducted and what the results mean. If you're looking for a rig you should have a copy of the *Sourcebook.*

Take your time in deciding which gear to get, and even consider home brewing your gear. There's a lot of satisfaction to be had in telling the operator on the other end of your QSO (contact) "the rig here is home brew." *The ARRL Handbook for Radio Amateurs* has extensive coverage of construction and basic-to-advanced electronics theory information.

QRP vs QRO

Many hams find the low initial cost of low power (*QRP*) equipment on HF very attractive. In the ham radio community,

After 23 hours of almost continuous contest operation, Torfinn, LA4OF, and Liv, LA4YW, found it a little difficult to smile. Look carefully for the life-sustaining bags of potato chips, a necessary accessory in many contest stations.

there are both QRO (power output up to the full legal limit) and QRP (5 watts output) enthusiasts. Most hams run around 100 watts—basically the power level of most typical transceivers. There are times when it is necessary to run the full legal power limit to establish and maintain solid communications or compete effectively in a DX contest on certain bands. Most often the 100-watt level or less is more than enough to provide excellent contacts.

On VHF the situation is slightly different. Unless you are trying to work several hundred miles in an opening or operate a weak signal mode, a few watts (5 to 10) are generally enough power when coupled with a good antenna.

VHF/UHF Gear

If you're interested in voice and maybe packet radio, all you need is an FM transceiver. FM transceivers are available for hand-held and mobile use; the mobile rig can be used at home if you have a 12-V battery or suitable power supply. Fig 3-1 illustrates the choices available for a VHF or UHF station and how portable, mobile and home (fixed) rigs and antennas can be used.

1) *Hand-held transceivers.* Hand-held transceivers put out from under a watt to 5 watts or more. If the hand-held you're considering is capable of high-power operation, you'll want to be able to switch to low power when possible, to conserve the battery. Most hand-helds offer a **HIGH/LOW** power switch. Good used hand-held transceivers sell from $150 and up. New transceivers start at $250. See *The ARRL Radio Buyer's Sourcebook* for more information. Remember: with higher power levels you will want to use a remote antenna. Radiation of a lot of power in front of your eyes or near your head is not a good idea. This important topic is discussed in the Safety chapter of *The ARRL Handbook for Radio Amateurs,* 1995 and later editions.

2) *Mobile transceivers* have power outputs ranging from 10 to 50 watts. In populated areas with many repeaters, any more than 10 watts is probably unnecessary. Mobile transceivers are available used from $150 and up. New transceivers start at about $250 for a 10-watt unit. A mobile transceiver can be used indoors, too, if you have an adequate power supply. Selection of a mobile antenna is just as important as selection of the rig. Most hams are very disappointed when they try to use a hand-held transceiver in a car with the rubber ducky antenna attached to the rig.

3) *Fixed rigs.* As shown in Fig 3-1, a mobile rig connected to a power supply or 12-V automotive battery makes a good fixed station. It can also be disconnected and moved into the car for mobile operation. Powering the rig from a battery has the added advantage of allowing emergency operation when the local power lines go down in a storm.

4) *VHF packet radio.* Most VHF packet-radio operation takes place on the 2-meter and 440-MHz bands. A mobile transceiver makes a good packet-radio station rig, too. You'll need a computer or terminal, and a terminal node controller (TNC). Used TNCs are available for $70 or so, but new ones are only slightly more expensive. Newer units may offer "mailbox" features, where other hams can leave messages for you when you aren't home.

HF Equipment

1) *Separate transmitter and receiver.* Before single-sideband caught on, most amateur stations featured a separate transmitter and receiver. In many cases, the transmitter was homemade and the receiver was a commercial model. Most of

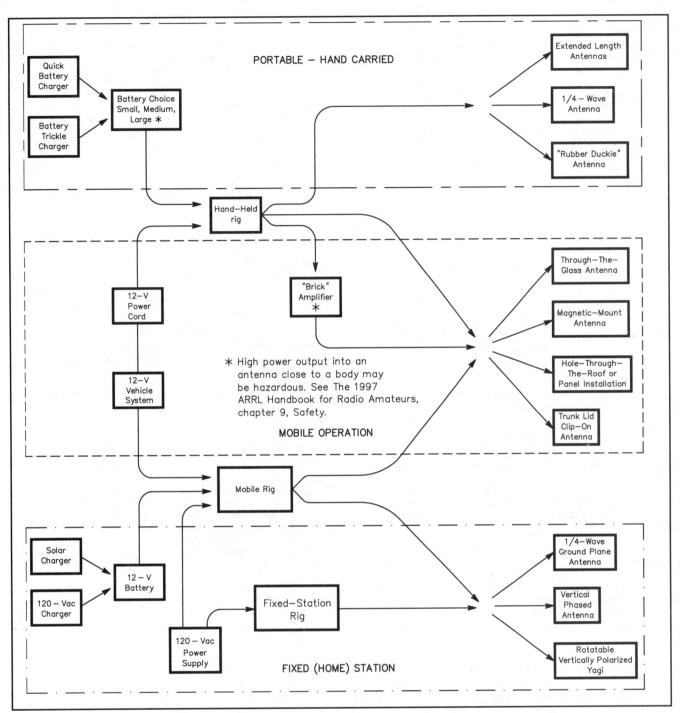

Fig 3-1—There are many choices of equipment for use with repeaters. A hand-held transceiver is the most versatile but it is usually low power and may require a number of accessories to use from a car or from home.

this equipment is over 30 years old now, and maintenance problems are common. There wasn't the emphasis on compact, lightweight equipment we see now, and the receiver alone was usually larger than most modern transceivers. Some hams who were licensed many years ago like to use this older gear for nostalgic reasons. It often appears on used equipment shelves and at hamfests. Our advice: Leave this equipment to people who enjoy working with older equipment or buy one of the receivers as a spare or for SWLing.

2) *Used separate* transmitter/receiver *combinations*. At one time all the major equipment manufacturers sold separate receivers and transmitters to "transceive" together. Examples are

the Heath SB401/SB303, Drake T-4/R-4, Kenwood T-599/R-599 and Collins 32S-3/75S-3. Each of these rigs was a classic in its time, but all are outdated now. They lack many refinements common on newer transceivers, including the ability to work on the WARC bands (30, 17 and 12 meters). All the transmitters mentioned, and most of the receivers, used vacuum tubes.

Years and high temperatures can really take its toll on electronic equipment. Tubes are expensive and harder to find. Capacitors tend to degenerate and finding replacements rated at the operating voltages of tubes can be a real chore. If you can find a good deal and know the equipment is in good working order (don't take the seller's word unless you know him or her

A Hobby with No Barriers

One of the outstanding characteristics of hamming is that anyone can enjoy it. Amateur Radio has a special appeal to persons with visual or physical disabilities: it provides a means of people-to-people contact on a basis of absolute equality. Ham radio has always been a natural way of making new friends, renewing old acquaintances and visiting—via radio—people and places all over the world. It is perfect for those with limited mobility. There are hundreds of individuals with disabilities active in Amateur Radio. They have shown that with the proper training, patience and determination, almost any physical disability can be overcome.

The HANDI-Hams

A nationwide group specializes in helping persons with disabilities become a part of the hobby—the HANDI-Hams. Bruce Humphrys, KØHR, a former director of the HANDI-Hams, described the program in *QST*: "Courage Center is a comprehensive rehabilitation facility for persons with physical, speech, hearing and visual handicaps, based in Golden Valley, Minnesota. Aside from the HANDI-Ham System, Courage Center offers a tremendously wide array of services for handicapped people . . . the HANDI-Ham system is one of the Center's services—serving more than 7000 people all over the world. The system provides three direct (and many indirect) services: (1) educational material, fraternity and close personal attention; (2) Amateur Radio equipment on loan; and (3) specially designed devices for ease of station control. The system relies on a trained cadre of radio amateurs to help provide these services. We not only help new students to get their first license, but also help handicapped licensed hams to upgrade."

Every year, the HANDI-Hams hold radio camps for students with disabilities who are studying for new or upgraded ham tickets. Amateur exams are conducted after each session.

Contact the Courage HANDI-Hams, 3915 Golden Valley Rd, Golden Valley, MN 55422, Telephone 612-520-0515, e-mail **handiham@mtn.org**, for further information.

Getting the Ticket

All candidates for an amateur license must prove their ability to send and receive the International Morse code, and must pass a test on radio regulations and theory.

It is possible to exempt individuals with severe handicaps or disabilities from the 13 or 20-wpm code tests, or to use special accommodation procedures for these tests.

Section 97.509(k) describes these procedures: "The administering Volunteer Examiners must accommodate an examinee whose physical disabilities require a special examination procedure. The administering VEs may require a physician's certification indicating the nature of the disability before determining which, if any, special procedures must be used." These special accommodation procedures, at the discretion of the Volunteer Examiner Team, could mean administering the examination at a place convenient or comfortable to the examinee (even at his or her bedside).

If the disability is so severe that the examinee is unable to pass the examination even with the special accommodation procedures, then a Morse code exemption for the 13 or 20-wpm code element may be obtained. For the purpose of obtaining an exemption for the higher speed code examination, the FCC defines a "severe handicap" as a disability that extends more than 365 days after the certification. (The 5-wpm code element cannot be waived.) The FCC states that an applicant for a 13 or 20-wpm Morse code exemption must submit to the VE Team an FCC Form 610 signed by a physician. The form (available from ARRL HQ) must be dated November 1993 or later.

ARRL Involvement

ARRL Headquarters helps enable persons with disabilities to prepare for FCC amateur exams. It serves as an information center and clearinghouse for this purpose, and maintains a database containing lists of operating aids and clubs, and other pertinent information. In addition, *QST* publishes technical and general-interest articles of special interest to persons with disabilities as well as a comprehensive booklet, *Source Book for the Disabled.* Contact ARRL HQ for information on obtaining a copy in printed or electronic format.

Membership in the League is available at a reduced rate without *QST* to persons who are blind. This gives already licensed blind members an opportunity to take an active role in ARRL activities, vote in ARRL Director and SM elections, run for League offices and otherwise share in the privileges of membership. Unlicensed blind persons, too, may show their support for the work of the ARRL through an associate membership. For more information, write the Program for the Disabled Coordinator at ARRL HQ.

QST is available on recorded disks from the National Library Service - Library of Congress. Contact your local library for details.

---

well!), you may be able to pick up a pair for under $200. Don't be surprised at some of the prices quoted. Some rigs, such as the Collins S-line, command high prices from collectors. If you're looking for a first rig, the cost of a used rig such as the S-line is such that you're better off considering a more modern, used transceiver.

3) *Transceivers.* HF transceivers have been common for over 25 years. Like all older gear, older transceivers are likely to have maintenance problems. Mobile use subjects a radio to a great deal of vibration and wide ranges of temperature. A rig showing signs of having been used for mobiling may not be a good choice for your first station.

The modern age of transceivers can be divided in half. Transceivers manufactured in the '70s were partly solid state. The transmitters usually had three tubes: two in the power amplifier and another serving as the driver stage. Some transceivers from this era had built-in ac power supplies. Because they are newer, there may be fewer maintenance problems. However, mechanical parts, such as those used in the tuning assembly, may be impossible to obtain. Make sure everything works properly

before you buy. Transceivers from this era probably won't operate on the 30, 17 and 12-meter bands.

The second era of modern transceivers feature solid-state construction. Tuning mechanisms have fewer mechanical parts to wear and all-band operation is common. Many feature general-coverage receivers that continuously tune from below 100 kHz to 30 MHz. With the general-coverage receiver you can listen to time- and frequency-standard stations like WWV and WWVH and enjoy the variety of short-wave broadcasting (see Chapter 1).

New transceivers don't usually offer many more features than the better-equipped used gear mentioned in the previous paragraph. Of course, new equipment comes with a factory warranty, and you know no one opened the case to make modifications.

4) *Homemade equipment.* Beginners usually lack the skill and experience to design or build their first stations, but if you have a background in electronics, don't pass up the chance to build ("home brew") your own. Many good circuits appear in *The ARRL Handbook for Radio Amateurs.*

Accessories

A few accessories found in almost all stations, are needed either to help set up and test the equipment or to operate it. Plans for a station should include some of these accessories.

SWR Indicator

This device is handy for testing an antenna and feed line when the antenna is first erected, and later to make sure the antenna is still in good shape. If a Transmatch is used with a multiband antenna system, such as a Windom or G5RV, an SWR indicator is essential for ensuring the Transmatch is adjusted for a reasonable SWR. Many modern HF transceivers include an SWR indicator. External SWR meters can cost as little as $30. If you plan to operate VHF and HF you may have to buy one for HF and a second for VHF/UHF.

Transmatch (Antenna Tuner/Antenna Coupler)

There are so many Transmatches available on the market that you might think you can't get by without one. Manufactured Transmatches cost between $75 and $1500. The circuits are pretty simple, though, and many first-time home-brew projects are Transmatches. *(The ARRL Handbook for Radio Amateurs* contains plans for building Transmatches.) By carefully shopping at hamfest flea markets you can often assemble your own for less than $50.

Keys, Keyers and Paddles

If you're interested in CW, you have to have some means of sending the code. You should start with a straight key until you feel you have the proper rhythm. Then you may wish to buy a keyer and paddles, or use your computer to send code. Some modern rigs have keyers built in, so all you need is a paddle. Paddles may be "standard" or "iambic." The iambic type requires less hand motion but usually a longer learning period.

Keyers cost from $50 to $250, and paddles cost about the same. Code-transmitting programs are available for the common computers; see the ads in *QST.*

Computers

Computers have become a very common part of the ham shack. Some are used "on-the-air"; that is, to send and receive Morse code, fax, slow-scan TV or digital modes. Others are

used for logging and record keeping. Hams use various types of computers, ranging from old *VIC-20s* to *UNIX*-based work stations. The most popular type—having the most software available—is the PC- or MS DOS-based unit.

When considering a computer for the shack remember inside the computer is an oscillator—the "clock"—and it has the capability of generating signals right in the middle of some ham bands. Other computers are very sensitive to the presence of RF such as generated by your transmitter a foot or so away. For help with this topic look at the ARRL book *Radio Frequency Interference—How to Find It and Fix It.* It contains an entire chapter on computer problems in the shack.

Test Meter

An inexpensive multimeter capable of measuring voltage, current and resistance is very helpful around the shack. High accuracy is not needed for most problems. A $15 to $30 unit will pay for itself the first time you need to check the integrity of a coax connector you've just installed.

Station Setup

How you set up your station is determined by how much space you have, and how much equipment you have to squeeze into it. The table should be about 30 inches high and 30 inches deep. An old desk makes a good operating table. Build shelves for your equipment. Radios stacked on top of one another can't "breathe" and may overheat. It's also easier to change cables and move equipment around when you use shelves.

Place your key far enough from the edge of the table so your entire arm is supported, to prevent fatigue. A microphone can be mounted on a stand placed on the table or on an extension that reaches in from the back or side. The best way to test the arrangement of your rig is to sit in the operating chair and operate the rig's controls. If the knobs are too low try placing spacers or blocks under the front feet of the rig. Too high? Try placing the spacers under the rear legs.

Check that you can see the frequency dial. While it may be nice to place a keyboard right in front of your position, it may cause considerable strain on your back, shoulder and arms if you have to reach across the keyboard and to operate the rig. In this case consider a "T" or "U" shaped operating position. If this is your first shack remember it won't be the last time you will rearrange the equipment.

Be sure to have an effective earth ground routed to your station. Use as many grounds as you can locate, bonding them together electrically and bringing the connection plate to the operating position. If you live on a higher floor and a ground is not available, use a cold-water pipe if you can.

At least one manufacturer makes an "artificial ground," which is really a Transmatch for ground radials. While this device probably won't let you have a stronger signal, it may remove RF from the equipment cases.

Connections from the various ground points to the shack should be as short as possible. Since RF currents flow mainly on the surfaces of conductors, ground conductors such as shield braid from RG-8 coaxial cable or wide strips of flashing copper are best.

Safety should be a prime consideration. Many hams will not plug in their equipment for the first time until all the cases have been tied to ground. Use a master switch and make sure other people in the house know where it is and how to use it. Consider running your shack through a *GFI* (ground fault interrupt) outlet. Unfortunately not all ham gear, especially older units, will operate with these devices. However, by detecting unbal-

The two most often used objects in the shack are the keyboard and the tuning controls. In this photo, Tim Jellison, W3YQ, reaches above the keyboard to the rig controls. The rig is slightly elevated to clear the keyboard. The computer display is eye-level, and directly in front of Tim.

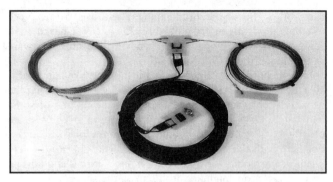

You can build a single-band dipole or buy one already assembled and ready to go. This MFJ unit uses low-loss leader line, and includes a balun to connect the balanced ladder-line to coax (pictured in the center of the coil of ladder-line). (*Photo courtesy of MFJ, Inc.*)

anced line currents and shutting off the voltage when unbalance occurs, they could save your life.

Antennas

There are so many varieties of amateur antennas you could write books about them. We did: *The ARRL Antenna Book* and *Your Ham Antenna Companion*. Whatever antennas you select, install them safely. Don't endanger your life or someone else's for your hobby!

Antennas are important. The best (and biggest) transmitter in the world will not do any good if the signal is not radiated into the air. A good rule of thumb to follow is "always erect as much antenna as possible." The better your antenna array, the better will be your radiated signal. A good antenna system will make up for inadequacies or shortcomings in station equipment. A less-sensitive receiver "hears better" with a good antenna system and a "bigger" antenna system will make a QRP (low power) station sound a lot louder at the receiving end.

If you are thinking of a tower talk to a few local hams before starting construction (or applying for a building permit). Local rules and ordinances may have a large impact on your plans.

Apply common sense to your antennas! Many hams act as though the world will end if they put up an antenna and measure an SWR greater than 1.5:1. For most purposes an SWR of 3:1 is perfectly acceptable at HF with good quality feed lines 100 feet or less in length.

First VHF/UHF Antennas

One of the nice things about VHF and UHF operation is that often the simplest antennas, if mounted high enough, will do an excellent job. Ground planes, "J-poles" and simple beams can either be purchased at a reasonable cost or often constructed in a few minutes. If you want to test a 2-element quad for 144-MHz, just take the design from the *ARRL Antenna Book*, build it from scrap wood and heavy gauge copper or aluminum wire, and run a few tests. The unit you built may not stay up for a long time in bad weather but it will be fine for determining if this is the sort of antenna you want to put up "permanently."

A simple ground-plane antenna is shown in Fig 3-2. It can be mounted by taping the feed line and bottom connector to a pole so that the antenna extends over the top of the pole. It can also be suspended by a cord by lengthening the vertical element and bending the extra length into a loop. One end of the cord is fastened to the loop and the other end run over a tree branch.

If you are willing to make a temporary antenna—which may last only a few months or a year—you can substitute ordinary wire from a wire clothes hanger. It may rust but as long as the connections are tight it will operate well.

Simple HF Antenna

The most popular "first" antenna is the half-wavelength dipole. It consists of a half wavelength of antenna wire with a feed line and an insulator at its center. See Fig 3-3.

For a new ham's activities, a resonant dipole system is a good selection. It is very easy to erect and has a low *SWR* (standing wave ratio—a measure of how well an antenna is tuned to the transmitted frequency) for its desired cut-to-length operating range. A single-band dipole fed with low-loss feed line can also be used on other bands. In fact with a Transmatch, a balun, and a random length center fed dipole you can actually operate on any HF band. The trick is to use a low-loss feed line and provide a match with a tuning network or Transmatch.

Where to Put the Antenna?

Think about your antenna location next. Remember: Your antenna should be as high and as far away from surrounding trees and structures as possible. Never put an antenna around or even near power lines! The dipole will require one support at each of its ends (perhaps trees, poles or even house or garage eaves), so survey your potential antenna site with this in mind. If you find space is so limited that you can't put up a straight-line dipole, don't give up. You can bend the ends of the dipole and still make plenty of contacts.

Some hams have been known to become very frustrated when they cannot put up a 135-foot long dipole that looks like it came out of an engineering design manual. Most dipoles are

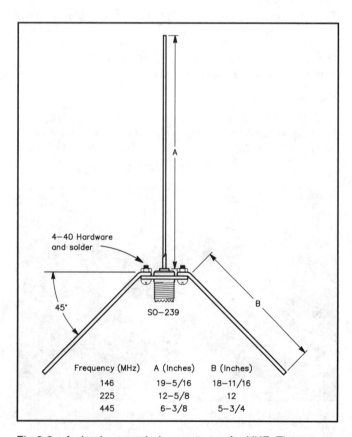

Frequency (MHz)	A (inches)	B (inches)
146	19–5/16	18–11/16
225	12–5/8	12
445	6–3/8	5–3/4

Fig 3-2—A simple ground-plane antenna for VHF. The elements are made from 3/32 or 1/16-in brass welding rod or #10 or #12 copper wire.

very close to the ground and still work well. Many hams successfully use a dipole suspended from the middle, called an "inverted V." The performance of this antenna is so satisfactory they are not even aware of the large losses this configuration may have as compared to the ideal dipole.

You can put up an antenna under almost any circumstances, but you may need to use your imagination. Look at *The ARRL Antenna Book, Now You're Talking!, W1FB's Antenna Notebook* and *QST* for many workable antenna ideas.

Antenna Parts

If this is the first time you have tried to put up a dipole by yourself the following parts list will give you some guidance.

1) *Antenna wire.* #12 or #14 hard-drawn Copperweld (copper-clad steel) is preferred, so the antenna won't stretch. It will be strong enough to support itself as well as the weight of the feed line connected at its center. Always buy plenty of wire. It never goes to waste!

2) *Insulators.* You need one center and two end antenna insulators for a simple dipole.

3) *Clamp.* Large enough to fit over two widths of your co-axial cable.

4) *Coax.* Feed line made of a center conductor surrounded by an insulating dielectric. This in turn is surrounded by a braid called the shield. You need RG-58/U (or /AU or /BU or polyfoam) or RG-8U. Look for coax with heavy braid shield such as Belden, Saxton or Times Wire and Cable. Stay away from cheap cable from unknown manufacturers, or too-good-to-be-true deals on "surplus cable."

A good alternative is balanced "open-wire line." This may be actually two parallel pieces of wire connected and spaced with plastic rods or enclosed in a plastic jacket, similar to TV 300-Ω wire but with pieces of the center plastic removed. If you are sure you want to build a single band dipole stick to the coax, otherwise consider the Transmatch, balun and open-wire configuration.

5) *Connector.* Connects the feed line (coax) to your rig. This will probably be a PL-259, standard on most rigs. If your radio needs another kind, check your radio's instruction manual for installation information. You also need connectors for the coax lines between your Transmatch, SWR-meter and your rig.

6) *Electrical tape and coax sealant.* This is needed to cover the antenna ends and joints to make them waterproof. Otherwise water can get into coax and "short" it, ruining your antenna system.

7) *Rope.* You need enough to tie the ends of the antenna to a supporting structure. Rope may be used; however, it does degenerate with time and weather. You can use conductive guy wire. Conventional wisdom says you should make sure the length of the guy wire is not resonant in a ham band. In practice, it's doubtful you'll see any difference in antenna performance with any length of guy wire.

8) *SWR meter.* This is essential for your station and especially for antenna adjustment. SWR meters are readily available and inexpensive, making them easier to buy than build.

Gather all the parts you'll need for your chosen antenna. Almost everything is available from your local electronics store or from suppliers advertising in *QST*. When this is done, the fun—actually putting together your antenna—can begin.

Assembly is quite simple. Your dipole consists of two lengths of wire, each approximately 1/4 wavelength long at your chosen operating frequency. These two wires are connected in the center, at an insulator, to the feed line. In our antenna, the feed line is coaxial cable, and it brings the signals to and from your radio. Calculate the length of the half-wave dipole by using this formula: antenna length in feet = 468/frequency in MHz. (The Novice/Technician Plus-band information below has approximate lengths already calculated.) Now measure the antenna wire, keeping it as straight as possible.

Antenna Lengths in Feet

(cut to center of Novice/Technician Plus band)

	1/2 wavelength	1/4 wavelength
80 m	126'6"	63'3"
40 m	65'8"	32'10"
15 m	22'2"	11'1"
10 m	16'7"	8'3"

Remember to add about 1 foot to each end of the dipole for tuning adjustment.

Putting It Together

Carefully assemble your antenna, paying special attention to waterproofing the coax. Don't solder the antenna ends until later, when retuning is completed. Just twist them for now. Route the coax to your station, remembering to keep it as unobtrusive as possible. Cut the coax to a length that will leave some excess for strain relief so your rig won't be pulled around during strong winds! Install the connector(s) according to the diagrams. Now connect your SWR bridge between the feed line (coax) and the transmitter.

One trick used by old-timers is the addition of a 10,000-Ω resistor, soldered directly across the center insulator of the dipole. An ohmmeter connected from the one side to the feed line to the other at the ground end should measure this value of 10,000 Ω. If it measures an open circuit it means the feed line is disconnected or broken. A short circuit means the feed line or the connector is shorted. The resistor has no effect on the antenna or SWR. Left in place it acts as a handy check on the antenna and feed line.

There are several ways to see if you built the antenna correctly. One very direct way is to buy or borrow an antenna noise bridge. Details of these handy devices are in the *ARRL Handbook*. The bridge will tell you the resonant frequency of the antenna. If the frequency is too low you should shorten the length of the antenna (cut equal amounts from both ends). Resonant frequency too high? The antenna must be lengthened.

The other common method is the use of an SWR meter. If it shows an SWR of 2:1 or less at your desired operating frequency, your antenna system is "tuned," and you can go ahead and operate. If your SWR is greater than 2:1, you must "retune" your antenna to obtain a lower SWR. Disconnect the transmitter, and try shortening the antenna a few inches on each end. Keep notes! Reconnect the transmitter, and check the SWR. If the SWR gets lower, continue shortening the ends until you get the lowest possible SWR reading. Be careful not to shorten it too much! As the last step, when you are sure the length of the dipole is correct, solder the wire wrapped around the end insulators.

Problems and Cures

A high SWR that does not change when you change the antenna length by a few feet (HF only) probably means something is wrong beyond merely length misadjustments of your simple, one-band dipole. Check to see if your coax is open or shorted. Make sure your antenna isn't touching anything and all your connections are sound.

When all systems are "go," get on the air and operate. As

Build Your Own Antenna

To connect the coax to the center insulator, cut a few inches of the outer covering off the coax. Next, separate the copper braid from the inner conductor and insulation. After you've done that, twist the strands of copper together to form a single wire. Remove about half the insulation covering the inner conductor and bend it away from the twisted strands of copper braid. Loop the cable over the insulator as shown and solder the braid to one half the antenna and the inner conductor to the other. Be sure to tape all connections securely for waterproofing; the braid can soak up water like a sponge or wick, making the coax useless after a while. B and C show how the wire is connected to various types of insulators at the ends, and D shows the connection of the feed line at the center.

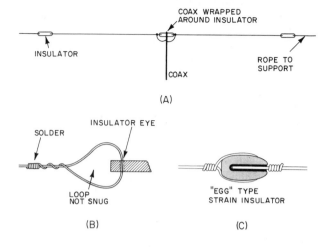

Examples of simple but effective wire antennas. A horizontal dipole is shown at A. The legs can be drooped to form an "inverted V" as shown at B. A sloping dipole (sloper) is illustrated at C. The feed line should come away from the sloper at a 90° angle for best results. If the supporting mast is metal, the antenna will have some directivity in the direction of the slope.

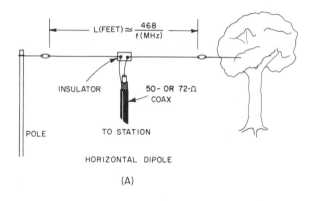

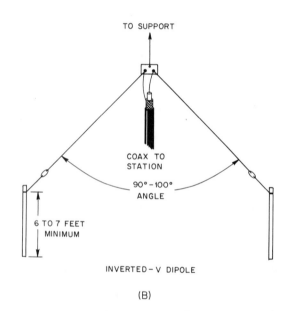

Using a PL-259 Connector

If you are using RG-58 or RG-59 with a PL-259 connector, you will need to use an adapter, as shown here. This material courtesy of Amphenol Electronic Components, RF Division, Bunker Ramo Corp.

1) Cut end of cable even. Remove vinyl jacket ¾″—don't nick braid. Slide coupling ring and adapter on cable.

2) Fan braid slightly and fold back over cable.

3) Position adapter to dimension shown. Press braid down over body of adapter and trim to 3/8″. Bare 5/8″ of center conductor. Tin exposed center conductor.

Figure 3-3—Building your own HF dipole antenna is a popular project.

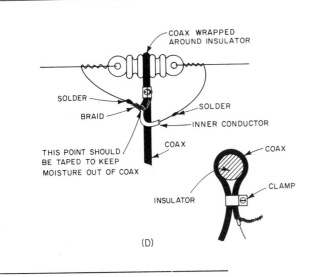

COAX WRAPPED AROUND INSULATOR

SOLDER

BRAID

SOLDER

INNER CONDUCTOR

THIS POINT SHOULD BE TAPED TO KEEP MOISTURE OUT OF COAX

COAX

COAX

CLAMP

INSULATOR

(D)

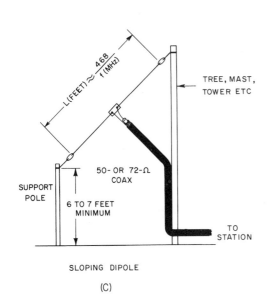

$L(FEET) \simeq \frac{468}{f(MHz)}$

TREE, MAST, TOWER ETC

SUPPORT POLE

50- OR 72-Ω COAX

6 TO 7 FEET MINIMUM

TO STATION

SLOPING DIPOLE

(C)

4) Screw the plug assembly on adapter. Solder braid to shell through solder holes. Solder conductor to contact sleeve.

5) Screw coupling ring on plug assembly.

you settle into that first QSO with your new antenna, enjoy those feelings of pride, accomplishment and fun that will naturally follow. After all, that's what Amateur Radio is all about.

OPERATING—WHAT, WHERE AND HOW

We're finally getting down to the real nitty-gritty. You have read a little about the basics of ham radio and have some ideas about selecting a rig and station equipment and erecting a decent antenna system.

Now it's time to learn a little about how to go about playing the ham radio game. For everyone to be able to communicate with everyone else effectively, reasonable operating procedures are needed.

A brief explanation follows, describing the major modes of ham radio communication and a few of the procedures and conventions in use on the air today. Other chapters of this manual discuss in great detail the operating procedures for many specialized modes of communication. Here we are going to look at voice and CW (Morse code) operating.

Spend as much time as possible listening to other operators on the bands. Find out what works and what doesn't. Copy the good habits of the good communicators and incorporate those habits into your own operating style. Good operating habits will make you an effective communicator for as long as you pound that CW key or speak into that microphone.

Where To Operate

One of the first questions you may have is where to operate? The ham bands are defined by FCC rules and these rules limit where you can operate. Fig 3-4 lists the frequencies each class of licensee can use for various modes. Within these rules hams have agreed to limit their operation to certain frequencies for certain modes. Table 3-1 shows these agreements.

If you're a new ham and hold a Technician or higher class license, you will probably do some operating on 2 meters, the most popular ham band. Although packet radio is also very common, the most popular mode is 2-meter FM using repeaters. Chapter 11 of this manual goes into great detail on this type of operation. The sidebar *Picking a Band* gives the characteristics of the various ham bands.

CW Operating

CW (continuous wave, Morse code) is the universal language, and a common bond of hamdom. Many hams take pride in their code proficiency. It takes practice to master the art of sending good code on a hand key, bug (semi-automatic key) or electronic keyer. You have to practice to get that smooth rhythm, practice to get that smooth spacing between words and characters, and practice to learn the sound of whole words and phrases, rather than just individual letters.

CW is an expedient mode of communications. CW transmitters are simpler devices than their phone counterparts, and a CW signal can usually get through very heavy QRM (interference on the band) much more effectively than a phone signal. Therefore, shortcuts and abbreviations are used during a CW QSO. Many of the abbreviations hams use have developed within the ham fraternity, while some are borrowed from the old-time telegraph operators. Q signals are among the most useful of these abbreviations. A list of the most popular Q signals is in Table 3-2.

You don't have to sit down and memorize this list. Copy the list and keep the copy on your operating table. You can also request Form FSD-218 from the ARRL. It contains a handy list of Q signals and abbreviations printed on cardboard for mount-

Table 3-1

The "Considerate Operator's Frequency Guide"

The following frequencies are generally recognized for certain modes or activities (all frequencies are in MHz).

Nothing in the rules recognizes a net's, group's or any individual's special privilege to any specific frequency. Section 97.101(b) of the Rules states that "Each station licensee and each control operator must cooperate in selecting transmitting channels and in making the most effective use of the amateur service frequencies. No frequency will be assigned for the exclusive use of any station." No one "owns" a frequency.

It's good practice—and plain old common sense—for any operator, regardless of mode, to check to see if the frequency is in use prior to engaging operation. If you are there first, other operators should make an effort to protect you from interference to the extent possible given that 100% interference-free operation is an unrealistic expectation in today's congested bands.

Frequency	Mode/Activity
1.800-1.830	CW, data and other narrowband modes
1.810	QRP CW calling frequency
1.830-1.840	CW, data and other narrowband modes, intercontinental QSOs only
1.840-1.850	CW; SSB, SSTV and other wideband modes, intercontinental QSOs only
1.850-2.000	CW; phone, SSTV and other wideband modes
3.560	QRP CW calling frequency
3.590	RTTY DX
3.580-3.620	Data
3.620-3.635	Automatically controlled data stations
3.710	QRP Novice/Technician CW calling frequency
3.790-3.800	DX window
3.845	SSTV
3.885	AM calling frequency
3.985	QRP SSB calling frequency
7.040	RTTY DX
	QRP CW calling frequency
7.080-7.100	Data
7.100-7.105	Automatically controlled data stations
7.110	QRP Novice/Technician CW calling frequency
7.171	SSTV
7.285	QRP SSB calling frequency
7.290	AM calling frequency
10.106	QRP CW calling frequency
10.130-10.140	Data
10.140-10.150	Automatically controlled data stations
14.060	QRP CW calling frequency
14.070-14.095	Data
14.095-14.0995	Automatically controlled data stations
14.100	NCDXF/IARU beacons
14.1005-14.112	Automatically controlled data stations
14.230	SSTV
14.285	QRP SSB calling frequency
14.286	AM calling frequency
18.100-18.105	Data
18.105-18.110	Automatically controlled data stations
21.060	QRP CW calling frequency
21.070-21.100	Data
21.090-21.100	Automatically controlled data stations
21.340	SSTV
21.385	QRP SSB calling frequency
24.920-24.925	Data
24.925-24.930	Automatically controlled data stations
28.060	QRP CW calling frequency
28.070-28.120	Data
28.120-28.189	Automatically controlled data stations
28.190-28.225	Beacons
28.385	QRP SSB calling frequency
28.680	SSTV
29.000-29.200	AM
29.300-29.510	Satellite downlinks
29.520-29.580	Repeater inputs
29.600	FM simplex
29.620-29.680	Repeater outputs

Note

ARRL band plans for frequencies above 29.680 MHz are shown in *The ARRL Repeater Directory* and *The FCC Rule Book*.

Table 3-2

Q Signals

These Q signals are the ones used most often on the air. (Q abbreviations take the form of questions only when they are sent followed by a question mark.)

QRG Will you tell me my exact frequency (or that of ___)? Your exact frequency (or that of ___) is ___ kHz.

QRL Are you busy? I am busy (or I am busy with ___). Please do not interfere.

QRM Is my transmission being interfered with? Your transmission is being interfered with ___ (1. Nil; 2. Slightly; 3. Moderately; 4. Severely; 5. Extremely.)

QRN Are you troubled by static? I am troubled by static ___. (1-5 as under QRM.)

QRO Shall I increase power? Increase power.

QRP Shall I decrease power? Decrease power.

QRQ Shall I send faster? Send faster (___ WPM).

QRS Shall I send more slowly? Send more slowly (___ WPM).

QRT Shall I stop sending? Stop sending.

QRU Have you anything for me? I have nothing for you.

QRV Are you ready? I am ready.

QRX When will you call me again? I will call you again at ___ hours (on ___ kHz).

QRZ Who is calling me? You are being called by ___(on ___ kHz).

QSB Are my signals fading? Your signals are fading.

QSK Can you hear me between your signals and if so can I break in on your transmission? I can hear you between signals; break in on my transmission.

QSL Can you acknowledge receipt (of a message or transmission)? I am acknowledging receipt.

QSN Did you hear me (or ___) on ___ kHz? I did hear you (or ___) on ___ kHz.

QSO Can you communicate with ___ direct or by relay? I can communicate with ___ direct (or relay through ___).

QSP Will you relay to ___? I will relay to ___.

QST General call preceding a message addressed to all amateurs and ARRL members. This is in effect "CQ ARRL."

QSX Will you listen to ___ on ___ kHz? I am listening to ___ on ___ kHz.

QSY Shall I change to transmission on another frequency? Change to transmission on another frequency (or on ___ kHz).

QTB Do you agree with my counting of words? I do not agree with your counting of words. I will repeat the first letter or digit of each word or group.

QTC How many messages have you to send? I have ___ messages for you (or for ___).

QTH What is your location? My location is ___.

QTR What is the correct time? The time is ___.

160 METERS

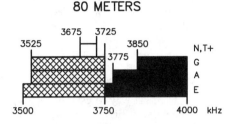

E,A,G

1800   1900   2000   kHz

Amateur stations operating at 1900–2000 kHz must not cause harmful interference to the radiolocation service and are afforded no protection from radiolocation operations.

80 METERS

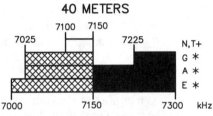

3675   3725
3525   3850
3775
N,T+
G
A
E

3500   3750   4000   kHz

5167.5 kHz (SSB only): Alaska emergency use only.

40 METERS

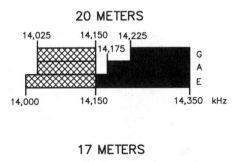

7100   7150
7025   7225
N,T+
G *
A *
E *

7000   7150   7300   kHz

* Phone operation is allowed on 7075–7100 kHz in Puerto Rico, US Virgin islands and areas of the Caribbean south of 20 degrees north latitude; and in Hawaii and areas near ITU Region 3, including Alaska.

30 METERS

E,A,G

10,100   10,150   kHz

Maximum power on 30 meters is 200 watts PEP output. Amateurs must avoid interference to the fixed service outside the US.

20 METERS

14,025   14,150   14,225
14,175
G
A
E

14,000   14,150   14,350   kHz

17 METERS

E,A,G

18,068   18,110   18,168   kHz

15 METERS

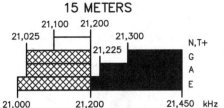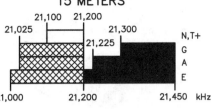

21,100   21,200
21,025   21,300
21,225
N,T+
G
A
E

21,000   21,200   21,450   kHz

12 METERS

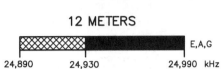

E,A,G

24,890   24,930   24,990   kHz

10 METERS

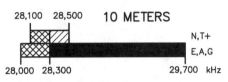

28,100   28,500
N,T+
E,A,G

28,000   28,300   29,700   kHz

Novices and Technicians are limited to 200 watts PEP output on 10 meters.

6 METERS

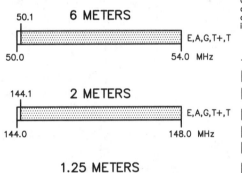

50.1
E,A,G,T+,T

50.0   54.0   MHz

2 METERS

144.1
E,A,G,T+,T

144.0   148.0   MHz

1.25 METERS

E,A,G,T+,T,N

222.0   225.0   MHz

Novices are limited to 25 watts PEP output from 222 to 225 MHz.

70 CENTIMETERS **

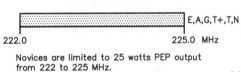

E,A,G,T+,T

420.0   450.0   MHz

33 CENTIMETERS **

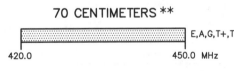

E,A,G,T+,T

902.0   928.0   MHz

23 CENTIMETERS **

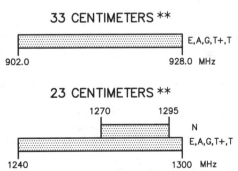

1270   1295
N
E,A,G,T+,T

1240   1300   MHz

Novices are limited to 5 watts PEP output from 1270 to 1295 MHz.

US AMATEUR BANDS

December 20, 1994

US AMATEUR POWER LIMITS

At all times, transmitter power should be kept down to that necessary to carry out the desired communications. Power is rated in watts PEP output. Unless otherwise stated, the maximum power output is 1500 W. Power for all license classes is limited to 200 W in the 10,100–10,150 kHz band and in all Novice subbands below 28,100 kHz. Novices and Technicians are restricted to 200 W in the 28,100–28,500 kHz subbands. In addition, Novices are restricted to 25 W in the 222–225 MHz band and 5 W in the 1270–1295 MHz subband.

Operators with Technician class licenses and above may operate on all bands above 50 MHz. For more detailed information see The FCC Rule Book.

——— KEY ———

▨ = CW, RTTY and data

░ = CW, RTTY, data, MCW, test, phone and image

■ = CW, phone and image

▨ = CW and SSB

▨ = CW, RTTY, data, phone, and image

☐ = CW only

E = EXTRA CLASS
A = ADVANCED
G = GENERAL
T+ = TECHNICIAN PLUS
T = TECHNICIAN
N = NOVICE

** Geographical and power restrictions apply to these bands. See The FCC Rule Book for more information about your area.

Above 23 Centimeters:

All licensees except Novices are authorized all modes on the following frequencies:
2300–2310 MHz
2390–2450 MHz
3300–3500 MHz
5650–5925 MHz
10.0–10.5 GHz
24.0–24.25 GHz
47.0–47.2 GHz
75.5–81.0 GHz
119.98–120.02 GHz
142–149 GHz
241–250 GHz
All above 300 GHz

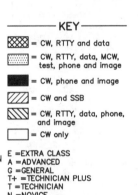

For band plans and sharing arrangements, see The ARRL Operating Manual or The FCC Rule Book.

Fig 3-4—Frequency Allocation Chart.

Table 3-3
Some Abbreviations for CW Work

Although abbreviations help to cut down unnecessary transmission, make it a rule not to abbreviate unnecessarily when working an operator of unknown experience.

AA	All after	OC	Old chap
AB	All before	OM	Old man
ABT	About	OP-OPR	Operator
ADR	Address	OT	Old timer; old top
AGN	Again	PBL	Preamble
ANT	Antenna	PSE	Please
BCI	Broadcast interference	PWR	Power
BCL	Broadcast listener	PX	Press
BK	Break; break me; break in	R	Received as transmitted; are
BN	All between; been	RCD	Received
BUG	Semi-automatic key	RCVR (RX)	Receiver
B4	Before	REF	Refer to; referring to; reference
C	Yes	RFI	Radio frequency interference
CFM	Confirm; I confirm	RIG	Station equipment
CK	Check	RPT	Repeat; I repeat; report
CL	I am closing my station; call	RTTY	Radioteletype
CLD-CLG	Called; calling	RX	Receiver
CQ	Calling any station	SASE	Self-addressed, stamped envelope
CUD	Could	SED	Said
CUL	See you later	SIG	Signature; signal
CW	Continuous wave (i.e., radiotelegraph)	SINE	Operator's personal initials or nickname
DLD-DLVD	Delivered	SKED	Schedule
DR	Dear	SRI	Sorry
DX	Distance, foreign countries	SSB	Single sideband
ES	And, &	SVC	Service; prefix to service message
FB	Fine business, excellent	T	Zero
FM	Frequency modulation	TFC	Traffic
GA	Go ahead (or resume sending)	TMW	Tomorrow
GB	Good-by	TNX-TKS	Thanks
GBA	Give better address	TT	That
GE	Good evening	TU	Thank you
GG	Going	TVI	Television interference
GM	Good morning	TX	Transmitter
GN	Good night	TXT	Text
GND	Ground	UR-URS	Your; you're; yours
GUD	Good	VFO	Variable-frequency oscillator
HI	The telegraphic laugh; high	VY	Very
HR	Here, hear	WA	Word after
HV	Have	WB	Word before
HW	How	WD-WDS	Word; words
LID	A poor operator	WKD-WKG	Worked; working
MA, MILS	Milliamperes	WL	Well; will
MSG	Message; prefix to radiogram	WUD	Would
N	No	WX	Weather
NCS	Net control station	XCVR	Transceiver
ND	Nothing doing	XMTR (TX)	Transmitter
NIL	Nothing; I have nothing for you	XTAL	Crystal
NM	No more	XYL (YF)	Wife
NR	Number	YL	Young lady
NW	Now; I resume transmission	73	Best regards
OB	Old boy	88	Love and kisses

ing at your operating position.

After using some of the Q signals and abbreviations a few times you will quickly learn the most popular of them without needing any reference. With time, you'll find that as your CW proficiency rises, you will be able to communicate almost as quickly on CW as you can on the voice modes. For more information on CW operating, see *Morse Code: The Essential Language,* published by ARRL. A list of common abbreviations is in Table 3-3.

It may not seem that way to you now, but your CW sending and receiving speed will rise very quickly with on-the-air practice. For your first few QSOs, carefully choose to answer the calls from stations that are sending at a speed you can copy (perhaps another first timer on the band?). Courtesy on the ham bands dictates that an operator will slow his or her code speed to accommodate another operator. Don't be afraid to call some-

one who is sending just a bit faster than you can copy comfortably. That operator will generally slow down to meet your CW speed. Helping each other is the name of the game in ham radio.

To increase your speed, you may continue to copy the code practice sessions from W1AW, the ARRL HQ station. It might be a good idea to spend some time sending in step (on a code-practice oscillator—not on the air, of course) with W1AW transmissions; it may help develop your sending ability.

Correct CW Procedures

The best way to establish a contact, especially at first, is to listen until you hear someone calling CQ. CQ means, "I wish to contact any amateur station." Avoid the common operating pitfall of calling CQ endlessly; it clutters up the air and drives off potential new friends. The typical CQ would go like this: CQ CQ CQ DE K5RC K5RC K5RC K. The letter K is an invitation for any station to go ahead. If there is no answer, pause for 10 or 20 seconds and repeat the call.

If you hear a CQ, wait until the ham finishes transmitting (by ending with the letter K), then call him, thus: K5RC K5RC DE K3YL K3YL $\overline{AR}$ ($\overline{AR}$ is equivalent to "over"). In answer to your call, the called station will begin the reply by sending K3YL DE K5RC R....

That R ("roger") means that he has received your call correctly. Suppose K5RC heard someone calling him, but didn't quite catch the call because of interference (QRM) or static (QRN). Then he might come back with QRZ? DE K5RC K (Who is calling me?).

The QSO

During the contact, it is necessary to identify your station only once every 10 minutes and at the end of the communication. Keep the contact on a friendly and cordial level, remembering that the conversation is not private and many others, including non-amateurs, may be listening. It may be helpful at the beginning to have a fully written-out script typical of the first couple of exchanges in front of you. A typical first transmission might sound like this: K3YL DE K5RC R TNX CALL. RST 599 599 QTH HOUSTON TX NAME TOM. HW? K3YL DE K5RC $\overline{KN}$. This is the basic exchange that begins most QSOs. Once these basics are exchanged the conversation can turn in almost any direction. Many people talk about their jobs, other hobbies, families, travel experiences, and so on.

Both on CW and phone, it is possible to be informal, friendly and conversational; this is what makes the Amateur Radio QSO enjoyable. During the contact, when you want the other station to take a turn, the recommended signal is KN, meaning that you want only the contacted station to come back to you. If you

Table 3-4
ARRL Procedural Signals

Situation	CW	Voice
check for a clear frequency	QRL?	Is the frequency is use?
seek contact with any station	CQ	CQ
after call to a specific named station or to indicate the end of a message	$\overline{AR}$	over, end of message
invite any station to transmit	K	go
invite a specific named station to transmit	$\overline{KN}$	go only
invite receiving station to transmit	BK	back to you
all received correctly	R	received
please stand by	$\overline{AS}$	wait, stand by
end of contact (sent before call sign)	$\overline{SK}$	clear
going off the air	CL	closing station

Additional RTTY prosigns
    SK QRZ—Ending contact, but listening on frequency.
    SK KN—Ending contact, but listening for one last transmission from the other station.
    SK SZ—Signing off and listening on the frequency for any other calls.

don't mind someone else signing in, just K ("go") is sufficient.

Ending the QSO

When you decide to end the contact or the other ham expresses the desire to end it, don't keep talking. Briefly express your thanks for the contact: TNX QSO or TNX CHAT—and then sign out: 73 $\overline{SK}$ WA1WTB DE K5KG. If you are leaving the air, add CL to the end, right after your call sign.

These ending signals, which indicate to the casual listener the "status of the contact," establish Amateur Radio as a cordial and fraternal hobby. At the same time they foster orderliness and denote organization. These signals have no legal standing; FCC regulations say little about our internal procedures.

Table 3-4 contains a brief list of these signals. You can also request Form FSD-220 from the ARRL. It contains this list plus other handy information for the operating position.

Phone Operating Procedures

These phone or voice operating procedures apply to operation on the HF bands as well as SSB on VHF/UHF. Procedures used on repeaters are different since the operation there is channelized—that is, anyone listening to the repeater will hear you as soon as you begin to transmit. Therefore there is no need to call CQ. Each repeater may use a slightly different procedure. There is a complete discussion of repeater operations in Chapter 11 of this book.

Learning phone procedures for HF is exactly like learning CW operating procedure. Listen to what others are doing, and incorporate their good habits into your own operating style.

Use common sense in your day-to-day phone QSOs, too:

1) Listen before transmitting. Ask if the frequency is in use before making a call on any particular frequency.

2) Give your call sign as needed, using the approved ITU (International Telecommunication Union) Phonetics. These phonetics are given in a table later in this chapter..

3) Make sure your signal is "clean." Do not keep your microphone gain turned up too high. If you have a speech processor use it only when you are sure it is properly adjusted. Don't take a chance of transmitting spurious signals.

4) Keep your transmissions as short as possible to give as many operators as possible a chance to use the frequency

spectrum.

5) Give honest signal reports. Chapter 17 has a list of what the various RST reports mean.

Whatever band, mode or type of operating some are undertaking, there are three fundamental things to remember. The first is that courtesy costs very little and is very often amply rewarded by bringing out the best in others. The second is that the aim of each radio contact should be 100% effective communication. The good operator is never satisfied with anything less. The third is that the "private" conversation with another station is actually in public. Keep in mind that many amateurs are uncomfortable discussing so-called controversial subjects over the air. Also, never give any information on the air that might be of assistance to the criminally inclined. As an example, never state when you are going to be out of town!

Correct phone operation is more challenging than it first may appear, even though it does not require the use of code or special abbreviations and prosigns. This may be because operators may have acquired some imperfect habits in their pronunciation, intonation and phraseology even before entering Amateur Radio! To these might be added a whole new set of clichés and mannerisms derived from listening to below-par operators.

Using the proper procedure is very important. Voice operators say what they want to have understood, while CW operators have to spell it out or abbreviate. Since the speed of transmission on phone is generally between 150 and 200 words per minute, the matter of readability and understandability is critical to good communication. The good voice operator uses operating habits that are beyond reproach.

It is important to speak clearly and not too quickly. This is particularly important when talking to a DX station who does not fully understand our language.

Avoid using CW abbreviations (including "HI") and Q signals on phone, although QRZ (for "who is calling?") has become accepted. Otherwise, plain language should be used. Keep jargon to a minimum. Some hams use "we" instead of "I," "handle" instead of "name" and "Roger" instead of "that's correct." These expressions are not necessary and do not contribute to better operating. No doubt you will hear many more.

Traffic handling (see Chapter 15 of this book) is an interesting feature of the annual ARRL Field Day. Often, casual visitors to the Field Day site are asked if they would like to send messages by Amateur Radio. Mary-Jo Giannusa, N2ZMZ, sent a number of messages on 2 meters for the Kings County Repeater Association Field Day group.

Procedure

There are two ways to initiate a voice contact: *call* (a general call to any station) or *answer a CQ*. If activity on a band seems low and you have a reasonable signal, a CQ call may be worthwhile.

Before calling CQ, it is important to find a frequency that appears unoccupied by any other station. This may not be easy, particularly in crowded band conditions. Listen carefully—perhaps a weak DX station is on frequency.

Always listen before transmitting. Make sure the frequency isn't being used *before* you come barging in. If, after a reasonable time, the frequency seems clear, ask if the frequency is in use, followed by your call: "Is the frequency in use? This is N2EEC." If, as far as you can determine, no one responds, you are ready to make your call.

As in CW operation, CQ calls should be kept short. Long calls are unnecessary. If no one answers you can always call again. If you do transmit a long call you may interfere with stations that were already on the frequency but whom you didn't hear in the initial check. In addition, any stations intending to reply to the call may become impatient and move to another frequency. If two or three calls produce no answer, there may be interference on the frequency. It's also possible that the band isn't open.

An example of a CQ call would be:

"CQ CQ Calling CQ. This is N2EEC, November-Two-Echo-Echo-Charlie, November-Two-Echo-Echo-Charlie, calling CQ and standing by."

When replying to a CQ both call signs should be given clearly. Use the ITU phonetic alphabet (see table later in this chapter) to make sure the other station gets your call correct. Phonetics are necessary when calling into a DX pileup and initially in most HF contacts but not usually used when calling into an FM repeater.

When calling a specific station, it is good practice to keep calls short and to say the call sign of the station called once or twice only, followed by your call repeated twice. VOX (voice operated switch) operation is helpful because, if properly adjusted, it enables you to listen between words, so that you know what is happening on the frequency. "N2EEC N2EEC, this is W2GD, Whiskey-Two-Golf-Delta, Over."

Once contact has been established, it is no longer necessary to use the phonetic alphabet or sign the other station's call. According to FCC regulations, you need only sign your call every 10 minutes, or at the conclusion of the contact. (The exception is handling international third-party traffic; you must sign both calls in this instance.) A normal two-way conversation can thus be enjoyed, without the need for continual identification. The words "Over" or "Go Ahead" are used at the end of a transmission to show you are ready for a reply from the other station.

Signal reports on phone are two-digit numbers using the RS portion of the RST system (no tone report is required). The maximum signal report would be "59"; that is, readability 5, strength 9. On FM repeaters, RS reports are not appropriate—when a signal has fully captured the repeater, this is called "full quieting."

Conducting the Contact

Aside from signal strength, it is customary (as in a typical CW QSO described elsewhere in this chapter) to exchange name, location and a brief description of the rig and antenna. Keep in mind that many hams will not know the characteristics of your *Loundenboomer 27A3* and your *SignalSquirter 4*. Therefore, you may be better off just saying that your rig runs 150 watts input and the antenna is a trap dipole. Once these routine details are out of the way, you can discuss virtually *anything* the two of you find appropriate and interesting.

DX can be worked on any HF band as well as occasionally on 6 meters. However as good conditions return to 10 meters, worldwide communication on a daily basis will again be com-

A warm sunny day, a comfortable tent, and what else do you need? Four large solar panels is the answer, charging a storage battery to power KH6WD. (*Photo courtesy of Kevin Bogan, WH6ML*)

monplace on this band. Ten meters is an outstanding DX band when conditions are right. A particular advantage of 10 for DX work is that effective beam-type antennas tend to be small and light, making for relatively easy installation.

Keep in mind that while many overseas amateurs have an exceptional command of English they may not be familiar with many of our colloquialisms. Because of the language differences, some DX stations are more comfortable with the "barebones" type contact, and you should be sensitive to their preferences. In unsettled conditions, it may be necessary to keep the whole contact short. The good operator takes these factors into account when expanding on a basic contact.

When the time comes to end the contact, end it. Thank the other operator (once) for the pleasure of the contact and say good-bye: "73, N9ABC THIS IS N2EEC, CLEAR." This is all that is required. Unless the other amateur is a good friend, there is no need to start sending best wishes to everyone in the household including the family dog! Nor is this the time to start digging up extra comments on the contact that will require a "final final" from the other station (there may be other stations waiting to call in). Please understand that during a band opening on 10 meters or on VHF, it is crucial to keep contacts brief so as many stations as possible can work the DX coming through.

Additional Recommendations

Listen with care. It is very natural to answer the loudest station that calls, but sometimes you will have to dig deep into the noise and interference to hear the other station. Not all amateurs can run a kilowatt.

Use VOX or push-to-talk (PTT). If you use VOX, don't defeat its purpose by saying "aaah" to keep the relay closed. If you use PTT, let go of the mike button every so often to make sure you are not "doubling" with the other station. Don't filibuster.

Talk at a constant level. Don't "ride" the mike gain. Try to keep the same distance from the microphone. It is a good idea to keep the mike gain down to eliminate room noise. Follow the manufacturer's instructions for use of the microphone. Some require close talking, while some need to be turned at an angle to the speaker's mouth. Speech processing (sometimes built in to contemporary transceivers or available as an outboard accessory) is often a mixed blessing. It can help you cut through the interference and static, but if too much is used, the audio quality suffers greatly. Tests should be made to determine the maximum level that can be used effectively, and this should be noted or marked on the control. Be ready to turn it down or off if it is not really required during a contact.

The speed of voice transmission (with perfect accuracy) depends almost entirely on the skill of the two operators concerned. Use a rate of speech that allows perfect understanding as well as permitting the receiving operator to record the information.

VHF/UHF SSB

"Weak-signal" VHF/UHF operating practices are not very different than those common on HF. Pileups can occur in good conditions, and because these openings can be short-lived, it is good procedure to limit the contact to exchange of signal reports or grid squares and location details. Other details can be left to the QSL card. As mentioned above, in this way more stations get a chance to work the DX station.

When conditions are good *call CQ sparingly,* if at all, unless you have an extremely potent signal. The old adage that "if you can't hear 'em, you can't work 'em" is just as true at VHF as HF. All too often, stations can be heard calling CQ DX on the same frequency as distant stations that they cannot hear.

One feature of VHF operating that's rarely found on the HF bands is the use of *calling frequencies.* The idea is that each frequency provides a "meeting place" for operators using the same mode. Once contact has been set up, a change to another frequency (the working frequency) is arranged so others can use the calling frequency. Table 3-5 contains the current VHF and UHF calling frequencies. See Chapter 12 for additional details.

Repeater Customs

A repeater is a device that receives a signal on one frequency and simultaneously transmits (repeats) it on another frequency. In a sense, it is a robot transmitter. Often located on top of a tall building or high mountain, VHF/UHF repeaters greatly extend the operating coverage of amateurs using mobile and hand-held transceivers.

Table 3-5
VHF/UHF/EFH Calling Frequencies

Band (MHz)	Calling Frequency
50	50.125 SSB
	50.620 digital (packet)
	52.525 National FM simplex frequency
144	144.010 EME
	144.100, 144.110 CW
	144.200 SSB
	146.520 National FM simplex frequency
222	222.100 CW/SSB
	223.500 National FM simplex frequency
432	432.010 EME
	432.100 CW/SSB
	446.000 National FM simplex frequency
902	902.100 CW,SSB
	903.1 Alternate CW, SSB
	906.500 National FM simplex frequency
1296	1294.500 National FM simplex frequency
	1296.100 CW/SSB
2304	2304.1
10000	10364

VHF/UHF Activity Nights

Some areas do not have enough VHF/UHF activity to support contacts at all times. This schedule is intended to help VHF/UHF operators make contact. This is only a starting point; check with others in your area to see if local hams have a different schedule.

Band (MHz)	Day	Local Time
50	Sunday	6 PM
144	Monday	7 PM
222	Tuesday	8 PM
432	Wednesday	9 PM
902	Friday	9 PM
1296	Thursday	10 PM

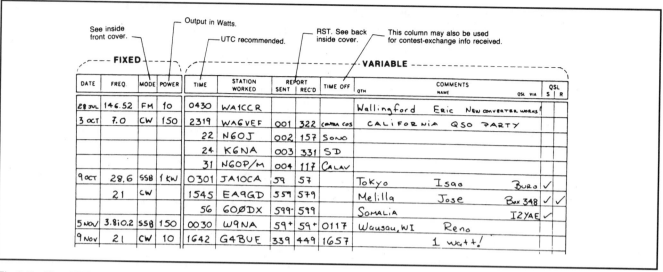

Fig 3-5—The ARRL Log Book is adaptable to all types of operating.

UTC Explained

Ever hear of Greenwich Mean Time? How about Coordinated Universal Time? Do you know if it is light or dark at 0400 hours? If you answered no to any of these questions, you'd better read on!

Keeping track of time can be pretty confusing when you are talking to other hams around the world. Europe, for example, is anywhere from 4 to 11 hours ahead of us here in North America. Over the years, the time at Greenwich, England, has been universally recognized as *the* standard time in all international affairs, including ham radio. (We measure longitude on the surface of the earth in degrees east or west of the Prime Meridian, which runs approximately through Greenwich, and which is halfway around the world from the International Date Line.) This means that wherever you are, you and the station you contact will be able to reference a common date and time easily. Mass confusion would occur if everyone used their own local time.

Coordinated Universal Time (abbreviated UTC) is the name for what used to be called Greenwich Mean Time.

Twenty-four-hour time lets you avoid the equally confusing question about AM and PM. If you hear someone say he made a contact at 0400 hours UTC, you will know immediately that this was 4 hours past midnight, UTC, since the new day always starts just after midnight. Likewise, a contact made at 1500 hours UTC was 15 hours past midnight, or 3 PM (15 − 12 = 3).

Maybe you have begun to figure it out: Each day starts at midnight, 0000 hours. Noon is 1200 hours, and the afternoon hours merely go on from there. You can think of it as adding 12 hours to the normal PM time—3 PM is 1500 hours, 9:30 PM is 2130 hours, and so on. However you learn it, be sure to use the time everyone else does—UTC. See chart below.

The photo shows a specially made clock, with an hour hand that goes around only once every day, instead of twice a day like a normal clock. Clocks with a digital readout that show time in a 24-hour format are quite popular as a station accessory.

UTC	EDT/AST	CDT/EST	MDT/CST	PDT/MST	PST
0000*	2000	1900	1800	1700	1600
0100	2100	2000	1900	1800	1700
0200	2200	2100	2000	1900	1800
0300	2300	2200	2100	2000	1900
0400	0000*	2300	2200	2100	2000
0500	0100	0000*	2300	2200	2100
0600	0200	0100	0000*	2300	2200
0700	0300	0200	0100	0000*	2300
0800	0400	0300	0200	0100	0000*
0900	0500	0400	0300	0200	0100
1000	0600	0500	0400	0300	0200
1100	0700	0600	0500	0400	0300
1200	0800	0700	0600	0500	0400
1300	0900	0800	0700	0600	0500
1400	1000	0900	0800	0700	0600
1500	1100	1000	0900	0800	0700
1600	1200	1100	1000	0900	0800
1700	1300	1200	1100	1000	0900
1800	1400	1300	1200	1100	1000
1900	1500	1400	1300	1200	1100
2000	1600	1500	1400	1300	1200
2100	1700	1600	1500	1400	1300
2200	1800	1700	1600	1500	1400
2300	1900	1800	1700	1600	1500
2400	2000	1900	1800	1700	1600

Time changes one hour with each change of 15° in longitude. The five time zones in the US proper and Canada roughly follow these lines.

*0000 and 2400 are interchangeable. 2400 is associated with the date of the day ending, 0000 with the day just starting.

Fig 3-6—Time is recorded both in log books and on QSL cards. UTC is always used.

To use a repeater, you must have a transceiver with the capability of transmitting on the receiver's *input frequency* (the frequency that the repeater listens on) and receiving on the repeater's *output frequency* (the frequency the repeater transmits on). This capability can be acquired by dialing the correct frequency and selecting the proper offset (frequency difference between input and output).

When you have the frequency capability (and most rigs today are fully synthesized), all you need do is key the microphone button and you turn on (access) the repeater. Some repeaters have limited access, requiring the transmission of a subaudible tone, a series of tones or bursts to gain access. However, most repeaters are open. Most repeaters briefly transmit a carrier after a user has stopped transmitting to inform the user that he is actually accessing a repeater.

After acquiring the ability to access a repeater, you should become acquainted with the operating practices of this unique mode. See Chapter 11 for specifics on operator practices.

Specialized Communications

Specialized communications, including satellite communications and the digital modes (including radioteletype, AMTOR and packet radio), all have distinct procedures. This manual thoroughly covers these modes in other chapters, while additional information can be found in *The ARRL Handbook for Radio Amateurs.*

Logging and QSLing

Although the FCC does not require logging except for certain specialized occurrences, you can still benefit by keeping an accurate log (see Fig 3-5). *The ARRL Logbook,* on sale at your local radio bookstore or directly from ARRL HQ, provides a good method for maintaining your log data at your fingertips.

The log entry should include:

1) The call sign of the station worked.

2) The date and time of the QSO. Always use UTC (sometimes also called GMT or Zulu time) when entering the date and time. Use UTC whenever you need a time or date in your hamming activities. The use of UTC helps all hams avoid confusion through conversion to local time. Fig 3-6 explains the UTC system.

3) The frequency or frequency band on which the QSO took place.

4) The emission mode of communication.

5) Signal reports sent and received.

6) Any miscellaneous data, such as the other operator's name or QTH that you care to record.

The FCC has promulgated a rather elaborate system of emission designators (see Chapter 2 on spectrum management). For logging simplicity, the following common abbreviations are often used:

Abbreviation	Explanation
CW	telegraphy on pure continuous wave
MCW	tone-modulated telegraphy
SSB	single-sideband suppressed carrier
AM,DSB	double-sideband with full, reduced or suppressed carrier
FAX	facsimile
FM	frequency- or phase-modulated telephony
RTTY	radioteletype
AMTOR	time diversity radioteletype
P	pulse
TV or SSTV	television or slow-scan television

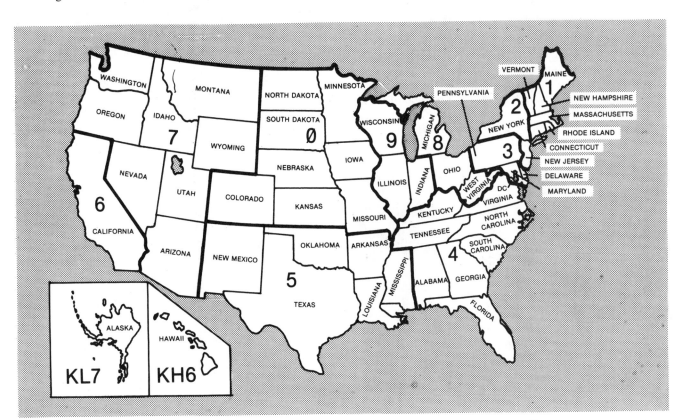

Fig 3-7—The 10 US Call Districts established by the FCC. In the past, the number in each ham call sign told you the district the ham was in when the call sign was issued. Today, under the vanity callsign program, the numbers are no longer a reliable guide.

Computer Logging

If you are into personal computing, there are many log-keeping programs that will take your input, format your log file and even give you a printed output. In addition, computer programs for contest applications are discussed in Chapter 7.

A well-kept log will help you preserve your fondest ham radio memories for years. It will also serve as a bookkeeping system should you embark upon a quest for ham radio awards, or a complete collection of QSL cards.

QSLing

The QSL card is the final courtesy of a QSO. It confirms specific details about your two-way contact with another ham. Whether you want the other station's QSL as a memento of an enjoyable QSO, or for an operating award, it's wise to have your own QSL cards and know how to fill them out. That way, when you send your card to the other station, it will result in the desired outcome (his card sent to you).

Your QSL

Your QSL card makes a statement about you. It may also hang in ham radio shacks all over the world. So you will want to choose carefully the style of QSL that you have printed.

There are many QSL vendors listed in the Ham Ad section of *QST* each month. A nominal fee will bring you many samples from which to choose or you may design and/or print your own style. The choice is up to you. See the accompanying sidebar for more information on QSLs.

That just about finishes our ever so slight foray into operating procedures. Your knowledge will grow quickly with your on-the-air experience. Enjoy the learning process. Ham radio is such a diverse activity that the learning process never stops—there is always something new to learn.

ON-THE-AIR ACTIVITIES

One of the advantages of Amateur Radio is the number of activities open to us. Often it's hard to decide which to do first. Many hams will spend a few months or even years on one of them—become proficient—and then jump to another to have a new experience. Among the most popular of these activities are DXing (talking to distant stations), contesting and awards hunting. Ragchewing (chatting), traffic handling, QRP (low power operation) and packet radio are just some of

If you feel you have been sitting around your shack for too long and need some exercise, consider the natural power source used by AA7QH.

the other popular activities. Most of these are covered in complete chapters of this manual. This section gives you a brief overview.

DXing

DX (ham shorthand for long distance) is one of those terms in ham radio that holds a special fascination for many hams. It can be correctly defined in different ways by different people. In the beginning, to most amateurs DX is the lure of seeing how far away you can establish a QSO—the greater the distance, the better. DX is a personal achievement, bettering some previous "best distance worked," involving a set of self-imposed rules.

DX can also be competing on a larger scale, trying to break the pileups on a rare DX (foreign) station and aiming for one of the DX-oriented awards. DXing can be a full-time goal for some hams and a just-for-fun challenge for others. No matter if you turn into a serious DXer or just have a little fun, DXing is one of the more fascinating aspects of Amateur Radio. See Chapter 5 for more particulars on DXing.

Contesting

Contesting is to Amateur Radio what the Olympic Games are to worldwide amateur athletic competition: a showcase to display talent and learned skills, as well as a stimulus for further achievement through competition. Increased operating skills and greater efficiency may be the end result of Amateur Radio contesting but fun is the most common experience.

The contest operator is also likely to have one of the better signals on the band—not necessarily the most elaborate station equipment, but a signal enhanced by constant experimentation and improvement. Contest operation encourages station and operator efficiency.

Nearly every contest has competitors vying to see which one can work the most stations (depending on the rules of the particular contest) in a given time frame. In some contests, the top-scoring stations have consistently worked 100 or more stations per hour for the entire 48-hour contest period.

The ARRL contest program is so diverse that it holds some appeal for almost every operator—the beginning contester and the old hand, the newest Technician and oldest Extra-Class veteran, "Top Band [160 meters] Buff" and microwave enthusiast. Complete contest entry rules and results appear in *QST* a few months before the contest. See Chapter 7 for further details on the exciting world of contesting. There is at least one contest for every interest.

Awards Hunting

Some hams spend most of their time in a contest that never ends—earning awards. The ARRL offers the awards listed in Table 3-6. Most other national amateur societies, private clubs and contest groups sponsor awards and certificates for various operating accomplishments. Many of the awards are very handsome certificates or intricately designed plaques very much in demand by awards chasers. Chapter 8 lists the most popular of these awards and how to earn them.

Rag Chewing

"Chewing the rag" refers to getting on the air and spending minutes (or hours!) in interesting conversation on virtually any and every topic imaginable. Without a doubt, the most popular operating activity is ragchewing. The rag chew may be something as simple as a brief chat on a 2-meter or 440-MHz FM repeater as you drive across town. It also may be a group of friends who have been meeting on 15-meter

The Final Courtesy

QSL cards are a tradition in ham radio. Exchanging QSLs is fun, and they can serve as needed confirmations for many operating awards. You'll probably want your own QSL cards, so look in *QST* for companies that sell them, or you may even want to make your own. Your QSL should be attractive, yet straightforward. All necessary QSO information on one side of the card will make answering a QSL a relatively simple matter.

A good QSL card should be the standard 3.5 × 5.5-inch size (standard post card size) and should contain the following information.

1) Your call. If you were portable or mobile during the contact, this should be indicated on the card.

2) The geographical location of your station. Again, portables/mobiles should indicate where they were during the contact.

3) The call of the station you worked. This isn't as simple as it sounds. Errors are very common here. Make sure it is clear if the call contains the numeral *1 (one)*, capital letter *I ("eye")* or lower case letter *l ("el".)*

4) Date and time of the contact. Use UTC for both and be sure you convert the time properly. It is best to write out the date *in words* to avoid ambiguity. Use May 10, 1995 or 10 May 1995, rather than 5/10/95. Most DX stations will use 8-2-95 to mean February 8, 1995. The day is written before the month in many parts of the world.

5) Frequency. The band in wavelength (meters) or approximate frequency in kHz or MHz is required.

6) Mode of operation. Use accepted abbreviations, but be specific. CW, SSB, RTTY and AM are clear and acceptable. FCC emission designations are not always understood by DX stations.

7) Signal report.

8) Leave no doubt the QSL is confirming a two-way contact by using language such as "confirming two-way QSO with," or "2 X" or "2-Way" before the other station's call. Other items, such as your rig, antenna and so on, are optional.

9) If you make any errors filling out the QSL, destroy the card and start over. Do not make corrections or mark-overs on the card, as such cards are not acceptable for awards purposes.

Now comes the problem of how to get your QSLs to the DX station. Sending them directly can be expensive, so, many amateurs use the ARRL's Overseas QSL Service. This is an ongoing service for ARRL members to send DX QSL cards to foreign countries at a minimum of cost and effort.

To receive QSL cards from DX (overseas) stations, the ARRL sponsors incoming QSL Bureaus, provided free for all amateurs throughout the United States and Canada. Each call area has its own bureau staffed totally by volunteers. To expedite the handling of your QSL cards (both incoming and outgoing), be sure to follow the bureau's requirements at all times. See Chapter 5 for further details on the ARRL Overseas QSL Service and the ARRL QSL Bureau System.

Table 3-6
ARRL Operating Awards

Award	Qualification
Friendship Award	Contact 26 stations with calls ending A through Z.
Rag Chewer's Club	A single contact ¹/₂ hour or longer
Worked All States (WAS)	QSLs from all 50 US states
Worked All Continents (WAC)	QSLs from all six continents
DX Century Club (DXCC)	QSLs from at least 100 foreign countries
VHF/UHF Century Club (VUCC)	QSLs from many grid squares
A-1 Operator Club	Recommendation by two A-1 operators
Code Proficiency	One minute of perfect copy from W1AW qualifying run
Old Timers Club	Held an Amateur Radio license at least 20 years prior
ARRL Membership	ARRL membership for 25, 40, 50, 60 or 70 years

SSB every Saturday afternoon for 20 years. The essential element is the same—hams talking to each other on any subject that interests them. The ARRL awards a certificate that declares membership in the *Rag Chewer's Club* earned by carrying on a conversation of half an hour or longer. Most hams will tell you that it is not difficult to chat back and forth for this period or longer!

Traffic Handling and Nets

Traffic handling involves passing messages to others via the amateur bands. Hams handle *third-party traffic* (messages for nonhams) in both routine situations and in times of disaster. Public-service communications make Amateur Radio the valuable public resource that it is. Traffic handling is covered in detail in Chapter 15.

Nets are regular gatherings of hams who share a mutual interest and use the net (short for "network") to further that interest. *Nets* most commonly meet to pass traffic or participate in one of the many other ham activities from awards chasing and DXing to just plain old ragchewing. County- and State-hunting nets are very popular since they provide a frequency to work that "49th" and "50th" state for the WAS (Worked All States) award. DX nets provide the same opportunity for those interested in working DX stations without the competition of contests or random calling.

There are several nets dedicated to Novice or slow speed CW operation that can help a newcomer sharpen operating skills. The *ARRL Net Directory*, a compilation of public service nets, is available from ARRL HQ for $2.

QRP Operation

QRP is the Q signal for low power. Sent with a question mark QRP? it means "Shall I lower my power?" This Q signal is also used to describe low power operation—some hams like the challenge of using lower power to talk to other hams.

Packet Radio and Digital Modes

BC (Before Computers) hams used RTTY (teletype) as their only digital mode. As computers became more popular the number of digital modes expanded. The most common and popular is packet radio.

Picking a Band

160 and 80 Meters

Eighty meters, and its single-sideband neighbor, 75 meters, are favorites for ragchewing. I frequently check out the 80-meter Novice/Technician CW subband. There I find both newcomers as well as old-timers trying to work the rust out of their fists! Around 3600 kHz you'll find the digital modes, RTTY, AMTOR and packet. The QRP frequency is 3560 kHz. If you hear a weak signal calling CQ near 3560, crank down your power and give a call! Another favorite frequency is 3579.5 kHz. If you live in the eastern half of North America, listen for W1AW on 3581 (CW), 3625 (RTTY/AMTOR/packet) or 3999 kHz (SSB). W1AW runs 1000 watts to a modest antenna—an inverted V at 60 feet. If you can copy W1AW, you can probably work the East Coast, even with low power. An outside antenna is a definite plus on these bands, however.

Even if you can't chase DX, you'll find plenty to do on either band. Ionospheric absorption is greatest during the day, thus local contacts are common. At night, contacts over 200 miles away are more frequent, even with a poor antenna. Summer lightning storms make for noisy conditions in the summer while winter is much quieter. You may also be troubled by electrical noise here. A horizontally polarized antenna, especially one as far from the building as possible, will pick up less electrical noise.

"Top Band," as 160 meters is often called, is similar to 80 meters. QSOs here tend to be a bit more relaxed with less QRM. DX is frequent at the bottom of the band. Don't let the length of a half wave dipole for 160 keep you off the band; a 25 or 50-foot "long wire" can give you surprisingly good results. One favorite trick is to connect together the center conductor and shield of the coax feed line of a 40 or 80-meter dipole and load the resulting antenna as a "T."

40 and 30 Meters

I must confess to being biased in favor of these bands, especially 40 meters. If I could only have a receiver that covered one band, it would be 40. Running 10 watts from my East Coast apartment (indoor antenna) I can work European hams, ragchew up and down the coast, check into Saturday morning QRP nets, and listen to foreign broadcast stations besides. Yes, 40 is a little crowded. Look at the bright side: you won't be lonely. I think it's possible to work someone on 40 any time of the day or night.

From a modest station, daytime operation on 40 is easier than at night. The "skip" (distance you can work) during the day is shorter, but that helps reduce interference. Most other countries have SSB privileges down to 7050 kHz. During the day, they, and the broadcasters, won't bother you much. Outside the US, QRPers hang around 7030 kHz; in the US we use 7040 kHz. RTTY/AMTOR/packet operators work around 7060 to 7100 kHz. Don't miss the Novice/Technician CW subband from 7100 to 7150 kHz. Even when the foreign broadcasters are booming in, you can usually find someone to work there. Forty is a good band for daytime mobile SSB operation, too. You'll find plenty of activity, and propagation condi-

tions tend to be stable enough to allow you to ragchew as you roll along.

The 30-meter band has propagation similar to 40 meters. Skip distances tend to be a little longer on 30 meters, and it's not so crowded. At present, stations in the US are limited to 200-watts output on this band. DX stations seem to like the low end of the band, from 10100 to 10115 kHz. Ragchewers often congregate above 10115. We share 30 meters with other services, so be sure you don't interfere with them. SSB isn't allowed on 30, but you can use CW and RTTY/AMTOR/packet. There is no agreed-upon QRP frequency on 30 meters.

20 Meters

As much as I like 40 and 30 meters, I have many fond memories of 20 meters as well. When I upgraded my license to General in 1963, I made a bee line to 20 meters. To this day, I can't stay away for long. A 20-meter dipole is only 33 feet long, and that doesn't have to be in a straight line. Many US hams have worked their first European or Australian contacts with a dipole and 100 watts.

While many hams think 20 meters is the workhorse band, it isn't always reliable. At the bottom of the sunspot cycle, 20 meters may be usable for a few hours a day. During years of higher solar activity though, 20 is often open all night.

There's plenty of room on the band. CW ragchewers hang out from 14025 to 14070 kHz, where you start hearing RTTY/AMTOR/packet stations. The international QRP frequency is 14060 kHz. The sideband part of the band is sometimes pretty busy and then it may be difficult to make a contact with low power or a modest antenna. Look above 14250 for ragchewers. Impromptu discussion groups sometimes spring up on 20 SSB, and make for interesting listening, even if you don't participate. If you like photographs, look around 14235 kHz for slow-scan TV. You'll need some extra equipment (as discussed in Chapter 16 of this book) to see the pictures.

17, 15 and 12 Meters

Except during years of high solar activity, you'll do most of your operating during daylight hours. Propagation is usually better during the winter months.

Seventeen and 12 meters are relatively new amateur bands, so they aren't as crowded as 15. Fifteen, though, is not nearly as crowded as 20. On 15, the QRP calling frequency is 21060 kHz. Don't overlook the Novice/Technician CW subband from 21100 to 21200 kHz. No special frequencies are used for QRP operation on 17 and 12. SSB operation is much easier on 17 and 12, because of lower activity. Low activity doesn't mean no activity; when those bands are open, you'll find plenty of stations to work. You only need one at a time, after all. RTTY/AMTOR/packet operation is found from 21070 to 21100 kHz, and around 18100 and 24920 kHz.

Some Russian RS-series satellites operate on 15 and 10 meters. Because of their low orbits, it isn't possible to work more than about 3000 miles on the RS satellites.

However, their low orbits make it possible to use these satellites with low power and simple antennas.

Practical indoor, outdoor, mobile or portable antennas for these bands are simple to build and install. It's even possible to make indoor beam antennas for the range of 18 to 25 MHz.

10 Meters

The 10-meter band stretches from 28000 to 29700 kHz. During years of high solar activity, 10 to 25-watt transceivers will fetch plenty of contacts. When the sun is quiet, there are still occasional openings of a few thousand miles. Ten meters also benefits from sporadic-E propagation. You'll find most sporadic-E openings in the summer, but they can happen anytime. Sporadic-E openings happen suddenly and end just as quickly. You won't be able to ragchew very long, but you'll be amazed at how many stations you can work.

SSB activity is heaviest in the Novice/Technician subband from 28300 to 28500 kHz. The Novice/Technician CW subband, 28100 to 28300 KHz, is a good place to look for activity, as is the QRP calling frequency at 28060 kHz. You can operate 1200-baud packet radio on 10 meters, whereas we're limited to 300 baud on the lower bands. RTTY/AMTOR/packet operation takes place from 28070 to 28100 kHz.

Higher in the band, above 29000 kHz, you'll find amateur FM stations and repeaters, and the amateur satellite subband.

Operating On 50 MHz and Above

The VHF/UHF/microwave bands offer many advantages to the low-power operator. The biggest plus is the relatively smaller antennas used. A good-sized 2-meter beam will easily fit in a closet when not in use. Portable and mobile operation on these bands is also easy and fun.

6 Meters

Six meters is perhaps the most interesting amateur band. When solar activity is high, worldwide QSOs are common. When solar activity is low, however, the opportunities for long-distance communication decrease. Sporadic-E propagation, which I mentioned earlier, is the most reliable DX mode during periods of low solar activity.

With small antennas, like 3-element beams, it's possible to work 1000 miles on sporadic E. Three-element 6-meter beams don't fit well inside houses or apartments, but you might be able to put one in an attic or crawl space. Even if you can only use a dipole, you'll be able to work locals, and snag some more-distant stations when the band opens.

Just about any mode found on the HF bands is used on 6 meters. CW and SSB operation take place on the lower part of the band. Higher up you'll find FM simplex and repeater stations. Another mode you'll sometimes find on 6 meters is radio control (RC) of model planes, helicopters, boats and cars.

2 Meters

Simply stated, 2 meters is the most popular ham band in North America. From just about any point in the US, you can probably work someone on 2 meters, 24 hours a day. Most hams know about 2-meter FM and packet radio operation, but CW and SSB are used here too. There's even an amateur satellite subband on 2 meters.

CW and SSB operation is done mostly with horizontally polarized antennas. FM and packet operators use vertical polarization, while satellites can be worked with either. A popular 2-meter antenna called a halo is perfect for indoor or mobile use on CW or SSB. The omnidirectional halo has no gain, but you'll be able to work locals, and up to 100 miles during band openings.

FM and packet usually require only a simple vertical antenna. The *ARRL Repeater Directory* will tell you what repeaters are available in your area. This book lists all repeaters reported to ARRL for the bands from 50 to 10,000 MHz.

Packet radio and VHF were made for each other. ARRL publishes several books on the subject including *Your Packet Companion* and *Practical Packet Radio*. Unlike FM repeaters, which have separate input and output frequencies, packet radio repeater stations use the same frequency for both.

The 222 and 430-MHz Bands

Every mode used on 2 meters is found on 222 except satellite communication. The 430-MHz or 70-cm band is becoming more and more popular, too. Multiband hand-held and mobile transceivers are available at prices only slightly higher than single-band rigs. If you think you'd like to try these bands in addition to 2 meters, look into a multiband rig.

One mode you'll find on 70 cm that isn't allowed on the lower frequencies is fast scan amateur television (ATV). Assuming you already have a broadcast TV set, all you need is a receive converter, transmitter, antenna and camera. Inexpensive cameras designed for home-video use are fine for ATV. ATV repeaters may be found in larger metropolitan areas. They're listed in the *ARRL Repeater Directory*.

23 Cm and Up

As you go higher in frequency, the relative size of antennas gets smaller. This fact allows you to use very high-gain antennas that aren't very big. Commercial equipment is available for the bands through 10,000 MHz (10 GHz). You'll also find easy-to-build kits (the tuned circuits are etched onto the circuit boards).

Because antennas are so small, it's possible to have 20 to 30-dB gain antennas that fit in your car's trunk. In comparison, a big 20-meter beam might offer only 10-dB gain. Operating from the field with battery-powered equipment is very popular, especially during VHF/UHF/microwave contests. Thanks to high-gain antennas, contacts over several hundred miles are possible with equipment running 1 or 2 watts.—*Jim Kearman, KR1S, from* Low Profile Amateur Radio, *published by the ARRL*

Less is More...Or Enjoying QRP

Are you a newly licensed radio amateur trying to get on the air with a minimum dollar investment? Are you a certified old-timer with 5BWAS, 5BWAC and 5BDXCC under your belt and have found hamming has lost some of its excitement? Are you somewhere in between Novice and veteran, and want to introduce a new spice to your ham palate? If you fit into any of these categories, read on.

Operating QRP (low power, defined by ARRL as 10 watts input or 5 watts output) is a popular modus operandi of thousands of hams. The thrill of communicating at low power levels is perhaps best described as being similar to the excitement experienced during your first QSO. Interestingly enough, the level of enjoyment proportionately goes up as your power goes down.

Many amateurs believe QRP is a relatively new avenue of hamming that started in the '70s with the introduction of the Ten-Tec PM series and the Heathkit HW-7 transceivers. In reality, QRP dates back at least 60 years before the birth of these rigs. In the early days of Amateur Radio, particularly during the Depression years in the '30s, amateurs didn't have a lot of money to invest in their hobby. "Make do with what you have" and professional scrounging were the order of the day to acquire the necessary parts to build up a station. This, more often than not, limited the station transmitter to one of the QRP variety.

A look at page 45 of May 1933 QST reveals an article entitled "Low Power Records," listing success stories hams of the day were enjoying using QRP. Actual QSOs not just hello-good-bye contacts, at 30-mW input, were being enjoyed by these pioneer QRPers. Low-power operating is, therefore, nothing new and might well be one of the few remaining common bonds with a simpler time.

The first thing anyone contemplating a jump into the QRP sport should do is cast off the notion you must run high power (QRO). A more enlightened attitude is needed. Consider your QRP operating as an adventure, a challenge, a unique and very personal voyage on the airwaves—riding a leaf instead of a supersonic jet. It's a gentle form of communication; think "heart and soul," not "blood and guts." Shoot down your DX prey with a pea-shooter rather than a double-barrel shotgun. A positive frame of mind will set the stage for an enjoyable time with QRP. Here are some important ground rules:

1) Listen, listen, listen.
2) Call other stations, don't call CQ.
3) Expect less-than-optimum signal reports.
4) Be persistent and patient.
5) Know when to quit.

Listen to the bands and try to figure out what the prevailing propagation is. Is the skip short or long? Who's working who? Is there much interference, static or fading (QRM, QRN or QSB respectively) present? A quick analysis of the band conditions should be the first thing you do when sitting down for a session of QRP operating. Listening will help you decide what band to operate. A thorough study of the radio-wave propagation information in this and other ARRL publications will help you understand what you hear. Always listen, listen, listen!

Call CQ as a last resort! Most hams prefer to answer a strong signal, which you probably will not have. You will be much better off answering a CQ. Try answering someone like this: WJ1Z DE KA1CV/QRP or WJ1Z DE KR1S/2W K. This tells WJ1Z why you aren't doing a meltdown on his headphones!

Don't be discouraged if you receive signal reports like RST 249 or RS 33. With less than 5 watts output, you can't expect to be overloading the receiver front ends out there in DX-land. Have faith, for you will get more than your fair share of very respectable reports. The ultimate ego gratification of a 599 or 59 will be yours if you keep at it!

If at first you don't succeed, then try again. And again. This QRP stuff is a game of persistence, so don't give up if you don't get an answer to your call on the first try. Don't be discouraged. Make up your mind—instant success just isn't part of the plan. That's what makes it so much fun.

If band conditions are inadequate, if your dipole has become the prey of a falling tree limb, if your neighbor has just started up a noise-generating appliance, if the dog is barking, the baby is crying and your spouse is out shopping, give it up! The last thing a QRPer needs is such distractions. Go tend the baby and see what the hound is upset about. Know when to quit. Sometimes you will have days like this, so know when to take a break from hamming. Try again tomorrow (or at least wait until your spouse returns from the store).

There is a relatively small selection to choose from if you plan to purchase a QRP-only transceiver. Only a limited amount of commercially made (or kit) equipment specifically designed for QRP operating is available. However, QST Ham Ads or flea markets are usually fruitful hunting grounds for good "preowned" radios at very reasonable prices.

Making its long awaited return to the amateur equipment scene is the Ten-Tec Argonaut II: a completely updated version of the popular 505, 509 and 515 Argonauts of the 1970s. With the addition of 160, 30, 17 and 12 meters, as well as 1990s' technology, the Argonaut II is a "big little rig" for the QRPer.

The earlier Argonauts still command respectable prices on the used market and are considered by some to be the rig of rigs for QRP. All featured full CW and SSB coverage on the "traditional" 80 to 10-meter bands. In addition, don't overlook Ten-Tec's old PM series. They were basic barebones units and are excellent for getting started.

The Heathkit HW-7, -8 and -9 transceivers were CW-only units covering the lower portions of various HF bands. Direct-conversion receivers were the weak links in the HW-7s and -8s. Heath recognized this deficiency and came out with an improved superhet receiver design for the HW-9.

Some of the Japanese multimode HF transceivers can be purchased with a low-power amplifier (Japan has a 10-watt amateur license class). These are full-feature transceivers, with all the advantages of modern technology, sans the full-power amplifier.

If you already have the typical amateur transceiver running 100 to 200 watts, never fear! All you have to do is reduce the output power by cranking down the drive or RF output control. (An accurate wattmeter also comes in handy.) With some rigs, minor tinkering with ALC circuitry can give you smooth level control well below 1 watt output.

Many amateurs have as much fun building QRP transmitters as they do operating them! Let's face it—nobody has the resources to build a deluxe, professional-quality transceiver. A two- or three-stage CW transmitter, on the other hand, is certainly within the technical ability of many of us. See the ARRL publication QRP Classics if you have the urge to roll your own!

A very useful tool for the QRP station is the wattmeter. A commercially built unit can be found for a reasonable price. A basic single meter unit with switchable forward and reverse power is a good way to start. In time you may want to add another meter to eliminate the need to switch back and forth between forward and reverse power. Save the switch and use it to change the power range of the meter. This way you can have one range for a 5-watt full scale, and the other a 1-watt scale. You can calibrate this wattmeter with a VTVM (vacuum-tube voltmeter), a simple home brew dummy load and an RF probe. Not only will you be saving some hard-earned bucks, but you will be gaining experience in designing, building, modifying and calibrating test equipment.

With QRP, your antenna is going to be much more important and instrumental in your success than if you run QRO. Running 100 watts into a random wire will net you plenty of solid contacts. But when you reduce power into the same wire, your signal effectiveness will decrease, too. As a result, the old axiom of putting up the biggest antenna you can muster, as high and as in the clear as possible, means more to the QRPer than someone running 100 or 1000 watts. The important thing is to optimize your antenna to your own personal circumstances. Many operators have reported amazing results using less-than-optimum antenna systems, but this is not to say you should be lax in your antenna installation. By running QRP, you are already reducing your effective radiated power (ERP); no need for a further (unintentional) reduction by cutting corners on your antenna system.

For 160 to 30 meters, dipoles generally will work well. Height is always nice, so do the best you can. Loops, end-fed wires and verticals are also used on these bands. The popular HF DX bands, 20 to 10 meters, deserve some serious thought as to rotatable gain antennas—Yagis or quads. Although this train of thought usually leads to a considerable outlay of cash, you will benefit in several ways. A 1-watt signal to a 10-dB gain Yagi will give you an effective radiated output of about 10-watt! That's just like having an amplifier that needs no power to run. A directional antenna is a reciprocal device as well. It is effective on both received signals and the transmitted signals. Listening to Europeans is so much more fun when you don't have to hear them along with signals from other unwanted directions.

The VHF and UHF bands are ripe for QRP operating and some serious antenna-design work. Amateurs with even very limited space can put up a high-gain Yagi for 6 or 2 meters or above. Many of the headaches of the HF bands are virtually nonexistent on VHF/UHF. The creative QRPer should be able to have a lot of success on these bands.

If you can't swing a big Yagi, either physically or financially, just get a decent wire antenna in the air. They all work, just some better than others. Again, the important thing is to put up the best system you possibly can.

Now that you're hooked, what to do about it? General operating and ragchewing are always fun, but do not limit yourself to just one kind of operating. Study the rest of this *Operating Manual,* and you should have enough info to attack several fronts with your mighty QRP signal! Here are a few suggestions.

Chase DX. Don't plan on busting a 250 station pileup, but very respectable DX can be worked with an extremely modest station and antenna. If you are cagey enough and have a few tricks up your sleeve (see Chapter 5 on DX operating techniques), then DX will indeed be yours.

Get in a contest. During the course of a contest weekend, activity on the bands is very high. What better time to try out a new transmitter or antenna? Go ahead and jump in—the water's fine. How do you compete against the super contest stations? You don't. You can compete with other QRPers or, better yet, compete against yourself. Set a personal goal—100 contacts, 30 states or countries or whatever—and keep a record of your results. Try to beat your previous record during the next contest. Once you get involved with QRP contesting, though, watch out—it's addicting!

Chase VHF/UHF grid squares. First, pack up your equipment and drive up to a mountain or hilltop location.

On VHF, if the band is open, your single watt will sound like 100 watts! With a couple of watts output to a gain antenna at a high elevation, you should be able to bend a few S-meter needles.

Go portable. Put the rig in your knapsack take a hike up a hill or mountain and set up your rig. Throw a wire over a tree limb and load 'er up. Working QRP from a scenic spot is hard to beat. Battery powered and away from all man-made electrical noise, you'll be able to hear what your receiver really can do. You can operate Field Day—not once a year, but every time you go for a hike!

Try HF digital modes, SSB or SSTV. Low-power operating does not mean you have to operate CW exclusively. If you like PacTOR, then you'll have it using QRP. The same goes for SSB, SSTV or other models. Further, operating SSB is an excellent way to spread the QRP word. More people seem to be "reading the mail" in the voice portions of the bands, so what better place to be a representative for the QRP movement?

You might be wondering how low can you go. Conduct minimum-power-level experiments and log your results. Get on the air with a fellow QRPer and, while monitoring your output power, see just how low in power you can go while still maintaining communications. You will be amazed at the possibilities. If the band conditions are right, there will be no difference between 1 watt and 5 watts; you may even be able to communicate at 10- or 20-mW. That's less power than some flashlights run. This type of threshold experimentation can be one of the most ego gratifying experiences in your Amateur Radio career.

On the bands, CW QRP activity centers around 1810, 3560, 7040 (7030 for QRP DX), 10, 106, 14,060, 21,060 and 28,060 kHz. Phone operation is around 3985, 7285, 14,285, 21,385 and 28,885 kHz. The WARC bands, 10/18/ 24 MHz, are also hotbeds of QRP activity. Novice and Technician Plus licensees should check 3710, 7110, 21,110 and 28,110 kHz.

QRP operating can result in earning some awards with a special QRP endorsement. There are many hams who have WAS and WAC under their belt using no more than 2 watts. DXCC has even been achieved by a few hardy QRP souls.

Several clubs cater to the QRPer. Most sponsor contests and awards. A number of clubs also publish newsletters. Joining one (or all!) of these clubs can give the interested amateur a much more rounded view of the QRP game.

The *QRP ARCI* (Amateur Radio Club International) offers membership for US amateurs at $12 per year and publishes the newsletter *QRP Quarterly.* Contact:

Michael Bryce
2225 Mayflower NW
Massillon, OH 44647

The *Michigan QRP Club* dues are $7 for US members. The club prints an informative newsletter, *The Five Watter.* Contact:

Membership Chairperson
564 Georgia Street
Marysville, MI 48040

When the bands are dead, the *real* QRPer either hits the workbench or sits down with a good book. Check out W1FB's QRP Notebook by Doug DeMaw, and *QRP Power.* Both are available from your local radio bookstore or ARRL HQ.

So what's holding you back? Join in on the fun and thrill of communicating with low power. Amaze your fellow amateurs and outdo the "big gun" across town with your operating accomplishments. Be a rebel; Quit Running Power!—*Jeff Bauer, WA1MBK*

AM: The Good Old Days are Now

Tim Walker, N2GIG, is quite proud of the waxed front panels of his 250-watt AM rig. Clyde, his 2nd operator and trusted assistant, can't seem to take his eyes off the 6-foot rack containing the recycled transmitter from commercial broadcast station WSK. The call letters WADP in the photograph were assigned to the station's remote pick-up transmitter.

Tim and many others enjoy restoring old *amplitude modulated* (AM) equipment for use on the ham bands. While it is not the sort of transmitter you would want to use during a DX contest, there are many AM fans who operate—some call it *broadcast*—on a regular basis in round tables with similarly minded friends. Transmissions tend to be a bit longer than SSB conversations, and the audio quality on many of the older rigs is often warmer and more realistic than a 300 to 2400-Hz system.

Favorite operating frequencies include the upper portions of the 160-meter band, 3.885, 7.290, 14.286, 21.390 and the range between 29.000 and 29.200 MHz. Occasionally you will hear an OT (*Old-Timer*) using a rig he saved from the late 1930s or 1940s. Other times the station operator is a *young squirt* who just enjoys using vacuum tube equipment which both heat and light the operating shack.

The table contains the names and addresses of several groups and publications that feature AM radio. Periodic meetings and flea markets make some of the older equipment available to anyone interested. One caution, however—if you manage to get your hands on a vintage receiver such as a Hammarlund SP-600, National NC-300, Collins 75A-1,2,3 or 4 or even a military surplus R-390 series unit—you might have to reinforce the operating table. The good news, however, is that compact, modern transceivers often include AM.

Publications:
Electric Radio
PO Box 57
Hesperus, CO 81326
  monthly publication

AM Press/Exchange
2116 Old Dover Road
Woodlawn, TN 37191
  bimonthly publication

Groups
AM International
9 Dean Ave
Bow, NH 03304
  active AM group

AM Radio Network
Box 73
West Friendship, MD 21794
  informal associations

Clyde, with his back to the camera, is admiring both his master, Tim Walker, N2GIG, and Tim's classic AM transmitter.

Packet radio requires either a full computer or a terminal (display and keyboard) at each end in addition to a packet radio controller and the radios themselves. Chapter 10 discusses this mode. If you are interested in packet, the ARRL publication *Your Packet Companion*, by Steve Ford, WB8IMY, is an invaluable guide for the newcomer.

ARRL On-The-Air Services

Up-to-the-minute information on everything of immediate interest to amateurs—from Federal Communications Commission policy-making decisions to propagation predictions, to news of DXpeditions—is transmitted on a timely basis in W1AW bulletins. W1AW is the Amateur Radio station maintained at ARRL HQ in Newington, Connecticut. W1AW bulletins are transmitted at regularly scheduled intervals on CW (Morse code), phone (voice) and RTTY (teletype) on various frequencies, as is a comprehensive Morse-code practice agenda. A detailed schedule of W1AW transmissions is in Chapter 17 and also appears regularly, with changes if any, in the ARRL monthly magazine *QST*. The schedule may also be obtained on request from ARRL HQ (please include a self-addressed, stamped envelope).

RULES AND REGULATIONS

The government of the United States grants us, as radio amateurs, certain privileges. In order to retain these privileges, we must conform to the rules and regulations in Part 97 of the Federal Communications Commission (FCC) rules.

This chapter will summarize only the most obvious rules as they pertain to day-to-day operation. This is by no means a complete discussion of all the FCC rules and regulations. For more complete information on the FCC rules as they pertain to Amateur Radio, along with the complete text of Part 97, see *The FCC Rule Book*, published by ARRL. (Contact ARRL HQ for ordering information.) See also Spectrum Management, Chapter 2 of this book.

Your License

You must have a license granted by the FCC to operate an Amateur Radio station in the United States, in any of its terri-

Table 3-7
Amateur Operator Licenses†

Class	Code Test	Written Examination	Privileges
Novice	5 WPM (Element 1A)	Novice theory and regulations (Element 2)	Telegraphy on 3675-3725, 7100-7150 and 21,100-21,200 kHz with 200 watts PEP output maximum; telegraphy, RTTY and data on 28,000-28,300 kHz and telegraphy and SSB voice on 28,300-28,500 kHz with 200 W PEP max; all amateur modes authorized on 222-225 MHz, 25 W PEP max; all amateur modes authorized on 1270-1295 MHz, 5 W PEP max.
Technician		Novice theory and regulations; Technician-level theory and regulations. (Elements 2 and 3A)*	All amateur privileges above 50.0 MHz.
Technician Plus	5 WPM (Element 1A)	Novice theory and regulations; Technician-level theory and regulations. (Elements 2 and 3A)*	All amateur privileges above 50 MHz. Telegraphy on 3675-3725, 7100-7150 and 21,100-21,200 kHz with 200 watts PEP output maximum; telegraphy, RTTY and data on 28,100-28,300 kHz and telegraphy and SSB on 28,300-28,500 kHz with 200 watts PEP max.
General	13 WPM (Element 1B)	Novice theory and regulations; Technician and General theory and regulations. (Elements 2, 3A and 3B)	All amateur privileges except those reserved for Advanced and Amateur Extra class; see Fig 3-4.
Advanced	13 WPM (Element 1B)	All lower exam elements, plus Advanced theory. (Elements 2, 3A, 3B and 4A)	All amateur privileges except those reserved for Amateur Extra class; see Fig 3-4.
Amateur Extra	20 WPM (Element 1C)	All lower exam elements, plus Extra-class theory (Elements 2, 3A, 3B, 4A and 4B)	All amateur privileges; see Fig 3-4.

†A licensed radio amateur will be required to pass only those elements that are not included in the examination for the amateur license currently held.

*If you hold a valid Technician class license issued before March 21, 1987, you also have credit for Element 3B. You must be able to prove your Technician license was issued before March 21, 1987 to claim this credit.

tories or possessions, or from any vessel or aircraft registered in the United States. The FCC sets no minimum or maximum age requirement for obtaining a license, nor does it require that an applicant be a US citizen.

An Amateur Radio license incorporates two kinds of authorization. For the operator, the license grants operator privileges. The license also authorizes an Amateur Radio station—operation of transmitting equipment.

Alien Amateur Radio operators, if they're licensed by a country with which the United States has signed a reciprocal-operating agreement and are citizens of that country, may apply to the FCC for permission to operate without having to pass the FCC amateur examinations.

The United States has six classes of operator licenses: Novice, Technician, Technician Plus, General, Advanced and Amateur Extra. There are two entry-level classes of license. The Novice license requires a minimum of technical knowledge and Morse code proficiency. The Novice license authorizes operation on six of the amateur frequency bands, with a variety of modes. The Technician license requires that applicants pass the Novice and Technician written tests, but a code examination is not required. The Technician license authorizes operation on all frequencies above 30 MHz. Upon passing the 5 wpm code test, a Technician becomes a Technician Plus, which includes all Novice HF privileges. Successive classes of license require more technical knowledge and/or Morse code proficiency. But each license rewards the licensee with more privileges, including expanded frequency allocations and more modes of operation. See Table 3-7.

License Renewal/Modification

Once you have a license, you can keep it for your lifetime,

but it must be renewed every 10 years.

1) To renew your license, complete FCC Form 610 (available from ARRL HQ; SASE, please) being careful not to miss any applicable boxes or blanks. Type or print neatly in pen. As of early 1997, only November 1993 and later editions of FCC Form 610 may be used.

2) Remember to attach a photocopy (or the original) of your license to the Form 610.

3) Mail your application to FCC, 1270 Fairfield Road, Gettysburg, PA 17325-7245. There is no fee for renewing an amateur license.

4) Application to renew cannot be made more than 90 days before expiration.

5) It's a good idea to retain a copy of your application as

How to Apply for an Amateur Radio License

If you are applying for any class of amateur license, you must appear before a team of three accredited Volunteer Examiners. If you want the latest list of test sessions near you, send an SASE to the Volunteer Examiner Department at ARRL HQ. The tests are typically given by local Amateur Radio clubs and are assembled from a standard list of questions issued by the FCC. These questions and answers appear in the ARRL *License Manuals* series. If you are physically unable to travel to an examination point, you may make special arrangements for taking your exam. Special examination procedures are available for persons with severe disabilities. See Chapter 3.

Complete details on the Volunteer Examining program are available from ARRL HQ. Also, see *The FCC Rule Book* published by the ARRL.—N1KB

proof of filing before the expiration date of your license. If your renewal application arrives at the FCC's Gettysburg office before the expiration date, you may continue to operate until the new license arrives. Otherwise you may not operate until you receive a new license.

6) If your license has already expired, it is still possible to renew. The FCC allows a 2-year grace period after the expiration date before *both* the license and the call sign expire.

7) If you are simply modifying your license (change of address, for example), you must also fill out Form 610. If you have any questions, contact the Regulatory Information Branch, ARRL HQ.

8) You are *required* to keep your address updated properly. The FCC may cancel or suspend your license when FCC mail is returned as undeliverable.

The FCC also issues licenses to bona fide clubs. The club station trustee, who must be a Technician class or higher, must complete an FCC Form 610B for a new (or modified) club station license. The license is for 10 years. Vanity call signs are also available to clubs.

Information about the status of your application can be obtained by calling the FCC Consumer Assistance line, 717-337-1212 or 1-800-322-1117.

If your license is lost, mutilated or destroyed, you must apply to the FCC for a duplicate. Send a letter to the FCC giving the circumstances and details under which the license was lost, mutilated or destroyed. After receiving your letter, the FCC will issue you a duplicate license bearing the same expiration date as the original license. If, after applying, you find the original license, either the original or the duplicate must be returned to the FCC.

US AMATEUR CALL SIGNS

The International Telecommunication Union (ITU) radio regulations outline the basic principles used in forming amateur call signs. According to these regulations, an amateur call sign must consist of one or two letters (sometimes the first or second may be a number) as a prefix, followed by a number and then a suffix of not more than three letters. Refer to Chapter 17 for a complete listing of international call-sign prefixes.

Every US Amateur Radio station call sign is a combination of a 1 or 2-letter prefix, a number and a 1, 2 or 3-letter suffix. The first letter of every US Amateur Radio call sign is always an A, K, N or W. For example, in the call sign W1AW, the W is the prefix, 1 is the number, and AW is the suffix.

For many, but not all, Amateur Radio stations located within the continental United States, the number in the call sign designates the geographic area in which the station was *originally* licensed. The call-sign districts for the continental US are illustrated in Fig 3-7. The prefix/ numeral designators for US Amateur Radio districts outside the continental US are shown in Table 3-8.

The FCC used to require the number (or in the case of US stations located outside the continental US, the prefix/number) to correspond to the call-sign district in which the station was located. In 1978, however, the FCC relaxed the rules and now allows hams to retain their present call sign even if the station location is moved permanently to another call-sign district. The Vanity Call Sign Program, begun in 1996, allows individual amateurs and club stations to pick a particular call sign for a $30 fee. It's routine to hear station call signs on the air that do not correspond to the call-sign district in which the station is located. It is no longer necessary to identify a station as

Table 3-8

FCC-Allocated Prefixes for Areas Outside the Continental US

Prefix	Location
AH1, KH1, NH1, WH1	Baker, Howland Is
AH2, KH2, NH2, WH2	Guam
AH3, KH3, NH3, WH3	Johnson I
AH4, KH4, NH4, WH4	Midway I
AH5K, KH5K, NH5K, WH5K	Kingman Reef
AH5, KH5, NH5, WH5 (except K suffix)	Palmyra, Jarvis Is
AH6,7 KH6,7 NH6,7 WH6,7	Hawaii
AH7, KH7, NH7, WH7	Kure I
AH8, KH8, NH8, WH8	American Somoa
AH9, KH9, NH9, WH9	Wake I
AHØ, KHØ, NHØ, WHØ	Northern Mariana Is
ALØ-9, KLØ-9, NLØ-9, WLØ-9	Alaska
KP1, NP1, WP1	Navassa I
KP2, NP2, WP2	Virgin Is
KP3,4 NP3,4, WP3,4	Puerto Rico
KP5, NP5, WP5	Desecheo

"portable," "mobile," "fixed" or "temporary."

Some amateurs choose to make their locations clear by including an indicator at the end of their call signs. For example, the voice identification, "This is W9KDR, permanent W1" or "fixed W1 " means that the station, though originally licensed in the ninth call district, is now licensed to an address in the first call district. When identifying on CW, W9KDR/1 could be used to clarify the station's location.

On-The-Air Identification Requirements

The amateur rules specifically prohibit unidentified transmissions, but what is required for proper station identification? An amateur station must be identified at the end of a transmission or series of transmissions. You must also iden-

---

Note to Canadian Readers

For purely practical reasons, this chapter was written to reflect the regulations established by the Federal Communications Commission for US amateurs. Canadian amateurs are regulated by Industry Canada (IC). Though the privileges of Canadian and US amateurs are generally similar, there are some differences.

There is one operator certificate available in Canada (the Radio Amateur Operator's Certificate) and it is available in four versions. The Basic Qualification is the entry level license and requires that the applicant pass a test composed of 100 questions. The Morse Code 5 wpm Qualification requires that the applicant pass the code test with fewer than 5 errors and the Morse Code 12 wpm has the same requirements, but at a higher speed. The Advanced Qualification is the most difficult to obtain and it consists of 50 questions. The Basic Qualification is mandatory and code is not required. The other three levels of qualification are voluntary. There is no minimum age requirement.

More information on licensing in Canada can be obtained from Radio Amateurs of Canada, Inc (RAC), 720 Belfast Road, Suite 217, Ottawa, ON K1G 0Z5, Canada, tel 613-244-4367. RAC publishes the *Canadian Amateur Study Guide* as well as the *Basic Question Bank*.

Guest Operating

An FCC rules interpretation allows the person *in physical control* of an Amateur Radio station to use his or her own call sign when guest operating at another station. Of course, the guest operator is bound by the frequency privileges of his or her own operator's license, no matter what class of license the station licensee may hold. For example, Joan, KB6MOZ, a Novice, is visiting Stan, N6MP, an Extra Class licensee. Joan may use her own call sign at N6MP's station, but she must stay within the Novice subbands.

When a ham is operating from a club station, the club call is usually used. Again, the operator may never exceed the privileges of his or her own operator's license. The club station trustee and/or the club members may decide to allow individual amateurs to use their own call signs at the club station, but it is optional. In cases where it is desirable to retain the identity of the club station (W1AW at ARRL HQ, for example), the club may require amateurs to use the club call sign at all times.

In the rare instance where the guest operator at a club station holds a higher class of license than the club station trustee, the guest operator must use the club call sign and his/her own call sign. For example, Billy, KR1R, who holds an Amateur Extra Class license visits the Norfolk Novice Radio Club station, KA4CVX. Because he wants to operate the club station outside the Novice subbands, Billy would sign KA4CVX/KR1R on CW and "KA4CVX, KR1R controlling" on phone. (Of course, this situation would prevail only if the club requires that their club call sign be used at all times.)

tify at intervals not to exceed 10 minutes during a single transmission or a series of transmissions of more than 10 minutes' duration. The identification must be given using the International Morse code or, when identifying on phone, in the English language. At the end of an exchange of third-party communications (that is, on behalf of anyone other than the control operator) with a station located in a foreign country, you must also give the call sign of the station with which third-party traffic was exchanged.

Temporary Designator

Temporary designators must be used after a call sign under some circumstances. These designators are used for "instant upgrading," an FCC rule that permits holders of an amateur license who successfully pass an examination for a higher class of license to use the new privileges immediately after leaving the test session. The accredited Volunteer Examiners issue the amateur a certificate of successful completion, (CSCE) which authorizes the new privileges for one year or until the issuance of the new, upgraded, permanent license (whichever comes first). The temporary certificate also sets forth a two-letter indicator for the upgraded class of license. For example, "temporary AA" means the operator has upgraded to Advanced.

When operating under the authority of a temporary amateur CSCE with privileges authorized by the certificate, but which exceed the privileges of the licensee's permanent station license, the station must be identified in the following manner:

1) On phone, by the transmission of the station call sign, followed by a word such as "temporary" followed by the special indicator shown on the certificate, appropriate to the newly earned class of license.

2) On CW and digital modes, by the transmission of the station call sign, followed by the slant mark, followed by the special indicator shown on the temporary amateur permit, for example, KB5NOW/AA.

Additional Identification Requirements

To meet the identification requirements of the amateur rules, the call sign must be transmitted on each frequency being used. If identification is made by an automatic device used only for identification by CW, the code speed must not exceed 20 words per minute. While the FCC does not require use of a specific list of words to aid in sending one's call sign on phone, it does encourage the use of a nationally or internationally recognized standard phonetic alphabet as an aid for correct phone identification. The ITU phonetic alphabet appears on the next page.

Permitted Communications

The US amateur rules clearly set forth the kinds of stations with which amateurs may communicate. These stations are:

1) other amateur stations;
2) stations in other FCC-licensed services during emergencies, and with RACES stations.
3) *any* station that is authorized by the FCC to communicate with amateurs.

Additionally, certain "one-way" transmissions are allowed, including those for tuning up your rig, calling CQ, while in remote control operation, during emergency communications, for code practice and bulletins, and for retransmission of space-shuttle communications.

Every amateur should operate with one eye on the basis and purpose of the Amateur Radio Service, as set forth in Section 97.1 of the amateur rules. Adhere to these fundamental principles and you can't go wrong!

§97.1 Basis and Purpose

The rules and regulations in this part are designed to provide an amateur radio service having a fundamental purpose as expressed in the following principles:

(a) Recognition and enhancement of the value of the amateur service to the public as a voluntary noncommercial communication service, particularly with respect to providing emergency communications.

(b) Continuation and extension of the amateur's proven ability to contribute to the advancement of the radio art.

(c) Encouragement and improvement of the amateur radio service through rules which provide for advancing skills in both the communication and technical phases of the art.

(d) Expansion of the existing reservoir within the amateur radio service of trained operators, technicians, and electronics experts.

(e) Continuation and extension of the amateur's unique ability to enhance international goodwill.

Third-Party Communications

Third-party traffic is defined in the US amateur rules as, "a message from the control operator (first party) of an amateur station to another amateur station control operator (second party) on behalf of another person (third party)." In other words, sending or receiving any communication on behalf of anyone other than yourself or the control operator of the other station. If you allow anyone to say "hello" into the microphone at your station, that is third-party traffic. Sending or receiving formal radiogram messages is third-party traffic (see Chapter 15). Autopatching and phone patching (interconnecting your radio to the telephone system) are also third-party traffic. You, the control operator of the station, are the first party; the control operator at the other station is the second party; and anyone else participating in the two-way communication from either station is a third party.

Certain kinds of third-party traffic are strictly prohibited. All amateurs should know that they must *not* handle the following kinds of third-party traffic:

1) international third-party traffic, except with countries that allow it or except where the third party is a licensed amateur eligible to be a control operator of the station;

2) third-party traffic involving material compensation, direct or indirect, paid or promised to any party.

Business Rules

The FCC rules were changed in 1993 giving amateurs greater flexibility in their public service and personal communications. Amateurs should look at Section 97.113 when trying to decide if a particular use of the ham bands is allowed:

1) Is is expressly prohibited in the rules?

2) Is there compensation—is anyone being paid?

3) Does the control operator or his employer have a financial interest in the activity using the ham bands?

4) Are communications transmitted on a regular basis, which could be reasonably furnished through other radio services?

Under the revised business rules, an amateur may conduct his or her personal business over Amateur Radio, including such activities as ordering a pizza or making a dentist appointment. Although this activity is legal, some repeater owners/trustees may ask you not to use their repeater for some of these transactions.

Third Party Traffic Outside the US

Third party traffic is permitted only with countries that have entered into third party traffic agreements with the US (see Chapter 5 for a list of countries that do permit third-party traffic handling). This is not through reluctance on the part of the US Government, but because of prohibitions on the part of the other governments. The amateur at the other end is commonly forbidden to handle any third-party traffic. US amateurs must abide by this restriction and must have no participation in the handling of third-party traffic in such cases.

Active participation by a third party: One of the best ways to get a nonham interested in Amateur Radio is to allow him or her to speak into a microphone and experience first-hand the excitement of amateur communication. While the US amateur rules allow active participation by third parties, the control operator must make certain that the communications abide by all the restrictions pertaining to third-party traffic. For example, the third party may not talk about business and may not communicate with any country that has not signed a third-party agreement with the US. Also, the control operator may not leave the controls of the station while the third party is participating in amateur communications. The amateur rules allow third-party participation only if a control operator is present at the control point—and if he or she is continuously monitoring and supervising the third party's participation.

Third-party participation is not the same as designating another licensed amateur to be control operator of a station. Control operators designated by a station licensee may operate the station to the extent authorized by the control operator's class of license.

In the 1980s, the FCC modified the rules to close a loophole that previously allowed amateurs who have had their licenses suspended or revoked to participate in Amateur Radio communications as third parties.

Logging

Although there is no FCC requirement that amateurs keep logs of routine operating activities, amateurs are encouraged to keep voluntary logs, and FCC Engineers-In-Charge may still mandate log-keeping in specific cases.

Authorized Power

Generally, an Amateur Radio transmitter may be operated with a PEP *(peak envelope power)* output of up to 1500 watts. However, there are circumstances under which less power must be used. These are:

1) Under all circumstances, an amateur is required to use the minimum amount of transmitter power necessary to carry out the desired communications.

2) Novices are limited to a maximum of 5 watts PEP output on the 1270-MHz band, 25 watts PEP on the 222-MHz band, and 200 watts PEP on the 80, 40, 15 and 10-meter bands. The 200-watt limitation also applies to any licensed radio amateur who operates in the 80, 40 and 15-meter Novice subbands and to any amateur operating on the 30-meter band. The rules limit the maximum transmitter output power in the Amateur Radio Service to 1500-watts PEP. Higher-class licensees may use 1500-watts PEP in the Novice sections of 28 MHz, 222 MHz and 1270 MHz.

3) Certain power limitations apply at VHF and UHF. See §97.313 of the FCC rules.

Amateur Operation in a Foreign Country

Operation may not be conducted from within the jurisdic-

The Phonetic Alphabet

When operating phone, a standard alphabet is often used to ensure call letters and other spelled-out information is understood. Thus Larry, WR1B, would announce his call as *Whiskey Romeo One Bravo* if he felt the station on the other end could misunderstand his call. Phonetics are not routinely used when operating VHF-FM.

A—Alfa (**AL** FAH)
B—Bravo (**BRAH** VOH)
C—Charlie (**CHAR** LEE)
D—Delta (**DELL** TAH)
E—Echo (**ECK** OH)
F—Foxtrot (**FOX** TROT)
G—Golf (GOLF)
H—Hotel (HOH **TELL**)
I—India (**IN** DEE AH)
J—Juliet (**JEW** LEE ETT)
K—Kilo (**KEY** LOH)
L—Lima (**LEE** MA)
M—Mike (MIKE)
N—November (NO **VEM** BERR)
O—Oscar (**OSS** CAR)
P—Papa (PAH **PAH**)
Q—Quebec (KEY **BECK**)
R—Romeo (**ROW** ME OH)
S—Sierra (SEE **AIR** AH)
T—Tango (**TANG** OH)
U—Uniform (**YOU** NEE FORM)
V—Victor (**VIK** TORE)
W—Whiskey (**WISS** KEY)
X—X-Ray (**EX** RAY)
Y—Yankee (**YANG** KEY)
Z—Zulu (**ZOO** LOU)

The boldfaced syllables are emphasized.

tion of a foreign government unless that foreign government has granted permission for such operation. Some countries have allowed US amateurs to operate while visiting; however, not all countries permit foreigners this privilege. Every country has the right to determine who may and who may not operate an Amateur Radio station within its jurisdiction. There are severe penalties in some countries for operating transmitting equipment without the proper authority. See Chapter 6 for details on overseas operating.

Canada-US reciprocal operating: The most liberal agreement between two countries allowing reciprocal operating for their radio amateurs is between Canada and the US. No formal application or permit is required for US amateurs to operate amateur equipment in Canada, nor is any formal paperwork required for Canadian amateurs to operate in the United States. It's automatic. US radio amateurs may operate in Canada with the mode and frequency privileges authorized them in the United States. Canadian radio amateurs may operate in the US with the mode and frequency privileges authorized to them in Canada, but not exceeding the privileges of a US Extra Class operator. In all cases, however, visitors must stay within the band and mode restrictions of the host country.

Other reciprocal operating agreements: Except for Canada, US amateurs must obtain a written permit to be allowed to operate in any foreign country. The administrations of some countries have signed reciprocal operating agreements with the US. Such agreements facilitate the application procedures for getting permission to operate. These agreements, because they are reciprocal, also allow foreign amateurs the opportunity to receive permission to operate while in the US. If you want to operate a station in a foreign country, ARRL HQ can help you obtain forms and information. Send an SASE to ARRL HQ's Regulatory Information Branch for details. Include one unit of postage for every country requested. A few countries require a US licensee to hold a General or higher-class license to qualify for a permit to operate. You will be required to pay a licensing fee in order to obtain a permit, which must be obtained before equipment is taken into a country. Plan as far ahead as possible. See Chapter 6 for a current list of the countries that have signed reciprocal operating agreements with the US.

Radio Frequency Interference

The FCC authorizes radio amateurs to use up to 1500 watts PEP output and many different bands of the radio spectrum. However, these privileges do not come without some effort on the part of the license applicant. A prerequisite for getting on the air is passing an examination that tests the applicant's technical knowledge. One area of special importance is knowing about radio frequency interference (RFI), its causes and its cures. The FCC expects radio amateurs to identify and solve most RFI problems, but RFI is not always a simple matter.

Much has been written about RFI. For example, the material in the ARRL book *Radio Frequency Interference: How to Find It and Fix It* examines this subject from both a legal and technical standpoint. Also, *The ARRL Handbook for Radio Amateurs* devotes an entire chapter to solving radio frequency interference problems. If you are the recipient of interference complaints, these two books will be of interest to you.

A growing number of amateurs are unjustly accused of causing RFI. This occurs when an amateur transmitter is operating properly, and the responsibility for the interference falls with the electronic device experiencing the interference. For example, some television sets receive signals they are not supposed to because the manufacturer decided that reducing the per-unit cost was more important than incorporating adequate shielding in the circuit design. In these cases, the transmitter is operating with a "clean" signal; it is the television set that must be brought up to today's engineering standards. This usually means that the only effective cure must be performed by adding a filter at the television set. Trying to make your neighbor understand that it is his equipment that is at fault is usually no easy task, however.

Public Law 97-259 allows the FCC to require standards for unwanted-signal-rejection filtering in electronic home-entertainment devices. Also, some manufacturers have made some progress in dealing with this problem by providing free assistance to owners of their products who experience RFI. However, the onus for preventing the interference continues to fall on amateurs.

Today, the interference problem is of huge proportions. This can be attributed to the rapid growth of the consumer electronics industry and the failure of manufacturers to include proper shielding or filtering in the home-entertainment equipment.

If you are accused of causing RFI:

1) Check your log. Were you operating at that time? (A complete log, although no longer required by the FCC, is very useful in interference situations.)

2) Check with your own nonamateur equipment. If you are not interfering with your own TV set, chances are the problem lies with your neighbor's receiver and not with your transmitter.

3) Solicit the cooperation of your neighbor in testing to determine the exact cause of the interference.

4) Check with your local radio club for a TVI committee or other assistance or contact your ARRL Section Manager for a referral to an ARRL Technical Coordinator or Technical Specialist in your area.

5) Request RFI assistance from the manufacturer of the home-entertainment device.

6) Read the ARRL *Radio Frequency Interference* book and the Interference chapter in *The ARRL Handbook for Radio Amateurs*.

7) Be prompt, courteous and helpful; Amateur Radio's reputation is at stake, as well as your own.

(A)

(B)

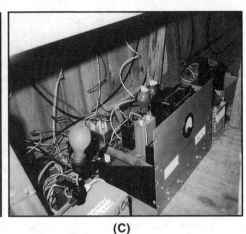
(C)

On the 75th anniversary of the first complete transatlantic Amateur Radio message, station W1BCG was reactivated by Bob Raide, W2XM, members of the AWA (Antique Wireless Association) and several Connecticut ham clubs. The transmitter consisted of a 204A oscillator, driving three 204As in parallel (Photo A). Al Brogdon, K3KMO, is in Photo B pounding brass through a floor-shaking keying relay. The relay, not part of the original Armstrong design, was used to keep the 1700-V plate supply away from the operator's fist. In place of a motor-generator power supply, the AWA constructed the more up-to-date supply in Photo C. A pair of 866 mercury vapor rectifiers cast a pleasant blue glow over the station, located behind the home of George Wells, KA1JUV, in Greenwich, Connecticut—very close to the 1921 site! (*Photos courtesy of N1II*)

8) For further assistance, write ARRL HQ. Include as much detail as you can about the symptoms and circumstances, but avoid emotional commentary.

On the Horizon

A basic appreciation of the rules will provide amateurs with increased levels of enjoyment. The rules have evolved over a long time and will continue to evolve. The deregulation of the Amateur Radio Service has had a desirable impact. The FCC proposes additional rules changes periodically; check *QST* for news of late-breaking events.

The rules are dynamic because Amateur Radio is dynamic—constantly changing to meet and create new communications technologies to better serve the public and society.

CHAPTER 4

Antenna Orientation

CHUCK HUTCHINSON, K8CH
ARRL HEADQUARTERS

It is probably no news to most people nowadays that true direction from one place to another is not what it appears to be on the old Mercator school map. On such a map, if one starts "west" from Wichita, Kansas (the approximate center of the continental US) he winds up in the neighborhood of Beijing, People's Republic of China. Actually, as a minute's experiment with a strip of paper on a small globe will show, a signal starting due west from Wichita never hits China at all but rather passes near Perth, in Western Australia.

"The shortest distance between two points is a straight line" is true only on a flat surface. The determination of the shortest path between two points on the surface of a sphere is a bit more complicated. Imagine a plane that intersects two points on the surface and the center of the sphere. The intersection of the plane and the sphere describes a circle on the surface of the sphere that is defined as a great circle. The shortest distance between the points follows the path of the great circle. The direction or bearing from your location to another point on the Earth is the direction of a great circle as it passes through your location on its way to the other point.

If, therefore, we want to determine the direction of some distant point from our own location, the ordinary Mercator projection alone is utterly useless. True bearing, however, may be found in several ways: by using a special type of world map that does show true direction from a specific location to other parts of the world, by working directly from a globe or by using mathematics.

DETERMINING TRUE NORTH

Determining the direction of distant points is of little use to amateurs erecting a directive array unless they can put up the array itself in the desired direction. This, in turn, demands a knowledge of the direction of true north (as against magnetic north), since all our directions from a globe or map are worked in terms of true north.

A number of ways may be available to amateurs for determining true north from their location. Frequently, the streets of a city or town are laid out, quite accurately, in north-south and east-west directions. A visit to the office of your city or county engineer will enable you to determine whether or not this is the case for the street in front of or parallel to your own lot. Or from such a visit it is often possible to locate some landmark, such as a factory chimney or church spire, that lies true north with respect to your house. If you cannot get true north by such means, three other methods are available: compass, pole star and sun.

By Compass

Get as large a compass as you can; it is difficult, though not impossible, to get satisfactory results with the "pocket" type. In any event, the compass must have not more than 2 degrees per division.

It must be remembered that the compass points to magnetic north, not true north. The amount by which magnetic north differs from true north in a particular location is known as variation. Your city engineer's office or the flight office at a nearby airport can tell you the magnetic variation for your locality. The information is also available from US Geological Survey topographic maps for your locality. These may be available in the orienteering equipment section of your local sporting goods store, or may be on file at your local library.

If you have access to the World Wide Web, you can determine the magnetic variation for your area. Point your browser to **http://www.airnav.com/airports/** and you'll have access to data, including magnetic variation, for a nearby airport or heliport. Another web site that will give you your magnetic variation (they call it "declination" at this site) is at **http://www.ngdc.noaa.gov/cgi-bin/seg/gmag/fldsnth1.pl**. You'll need to know your latitude, longitude and elevation. The same site will offer to let you download a copy of *pgeomag3.exe*, a self-extracting compressed file. Run it, and you will be able to get your variation off-line.

When correcting your "compass north," do so *opposite* to the direction of the variation. For instance, if the variation for your locality is 12° west (meaning that the compass points 12° west of north), then true north is found by counting 12° east of north as shown on the compass.

When taking the bearing, make sure that the compass is located well away from ironwork, fencing, pipes, etc. Place the instrument on a wooden tripod or support of some sort, at a convenient height as near eye level as possible. Make yourself a sighting stick from a flat stick about 2 feet long with a nail driven upright in each end (for use as "sights") and then, after the needle of the compass has settled down, carefully lay this stick across the face of the compass—with the necessary allowance for variation—to line it up on true north. *Be sure you apply the variation correctly.*

This same sighting-stick and compass rig can also be used in laying out directions for supporting poles for antennas in other directions—provided, of course, that the compass dial is graduated in degrees.

By the Pole Star

Many amateurs in the Northern Hemisphere use the pole star, Polaris, in determining the direction of true north. An advantage is that the pole star is never more than 0.8° from true north, so that in practice no corrections are necessary. Disadvantages are that some people have difficulty identifying the pole star, and that because of its comparatively high angle above the horizon at high northerly latitudes, it is not always easy to "sight" on it accurately. Further, Polaris is not a very bright star. Once you've sighted it, you can use a string with a weight tied on the end and held high at arm's length to look along and to identify a landmark that is north of your position. Polaris is not visible in the Southern Hemisphere.

By the Sun

The sun can be used for determination of true north. The method is based on the fact that exactly at noon, local time (not Standard Time), the sun bears due south (in the northern latitudes), so at that time the shadow of a vertical pole or rod will bear north. The resulting shadow direction is true north.

Clock or Standard Time for local noon is halfway between calculated sunrise and sunset times for your location. For example, calculations show that the sun rises in Newington, Connecticut, at 1144 UTC on February 20, 1997. It sets at 2225 UTC. The day is 10 hours and 41 minutes long, and local noon is 5 hours and 20.5 minutes after sunrise. This is 17:04:30 UTC or four and a half minutes after noon EST.

Many local newspapers publish sunrise and sunset times. A number of popular Amateur Radio oriented software packages also include this information. On the Internet you can point your Web browser to: **http://www.almanac.com/rise/ rise.html**.

Here's another way to use the shadow of a vertical pole or rod. Mark the end of the shadow at some convenient time around mid-morning. When the shadow is the same length in the afternoon, mark that spot. The line between the marks runs east/west.

An Alternative Method for Antenna Orientation

It is not necessary to use true north for orienting your antenna. Any convenient landmark at a known bearing can be used for this purpose. The method is explained in the following example.

W1AW has four towers with rotating antennas—three at 60 feet, and one at 120 feet. A neighbor's chimney serves as a south reference for the 120 foot tower. (It's a bit too close, but it works.) There are no north/south reference landmarks for the three 60 foot towers.

There is a landmark over a mile away—a small structure atop a tall building on a hill. From a US Coast and Geodetic Survey Map, staff determined the bearing to be 118 degrees true. Corrected compass sightings verified the bearing.

Today it's a simple matter to set a rotator indicator at 118 degrees and align the antenna to point at the reference structure. Because the reference point is over a mile away, the same bearing works for all four towers. Any errors are small enough to be insignificant—especially for HF antennas.

AZIMUTHAL MAPS

While the Mercator projection does not show true directions, it is possible to make up a map that will show true bearings for all parts of the world from any single point. Three such maps are reproduced in this chapter. Fig 4-1 shows directions from Washington, DC, Fig 4-2 gives directions from San Fran-cisco and Fig 4-3 (a simplified version of the ARRL Amateur Radio map of the world) gives directions from the approximate center of the United States—Wichita, Kansas.

For anyone living in the immediate vicinity (within 150 miles) of any of these three reference points, the directions as taken from maps will have a high degree of accuracy. However, one or the other of the three maps will suffice for any location in the United States for all except the most accurate work; simply choose the map whose reference point is nearest you. Greatest errors will arise when your location is to one side or the other of a line between the reference point and the des-

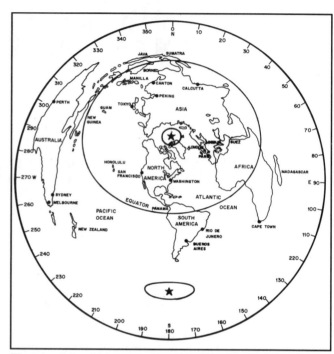

Fig 4-1—Azimuthal map centered on Washington, DC.

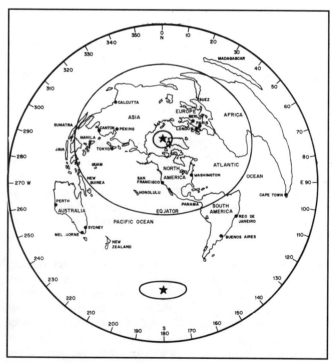

Fig 4-2—Azimuthal map centered on San Francisco, California.

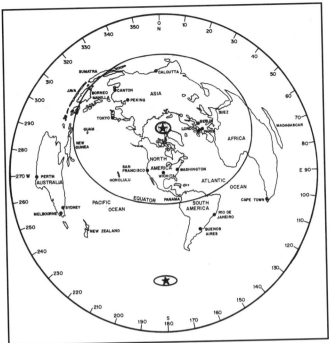

Fig 4-3—Azimuthal map centered on Wichita, Kansas. Copyright by Rand McNally & Co, Chicago. Reproduction License No. 394.

tination point; if your location is near or on the resulting line, there will be little or no error.

By tracing the directional pattern of the antenna system on a sheet of tissue paper, then placing the paper over the azimuthal map with the origin of the pattern at one's location, the "coverage" of the antenna will be readily evident. This is a particularly useful technique when a multi-lobed antenna, such as any of the long single-wire systems, is to be laid out so the main lobes cover as many desirable directions as possible. Often a set of such patterns will be of considerable assistance in determining what length antenna to put up, as well as the direction in which it should run.

The current edition of the ARRL Amateur Radio Map of the World, entirely different in concept and design from any other radio amateurs' map, contains a wealth of information especially useful to amateurs. A special type of azimuthal projection made by Rand-McNally to ARRL specifications, it gives great-circle bearings from the geographical center of the continental United States, as well as the great-circle distance measurement in miles and kilometers, within an accuracy of 2%. The map shows principal cities of the world, local time zones, WAC divisions, index of DXCC countries and amateur prefixes throughout the world. The map is large enough to be easily readable from the operating position, 31 × 41 inches, and is printed in six colors on heavy paper. The map is available from ARRL HQ; write for details.

Bill Johnston, N5KR, offers computer-calculated and computer-drawn great-circle maps; an extensive selection of these fine maps for various areas of the world appears in Chapter 17. An 11 × 14-inch map can be custom made for your location. Write to PO Box 370, White Sands, NM 88003.

DIRECTION AND DISTANCE BY TRIGONOMETRY

The methods to be described will give the bearing and distance as accurately as one cares to compute them. All that is required is a table of latitude and longitude information, such as Table 4-2, and a calculator or computer with trigonometric functions. The latitude and longitude for any other location can be taken from a map of the area in question.

Fig 4-4 will help you to visualize the nature of the situation. That sketch represents the path between points situated relatively such as Wichita, Kansas, USA (at point A), and Perth, Western Australia (at point B). In using these equations, northerly latitudes are taken as negative. Also, westerly longitudes are taken as positive, and easterly longitudes are taken as negative. *In all calculations*, the appropriate signs are to be retained. *All additions and subtractions throughout the procedure are to be made algebraically.* Thus, if a negative-value number is subtracted from a positive-value number, the resultant will be positive, and it will be the sum of the two absolute values, and so on.

The Calculations

The two equations we'll be using for these calculations are:

$$\cos D = \sin A \sin B + \cos A \cos B \cos L \qquad \text{(Eq 1)}$$

$$\cos C = \frac{\sin B - \sin A \cos D}{\cos A \sin D} \qquad \text{(Eq 2)}$$

where

A = *your* latitude in degrees
B = latitude of the other location in degrees
L = *your* longitude minus that of the other location (algebraic difference)
D = distance along path in degrees of arc
C = true bearing from north if the value for sin L is positive. If sin L is negative, true bearing is 360 − C.

The term *cos* is an abbreviation for cosine, and the term *sin* is an abbreviation for sine. A knowledge of the meanings of these terms isn't necessary for their use here.

The actual calculating procedure uses, first, Eq 1 to determine the angular value for D, in degrees. From this value the path-length distance may be determined in miles or kilometers. Next, Eq 2 is used to determine the bearing angle.

Using the Wichita-to-Perth example mentioned earlier, refer to Fig 4-4 to see how the equations are used. From

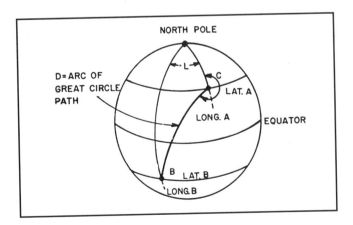

Fig 4-4—The various terms used in the equations for determining bearing and distance. North latitudes and west longitudes are taken as positive, while south latitudes and east longitudes are taken as negative.

Table 4-1, it can be seen that the location of Wichita is 37.7° north latitude, 97.3° west longitude. Similarly, Perth is located at 31.9° south latitude, 115.8° east longitude. Your location is in Wichita. Values for use in the equations are as follows:

A = lat A = + 37.7°
B = lat B = –31.9°
L = long A – long B
  = 97.3° – (–115.8°) = 213.1°

Solving Eq 1, cos D = sin 37.7° × sin (–31.9°) + cos 37.7° × cos (–31.9°) × cos 213.1°. D = 152.4°. Each degree along the path equals 60.0359 nautical miles. Therefore, 152.4° of arc is equivalent to 60.0359 × 152.4 = 9,147 nautical miles. To convert to statute miles, multiply degrees by 69.0826. If the distance is desired in kilometers, multiply degrees by 111.1775. This means that the Wichita-to-Perth distance is 10,525 miles or 16,939 kilometers.

Solving Eq 2

$$\cos C = \frac{\sin (-31.9°) - \sin 37.7° \cos 152.4°}{\cos 37.7° \sin 152.4°}$$

C = 87.9°. Because the sin of L (213.1°) is negative, however, the correct value for C is 360° – 87.9° = 272.1°. Thus, the true bearing from Wichita to Perth is 272.1° and the distance is 10,525 statute miles. If the bearing from Perth were desired, it would be necessary only to work through Eq 2, interchanging latitude values for A and B. Because of the way L is defined (now –213.1°), sin L will be positive, and it will not be necessary to subtract from 360° to get the true bearing at Perth, which is 68.6°.

These equations give information for the great-circle bearing and distance for the shortest path. For long-path work, the bearing will be 180° away from the answers obtained.

The equations described above may be used for any two points on the Earth's surface—both locations in the Northern Hemisphere, both locations in the Southern Hemisphere, either or both on the equator, and so on. The equations themselves are exact, not being based on any approximations. However, there are some cases where practical limitations exist in the accuracy of the results obtained from Eq 2, in relation to the number of significant figures used during calculations. (Round-off errors in calculators and computers during computations will effectively reduce the number of significant figures in the resulting answers.) These cases are where both locations are near or at exact opposite points on the Earth (antipodes), where the locations are close together, or where your location is at or near one of the poles. (At the poles, all directions are either south or north, anyway.) More specifically, these situations exist when lat A is near +90°, or where D is near 0° or 180°.

Other Computer Resources

GeoClock was one of the first shareware programs widely used by hams. It is currently available both as shareware and as an enhanced version from GeoClock, 2218 N. Tuckahoe Street, Arlington, VA 22205. Telephone 703-241-2661, e-mail **geoclock@compuserve.com**. Their web page at **http://www.clark.net/pub/bblake/geoclock** contains descriptions and information on the versions available. When ordered, it comes in both DOS and Windows versions.

The basic program generates maps with the night/dark profile superimposed—in other words a gray-line plot. The sunlight and twilight settings are adjustable, with an explanation included in the program on-line help. If you want the gray-line

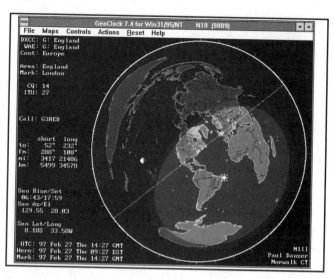

Fig 4-5—This screen capture of the *GeoClock* program gives the bearing from Connecticut to a G3. Any two spots on the globe may be selected.

plot for a selected date and time, simply change the clock setting and the new map is generated. Various display types are accommodated by changes made in a word processor to a configuration file. The ham add-in generates distance, short-path bearing and long-path bearing to any prefix you enter. In Fig 4-5 G3RED was entered (left center of the screen). The resulting bearing line running from Connecticut to England corresponds to a short path bearing of 52° with a distance of 3417 miles. Long path bearing and direction, contest and certificate zones and sunrise/sunset times are also displayed on the screen.

In this screen picture, the gray-line runs through Alaska, Western Australia and around through Eastern Europe. If you are a US ham, you can see in one glance you are too late today for any gray-line propagation!

MiniProp Plus was developed by Sheldon Shallon, W6EL (Telephone 310-473-7322, e-mail **ad363@lafn.org**). It is a full-featured propagation program that includes gray-line and bearing/distance calculations. It installs under DOS, Windows 3.X or Windows 95, and displays a number of maps. The map shown in Fig 4-6 is centered on Connecticut, but illustrates the

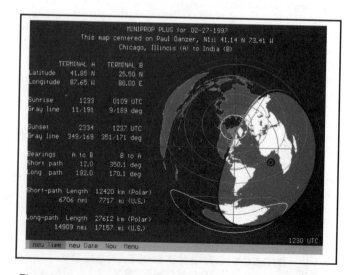

Fig 4-6—In addition to propagation forecasts, MINIPROP PLUS will give gray-line plots and path information. Here the gray-line falls between the selected Chicago QTH and India.

path on February 27, 1997 from Chicago to India. In this case, the gray-line falls just about on this path, and so a W9 who was up and operating around 1230Z stood a good chance of working India on the gray-line.

A wide variety of software is available for producing antenna bearings, distances and gray-line predictions. From time to time *QST* announces new offerings and prints reviews of them. The ARRL book *Personal Computers in the Ham Shack* summarizes a number of PC-based ham software packages. Frequent searches on the Internet will probably turn up new programs every few months.

Table 4-1
Latitude and Longitude of Various US/Canadian Cities and DX Locations with Bearing from East, Central and Western USA

Prefix	State/Province/Country/City	Lat	Long	fm E USA	fm C USA	fm W USA	Prefix	State/Province/Country/City	Lat	Long	fm E USA	fm C USA	fm W USA
VE1	New Brunswick, Saint John	45.3N	66.1W	58.0	62.6	64.8		Idaho, Boise	43.6N	116.2W	289.6	297.9	357.2
	Nova Scotia, Halifax	44.6N	63.6W	63.8	64.5	65.4		Pocatello	42.9N	112.5W	287.5	298.5	41.0
VE2	Quebec, Montreal	45.5N	73.6W	38.4	59.7	65.7		Montana, Billings	45.8N	108.5W	295.0	318.4	40.9
	Quebec City	46.8N	71.2W	40.3	57.3	63.0		Butte	46.0N	112.5W	295.1	311.3	21.9
VE3	Ontario, London	43.0N	81.3W	342.3	63.2	71.9		Great Falls	47.5N	111.3W	298.9	318.6	22.8
	Ottawa	45.4N	75.7W	28.9	58.8	66.1		Nevada, Las Vegas	36.2N	115.1W	273.5	267.7	169.0
	Sudbury	46.5N	81.0W	353.9	50.4	63.9		Reno	39.5N	119.8W	282.1	281.9	261.5
	Toronto	43.7N	79.4W	6.9	62.0	70.1		Oregon, Portland	45.5N	122.7W	294.4	300.1	320.4
VE4	Manitoba, Winnipeg	49.9N	97.1W	315.1	2.8	47.0		Utah, Salt Lake City	40.8N	111.9W	282.3	289.0	74.3
VE5	Saskatchewan, Regina	50.5N	104.6W	309.6	341.5	33.5		Washington, Seattle	47.6N	122.3W	298.4	306.3	331.2
	Saskatoon	52.1N	106.7W	312.4	339.5	24.8		Spokane	47.7N	117.4W	298.6	310.7	353.0
VE6	Alberta, Calgary	51.0N	114.1W	306.4	324.0	6.2		Wyoming, Cheyenne	41.1N	104.8W	281.4	302.7	79.0
	Edmonton	53.5N	113.5W	312.1	330.5	6.3		Sheridan	44.8N	107.0W	292.5	318.1	51.2
VE7	British Columbia, Prince George	53.9N	122.8W	310.4	321.2	344.0	W8	Michigan, Detroit	42.3N	83.0W	316.5	64.7	73.8
	Prince Rupert	54.3N	130.3W	310.4	317.2	330.9		Grand Rapids	43.0N	85.7W	307.0	58.0	72.5
	Vancouver	49.3N	123.1W	301.7	310.2	333.9		Sault Ste. Marie	46.5N	84.4W	335.2	45.4	63.6
VE8	Nwest Territories, Yellowknife	62.5N	114.4W	329.0	343.0	1.9		Traverse City	44.8N	85.6W	321.1	49.8	67.8
	Resolute	74.7N	95.0W	353.2	1.3	9.3		Ohio, Cincinnati	39.1N	84.5W	256.9	79.9	81.9
VY1	Yukon, Whitehorse	60.7N	135.1W	320.6	326.6	336.6		Cleveland	41.5N	81.7W	319.8	69.3	75.4
VY2	P.E.Island, Charlottetown	46.2N	63.1W	57.7	61.0	62.7		Columbus	40.0N	83.0W	271.0	75.6	79.2
VO1	Newfoundland, St. John's	47.6N	52.7W	59.7	58.8	58.4	W9	W. Virginia, Charleston	38.4N	81.6W	218.4	83.1	82.3
VO2	Labrador, Goose Bay	53.3N	60.4W	38.5	47.0	51.2		Illinois, Chicago	41.9N	87.6W	290.8	60.7	75.6
W1	Connecticut, Hartford	41.8N	72.7W	69.6	70.9	72.4		Indiana, Indianapolis	39.8N	86.2W	269.6	75.2	80.8
	Maine, Bangor	44.8N	68.8W	56.2	63.3	66.2		Wisconsin, Green Bay	44.5N	88.0W	309.9	45.9	68.5
	Portland	43.7N	70.3W	59.7	65.9	68.4		Milwaukee	43.0N	87.9W	299.5	53.7	72.6
	Massachusetts, Boston	42.4N	71.1W	67.4	69.3	70.9	W0	Colorado, Denver	39.7N	105.0W	277.2	289.5	88.5
	New Hampshire, Concord	43.2N	71.5W	60.5	67.0	69.6		Grand Junction	39.1N	108.6W	276.9	280.8	96.6
	Rhode Island, Providence	41.8N	71.4W	71.7	71.0	72.1		Iowa, Des Moines	41.6N	93.6W	283.2	41.8	79.3
	Vermont, Montpelier	44.3N	72.6W	49.5	63.6	67.8		Kansas, Pratt	37.7N	98.7W	267.0	242.1	94.2
W2	New Jersey, Atlantic City	39.4N	74.4W	96.1	78.3	77.4		Wichita	37.7N	97.3W	265.8	117.9	93.0
	New York, Albany	42.7N	73.8W	57.9	68.0	71.0		Minnesota, Duluth	46.8N	92.1W	311.8	24.4	60.6
	Buffalo	42.9N	78.9W	15.6	65.3	71.7		Minneapolis	45.0N	93.3W	301.4	25.2	65.8
	New York City	40.8N	74.0W	78.1	73.9	74.7		Missouri, Columbia	39.0N	92.3W	267.9	75.6	85.5
	Syracuse	43.1N	76.2W	41.3	65.9	70.8		Kansas City	39.1N	94.6W	270.1	66.5	86.2
W3	Delaware, Wilmington	39.7N	75.5W	93.5	77.4	77.2		St. Louis	38.6N	90.2W	263.2	82.0	85.7
	District of Col., Washington	38.9N	77.0W	114.4	80.3	79.3		Nebraska, North Platte	41.1N	100.8W	280.7	326.0	79.6
	Maryland, Baltimore	39.3N	76.6W	103.9	78.9	78.4		Omaha	41.3N	95.9W	281.3	25.5	78.6
	Pennsylvania, Harrisburg	40.3N	76.9W	81.8	75.4	76.6		North Dakota, Fargo	46.9N	96.8W	305.1	5.3	57.2
	Philadelphia	39.9N	75.2W	90.0	76.7	76.7		So. Dakota, Rapid City	44.1N	103.2W	291.0	328.9	62.5
	Pittsburgh	40.4N	80.0W	15.4	74.6	77.3	1A0	S.M.O.M.	41.9N	12.4E	54.4	45.5	35.8
	Scranton	41.4N	75.7W	65.4	71.8	74.0	1S	Spratly Is.	8.8N	111.9E	344.6	322.8	306.5
W4	Albama, Montgomery	32.4N	86.3W	215.7	116.9	98.3	3A	Monaco	43.7N	7.4E	54.9	46.4	37.6
	Florida, Jacksonville	30.3N	81.7W	188.7	114.9	98.4	3B6	Agalega	10.4S	56.6E	64.6	46.2	14.5
	Miami	25.8N	80.2W	180.9	123.9	104.5	3B7	St. Brandon	16.3S	59.8E	67.5	48.1	10.0
	Pensacola	30.4N	87.2W	213.7	127.2	103.3	3B8	Mauritius	20.3S	57.5E	74.2	57.1	17.7
	Georgia, Atlanta	33.8N	84.4W	210.9	106.8	93.8	3B9	Rodrigue Is.	19.7S	63.4E	67.9	46.6	1.7
	Savannah	32.1N	81.1W	186.8	108.1	94.7	3C	Equ. Guinea, Bata	1.8N	9.8E	88.7	77.4	63.7
	Kentucky, Lexington	38.0N	84.5W	241.7	85.8	84.5		Malabo	3.8N	8.8E	87.9	76.5	63.0
	Louisville	38.2N	85.8W	250.1	85.0	84.6	3C0	Annobon Is.	1.5S	5.6E	94.0	82.7	69.6
	North Carolina, Charlotte	35.2N	80.8W	187.9	96.2	88.5	3D2	Fiji Is., Suva	18.1S	178.4E	263.0	251.8	240.5
	Raleigh	35.8N	78.6W	164.8	92.1	86.1	3D2	Conway Reef	21.4S	174.4E	262.5	251.5	240.8
	Wilmington	34.2N	77.9W	163.2	97.1	86.6	3D2	Rotuma Is.	12.3S	177.7E	268.2	256.8	245.1
	South Carolina, Columbia	34.0N	81.0W	188.0	101.1	91.0	3DA	Swazilnd, Mbabane	26.3S	31.1E	99.0	90.1	73.5
	Tennessee, Knoxville	36.0N	83.9W	218.8	95.8	88.7	3V	Tunisia, Tunis	36.8N	10.2E	60.1	50.5	40.3
	Memphis	35.1N	90.1W	241.7	112.2	95.2	3W, XV	Vietnam, H C Minh City (Saigon)	10.8N	106.7E	351.5	329.5	312.4
	Nashville	36.2N	86.8W	236.8	98.0	90.1		Hanoi	21.0N	105.8E	353.8	335.0	319.2
	Virginia, Norfolk	36.9N	76.3W	135.7	87.0	82.8	3X	Guinea, Conakry	9.5N	13.7W	98.1	85.9	74.8
	Richmond	37.5N	77.4W	140.1	85.4	82.2	3Y	Bouvet	54.5S	3.4E	139.1	135.1	131.0
W5	Arkansas, Little Rock	34.7N	92.3W	245.4	124.0	98.2	3Y	Peter I Is.	68.8S	90.6W	184.0	177.2	170.5
	Louisiana, New Orleans	29.9N	90.1W	222.4	138.7	107.5	4J, 4K	Azerbaijan, Baku	40.4N	49.9E	35.8	24.0	10.8
	Shreveport	32.5N	93.7W	240.0	146.2	105.7	4L	Georgia, Tbilisi	41.7N	44.8E	38.0	26.9	14.3
	Mississippi, Jackson	32.3N	90.2W	230.0	129.5	102.2	4S	Sri Lanka, Colombo	7.0N	79.9E	26.2	2.9	339.0
	New Mexico, Albuquerque	35.1N	106.7W	265.4	250.1	120.7	4U	ITU Geneva	46.2N	6.2E	52.8	44.9	36.6
	Oklahoma, Oklahoma City	35.5N	97.5W	257.5	170.5	101.3	4U	United Nations Hq.	40.8N	74.0W	78.1	73.9	74.7
	Texas, Abilene	32.5N	99.7W	250.8	194.7	114.7	4X, 4Z	Israel, Jerusalem	31.8N	35.2E	50.4	38.7	24.9
	Amarillo	35.2N	101.8W	261.4	228.6	108.6	5A	Libya, Tripoli	32.5N	12.5E	62.7	52.5	41.4
	Dallas	32.8N	96.8W	247.2	168.9	109.0		Benghazi	32.1N	20.0E	59.0	48.4	36.4
	El Paso	31.8N	106.5W	257.3	230.9	133.9	5B	Cyprus, Nicosia	35.2N	33.4E	49.1	37.9	25.0
	San Antonio	29.4N	98.5W	240.7	183.0	121.1	5H	Tanzania, Dar es Salaam	7.0S	39.5E	75.7	62.1	40.2
W6	California, Los Angeles	34.1N	118.2W	271.3	262.7	197.3	5N	Nigeria, Lagos	6.5N	3.4E	89.2	77.8	65.2
	San Francisco	37.8N	122.4W	280.1	277.0	248.2	5R	Madagascar, Antananarivo	18.9S	47.5E	80.8	67.2	38.7
W7	Arizona, Flagstaff	35.2N	111.7W	269.3	259.9	143.3	5T	Mauritania, Nouakchott	18.1N	16.0W	92.0	80.2	69.8
	Phoenix	33.5N	112.1W	266.0	252.8	153.1							

Prefix	State/Province/Country/City	Lat	Long	fm E USA	fm C USA	fm W USA
5U	Niger, Niamey	13.5N	2.0E	84.6	73.3	61.2
5V	Togo, Lome	5.8N	1.2E	91.2	79.8	67.3
5W	Western Samoa, Apia	13.5S	171.8W	260.7	249.4	236.6
5X	Uganda, Kampala	0.3N	32.5E	74.9	62.0	43.4
5Z	Kenya, Nairobi	1.3S	376.8E	86.6	75.2	60.3
6W	Senegal, Dakar	14.7N	17.5W	96.2	83.9	73.3
6Y	Jamaica, Kingston	18.0N	76.8W	171.9	131.3	111.5
7O	Yemen, Aden	12.8N	45.0E	56.5	41.9	22.6
	Sanaa	15.4N	44.2E	55.3	41.1	22.5
7P	Lesotho, Maseru	29.3S	27.5E	103.9	95.9	81.7
7Q	Malawi, Lilongwe	14.0S	33.8E	85.7	74.0	54.1
	Blantyre	15.8S	35.0E	86.5	74.9	54.5
7X	Algeria, Algiers	36.7N	3.0E	63.6	54.3	44.7
8P	Barbados, Bridgetown	13.1N	59.6W	140.6	115.7	102.0
8Q	Maldive Is.	4.4N	73.4E	35.2	12.6	346.7
8R	Guyana, Georgetown	6.8N	58.2W	143.7	120.6	106.5
9A	Croatia, Zagreb	45.8N	16.0E	49.3	40.7	31.4
9G	Ghana, Accra	5.5N	0.2W	92.3	80.9	68.6
9H	Malta	36.0N	14.4E	58.7	48.9	38.0
9J	Zambia, Lusaka	15.4S	28.3E	90.6	79.6	61.9
9K	Kuwait	29.5N	47.8E	43.7	30.5	14.9
9L	Sierra Leone, Freetown	8.5N	13.2W	98.6	86.4	75.2
9M2	W Malaysia, K. Lumpur	3.2N	101.6E	357.7	331.7	312.1
9M6,8	E Malaysia, Sabah(9M6)	5.8N	118.1E	335.7	314.4	299.3
	Sarawak, Kuching (9M8)	1.6N	110.3E	344.7	320.0	302.8
9N	Nepal, Kathmandu	27.7N	85.3E	13.9	356.8	340.1
9Q	Zaire, Kinshasa	4.3S	15.3E	89.9	78.7	64.0
	Kisangani	0.5N	25.2E	79.7	67.7	51.0
	Lubumbashi	11.7S	27.5E	87.9	76.6	59.0
9U	Burundi, Bujumbura	3.3S	29.3E	79.9	67.6	49.6
9V	Singapore	1.3N	103.8E	354.3	327.8	308.6
9X	Rwanda, Kigali	2.0S	30.1E	78.3	65.9	47.7
9Y	Trinidad & Tobago , P. of Spain	10.5N	61.3W	145.5	120.2	105.7
A2	Botswana, Gaborone	24.8S	25.9E	100.5	91.4	76.2
A3	Tonga, Nukualofa	21.1S	175.2W	256.6	245.7	234.0
A4	Oman, Masqat	23.6N	58.6E	39.0	23.6	5.5
A5	Bhutan, Thimpu	27.3N	89.4E	10.2	352.8	336.3
A6	U. A. E., Abu Dhabi	24.5N	54.2E	41.9	27.3	9.8
A7	Qatar, Ad-Dawhah	25.3N	51.5E	43.5	29.3	12.3
A9	Bahrein, Al-Manamah	26.2N	50.6E	43.7	29.7	13.0
AP	Pakistan, Karachi	24.9N	67.1E	31.2	15.0	356.9
AP	Islamabad	33.7N	73.2E	22.6	7.7	352.1
BS7H	Scarborough Reef	15.1N	117.8E	339.5	320.6	306.0
BV	Taiwan, Taipei	25.1N	121.5E	339.0	323.0	309.9
BV9P	Pratas Is.	20.7N	116.7E	342.4	324.7	310.5
BY	P.R. of China, Beijing	40.0N	116.4E	347.4	334.2	322.6
	Harbin	45.8N	126.7E	341.7	330.6	320.6
	Shanghai	31.2N	121.5E	341.0	326.3	313.8
	Fuzhou	26.1N	119.3E	341.4	325.3	312.0
	Xian	34.3N	108.9E	352.4	337.4	324.2
	Chongqing	29.8N	106.5E	354.0	337.7	323.5
	Chengdu	30.7N	104.1E	356.3	340.1	325.8
	Lhasa	29.7N	91.2E	8.1	351.4	335.6
	Urumqi	43.8N	87.6E	9.0	355.9	343.2
	Kashi	39.5N	76.0E	18.5	4.7	350.6
C2	Nauru	0.5S	166.9E	284.9	272.8	261.2
C3	Andorra	42.5N	1.5E	58.5	50.1	41.6
C5	The Gambia, Banjul	13.5N	16.7W	96.7	84.4	73.7
C6	Bahamas, Nassau	25.1N	77.4W	170.9	120.5	103.0
C9	Mozambique, Maputo	26.0S	32.6E	97.8	88.8	71.4
	Mozambique	15.1S	40.7E	82.0	69.3	45.9
CE	Chile, Santiago	33.5S	70.8W	172.0	156.9	143.5
CE0Y	Easter Island	27.1S	109.4W	207.3	191.1	173.6
CE0Z	Juan Fernandez	33.6S	78.8W	179.0	163.4	149.4
CE0X	San Felix	26.3S	80.1W	180.1	162.5	146.9
CM,CO	Cuba, Havana	23.1N	82.4W	187.6	133.7	110.7
CN	Morocco, Casablanca	33.6N	7.5W	71.7	62.1	53.1
CP	Bolivia, La Paz	16.5S	68.4W	166.8	147.3	131.8
CT	Portugal, Lisbon	38.7N	9.2W	66.9	58.3	50.1
CT3	Madeira Isl., Funchal	32.6N	16.9W	77.4	67.5	59.1
CU	Azores , Ponta Delgada	37.7N	25.7W	75.0	66.1	59.2
CX	Uruguay, Montevideo	34.9S	56.2W	160.2	146.5	134.8
CY0	Sable Is.	43.8N	60.0W	69.1	66.3	65.7
CY9	St. Paul Is.	47.2N	60.1W	56.8	59.3	60.6
D2	Angola, Luanda	8.8S	13.2E	94.7	83.8	69.6
D4	Cape Verde, Praia	14.9N	23.5W	100.3	87.3	76.9
D6	Comoros, Moroni	11.8S	43.7E	76.7	62.8	38.1
DA-DL	Fed. Rep. of Germany, Bonn	50.7N	7.0E	47.9	40.7	33.1
	Berlin	52.5N	13.4E	43.9	36.6	28.8
DU	Philippines, Manila	14.6N	121.0E	335.9	317.4	303.2
E3	Eritrea, Asmara	15.3N	38.9E	59.3	45.8	28.2
EA	Spain, Madrid	40.4N	3.7W	62.8	54.3	45.9
EA6	Balearic Is., Palma	39.5N	2.6E	61.0	52.2	43.1
EA8	Canary Is., Las Palmas	28.4N	14.3W	80.7	70.2	60.9
EA9	Ceuta & Melilla, Ceuta	35.9N	5.3W	68.2	59.0	50.1
	Melilla	35.3N	3.0W	67.8	58.5	49.2
EI	Ireland, Dublin	53.3N	6.3W	48.6	42.9	37.1
EK	Armenia, Yerevan	40.3N	44.5E	39.1	27.8	14.9
EL	Liberia, Monrovia	6.3N	10.8W	98.7	86.7	75.3
EP	Iran, Tehran	35.8N	51.8E	37.2	24.6	10.1
ER	Moldova, Kishinev	47.0N	28.8E	42.6	33.4	23.2
ES	Estonia, Tallinn	59.4N	24.8E	33.5	26.8	19.4
ET	Ethiopia, Addis Ababa	9.0N	38.7E	63.9	50.1	31.3
EU,EV,EW	Belarus, Minsk	53.9N	27.6E	37.4	29.5	20.7
EX	Kyrgyzstan, Bishkek	42.9N	74.6E	18.4	5.5	352.2
EY	Tajikistan, Samarkand	39.7N	66.8E	25.0	11.9	357.8
	Dushanbe	39.1N	68.8E	23.8	10.4	356.2

Prefix	State/Province/Country/City	Lat	Long	fm E USA	fm C USA	fm W USA
EZ	Turkmenistan, Ashkhbd	38.0N	58.4E	31.6	18.7	4.5
F	France, Paris	48.8N	2.3E	51.5	44.2	36.7
FG	Guadeloupe	16.0N	61.7W	141.2	114.5	100.8
FJ, FS	St. Martin	18.1N	63.1W	141.4	113.4	99.8
FH	Mayotte	13.0S	45.3E	76.6	62.5	36.6
FK	New Caledonia, Nouma	22.3S	166.5E	266.4	255.1	245.1
FM	Martinique	14.6N	61.0W	141.4	115.5	101.7
FO	Clipperton Is.	10.3N	109.2W	229.1	202.9	166.7
FO	Fr. Polynesia, Tahiti	17.6S	159.5W	249.9	238.3	224.1
	Rurutu, (Austral Is.)	22.5S	151.3E	276.2	263.4	254.0
	Hiva Oa, (Marquesas Is.)	9.9S	139.0W	241.5	227.5	208.3
FP	St. Pierre & Miquelon, St. Pierre	46.7N	56.0W	61.0	60.5	60.4
FR/G	Glorioso	11.5S	47.3E	73.6	58.7	32.0
FR/J,E	Juan de Nova	17.0S	42.8E	82.3	69.5	44.7
	Europa	22.3S	40.4E	89.3	78.3	55.5
FR	Reunion	21.1S	55.6E	76.8	60.9	23.2
FR/T	Tromelin	15.9S	54.4E	72.0	55.3	21.9
FT5W	Crozet	46.0S	52.0E	116.0	119.5	128.4
FT5X	Kerguelen	49.3S	69.2E	123.5	144.9	199.9
FT5Y	Antarctica, Dumont D'Urville	66.6S	140.0E	206.7	209.6	211.0
FT5Z	Amsterdam & St. Paul Is.	37.7S	77.6E	89.7	86.4	277.9
FW	Wallis & Futuna Is., Wallis	13.3S	176.3W	263.7	252.4	240.1
FY	Fr. Guiana, Cayenne	4.9N	52.3W	137.3	116.9	103.7
G	England, London	51.5N	0.1W	49.2	42.6	35.9
GD	Isle of Man	54.3N	4.5W	46.9	41.3	35.6
GI	No. Ireland, Belfast	54.6N	5.9W	46.8	41.5	35.9
GJ	Jersey	49.3N	2.2W	52.3	45.5	38.5
GM	Scotland, Glasgow	55.8N	4.3W	45.0	39.8	34.2
	Aberdeen	57.2N	2.1W	42.9	37.7	32.2
GU	Guernsey	49.5N	2.7W	52.2	45.5	38.6
GW	Wales, Cardiff	51.5N	3.2W	50.0	43.7	37.3
H4	Solomon Isl., Honiara	9.4S	160.0E	282.7	269.9	259.0
HA	Hungary, Budapest	47.5N	19.1E	46.5	38.0	28.7
HB	Switzerland, Bern	47.0N	7.5E	51.5	43.7	35.3
HB0	Liechtenstein	47.2N	9.6E	50.6	42.6	34.1
HC	Ecuador, Quito	0.2S	78.0W	176.9	149.5	129.6
HC8	Galapagos Is.	0.5S	90.5W	195.9	168.1	143.7
HH	Haiti, Port-Au-Prince	18.5N	72.3W	160.6	123.9	106.8
HI	Dominican Rep., Santo Dmingo	18.5N	70.0W	155.3	120.8	104.8
HK	Colombia, Bogota	4.6N	74.1W	169.9	141.0	122.0
HK0	Malpelo Is.	4.0N	81.1W	181.9	151.4	129.6
HK0	San Adreas	12.5N	81.7W	183.6	146.0	122.7
HL	Korea, Seoul	37.5N	127.0E	338.6	325.8	314.6
HP	Panama, Panama	9.0N	79.5W	179.0	145.3	123.6
HR	Honduras, Tegucigalpa	14.1N	87.2W	195.7	155.3	127.6
HS	Thailand, Bangkok	13.8N	100.5E	359.4	337.8	319.8
HV	Vatican City	41.9N	12.5E	54.4	45.4	35.7
HZ, 7Z	Saudi Arabia, Dharan	26.3N	50.0E	44.1	30.2	13.6
	Mecca	21.5N	39.8E	54.4	41.2	24.8
	Riyadh	24.6N	46.7E	47.5	33.8	17.1
I	Italy, Rome	41.9N	12.5E	54.4	45.4	35.7
	Trieste	45.7N	13.8E	50.3	41.9	32.7
	Sicily	37.5N	14.0E	57.6	48.0	37.4
IS	Sardinia, Cagliari	39.2N	9.1E	58.4	49.2	39.4
J2	Djibouti, Djibouti	11.6N	43.2E	58.7	44.3	25.1
J3	Grenada	12.0N	61.8W	145.1	119.1	104.7
J5	Guinea-Bissau, Bissau	11.9N	15.6W	97.3	85.1	74.2
J6	St. Lucia	13.9N	61.0W	142.1	116.2	102.3
J7	Dominica	15.4N	61.3W	141.1	114.8	101.1
J8	St. Vincent	13.3N	61.3W	143.1	117.2	103.1
JA-JS	Japan, Tokyo	35.7N	139.8E	328.5	316.6	306.1
	Nagasaki	32.8N	129.9E	334.6	321.1	309.6
	Sapporo	43.1N	141.4E	331.1	320.7	311.3
JD1	Minami Torishima	24.3N	154.0E	311.8	299.9	289.1
JD1	Ogasawara, Kazan Is.	27.5N	141.0E	323.4	310.3	299.1
JT	Mongolia, Ulan Bator	47.9N	106.9E	355.4	343.6	332.7
JW	Svalbard, Spitsbergen	78.8N	16.0E	14.2	12.2	9.8
JX	Jan Mayen	71.0N	8.3W	25.1	23.6	21.5
JY	Jordan, Amman	32.0N	35.9E	49.8	38.1	24.2
KC4	Antarctica, Byrd Station	80.0S	120.0W	187.6	184.2	180.8
	McMurdo Sound	77.7S	166.7E	195.7	195.6	194.9
	Palmer Station	64.8S	64.0W	173.0	165.6	158.7
KC6	Belau, Yap	9.5N	138.2E	315.7	299.7	287.5
	Koror	7.3N	134.5E	317.9	300.9	288.4
KG4	Guantanamo Bay	19.9N	75.2W	167.0	126.1	107.8
KH0	Mariana Is., Saipan	15.2N	145.8E	312.5	298.4	286.9
KH1	Baker, Howland Is.	0.5N	176.0W	274.2	262.9	250.0
KH2	Guam, Agana	13.5N	144.8E	312.2	297.9	286.2
KH3	Johnston Is.	17.0N	168.5W	282.3	272.2	258.8
KH4	Midway Is.	28.2N	177.4W	296.5	287.5	276.7
KH5	Palmyra Is.	5.9N	162.1W	269.5	258.2	243.0
KH5K	Kingman Reef	7.5N	162.8W	271.2	260.1	245.0
KH6	Hawaii, Hilo	19.7N	155.1W	276.4	266.4	250.4
	Honolulu	21.3N	157.9W	279.5	269.9	254.9
KH7	Kure Is.	28.4N	178.4W	297.2	288.2	277.5
KH8	American Samoa, Pago Pago	14.3S	170.8W	259.4	248.2	235.3
KH9	Wake Is.	19.3N	166.6E	299.7	288.5	277.5
KL7	Alaska, Adak	51.8N	176.6W	316.5	311.9	307.1
	Anchorage	61.2N	150.0W	321.3	323.2	327.0
	Fairbanks	64.8N	147.9W	326.2	329.1	334.0
	Juneau	58.3N	134.4W	316.8	322.7	333.4
	Nome	64.5N	165.4W	327.3	326.9	327.4
KP1	Navassa Is.	18.4N	75.0W	167.3	127.9	109.3
KP2	Virgin Isl., Charlotte Am.	18.3N	64.9W	144.6	115.0	100.9
KP4	Puerto Rico, San Juan	18.5N	66.2W	147.0	116.2	101.7
KP5	Desecheo Is.	18.3N	67.5W	150.0	118.0	103.0
LA-LJ	Norway, Oslo	60.0N	10.7E	36.8	31.2	25.1

Prefix	State/Province/Country/City	Lat	Long	fm E USA	fm C USA	fm W USA
LU	Argentina, Buenos Aires	34.6S	58.4W	161.9	147.9	136.0
LX	Luxembourg	49.6N	6.2E	49.3	42.0	34.3
LY	Lithuania, Vilna	54.5N	25.5E	37.7	30.0	21.5
LZ	Bulgaria, Sofia	42.7N	23.3E	48.7	39.2	28.6
OA	Peru, Lima	12.1S	77.1W	176.4	154.4	136.6
OD	Lebanon, Beirut	33.9N	35.5E	48.8	37.3	23.8
OE	Austria, Vienna	48.2N	16.3E	47.0	38.8	29.8
OH	Finland, Helsinki	60.2N	25.0E	32.7	26.1	18.9
OH0	Aland Is.	60.2N	20.0E	34.2	27.9	21.1
OJ0	Market Reef	60.3N	19.0E	34.4	28.2	21.5
OK,OL	Czech Rep., Prague	50.1N	14.4E	45.9	38.1	29.7
OM	Slovak, Rep., Bratislava	48.0N	17.0E	46.9	38.6	29.6
ON	Belgium, Brussels	50.9N	4.4E	48.5	41.6	34.2
OX	Greenland, Godthaab	64.2N	51.7W	25.0	31.0	34.6
	Thule	76.6N	68.8W	4.3	10.0	14.8
OY	Faroe Islands, Torshavn	62.0N	6.8W	37.3	33.8	29.8
OZ	Denmark, Copenhagen	55.7N	12.6E	40.9	34.3	27.1
P2	Papua New Guinea, Madang	5.2S	145.6E	298.0	282.7	271.4
	Port Moresby	9.4S	147.1E	293.0	278.1	267.1
P4	Aruba, Oranjestad	12.1N	69.0W	157.8	127.3	110.5
P5	North Korea, Pyongyang	59.0N	125.8E	346.8	338.5	330.8
PA-PI	Netherlands, Amsterdam	52.4N	4.9E	46.7	40.0	32.9
PJ2,4,9	Netherlands Antilles, Willemstd	12.6N	70.1W	159.6	128.2	111.0
PJ5-8	St. Maarten and Saba	17.7N	63.2W	142.1	114.0	100.3
PY	Brazil, Brasilia	15.8S	47.9W	145.1	128.8	116.2
	Rio De Janeiro	23.0S	43.2W	144.5	130.2	118.3
	Natal	6.0S	35.2W	127.3	112.3	100.5
	Manaus	3.1S	60.2W	152.3	130.8	116.0
	Porto Alegre	30.1S	51.2W	154.4	140.3	128.4
PY0	Fernando De Noronha	3.9S	32.4W	123.3	108.7	97.1
PY0	St. Peter & St. Paul Rcks	1.0N	29.4W	117.1	102.8	91.4
PY0	Trindade & M. Vaz Is., Trindade	20.5S	29.3W	131.9	119.1	107.9
PZ	Suriname, Paramaribo	5.8N	55.2W	140.3	118.7	105.1
R1FJ_	Franz Josef Land	80.0N	53.0E	8.7	5.5	2.2
R1MV_	Malyj Vysotskij Is.	60.6N	28.6E	31.2	24.5	17.1
S0	Western Sahara, Smara	26.4N	11.4W	81.1	70.4	60.7
S2	Bangladesh, Dacca	23.7N	90.4E	9.8	351.3	334.0
S5	Slovenia, Ljubljana	46.0N	14.5E	49.8	41.3	32.1
S7	Seychelles, Victoria	4.6S	55.5E	60.5	42.5	14.4
S9	Sao Tome	0.3N	6.7E	91.9	80.6	67.3
SM	Sweden, Stockholm	59.3N	18.1E	35.6	29.3	22.5
SP	Poland, Krakow	50.0N	20.0E	43.9	35.7	26.8
	Warsaw	52.2N	21.0E	41.5	33.6	25.1
ST	Sudan, Khartoum	15.6N	32.5E	63.5	50.8	34.4
ST0	Southern Sudan, Juba	5.0N	31.6E	71.9	59.1	41.3
SU	Egypt, Cairo	30.0N	31.4E	54.0	42.4	28.7
SV	Greece, Athens	38.0N	23.7E	52.3	42.2	30.7
SV/A	Mount Athos	40.2N	24.3E	50.2	40.4	29.2
SV5	Dodecanese, Rhodes	36.4N	28.2E	51.1	40.5	28.3
SV9	Crete	35.4N	25.2E	53.5	43.0	30.9
T2	Tuvalu, Funafuti	8.7S	178.6E	270.6	259.1	247.1
T30	West Kiribati, Bonriki	1.4N	173.2E	282.1	270.4	258.5
T31	Central Kiribati, Kanton	2.8S	171.7W	268.9	257.6	244.2
T32	E. Kiribati, Christmas Is.	1.9N	157.4W	263.3	251.5	235.3
T33	Banaba Is.	0.5S	169.4E	283.2	271.2	259.6
T5	Somalia, Mogadishu	2.1N	45.4E	63.8	48.7	26.6
T7	San Marino	43.9N	12.3E	52.6	44.0	34.6
T9	Bosnia-Herz., Sarajevo	43.9N	18.4E	50.0	41.0	31.0
TA	Turkey, Ankara	39.9N	32.9E	46.0	35.4	23.4
	Istanbul	41.2N	29.0E	47.1	37.0	25.6
TF	Iceland, Reykjavik	64.1N	22.0W	34.4	33.4	31.6
TG	Guatemala, Guatmla City	14.6N	90.5W	202.9	162.1	131.5
TI	Costa Rica, San Jose	9.9N	84.0W	187.8	152.3	127.8
TI9	Cocos Is.	5.6N	87.0W	192.2	160.1	135.1
TJ	Cameroon, Yaounde	3.9N	11.5E	86.0	74.6	60.8
TK	Corsica	42.0N	9.0E	55.8	47.1	37.7
TL	Cen. Afr. Rep., Bangui	4.4N	18.6E	81.1	69.4	54.4
TN	Congo, Brazzaville	4.3S	15.3E	89.9	78.7	64.0
TR	Gabon, Libreville	0.4N	9.5E	90.0	78.7	65.1
TT	Chad, N'Djamena	12.1N	15.0E	77.5	66.0	52.2
TU	Ivory Coast, Abidjan	5.3N	4.0W	95.0	83.4	71.4
TY	Benin, Porto Novo	6.5N	2.6E	89.7	78.3	65.8
TZ	Mali, Bamako	12.7N	8.0W	91.6	79.9	68.7
UA	Russia, St Ptrsbrg (UA1)	59.9N	30.3E	31.3	24.3	16.6
	Archangel (UA1)	64.6N	40.5E	24.0	17.4	10.3
	Murmansk (UA1)	69.0N	33.1E	22.3	17.1	11.4
	Moscow (UA3)	55.8N	37.6E	31.9	23.6	14.6
	Samara (UA4)	53.2N	50.1E	28.0	18.5	8.3
	Rostov (UA6)	47.5N	39.5E	37.1	27.2	16.3
UA2	Kaliningrad	55.0N	20.5E	39.1	31.8	23.8
UA9,0	Russia, Nvsibirsk (UA9)	55.0N	82.9E	9.8	359.5	349.2
	Perm (UA9)	58.0N	56.3E	22.2	13.4	4.1
	Omsk (UA9)	55.0N	73.4E	15.0	4.9	354.6
	Norilsk (UA0)	69.3N	88.1E	4.4	357.7	351.1
	Irkutsk (UA0)	52.3N	104.3E	357.4	346.6	336.4
	Vladivostok (UA0)	43.2N	131.9E	337.3	326.1	316.1
	Petropavlovsk (UA0)	53.0N	158.7E	327.7	320.7	314.0
	Khabarovsk (UA0)	48.5N	135.1E	337.6	327.6	318.6
	Krasnoyarsk (UA0)	56.0N	92.8E	4.0	354.0	344.2
	Yakutsk (UA0)	62.0N	129.7E	346.1	338.7	331.9
	Wrangel Island (UA0)	71.0N	179.5W	337.1	335.8	335.2
	Kyzyl (UA0Y)	51.7N	94.5E	3.4	352.3	341.6
UJ-UM	Uzbekistan, Bukhoro	39.8N	64.4E	26.6	13.7	359.7
	Tashkent	41.2N	69.3E	22.7	9.7	356.0
UN-UQ	Kazakhstan, Alma-Ata	43.3N	76.9E	16.6	3.8	350.6
UR-UZ,EM-EO	Ukraine, Kiev	50.4N	30.5E	39.1	30.4	20.7
V2	Antigua & Barbda, St. Jhns	17.1N	61.8W	140.2	113.3	99.8

Prefix	State/Province/Country/City	Lat	Long	fm E USA	fm C USA	fm W USA
V3	Belize, Belmopan	17.3N	88.8W	201.1	156.2	126.1
V4	St. Kitts & Nevis	17.3N	62.6W	141.4	113.9	100.2
V5	Namibia, Windhoek	22.6S	17.1E	103.6	94.2	80.6
V6	Micronesia, Ponape	6.9N	158.3E	296.8	283.9	272.6
V7	Marshall Islands, Kwajalein	9.1N	167.3E	291.9	280.1	268.5
V8	Brunei, Bandar Seri Bgwan	4.9N	114.9E	339.5	317.1	301.3
VK	Australia, Canberra (VK1)	35.3S	149.1E	260.9	250.9	243.9
	Sydney (VK2)	33.9S	151.2E	261.8	251.6	244.2
	Melbourne (VK3)	37.8S	145.0E	258.8	249.6	243.4
	Brisbane (VK4)	27.5S	153.0E	269.1	257.5	248.8
	Adelaide (VK5)	34.9S	138.6E	267.1	255.8	249.3
	Perth (VK6)	31.9S	115.8E	297.5	272.2	264.2
	Hobart, Tasmania (VK7)	42.9S	147.3E	249.4	242.4	237.3
	Darwin (VK8)	12.5S	130.9E	306.6	287.0	275.1
VK0	Heard Is.	53.0S	73.4E	134.6	161.1	203.1
VK0	Macquarie Is.	54.7S	158.8E	228.8	225.0	221.3
VK9C	Cocos-Keeling Is.	12.2S	96.8E	6.7	329.0	304.7
VK9L	Lord Howe Is.	31.6S	159.1E	260.7	250.4	242.1
VK9M	Mellish Reef	17.6S	155.8E	278.1	265.3	255.2
VK9N	Norfolk Is.	29.0S	168.0E	258.8	248.5	239.1
VK9W	Willis Is.	16.3S	149.5E	284.3	270.3	260.1
VK9X	Christmas Is.	10.5S	105.7E	348.7	316.1	296.9
VP2E	Anguilla	18.3N	63.0W	141.0	113.0	99.5
VP2M	Montserrat	16.8N	62.2W	141.2	114.1	100.4
VP2V	Brit. V. Is., Tortola	18.4N	64.6W	144.0	114.6	100.6
VP5	Trks & Cacos Is., Gr. Trk	21.4N	71.2W	155.5	118.1	102.5
VP8	Falkland Isl., Stanley	51.7S	57.9W	166.5	156.3	147.0
VP8	So. Georgia Is.	54.3S	36.8W	156.0	147.8	140.4
VP8	So. Orkney Is.	60.6S	45.5W	163.3	155.9	149.1
VP8	S. Sndwch Is., Snders Is.	57.8S	26.7W	153.4	146.8	140.8
VP8	S. Shetlnd Is., K Grge Is.	62.0S	58.3W	169.7	161.9	154.7
VP9	Bermuda	32.3N	64.7W	117.2	91.7	84.0
VQ9	Chagos, Diego Garcia	7.3S	72.4E	44.5	18.2	344.8
VR2,VS6	Hong Kong	22.3N	114.3E	345.2	327.6	313.3
VR6	Pitcairn Is.	25.1S	130.1W	224.9	210.8	193.9
VU	India, Bombay	19.0N	72.8E	28.7	10.3	350.3
	Calcutta	22.6N	88.4E	12.0	353.2	335.4
	New Delhi	28.6N	77.2E	21.0	4.6	347.7
	Bangalore	13.0N	77.6E	26.3	5.5	343.7
VU	Andaman Is., Pt. Blair	11.7N	92.8E	8.9	346.3	326.3
VU	Laccadive Is.	10.0N	73.0E	32.7	11.8	348.5
XE	Mexico, Mexico City (XE1)	19.4N	99.1W	224.1	183.3	139.9
	Chihuahua (XE2)	28.7N	106.0W	250.1	218.0	140.9
	Merida (XE3)	21.0N	89.7W	206.4	154.8	122.5
XF4	Revilla Gigedo	19.0N	111.5W	241.4	215.5	168.2
XT	Brkina Faso, Ogadougou	12.4N	1.6W	87.7	76.3	64.5
XU	Cambodia, Phnom Penh	11.7N	104.8E	354.0	332.1	314.6
XW	Laos, Viangchan	18.0N	102.6E	357.1	337.1	320.3
XX9	Macao	22.2N	113.6E	345.9	328.2	313.8
XZ	Myanmar, Yangon	16.8N	96.0E	4.6	343.8	325.7
YA	Afghanistan, Kandahar	31.0N	65.8E	29.5	14.7	358.4
	Kabul	34.4N	69.2E	25.4	11.0	355.6
YB-YD	Indonesia, Jakarta	6.2S	106.8E	348.0	318.5	299.8
	Medan	3.6N	98.7E	1.9	335.8	315.4
	Pontianak	0.0	109.3E	345.7	320.0	302.5
	Jayapura	2.6S	140.7E	304.8	288.4	276.6
YI	Iraq, Baghdad	33.0N	44.5E	43.7	31.3	16.8
YJ	Vanuatu, Port Vila	17.7S	168.3E	269.6	258.0	247.4
YK	Syria, Damascus	33.5N	36.3E	48.5	36.9	23.4
YL	Latvia, Riga	57.0N	24.1E	36.0	28.8	21.0
YN	Nicaragua, Managua	12.0N	86.0W	192.4	154.4	128.2
YO	Romania, Bucharest	44.4N	26.1E	46.0	36.6	26.0
YS	El Salvador, San Salvdor	13.7N	89.2W	199.7	159.8	130.6
YU	Yugoslavia, Belgrade	44.9N	20.5E	48.2	39.2	29.2
YV	Venezuela, Caracas	10.5N	67.0W	155.1	126.5	110.3
YV0	Aves Is.	15.7N	63.7W	145.1	117.0	102.7
Z2	Zimbabwe, Harare	17.8S	31.0E	91.0	80.2	61.7
Z3	Macdnia, (ex Yugoslav), Skpje	42.0N	21.4E	50.2	40.7	30.2
ZA	Albania, Tirane	41.3N	19.8E	51.6	42.1	31.6
ZB2	Gibraltar	36.1N	5.4W	68.1	58.9	50.0
ZC4	British Cyprus	34.6N	33.0E	49.7	38.5	25.5
ZD7	St. Helena	16.0S	5.9W	112.4	101.5	89.9
ZD8	Ascension Is.	8.0S	14.4W	112.3	100.2	88.7
ZD9	Tristan da Cunha	37.1S	12.3W	131.7	122.8	113.7
ZF	Cayman Is.	19.5N	81.2W	183.2	137.0	114.2
ZK1	No. Cook Is., Manihiki	10.4S	161.0W	256.3	244.5	229.9
ZK1	S. Cook Is., Rarotonga	21.2S	159.8W	247.3	235.9	222.3
ZK2	Niue	19.0S	168.9W	254.6	243.4	230.8
ZK3	Tokelaus, Atafu	8.4S	172.7W	265.3	253.9	240.9
ZL	New Zland, Acklnd (ZL1)	36.9S	174.8E	247.2	238.1	229.3
	Wellington (ZL2)	41.3S	174.8E	242.4	234.1	226.0
	Christchurch (ZL3)	43.5S	172.6E	240.7	233.0	225.5
	Dunedin (ZL4)	45.9S	170.5E	238.6	231.4	224.6
ZL5	Antarctica, Scott Base	77.9S	166.4E	195.4	195.4	194.7
ZL7	Chatham Is.	44.0S	176.5W	236.1	228.0	219.5
ZL8	Kermadec Is.	29.3S	177.9W	251.0	240.8	230.3
ZL9	Acklnd & Cmpbll Is., Acklnd	50.7S	166.5E	233.5	227.8	222.2
	Campbell Is.	52.5S	169.1E	230.5	225.1	219.6
ZP	Paraguay, Asuncion	25.3S	57.7W	158.4	142.5	129.4
ZS	S. Africa, Cpe Twn (ZS1)	33.9S	18.4E	113.0	105.9	95.2
	Port Elizabeth (ZS2)	34.0S	25.7E	109.6	102.9	91.1
	Bloemfontein (ZS4)	29.2S	26.1E	104.6	96.6	82.7
	Durban (ZS5)	29.9S	30.9E	102.7	94.9	79.8
	Johannesburg (ZS6)	26.2S	28.1E	100.6	91.8	76.2
ZS8	Prnce Edwrd & Mrn Is	46.8S	37.8E	120.3	119.7	118.4

CHAPTER 5

DXing

BILL KENNAMER, K5FUV
ARRL HEADQUARTERS

D X Is...the essence of Amateur Radio. If it were not for the desire to send a signal over the hill, then over the water, then across the oceans, and around the world, Amateur Radio, or even wireless itself would probably have been dismissed as a useless laboratory phenomenon. But that first DXer, Guglielmo Marconi, put the signals over the hill, across the oceans, and around the world, and the world hasn't been the same since. It isn't an overstatement to say that worldwide communication owes everything to DXing, and to Marconi, the original DXer and DXpeditioner.

The history of DXing is long and varied. Of course it starts with those early efforts of Marconi, who very early turned to commercial development, and continued with amateurs who continued to work stations farther and farther away. But the real dream was to bridge the oceans. The ARRL had a part in that dream. At the Board of Directors meeting at the first National Convention in Chicago, 1921, then Traffic Manager Fred Schnell presented a plan that would give the best possible chance for radio signals being heard across the Atlantic. For the transatlantic receiving test scheduled in the late fall, Paul Godley, 2XE, considered to be probably the foremost receiving expert in the United States at the time, and a member of the ARRL Technical Committee, was dispatched across the Atlantic with his best receiving equipment. Setting up in a tent on the coast of Scotland, Godley began his tests. By December 7, 1921, he was ready. Tuning across the bands, he began to hear a 270-meter spark. The operator's call sign, 1AAW, was clearly heard! (But 1AAW turned out to be a pirate! Even at the beginning pirates were one of the hazards of DXing.) The signals of more than 30 American amateurs were heard during this series of tests, waiting only for a two-way crossing to be completed.

Even then, the experimenting and modernization spurred by a desire for better DX performance was apparent. Godley reported that of the signals heard, over 60% had used CW rather than spark, and with less power. After these tests, the death knell for spark had been sounded, and the future of tube transmission was ensured. All because of the desire for DX.

In late 1921 and early 1922, Clifford Dow, 6ZAC, who was located in Hawaii, heard signals from the western United States. He announced this in a letter to *QST*, and said if he could get some transmitting equipment, he believed he could make it across the Pacific. A group from the West Coast sent him the needed transmitter, and two-way contact was established on April 13, 1922, between Dow and 6ZQ and 6ZAF in California. Thus was yet another tradition of DXing established, that of providing equipment to activate a "new one."

Léon Deloy, 8AB, of France, had participated in the transatlantic tests of early 1923. His signal had been one of several heard by US amateurs, but no two-way communication resulted. During the summer, he studied American receiving methods, even came to Chicago for the National Convention. He returned to France with American receiving equipment, determined to be the first to make the transatlantic crossing.

After setting up and testing, by November Deloy was ready. He cabled ARRL Traffic Manager Schnell that he would transmit on 100 meters from 9 to 10 PM, beginning November 25. He was easily heard. By November 27, Schnell had secured permission for use of the 100-meter wavelength at 1MO and the station of John Reinartz, 1XAM. The two of them waited for Deloy.

For an hour Deloy called and sent messages. Then he signed. The first DX pileup began as Schnell and Reinartz both called. Deloy asked Reinartz to stand by, and worked first Schnell and then Reinartz for the first transatlantic QSOs. The age of DXing had finally truly begun.

From these small beginnings, to the later exploits of such as Bill Huntoon, Danny Wiel, Gus Browning, and the latter day exploits of Martti Laine, and such as the VK0IR group, DXing has grown. While the early days were marked by exploration and discovery, we now recognize that we have the equipment, power and antennas to hear a pin drop on the other side of the world and respond to it. While some discovery is still involved, and that mostly on the VHF/UHF bands, for most of us DXing takes the form of Radiosport at its Zenith, a ritual of great importance to some, and a source of friendship and occasional enjoyment to others. DX IS. . .no doubt about it, the essence of Amateur Radio.

WHAT IS THIS THING CALLED DX?

In the early days of radio, DX became the acronym for "distance." At the time, DX could well have been determined by a new town, a new county, or even a new state. With the passage of time, DXers became more proficient, and DX is now generally accepted, at least at HF, as being contacts outside your own country. VHFers will consider it to be a new and distant grid square, while the microwave DXer will consider his DX in terms of miles. Our discussion will be primarily concerned with HF DXing.

A DXer most often pursues his hobby by chasing DX contacts that will bring him new credits for one of the popular award programs that he might be chasing. This may be an award such as the ARRL DX Century Club (DXCC), *CQ Magazine's* Worked All Zones Award (WAZ), or the Radio Society of Great Britain's Islands on the Air program (IOTA). Each of these programs has a different objective. All, however, have in common the pursuit of DX, no matter the form it may take. A further discussion of US awards is in the Operating Awards chapter of this book.

Almost every ham has a bit of DXer ingrained in his system. It's only human nature to want to see how far away your equipment will work. The VHF operator may want to see how far away he can be from the repeater and still make it back with an HT. Listen to a repeater pileup on a tropo opening someday. You'll quickly see that the DX spirit exists in all of us. However, those who feel a stronger pull, a desire to achieve more, will soon hear the siren's call of DXing, and all manner of new transceivers, amplifiers, and antennas will appear. In some, the bug has lain dormant for many years, only to break though with a sudden, passionate surge that takes hold of life's direction itself, and guides him to new altitudes in pursuit of his own particular dream.

In this chapter, we will look at the various stages a DXer may go through, and see how his operating techniques, equipment and need for knowledge varies. We will see what resources are needed at each stage, and will explain several of the different facets of the DX hobby.

DXing 101: FROM THE BEGINNING

Few individuals start out in Amateur Radio just to DX. Many were interested in just getting on the air, and suddenly bumped into a DX station. The fascination began at that point,

and grew over time. Yet it is likely to be the beginning DXer who enjoys the hobby more. There's a lot to be said for just beginning, not knowing all the tricks, not knowing what to expect. The simple fact is that an experienced DXer is what he is because of one thing. . . experience! Experience doesn't come overnight, but through both the passage of time and by practicing. The beginning DXer certainly can get some practice, because everything is new. By the time he's completed his DXCC, he's learned a lot, and will be ready to go forward. And, while an experienced operator with a good station might work DXCC in a weekend, it's likely to take even a highly motivated beginner a year or more. So, let's examine a few things that should be developed over that first year.

Equipment

What to start with? Actually the beginning DXer can start with almost anything. Having said that, there are a few things the beginning DXer *shouldn't* do. One of them is QRPp. Although some may have had some success with it in the past, the fact is that the beginning DXer doesn't need the additional frustrations that being inexperienced and operating with 5 watts will bring. So the first worthwhile piece of equipment for the DXer is a rig with at least 100 watts output. Rig selection is a complex subject, best left for a much longer discussion. In general, here are some suggested features for a starting point:

- 100 watts minimum output.
- Adequate receiver design. The better you can afford, the better you can perform. The old adage "You can't work 'em if you can't hear 'em" applies almost as much to rig selection as to hearing ability.
- Adequate filters. Narrow filters are needed for CW and SSB—a minimum of 500 Hz for CW, while 2.4 kHz is a desirable figure for SSB. These should be mechanical or crystal filters. Generally, outboard audio filters are not adequate for DXing on a crowed band.
- A good set of headphones. Usually this means headphones with a response between 200 to 3500 cycles. Remember, we're going DXing, not listening to hi-fi.
- A second VFO or separate receiver. These will be needed if you should encounter any split frequency pileups.

With this type equipment, worldwide DX can be worked. In fact, many of the DX stations the DXer will work over his career are using nothing more than this minimum list themselves. It is how the equipment is set up and operated that will make the difference.

There are some tools the DXer should have in his possession before beginning. One is *The ARRL DXCC Countries List* publication. Most of the information found there is also contained in the back of this publication. Another important tool is a good world map, such as the *ARRL World Map*. Mount it on the wall within sight of the operating position. It's a good idea to be able to see where countries are in relationship to one another. For example, if you wanted to work Japan, it helps to know that it's around the world from Europe, so that if you're hearing many loud European signals on a band, you're not as likely to hear Japanese stations coming from the same direction at the same time. Over time,

The birthplace of radio and DXing, Sasso Marconi, the home of Guglielmo Marconi, is now the home of the Guglielmo Marconi Foundation. IY4FGM, the Foundation's station still provides DX for this, the original DX QTH.

you'll learn how to use this to your advantage. For example, experience will tell you what time and band it would be expected to hear the Philippines (DU). So, if a DXpedition to very rare Scarborough Reef (BS7H) was known to be on the air, it could be expected that it could be found at the same time and frequency band as the Philippines, since it's only about 150 miles from Subic Bay. To use this for any of the rare countries, find another more common one that's very close. The times and frequencies won't vary that much, and you'll be prepared.

The ARRL DXCC Countries List contains not only the list itself, but also a list of the ITU call-sign allocation series. Between the two lists, a DXer should be able to identify any call sign heard on the air. In fact, the DXer should begin early on memorization of the entire DXCC list, and the ITU prefix allocation table. This should be coupled with memorizing the short path beam headings as well. The DXer can then identify any call sign and be prepared to point a directional antenna.

Antennas for DXing are many, and varied. While most anything will work somewhere at some time, the secret to DXing is making sure that whatever antenna system is in use is installed properly. This means using the right feed line, having the connections properly prepared, and simple things such as making sure the coaxial cable and connectors are dry. If you are a new ham, or new to the HF bands, you might want to get a copy of *Your Ham Antenna Companion*, published by the ARRL. It will give you an overview of HF antennas and some suggestions for antennas you might build yourself. Nothing quite matches the thrill of working DX with a home-brewed antenna!

One of the first antennas a DXer might use is a dipole. Properly cut and installed, a dipole is a suitable antenna for DXing on any band. A minimum of 35 to 40 feet would be a good height for 10 to 30 meters. Of course, for most installations the higher, the better. The dipole may be mounted as an Inverted V, but would generally perform better for DX work with both ends at the same height. The dipole may also be mounted in a sloping position, with one end much higher than the other. This is often done for DXing on 40, 80, and 160 meters. There are other configurations for the dipole, These may be found in the ARRL's *Low-Band DXing* by Devoldere, ON4UN. The configurations are as valid for the higher frequencies as the low bands. All things considered, a full size dipole will usually outperform a trapped single element antenna, especially if that single element antenna isn't properly installed.

Vertical antennas are also suitable for DXing. They can provide the lower takeoff angles that are often superior for DXing. However, installation is more critical for the vertical than the horizontal. Radials should be used with any vertical. If ground mounted, at least 40 should be used, and the more the merrier. If elevated, it is best to provide at least four full size radials for each desired band. This will provide increased performance over an installation without radials. Excellent references for installation of radial systems for verticals may be found in *The Antenna Compendium, Vol. 5* and *The Antenna Anthology*.

If the space is available, a small tribander is also a good antenna system. Properly mounted, it will provide superior performance over a unity gain antenna system.

Listening

Remember this if you remember nothing else about DXing: Listening carefully is the best thing one can possibly do to improve his DXing ability. It's sometimes hard to remember that, with the excitement of the chase, but over a period of time, the DXer will realize this is very true. It's the difference between having, for example, 10 years of experience, or having one year's experience 10 times.

As an example, suppose it is morning, during a period when the sunspots are low to moderate. For these conditions, the band of choice would be 20 meters. Begin on SSB, near the bottom of the band. You do have your headphones on don't you? Remember now, we're just listening, so it doesn't matter where we are. Of course when transmitting, it will be necessary to be in the part of the band equivalent to your license class. Tune slowly up the band. Listen to the QSOs going. If the signals are steady, and the accent sounds like your own countryman, pass it up now, and move slowly up the band. A signal is heard speaking in accented English. Stop here and listen. The signal may be strong and steady, or it may be strong but occasionally dipping in strength. This may well be a DX station. Stick around long enough for him to ID, or otherwise give a clue as to his location. See what style of operating he may be using. If he's in a QSO, mark the frequency and his call down in your notebook. If it is a European, then you know the band is open in that direction. But if he's in QSO, don't call him until he is definitely finished. To do otherwise would be rude, and usually only LIDs are rude. You can look at your notebook in a few minutes, and check back to see if he's still there.

Keep moving up slowly. Be sure to pay attention to weak signals. Many times they are passed over because people either can't hear them through their speakers (but you're wearing headphones, remember?). More than once a rare DX station has appeared on the band, called a few CQs with no response, and gone away. With 100 watts and a simple antenna, it's to your advantage to try to find stations before anyone else is calling and the pileup starts. It's likely

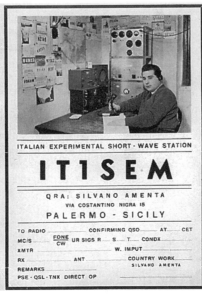

Metairie, Louisiana, DXer Silvano Amenta, KB5GL, brought his love of DXing with him from Sicily, where he began DXing as IT1SEM. (*Photo courtesy of KB5GL*)

Pirates and Policemen, LIDs and Jammers

The DXer does encounter a few problems in the pursuit of his avocation. Those nefarious denizens above are not there to make life easier for the DXer. However, all DXers will ultimately have in common the fact that they have overcome and persevered through the jungle created by the antics of these creatures of the ether. These cretinous individuals should not diminish the pleasure of the DXer, but allow him to puff out his chest with pride, as he overcomes the obstacles put into his path by such miscreants.

Pirates have been with us always. We can't be sure if Marconi heard one when he first fired up, but certainly the first signal received across the Atlantic by Godley was one. Probably the most famous one was named Slim. At a time when a volcanic island had popped up out of the sea near Iceland, Slim turned up from Cray Island as 8X8AA, claiming that the island had just popped up in the North Atlantic, and would qualify for a new country. He held forth for several days, then disappeared forever. Since then, many pirates over the years have been tagged as Slim. So, if an operation comes onto the air, and it's so improbable as to be suspect, it may be Slim, back for another run.

One thing the DXer must do when encountering a suspected pirate: *Work Him!* Yes, always work him, because sometimes the improbable is true. The rule is *Work 'em First, Worry Later (WFWL)*. Two things are accomplished by doing so: first, the DXer can practice his pileup technique, and second, if it is for real, it's in the log. If it's suspect, the QSL need not be sent until later. But if it's not in the log, it's hard to get a QSL. Some pirates and bootleggers do QSL, so it's not unusual to get cards rejected for awards credit. Everybody does, so it's best to continue working stations until one sticks.

Every pileup will have Policemen and Jammers. The Policemen may be well meaning souls, but they really are in the same category as the Jammers. They just do it in a less sophisticated way. The DX is calling, working a few hundred an hour by split frequency. Meanwhile, on his transmit frequency, the LIDs, Policemen, and Jammers may be all heard at once, each pursuing his own route to DX infamy. The LID will start with "Who's the DX?", "Where's he listening?" or the inevitable, "Is the frequency in use?" All these questions could be answered by listening a little, but either the LID's time is too valuable, or his intelligence level is too low to think of that.

This is, of course, an excuse for the Policemen to jump in, with varying responses. Ten Policemen will, one after the other (never in unison) give them all the information anyone would care to know about the DX station, meanwhile totally obliterating the DX station and stopping the pileup in its tracks. This is immediately followed by 10 more who attack the LID, questioning his parentage, or alluding to his intelligence level. Finally come the last 10, who are telling the previous ones and the LID to just shut up. Ah yes, the musical chaos of the pileup!

Did we say music? That must be the cue for the Jammer to appear, as music is one form of jamming. The Laughing Box is also good, although not in much use in recent years. Finding someone calling in the pileup and recording it in the DVP and replaying it on the DX frequency is also a frequently used technique. Sometimes the call of their favorite DX net control is used this way. Most Jammers these days are not that sophisticated. Now it's mostly just tuning up or calling CQ on the DX frequency, or running a couple of minutes of unsquelched 2 meters onto the air. One of the easiest ways for the Jammer to enjoy himself is to just ask who the DX is, and get the Policemen started.

Meanwhile, the DXer can rejoice in the fact that, although all these obstacles are placed in the way, he's still working the DX! So let the others moan and groan, let them complain to the utmost. The fact remains, those who learn their skills well will succeed.

---

that the DX station may be running 100 watts to a dipole, sometimes even an indoor dipole. If you can hear him, he can most likely hear you too, especially if no one else is calling.

The important thing while you're tuning is to listen to as many different DX stations as possible. Notice the sound, the accents, audio quality. With a little practice, you can tune the band quickly, and immediately identify the stations that are DX. Later, you will be able to do the same thing on CW.

Operating

Now it's time to begin calling some of the DX stations you've been hearing. You hear the DX station sign. You know his call. Now call him. If it's a station who has been working short QSOs, give his call one time only (he *knows* his call!), and yours twice, using standard phonetics. Then wait. If you're lucky, he'll come back to you! If he comes back to someone else, wait patiently until he's finished. DXing is a game of patience. But turn off your VOX while waiting, Tripping your VOX while waiting on the frequency is the behavior of a LID.

You're lucky!! He's come back to you! You're now in QSO with the DX station. The first QSO with a DX station can be somewhat like dancing. He leads, you follow. As a rule, on your first transmission work by a formula. State your name, state, and signal report. A good form would be, "My name is Bill, in the state of Connecticut. Your report is 5 by 9, over." The DX station's English may be limited, he may be working many stations, or maybe is waiting on frequency for a schedule. In any case, what he does on his next transmission will determine whether he will say 73, or want to have an hour's discussion of the state of the fish in the river in your state. Just follow his lead and enjoy the QSO, no matter how long or short.

Sylvie, JP1LAB, operates 3D2LA from Fiji. Sylvie and OM Mike, JH1KRC, often take DX vacations in the Pacific. (*Photo courtesy of JH1KRC*)

What to Look For

Your receiver is a critical element in being an effective DXer. The following description of receiver characteristics was written by ARRL Lab Test Engineer Mike Gruber, W1DG. Mike performs the tests for each piece of equipment reviewed in *QST*. This material was taken from *QST Product Reviews: A Look Behind the Scenes*, in the October 1994 issue of *QST*, page 35. A full description of receiver performance tests is included in recent editions of *The ARRL Handbook For Radio Amateurs*, Chapter 26.

The ARRL Technical information Service (TIS) offers a package titled *RIG*–a set of reprints from *QST*, answering the question: "Which rig should I buy?" See the back of this book for information on receiving TIS packages. *The ARRL Radio Buyer's Sourcebooks* are another good source of information on ham gear.

CW and SSB Sensitivity:

One of the most common sensitivity measurements you'll find for CW and SSB receivers is Minimum Discernible Signal (MDS). It indicates the minimum discernible signal that can be detected with the receiver (although an experienced operator can often copy somewhat weaker signals). MDS is the input level to the receiver that produces an output signal equal to the internally generated receiver noise. Hence, MDS is sometimes referred to as the receiver's "noise floor."

You'll find MDS expressed in most spec sheets as µV or dBm. The lower the number in µV, or more negative the number in dBm, the more sensitive the receiver. (For example, a radio with an MDS of −139 dBm (0.022 µV) is more sensitive than one with an MDS of −132 dBm (0.055 µV).)

When making sensitivity comparisons between radios, keep in mind that more is not always better. The band noise (not the receiver noise) floor often sets the practical limit. Once you've reached that point, greater sensitivity simply amplifies band noise. Also, too much sensitivity may make the receiver more susceptible to overload and decrease dynamic range.

A typical modern HF transceiver has an MDS of between −135 and −140 dBm (or 0.0398 to 0.0224 µV). You can easily make a dB comparison between two radios if the MDS is expressed in dBm. Simply subtract one MDS from the other. If, for example, one has an MDS of −132 and the other −139 dBm, the later radio has a sensitivity that is 7 dB better than the first–if both measurements are made at the same receive bandwidth.

Dynamic Range

Dynamic range is a measure of the receiver's ability to tolerate strong signals outside of its band-pass range. Essentially, it's the difference between the weakest signal a receiver can hear, and the strongest signal a receiver can accommodate without noticeable degradation in performance. Two types are considered in *QST* product reviews: blocking dynamic range (blocking DR) and third order intermodulation distortion dynamic range (IMD DR).

Blocking dynamic range describes a receiver's ability to maintain sensitivity (or not to become desensitized) while tuned to a desired signal due to a strong undesired signal on a different frequency. IMD dynamic range, on the other hand, is an indication of a receiver's ability to not generate false signals as a result of the two strong signals on different frequencies outside the receiver's passband. Both types of dynamic range are normally expressed in dB relative to the noise floor.

Meaningful dynamic range comparisons can only be made at equal frequency spacings. Also, as in the case of our hypothetical receiver, the IMD DR is usually 20 dB or more below the Blocking DR. This means false signals will usually appear well before sensitivity is significantly decreased. It's not surprising that IMD DR is often considered to be one of the more significant receiver specifications. It's generally a conservative evaluation for other effects that may or not be specified.

Third Order Intercept Point

Another parameter used to quantify receiver performance in *QST* product reviews is the third-order intercept point (IP3). This is the extrapolated point at which the desired response and the third-order IMD response intersect. Greater IP3 indicates better receiver performance.

Second Order IMD Dynamic Range and Intercept Point

We've recently added a new test to the battery–second-order IMD distortion dynamic range. These products, like third-order products, also are generated within a receiver. Their relationship to the offending signals that cause them is f1 +/- f2. In today's busy electromagnetic environment, they can become truly offensive under certain conditions. Consider the case of two strong shortwave stations, on two different bands, that sum to the frequency of the weak DX you're trying to copy. If their intermod product is strong enough, you may not be able to copy the station through the interference. As an example, a high seas coast telephone station on 4400 kHz might sum with a short wave broadcast station on 9800 kHz to produce and intermod product on 14200 kHz.

IF and Image Rejection

A station at a receiver's IF frequency can be truly offensive if its signal is strong enough and the radio's IF rejection is insufficient. The station appears across a radio's entire receiving range! Be sure to consider IF rejection when considering a receiver and a nearby transmitter is at its IF frequency.

We're now testing both image and IF rejection in the Lab. We first measure the level that causes the unwanted signal to be at the MDS. Then, we reference this level to the MDS in order to express the measurement in dB.

Yes, you'll want to QSL. We'll talk about that later.

Many times you'll hear DX stations who are running stations at a fairly high rate of speed. They'll just answer with a call sign and signal report, then go on to the next station. When calling a running station, sign your call sign *one time*, using phonetics. If you're beaten in the pileup, wait until the next over, and call again. Don't ask the station for his QSL information in this type of pileup. If it's a European who has a QSL bureau in his country, he will most likely answer cards by the bureau. If he's somewhat rare, he will announce his QSL information on frequency, or it may be found from several sources off the air.

For the beginning DXer (or any DXer), contests are a very good way to pad your DX score. Contacts are often easier due to the number of pileups on the bands. There are a great many stations on the air, so it is often easy to work many stations from a given country. The more stations you work, the more QSLs you will receive, and the best way to QSL many of these contest operators is via the bureau. Either they are happy to QSL as the price they pay for your contest QSO, or they wouldn't QSL anyway, direct or otherwise. So go for the quantity.

THE INTERMEDIATE DXer

The pure pleasure of DXing may be best enjoyed by those who have worked a DXCC level of between 100 to 250. It is at

Massimo, IK1GPG, and Betty, IK1QFM, at their beautiful station in Mondovì. (*Photo courtesy of IK1GPG*)

this point that The DXer has become confident in what he is doing, has developed operating proficiency, and can still find plenty of exciting DX to chase. This Golden Time should best be enjoyed to the utmost, for once it passes, it cannot be reclaimed.

Equipment

While some of the requirements for equipment are the same, some areas may be upgraded for even better results.

- The transceiver can remain the 100 watt unit, but may be upgraded to one with a few more features. Cascaded filters are more desirable, and DSP, either outboard or inboard, has some advantages. Nothing like that DSP autonotch to remove those heterodynes–but be prepared to spend many hours learning to use it quickly and correctly!
- The headset is still essential, but now with an added boom microphone. It's nice to have both hands free for something else, like moving the antennas while sending your call on CW.
- A footswitch becomes a very handy thing to have, too. Not only are your hands left free, but there's no sneezing and tripping your VOX just as the 7O is coming back to you!
- A good keyer is nice, because the DXer has to be versatile. If the needed DX works only CW, the DXer must either work CW or maybe wait another five to ten years for him to come up again.
- An amplifier is a good purchase at this level. The rarer the DX, the deeper the pileups. With an amplifier, there's a reasonable chance of getting through. Without, better wait until the fourth or fifth day to begin calling.
- Desirable before, the second receiver or VFO is now essential. You will need to be capable of operating split frequencies most of the time.

By now, the DXer would have considered some sort of tower and beam antenna. This is a great investment, one of the best the DXer could ever make. A small tribander at 40 to 50 feet will provide a noticeable increase in results on both transmit and receive. Wires may still be used for the low bands, or the tower itself may be fed as a vertical.

Listening

Listening will always be the most important operating strategy the DXer can practice. The DXer has already learned what a DX station sounds like. The over the pole flutter is recognizable, as well as the QSB from a station far away. Weak signals stand out from the crowd for the DXer, who sees this as the sign of rare DX on the band. Now it's time to progress to listening for more than just the DX itself. It's time to listen to the pileup, and capture the dynamics in memory, put back for later use.

There are two kinds of pileups, transceive and split frequency. These are exactly what their names imply— transceive on the same frequency, and split by using one frequency for transmitting and another for listening. Both are easily manageable if proper technique is used. So, it's important to see how other operators apply that technique.

Tuning the band, a DX station is heard passing out rapid fire QSOs, signal reports only. First, as a rule, listen for a moment or two to see who it is. If needed, great, it's time to call. If not, stick around for a minute to listen. The pileup builds, almost to the point of getting unmanageable. Notice who's getting through and who's not. The guys transmitting their last two letters on phone seem to be having a more difficult time. That's because many DX stations these days won't work anyone sending the last two as long as they can hear a full call sign. So, even if they have a good signal, the guys doing that are spinning their wheels to some degree. Full calls are always best. On CW this isn't much of a problem. . . .yet. But a few have been heard in a CW pileup, too.

Notice that occasionally you will hear a station signing his call sign while extraneous information is being given to the DX station. This may be something like the second station signing his call on top of the signal report. This is a technique called *tailending*. Although it is a good technique, one needs to listen to a few good practitioners of the art, for it is an art, before trying it. Otherwise it can blow up in your face. Notice that often the calling station is on a slightly different frequency, and different speed on CW. Sometimes he's weaker. Some practitioners of the art actually turn off their amplifier before trying this. Doing so avoids being obnoxious. Notice also that if the station trying the technique tries it once or twice and it doesn't work, it isn't tried again. To continue could annoy the DX station, and he may be deliberately not taking tailenders when they call, but one or two calls later just to keep everyone from trying it at once. A pileup can quickly turn into bedlam if two or three tailenders are taken in a row.

Continue listening to the pileup, and see where callers are positioning themselves in the pileup. On CW, they may have their carrier up a hundred cycles or two from where the last caller was to make their call stand out. Sometimes slower speed is better if everyone else is fast. This also works somewhat on Phone. By varying the RIT, it's possible to make the voice on SSB take on a different pitch that may cut through the pileup.

Tuning up the band, somewhere around the "usual DX frequencies," a signal is heard, again passing out rapid signal reports. This time, however, no one is heard coming back to him, yet the signal reports keep on coming. He is obviously working DXers on a split frequency. Now is when that second receiver pays for itself. Split frequency operation can be, and is, done most frequently with two VFOs, but can be even more effective when done with two receivers. On CW, start looking for the stations calling about one kHz up. Keep tuning up until the calling stations are found. The pileup may be as much as 5 to 10 kHz away. Then begin to look for the calling stations. Start tracking the stations that are working the DX station to see how far, and in which direction, he's moving between QSOs. See if he's taking tailenders (easier to do split than transceive). Do the same thing on SSB if it's a Phone pileup. *Listen carefully to what the DX station is saying. If you can't hear, wait until you can.* Surprisingly, when these pileups get really loud and raunchy, the DX station with often call out a specific spot frequency where no one is calling. Those who are listening will catch it. Those who don't will continue to call in vain. Look also for operators who operate at or below the edge of the announced calling frequency range. Sometimes the DX stations will announce something like "200 to 210." At that moment, he may have moved down to 200 to start tuning up again. Often a caller at 199.5 can get the QSO because of the difference in pitch, or because his sidebands make it difficult to hear on 200. This only works once. If he moves away from 200, then it's back to square one. Find and follow, then get ahead.

Have you heard the guy at the DX club who comes into the discussion on the latest big DXpedition? "Yep, worked him with one call," he says. You'll know that he may have spent half an hour or so using just the techniques above to set up that one call.

Information Systems

The DXer thrives on information. When is the next DXpedition? Who are the operators? Where are they going? All these questions can be answered from information sources available to the DXer, either by print, PacketCluster, or over the Internet. Frequently, in a pileup you might hear someone asking who the DX is or how to QSL. If that DXer had proper information sources, he would know without asking, and wouldn't be adding QRM to a pileup. Gathering information is as important to a DXer as any other facet of operating, because knowledge really is power.

The weekly DX newsletter, such as *QRZ DX* or other publications of that type, have been around for almost as long as there have been DXers. While some say their time is past, actually it isn't. They provide a hard copy record of events upcoming as well as events past, a record of QSL routes and corrections, lists of stations worked (very important if you're looking for specific resident DX stations), and provide this all in short factual tidbits. These publications are best used for upcoming events that have been announced well in advance. They also have the ability to provide chronicles of ongoing events, especially with resident operator situations.

Most recently, PacketCluster has been added as a tool for the DXer. In fact, it has been said that many owe their Honor Roll plaques to the PacketCluster network. Certainly it's likely that, coupled with spot filtering software that call the operator to the radio if a needed one comes in, it's made TV advertisers get more for their money!

The PacketCluster network now stretches around the world. Through the use of Internet connections, it's possible to see what's being worked in Italy or Japan. This is actually more useful than one may think. First, it's possible to know that stations from certain areas are actually on the air. It's possible to identify these stations, and even write for schedules. There have been some who have found that spots from another area of the country actually can be used. It's sometimes possible to hear spotted stations where previously no one would have even listened. And it has tipped band openings not thought possible

DXpeditioners often travel to rare and remote islands, such as Mellish Reef, where the crew of VK9MM provided DXexcitement in 1993. (*Photo courtesy of V73C*)

before. Real time, PacketCluster is probably the most superior method of intelligence gathering for ongoing operations currently in use. It is a truly amazing system, and the hybrid radio/Internet system is one of the best forms of useful ham ingenuity.

The Internet has some capabilities for information transfer as well. However, one really has to separate the wheat from the chaff with the various reflectors and newsgroups available. There is a lot of information, but a greater amount of misinformation, available. This is because everyone with a keyboard and a connection can put in their opinion on anything. It's best used for specific information posted to the various reflectors by those who are involved with DXpeditions, or by resident stations. It's been excellent for stations such as XV7SW, XU6WV, or 9X4WW, who have often provided information on what they're doing or what their times and conditions are on 160 meters. *The DX Reflector*, *The Top Band Reflector*, and others provide much in the way of information. See Chapter 6 of this book, The Internet, for more information on electronic newsletters and reflectors.

Propagation

Here we can start by breaking one of the paradigms of DXing: Whatever the propagation is, *It Doesn't Matter!* That's right, it doesn't matter at all. Because if the DX station comes on the air on his chosen schedule, if the contest weekend falls at the worst propagation time of the year at the bottom of the sunspot cycle, the DXer still has to be there to work it! So rather than learning about corona holes and solar winds, it's better to concentrate on what you need to know about practical uses of propagation no matter what the solar conditions are. The basic difference between high sunspots and low ones is simply the choice of bands in use. The higher the spots, the higher the band.

So let's start a typical day on the bands in fall and winter. At sunrise, you have many choices. From the US, you can beam east on the high bands, or work west on the lower bands. In general, your antenna can follow the sun, starting southeast, then northeast, and by noon down to the south. In the early afternoon, look either east or southwest, moving up to northwest by late afternoon to early evening. From noon on, the southerly propagation should always be there, even into the late evening.

Another type of propagation is the *long path*, where propagation comes from an almost opposite direction. For example, in the morning especially in fall and spring, it's possible to work African stations by beaming west rather than east. The path seems to fall near the terminator (the area where darkness and light meet), and can sometime be more effective than the short path. Long path may even be used on 10 meters during times of high sunspot activity, with morning openings to Asia from the southeast.

For the low bands, follow the darkness as you would follow the sun. Look toward the east to southeast at sunset, the south anytime after dusk, and the west from midnight on. Long path also works here, as Asia and Australia may be heard on the East Coast at sunset, and Europe on the West Coast in the mornings.

The 80,000+ QSOs made by the VK0IR team from Heard Island in the winter of 1997 merely illustrates the principle that when the DX is out there, it will be worked, regardless of propagation. It's just a matter of band selection.

ADVANCED DXing

The Advanced DXer will be nearing the DXCC 300 level. By now, needed "New Ones" may be coming largely from DXpeditions, as only perhaps one or two will still be available from resident operators. By now, the DXer may have run into an unusual phenomenon, wherein a Needed One that seems quite common to other DXing friends becomes almost impossible to find. The DXer may even have found one or two others on his way up, countries where there are resident operators that seem to elude him. Sometimes his friends report working two or three in a contest, or sometimes he happens to wander onto the frequency of one who suddenly goes QRT before he can even give a call. One day the spell will be broken, however, and suddenly the DXer will be deluged with contacts from the former Elusive One, as they become QRM.

The Advanced DXer will find he spends less time transmitting than listening. He will find himself sometimes jumping into large pileups just to maintain those all important pileup skills. Now is the time to develop the analytical skills to climb to the top, or develop new interests in news bands and modes.

Equipment

The Advanced DXer now can depend more on the skills he's developed rather than on the equipment. Still, equipment is important. It won't be so different from the Intermediate

Dudley Kaye-Eddie, Z22JE, relaxing at his shack in Harare, Zimbabwe. (*Photo courtesy of Z22JE*)

Gene Neill, ZA5B, operates from his shack in Shkodra, Albania. Gene, whose US call is WA7NPP, previously operated as TL8NG from the Central African Republic. (*Photo courtesy of WA1ECA*)

DXer's rig, and in fact there are some top DXers who are reluctant to change from a tried and true old rig because they don't want to miss an opportunity. However, the top of the line transceivers now are offering dual receive, which is almost a license to steal in a split frequency pileup. A receiver with a strong front end, able to stand nearby strong signals without creating internal distortion products is highly desirable, as he will be in pileups with everyone. After all, what he needs is most likely needed by everyone in the entire world, and the pileups will usually be deepest at this level.

If not using a stereo headset already, the DXer should add one with a boom mike. This makes it easier to have the DX in one ear, the pileup in the other while using two receivers.

While it's possible to work DX without an amplifier at all, or one with only 500-1000 watts, a full power amplifier is desirable. While some may say it represents only a decibel, remember that a decibel is sometimes defined as the smallest difference one can hear. Thus, in a competitive situation, it can make a difference. However, there are some who could have twice the legal power and still wouldn't get through. Still, it's nice to have strength and skill.

Antennas may range from the ubiquitous tribander at 50 feet to 200 pounds of aluminum at 200 feet. Usually the bigger the better. A height preference would run to at least a minimum of one wavelength at 20 meters or higher, and at least a half wavelength or a full size vertical for the lower bands. Local ordinance or deed restrictions may preclude this. DX can still be worked, but it does become a little harder. The Truly Serious will most likely have a multi-element monobander on 20 meters, and perhaps a two-element 40-meter beam. There are more than a few DXers who have multiple towers and monoband beams. From a city lot, the DXer will have to have good skills to beat these setups with a good operator on even a semi-regular basis, but he should know they're out there, and those guys will win more than their share of the pileups. It's best to be one of those guys, but not totally necessary.

Listening

Listening will always be an important part of the DXer's strategy at any level. In fact, the more experience the DXer gets, the more his listening skills will truly develop. Experience also brings a different perception of what he hears. For example, weak, watery signals may bring the thought that he should try another band. Or, hearing an expected station from an unexpected direction may signal an unusual opening, and tuning the band may find more stations from that part of the world, but maybe an even more rare country. The rule that most people would rather work loud stations rather than weak ones always holds true, and sometimes tuning away from the big pileup may bring another station from the same country, but weaker, and with no pileup. For example, in 1992 the YA5MM operation from Afghanistan was very successful. Big signals at seemingly all times of the day or night, good operators, and most especially large pileups. Meanwhile, OK1IAI/YA held forth each evening near 14.036 MHz, with a weak but workable signal. Seldom was there any pileup at all. The listeners were finding and working this station, while many others were waiting on a DXpedition with a loud signal.

Listening at this level now also includes surveillance activities. The DXer has learned the players by now, so listening on the bands should include some conversations. You'll hear such as "Our friend from the East is planning to go south for the summer." Does this mean a major Japanese DXpedition to Bouvet? Perhaps, if the DXer knows who he's listening to, and who the speaker may have close contacts with. More on intelligence gathering later.

Listening also becomes important while in the dynamics of the pileup. Often the DX station will say or do something that will give you a clue as to what will happen next. For example, in 1991, XU1JA would suddenly disappear from 15-meter CW with no warning if the pileup got too large. A quick search of 10 or 20 meters right after that happened would net a QSO before the pileup got there and he returned to his original band. Many times the pileup is told that the operator is going QRT. Sometimes that's true. Sometimes it also means he's running to a different band or frequency for a schedule. The DXer hasn't anything to lose, so going to look can't hurt. Trying the higher end of the band may turn him up. To use this technique, the DXer *must* be able to identify his quarry by the sound of his signal, as call signs will be used as little as possible. Also, the DXer must avoid calling the DX station until the schedule is over, or risk the wrath of never appearing in the log.

The DX station quite often gives the instructions on his next band move over the air, yet the pileup never hears. The DXer who listens will. In short. For his entire DXing career, the DXer needs to be aware of everything that's going on around him on the air. He must hear everything, and process the information at high speed to make his decision about how to do the most important thing, which is to get into the DX station's log with a good contact.

Operating

By now, the DXer has experienced tailending. He's worked many successful contacts using split frequency techniques. Yet he will continually find new ways to pick his way through the pileups, for as one DXer's successful techniques are propagated to others, then that technique, for a time, becomes less usable. So, a new technique, or dusting off an old one, may be necessary with any given pileup. The DXer will become adaptable, and will do so almost instinctively.

One of the most important techniques is to make short calls and listen after each one. For example, the DXer signs his call once, either Phone or CW, and listens for 1.5 to 2 seconds before signing his call again. He should not sign his call more than three times for each DX station's over. He remembers that if he's not the QSO, he's just QRM. The DXer will get beat in a pileup, and needs to remember that and not get frustrated about it. Sometimes it's just impossible to beat better propaga-

tion or better antennas. It is better to let the pileup proceed at as fast a pace as possible. Strangely, sometimes the DXer can actually improve his chances of getting through by not transmitting. Letting the pileup proceed at a faster pace will get more people through. Less QRM means a faster pace.

The DXer can also take advantage of a knowledge of propagation. It's pointless to waste kW hours of electricity while getting totally frustrated when there's no possible way to break a pileup. Since it's reasonable to expect propagation to flow from east to west as it follows the sun, then it's also reasonable to expect to take advantage of that knowledge. A 1981 pileup on 5H3KS on 20-meter SSB illustrates the point. He was somewhat audible, with an East Coast pileup making it through with some difficulty in the early afternoon. It was evident from the call sign prefixes that only the east coast was making it through. Slowly over time, he became louder, and stations from the midwest began joining in the calling. Still, only East Coast stations were getting through. Finally, he reached over S9. Mostly East Coast stations were making through still. However, as his signal peaked and began to fall off, stations from the 0 and 5th call districts were the dominant stations getting through the pileup. This demonstrates two concepts: first, that patience is rewarded, and second, that an easterly station is often best worked *after* his signal peaks.

Advanced Intelligence Gathering

As the DXer nears the 300 zone, the coveted "New One" seems to be harder and harder to find. In some cases, a DXpedition is required. In others, a little research into the operating habits of the station is required. In both cases, sometimes more is needed than just an occasional foray to the bands in the hopes that they may be there (although that sometimes happens too!). So, a little intelligence gathering is necessary.

The DX weeklies, such as *QRZ DX*, begin to pay off in the search for the operations of resident DX operators. The DXer may get a bulletin board, and begin keeping a chart of the activities of these operators by using the QSN (I heard on __kHz) reports. For example, it's noted that a VKØ in Macquarie is beginning to show up on a certain band, on or near a certain frequency, and usually around the same time of day. Further charting may observe that this occurs on the same day, but every other week. Thus, there's a high probability that

Bill, EA9EO, attracts a crowd of onlookers as 3C0AB from Annobon Island in 1979. *(Photo courtesy of EA9EO).*

if the DXer is there at the right time on the next likely day, he will get a shot.

The bulletin board may also be used to post notes about DXpeditions that will occur on certain days in the future. Sometimes the announcements are made weeks ahead of time, and not repeated. So, it's a good idea for the DXer to keep his own notes. That way, he checks his bulletin board daily for anything that may come up that day. Even information gleaned from over the air conversations or other not readily available sources may be tracked this way.

Earlier, it was mentioned that listening to conversations could pay off. Sometimes these are just rumors, but sometimes there's real information passed over the air about setting up DXpeditions. When this is done, it's often in seemingly meaningless conversation. Make no mistake, however, those in conversation know what they're talking about. With a little practice and thought, the DXer will too.

Many of these type conversations have now moved to e-mail. Private e-mail has greatly facilitated the logistical support of international DXpeditions. Private e-mail has also helped those DXers with friends who want him to know what's going on. The DXer will have acquired plenty of those on his way up. While not yet totally replacing the telephone, e-mail has made big gains for intelligence gathering.

Finally, keep track of national holidays and customs in the area of interest. Know when the DXer from other lands might be taking his cup of coffee (or whatever) into the shack to enjoy a few minutes of radio before work. Read a newspaper with a good international section to keep track of world events that may affect DXing.

BASIC DXpeditioning

By now the DXer may be thinking of the pileups, of what it may be like to be on the other side. He has seen the best and worst of DXpeditions, and would like to attempt something of his own. Great!! The world will be waiting for him, because somebody needs anyplace he might go, and as for the rest, well, who can resist the siren's call?

To start planning a DXpedition, there must first be an objective in mind. For the first time DXpeditioner, it would be best to start with something with a minimum of hassle. The DXer wouldn't want to look up someplace like Heard Island and try to go there by himself to get started! Far better to begin by picking a destination where licenses are obtainable, where a hotel room can be arranged, and where airlines fly. A location with propagation to population centers is best, as part of the mission is to make some contacts in quantity as practice.

After site selection (and that includes finding a place friendly to antennas and transmitters), inquiries need to be made about the license. Unless the location selected is under FCC jurisdiction, the DXer will need to obtain a license from the governing body of the DX location. In some cases, this may take months, especially if he is not on site. Check with the Regulatory Information Branch at ARRL HQ for reciprocal licensing information. Most forms are on file there, along with a recap of necessary procedures to follow.

The DXer should have a passport. While not strictly necessary in some countries he may visit, it's always a good idea to have it handy. Also, a photocopy should be carried on his person at all times. In the event the passport is lost, or the DXer is stopped by someone in authority, it may be necessary to produce satisfactory papers instantly.

The next step would be to determine what to take. Transceiver selection in many cases boils down to what is owned.

In 1977, DXers focused their attentions on tiny Kingman Reef, where KP6BD, operated by K4SMX, WB9KTA, N9MM and K6NA made 11,000 QSOs in three days. (*Photo courtesy of K4SMX*)

However, the DXer will find that the transceiver needed for the average DXpedition need not be quite as full featured as what is owned. Dual receive is not a requirement, but a built-in keyer is. RIT or receiver offset is necessary to work split. Light weight and small size is important, as it is best to carry the transceiver on board. As checked baggage it may never reach the destination.

Antennas will be determined by what can be carried. The DXer would never go wrong by selecting an all band vertical to start. Wire dipoles are also good, especially if one or both ends can be placed on high supports. Carry two or three hundred foot lengths of small coax with connectors installed, along with several barrel connectors. Some spare connectors may also be handy. Take nylon rope of some type to use for antenna installation.

It is very important to check the line voltage available at the chosen location. More countries use 220 Vac than the usual 115 Vac found in the US. The DXer must have a power supply available for whatever voltage is found.

By packing the carry-on baggage correctly, the DXer will be assured his DXpedition will come off even if his baggage is lost. The transceiver, power supply, headset with boom mike, and wire antenna and coax should all go in the carry-on. A lap-top computer, if available, should go too. Everything else, including clothes should go in the checked baggage. If the checked baggage doesn't make it, at least the station is there. Log sheets, etc., may be obtained locally (most parts of the world have blank sheets of paper and writing instruments), rudimentary clothes will be obtainable, and most hotels will have laundry services.

Upon arriving, it may be necessary for the DXer to make some arrangements with customs for admitting the radio equipment. This may in some cases require a large deposit, although sometimes the gear can be talked through. The DXer should try to find out as much about customs procedures in advance as possible.

When possible, the DXer should try to find an accommodating high rise hotel overlooking the ocean. The antennas should go as near to the edge of the roof as possible, to the DXer's top floor room. Of course it's possible that those kinds of accommodations will be impossible to obtain. In that case, the DXer will have to work with obtaining the best compromise he can.

Once the station is set up and on the air, it's time for the DXer to enjoy the pileups he came to create. After a few minutes of operation, the pileup will build. It is here the DXer should make a transition in operating style that will let him make the maximum number of contacts in the minimum amount of time. On SSB, try to operate transceive if the pileup isn't too deep. If it begins to get large and continues for a long time, it may be time to listen up. On CW, as soon as there are multiple stations calling, it's time to start listening about two up. Working split allows the pileup to hear the DX.

The DXer should be very consistent in working the contacts. The call of the station worked should be given, and a report, as in "K5FUV, five-nine." Every contact should be ended uniformly as in "Thanks, XZ1A." A minimum of verbiage should be used, only enough to make the contact and get on to the next one. On CW, it's similar, as in "K5FUV 5NN" and "TU XZ1A." QSL information may be given about every 10 minutes.

The DXer will find that practice helps. Rates will increase, and so will his enjoyment. DXpeditioning is a fever that may even be worse than the regular DX Bug.

ADVANCED DXpeditioning

At this point the DXer is advised to seek company, for DXpeditioning at this level requires more work in all areas, both in logistics and in political savvy. These are the kinds of places where one doesn't just waltz in and request a license. In some cases, two or three trips may be required before even getting a decent hearing from the guy who can say yes. The DXer who wants to play at this level must be prepared to spend some bucks.

First, letter writing is required. If there is someone the DXer knows who might know someone with some influence in gov-

William Wu, BV2VA, running BV9P from Pratas Island in 1995. Operators from the CTARL operated for 10 days. (*Photo courtesy of BV2VA*)

DXers may be found anywhere, including the Great Wall of China. DXers Kan Mizoguchi, JA1BK, Petri Laine, OH2KNB, and Martti Laine, OH2BH, are ready for operation from BT1X. (*Photo courtesy of JA1BK*)

ernment in some country, it may be possible that they could help facilitate contacts. In many cases, the hardest job is getting to the right person. Letters should be written. In many cases, the DXer will be rejected. Perhaps after an exchange of letters, it may be possible to go and meet with officials. Sometimes this can take months or years. The techniques or responses for doing this are not well defined, and nothing ever works the same way twice. Sometimes a check with the DXCC Desk may give some insight as to what's required. In most cases, getting the permission is the hardest part. Sometimes, it never happens.

The logistics of such a trip are more difficult as well. Since this is likely a trip to a highly needed country, it is necessary to take enough equipment to put a loud signal on the bands, possibly on two or three bands at the same time. This means amplifiers and beam antennas. More operators allow for more equipment. Some things may be purchased locally. A 6-meter-long steel tube provided the mast for a beam in Myanmar. From the top of a hotel, that's usually enough. The DXer should heed the Scout's motto: "Be Prepared!"

Being prepared also means checking into State Department travel advisories for the area, as well as checking with the Centers for Disease Control in Atlanta for any health concerns in the region. The DXer should be sure he has any required inoculations before going. Some are for his health; some are required by the country visited. Also, the DXer may want to take a suit. One can never tell when he may need to go meet a high government official!

Arriving is sometimes exciting. After all, the DXer may be bringing several boxes of otherwise contraband equipment into the country! Many countries don't look at people with small radio transmitters in quite the same way that they do in the US. If at all possible, the DXer should be met at the port of entry by someone who can help get him through.

Operating from a country such as this will be different. The DXer should determine from a Great Circle map centered on the location the areas of the world that will be most difficult to work. These will normally be those where signals must pass through the auroral zones. An operating strategy should be planned that will allow for working only those areas when the band is open. Other areas may be worked at a later time. This technique is called *targeting*. For a more extensive discussion of targeting, see *DXpeditioning Basics* by Wayne Mills. It's available from INDEXA, PO Box 607, Rock Hill, SC 29731, for $5 postpaid.

With several operators, scheduling should be arranged so that the prime station is on the air at all times, and if available, a second station should be manned as well. The prime objective of a DXpedition such as this should be to maximize the number of different call signs in the log, since it may be a long time before anyone goes back.

The DXer should plan to operate split frequency almost as soon as he hits the band. By announcing likely frequencies, he will find people waiting for him to arrive, and may need no more than one CQ for as long as he remains on the frequency. However, while operating split, efforts should be made to listen on only 10 to 20 kHz at most. It is tempting to keep going up looking for a clear receive frequency, but it isn't really necessary. Just move back to the bottom and start back up. Remember, by the time the DXer gets to the operating part of the project, most of the real work has already been done. The operating is the fun part.

ULTIMATE DXpeditioning

There are many obstacles in the path of the Ultimate DXpedition. First, permission must be obtained. This is sometimes easy, sometimes difficult. Often, the Permittor wants to be sure the DXpeditioner is capable of pulling this off without being killed or stranded, a very real possibility. Once that is assured, there's always the transportation problem, Then, equipment, food, survival gear, and other necessities. Oh yes, there's radio stuff too. The costs for the Ultimate Class DXpedition can run into hundreds of thousands of dollars. So, funding is needed too. The Ultimate DXpedition is not a task to be taken lightly, and certainly not without some experience.

Operating Permission and Documentation

Operating permission is most certainly needed for these destinations. In many cases it consists only of landing permission. For example, some French and American possessions have limited access, but no need for a special license. These destinations will require special landing permission, and it must be in writing from the proper agency. This may be a difficult task, but it is not insurmountable, as proven by the fact that it has been done before. But it may take many letters and

DXpedition landings aren't always as easy as this one in the Glorioso Group in 1980. (*Photo courtesy of DK9KX*)

a few visits before the goal is accomplished. Nothing difficult is ever easy.

The copies of operating and landing permissions will document that the DXer had permission to be there for the purpose of Amateur Radio. There is a need, however, to further document presence at the location. Documents may consist of transportation receipts, ship's logs or certifications signed by the captain, and pictures of the operators at known locations. Postcards mailed from the location, or, if that's impossible, from the nearest port to the DXCC Desk also help document the operation.

Advanced Transportation

The DXer will find that getting to the Ultimate DX location requires a bit more ingenuity than a trip to Aruba. These locations are mostly exotic because they're hard to reach! They will require special charter flights or boat transportation. The DXer should use care in choosing the provider. It's best to find someone who has used transportation in the area before, and ask them about a provider and the quality of service provided. In reading the details of past DXpeditions, the DXer may find that the same boat was used by several different DXpeditions. Charter flights are usually very difficult to arrange. Distances involved and lack of landing usually preclude the use of an airplane at all, although DXpeditions to both Willis Island and Annobon have on occasion used airplanes. But in reality, the DXer should plan a long boat trip.

Life Support

Whether in the Antarctic or a desert island in the Pacific, life support for the DXpedition team is very important. The shelters chosen should provide shelter from whatever elements may be found at the DXpedition site. For Antarctic regions, this means shelters that can protect against cold without being blown apart by high winds. Some form of heating may be required, and exposure suits for the operators may also be essential.

In hot, sunny areas, it's necessary to protect against wind, rain, and sun. It should be stressed that long term exposure to the sun is harmful, and so the operators should take precautions to prevent permanent damage. Prevailing winds may bring salt spray to damage equipment, and provisions should be made to block any spray that may appear.

Food is always a consideration. Meals should be planned to make sure that there are adequate provisions not only for the DXpedition period itself, but also for the trip both ways, and a several day safety factor should be allowed, in the event the group is trapped on the island due to weather conditions.

The Ultimate DXpedition will take place days from civilization. Medical aid will not be available for the team members should they need it. Exposure to hazardous plants, fish, and animals is highly likely. Therefore it is a good idea to choose a doctor as part of the operating crew when possible. He will know what medications to take for the area and potential health threats involved.

Equipment

The Ultimate DXpeditioner should do as much as he possibly can to ascertain the conditions he may find at the operating site, and plan for any possibility. This is especially true in the area of antennas, where selections should be made based upon what may be carried and landed for their support.

Antennas at this level should consist of beams, as many as can be found for the number of stations, along with a spare. Masts will have to be taken to get them into the air. Often this can consist of heavy duty TV masts with multiple guy wires. A TV mast will allow up to a 20-meter beam to be placed 30 to 40 feet high when properly installed. Taking an aluminum step ladder is necessary, as the masts need to be pushed up vertically. A rotator is not necessary, as the antenna may be rotated by a rope tied to the director end of the boom, and tied off in the desired direction. A DXpedition isn't required to rotate an antenna as often, since they *are* the DX, and don't have to look for it. Remember that all peripheral equipment, such as coax, connectors, and soldering equipment will have to be carried.

Amplifiers will be needed. The Ultimate DXpedtion won't be back here for five to ten years, so needs must be satisfied. The only way to do that satisfactorily is to be *loud*. The amplifier chosen should provide the necessary output, and be flexible about running on the varying voltages sometimes found with a generator. Sophisticated control systems sometimes won't take this type of service. It is recommended that an amplifier taken on a DXpedition be tested with its generator. Any other electrical devices to be used with that generator should also be operating during the test. Simpler is often better when it comes to DXpedition choice.

Since the DXpedition will use generators, it is best to select those with adequate capacity for the job at hand. Figure maximum current requirements for the equipment to be used with a particular generator, and add a safety factor. Generators should use diesel fuel, as the transportation operator will feel better about carrying this than more volatile gasoline.

Packing

The Ultimate DXpeditioner may decide that the best way to get the equipment to the embarkation point is by shipping. The equipment should be well packed so that it won't be destroyed by shipping, and sent to an agent at the embarkation point. It should be sent well in advance so that it will be in place before the DXpeditioner gets there. It is possible to take smaller loads as checked baggage, but usually only if one can get the gear in hand between airplane transfers. Checking through to final destination is sometimes risky. It's hard for an airline to bring lost baggage out to Heard Island if it should go astray.

Clothes should be taken with a view of the needs at the DXpedition site. For the Pacific, bathing suits and shorts may be the order of the day. For cold regions, one may change clothes every several days rather than daily. Pack accordingly.

If any equipment is to be carried onto an airplane, it should be items such as lap top computers, keyers, headsets, etc. that would be needed by the individual operator. Don't forget the camera, and books are nice for airplanes and boats.

QSLING

For DXers who are pursuing awards, QSL cards are almost as important as the contact itself. The QSL has been a part of Amateur Radio and DXing for as long as radio itself has existed. It is the QSL that actually proves the contact. As many have found, pirates and bootleggers enjoy having a good time, and the DXer would never know if the station was legitimate without the attempt to get a QSL. While electronic QSLs may be on the horizon, the majority of DXers currently show strong support for the old fashioned paper QSL, so it seems likely that QSL cards will be with us for a while longer.

Outgoing

Yes, the DXer wants to receive that all important piece of cardboard. However, in order to receive, one must first send.

XF4CI took place from this small shack on Soccoro Island in the Revillagigedo group. This is the second operating location. (*Photo courtesy of XE1CI*)

Strangely enough, most Rare DX stations are not eagerly waiting for the DXer's QSL card. They already have a few, all bands, all modes. So, it is up to the DXer to present his card in such a way that it will get returned. This is very important, as otherwise the card might as well be going into a Black Hole.

First, the card itself should be designed for maximum ease of the DX station or his manager. After processing 20 to 30,000 QSLs, it's easy to understand how a manager could get a little upset with having to check around the card, or get a case of carpel tunnel syndrome from turning a card over to check out both sides. QSL card design, while not making the chore totally painless, can at least prevent the manager or DX station from learning the DXer's call sign in a negative context! So, the QSL card should have all the information on one side only. It should be easy to read, and in a logical order.

There is certain information required on a QSL card that is to be submitted for DXCC credit. Confirmation for two-way communication must include the call signs of both stations, the DXCC country, mode, date, time and the frequency band used. Desirable information includes the county and grid square. If the DXer lives on an island, an Islands on the Air (IOTA) identifier number is also desirable. This way, if the DX station happens to be pursuing awards of his own, he will be able to use the DXer's QSL card for his own purposes.

The card should have the information printed plainly on the card. After seeing a few of the optical illusion cards, or ones with overly embellished lettering, it's easy to understand how confusing this can be. Plain lettering on a plain card is much better. If a picture card is desirable, it may be better to have a plain card printed on one side, and the picture card on the other. Print your call on both sides of the card. That way the card can be displayed by the DX station, but processed rapidly as well.

There are three ways to get a card to its destination: By bureau, by QSL service, and direct mail. Each has its advantages, although there are some differences in the speed. There's also some differences in how a DX station will handle them. The method the DXer uses depends upon his personal requirements.

The ARRL Outgoing QSL Bureau (see the end of this chapter) provides economical service to the countries who have incoming bureaus. At $4 per pound, this is indeed a bargain for those who don't mind waiting for a QSL. While slow, with turnaround time sometimes exceeding a year or more, still this may be considered an efficient and cost effective method for

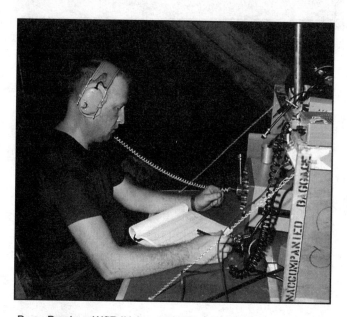

Dave Bowker, WØRJU (now K1FK), operates the night shift from WØRJU/KP1 on Navassa Island. He's using a unique shipping crate which turns into an operating desk. (*Photo courtesy of NØTG*)

The envelopes themselves are important. Other countries do not use the same envelope sizes that are commonly found in the United States. This means that using a standard US #10 envelope will create the undesirable situation of sending something through the mail that's crying "Steal me! Steal me!." It is far better to obtain the proper size envelope for the job at hand rather than trying to make do. The best size is a $4^{3}/_{4} \times 6^{1}/_{2}$ outer Air Mail envelope, and a $4^{1}/_{2} \times 6^{1}/_{4}$ inner Air Mail envelope for returns. If possible, the addresses should be typed directly on the envelope, or printed labels. It is important to remember that *no* call signs should go on the outside of any envelope.

Inside the flap of the return envelope the DXer should put his call and date, time, and mode of the QSO. In the event the card gets separated from the return envelope, it is still possible for the manager to look up the QSO and send the card. Also, either cards or a list of all QSOs made with the station, if a DXpedition, should be sent so that the manager won't have to guess about whether to provide cards for all QSOs he finds in the log. Chances are he *won't* if he's not asked, as he would then have to determine if the DXer worked him or if what's in the log is the result of a miscopied call sign.

The old saying goes "The final courtesy of a QSO is a QSL." That may have been so years ago when postcards were a penny, but that time is long past. Perhaps the saying should be changed to "It is discourteous to send a QSL card *without* return postage provisions." QSL cards aren't cheap these days, with even the cheapest around 3 1/2 cents each. As this is written, inside the United States it costs 32 cents for mailing in an envelope or 20 cents without. Multiply this by a thousand, and you'll find that QSLing could easily cost $225 to $355! Make it air mail, and the price jumps even further. Then consider that United States postal rates are among the cheapest in the world, which could put the tab for a thousand QSLs without return postage well over $1,000! That's an awful lot of money for something the DX station likely doesn't need. So it's easy to see why there is poor or no response to cards received without return postage. This even applies to Stateside cards these days. A word to the wise: if you really want that exotic QSL card, be sure that return postage is provided in whatever form is necessary.

The best way to provide return postage these days is the International Reply Coupon (IRC). They may be obtained from a Post Office, and are redeemable for the lowest air mail rate in any UPU country. IRCs can often be obtained from a QSL manager in the DXer's own country. IRCs purchased this way trade somewhere between the value of new IRCs and the redemption value. At the present, in the US this would mean somewhere between 60 cents and $1.05. Usually 75 cents will get an IRC.

A caveat is that in some countries, the lowest Air Mail rate won't provide enough postage to return an envelope and QSL card. Sometimes it takes two. For example, until recently, Germany's lowest Air Mail rate provided for less weight than the average QSL card and envelope. So it became necessary to go to the next highest rate, which was 3 Marks. One IRC exchanged for the lowest amount of postage, 2 Marks. So it took two IRCs to provide enough postage. At the same time, it took about $1.25 US to exchange for 2 Marks. $1 US was not enough for return postage.

Another way to provide return postage is through the use of mint stamps from the DX station or QSL manager's country with sufficient value to provide return postage. This allows the manager to fill out the card and put it in the envelope with no trip

QSLing. This is especially true when compared with pouring money down black holes with direct QSLs to some parts of the world. The disadvantages are that in some countries served by bureaus, only cards to members are delivered. This method is highly recommended when QSLing North and South America, Japan, Europe, and the CIS countries in particular.

To use the bureau, the cards should be sorted in country order. Cards to managers should have the call of the manager on both the front and back side of the card. Turn the card over and put the call sign of the station you want it to go to on the back in block letters. This makes it easier for the sorters at either end.

Next up the scale is a QSL service, such as the WF5E QSL Service. For only a small amount per card, the QSL service will send the cards by the bureau, direct, or to a manager. Cards are returned via the DXer's own incoming bureau unless special arrangements are made. The DXer will notice the cards coming back this way, as they are usually marked with the service's stamp. Turnaround time is improved as cards are sent to bureaus in smaller quantities, and cards are sent to managers with return postage. In some cases, it isn't even necessary to know the manager, as the QSL service keeps track of managers, and will get the card to the right place. The success rate over time with this method is very good.

Direct QSLs are used by many. The caveat here is that QSLing direct isn't cheap, so be prepared to shell out some money if this method is used. Presentation is worth a lot too, so one must be prepared to go the whole way for any chance of a return. This method is generally best not used for resident amateurs if a bureau is available. The bureau is often more convenient for resident amateurs.

Direct QSL returns start with a good address. Sometimes the station will give his complete address over the air. If not, then *The Radio Amateur Callbook*, the Buckmaster CD ROM (available from the ARRL), or *QSL Routes* may have the information. *QRZ DX* often has newer addresses for stations who have recently been active. It is best to get the complete name of the operator, as it is always best that call signs *not* be placed on the envelope.

to convert money or IRCs, purchase postage, and affix it to the envelope. It also removes the temptation to keep the IRCs or money and think about returning the card later. If this method is used, it's best to put the postage on the return envelope so that it may only be used for returning the DXer's cards.

The least effective, but most often used method of providing return postage is the ubiquitous Green Stamp, the United States one dollar bill. There are several problems associated with using Green Stamps, not the least of which is mail pilferage. Even mail passing through the US is not immune to it, and in some countries if the envelope is identified as going to an Amateur Radio operator, it is almost certain to disappear. One card sent as registered mail to the Middle East was returned as undeliverable. It was still sealed, but the $5 bill enclosed had been removed! This is often the fate of the Green Stamp. In addition to exchange problems at the country of destination, there are some countries around the world where hard currency is so controlled it is just flat illegal to have it. Two countries that quickly come to mind where this is so are South Africa and India. There are many others.

In other countries, Green Stamps trade much like IRCs; that is, they are not exchanged at all, but traded for the Amateur's own QSLing needs. They are often sent with visitors from the US back to the States to buy US postage and forward the cards on from there. One Green Stamp no longer will buy a sufficient amount of postage in many countries. So two are required if the DXer expects to get a return. However, they are convenient. For most DXers, a reach into the pocket is all it takes. So, the Green Stamp is used mostly for convenience. The price paid may well be loss of the card.

With the QSL card, return postage in whatever form, and envelopes, the DXer is ready to prepare the QSL for mailing. First be sure the QSL is completed properly, with all QSOs made with the station listed on the card. The card should then be inserted into the envelope, which should have the DXer's return address typed or printed on it. Be sure that the country name is the bottom line of the return address. If a Green Stamp or IRC is used, insert that into the return envelope. The idea is to make as flat a package as possible. If the card is going to an area of high humidity, putting a piece of waxed paper under the return envelope flap can prevent it from sticking together before it is used. All this should then be placed into the outer envelope and sealed. Remember, no call signs or references to Amateur Radio should be on the outer envelope. With some luck, the DXer will see his desired rare QSL coming back to him in a few months.

One word about QSL turnaround: Many DXers have much too high an expectation about how fast they should receive QSLs. It depends upon the operation and whether a special effort is put into quick turnaround, but six months is very reasonable for QSL return. Whereas the MegaDXpedition may have a small army of helpers to spring into action to complete the QSLing, most smaller DXpeditions depend upon one individual. Looking at an operation that made 10,000 QSOs, and allowing 3 minutes per card for opening, finding and checking the log entry, writing or attaching a label to the card, inserting the card and sealing the envelope, and obtaining and affixing stamps, figure on the manager handling 100 cards per day. That's five hours per day for 100 days without a break! If the manager also had a full time job and family, the DXer can see where he might want a little time off from this routine. The DXer should allow a minimum of at least six months before even thinking about a second QSL, even if his friends are receiving theirs. His card may have been near the bottom of the pile. Also, the DXer shouldn't send a request via the bureau at the same time as sending the direct card. QSL managers note all cards received and sent in the log. When they find a DXer using this practice, they make a note of the call sign. The next time, that DXer's card often goes out with the last batch mailed, or through the bureau.

Incoming cards are easy: the DXer just needs to have a mailing address! However, he also needs to be prepared to receive cards by the incoming bureau. Each call district in the US has its own Incoming QSL Bureau. The DXer should send cards or envelopes with postage to his own bureau, according to the rules established by the bureau in his district. DXers should also note that the bureau depends upon their call sign, not upon where they live. *With the new Vanity Call system allowing choices of calls anywhere within the 48 states, some DXers living in Arkansas will now find that they should be receiving cards from the W9 Bureau in Illinois.* The latest addresses of the various bureaus are printed periodically in *QST*.

Coming back from the mailbox with a new batch of QSL cards is one of the joys of DXing. Following these instructions should assure the DXer that his chances of making many happy trips are increased.

Propagation

While this was stated before, it is worth saying again: Whatever the propagation is, *It Doesn't Matter!* The DXpedition will still start on time, the contest will arrive on its appointed weekend, and the DXer will be there to work it, whether there are 50 spots on the sun or none. Yet it is good to know how each band works in order to take advantage of what is offered. Herein is offered a band-by-band breakdown of what to expect.

160 Meters

Often called the "Top Band," 160 offers one of the great DX challenges. Openings are sometimes very short in duration, and usually provide shorter distance openings than other bands. During high sunspot levels, the band is open, often with greater intensity. But the sun's activity causes an increase in noise level, and absorption of signals increases as the D layer, which forms at sunrise and begins to dissipate at sunset, is more intense in years of high solar activity. This effect is somewhat lessened during times of low sunspot activity, but for the most part, 160 meters is still a night-time band.

Propagation on 160 begins in the late afternoon to early evening, in the direction of the approaching darkness. As stations in the east approach their sunrise, there will likely be a short but noticeable increase in their signal strength. This is the opportunity to work the longest distances, and the greatest effect (*gray line*) occurs when it is sunrise at one end of the path, and sunset at the other. Propagation continues to flow east to west, as the DXer follows the sun. As the sun rises, the DXer can expect to work stations to his west until they finally sink into the noise as the D layer begins to absorb the signal again. See Chapter 4 of this book, Antenna Orientation, for information on software that generates gray line path plots.

Also worth more than a little mention are the aurora belt zones that occur around the magnetic poles. Charged particles discharged by the sun are attracted by the magnetic poles, and create a dense layer that attenuates any signals passing through it. Of course, the longer the wave, the greater the attenuation. Thus, low band signals are affected more than others by the aurora belts. This will be apparent on northerly paths, as stations in the Yukon and Alaska often report one way propagation where they can hear stations to the south, yet are not able to communicate with them.

Due to longer periods of sunlight and thunderstorm activity, 160 meters provides better propagation in the fall and winter. However, one should not overlook possible openings to the Southern Hemisphere in the summer, as the seasons are reversed, and those in the Southern Hemisphere are experiencing their best propagation of the year.

80 Meters

Many of the same phenomena that affect 160 meters also affect 80 meters. The aurora belts, D layer absorption, and thunderstorm activities also have limiting effects on this band. It usually begins to open in the early afternoon. The gray line is especially useful at the beginning of the opening, but will require a good antenna system and power to be able to take full advantage of it. The band opens to the east, and the gray line stations will often come in from the southeast. The opening will continue toward the east until after sunrise at the eastern end of the path, then to the south, finally ending as sunrise approaches with westerly openings.

Eighty meters is seasonal, with better conditions in the wintertime, but with DX still available in the summer. In times of low sunspots, 80 (or 75, as the phone part of the band is commonly called) becomes the prime nighttime phone band, as 20 closes and 40 turns into a morass of broadcast stations. While less used in times of high sunspots, it's still open, although usually later in the evening.

40 Meters

If 20 meters is the King of DX bands, then 40 meters surely must be the Queen. While affected by many of the same conditions as 80, the D layer absorption problem is reduced because of the shorter wavelength, and it becomes possible to work intercontinental DX during the late afternoon. In fact, at the sunspot minimum, it's possible to hear DX all day long, and some of it may be worked if the signals are not absorbed too badly.

In the late afternoon, the band is likely to provide good long path openings to the southeast. Openings are again basically toward the east until well after eastern sunrise. Afterward, the opening will swing south, then toward the west as sunrise approaches. An opening to the southwest is very common in the morning, and stations that would normally be found with a northwest Great Circle path will be found there, with no opening at all to the northwest! An example of this is the USA to Hong Kong path, which will often be open at a heading of 210-225°, rather than the expected 330°.

30 Meters

If ever there was a DX band that provided 24 hour coverage at most stages of the sunspot cycle, 30 meters would be it. It shares many of the characteristics of both 20 and 40 meters, with long nighttime openings, while providing DX throughout most of the daytime as well. Propagation is hard to describe because of this, as 30 meters can almost be open to anywhere at anytime as the seasons change. The DXer should check this band frequently for some pleasant surprises.

20 Meters

The undisputed King of the DX bands is 20 meters. For long distance propagation at any time of the sunspot cycle, 20 will be the band of choice. During low sunspots, it's likely to be the

only band open during the daytime. At sunspot maximums, 20 will be open to somewhere almost 24 hours each day. Almost every form of propagation can be found here.

During the winter, 20 will open to the east at sunrise. The opening will last until stations at the eastern end of the path are past sunset. In the early afternoon, the opening will extend to the south and to the southwest. Early evening should find the band opening to the west to northwest. In the spring and fall, long path propagation exists, with opening to the east coming from a southwesterly to western path just after sunrise, and lasting for several hours. In the evening, the path opens to the southeast. It is not unusual to find long path openings to areas of the world that are actually stronger than the short path opening.

Summertime often finds openings to the Far East in the mornings, while European openings continue through the early evening and on into the night. There is always a southern path, extending well into the night after other paths have closed. The band may seem to close, then open up again later, around midnight, on certain paths.

This single band is so productive that more than a few serious DXers restrict their antenna choice to one large monoband 20-meter antenna. All DX passes through 20 meters at some point, it's impossible to be a truly effective DXer without it.

17 Meters

Another of the newer bands, 17 meters hasn't been with us for a full sunspot cycle yet. It has many of the characteristics of 15 meters, but due to its lower frequency it's open more often, especially during times of low sunspots. Again, look for morning openings from the east, and evening openings to the west.

15 Meters

This is somewhat of a transition band. It has many of the characteristics of 20 meters at times, including possible long path openings, and in addition some of the characteristics of 10 meters. It often provides very long distance openings, and would be a good choice for working Africa in the afternoon during times of moderate to good solar activity. At times of

The beauty of a summer day in the South Orkney Islands greets the operators of LU6Z in 1996. (*Photo courtesy of GACW*)

low sunspots, it may not open at all on east-west paths, but is usually open to the south.

Typically 15 opens after sunrise, if it opens at all. At the bottom of the sunspot cycle, these openings are likely to be of very short duration, but during cycle peaks are likely to last from sunrise to long after sunset. Propagation begins to move noticeably from east to west, with absorption around local noon often making signal levels go down for a while, only to return later in the afternoon. The band will begin to open to the west as the sun moves across the Pacific.

At times of high solar activity, 15 meters will provide openings to Asia as late as local midnight. However, at times of low solar activity, the band isn't likely to open to Asia at all, unless the DXer happens to live on the West Coast. When the bands are open, no band provides more access to exotic DX than 15 meters.

12 Meters

As you might expect, 12 meters closely resembles 10 meters. However, because of the lower frequency, it is open more often than 10, but slightly less than 15. If 15 is open, it is worthwhile to try 12. Another full sunspot cycle will allow its characteristics to be explored more fully.

10 Meters

Newer hams may never have heard 10 meters in all its glory. Yet during a strong opening during a contest, it may well be impossible to find a clear frequency from 28.300 MHz all the way to 29.000 MHz! Band openings here are incredible, although worldwide band openings are somewhat more rare. Yet it is possible to cover the world, and with less power, when conditions are right.

Unfortunately, when sunspots are low, 10 meters stays in hibernation for years, with only an occasional flash of its greatness. That's why it is necessary to know all its propagation well. Of all the HF bands, 10 meters has more propagation modes than any other. Each has its place, depending upon the time of year, time of the sunspot cycle, or time of day.

During low sunspots, 10 meters often opens in the *sporadic E* mode. E layer clouds form rather quickly, and often break up just as fast. While sometimes intense enough to even provide propagation on 20 meters, it is most noticeable on 10 and 6 meters. Openings can be from 1500 to 3000 miles, depending upon whether the cloud is formed well enough to permit one or two hops. Sporadic E often occurs during the summer, and sometimes in mid-winter. Openings can hold in as long as 24 hours. When this occurs, it's likely to be open all night long. Summertime openings can provide openings to Europe or the Pacific, and can occur at either low or high sunspot numbers. The best seasons are May through early August, and December.

Scatter is another form of 10-meter propagation. Using high power and high antennas, stations from 400 to 800 miles away can be worked, albeit with weak signals. Occasionally the ping of a meteorite can be heard, as signal strengths jump suddenly, then fall. Scatter occurs during high sunspot or low sunspot periods.

Another form of scatter involves beaming away from the desired direction. Often called *side scatter*, this propagation mode can be used by beaming southeast in the morning, and southwest in the afternoon. At the early stages of a sunspot cycle, this is a main propagation path to Europe, open when the direct path never does. It will sometimes seem that all signals come from this one direction. When in doubt about

The gang's all here on Sable Island, CYØXX, in 1996. (*Photo courtesy of WA4DAN*)

which way to point the beam, this is the mode to try.

Back scatter is still another of the scatter modes that can be useful. Sometimes 10 meters is an extreme long distance band. At times like that, it's often difficult to work close-in stations. Back scatter can be used for these situations. To use this mode, *both* stations should point their antennas in the *same* direction. The stations then receive each other by bouncing signals back to each other. The band needs to be somewhat quiet for this to occur. As an example, suppose a station in Texas wants to work a station in Mexico. The Mexican station is working the East Coast or Europe. The Texan's best chance is to turn his antenna in the direction of the stations being worked by the Mexican station.

During high sunspots, the best propagation mode is by the F layer. This produces the best long distance propagation of all. Due to the nature of 10 meters and the long skip involved, there is less interference from stations on the same frequency. This can be deceptive, as the DXer is often beaten in pileups by stations he can't hear at all! Still, this is preferable to hearing the pileup hammer his ears.

North-south trans-equatorial propagation is available at almost any time in the sunspot cycle. It is possible to work Africa, New Zealand, Australia, or South America at almost any time of the sunspot cycle. Paths south of 90° or 270° are open weakly throughout most of the sunspot cycle.

With its very unique characteristics, 10 meters gets attention from amateurs of all interests. When 10 meters is open again, most of the DX will be here most of the time.

6 Meters

A DX band? Yes! It can provide some of the most fun DX of any band. Again, many modes of propagation are available, but sporadic E is probably the most popular. Occurring in the spring and early summer, and again in December, sporadic E can provide DX contacts up to 4000 miles, and sometimes more.

This band is the practical lowest frequency for EME (moonbounce) operations. Stations utilizing four or more high gain antennas and high power have been able to communicate internationally even though the band is closed for other propagation modes.

At present over 200 DXCC awards have been issued for 6 meters. No longer a curiosity, 6 meters is a legitimate DX band.

2 Meters

Two meters is also a DX band. Through the use of meteor scatter, many Europeans have earned credits for their Mixed DXCC on 2 meters before upgrading their license and finishing up on the HF bands. However, most DXers who have earned DXCC on 2 meters did so through the use of EME. Over 130 countries have been worked on 2 meters.

Pick a Band, from 160 to 2

This is not a scientific guide to propagation on the DX bands. It is only to alert the DXer to the possibilities of the DX bands and how to use them. The DXer should note the time of day, solar activity, and the direction of the DX, and make a band choice that may give him an opportunity to find the DX. No attempt has been made to fully explain any phenomenon, only to point out that it does exist, and the DXer should adjust his plans to the possibility. For example, he may have to calculate the sun's position at the other end of the desired path in order to make a band choice. The important thing to remember is that the DXpedition doesn't usually sit at home waiting for the sun. So propagation doesn't matter; how to use it does.

RESOURCES FOR THE DXer

The following resources are available to the DXer from the Internet:

The Arkansas DX Association home page: **http://www.seark.net/adxa/**. This page, maintained by Wayne Beck, K5MB, has links to many DX resources online. The DXer may also make Telenet connections to many PacketClusters worldwide. This may be the only url the serious DXer will ever need.

ARRL Home Page: **http://www.arrl.org/**. Source of information about ARRL award programs, including DXCC Rules and forms.

WF5E DX QSL Service: **http://www.seark.net/adxa/wf5e.html**. The story of the WF5E QSL Service, and how to use it.

WA5YKO's DX Notebook: **http://www.dxer.org/**. Links to much DX related information. If it wasn't on the ADXA page, it'll likely be here.

The VE7TCP DX Reflector: **dx@ve7tcp.ampr.org**. This remailer contains some DX information and a few tips as well. Sometimes it requires separating the wheat from the chaff, but overall it's worthwhile for the serious DXer to follow.

The Top Band Reflector: **topband@contesting.com**. This reflector deals with 160 meter information, including DX, antennas, propagation, and conditions. Moderator W4ZV does well in keeping threads under control.

The SM7PKK DXpedition Home Page: **http://home1.swipnet.se/~w-17565/**. This URL provides information about DXpeditioning from one of the experienced Pacific travelers. Tips on how to do it and how to prepare.

QSLing resources:
QSL Routes
Theuberger Verlag GmbH
PO Box 73
10122 Berlin
Germany
20 IRC or US $20 for 1997 edition
Air mail 30 IRC or US $30.

One of the best resources for the serious QSLer. This book contains routes for many of the very old QSL cards. An ideal source for tracing operators from long ago.

Envelopes and mint stamps:

James E. Mackey, PO Box 270569, West Hartford, CT 06127-0569

William Plum, 12 Glenn Road, Flemington, NJ 08822 Fax 908-782-2612

Both of these sources provide mint stamps and European Air Mail inner and outer envelopes.

THE ARRL OUTGOING QSL SERVICE

Note: The ARRL QSL Service should not be used to exchange QSL cards within the 48 contiguous states.

One of the greatest bargains of League membership is being able to use the ARRL Outgoing QSL Service to conveniently send your DX QSL cards overseas to foreign QSL Bureaus. Your ticket for using this service is proof of ARRL Membership and just $4.00 per pound. For those not quite so DX active (sending 10 cards or fewer), enclose $1.00. You can't even get a deal like that at your local warehouse supermarket! And the potential savings over the substantial cost of individual QSLing is equal to many times the price of your annual dues. Your cards are sorted promptly by the Outgoing Service staff, and cards are on their way overseas usually within a week of arrival at HQ. Approximately two million cards are handled by the Service each year!

QSL cards are shipped to QSL Bureaus throughout the world, which are typically maintained by the national Amateur Radio Society of each country. While no cards are sent to individuals or individual QSL managers, keep in mind that what you might lose in speed is more than made up in the convenience and savings of not having to address and mail QSL cards separately. (In the case of DXpeditions and/or active DX stations that use US QSL managers, a better approach is to QSL directly to the QSL manager. The various DX newsletters, the GOLIST QSL manager directory, and other publications, are good sources of up-to-date QSL manager information.)

As postage costs become increasingly prohibitive, don't go broke before you're even halfway towards making DXCC. There's a better and cheaper way — "QSL VIA BURO" through the ARRL Outgoing QSL Service!

How To Use The ARRL Outgoing QSL Service

1) Presort your DX QSLs alphabetically by parent call-sign prefix (AP, C6, CE, DL, ES, EZ, F, G, JA, LY, PY, UN, YL, 5N, 9Y and so on). NOTE: Some countries have a parent prefix and use additional prefixes, i.e., CE (parent prefix) = XQ, 3G,.... When sorting countries that have multiple prefixes, keep that country's prefixes grouped together in your alphabetical stack.

Addresses are not required. DO NOT separate the country prefixes by use of paper clips, rubber bands, slips of paper or envelopes.

2) Enclose proof of current ARRL Membership. This can be in the form of a photocopy of the white address label from your current copy of QST. You can also write on a slip of paper the information from the label, and use that as proof of Membership. A copy of your current Membership card is also acceptable.

3) Members (including foreign, QSL Managers, or managers for DXpeditions) should enclose payment of $4.00 per each pound of cards or portion thereof—approximately 150 cards weigh one pound. A package of only **ten (10)** cards or fewer sent in a single shipment costs only $1.00. **Eleven (11)** to **twenty (20)** is $2.00. **Twenty-one (21)** to **thirty (30)** is $3.00, etc. Please pay by check (or money order) and write your callsign on the check. Send "green stamps" (cash) at your own risk. **DO NOT** send postage stamps or IRCs. (DXCC credit **CANNOT** be used towards the QSL Service fee.)

4) Include only the cards, proof of Membership, and fee in the package. Wrap the package securely and address it to the ARRL Outgoing QSL Service, 225 Main Street, Newington CT 06111.

5) Family members may also use the service by enclosing their QSLs with those of the primary member. Include the appropriate fee with each individual's cards and indicate "family membership" on the primary member's proof of Membership.

6) Blind members who do not receive QST need only include the appropriate fee along with a note indicating the cards are from a blind member.

7) ARRL affiliated-club stations may use the service when submitting club QSLs by indicating the club name. Club secretaries should check affiliation papers to ensure that affiliation is current. In addition to sending club station QSLs through this service, affiliated clubs may also "pool" their members' individual QSL cards to effect an even greater savings. Each club member using this service must also be a League member. Cards should be sorted "en masse" by prefix, and proof of Membership enclosed for each ARRL member.

Recommended QSL Card Dimensions

The efficient operation of the worldwide system of QSL Bureau requires that cards be easy to handle and sort. Cards of unusual dimensions, either much larger or much smaller than normal, slow the work of the Bureaus, most of which is done by unpaid volunteers. A review of the cards received by the ARRL Outgoing QSL Service indicates that most fall in the following range: Height = 2-3/4 to 4-1/4 in. (70 to 110 mm), Width = 4-3/4 to 6-1/4 in. (120 to 160 mm). Cards in this range can be easily sorted, stacked and packaged. Cards outside this range create problems; in particular, the larger cards often cannot be handled without folding or otherwise damaging them. In the interest of efficient operation of the worldwide QSL Bureau system, it is recommended that cards entering the system be limited to the range of dimensions given. [Note: IARU Region 2 has suggested the following dimensions as optimum: Height 3 1/2 in. (90 mm), Width 5 1/2 in. (140 mm).]

Countries Not Served By The Outgoing QSL Service

Approximately 260 DXCC countries are served by the ARRL Outgoing QSL Service, as detailed in the ARRL DXCC Countries List. This includes nearly every active country. As noted previously, cards are forwarded from the ARRL Outgoing Service to a counterpart Bureau in each of these countries. In some cases, there is no Incoming Bureau in a particular country and cards therefore cannot be forwarded. However, QSL cards can be forwarded to a QSL manager, e.g., 3C1MB via (EA7KF). The ARRL Outgoing Service cannot forward cards to the following countries:

A5	Bhutan
A6	United Arab Emirates
D2	Angola
EP	Iran
J5	Guinea-Bissau
KC6	Belau
KHØ	Mariana Is.
KH1	Baker and Howland Is.
KH4	Midway I.
KH5	Palmyra and Jarvis Is.
KH7	Kure I.
KH8	Am. Samoa
KH9	Wake I.
KP1	Navassa I.
KP5	Desecheo I.
P5	North Korea
S7	Seychelles
T2	Tuvalu
T3	Kiribati
T5	Somalia
TJ	Cameroon
TL	Central African Republic
TN	Congo
TT	Chad
TY	Benin
V6 (KC6)	Micronesia
VP2M	Montserrat
XU	Kampuchea
XW	Laos
XX9	Macao
XZ (1Z)	Myanmar (Burma)
YA	Afghanistan
ZD9	Tristan da Cunha
3CØ	Pagalu I.
3C	Equatorial Guinea
3V	Tunisia
3W, XV	Vietnam
3X	Guinea
5A	Libya
5R	Madagascar
5T	Mauritania
5U	Niger
7O, 4W	Yemen
7Q	Malawi
8Q	Maldives
9N	Nepal
9Q	Zaire
9U	Burundi
9X	Rwanda

Additional information:

We no longer hold cards for countries with no Incoming Bureau. Only cards indicating a QSL manager for a station in these particular countries will be forwarded.

When sending cards to Foreign QSL Managers, make sure to sort these cards using the Manager's callsign, rather than the station's callsign.

SWL cards can be forwarded through the QSL Service.

The Outgoing QSL Service **CANNOT** forward stamps, IRCs or "green stamps" (cash) to the foreign QSL bureaus.

Please direct any questions or comments to the ARRL Outgoing QSL Service, 225 Main Street, Newington CT 06111-1494. Inquiries via email may be sent to buro @arrl.org.

THE ARRL INCOMING QSL BUREAU SYSTEM

Purpose

Within the U.S. and Canada, the ARRL DX QSL Bureau System is made up of numerous call area bureaus that act as central clearing houses for QSLs arriving from foreign countries. These "incoming" bureaus are staffed by volunteers. The service is free and ARRL membership is not required.

How it Works

Most countries have "outgoing" QSL bureaus that operate in much the same manner as the ARRL Outgoing QSL Service. The member sends his cards to his outgoing bureau where they are packaged and shipped to the appropriate countries.

A majority of the DX QSLs are shipped directly to the individual incoming bureaus where volunteers sort the incoming QSLs by the first letter of the call sign suffix. One individual may be assigned the responsibility of handling from one or more letters of the alphabet. Operating costs are funded from ARRL membership dues.

Claiming your QSLs

Send a $5 \times 7^1/_2$ or 6×9 inch self-addressed, stamped envelope (SASE) to the bureau serving your callsign district. Neatly print your call-sign in the upper left corner of the envelope. Place your mailing address on the front of the envelope. A suggested way to send envelopes is to affix a first class stamp and clip extra postage to the envelope. Then, if you receive more than 1 oz. of cards, they can be sent in the single package.

Some incoming bureaus sell envelopes or postage credits in addition to the normal SASE handling. They provide the proper envelope and postage upon the prepayment of a certain fee. The exact arrangements can be obtained by sending your inquiry with a SASE to your area bureau. A list of bureaus appears below.

Helpful Hints

Good cooperation between the DXer and the bureau is important to ensure a smooth flow of cards. Remember that the people who work in the area bureaus are volunteers. They are providing you with a valuable service. With that thought in mind, please pay close attention to the following DOs and DON'Ts.

DOs

- DO keep self-addressed $5 \times 7^1/_2$ or 6×9 inch envelopes on file at your bureau, with your call in the upper left corner, and affix at least one unit of first-class postage. * DO send the bureau enough postage to cover SASEs on file and enough to take care of possible postage rate increases.
- DO respond quickly to any bureau request for SASEs, stamps or money. Unclaimed card backlogs are the bureau's biggest problem.
- DO notify the bureau of your new call as you upgrade. Please send SASEs with new call, in addition to SASEs with old call.
- DO include a SASE with any information request to the bureau.
- DO notify the bureau in writing if you don't want your cards.
- DO inform the bureau of changes in address.

DON'Ts

- DON'T send domestic US to US cards to your call-area bureau.
- DON'T expect DX cards to arrive for several months after the QSO. Overseas delivery is very slow. Many cards coming from overseas bureaus are over a year old.
- DON'T send your outgoing DX cards to your call-area bureau.
- DON'T send SASEs to your "portable" bureau. For example, NUØX/1 sends SASEs to the WØ bureau, not the W1 bureau.
- DON'T send SASEs to the ARRL Outgoing QSL Service.
- Don't send SASEs larger than 6×9 inches. SASEs larger than 6×9 inches require additional postage surcharges.

ARRL Incoming DX QSL Bureau Addresses

First Call Area: All calls*

W1 QSL Bureau
YCCC
P.O. Box 80216
Springfield, MA 01138-0216

Second Call Area: All calls*

ARRL 2nd District QSL Bureau
NJDXA, P.O. Box 599
Morris Plains, NJ 07950

Third Call Area: All calls

Pennsylvania DX Association
P.O. Box 100
York Haven, PA 17370-0100

Fourth Call Area: All single-letter prefixes (K4, N4, W4)

Mecklenburg Amateur Radio Club
P.O. Box DX
Charlotte, NC 28220

Fourth Call Area: All two-letter prefixes (AA4, KB4, NC4, WD4, etc.)

Sterling Park Amateur Radio Club
Call Box 599
Sterling, VA 20167

Fifth Call Area: All calls*

ARRL W5 Incoming QSL Bureau
P.O. Box 50625
Midland, TX 79710

Sixth Call Area: All calls* +

ARRL Sixth (6th) District DX QSL Bureau
P.O. Box 1460
Sun Valley, CA 91352

Seventh Call Area: All calls*

Willamette Valley DX Club, Inc.
P.O. Box 555
Portland, OR 97207

Eighth Call Area: All calls

> 8th Area QSL Bureau
> P.O. Box 182165
> Columbus, OH 43218-2165

Ninth Call Area: All calls*

> Northern Illinois DX Assn.
> Box 1450
> Woodstock, IL 60098

Zero Call Area: All calls*

> WØ QSL Bureau
> P.O. Box 4798
> Overland Park, KS 66204

Puerto Rico: All calls*

> QSL Bureau de Puerto Rico
> P.O. Box 9021061
> San Juan, PR 00902-1061

U.S. Virgin Islands: All calls

> Virgin Islands ARC
> GPO Box 11360
> Charlotte, Amalie
> Virgin Islands 00801

Hawaiian Islands: All calls*

> Wayne Jones, NH6GJ
> P.O. Box 788
> Wahiawa, HI 96786

Alaska: All calls*

> Alaska QSL Bureau
> 4304 Garfield St.
> Anchorage, AK 99503

Guam:

> MARC
> Box 445
> Agana, Guam 96910

SWL:

> Mike Witkowski, WDX9JFT
> 4206 Nebel St.
> Stevens Point, WI 54481

QSL Cards for Canada may be sent to:

> RAC National Incoming QSL Bureau
> Loyalist City Amateur Radio Club
> P. O. Box 51
> Saint John, NB E2L 3X1
> Canada

QSL cards for Canada may also be sent to the individual bureaus:

> VE1, VE9, VEØ, VY2*
> Brit Fader Memorial QSL Bureau
> P. O. Box 8895
> Halifax, NS B3K 5M5

> *VE2*
> J. Dube, VE2QK
> 875 St. Severe St.
> Trois-Rivieres, PQ G9A 4G4

> *VE3*
> The Ontario Trilliums
> PO Box 157
> Downsview ON M3M 3A3

> *VE4*
> Adam Romanchuk, VE4SN
> 26 Morrison St.
> Winnipeg, MB R2V 3B4

> *VE5**
> Bj. Madsen, VE5FX
> 739 Washington Dr.
> Weyburn, SK S4H 2S4

> *VE6**
> Larry Langston, VE6LLL
> Box 3364
> Ft. Saskatchewan, AB T8L 2T3

> *VE7**
> Dennis Livesey, VE7DK
> 8309 112th St.
> Delta, BC V4C 4W7

> *VE8**
> Rolf Ziemann, VE8RZ
> 2 Taylor Road
> Yellowknife, NWT X1A 2K9

> *VO1, VO2*
> Roland Peddle, VO1BD
> P.O. Box 6
> St. John's, NF A1C 5H5

> *VY1*
> W.L. Champagne, VY1AU
> P.O. Box 4597
> Whitehorse, YU Y1A 2RB

*These bureaus sell envelopes or postage credits. Send an SASE to the bureau for further information.

+These bureaus can only accept specific sized envelopes. Send an SASE to the bureau for further information.

CHAPTER 6

The Internet

STAN HORZEPA, WA1LOU
ONE GLEN AVENUE
WOLCOTT, CT 06716-1442
e-mail: **stanzepa@ct2.nai.net**
URL: **http://www.tapr.org/~wa1lou**

WHY THE INTERNET?

What is a chapter about the Internet doing in a book about operating an Amateur Radio station? The Internet is a mode of communications, but it is not an Amateur Radio mode of communications, so why the Internet?

The computer has revolutionized Amateur Radio. The first home computers became available in the late 1970s and hams, being the gadgeteers we are, were among the first buyers, users, and hackers of home computers. Build a gadget and hams will find a use for it. And that we did.

Today, computers are an integral tool in almost every aspect of Amateur Radio. Let me emphasize the word "tool." Like other tools we use in this hobby, computers have not replaced Amateur Radio, they have augmented Amateur Radio. And, like the computer, the Internet is another tool that augments our hobby.

Hams require a lot of information to get the most out of their hobby. Hams ask, "How do I modify my transceiver for 9600 packet? Where is the gray line? How do I QSL WA1LOU? What is my grid square? Where can I download the latest version of APRS? How can I make a schedule to work Clipperton on Christmas Eve?"

You can look it up if you have the right book or magazine. You can figure it out with your calculator if you have the correct formula. You can mail a blank disk and an SASE to the guy who wrote the software. You can write a letter to the Clipperton DX Club to set up a sked for December 24.

Or you can use the Internet!

The Internet is an information tool, perhaps the most powerful information tools in existence. And it is getting more powerful all the time.

It is powerful because it is so accessible. You can access the Internet from a computer sitting on your desk (right next to your radio).

It is powerful because it accesses the world. The Internet literally puts the world at your fingertips. If the information you seek is stored on a computer anywhere in the world that is interfaced to the Internet, then that information is accessible to you. Similarly, if the information you seek is stored in the brain of a computer user anywhere in the world who has access to the Internet, then that information is accessible to you, too (assuming the user is willing to share that information).

The Internet has Amateur Radio applications because there is a huge amount of Amateur Radio information available from those same computers (and computer users) that are interfaced to the Internet.

WHAT IS THE INTERNET?

The Internet is a network that interconnects computers all over the world for the purpose of freely exchanging information between the interconnected computers.

The Internet started out as something completely different. It was originally a network called ARPANET, which was devised by the U.S. Defense Department Advanced Research Projects Agency (DARPA) in the late 1960s. It interconnected computers belonging to the military, its contractors and educational institutes doing its research. Its purpose was to provide interconnectivity in the event that the Soviets dropped *the big one* and wiped out all other means of communications.

As time went by, non-military interests were permitted to use ARPANET as well. By the late 1980s, most corporations and institutes of higher learning were using the Internet, primarily for exchanging, software, data, and electronic mail (or e-mail, for short).

At this time, the *graphical user interface* (GUI) of the Macintosh computer and its *hypertext*-linked software was getting a lot of attention. The Macintosh GUI and hypertext links, as embodied in the Macintosh software called *Hypercard*, were very user-friendly as compared to the command line interface that was the run of the mill of the majority of computers up until then. Hypertext links were the epitome of user-friendliness.

Reread the previous paragraph pretending that what you are reading is displayed on a Macintosh computer and the term *graphical user interface* has a hypertext link. Suppose you found that you were unsure as to what GUI meant or that you wanted more information about it. You could use your mouse to click on the term GUI and the hypertext link would bring up a window that provided more information, maybe even a diagram, about graphical user interfaces.

This was magic. In 1989, a Swiss physicist, Tim Berners-Lee, wrote a paper proposing a protocol that used a graphical user interface and hypertext links to transfer, navigate, and display the information stored in computers. His protocol became the *HyperText Transfer Protocol* (http, for short) and within the following year, the term *World Wide Web* was coined and the first Web browser software was written. It had a GUI and used hypertext links for navigation. Web pages began to appear here and there on the Internet.

More Web pages created a demand for better Web browsing software. Better Web browsing software resulted in even more Web pages. ARPANET became the Internet and the rest is history.

GETTING ON THE INTERNET

Getting on the Internet is getting easier all the time. Back in the "good" old days (way back in the early 1990s), getting on the Internet required having access to a computer at work or school that was connected to the Internet or having deep pockets in order to afford the cost of connecting your computer directly to the Internet yourself.

The first option had drawbacks because your employer or school often put limits on Internet access. ("Mr. Employee, please justify logging into alt.fun.fooling.around for 20 hours this past week... on company time!") The second option, the direct connection, had no drawbacks except that it was too darn expensive for the average computer user. Unless you were using the Internet for your business, this option was not an option.

Then, things began to change. For one thing, the price of high speed modems fell.

When someone says *Internet,* most people who know anything about the subject think of those beautiful, graphical Web pages, those www-dot-something-dot-somethings that we see everywhere in the media yesterday and today. Beautiful www-dot-something-dot-somethings pack a lot of data. In order to display those Web pages *on* your computer, all that data must be transferred *to* your computer. The faster you can transfer that data, the faster your computer displays a Web page.

To transfer data fast requires a high speed modem. Whereas modems operating at 9600 are fine for transferring plain text (7-bit ASCII characters), these modems are excruciatingly slow when it comes to transferring all the data contained in your typical Web page. Unless you have the patience of Job, you would not consider using such a modem for Internet surfing.

Until recently, the price of high speed modems was prohibitively high and, like an Internet direct connection, not affordable for the average computer user. But prices for high speed modems did fall and suddenly, Web page speed requirements were within the reach of the average user. The problem was that even though you could now afford a high speed modem, there was nothing to hook up at the Internet end of your modem. The best you could do was dial-in to the computer at work or school and use that computer as a gateway to the Internet. Now you are back where you started from, faced with the limitations imposed by your employer or school. The only difference being that now you were at home!

Things continued to change for the better. Inexpensive high speed modems begot a demand for Internet access which begot the proliferation of *Internet Service Providers* (ISPs). As its name implies, an ISP is in the business of providing access to the Internet. The ISP pays for a direct connection to the Internet, and sets up a bank of telephone lines connected to this direct connection. The user pays the ISP for access to the telephone line connection to the Internet. The ISP now provided you with a place to use your high speed modem.

And that is basically where we are today!

It All Comes Down To Collecting Parts

You need a few parts to get on the Internet. You probably have some of these parts lying around already. Everybody has a phone line and if you are reading this, you probably have a computer, too.

Maybe you also have a modem, but if it is 9600 or slower, you should seriously consider buying something faster if you are serious about using the Internet. High speed modems are very affordable today. In the early '90s, 2400 baud modems sold at the Dayton HamVention for what was then a bargain price of $125. Today, "33.6 kbps as low as $125.95" stares you in the face from the page of a computer-parts mail order catalog. At Dayton or your local ham radio/computer flea market $100 or less could be the "show price." And prices are still falling!

You also need an ISP, preferably one that provides local access in order to avoid long distance telephone toll charges. Check the yellow pages of your telephone book under *Internet Service Provider.* Another source for ISPs is the ISP listings and advertisements appearing in the computer/Internet section that a lot of newspapers publish on a regular basis these days. The sidebar contains the contact information of a number of national ISPs.

Do some comparative shopping. The ISP business is very competitive, so get the best price and best service. Today, you should be able to obtain local 28.8 kbit/s Internet access with no time restrictions (there is no meter running) for well under $20 per month.

Typically, when you begin doing business with an ISP, there is a sign-up fee for setting up your account and providing you with the software you need to access the Internet from your computer. The quantity and quality of the ISP-provided software varies, but at a minimum, it should get you up and running on the Internet. If need be, you can always upgrade the software later. The ISP should also provide you with the information you need for configuring the software (e.g., IP number, host name, initial password, POP server, SMTP server) with directions on how to plug this information into the software. Even on user-friendly computer systems like the Macintosh and Windows 95, this process is daunting to the first-time user. So, if you are having trouble, call the technical support department of your ISP and have them step you through the software configuration process.

Depending on how you plan to use the Internet, you may not need all the software your ISP provides. The following lists the type and function of software that your ISP may provide. Often, the software supplied by an ISP is multi-purpose.

Don't get overwhelmed by this list. It is often simpler than it appears. Most net browsers include mailers and news readers. Major ISPs supply *front-end* software, that let you log onto their service. This front-end software usually includes a Web browser, mailer, news reader and FTP capability. Many Web

National Internet Service Providers (ISPs)

No guarantee is made to the accuracy (especially pricing) because this is a very competitive and volatile market. By the time you read this, prices will likely be lower for many of these providers. Many new local providers are coming on-line every day. Check with the hams in your area to see who they are.

Uunet Technologies, Inc.
Product: AlterDial
Web Site: **http://www.uu.net/alterdia.htm**
Points of Presence: 100 U.S. cities
Pricing: $25 setup fee, $30/month for 5 hours, $2/each additional hour
Customer Service: 800-900-0241

America Online Inc.
Product: America Online
Web Site: **http://www.aol.com/**
Points of Presence: 166 U.S. cities, AOL users also can connect via SprintNet
Pricing: no setup fee, $9.95/month for 5 hours, $2.95/each additional hour or $19.95/month unlimited access
Customer Service: 800-827-3338

AT&T WorldNet
Product: AT&T Business Network
Web Site: **http://www.att.com/worldnet/**
Points of Presence: Major Cities
Pricing: Free software provided, over 2500 business information offerings, $35.95/month for 10 hours, $2.95/each additional hour
Customer Service: 800-394-8840

CompuServe Inc.
Product: CompuServe Information Service
Web Site: **http://www.compuserve.com/**
Points of Presence: 420 U.S. cities
Pricing: no setup fee, $9.95/month for 5 hours, $2.50/each additional hour, or $24.95/month for 30 hours plus $1.95/each additional hour

Concentric Network Corporation
Product: Concentric Network
Web Site: **http://www.concentric.net/**
Points of Presence: 197 U.S. and 7 Canadian cities
Pricing: no setup fee, $7.95/month for 5 hours, $1.95/each additional hour, or $29.95/month for unlimited access plus free access to over 35 top US bulletin board systems via *BBS Direct* (a separate service of the company); discounts given for six-month and annual prepays

Global Network Navigator (An America Online Subsidiary)
Product: GNN
Web Site: **http://gnn.com/gnn/**
Points of Presence: more than 700 U.S. cities (most major ones; SprintNet & GNNNet)
Pricing: no setup fee, $14.95/month for 20 hours, $1.95/each additional hour
Customer Service: 800-819-6112

IBM Global Network
Product: IBM Internet Connection Service
Web Site: **http://www.ibm.com/globalnetwork/**
Points of Presence: 111 U.S. cities, over 400 access numbers world wide
Pricing: $19.95 setup fee, $29.95/month for 30 hours, $2.00/each additional hour
Customer Service: 800-426-3333

MCI Telecommunications
Product: InternetMCI
Web Site: **http://www.internetmci.com/**
Points of Presence: 60 U.S. cities
Pricing: $3.00/month for 3 hours, $1.80/each additional hour or $19.95/month for unlimited access
Customer Service: 800-550-0927

Netcom On-Line Communications Services Inc.
Product: NetCom Internet Services
Web Site: **http://www.netcom.com/**
Points of Presence: 175 U.S. cities
Pricing: $25 setup fee, $19.95/month includes 40 prime time hours (9 AM-midnight M-F) with unlimited access during non-prime hours, $2/hour for additional prime time hour
Customer Service: 800-353-5600

Performance Systems International Inc.
Product: InterRamp
Web Site: **http://www.psi.net/internet/**
Points of Presence: 150 U.S. cities
Pricing: $9 startup, $9/month for 9 hours or $29/month for 29 hours, $1.50 each additional hour (weekends and 11 PM to 8 AM weekdays are free)
Customer Service: 800-774-0852

Performance Systems International Inc.
Product: Pipeline USA
Web Site: **http://www.psi.net/internet/**
Points of Presence: 140 U.S. cities
Customer Service: 800-395-1056

Prodigy Services Co.
Product: Prodigy Network
Web Site: **http://www.prodigy.com/**
Points of Presence: 325 U.S. cities; world's largest dial-up access provider to the Web
Pricing: $9.95/month for 5 hours or $30/month for 30 hours, $2.95/each additional hour
800-PRODIGY

Microsoft Corp.
Product: The Microsoft Network
Web Site: **http://www.msn.com/**
Points of Presence: Over 100 U.S. cities
Pricing: No setup fee but can only be accessed from Windows 95, $19.95/month (frequent users monthly plan)—20 hours, $2.00/each additional hour

(continued on next page)

sites allow downloading files through your browser, so separate FTP software is not needed. Telnet, an important Internet capability in the past, is less so these days as fewer sites allow casual net occupants to really do very much on their computers. You can probably live, and live very well, without Telnet capability!

Web Browser—A Web browser allows you to locate and display Internet Web pages. Most Web browsers include some or all the functions of the other software types in this list.

Mailer—A mailer allows you to send and receive e-mail via the Internet. Most Web browsers include a mailer function.

News Reader—A news reader allows you to read the messages that are published in the various special interest newsgroups that proliferate on the Internet. Pick a subject and there is likely to be a newsgroup for it. Most Web browsers include a news reader function.

FTP Software—FTP is the acronym for *file transfer protocol*. FTP software permits you to transfer files to or from another computer on the Internet. Most Web browsers include an FTP function.

Telnet Software—Telnet is a terminal-emulation protocol and Telnet software permits you to access the Internet with a command line interface. Using UNIX commands, you may perform many of the functions of the software types listed above. Telnet is recommended for the advanced Internet user. Few Web browsers include a Telnet function.

SURFING WITH A BROWSER

In order to surf the Web pages of the Internet, you need a Web browser. If your ISP provides you with software, that software, at a minimum, likely includes a Web browser. The basic function of a Web browser is to connect your computer to a remote computer site and transfer the contents of a selected Web page stored at that site to your computer.

Today, the browsers of choice are Netscape Navigator and Internet Explorer. You can download copies of these browsers from **http://www.netscape.com** and **http://www.microsoft.com** respectively. The big two browsers are very similar (see Fig 6-1). They usually can be used interchangeably for day-to-day Web surfing.

Let's Go Surfin' Now (Everyone Is Learnin' How)

To command a Web browser to connect your computer to a remote computer site and transfer the contents of a selected Web page stored at that site to your computer, you input the location of the Web page into the browser and it does the rest. The location of a Web page is known as its *uniform resource locator* or *URL,* for short. For example, the URL for the homepage of the author of this chapter is **http://www.tapr.org/~wa1lou**.

The letters preceding the colon in the URL indicate the protocol used to transfer data to your computer from the remote computer. In this example, **http** (HyperText Transfer Protocol) is the protocol for transferring, navigating, and displaying Web page data.

The twin slash bars (*//*) indicate that the *domain name* follows. The domain name is the name of the remote computer where the Web page is stored.

The start of the address, **www.tapr.org**, is the name of that remote computer. This name typically starts with www, the acronym for World-Wide Web, and ends with one of the following: **org** for nonprofit organizations, **com** for commercial enterprises, **edu** for educational institutes, **gov** for governmental organizations, **mil** for military organizations, and **net** for communication networks. However, just as the telephone companies are issuing new area codes, you can expect new URL suffixes to pop up any day now. While **www** is very common at the start of an address, it is not required, and other names or set of letters are often used.

The centerpiece of the domain name is unique and usually an acronym of the organization that owns the remote computer. In this case it is **tapr,** which is the acronym for Tucson Amateur Packet Radio, Inc.

On many remote computers, everything after the first solo slash bar (*/*) indicates the path to the Web page on the remote computer. A tilde (~) indicates a home directory. In this case, **/~wa1lou** indicates that the path to the Web page is the home directory named **wa1lou**.

After your Web browser navigates the Internet and locates the Web page you wish to view, it begins transferring the data that composes the Web page to your computer. Some Web pages appear almost instantly. Others seem like they take an eternity to appear.

The data rate that your modem uses to communicate with your ISP is the most obvious factor affecting the display speed of a Web page. The faster the data rate, the faster the Web page appears. Note that just because you have a high speed modem, there is no guarantee that it is operating at its top speed. Usually it isn't. Instead, it adapts its data rate for the current telephone line conditions and the data rate of the modem at the ISP.

The more graphical the Web page, the longer it takes to transfer its contents and display it on your computer. Web pages that contain only text appear very quickly. Web pages that contain graphics make you wait.

The configuration of your computer also affects the display speed of a Web page. The clock speed of your computer is certainly a factor, but other factors also come into play. Some computers handle the video display functions more efficiently and quickly than other computers. A computer that is advertised as *optimized for multimedia* is likely to display a Web

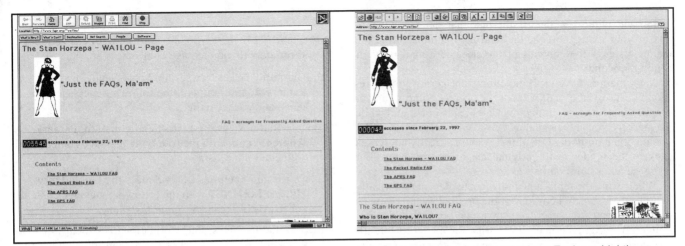

Fig 6-1—The similarities of the two most popular Web Browsers, Netscape Navigator (left) and Internet Explorer (right) are evident as they display the same Web page.

page faster than a plain vanilla computer. And so it goes.

Underwater Golf (Or Surfing The Links)

Most Web pages are larger than the default window provided by your Web browser, so open the window as much as possible (both vertically and horizontally) to permit the Web page to fill your monitor. Even after doing this, the Web page likely extends below the window. To display the hidden parts of the Web page you can use the scroll function of your browser to reveal the rest of the page or you may be able to use hypertext links to get around the Web page (assuming that such links are included in the page).

Both text and graphics can be a hypertext link. Text links are easy to spot because they are typically colored blue, while the rest of the text surrounding them is another color, typically black. Graphic links are less obvious. Moving the mouse pointer over a graphic determines if a graphic has a link, because when you do so, the pointer changes shape, from a narrow arrow to a hand with raised forefinger.

There are two types of hypertext links that you can use for navigation. There are navigational hypertext links that permit you to navigate within the currently displayed Web page. These work faster than the scroll function of your browser and are often contained in the Web page list of contents, if any, (to move you to a specific area of interest that is listed in the contents) or at the bottom of the page (to get you back to the top).

The other type of navigational hypertext link is one that replaces the currently displayed Web page with a different Web page, which may be located at the same computer site or at the other end of the Earth. Such links are used to display Web pages that are related to the subject matter of the Web page you were previously viewing.

Another type of hypertext link you are likely to find on most Web pages is an e-mail link. An e-mail link invites you to send e-mail to whoever or whatever is responsible for the Web page. You can use this link to comment or ask questions about the Web page. When you select an e-mail link, your Web browser opens a window already addressed to the party responsible for the Web page, waiting for you to enter your comments, questions, etc.

Those are the highlights of a typical Web browser. Use one for a while and soon you too will be surfing with the best of them.

SENDING AND READING THE MAIL

In order to send e-mail to other users via the Internet, you use a mailer. Netscape Navigator, Internet Explorer, and other Web browsers have built-in mailers, however, if you are like me and do a lot more e-mailing than Web surfing, you may prefer to use a standalone mailer like Eudora for e-mailing. Standalone mailers are much smaller than your typical Web browser and, as a result, they load into computer memory faster than a browser and often function faster, too.

When you write a letter and mail it via the postal system, you must address it properly, otherwise it is not delivered. Similarly, when you compose e-mail, you must address it properly in order to ensure delivery. E-mail addresses have some similarities to URLs, but are typically much shorter, for example, the author's e-mail address is **stanzepa@ct2.nai.net**

The at-symbol (@) divides the e-mail address in two. The first half of the address contains the name of the user located on the computer whose domain name appears in the second half of the address.

The user name may be some combination of the first and last name of the user (e.g., stanzepa for Stan Horzepa). There is often a limitation (eight characters) to the length of the user name. Thus, in this example the user name is **stanzepa,** not **stanhorzepa**. Some systems use random assignments of letters or numbers. Often *aliases* are permitted. At the ARRL Web site most Headquarters staff members can be reached by either a combination of initial and last name (**jkleinman @arrl.org**) or call sign (**n1bke@arrl.org**).

The domain name is typically an acronym of the entity that owns the computer handling the user's e-mail (usually your ISP) followed by a period and the same **com**, **edu**, **gov**, **mil**, **net** or **org** convention used in URLs.

With e-mail address in hand, using a mailer is easy. Enter the address in the address field, optionally enter the subject of the e-mail in the subject field, and enter the contents of the e-mail in the wide open field intended for the contents. After you compose the e-mail, you may send it immediately or save it for later, that is, the next time you connect to the Internet.

Reading e-mail is even easier than sending it. Invoke the check mail command and if your computer is not connected to the Internet, the mailer initiates a connection. Once connected, the mailer polls the mail server of your ISP for any new e-mail addressed to you. If new e-mail exists, it is transferred to your mailer and included in the list of e-mail that appears in the in-box of your mailer. Invoke the read mail command and the mailer displays the contents of the selected e-mail on your computer display.

Most mailers have reply functions that permit you to re-spond to received e-mail without the need of entering an ad-dress or subject. You simply invoke the reply command while the message you are responding to is displayed, and the mailer sets up blank outgoing e-mail that is automatically addressed back to the sender of the original message. Just enter your response and send it on its way.

When using the reply function, most mailers automatically include all or selected portions of the original e-mail contents in your reply message in order that you may comment on some-thing in the original e-mail. For clarity, the mailer differenti-ates the original e-mail contents from your response, typically by placing the greater-than symbol (>) at the beginning of each line of the original message.

Thus,

what's for dinner?
in the original e-mail becomes
>*what's for dinner?*
in your response.

Using e-mail is easy and fast. I recently composed e-mail to a friend who lives at the other end of the state. After composing the message, I selected the send function of my mailer and faster than a speeding bullet, my telephone started ringing. My friend was calling to ask me a question about the contents of the e-mail I had just sent him. I was impressed! But when your ISP is very busy, or the Internet is loaded with traffic, a brief e-mail can take several hours to go 50 miles.

E-MAIL LISTS

Related to e-mail, the e-mail list is an ongoing discussion group that uses e-mail to exchange messages concerning a particular topic of interest. The power of the e-mail list is that each message addressed to the list is disseminated to everyone who is a subscriber to the list. The message dissemination is performed automatically by the list server, which is software running on a computer. Whenever you want to send a message to the list, you send it to the listserver, and the server resends it to all the list subscribers.

Note that you usually don't have to be a subscriber to a list to send a message to a list. One way to find out how to sub-scribe to a list is to send a message addressed to the list asking how to subscribe. In general, subscribing to an e-mail list re-quires that you send e-mail addressed to the list server request-ing a subscription. The exact subscription process varies with each e-mail list and since you are dealing with a computer, you must follow the subscription process exactly; otherwise, the computer rejects your request.

A list of Amateur Radio related e-mail lists follows along with information you need to subscribe to each. To subscribe to a list, you send e-mail to the specified address, leave the subject field of your message blank, and include only the speci-fied one-line command in the body of your message. Some mailers will not allow you to send an e-mail with a blank sub-ject field—in that case enter a single space in the blank field.

After you e-mail your subscription request, the list server sends you a confirmation of your subscription along with in-formation on how to unsubscribe from the list, and how to send (*post*) messages to the list. Save this file for future reference.

AEA DSP 2232 List—Send e-mail to: **dsp2232-request @rmi.de** with no message contents.

alt.ham-radio.packet Newsgroup List—Send e-mail to: **listserv@ucsd.edu** with the following message con-tents: add packet-radio.

Amateur Radio Newsline List—Send e-mail to: **listserv @netcom.com** with the following message contents: sub scribe newsline-list.

Amateur Radio Satellites List—Send e-mail to: **listserv @PLEARN.BITNET** with the following message contents: subHAM-SAT *your first and last name*.

ARRL Bulletins, News, and Information List—Send e-mail to: **w1aw-list-request@arrl.org** with the following message contents: subscribe *your e-mail address*.

ARRL Computer Aided Design List—Send e-mail to: **listproc @tapr.org** with the following message contents: subscribe arrlcad *your first and last name*.

Other ARRL Information—The following e-mail lists are dis-tributed by Netcom internet services and are sponsored by Mike Ardai, N1IST, and the Boston Amateur Radio Club. If you have any questions about these lists, please direct them to Mike Ardai, N1IST at **n1ist@netcom.com**. For more information on the Boston ARC, subscribe to barc-list.
arrl-exam-list—Amateur Radio license examinations scheduled in the US and in some foreign areas.
arrl-nediv-list—Bi-monthly bulletins from the ARRL NE Division director
arrl-ve-list—Announcements to VEs and VE teams.
fieldorg-l—ARRL field organization discussions
letter-list—Redistribution of the *ARRL Letter*

These lists are automatically maintained by Majordomo soft-ware. To sign up or inquire about these lists, send an e-mail to **listserv@netcom.com** with the following in the body (subject is ignored) of the message. <listname> is the name of the list (exactly as printed above).
To subscribe: subscribe <listname>
To unsubscribe: unsubscribe <listname>
For more information about a list: info <listname>
For more information about Majordomo: help
To post (to the two-way lists), send your message to
<listname>@netcom.com

Bitbucket Mailing List (specialized Amateur Radio communi cation techniques)—Send e-mail to: **majordomo@ primenet.com** with the following message contents: SUBSCRIBE bitbucket.

Boatanchors List—Send e-mail to: **listproc@ThePorch.com** with no message contents.

Boston Amateur Radio Club List—Send e-mail to: **listserv @netcom.com** with the following message contents: subscribe barc-list.

Boston Amateur Radio Club RACES and Emergency Manage ment List—Send e-mail to: **listserv@netcom.com** with the following message contents: subscribe barc-races.

CQ Contesting List—Send e-mail to: **cq-contest-request**

@tgv.com with the following message contents: SUBSCRIBE.

CT (contesting software) List—Send e-mail to: **ct-user-request@ve7tcp.ampr.org** with the following message contents: subscribe.

Direction-Finding/Fox-Hunting List—Send e-mail to: **listserv @netcom.com** with the following message contents: subscribe fox-list.

DX Mailing List—Send e-mail to: **dx-request@ve7tcp. ampr.org** with the following message contents: subscribe.

Eastern ARRL Massachusetts Section List—Send e-mail to: **listserv@netcom.com** with the following message con tents: subscribe ema-arrl.

Experimental Digest from Ham Radio Related Newsgroup List—Send e-mail to: **listserv@ucsd.edu** with the following message contents: addham-radio.

Georgia Tech Radio Club Mailing List—Send e-mail to: **listserv@GITVM1.BITNET** with the following message contents: sub GTRADIO *your first and last name*.

Glowbugs List (homebrewing with tubes)—Send e-mail to: **listproc@theporch.com** with the following message contents: subscribe glowbugs *your first and last name*.

Ham List—Send e-mail to: **listserv@vm.ege.edu.tr** with the following message contents: SUB HAM-L *your first and last name*.

Ham Policy List—Send e-mail to: **listserv@ucsd.edu** with the following message contents: add ham-policy.

Ham Radio Applications of the Jolitz 386BSD Software List—Send e-mail to: **listserv@ucsd.edu** with the following message contents: addham-bsd.

Ham Radio Operation List—Send e-mail to: **listserv @ucsd.edu** with the following message contents: add info-hams.

KA9Q NOS Amateur Radio Networking List—Send e-mail to: **listserv@ucsd.edu** with the following message contents: add noose-hacks.

Microwave List—Send e-mail to: **listserv@ucsd.edu** with the following message contents: add ham-wave.

Military Affiliated Radio Service List—Send e-mail to: **listserv@stat.com** with the following message contents: subscribe mars-list.

New England VE Exams List—Send e-mail to: **listserv @netcom.com** with the following message contents: subscribe ky1n-list.

Packet Radio Internet Extension List—Send e-mail to: **listserv@UCSFVM.BITNET** with the following message contents: subPRIE-L *your first and last name*.

Polish Radio Amateur List—Send e-mail to: **listserv @PLEARN.BITNET** with the following message contents: subHAMS-PL *your first and last name*.

QRP List—Send e-mail to: **listserv@lehigh.edu** with the following message contents: subscribe QRP-L *your first and last name and call sign*.

Radio BBS Bulletin Distribution List—Send e-mail to: **listserv@ucsd.edu** with the following message contents: add bull-fwd.

rec.radio.amateur.antenna Newsgroup List—Send e-mail to: **listserv@ucsd.edu** with the following message contents: add ham-ant.

rec.radio.amateur.digital.misc Newsgroup List—Send e-mail to: **listserv@ucsd.edu** with the following message contents: addham-digital.

rec.radio.amateur.equipment Newsgroup List—Send e-mail to: **listserv@ucsd.edu** with the following message contents: add ham-equip.

rec.radio.amateur.homebrew Newsgroup List—Send e-mail to: **listserv@ucsd.edu** with the following message contents: addham-homebrew.

rec.radio.amateur.space Newsgroup List—Send e-mail to: **listserv@ucsd.edu** with the following message contents: add ham-space.

rec.radio.info Newsgroup List—Send e-mail to: **listserv @ucsd.edu** with the following message contents: add radio-info.

TAPR Automatic Packet Reporting System (APRS) Special Interest Group—Send e-mail to: **listserv@tapr.org** with the following message contents: subscribe APRSSIG *your first and last name*.

TAPR BBS Special Interest Group—Send e-mail to: **listserv @tapr.org** with the following message contents: subscribe TAPR-BB *your first and last name*.

TAPR DSP-93 Special Interest Group—Send e-mail to: **listserv@tapr.org** with the following message contents: subscribe DSP-93 *your first and last name*.

TAPR DTMF Accessory Squelch (DAS) Special Interest Group—Send e-mail to: **listserv@tapr.org** with the following message contents: subscribe DAS *your first and last name*.

TAPR HF Special Interest Group—Send e-mail to: **listserv@tapr.org** with the following message contents: subscribe HFSIG *your first and last name*.

TAPR Networking Special Interest Group—Send e-mail to: **listserv@tapr.org** with the following message contents: subscribe NETSIG *your first and last name*.

TAPR Spread Spectrum (SS) Special Interest Group—Send e-mail to: **listserv@tapr.org** with the following message contents: subscribe SS *your first and last name*.

TAPR-TNC Special Interest Group—Send e-mail to: **listserv@tapr.org** with the following message contents: subscribe TAPR-TNC *your first and last name*.

TCP/IP Ham Packet Radio List—Send e-mail to: **listserv @ucsd.edu** with the following message contents: add tcp-group.

TCP/IP Ham Packet Radio List Digest—Send e-mail to: **listserv@ucsd.edu** with the following message contents: addtcp-digest.

Technical Discussion List—Send e-mail to: **listserv @netcom.com** with the following message contents: subscribe ham-tech.

University Ham Radio Clubs Mailing List—Send e-mail to: **ham-univ-request@listserver.njit.edu** with the following message contents: subscribe ham-univ *your first and last name.*

University of Kentucky Amateur Radio Club List—Send e-mail to: **listserv@UKCC.BITNET** with the following message contents: sub UKARC *your first and last name.*

VHF List—Send e-mail to: **vhf-request @w6yx.stanford.edu** with the following message contents: subscribe.

Weather List—Send e-mail to: **listserv@netcom.com** with the following message contents: subscribe ham-wx.

Yaesu 990 Mailing List—Send e-mail to: **990-request @xyzoom.info.com** with no message contents.

HAVE YOU READ THE NEWS?

Newsgroups proliferate on the Internet. At last count, there were 13,538 newsgroups. There is likely to be at least one newsgroup covering any subject matter you can think of and a lot more that you'd never think of!

You need news reader software to read the messages that are posted in the newsgroups (most Web browsers include a news reader function).Typically, the news reader obtains a list of all the newsgroups that exist from the news server of your ISP and then you select the newsgroups you are interested in reading from that list. After making your selection, the newsreader obtains a list of all the messages posted in those newsgroups and displays the list to you. You pick and choose the messages you want to read.

The news reader software also allows you to post your own messages to a newsgroup and to reply to other messages already posted on the newsgroup. The posting and replying function operates in the similarly to the send and reply function of a mailer.

The following is a list of newsgroups related to Amateur Radio:

alt.ham-radio.packet
alt.radio.digital
alt.radio.pirate
alt.radio.scanner
alt.radio.scanner.uk
aus.radio.amsat
aus.radio.packet
francom.radio_amateur
in.ham-radio
rec.antiques.radio+phono
rec.radio.amateur
rec.radio.amateur.antenna
rec.radio.amateur.digital.misc
rec.radio.amateur.dx
rec.radio.amateur.equipment
rec.radio.amateur.homebrew
rec.radio.amateur.misc
rec.radio.amateur.policy
rec.radio.amateur.space
rec.radio.amateur.swap
rec.radio.info
rec.radio.scanner
rec.radio.shortwave
rec.radio.swap
rpi.union.ham-radio
sanet.radio.packet
slac.rec.ham_radio

su.org.ham-radio
tnn.radio.amateur
uk.radio.amateur
uwarwick.societies.amateur-radio

Lists Vs. Newsgroups

E-mail lists and newsgroups are similar in that they are both ongoing discussion groups exchanging ideas about a particular topic of interest. They differ in that the e-mail list, by virtue of its subscription feature, offers more privacy than a newsgroup. There is much less intentionally disruptive traffic (*spamming* and *flaming*) on an e-mail list than on a newsgroup. The privacy of the list deters transients, who intentionally spam and flame newsgroups, from doing the same on a list. Unless the spammer or flamer subscribes to a list, he cannot read the result of his spamming/flaming and thus has little incentive to intentionally spam and flame.

FETCHING FILES

There are a lot of files and software stored on the Internet that you can obtain if you have the software to go and fetch those files. The fetching file software uses the file transfer protocol (FTP) to transfer files to or from another computer on the Internet. Most Web browsers include this function.

To use FTP software, you simply command the software to connect with the computer site where the software you desire is stored, enter the path to the directory where the software is located, and tell the software to get the software.

The following is a list of Internet sites that have loads of Amateur Radio software for your FTP'ing pleasure:

ftp.amdahl.com
ftp.cs.buffalo.edu
ftp.tapr.org
ftp.uni-paderborn.de
ftp.usr.com
ftp.uu.net
grivel.une.edu.au
nic.funet.fi
oak.oakland.edu
ucsd.edu
vixen.cso.uiuc.edu
wolfen.cc.uow.edu.au

TELNET THE INTERNET

Telnet software predates the Web browsers that everyone seems to use today. Before browsers, Telnet software was one of the primary tools for surfing what we now call the Internet.

Using a command line interface, Telnet software is strictly text-based and, as a result, allows you to get around the Internet and perform tasks faster than the typical GUI-intensive Web browser. However, this command line interface uses UNIX commands and thus turns off people who have no need to learn a computer language when they can use a relatively user-friendly Web browser instead.

FINDING THE GOODIES

There is lots of stuff stored on the Internet. The problem is finding it! There are two ways: *search engines* and *directories*.

Search engines prompt you to enter one, two, or more words into a search field. In response, the engines return a list of the hypertext-linked Internet locations where the words you entered were found. Each engine has unique options for more intelligent searches, so check out the search help information for each engine before you initiate a search (and end up with a lot of useless information).

One very popular search engine is *Wired* magazine's Hot Bot at **http://www.hotbot.com/index.html** (see Fig 6-2). Two others frequently used are Digital Equipment Corporation's Alta Vista (**http://altavista.digital.com**) and Lycos (**http://www.lycos.com**).

Directories present you with a hierarchy of topics. You begin a search by selecting a general topic and working down through a continuing narrower hierarchy of topics. At the end of the search, you find a list of the hypertext-linked Internet locations related to the topic. Yahoo!(**http://www.yahoo.com**) is the most popular directory on the Internet.

Amateur Radio's Presence On The Internet

Considering the number of hams in the computer and communications fields, it is no surprise that Amateur Radio has had a presence on the computer communications network (the Internet) for a long time. There are thousands of ham radio Web sites, e-mail lists, newsgroups, software depositories, and links on the Internet (the Hot Bot search engine comes up with 52,301 hits for *amateur radio,* and 31,077 hits for *ham radio*).

Ron Klimas, WZ1V, has put together an extensive compilation of Amateur Radio's presence on the Internet at **http://uhavax.hartford.edu/~newsvhf/ham-www.html**. A list of Amateur Radio topics in WZ1V's list follows (an example follows each topic in brackets).

Businesses

Sales [RF Components, Inc.]
Manufacturers [Kantronics/RF Concepts]
Repair [Radio Doctor]
Software/Computer [NuMorse/NuTest for Windows]
Hamfests & Online Swap [Delaware Valley Radio Assn]
Publications [Klingenfuss Publications]
Non-Ham [ABSNET Internet Services]

Clubs (Non-specific)

National/International [The Finnish Amateur Radio League]
Local (USA) [San Bernardino Microwave Society]
Local (Canada) [The White Rock ARC]

Fig 6-2—Using a search engine to find Amateur Radio on the Internet results in 52,301 hits! The hot-link advertisement (What do Laws of Gravity...) is the *price* you pay for using this *free* search engine service.

Local (non-North America) [Sydney Waverley ARS]
University (North America) [Columbia University ARC]
University (non-NA) [Technical Univ. Warsaw Radio Club]
Company Related [Compaq's Amateur Radio Club]
Non-Ham Common Interest [Lambda ARC]

Special Interest Groups & Clubs

Antennas [Copper Cactus]
Antique or Vintage [Boat anchors!]
ATV & SSTV [TX ATV Society]
Contesting [Kentucky Contest Group]
County Hunting [The County Hunter Web]
CW Enthusiasts [FISTS]
Digital Radio/Packet [Colorado Digital Working Group]
Direction Finding [Southern California Transmitter Hunts]
DX [Northern Ohio DX Association]
Emergency, Service & Special Event [San Diego Co. RACES]
Handicapped or Disabled [Courage HANDI-HAM System]
Homebrew & QRP [QRP-L]
Linux [Linux Amateur Radio Software List]
Microwave [Microwave FAQ]
Nets, Non-Emergency [Ten-Ten International Net]
Space [SAREX - Shuttle Amateur Radio EXperiment]
Spread Spectrum [Spread Spectrum Communications]
Weak Signal (EME, Meteor, etc.) [North East Weak Signal]

Information

Amateur Radio practice exams [Ham Exam]
Aurora [Auroral Forecast Maps]
Call Sign Servers [Call Sign server at QRZ]
General Info [Radio Mods at OAK]
Governmental Bodies [FCC home page]
Repeaters [Colorado Repeater List by city]
Scanner Enthusiast Resources [NF2G Scannist Pages]
Software Archives & Shareware [*QST* related binaries]
Solar Flux Info / MUF / etc. [WWV last 25 reports]
SWL (Short Wave Listening) Resources [Listening Post]
Technology Resources [Motorola Data]
Time [US Naval Observatory Master Clock]
Web Resources and Search Mechanisms [HTML Access Counter]

Miscellaneous Things

Individual's Home Pages [WD1V home page]
Other Index Pages and Ham Related Lists [NZ1M's Links]
Misclass (But Ham related) [Hams on Internet]
BBS's and Internet Ties [N0ARY/BBS]
Links That Don't Relate to Radio [WeatherNet]

If you can't find what you are looking for on WZ1V's list, check out his links to other Amateur Radio lists, "Other Index Pages and Ham Related Lists."

POINTS OF INTEREST ON THE INTERNET

Now that you know what Amateur Radio is doing on the Internet, let us take a tour and sample some actual ham radio Internet sites.

javAPRS: Amateur Radio In Living Color

One of the most interesting marriages of Amateur Radio and the Internet is javAPRS. *Java* is a computer language for

Web browsers that permits a programmer to create a mini-application, known as an *applet,* that runs within a Web browser window. Steve Dimse, K4HG, wrote an applet that displays a live, interactive APRS window on a Web browser page as illustrated in Fig 6-3.

javAPRS is live because it is actually connected to a radio station monitoring an APRS channel and it is interactive because the user can control the APRS display. The user can zoom in and out of the display, center the display on a mouse-clicked location, and obtain lists of stations heard, weather reports monitored, beacons received, identification reports heard, and the last 25 messages monitored in the Java console.

javAPRS is popping up on Web pages all over North America. Currently, you can view javAPRS that display APRS activity in California, Georgia, Illinois, Maryland, Michigan, New York, and Ontario. Check out K4HG's javAPRS page at **http://www.bridge.net/~sdimse/javAPRS.html** to learn more about javAPRS and to find links to active javAPRS sites.

AMPRNET/Internet Gateways: Ham Transceivers On The Internet

Whereas javAPRS allows you to *monitor* live Amateur Radio activity, packet radio gateways allow you to receive *and* transmit live Amateur Radio activity by means of interfaces between the Internet and the amateur packet radio TCP/IP network (AMPRNET).

The Internet and AMPRNET use the same protocols, the suite of protocols commonly known as *TCP/IP,* for communications over their respective networks. This protocol compatibility makes interfacing the two networks a relatively simple matter (a lot simpler than hurdling the legal issues involved) and, as a result, AMPRNET/Internet gateways have sprung up all over the world. For a current list of these gateways and their capabilities, check out **http://www.ccnet.com/~rwilkins/gateways.html**.

The AMPRNET/Internet gateways provide a lot of services. Some services are apparent to the user. For example, your local gateway may permit you to establish SMTP (mail), Telnet, FTP, and finger (user information) connections directly with other amateur TCP/IP stations in the US and other countries. Your local gateway may also permit you to connect directly to the gateways (or to neighboring machines set up specifically to provide end-user services) to access weather data, call sign information, chat servers, etc. Most of the gateways also offer these services to AX.25 users and sometimes to NET/ROM users as well.

The AMPRNET/Internet gateways also provide services that are not apparent to the user. In the background, gateways enhance AX.25 and AX.25 network node users' enjoyment of packet radio without those users having to connect to a gateway or even being aware that a gateway exists. For example, some regional packet cluster networks use gateways to connect to each other. Other common *hidden* uses are carrying PBBS mail and connecting geographically dispersed AX.25 network nodes. AX.25 network node users who see dozens of nodes from all over the world are often seeing the results of an AMPRNET/Internet gateway in action.

There is misinformation and misconceptions about how to use the AMPRNET/Internet gateways because no two gateways are exactly alike. So, before using a gateway, familiarize yourself with its unique operating procedures. Do not assume that the procedures that worked at one gateway work at another.

Call Sign Lookup

Can't find your copy of the call sign directory? Did you lose your call sign directory compact disk? Well, don't worry—because you can dial up the Internet and use one of the Amateur Radio call sign servers, like the one shown in Fig 6-4, that are interfaced to the current FCC license information.

There are a number of such servers that you can access and most of them allow you to do more than just look up a call sign. You can perform searches using keywords such as surnames, city and state names, or ZIP Codes to obtain lists, for example, of all the hams with the surname of *Maxim* or all the hams who live in the ZIP Code of *06111.* Try and do that with a printed directory!

The Mother And Father Of All Amateur Radio Web Sites

There are two baseline Amateur Radio Web sites used by most hams: the ARRL and TAPR home pages. They are the Mother and Father of all Amateur Radio Web sites and most hams who are active on the Internet bookmark these pages with their Web browser.

As you would expect, the ARRL home page, *ARRL Web,* at **http://www.arrl.org**, contains a wealth of information that reflect the services the organization provides to the Amateur Radio community. Look at Fig 6-5 and you see an extensive selection of links that represent these varied services.

Similarly, the Tucson Amateur Packet Radio (TAPR) Home Page at **http://www.tapr.org** reflects the activities of the orga-

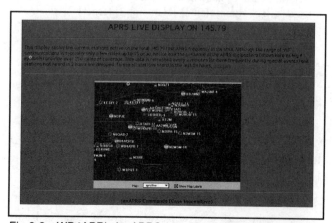

Fig 6-3—WB4APR's javAPRS page (**http://web.usna.navy.mil/~brununga/radio.html**) displays live APRS activity on 145.79 MHz in the Middle Atlantic States.

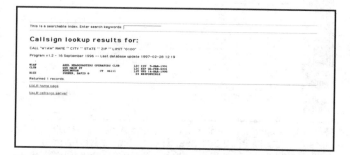

Fig 6-4—Servers like this one at **http://www.mit.edu:8001** allow you to search an on-line call sign directory by call, name, and location.

ARRLWeb:
The American Radio Relay League's World Wide Web Service

What's New? · **Announcements** · *The ARRL Letter* · **ARRL (W1AW) Bulletins**

Breaking news: Little LEO industry now wants 219-225 MHz, too! You can help!

Flash! ARRL Finds Cure for Cabin Fever in these Hot New Titles!

Advertising □ ARRL FTP | BBS | /VEC | Foundation □ Awards □ Band Threat News □ Clubs □ Contests □ Divisions & Sections □ Educational Activities □ Employment □ Exams □ FCC □ Government Relations □ Hamfests/Conventions □ IARU □ Information □ Links □ Product Notes □ Public Relations □ Public Service □ Publications Catalog □ Regulatory □ RF Exposure News □ SAREX □ Software □ Special Events □ Technical □ Vanity Calls □ W1AW Schedule □ *ARRLWeb* content search

7:39 PM ET 2/14/97 □ ARRL email: **hq@arrl.org** □ Webmaster: **webmaster@arrl.org**

Ham radio spoken here: Welcome to *ARRLWeb*, **the American Radio Relay League's home on the World Wide Web!** The League (email hq@arrl.org, telephone 860-594-0200, fax 860-594-0259), a membership service organization headquartered at 225 Main St, Newington, CT 06111, USA, serves the over 600,000 Amateur Radio operators, enthusiasts, experimenters and hobbyists in the United States, its territories and possessions.

The *ARRLWeb* Information Desk: What's new, how to contact us, links, *ARRLWeb* stats.

Your ARRL Membership: What we do for you, and how to join us.

In Your Neighborhood: Divisions, Sections, clubs, license exams, hamfests and conventions.

On the Air: Awards, contests and operating events for you and your radio, including special events, the latest ham-radio-in-space events in SAREX, the Shuttle Amateur Radio Experiment.

How to Get Started in Ham Radio: Your journey of ten thousand frequencies starts here.

Recruitment and Education: Help for instructors, school teachers and Scouting programs.

Delivering Amateur Radio's Message: ARRL's Public Relations and Government Relations activities.

Amateur Radio in the World: National ham radio societies the world over, and how they come together in the International Amateur Radio Union (IARU).

ARRL Products and Services: Ads, books, magazines, software, supplies, QSL bureaus, technical information service, regulatory information.

The ARRL Foundation: Funding for scholarships and projects that brighten Amateur Radio's light in the world.

The American Radio Relay League is the principal representative of the Amateur Radio Services, serving members by protecting and enhancing spectrum access and providing a national resource to the public.

Fig 6-5—The many links of ARRL Home Page represent the services that the League provides to the Amateur Radio community.

Fig 6-6—The Amateur Radio world refers to the TAPR home page to check the progress of the digital revolution.

nization that is spearheading the Amateur Radio digital revolution. The many links of the TAPR Home Page, as shown in Fig 6-6, indicate what TAPR is doing to make the revolution a reality.

KEEPING CURRENT

The Internet is evolving all the time. The Internet of yesterday is not the same as the Internet of today and it certainly will change by the time you check it out tomorrow. Unfortunately, this also means interesting and valuable Web pages come and go overnight—sometimes it seems as soon as their addresses are printed the page changes location!

One way to stay current is to keep an eye out for new books and reference materials. *Personal Computers In The Ham Shack,* published by the ARRL, contains not only information on the Internet for hams but also on all aspects of PCs—including circuit design, connecting PCs with your rig or test equipment, and antenna analysis. On a monthly basis, you can read the Digital Dimension column that appears in *QST*. Frequent checking of the ARRL Web site will alert you to new links and new information as soon as the ARRL staff finds them and posts them.

CHAPTER 7

Contests

CLARKE GREENE, K1JX
92 CYNTHIA LN, B2
MIDDLETOWN, CT 06457

The word "contest" brings to mind a number of images. Extravaganzas sponsored by various magazine publishers, where "you may have already won $100,000" may be most familiar. If you've ever had any dealings with the legal system or the legal profession, the term "contest your claim" may hold special significance (particularly depending on the outcome of that contest). Many sporting events are zealously described as "titanic contest(s) between close rivals."

To many radio amateurs, this same word describes an event encompassing none of these images, but at the same time, all of them. An Amateur Radio contest is an operating event, held over a predefined time period where the goal is…to enjoy yourself. Of course everyone has their own definition of enjoyment; that diversity of definition only adds to the appeal.

The first Amateur Radio contests were operating periods set aside specifically to attempt transoceanic communication between North America and Europe, just after World War I. Through this coordination effort, many of the participants successfully made some of the very first intercontinental contacts. Ultimately, this activity evolved into the ARRL International DX Contest. Although the technology and results have changed "somewhat," the object is still the same: Contact as many stations in as many areas of the globe as possible during the contest period.

Two other early operating activities were conceived as intense exercises to train radio amateurs in traffic handling and emergency communications. These are now known as the ARRL Sweepstakes and Field Day. The only real difference between the two is that Field Day emphasizes portable operation as might be required at a disaster site.

These examples demonstrate the two primary motivations for establishing radio contests: advancing the state of the art in radio amateur communications and advancing the expertise level of the Amateur Radio operator. Today, these same two goals provide the inspiration for sponsoring organizations to continue support of Amateur Radio contests. Many of the hams who take part in contests use them as opportunities to work a few new states or collect a few new countries. Others test the effectiveness of a new antenna or other piece of gear. Some work at increasing code speed. There are even some who take all of this very seriously and are out to *win*. In the final analysis, each participant gains expertise, learns a little more about propagation, and becomes a better operator, more equipped to perform effectively in the case of an emergency or to explore some new area of the Amateur Radio art, all while having a good time.

PHILOSOPHY OF EFFECTIVE CONTESTING

For many hams, participation in contests is very casual. If they happen to be on the air and hear a flurry of activity that can only be the result of a contest or big DXpedition, they'll figure out what's going on and pass out some contacts to the hams who are seriously going at it. Indeed, if it were not for these casual operators who are willing to get on the air and work a few contesters, most contests would be extremely slow and boring after the first few hours (see accompanying sidebar "Contesting for Non-Contesters"). For casual contesters, who may not have the time or inclination to compete (at least on that particular weekend), little forethought or planning is required. Getting on the air and knowing the contest exchange (or at least be willing to have someone tell you the exchange) is adequate preparation.

By contrast, serious contesting, at any level, requires some advance preparation. Preparation for a contest entails a number of considerations, and truly is an interdisciplinary subject. Unlike many forms of competition, the success of a radio amateur contest effort is a synergistic product of the station, the operator and a dose of luck as well. An inefficient operator at the controls of a giant station at a great location may not fare as well as a superior operator at a mediocre station. An experienced, competent operator at an outstanding station might completely dominate his or her class, or may have an electrical outage three hours into the contest and have to give up. One

A full crew on Field Day—From left to right: Willie, KA2ERE; Tom, KB2ORQ; Charles, N2JZA; Colonel Scuder, KB2PII and Lenny, KB2HQE. Hugh Gallagher, KA2SKO watched in the background. This table, one of several, held the packet and sideband station.

For Novices, Techs and New Hams

One Saturday evening you turn on the rig, and find the band jammed packed with signals. What's this, you say? Some national emergency? A convention of crazies? You may be close on the second guess—it's probably a contest.

What is it about contests that attracts people to them? For many it's a chance to work stations in rare places, whether it be DX locations, states or VHF grid squares. For others, it's the competition—participation in a sport where it's you against the other guy, whether it's across the street, the state, or the world. How well can you, your mind, and your equipment hold up against someone else? It's the mental equivalent to chess, yet can require the physical endurance of a marathon. Others use contests as a chance to try out new equipment, experiment with antennas or participate in a group effort with their radio club. On VHF, you may work stations on bands in places that would never see any activity apart from the contest weekends.

If you ask a die-hard contester, they'll probably boil it all down to one reason, though—excitement. Whether you run with the big boys, or just get on for an hour or two without even keeping track of points, contesting can be exciting. If you've never been bitten by the bug, perhaps we can strip away some of the mystique about this segment of the hobby for you.

What is a contest? Simply put, a ham radio contest is an event where hams try to contact as many other hams as possible in a given time under certain conditions. These rules can restrict the bands you can work (such as VHF/UHF QSO Parties); the places you can work (in the ARRL DX Contest, DX stations work US and Canadian stations *only*, and vice versa) or how many times you can work them (such as the November Sweepstakes, where you can only work a station once.) There about as many different contests as you can imagine, ranging from "major" events, such as ARRL DX, Field Day, November SS, and *CQ WW*, to smaller events such as state QSO parties. Each one has its own charms and personality.

How do you start? If you're like most folks, you've probably ran across a contest and wanted to make a contact but did not know what to do. First, find out which contest it is! On any given weekend there can be several small contests. The Contest Corral column in *QST* gives a good list of upcoming events and a thumbnail sketch of the rules. **Table 7-1** is a list of the major contests and when they are held.

The rules will have the next thing you're looking for—the *exchange*. Simply put, this is what the stations will be sending to one another. These can range from serial numbers, (in other words, a count of how many contacts you've made so far), to your state, county, or name. Some contest exchanges include your grid square (a letter-number combination used to determine the 1°×2° area of the world you live in—used in VHF/UHF Contests) or your zone in some DX contests. The VHF Operating chapter of this manual contains a grid square map you can use to find your own grid square or the location—and therefore the beam heading—of other stations. The DX zone map is in the reference section of this manual.

Once you think you have a good idea of what information is included in the exchange, it's time to make some contacts. There are two techniques folks use to do this: *Hunt and Pounce*, and *Running*.

Hunt and Pounce is just like its name—you hunt up and down the band looking for stations to work who are calling CQ. Then you pounce on them (call them quickly.) There's not much to it, on the face of it. Almost everyone starts at the bottom of the band, and works their way up to the top. If you're new at this, you will find you need to tune slowly. It takes a while to get used to hearing many signals close together. If you hear a station calling, just zero-beat it and send your call sign. Just once is sufficient—more than once and you may miss who he comes back to, and will probably QRM the QSO. When he comes back to you, don't expect a rag-chew! In contesting, the name of the game is speed, and they'll be looking to work as many stations as they can.

The *run* is where you stay in one spot and call CQ. This can have the advantage of working large amounts of stations in a short amount of time. Be careful though! Handling a pileup is not without its dangers and troubles, and it takes an experienced hand to be able to do it right. If you're just starting out, you may want to take some time and listen to how some other stations on the air do it, and perhaps pick up some helpful hints. At the same time, though, don't be afraid to call CQ! Many times a band may seem quiet, when in actuality everyone else on the band is listening.

What are the best contests to start with? Many people get their start during Field Day. Others try the relatively low-intensity state QSO parties to get their feet wet. The 3 major VHF contests (the January VHF Sweepstakes, and the June and September VHF QSO Parties) are perfect opportunities for you to check things out. In the more populated areas of the country, you can make lots of QSOs with just a hand held!

Don't hesitate to jump into the most popular of the HF contests. There are usually a few spots on one band or another offering you a chance to participate without being buried in the fray. The November Sweepstakes is a great chance to start as it doesn't take much of a station to be able to make contacts. You can even get a nice pin for making 100 contacts, and there's slow-speed segments set aside for ops who can't roll along at 35 wpm on CW.

If you've never been to Field Day, you're missing some of the best contest and camaraderie ham radio has to offer! When sunspots are at their peak, the 10-Meter Contest in December brings stations from all over the country and the world on the air for the weekend, and hams trying to make DXCC or WAS can find their totals taking a big jump.

Last, but not least, are the smaller contests and QSO parties. These range from ones sponsored on a state level (such as the California or Pennsylvania QSO Parties), to ones run by a country's ham society (such as the Worked All Europe, sponsored by DARC in Germany, or the All-Asian DX Contest, sponsored by JARL). These events offer a chance to make a large number of contacts in an afternoon, and may provide activity from places you wouldn't hear on the air normally.

Whatever contest strikes your fancy, remember these two maxims—get on the air, and have fun! Once you try it, you may find yourself a contest addict in no time. See you in the pileups!—NF1J

---

thing is certain, though; an inexperienced operator at a poorly prepared station not only should expect dismal results, but shouldn't expect to have fun, either.

Setting Goals

The very first step when entering competition is to set a goal for yourself. As mentioned earlier, a goal might be to improve your code speed by 10 words per minute or to work the last two states you need for the Worked All States award. These would certainly be very worthwhile achievements, and a contest is an excellent arena in which to attain them.

If you're more oriented toward the competitive aspects of contests, then you should establish a competitive goal for yourself. The goal should be reasonably attainable, or you'll just get discouraged when you don't achieve it. Don't set the goal too low or you won't progress, or, more importantly, draw any

Table 7-1
Major Contests

Month	Contest	Scope	Exchange	For more information
Jan	ARRL VHF Sweepstakes	Primarily W/VE	Grid-square locator	Dec *QST*
Jan	*CQ* Worldwide 160-Meter Contest (CW)	International	W/VE; signal report and state/province; DX; signal report and country	Dec *CQ*; Contest Corral Jan *QST*
Jan	ARRL RTTY Roundup	International	W/VE; Signal report and state/province; DX; signal report and serial number	Dec *QST*
Feb	ARRL International DX Contest (CW)	W/VE stns work DX stns	W/VE; signal report and state/province; DX; signal report and power	Dec *QST*
Feb	*CQ* Worldwide 160-Meter Contest (phone)	International	See above	See above
Mar	ARRL International DX Contest (phone)	International	See above	Dec *QST*
Mar	Spring RTTY Contest	International	UTC, signal report and consecutive QSO serial no.	Contest Corral May *QST*
Mar	*CQ* WPX Contest (phone)	International	Signal report and consecutive QSO serial no.	Jan *CQ*; Contest Corral Feb *QST*
May	Russian CQ-M Contest (phone and CW)	International	Signal report and consecutive QSO serial no.	Contest Corral, Apr *QST*
May	*CQ* WPX Contest (CW)	International	See above	See above
Jun	ARRL June VHF QSO Party	International	See above	May *QST*
Jun	All Asian DX Contest (CW)	Asian stns work	Signal report and age	Contest Corral Jun *QST*
Jun	ARRL Field Day	Primarily W/VE	Transmitter "class" & ARRL section	May *QST*
Jul	IARU HF World Championship (phone and CW)	International	Signal report and ITU zone	April *QST*
Jul	*CQ* VHF WPX Contest	International	Consecutive QSO serial no. and call sign	Feb *CQ*; Contest Corral Jul *QST*
Aug	Worked All Europe (CW)	EU stns work others	Signal report and consecutive QSO serial no.	Contest Corral May *QST*
Sep	Worked All Europe (phone)	See above	See above	See above
Sep	ARRL Sept. VHF QSO Party	International	Grid-square locator	Aug *QST*
Sep	All Asian DX Contest (phone)	Asian stns work	Signal report and age	Contest Corral Sep *QST*
Oct	*CQ* Worldwide DX Contest (phone)	International	Signal report and *CQ* zone	Sep *CQ*; Contest Corral Oct *QST*
Nov	ARRL Sweepstakes (CW)	W/VE	Consecutive QSO serial no., power level designator, call sign, last 2 digits of the year you were licensed, ARRL section	Oct *QST*
Nov	ARRL Sweepstakes (phone)	W/VE		Oct *QST*
Nov	*CQ* Worldwide DX Contest (CW)	International	See above	See above
Dec	ARRL 160-Meter Contest (CW)	International	W/VE; signal report and ARRL Section; DX; signal report	Nov *QST*
Dec	ARRL 10-Meter Contest (phone)	International	W/VE; signal report and state/province; DX; signal report and consecutive QSO serial number	Nov *QST*

The *Armadillo Gang*, N5MIV, operated for two days from this 24-foot air-conditioned travel trailer. The trailer is normally used as a mobile emergency operating post.

satisfaction out of achieving it.

The best types of goals are relative goals; that is, how well you can do compared to a particular station in your area who might have similar capabilities to your own. Absolute goals, such as making 250 contacts on 28 MHz during a DX contest, may prove to be totally unreasonable; the band might not even open during periods of low sunspot activity or might be so good that stations using a converted CB set and a mobile whip antenna are capable of making 500 contacts. By setting your sights on an absolute goal, you probably are preventing yourself from doing the best that you can.

Another reason for setting a relative goal as opposed to an absolute goal is the wide disparity between different geographical areas. For example, during a DX contest, because of the vast differences in propagation to population centers around the world, the only real relationship between a station competing from the East Coast of the United States and one from the West Coast during a DX contest is the coincidence that both stations happen to be active during the same time period.

The immense Amateur Radio population in Japan is readily accessible to the average western United States station, so typically the West Coast results show a higher number of contacts than do East Coast results. The East Coast stations, who may only have a short period of useful propagation to Japan, usually have good propagation to Europe. While the amateur population in Europe is less than that of Japan, the population is distributed among a large number of countries, each counting as separate multipliers.

Since the final score is the product of the number of contacts and the number of multipliers, the bottom-line score (with a lot of multipliers and a fair number of contacts) of an East Coast station may be similar to the bottom-line score of a West Coast station (with a lot of contacts, but relatively few multipliers). However, the scores are not comparable, since the conditions are so different. Are football championships comparable to baseball championships? The obvious answer is no; each is an achievement unto itself. By setting your contest goals relative to a nearby station, you eliminate the difficulty of competing with something you have no control over—propagation.

The ideal goal is to compete with a station similar in capability to yours, operated by someone who has similar experience and talent to your own. The absolute optimum goal would be to compete with someone you know personally; after the contest is over you can then compare notes, learn from each other, and spur each other on to new levels of achievement in the next contest.

Compete in a class that is commensurate with your own abilities as well as your station's. For example, if you live on a suburban lot and at present only have a three-element 15-meter monoband beam antenna, it is pointless to enter the all-band category in a DX contest. Similarly, if you don't own (or wish to own) a high-power linear amplifier for the HF bands, compete in a low-power category.

The contest sponsors make it easy for you to be judged only against your "peers." Not only are there multiple classes (such as monoband 15 meters, or transmitter power below 150 watts) that you can enter, but results are geographically processed. You only compete against other stations in your own geographical area (ARRL section, state, or call area, depending on the contest). The contest sponsors give awards based on these classes and geographical distributions.

Finally, be realistic. Don't try to set the contest world afire in your first attempt. Most of the top competitors not only share your drive and desire, but also have vast experience coupled with that desire. After all, there is a reason they got to the top. Don't forget, as you yourself advance, that the competition isn't standing still either. In the event you don't find yourself challenged enough, set higher goals or choose a new station to compete with.

If you find that you consistently win all the contests in every category you enter by a wide margin, please contact the editor of this book for possible employment in writing the contest chapter of the next edition!

The Fundamentals

Once you have set your goal, the next step is to decide exactly what tools you need to attain it. First, the fundamentals.

What kind of contest are you going to enter? Most contests fall into four somewhat overlapping categories:

Domestic—contact as many stations as possible during the contest period in the United States and Canada. The ARRL Sweepstakes (SS) is an example of a domestic contest.

DX—contact as many stations during the contest period outside your own country. The ARRL International DX Contest and the *CQ* Magazine World Wide DX Contest are examples of DX contests.

VHF—contact as many stations as possible on the amateur bands above 50 MHz during the contest period. The ARRL VHF QSO Parties and the *CQ* Magazine VHF WPX Contest are examples of VHF contests.

Specialty—contact as many stations as possible during the contest period. Usually, all contacts must be made with operators who either are a member of a particular group or are using a particular transmission medium. The ARRL 10-Meter Contest and the *CQ* Magazine 160-Meter Contest are examples of contests in which operation is restricted to a particular transmission medium (in these two cases, a single amateur band). The RTTY Roundup requires all contacts to be made using radioteletype communications.

Once you have analyzed the format of the contest (and have read and understood the rules), you are in a position to decide how to maximize points for that contest. For example, in the ARRL Sweepstakes, a maximum of 75 multipliers can be worked, and each station may be worked only once, regardless of band. Clearly the emphasis is on making contacts in the SS. Where are these contacts most likely to come from?

An examination of the *Radio Amateur Callbook* for the United States gives a good clue. The licensed amateur population per state is presented in tabular form. As might be expected, the amateur population is densest around the major population centers like the Northeast corridor (Washington, DC through Boston) and Southern California. From this information, you might conclude that a percentage of your contacts could come from these population centers. The *QST* summary of the 1990 SS results confirms this, and adds the information that there are some hotbeds of Sweepstakes activity not apparent from the amateur census figures. For example, the Northern California Contest Club over the past decade has consistently taken part in the SS as a major club activity. As a result, the number of Sweepstakes participants from Northern California is proportionally much higher than almost any other area of the United States and Canada.

Now that you have an idea where to geographically find the stations you will be attempting to work in bulk, the next subject to examine is how to go about reaching them. Here, a knowledge of propagation is essential. If you aren't already reasonably well versed in the fundamentals of propagation, you might want to stop now and consult *The ARRL Antenna Book* or *The ARRL Handbook for Radio Amateurs* (this suggestion applies regardless of whether you are interested in DX or domestic contests, or contesting on the HF bands or on VHF and above).

From your knowledge of propagation, you know that the higher frequency bands—typically 20 through 10 meters— are generally the best bands for long-distance contacts (over 1000 miles), particularly during the day. From a location in Connecticut (that is, if you're from Connecticut), this means that the higher frequency bands will be good for working the Midwest and the West Coast during the sunlight hours. Since your demographic study indicates that the largest percentage of stations available for working during the SS are in the Northeast, 20 through 10 meters aren't going to be as productive (particularly for long periods of time) as 80 and 40 meters where the optimum working range is shorter. This suggests that a station in Connecticut probably would want to emphasize 80 and 40 meters as much as possible to maximize the contact total. Because of ionospheric absorption during daylight hours, 80 meters is pretty much restricted to nighttime use; 40 meters becomes the daytime band of choice by default. At night, both 40 and 80 meters have good propagation and should be covered. The remainder of the time, particularly during the day, should be spent on the higher bands. The ideal mix (in theory, anyway) is to spend an equivalent amount of time on each band in proportion to the active amateur population workable on that band from your location (eg, if 38% of the potential contacts workable from your station are from the West Coast, you should spend 38% of your time on bands with propagation to the West Coast).

Of course, this analysis assumes that your station is equally effective on all bands and that there are no factors other than propagation involved. Unfortunately, neither is usually the case. Particularly on phone, the 40-meter band is flooded with foreign broadcast stations after European sunset. Very few amateur stations are loud enough to be heard through these broadcast signals, and even fewer have the ability to effectively hear stations responding through the raucous QRM (interference). This makes 40 meters a tough row to hoe. Fortunately (at least from the East), the skip zone on 40 meters stretches out dramatically by the time the broadcasters become unbearably loud. Late at night, stations in the Northeast are no longer reliably workable, and the exodus to 80 meters begins. On 80 meters, the QRM is ferocious, but at least none of the stations there run megawatts, play music or broadcast a carrier.

On 80 meters late at night, the goal is to work as many East Coast stations as possible, because during the day that feat will be more difficult.

In summary then, based on the analysis, if you live in the Northeast, maximize your effectiveness on 80 and 40 meters so that you can penetrate into the area with the greatest contact potential—the Northeast.

A station on the West Coast has a different perspective. For him (or her), the prime bands are 20 through 10 meters where propagation is good to the East Coast. This is a blessing in a number of ways. First, hot action during the day and little or no action late at night means that the West Coast operator can probably sleep when most people normally sleep— at night. Second, the higher frequency bands are generally quieter and certainly don't have the interference problems that 40 meters does. Third, antennas for the higher frequency bands are physically smaller than they are for 80 and 40 meters; the average ham in an urban or suburban area has a better chance to install an effective antenna on 20/15/10 meters. Since DXing is one of the more popular amateur activities, and DX is generally easier to work on the higher bands, the average ham probably has his or her most effective antenna system on some combination of these three bands. Therefore, the average West Coast ham's most effective bands are the bands with the greatest point potential. By the same reasoning, there are more active hams with good stations in the Northeast on the higher bands ripe for the plucking by West Coast stations. Fourth, there is less chance for TVI and other interference problems during the day than at night, so more "casual contesters" get on the air. The astute West Coast station, who may have made a similar study to the one you made, makes use of this advantage to get a higher score.

Are the odds heavily weighted in favor of the West Coast (never mind the South or the Midwest, which we didn't even examine)? Yes and no. From the *QST* SS summary, it's obvious that the East Coast typically fares poorly compared to the competition at points west. The odds become a little more favorable to the east when sunspot conditions deteriorate and the higher bands aren't open for as many hours coast to coast. But, so what? Earlier it was decided that it's impossible to compete with everybody everywhere, and it is preferable to set your goal to compete with someone in our own area. This is just another case in point.

Up the mast once more, and NC7K will be ready for the sweepstakes. The shiny spots are reflections of the Nevada sunshine, not RF.

An analysis of this type can be (and should be) made for each contest in which you plan to participate. If you're really serious in your own way, your station design (especially your antenna system), your operating "game plan," and even your choice of contests will be defined by an analysis of the fundamentals of the contest. In engineering and scientific terms, this is known as "defining the problem."

Now that the first stage is complete, and you know what you want to do, you can try to figure out how to go about doing it.

STATION PREPARATION

If you are serious about competing in contests, chances are that your station and antenna system already are (or will soon be) oriented toward producing higher contest scores. Unfortunately, very few of us have the luxury of being able to install the ideal contest antenna system. The key to building the most effective contest station within your own means is to analyze your objectives (see the previous section of this chapter), and take advantage of whatever resources you might have at hand to achieve these objectives.

The prime purpose in optimizing your station is to maximize your station's effectiveness into those areas where there is the greatest point potential. This may sound too simple to be accurate, but the concept really does apply, no matter what rung of the contest career-ladder you may be standing on. Station effectiveness is merely a relative term for describing your ability to hear and be heard (in that order of importance).

Antennas

The subject of engineering and installing antenna systems is far too complex to cover in a book like this one; the best we can do here is examine what characteristics might be desirable for a contest-oriented antenna farm. Antenna design is a very popular Amateur Radio topic, and antenna-related articles appear almost every month in the various amateur magazines. For a solid basis in the fundamentals of this subject, *The ARRL Antenna Book, The ARRL Handbook for Radio Amateurs,* and *Your Ham Antenna Companion* are offered by the ARRL. They all deserve a place on your bookshelf.

Simply stated, the idea in designing an antenna system for

The part of a Field Day station not usually seen in photographs. This is N5FD. In the foreground is the solar-charged battery. If any of the connecting wires develop a problem, it will probably be at 4 AM.

successful contesting is to obtain the greatest antenna gain possible over the geographical area where there is the greatest point potential. Determining the means to this end requires knowledge of propagation and how a signal might travel from your location to the target location. From the bottom of 160 meters up through at least 6 meters, signals propagate beyond the line of sight by means of high-altitude refraction of the transmitted electromagnetic wave front. Above 6 meters, most long distance propagation is the result of relatively low-altitude refraction. The causes for the refraction vary from band to band and between the various propagation modes. For a more detailed description of the processes involved and their effects, see the propagation references mentioned earlier.

Once the wave propagation path from you to the target area is understood, the best antenna system can be chosen by identifying a system whose characteristics produce a radiation pattern that matches the propagation path. The closer the transmitted wave follows the propagation path, the greater the signal strength that arrives at the target site. The reciprocal holds true as well for reception; the greater the energy captured from the propagation path, the greater the received signal strength. Like a lot of other subjects discussed so far, this one sounds very simple. In fact, given a large enough budget (for both antenna hardware and real estate), it is possible to build an ideal antenna for any given propagation path.

Unfortunately, propagation paths change hourly, daily, monthly and yearly. Some of the changes are subtle; some are manifest. The real challenge in building the optimum antenna system is to be able to accommodate these changes as they occur, rather than having to string new wires or plant new towers as the band changes. As the ionosphere goes through its changes, the geometric path that will support propagation changes. If the propagation path is no longer optimum for your antenna system, signals will fade (the old familiar QSB), perhaps until the "match" between your antenna system's pattern and the propagation path is so bad that the band will effectively close between you and that area. This effect explains why a station only a few miles from you might be able to carry on a conversation with another station that to you is inaudible. Perhaps your neighbor's antenna system has a different vertical radiation pattern than your system does, and the distant station's signals are arriving at a radiation angle at which your antenna system has a severe null. This might be because your neighbor's antenna is at a different height above ground, or because the ground beneath his or her antenna falls away at a different angle from your own, effectively modifying the vertical radiation pattern.

With the ionosphere performing its undulation act, no one "ideal" antenna is ever really optimum. The situation is similar to trying to shoot a moving target from a rocking boat; you know approximately the direction in which to aim, but not much more. In fact, if you are considering the situation up through 6 meters, where almost all antenna systems are at best only a few wavelengths above ground, an antenna's vertical radiation pattern is a series of peaks and nulls, as shown in the various antenna texts. This effect is caused by wave interference, both constructive and destructive, between the portion of the antenna's radiation lobes that are directed up away from the ground and the portion of the antenna's radiation lobes that reflect from the ground's surface. When one of the peaks in the antenna system's vertical radiation pattern matches the propagation path to a particular point, then signals will propagate well, with minimum loss. When the propagation path falls into one of the antenna system's nulls, where the antenna system response is very low, the transmission path loss is high, perhaps to the point that the band will appear "closed." Usually, neither of these extremes is the case, and the transmission path loss (and resultant signal levels) falls somewhere between the two.

Your goal is to be as loud as you can into your target area (to attract attention and get people to hear you) for as long as you can (to have exposure to the maximum number of stations). Thus, your antenna-system effectiveness is determined not by how loud signals are at their peak (when the propagation path and an antenna lobe match well), but rather by signal-strength consistency over a period of time. An antenna system that produces signal peaks of S9 + 30 dB for 20 percent of the time and signal fades that drop into the noise the rest of the time isn't nearly as effective as a system that produces consistent S9 signals all the time. It's obvious that you can make contacts only as long as the band is open; long band openings produce a lot more contacts than loud band openings. An effective antenna system might not make you any louder, but it should let you take advantage of any band opening, no matter how poor the propagation.

There are a number of ways to accommodate the changing ionosphere and maximize band-opening time. The general strategy is to use an antenna system that employs some sort of radiation-pattern steering mechanism to match the antenna pattern to the propagation path. Almost everyone who uses a "beam" antenna of some type has either an electronic or mechanical rotation facility to turn the beam in azimuth to favor the desired transmission path. However, to match an antenna system's radiation pattern with the propagation path, control of the vertical radiation pattern is needed. The most common way to achieve this is to employ antennas at different heights. Since the ground reflection patterns, for even identical antennas, change with height above ground, a variety of antennas at different heights produce a wide selection of vertical radiation patterns. The appropriate antenna for a given propagation path can then be chosen experimentally.

Another vertical-pattern steering technique is to employ a phased multiple-antenna system (such as "stacked" dipoles or beams), and electrically vary the phase relationship between the antenna elements. The actual shape of the radiation pattern then can be changed as needed. While this approach is very effective, the space required and the mechanical enormity of stacked antennas forces most station owners to employ the simpler arrangement of switched multiple antennas.

Keeping Your Ears Clean

So far, all the discussion has been aimed toward being as loud as you can into your chosen target areas, for as long as possible. Fortunately (thanks to Mother Nature), an antenna system that provides the best transmit performance also provides the best receive performance, as far as signal strength is concerned. Unfortunately, the reality of the amateur bands is that the signals you want to hear occupy the same RF spectrum as noise and other interfering signals. In fact, you are often limited in receive capability not by low signal levels, but rather by interference and noise. As far as your receiver is concerned, it doesn't know the difference between the signal you want to hear and the other junk on the same frequency. Once noise or interference makes it as far as your receiver's input terminals, there is often no way to separate the signal you want to hear from everything else on frequency (more on this later). The only panacea for interference ills is to prevent the interference from reaching your receiver in the first place. The most effec-

tive way to do this is through your antenna system.

The basic physical process that allows your antenna system to produce gain is a very simple one: Energy from one direction in an antenna's pattern is redirected into another direction. The total energy in the antenna pattern is fixed; it is merely redistributed. A gain is developed in one direction, "negative gain" is developed in another. This process is not only useful in making you louder toward one direction, but it is also the means to reduce undesired signals and noise from another direction. Most every beam antenna has a null directly off the "back" that is between 10 and 30 decibels below the strength of the main front lobe. Response off the "side" of the antenna is even less, as compared to the main front lobe. The rejection of signals off the main "beam" lobe is one of the more compelling reasons for using a directional antenna. Not only are you louder in the preferred direction, but you cause less interference and are interfered with less in other directions. The cliche "you can't work 'em if you can't hear 'em" really is true. The use of a directional antenna is a very effective step in beating the implications of the cliche.

Particularly on the lower frequencies, where antennas are very large to begin with, it is often very difficult, if not impossible, to install beam antennas (imagine a three-element 160-meter beam at a height of a half wavelength!). Reduced size antennas work reasonably well, but as an antenna's size is reduced, so is its radiation resistance. The antenna's losses increase as the radiation resistance drops. In most practical situations, the theoretical gain offered by a dramatically "shrunken" antenna is more than offset by the increased losses; as far as transmitting is concerned, the simpler dipole is at least as effective as the more complicated directional array. In addition, the simpler antenna probably can be installed at a greater height where it also might be more effective.

Even in a situation where a large directional antenna is out of the question, it is still possible to effectively employ a directional antenna to cut down received interference. The solution is quite simple; use the directional antenna for receiving only. As pointed out earlier, signal strength is usually not the limiting factor in hearing well; what really matters is the signal-to-noise/interference ratio. There are a number of directional antenna designs that are not very efficient and offer "negative gain" compared to a dipole. While these particular antennas would be poor transmitting antennas, they can be very effective for receiving. For example, signals received on a direc-

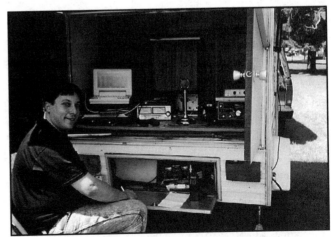

This neat set-up belongs to the Quaboag Valley Amateur Radio Club, KD1RW. Mark, KA1GMS is sitting in front of the packet and VHF station.

Field Day is not just operating. This is the visitor welcome table with Phil, NH6SA and Clyde, KH6NK ready for all visitors. Don't you wish you could visit the KH6WO Field Day station? *(Photo courtesy of Kevin Bogan, WH6ML)*

tional loop antenna might be 20 decibels weaker than the same signals as received on a dipole. However, by orienting the loop properly, an interfering signal might be reduced in strength by 40 decibels. The overall improvement in signal-to-noise/interference ratio is then 20 decibels, a substantial margin. In addition, many of these directional antennas also are capable of vertical radiation pattern-steering, giving further rejection of undesired signals. Some designs that fall into this category are the Beverage antenna, the Adcock antenna and the tuned loop.

Many of the same principles apply for both HF work and VHF work. On the VHF bands, where the wavelengths are physically short, even modest-size antenna systems develop prodigious gain compared to their HF counterparts. The effective radiated power (ERP) transmitted by most VHF contest stations is usually several times greater than even the largest HF amateur contest stations. That, coupled with the higher gain needed for best sensitivity in a VHF receiver, provides the breeding ground for severe receiver-overload conditions. Aside from improving receiver dynamic range, the most effective way to reduce receiver overloading is simply to reduce received signal strength (selectively, of course). An antenna system with a "clean" pattern (that is, few lobes aside from the desired main lobe) can offer effective rejection of an overpowering signal—when the antenna isn't pointed in the direction of the offending station!

Station Electronics

Today's Amateur Radio market offers a wide variety of electronic marvels for HF, VHF, UHF, CW, SSB, ASCII, RTTY, FM, SSTV, ATV and every other transmission mode and frequency range imaginable. The rigs scan, autotune and operate break-in. They offer passband tuning, notch filters, digital readout, speech processing, computer control, RIT, XIT, VOX and audio filters either built-in or as options. The combinations and possibilities are endless and can be very confusing even to the knowledgeable and experienced radio veteran. Furthermore, how any of these features find utility in contest work is a very confusing issue. Well, don't expect to find any definitive answers here. Most of the answers are a matter of your own personal taste and your pocketbook (although high priced doesn't necessarily mean better). There are a few areas that you should especially examine, however, if you are purchasing a radio specifically for contest use.

Unlike even a decade ago, virtually the only form of transmission equipment commercially available is the transceiver. In the past, transmitter and receiver "separates" were the standard bill of fare in the amateur station. When transceivers first became available, they were generally viewed as compromise units, primarily intended for portable or mobile use, mostly on phone. Many didn't have the capability for optional narrow CW receive filters. Few offered any facility for adjusting the receive frequency independently of the transmit frequency; in fact, on many designs, CW operation was included only as a poorly conceived afterthought. Most of the early HF transceivers were made for operation on a few of the amateur bands, perhaps only one. Among contesters, transceivers were a concession to portability or finances.

Earlier, the cliche "you can't work 'em if you can't hear 'em" was mentioned. The wisdom in this is pretty obvious. For most Amateur Radio work, the need to hear what is on frequency right now doesn't usually exist. In most activities, imminence isn't a priority. For example, if a DX station happens to fade for a few minutes, normally even a ravenous DXer can stand by for a couple of minutes until the band peaks up again. Since a contest is a timed event, a contester can't afford to spend a lot of time trying to dig through the QRM to hear the other station's information. The use of a high-performance receiver can go a long way toward helping the contester dig the weak ones out of the noise and QRM.

The most basic function of a receiver is to allow the operator to hear what is being transmitted. Again, this is one of those ultra-fundamental statements that is so obvious that little is made of it; the real underlying meaning does have particular significance here. The concept of hearing implies not only detecting the presence of sound, but also being able to interpret the sound and understand what was sent. This is because detecting a signal is only half the battle; you have to understand what is being sent for the transmission to be of value. A more descriptive version of our cliche might be "you can't work 'em if you can't hear 'em *and* understand 'em." Keep this in mind when evaluating a receiver.

One of the most audible indications of contest activity is just that, contest activity. The bands become very crowded, and signal levels are ferociously high. The combination of high signal-levels and lots of them creates a lot of QRM. The same combination causes receiver overload and exacerbates otherwise relatively subtle problems like reciprocal mixing. Clearly, there is enough interference to contend with on the bands without generating a dose of interference within your own station. Strive to get the very best receiver that you possibly can.

Many features such as multiple internal VFOs, receiver passband tuning and frequency "memory," just to name a few, are primarily the beneficiary of one major change in the commercial design of Amateur Radio equipment, the use of digital circuits. Features unknown a few short years ago are now within reach of almost every contester.

There are a number of parameters that require study when evaluating the performance of a particular rig. Receiver intermodulation distortion performance, receiver sensitivity and receiver selectivity performance are all characteristics of importance to contest operators. Unfortunately, manufacturer's specifications aren't all made to the same test standards, and can't be conveniently compared. Again, perhaps the best source of information is the Product Review column in *QST*; all the equipment is measured to the same set of standards using the same test equipment. Other specifications of interest to the contest operator are received audio distortion, AGC (automatic gain control) threshold, AGC range, AGC audio output vs RF input level performance, and frequency readout accuracy. Most of these characteristics are not specified anywhere, and many can be subjective; if you can, try out a piece of equipment yourself before you decide to purchase it.

The same cliche can be modified to describe the philosophy you should embrace with regard to your transmitting equipment; "you can't work 'em if you can't be understood." This particularly applies to speech processing on phone. Forget for a moment the effect of an improperly adjusted speech processor on your transmitted bandwidth; think of how you sound to the station at the other end. It is currently in vogue to use speech processing—lots of it—in an attempt to sound loud. Unfortunately, most of this sophisticated processing does raise the average level of a transmitted signal, but primarily by adding distortion products (which does make the wattmeter hang higher). It can be scientifically demonstrated that, except under very weak signal conditions when the signal-to-white noise ratio approaches unity, speech processing adds nothing to the readability of an SSB signal. One of the main reasons amateurs are deluded into using a lot of processing is the on-the-air reports they receive; they hear how "loud" they sound. Indeed, if high blower-noise levels and lots of distortion constitutes loudness, then cranked-up speech processors make you sound loud.

In fairness, speech processors do have their place. Very few of us have ideal radio voices, complete with an optimum distribution of audio energy. This, coupled with less than optimum microphones, can make you sound somewhat anemic on phone. A speech processor used with discretion can help equalize the energy distribution of your voice as transmitted. Overzealousness with the processing equipment does make the wattmeter needle stand straight up, but it actually detracts from your ability to be heard and understood at the other end. Particularly in a DX contest, when there is interference galore, and the operator at the other end may not understand English too well, you should try to make yourself as understandable as you possibly can.

As far as amplifiers are concerned, your choice is determined by personal preference and budget. Most HF contests have categories reserved specifically for stations running low power. Many operators choose to run less than the legal limit (either "low power," which is usually defined as approximately 150 watts output, or QRP, which is usually defined as either 10 watts input or 5 watts output) to add an additional challenge to their contesting. The subtleties of operating using a few watts vs a thousand appeals to a large segment of the contest community.

On the VHF bands, where signals are often weak, lowpower operation is truly a handicap. That doesn't mean you can't have fun operating a VHF contest using low power (particularly from a mountaintop). But if making a high score is a priority for you, you'd better consider the reality of highpower operation. Many of the propagation modes at VHF and above have formally and collectively been given the title "weak signal modes." To take advantage of these modes on a consistent basis, you need high power in addition to large antennas. For example, on 6 meters, one of the more useful propagation modes for communicating beyond 300 miles is generically known as "scatter." Scatter signals peak around dawn when the Earth collects its morning breakfast of meteors. It is quite possible for even a very modest 6-meter station (ie, 10 watts to a small beam) to contact some of the larger 6-meter contest stations through clever operating. The system-gain capabilities at the large station tend to overcome the lower system gain found at

the smaller station. Since there is a relatively modest number of "large" 6-meter stations, this does tend to limit the number of stations workable by a small station.

Station Accessories and Productivity Boosters

During a contest, your main activity should be making contacts, not performing mental gymnastics. There are a number of station accessories that can relieve you from having to *think* during a contest (if you don't understand why you don't want to think during a contest, just wait; you'll quickly learn from experience!).

Before worrying about these brain-conserving gadgets, however, you should give the rest of your body some consideration. First on your priority list should be a comfortable chair; well, not too comfortable. You'll be sitting for long periods, and if you value your physical well-being, the investment in a good chair is small compared to the benefits. A chair designed for use by a secretary, who also sits for hours at a work station, is an excellent choice for an operating-position chair. Since your major activities during a contest are tuning your receiver and writing in the log, it's obvious that you'll be spending a lot of time in an erect sitting position. For best comfort over long periods, the chair you select should offer good back support (primarily in the lumbar region) and good leg support when you are sitting upright. Also, the chair shouldn't be too confining.

It may sound like this subject is being overly stressed, but there are few things more debilitating to a contest effort than being physically uncomfortable.

Since you will be looking at your station clock every time you make a log entry, a digital clock set to UTC should be placed in a prominent position where you won't have to crane your neck to view it. The digital readout aspect is important, since you won't have to interpret the clock reading as you would with a normal clock face. Although telling time may seem second nature to you normally, most people don't actually discern the exact time when reading their watch (as you have to when making a log entry). Rather, they view the clock face in a relative manner, relating their daily schedule to the position of the hands on the clock face (this phenomenon may explain the current trend away from digital watches). In a contest, you don't necessarily want to know the time relative to the world; you want only a bookkeeping reference for your log. In fact, it's quite common to lose your sense of time during a contest;

Ed, WV8L, looking puzzled at the CW setup of K8ZFR Field Day. He seems to be saying: "Why does this guy keep calling me a dupe?"

Computers and Contesting—The Wave of the Present

Computers and contesting have come a long way in a short time. Programs that came out 10 years ago are already antiques when compared to the software packages on the market now. What do these programs do? How do they work? Which one is right for you? Let's try to answer some of these questions, which can confound even the experienced contester. For a full discussion of the uses of a computer in your shack, see the ARRL publication *Personal Computers in the Ham Shack*.

What do they do?

The first program function is to check whether you have worked a particular station. In the old days of paper logging, everyone's biggest fear was working too many duplicate calls. This could even mean disqualification from the contest. All the current programs on the market have some form of tracking to let you know what stations you have worked on what bands. Most even have what is called a "check partial." This function allows you to enter part of a call sign. The computer then checks to see if you have worked anyone on the current band with a call sign similar to your entry.

By identifying duplicates you can save a lot of time calling someone in a pileup whom you have already worked. It also can prove handy when *passing multipliers*—knowing instantly on what bands you've worked a station. This gives you the ability to ask the station to work you on other bands open at that time. Thus the dreaded "How could I miss a 6Y5 on 15 the entire contest?" syndrome can be prevented.

Another handy function is automatic score keeping. Some contests have extremely complicated rules and point schemes. It can be a real head-scratcher trying to figure out what's what. Nowadays you can just look at a window on your screen and automatically know exactly how you're doing.

The biggest advantage of these programs is the elimination of paperwork. No longer do you have to recopy messy log sheets or puzzle over those strange forms. Contesting software will automatically generate just what you need, whether it be log sheets, dupe sheets, or even the files needed for electronic submission. This allows you to eliminate all the paperwork altogether! (See the box

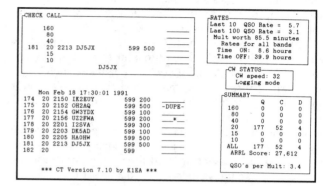

The *CT* software system configured for typical contest operation.

elsewhere on electronic submissions for ARRL contests.)

All of these packages give you additional features. Most show you how fast you're working stations by providing some kind of rate function—either a box, a chart or both. Virtually all of them have a QSL function to print QSL labels on command at the end of the contest! If you're a statistics lover, contesting software packages can generate enough figures to keep you happy until the next contest rolls around.

There are three other features of on-line logging software that should not be overlooked. The most popular is the ability to send and receive messages to and from your local *PacketCluster* via your TNC and 2-meter radio. If you're running in a multioperator or one of the various single operator-assisted classes, this will allow you to see and share spots at a moment's notice.

The second handy function is the ability to control your radio from your computer. Given the right connections and equipment, you can instantly change bands, work spots and monitor pileups from the keyboard. No more working someone and logging them on the wrong band.

The third (and most widely used) feature is the ability to send CW with your computer. The software acts as a memory keyer, and automatically sends all the information

time relates only to band propagation, schedules, strategic planning and (thank goodness) the end of the contest.

As a matter of practice, *always* use headphones for receiving. Aside from sparing the rest of the household from the contest din, headphones allow you to concentrate better than would a speaker. The frequency response of headphones is usually superior to the normal speaker placed in the typical acoustic environment found in an amateur station, and noise emanating from outside the station is sealed out. Many experienced operators claim that the use of headphones offers at least 10-dB signal-to-noise improvement over the use of a speaker. On phone, the use of headphones also eliminates most, if not all, of the problems associated with VOX operation.

The important feature to look for in headphones is comfort. Many modern stereo headphones are quite comfortable and can be worn for hours without causing pain or fatigue. The sound quality of the better units far surpasses any speaker that you might consider using. One note: Most amateurs seem content to use inexpensive ($10) headphones for contesting. These cheap units don't sound very good, aren't very durable, and certainly aren't too comfortable. It seems foolish to invest

hundreds, if not thousands, of dollars in electronic equipment (not even counting the cost of an antenna system) and then spend very little on inadequate headphones. A station is only as effective as its weakest link.

For CW operation, a computer or a memory keyer is a sound investment. Many operators find that their physical coordination tends to go awry late in the contest, and their sending skills trail off. Since you want the operator at the other end to understand you clearly, a device to form the letters and words automatically will provide an edge to the tired contester. It also frees you from sending redundant information. Most contest exchanges can be preprogrammed. When it's time for you to send your information, you merely have to press a key or button and you're able to sit back and catch up on the paperwork of dupe sheets, multiplier sheets and logs. Life is less frantic (and less tiring) when you use an electronic helper.

Along the same lines, choose a paddle that you find comfortable, and somehow prevent it from wandering around the operating table as you send. Taping a piece of sandpaper, rough side up, where you want the paddle to sit is a good measure to prevent paddle creep.

with just the touch of a key. The computer can also be configured to automatically tell a station if it's a dupe. For phone operators, two recent innovations have been the *Digital Voice Processor* or DVP from K1EA and the *Voice Blaster* software from LTA. The DVP is a plug-in board for your computer. It functions as a voice keyer and recorder, allowing you to send and receive audio from your keyboard. The LTA software fills the same purpose by using your computer's sound card.

How do they work?

Virtually all the programs used by contesters today were written for PCs. Some packages have different hardware requirements. It is always a good idea to check with the software supplier before plunking down your hard-earned money on a program that requires more computer than you own. You almost certainly need a hard drive. The size of these packages and all the different contests they cover take up a lot of disk space. The files they generate during a contest require even more!

Once you have installed your program, the rest is very simple. The first screen is the start-up screen where you enter all the pertinent information such as contest, call sign, class and address. At this point you are all set up and ready to contest. Most of your functions (such as changing a band or sending a CW message) are controlled by your computer's function keys.

You log stations by moving from field to field and typing in the information as you go. Hit **<enter>**, and you are on to the next contact. On CW (if you've set up to key the rig), typing the call and hitting the **<insert>** key will automatically send the station you're working the full exchange. All programs have some form of on-line help, and there are keyboard overlays for some of the most popular programs. It is a good idea to try the program for the first time by running your own mini-contest (off the air) before the day of the big contest.

Which one is right for me?

This depends on your preferences. Some folks are happy with software they have used for years, that just covers the few contests they always operate. Others want the latest and greatest offering. Any package you buy should allow you to

June, the weather is fine, and WB1ARU seems to be enjoying herself as she operates 20-meter SSB at K1BU.

at least work the major (ARRL DX, CQ WW, SS, WPX, Field Day and VHF) contests; most also cover smaller events like state QSO parties. You should make certain your software allows you to send CW from your keyboard, and it will talk to your TNC. (Not all do!) You want to have a package with a "post contest" function. This allows you to go in and correct mistakes you may notice after the heat of the battle has faded. Any good software program should also allow you to submit your log electronically by generating the proper files.

What do the "Big Gun" contesters use? There are three software packages that are the most popular with contesters. *CT*, written by Ken Wolff, K1EA, has grown to be extremely popular. Contact XX Towers Inc., 814 Hurricane Hill Rd., Mason, NH 03048 for details. LTA, PO Box 77, New Bedford, PA 16140 is the source of the second package called *NA*. The third program gaining favor is the *TR Logging Software*, written by Larry "Tree" Tyree, N6TR. It's available from George Fremin III, K5TR, 913 Ramona St, Austin, TX 78704. Any one of these should serve you quite well. You also want to check the advertisers in *QST* for other packages that might serve your purpose.— *Warren Stankiewicz, NF1J*

For phone operation, use a microphone that works effectively in combination with your own voice and your transmitter. Close matching of microphone characteristics with your voice can provide far more "punch" in your audio than any speech processor. If you can, mount the microphone on a boom so that the microphone base doesn't take up valuable writing space. Be sure to route the microphone cable in such a way as to avoid the writing surface. Most operators prefer to use the VOX (voice-operated switch) feature on their rigs, rather than PTT (push-to-talk), since VOX permits both hands to be free for logging and so on.

Perhaps the ultimate solution is the use of a headset mounted boom microphone. The advantages to this arrangement are numerous. The microphone is always optimally placed relative to your mouth, but never obstructs your view of the log materials or the rig. Since the microphone is in a fixed position relative to your head, you are free to move about and stay comfortable. Finally, the use of a headset-type microphone eliminates the possibility of bashing yourself in the mouth with the microphone when you're not paying close attention (don't laugh—it happens).

A very good accessory to use with a microphone is an equalizer. Within limits, an equalizer can compensate for an imperfect match between the energy distribution in your voice and the frequency response of the microphone.

Some operators prefer to use a foot switch to control transmit/receive switching (a sort of "STT"—step to talk). Footswitch operation is very fast, and of course doesn't tie up the use of your hands. The foot switch can be wired in parallel with the PTT connections within the transmitter microphone connector. Very often this arrangement allows control on CW as well as on phone.

Since a large part of a contest effort is devoted to logging and other paperwork, having an electric pencil sharpener nearby can be quite handy. Similarly, a pad of scrap paper finds numerous uses over a contest weekend.

There are lots of other station accessories that you might find useful. One item that should not be considered an optional accessory is good station lighting. Good lighting not only keeps your eyes healthy, but it also reduces fatigue. Fluorescent lighting, when used alone, causes many people to get headaches when they concentrate. High-level incandescent lighting is

More Operators Means More Fun

In contests, there seem to be two types. First there is the lone wolf who struggles through the weekend alone, fighting single handedly against QRM, equipment failures and the demands of sleep and family. Next there is the group operator, who joins with his friends for a multioperator effort. If you've never tried a multiop, as they are known, or even thought about it, then perhaps this will open your eyes to an entirely different way of contesting.

A good example of a multiop effort is a club during Field Day. It's more than just a bunch of people trying to make contacts together—it's an *atmosphere*, a camaraderie that comes with a team coming together trying to accomplish a specific purpose. There are two basic types of multioperator efforts—*single transmitter*, where you use only one transmitter at a time and *multitransmitter*, where you use more than one at a time. There are numerous rules and restrictions for each contest.

How hard (or easy) is multiop? It depends on what you're setting out to do. In contesting there are advantages and disadvantages to everything. How you approach it, and how you make the best use of the resources at hand, spell out the difference between a good time and the weekend of unmemorable frustration. One thing is certain; with a group even the biggest mistakes become the basis for stories to be told over and over!

First, let's cover the good part of operating multiop. Top on anyone's list is *sleep*. With more than one operator, you can usually try to schedule things so everyone gets a break when they need one, and no one gets so bored or tired they begin to make mistakes. Having more operators allows you to work stations on other bands more easily, or listen to multipliers. In multi-two or multi-multi, you can have signals on as many bands as you have rigs and antennas. You can also use a technique known as the *pass*: asking a station to move from one band to allow you to make certain you work that multiplier on every band possible. The advent of *PacketCluster* has enabled groups to find (and give out information on) DX stations to an extent unheard of in the old days of 2-meter FM local nets.

Best of all, it offers a great opportunity for you or other members of your club to cooperate in a contest. If you're new to the hobby, or aren't able to put up much of a station, a good multiop effort can get you some experience to pick up some important contest skills, or the chance to operate better equipment than you normally use. Besides, it's a lot of fun! If you've have a good station, but can't get interested in working through an entire contest (48 hours sometimes seems to get longer and longer) then perhaps you ought to consider opening up your shack to your local group. After all, any old-timer will tell you it takes time, skill and patience to be a good contester—this is your golden opportunity to show someone the *right* way to do things! We're sure there are plenty of hams in your community who would jump at the chance to operate from a station with a nice antenna farm, or with one of the latest rigs or amps.

How hard can it be? Well, at times, very. The challenges of running or operating a multioperator station can be formidable, and it takes a lot of time and preparation to make things flow smoothly. Here's some of the common pitfalls:

Logging: Almost everyone uses computers these days. When it comes to multioperating, though, nightmares crop up unexpectedly. Does everyone know how to use the software? If not, are there handy "cheat sheets" nearby the operating positions? Have you checked to make sure the software works on all the computers in the shack? More than one effort has been stymied when the PC someone brought wouldn't run the contest software on hand. (This usually happens when the clock shows 2354Z—6 minutes to contest start time.) Some computers generate noise in your receivers; others will lock up the first time you key-down the amplifier. If you're linking computers, or connecting them to radios, packet TNCs or keyers, the cabling between different ports and connectors can be a complicated nightmare! (Hint: make certain you have *plenty* of adapters of all shapes and sizes!)

Creature comforts. Do you have a place for people to sleep or relax, if need be? Operators in the shack should be left alone to operate—have a separate room for people to rest, eat, sleep or tell war stories. Make certain to stock enough

very tiring. A good compromise is a combination of fluorescent and incandescent lighting. Many operators find that local lighting at the operating position, with the rest of the room held at reduced lighting levels, allows them to concentrate more on operating. Additionally, the use of a dimmer to control light levels can reduce fatigue tremendously (try to avoid the solid-state variety; they tend to generate noise in the RF spectrum).

PREPARING FOR THE CONTEST

In addition to preparing your station for a contest effort, you must also prepare yourself for the main event. Practicing contest operating is quite a challenge; the only realistic simulation for a contest (until recently, anyway—see accompanying sidebar) *is* a contest. There is never the QRM level or amount of activity during non-contest periods that there is during a contest. Most of the skills you are interested in developing really can be honed only under fire.

To maintain their already substantial operating skills, many successful contesters put in at least token efforts in the less-active contests that seem to pop up every weekend. They gain valuable practice by participating, their call becomes more familiar to the rest of the contest fraternity, and they have fun. These "minor" contests are the best training ground for the "major" contests. Additionally, don't forget that just because

you consider the East Podunk QSO Party to be a minor event doesn't mean that there aren't some operators out there who treat the East Podunk QSO Party like the Olympics. Get on and give them a contact; they might return the favor some day.

There are some exercises that you can perform to sharpen your contest skills during non-contest periods. One of the greatest skills a contest operator can develop is the ability to copy weak signals through heavy interference. Not many amateurs spend their time looking for QRM, but you might actually benefit from doing just that. If you hear a big pileup somewhere, spend some time listening. Try to pick calls out. Listen to see who gets through the pileup first, and how he or she does it. The benefits here are threefold. First, you get practice at pulling signals through QRM. Second, you learn about some of your potential contest competition and their effectiveness in breaking through pileups. Third, you get to examine firsthand the various techniques people use to get through. Try to analyze and remember the *successful* approaches for your own use later. Take notes if you have to.

Once you reach the point where you have no difficulty copying all the stations calling in a pileup, try adding your own QRM—but not by getting on the air and adding to the chaos! Increase your receiver filter bandwidth a step wider than you normally use. For example, don't listen to a CW pileup using

food and beverages to last the weekend, or your operators may desert you! Of course, the need for well-developed toilet facilities has wiped out more than one operation. Operators at the famed top multi-multi station N2RM may be content to use an outdoor or porta-potty in February; your operators may not be such dedicated contesters.

How well lit is your shack? Are the operating positions comfortable? Is the ambient noise level good? Remember, in a phone contest, this room could have six people calling CQ at the top of their lungs.

The biggest hurdles to overcome are technical. Multiple stations means multiple interference, and it takes quite a bit of work to overcome some of these problems. To operate on more than one band, you need to be able to *hear* on more than one band. Harmonics from your 40-meter transmitter can trash 15; 80 can wipe out 20 or 10. It's not easy! The solutions are too complex to cover in any detail here, but let's cover some basics.

To start off, ground everything. Even if you don't think it needs to be. Use hefty grounds, too—a strap made of tinned braid an inch wide will work a lot better than #22 AWG wire. Ground everything in the station together, and to the main ground for your shack. Second is the physical placement of your antennas—two different antennas on one tower can lead to various problems if they're not done right. Most of the biggest multitransmitter stations have at least two towers for their farm, with adequate spacing between any two antennas for any given two bands in order to reduce the interference.

Inside the station there is a vital accessory you should not overlook. The well-equipped multiop uses a device known as a *passive receive filer*. These go in your receiver's line to eliminate unwanted signals from other bands.

Most widely known are the famed W3LPL filters, used at Frank Donovan's Maryland megastation, based on an LC design. Other stations use ones they're developed themselves, or one of the many commercial versions now appearing on the open market. These improvements can go a long way in turning a weekend of potential nightmares into a lean, mean, contesting machine.

The new guys take over. Brad Parker (in the rear) and Patrick Conroy were still waiting for their call letters to come in the mail when they operated N1II at Field Day for the Greater Norwalk Amateur Radio Club. A few weeks later the mailman told Brad he was KB1AKA and Patrick he was N1TFA. Peter Braun, N1RRA, took the photograph.

Whatever your aim—whether to be the king of the mountain, tops in your club, or to train some new operators, enjoy yourselves! Many hands can make light work—and for a bigger score!—*Warren Stankiewicz, NF1J*

your narrowest code filter. Listen on the SSB filter. If you're listening to an SSB pileup, switch over to the AM filter. Using a wider filter forces you, the operator, to separate the signals in your mind, instead of letting the receiver provide the interference rejection. After enough practice, you'll find that your brain is a very effective filter. When contest time comes along and you go back to the narrower filters, you'll marvel at the selectivity. You may find that your *cerebral* filter scheme works so well that, particularly on CW, you don't need as narrow an electronic filter as previously. If you can attain this level of proficiency, you gain an added advantage over most of the competition. Since your receive filter is wider, "chasing" contacts is much easier because you can effectively tune the band faster than someone with a narrow filter.

Strategy plays an important role in successful contest operating. The decision to be on a particular band at a particular time can be crucial to the outcome of a competition, particularly when you make the right move and the competition doesn't. Your strategic decisions should be made on the basis of hard facts and educated choices (in reality, a lot of luck will be involved, too). To have a full complement of facts available during the contest, you have to do yet more research before the contest. There are two areas where you can gather information prior to the contest that will be of value during the contest. The first is propagation.

During the weeks prior to the contest, study the bands. At what times do they open to where? From what direction? For how long? Keep records of your observations. The most significant times to observe are the few days just prior to the contest, and 28 days before the contest period (the sun rotates once every 28 days; propagation trends tend to follow the same pattern). Once you have completed your study, plot your results in graphical form. Use the resulting presentation in preparing a tentative operating plan for the contest. Of course, like the weather, you truly can't predict band openings with a high degree of reliability, so be prepared to modify your strategy according to the propagation you actually encounter. Be flexible.

Speaking of the weather, you should also follow the weather forecasts for the days immediately preceding the contest. Most enhanced VHF propagation conditions are the result of slowly moving weather systems; weather-map study can help you predict potential favorable directions in a VHF contest. HF contest enthusiasts should also follow the weather; passing storm fronts cause QRN (static), particularly on the lower HF bands. If you expect a front to pass through your area during the contest, plan to emphasize the higher frequency bands during that period to avoid the QRN. Plan your lower frequency op-

Contest Entries and the Electronic Age

Since 1989, the ARRL Contest Branch has accepted entries electronically. Since there are so many software packages out there, and different methods of transmission—from diskette to BBS to e-mail—it can be confusing, or at least a little intimidating, to figure out how to do it.

To submit a log electronically for an ARRL contest, you need two things. First is a log file. It must have the name CALLSIGN.LOG (such as K1ZZ.LOG). The second thing needed is a summary sheet. This can be the standard piece of paper if you are mailing in a disk. If you're sending it to HIRAM (the ARRL BBS) or by e-mail, you need a file with the same summary information.

Your .LOG file must be an MS-DOS file and follow a certain format. It has to be in the "Standard File Format for Submission of Contest Log Data," as shown in the accompanying sidebar. It can't be a binary (.BIN) or "all" (.ALL) file; nor should you name another file CALLSIGN.LOG. Fortunately, almost all of the major contesting software packages (see the sidebar "Computers and Contesting—the Wave of the Present" for details) will automatically generate the proper files for you—check your software manual.

Once you have the summary and the log file, four options are open for you. You can mail the disk in with the summary sheet; you can upload them as separate files to the ARRL BBS (860-594-0306), anonymous ftp to **ftp.arrl.org** or you can send them by e-mail to the Contest Branch Internet address: **contest@arrl.org**. (If you use the Internet, you can send your log and summary files as one file; on the ARRL BBS you can also ZIP them together.) Any of these methods should work just fine.

But wait! What if something happens to my disk?—the ARRL Contest Branch is not a government bureaucracy—if we hit a snag with a file, we won't disqualify you for it. What if my file gets lost? All e-mail submissions are acknowledged with a short message. If you don't get this message in about 2 weeks or so, try again! Many people who mail entries in to ARRL have also found a self-addressed, stamped postcard is a good way to keep tabs on their log. As soon as your entry is processed, we mail the card back to you, attenuating any worries by several tens of dBs.

Where's the payoff in all this? Time. You no longer have to work your way through reams of paper. This saves you time and effort. It also reduces the massive job the Contest Branch has to do every year in processing the 16 contests it sponsors. These contests generate roughly 16,000 pieces of mail in a year—a lot of paper for two people to sort through. The more time (and money) we can save, the more you, the members, save as well.

And from all of us, thank *you* for helping!—*NF1J*

station located in a rare multiplier habitually get on the air just before his or her bedtime every evening, perhaps way up in frequency on the band? Is a particular station willing to change bands with you for a new multiplier? Is someone going to a rare area specifically for the contest? All this information can be put to good use if you document it carefully and put the data in a form that you can understand well into the contest.

Sometime before the contest, prepare your logging materials for contest use. Obviously you can't fill in contact data in advance, but you can put in some constant information like your own call sign, the contest starting date, a consecutive number for each log page, the dupe sheet frequency band information and the like. Not only will this step save time during the contest, it will also lead to less confusion during the contest, allowing you to concentrate on making contacts.

It's also a good idea before the contest to have your station all set up for the event. Adjust the VOX controls on the transmitter, set your clock, calibrate your receiver, replace the burned-out light bulb over the operating position, program the memory keyer, note the optimum band markings for quick tuneup on each band, and any other maintenance tasks that might need doing. You certainly don't want to perform maintenance during the contest; you have enough to do just in operating and fixing the inevitable visits from Murphy.

Be sure to read and thoroughly understand the latest contest rules for the contest in question. Perhaps the rules have changed since last year or your recollection of the existing rules is rather fuzzy. In either case, being clear on the rules might prevent you from making a fatal mistake that might make your hard work during the contest weekend for naught. Finally, prepare yourself mentally for the contest. As most sports psychologists will tell you, your mental attitude often is the determining factor in how well you perform in a competitive activity. Convince yourself that you can do well by putting in the effort. Picture in your mind what you're going to do in each situation you might encounter during the contest. Run through it over and over until it becomes second nature. You have to determine for yourself how to get motivated, since everyone is different. One thing is for sure, though; if you aren't prepared to do your best, you won't.

OPERATING THE CONTEST

The big moment finally arrives. Now what? Regardless of whether you are operating a VHF contest, an HF DX contest or whatever, there is one fundamental guideline that you should follow in a contest: Do what gains you the most points. "Another obvious suggestion," you're probably thinking. It may be obvious, but if you truly follow it, you'll end up with the highest score possible for your station on that weekend.

For example, early in the contest you come across a rare multiplier. He is just casually working a few stations at a rate of about 12 an hour. He is on an "oddball" frequency, away from most of the contest activity. What should you do? If you decide that the only way you'll work the multiplier during the whole weekend is to stick it out and spend the time, you'll be missing the highest activity point in the contest. The time might be well spent. Of course, you could also spend 45 minutes and never work him; think of the contacts you lost and will never recover.

Decisions like this need to be made almost constantly during a contest. Your decision should be based on considering that simple fundamental. If you feel that propagation might be favorable right then, that your station is loud enough, and that

eration around the QRN. Some television stations now show foreign weather conditions as well as local forecasts; for DX contest operating, use the foreign weather information to plan your lower frequency operation around QRN at the DX end, so the DX stations can hear you.

The second area where you can gather useful contest information before the contest is in the study of activity patterns. Having a knowledge of certain stations' operating habits can truly be a blessing with regard to finding rare contest multipliers. Is there a particular gathering frequency for groups of stations in a particular geographical area (such as for severe weather reporting in tropical areas)? Does a

the addition of another multiplier is more beneficial to the final score than running up your contact total at that point, then by all means try to work the multiplier. Your gamble may or may not work. You might decide that you can more easily work a number of stations in the same time that it might take to work the multiplier, more than enough contacts to compensate for the "lost" multiplier. Your best choice then probably is to forget about the multiplier; he might be easier to work later. Decisions like these are the very essence of contesting, and only you can decide based on your knowledge of propagation, your station's capabilities, contest strategy and your own capabilities. Indeed, preparation, experience and raw talent determine your ability to make the right decision when needed.

There are some operating hints that might make your strategic decisions easier to make. First, learn the relative point value of a new multiplier compared to a new QSO. Simply dividing the number of QSOs made at that point in the contest by the number of multipliers worked at that point in the contest gives a ratio that represents the number of contacts a multiplier is worth at that point. You can do this during the contest with a calculator or slide rule, but arithmetic exercises when you're already fatigued likely will be very time consuming, and may not yield accurate results. Instead, a prepared chart, showing the number of QSOs along one axis, the number of multipliers along the other axis and the QSO/multiplier ratio at the intersections, is much easier and more reliable to use during a contest (in fact, add that to your list of items to prepare before the contest).

At each point in a contest, the QSO/multiplier ratio changes depending on the contest format and your own performance. For example, during a DX contest, where there is a relatively high number of multipliers available, the QSO/multiplier ratio might be around three. That means each new multiplier worked at that point is equivalent to making three (nonmultiplier) QSOs, as far as your point total is concerned. In the November Sweepstakes, in which there is a finite number of multipliers available (77), the QSO/multiplier ratio might be over 10. In that circumstance, it would require making 10 QSOs to equal the point value of one new multiplier.

By applying your knowledge of propagation, your intimate familiarity with your station's capabilities, and your best estimate of how the propagation and activity is evolving in the contest, you can use the QSO/multiplier ratio information to decide whether to look for contacts or to try jumping into a pileup for a new multiplier. During high activity and good propagation periods, making contacts is probably a better tack, since you might never be able to recover the band opening time later on. Spending 10 minutes in a pileup when the band is hot and you have the potential to make a lot of contacts is a poor choice compared to spending 10 minutes in a pileup when the activity or propagation has slackened, and you only might be able to make two contacts during that same 10 minutes.

The key principle is to utilize your band-opening time wisely, as you have little or no control over propagation and have to work what you can when you can. This strongly implies that any unusually good propagation should be taken advantage of immediately, since it may never return. At VHF especially, where band openings tend to be short-lived (anywhere from a second to a couple of hours), enhanced propagation is a sure sign to stop whatever you're doing and work what you can during the opening. You may not get another chance. The value of band-opening time is inversely proportional to its duration.

As an example, consider 6-meter scatter propagation during a VHF contest morning. A meteor burst might make a 1000-mile-distant station audible (and potentially workable) for 20 seconds. Unless you make the contact during that time period, you're out of luck. You should call quickly and succinctly and exchange contest reports the same way. If you make yourself understood the first time, there is no need for repeats that you might not have propagation for. There is no time to give the other station's call (which he or she presumably already knows) five times, your call eight times and the instruction to go ahead six times. By the third "go ahead," the burst is over, the contact is lost, and the operator at the other end is frustrated over the failure to complete the contact successfully. The next time there is a meteor burst, that other operator might tell you to "go away," and then try to work someone quicker on the draw than you were.

HF contesters are just as guilty of bad (or no) timing as VHF contesters. How many times have you heard the din of a pileup go on for two or three minutes after the station being called has already come back to someone? Who gained there? The moral: Don't repeat unless asked to.

There are two general approaches to making contacts—calling CQ (and "running" people/getting responses) and hunting. Each has its own advantages and disadvantages, as you might expect. Calling CQ effectively requires getting the attention of other stations and motivating them to call you. Getting attention at the other end is most easily accomplished by wielding a big signal. Big signals always seem to get answered (as do signals emanating from rare multipliers). Of course, you may not be fortunate enough to possess a big signal, so calling CQ may not be as productive for you. Here again, knowledge of your station's capabilities is required to make that decision.

Motivating respondents to call is usually easy early in the contest while you still haven't drained the pool of active contesters. The contesters are already motivated to call; they want the QSO. The strain begins *after* you have worked the serious competitors, and the casual operator is all that remains. Here you have to be prepared to answer questions about the contest, and perhaps tell the operator at the other end what information you need for a valid contest QSO.

There is one operating strategy to keep in mind here: Don't scare anyone off. Try to be as polite as you can, and give the information at a reasonable pace. Sending an exchange at 50

What does a contest station really look like at 3 in the morning? This is VE7NSR, looking abandoned as the operator and logger went to the food table for fresh coffee. A fast look at the table and floor, and you can see why many people prefer computer logging to using paper and pencil!

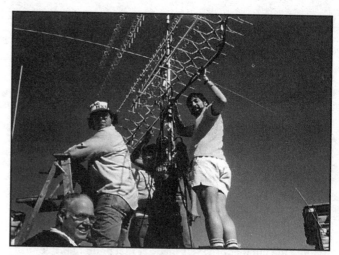

Inhabiting grid square DM04 was the gang at W6YLZ. Shown are WA6HXD, KA6ZVP, WA8MXA and W6YLZ. The array of antennas covered 440 and 1296.

words per minute or speaking as fast as you can may be fine when you are in contact with another contest devotee, but is totally lost on most everyone else. Not only will they not be able to acknowledge your transmitted information, they (as well as anyone else listening) might become discouraged or even incensed, disappearing without giving you your QSO.

There seemingly are as many hunting techniques as there are contest operators. While everyone has a favorite approach, there is one generalization about the successful versions; they are all organized. The operator starts at one end of the band and methodically scans to the other end, while trying to contact all the stations not already worked in between. Maintenance of an updated duplicate check sheet (dupe sheet) is a necessity for hunting. It is not only embarrassing but a waste of time to call someone already worked. Some operators maintain a "map" of the band that lists the call signs of the stations holding forth on the various frequencies. A map such as this allows the hunting operator to skip over frequencies that he or she knows are occupied by someone already in the log. This can save considerable time, especially on bands that are clogged with big pile-ups; spending 10 minutes to work a station that turns out to be a duplicate contact doesn't add much excitement to the contest.

Successful contest operating can be summarized very easily: Always spend your time where you can produce the most points and treat that time as if it is precious—because it is! Always strive to make your log entries an accurate and complete representation of what was sent to you. Accurate communication is the very basis for the contest; inaccurate communication can be the basis for disqualification.

AFTER THE CONTEST

Afterwards, every participant faces the same task, regard-less of his or her performance in the contest—preparing the log for submission. The proficient and ambitious contester adds a second task—analyzing the results. The first task shouldn't be taken lightly. Many contests have been won or lost as a result of accurate or inaccurate log checking before submission. Log preparation is a lot like preparing income tax returns. The accuracy of your submission has a lot to do with the outcome. While you may be able to get away with "creative logging" for a year (or contest) or two, eventually you will be revealed. From that point on, your return (or log) will be carefully scrutinized on a routine basis. What is probably worse about contest logs than income-tax returns is that once you are disqualified, you will never have any credibility with your fellow contesters. That in itself might be a more severe punishment than a few measly years in prison for defrauding the government!

A proper log submission consists of a photocopy of the original log sheets, with invalid and duplicate contacts so marked. Don't erase these contacts, just mark them so that the log checkers can identify and verify them by cross-checking with other logs. It is virtually imperative to "re-dupe" a log before submission. Any duplicate check-sheet made during the contest is bound to be inaccurate and should only be considered an operating aid. In these modern times, computers are increasingly being used to eliminate duplicate contacts from contest logs prior to submission (be sure to double check for keyboarding errors, however).

A contest log might be considered the history of a "lost" weekend. While there is no explicit description whatsoever of the events of the weekend, a clear record is available for the experienced to interpret. Your best source of information to use while preparing for the next contest is your record of the last contest. What mistakes did you make? What bands were the most or least productive and why? What equipment worked well? What equipment failed? Responding to these questions should keep you busy until the next contest. And you can be sure that the next contest will be more productive and satisfying than the last.

CHAPTER 8

Operating Awards

STEVE FORD, WB8IMY
ARRL HEADQUARTERS

Awards hunting is a significant part of the life-support system of Amateur Radio operating. It's a major motivating force of so many of the contacts that occur on the bands day after day. It takes skillful operating to qualify, and the reward of having a beautiful certificate or plaque on your ham-shack wall commemorating your achievement is very gratifying. (On the other hand, you don't necessarily have to seek them actively; just pull out your shoebox of QSLs on a cold, winter afternoon, and see what gems you already have on hand.) Aside from expanding your Amateur Radio-related knowledge, it's also a fascinating way to learn about the geography, history or political structure of another country, or perhaps even your own. This chapter provides detailed information on awards sponsored by ARRL and other major US organizations.

There are some basic considerations to keep in mind when applying for awards. Always carefully read the rules, so that your application complies fully. Use the standard award application if possible, and make sure your application is neat and legible, and indicates clearly what you are applying for. Official rules and application materials are available directly from the organization sponsoring the particular award; always include an SASE (self-addressed, stamped envelope) or, in the case of international awards, a self-addressed envelope with IRCs (International Reply Coupons, available from your local Post Office) when making such requests. Sufficient return postage should also be included when directing awards-related correspondence to Awards Managers, many (if not most) of whom are volunteers. So above all, be patient!

As to QSL cards, if they are required to be included with your application, send them the safest possible way and always include sufficient return postage for their return the same way. (A return-postage chart for ARRL awards appears in Chapter 17 of this book). It is vital that you check your cards carefully before mailing them; make sure each card contains your call and other substantiating information (band, mode, and so on). Above all, don't send cards that are altered or have mark-overs, even if such modifications are made by the amateur filling out the card. Altered cards, even if such alterations are made in "good faith," are not acceptable on this no-fault basis. If you are unsure about a particular card, don't submit it. Secure a replacement.

None of the above is meant to diminish your enthusiasm for awards hunting. Just the opposite, since this chapter has been painstakingly put together to make awards hunting even more enjoyable. These are just helpful hints to make things even more fun for all concerned. Chasing awards is a robust facet of hamming that makes each and every QSO a key element in your present or future Amateur Radio success.

ARRL AWARDS

To make Amateur Radio QSOing more enjoyable and to add challenge, the League sponsors awards for operating achievement, some of which are the most popular awards in ham radio. Exception for the RCC and Code Proficiency awards, applicants in the US and possessions, Canada and Puerto Rico must be League members to apply.

Rag Chewers Club (RCC)

New hams often go for RCC as their first operating award. This award is designed to encourage friendly contacts of more substance than the hello-goodbye type QSO. RCC has just one requirement: "Chew the rag" over the air for at least one solid half hour. If you want to obtain the RCC certificate, report the QSO to ARRL HQ, with a fee of $2, and you'll soon be issued the distinctive award. If you want to nominate someone else for membership, send the nomination

to him or her, not to ARRL HQ. This way, no one gets an unwanted certificate, and confirmed rag chewers can still nominate those they think are qualified. RCC is available to all amateur licensees.

Friendship Award

The purpose of this award is to encourage friendly, conversational contacts between radio amateurs. The Friendship Award is available to any ARRL member who submits log copies showing two-way communication with 26 stations whose call signs end with each of the 26 letters of the alphabet. (For example: W4RA, KØORB, W3ABC, K1ZZ.) QSL cards are not required. Applications must include a fee of $3.

Logs must indicate the contact date, call sign, name, location *and some other fact about the person you contacted.* (For example: age, other interests, occupation and so on.) Contacts can be made on any frequency or mode. All contacts must be made after November 1, 1993. For a friendship Award application, send a self-addressed envelope with one unit of First-Class postage to: ARRL, 225 Main St., Newington, CT 06111.

Code Proficiency (CP)

You don't have to be a ham to earn this one. But you do have to copy one of the W1AW qualifying runs. (The current W1AW operating schedule is printed in *QST,* listed on the ARRL Internet site (**http://www.arrl.org**) or from ARRL HQ for an SASE.) Twice a month, five minutes worth of text is transmitted at the following speeds: 10-15-20-25-30-35 WPM. For a real challenge, W1AW transmits 40-WPM twice yearly. To qualify at any speed, just copy one minute solid. Your copy can be written, printed or typed. Underline the minute you believe you copied perfectly and send this text to ARRL HQ along with your name, call (if licensed) and complete mailing address. Your copy is checked directly against the official W1AW transmission copy, and you'll be advised promptly if you've passed or failed. If the news is good, you'll soon receive either your initial certificate or an appropriate endorsement sticker. Please include an SASE with your submission; 9 ×12 with two units of First-Class postage for a certificate or a business-size envelope for an endorsement.

Worked All States (WAS)

The WAS (Worked All States) award is available to all amateurs worldwide who submit proof of having contacted each of the 50 states of the United States of America. The

WAS program includes the numbered awards and endorsements listed below.

Two-way communication must be established on amateur bands with each state. Specialty awards and endorsements must be two-way (2x) on that band and/or mode. There is no minimum signal report required. Any or all amateur bands may be used for general WAS. The District of Columbia may be counted for Maryland.

Contacts must all be made from the same location or from locations no two of which are more than 50 miles apart, which is affirmed by signature of the applicant on the application. Club-station applicants, please include clearly the club name and call sign of the club station (or trustee).

Contacts may be made over any period of years. Contacts must be confirmed in writing, preferably in the form of QSL cards. Written confirmations must be submitted (no photocopies). Confirmations must show your call and indicate that two-way communication was established. Applicants for specialty awards or endorsements must submit confirmations that

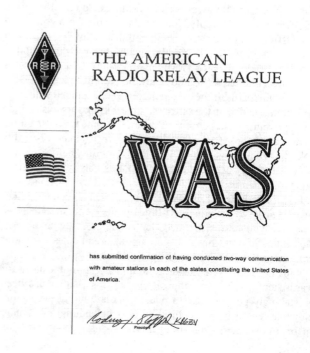

clearly confirm two-way contact on the specialty mode/band. Contacts made with Alaska must be dated January 3, 1959 or later, and with Hawaii dated August 21, 1959 or after.

Specialty awards (numbered separately) are available for OSCAR satellites, SSTV, RTTY, 432 MHz, 222 MHz, 144 MHz, 50 MHz, and 160 meters. Endorsements for the basic mixed mode/band award and any of the specialty awards are available for SSB, CW, EME, Novice, QRP, packet and any single band except 30 meters. The Novice endorsement is available for the applicant who has worked all states as a Novice licensee. QRP is defined as 10-watts input (or 5-watts output) as used by the applicant only and is affirmed by signature of the applicant on the application.

Contacts made through "repeater" devices or any other power relay method cannot be used for WAS confirmation. A separate WAS is available for OSCAR contacts. All stations contacted must be "land stations." Contact with ships (anchored or otherwise) and aircraft cannot be counted.

Applicants must be ARRL members to participate in the WAS program. DX stations are exempt from this requirement.

HQ reserves the right to "spot call" for inspection of cards (at ARRL expense) of applications verified by an HF Awards Manager. The purpose of this is not to question the integrity of any individual, but rather to ensure the overall integrity of the program. More difficult-to-be-attained specialty awards (222 MHz WAS, for example) are more likely to be so called. Failure of the applicant to respond to such a spot check will result in nonissuance of the WAS certificate.

Disqualification: False statements on the WAS application or submission of forged or altered cards may result in disqualification. ARRL does not attempt to determine who has altered a submitted card; therefore do not submit any marked-over cards. The decision of the ARRL awards Committee in such cases is final.

Application Procedure (please follow carefully): Confirmations (QSLs) and application form (MSD-217) may be submitted to an approved ARRL Special Service Club HF Awards Manager. ARRL Special Service Clubs appoint HF Awards Managers whose names/addresses are on file at HQ. If you do not know of an HF Awards Manager in your local area, call a club officer to see if one has been appointed or contact HQ. If you can have your application so verified locally, you need not submit your cards to HQ. Otherwise, send your application, cards, and required fees to HQ, as indicated on the application form (reproduced in Chapter 17).

Be sure that when cards are presented for verification (either locally or to HQ) they are sorted alphabetically by state, as listed on the back of application form MSD-217.

All QSL cards sent to HQ must be accompanied by sufficient postage for their safe return, and the required fee of $5 for each WAS certificate (which includes any endorsements with the same 50 QSL cards), or $2 per endorsement application.

Five-Band WAS (5BWAS)

This award is designed to foster more uniform activity throughout the bands, encourage the development of better antennas and generally offer a challenge to both newcomers and veterans. The basic WAS rules apply, including cards being checked in the field by Awards Managers; in addition, 5BWAS carries a start date of January 1, 1970. Unlike WAS, 5BWAS is a one-time-only award; no band/mode endorsements are available. Contacts made on 10/18/24 MHz are not valid for 5BWAS. The $10 application fee includes the certificate and

lapel pin. In addition, a 5BWAS plaque is available at an additional charge.

Pins

WAS and 5BWAS pins may be purchased through our Publication/Sales Department.

Worked All Continents (WAC)

In recognition of international two-way Amateur Radio communication, the International Amateur Radio Union (IARU) issues Worked All Continents certificates to Amateur Radio Stations of the world. Qualification for the WAC award is based on examination by the International Secretariat, or a Member-Society, of the IARU, that the applicant has received QSL cards from other amateur stations in each of the six continental areas of the world (see the ARRL DXCC Countries List in Chapter 17 for a complete listing of continents). All contacts must be made from the same country or separate territory within the same continental area of the world. All QSL cards (no photocopies) must show the mode and/or band for any endorsement applied for.

WAC Certificates

The following WAC certificates are available:
Basic Certificate (mixed mode)
CW Certificate
Phone Certificate
SSTV Certificate
RTTY Certificate
FAX Certificate
Satellite Certificate
5-Band Certificate

WAC Endorsements

The following WAC endorsements are available:
6-Band endorsement
QRP endorsement
1.8-MHz endorsement
3.5-MHz endorsement
50-MHz endorsement

144-MHz endorsement

430-MHz endorsement

Any higher-band endorsement

Contacts made on 10/18/24 MHz or via satellites are void for the 5-band certificate and 6-band sticker. All contacts for the QRP endorsement must be made on or after January 1, 1985, while running a maximum power of 5-watts output or 10-watts input.

For amateurs in the United States or countries without IARU representation, applications should be sent to the IARU International Secretariat, PO Box AAA, Newington, CT 06111, USA. After verification, the cards will be returned, and the award sent soon afterward. There is no application fee; however, sufficient return postage for the cards, in the form of a self-addressed, stamped envelope or funds is required. US amateurs must have current ARRL membership. [A sample application form appears in Chapter 17.] All other applicants must be members of their national Amateur Radio Society affiliated with IARU and must apply through the Society only.

Note: The DXCC Countries List in Chapter 17 includes a continent designation for each DXCC country.

A-1 Operator Club (A-1 OP)

Membership in this elite group attests to superior competence and performance in the many facets of Amateur Radio operation: CW, phone, procedures, copying ability, judgment and courtesy. You must be recommended for the certification independently by two amateurs who already are A-1 Ops. This honor is unsolicited; it is earned through the continuous observance of the very highest operating standards.

Old-Timers' Club (OTC)

In recognition of amateurs who have held an amateur license 20-or-more years (lapses permitted), a suitable award is available—OTC. If you qualify as an "old-timer," you'll find the necessary paperwork pretty easy. Drop a note to HQ (with a fee of $2.00) with the date of your first amateur license and your call then and now. HQ will verify the information, and if you're eligible, you'll soon receive your OTC certificate by return mail.

VHF/UHF Century Club (VUCC)

The VHF/UHF Century Club (VUCC) Award is awarded for contacts with a minimum number of "Maidenhead" 2° × 1° grid square locators per band as indicated below. Grid squares are designated by a combination of two letters and two numbers. More information on grid squares can be found in January 1983 *QST*, pp 49-51 (reprint available from HQ on request), The ARRL World Grid Locator and the ARRL Grid Locator for North America (see latest *QST* for prices and ordering information). The VUCC certificate and endorsements are available to ARRL members in Canada, the US possessions, and Puerto Rico, and other amateurs worldwide. Only those contacts dated January 1, 1983 and later are credible for VUCC purposes.

The minimum number of grid squares needed to qualify for each individual band award is as follows:

THE AMERICAN RADIO RELAY LEAGUE

VUCC

VHF/UHF CENTURY CLUB

has submitted confirmation of having conducted two-way communication with amateur stations in 2° longitude X 1° latitude grid squares on the frequency band indicated. This outstanding achievement has earned the above membership in the exclusive VHF/UHF Century Club (VUCC).

50 -144 MHz and Satellite—100 credits
222 and 432 MHz—50 credits
902 and 1296 MHz—25 credits
2.3 GHz—10 credits
3.4 GHz, 5.7 GHz, 10 GHz, 24 GHz, 47 GHz, 75 GHz, 119 GHz, 142 GHz, 241 GHz and Laser (300 GHz)—5 credits

Certificates for 222 and 432 MHz are designated as Half Century, 902 and 1296 MHz as Quarter Century, and those above as SHF Awards. Individual band awards are endorseable in the following increments:

50 and 144 MHz—25 credits
222 and 432 MHz—10 credits
902 MHz and higher—5 credits

There are no specialty endorsements such as "CW only." Separate bands are considered as separate awards. No cross-band contacts are permitted except for Satellite. No contacts through active "repeater" devices or other power relay methods are permitted, except for Satellite. Contacts with aeronautical mobiles (while airborne) are not permitted. Contacts with shipboard stations are permitted.

Stations who claim to operate from more than one grid square simultaneously (such as from the intersection of four grid squares) must be physically present in more than one square to give multiple square credit with a single contact. This requires the operator to know precisely where the intersection lines are located and placing the station exactly on the boundary to meet this test. To achieve this precision work requires either current markers permanently in place or the precision work of a professional surveyor. Operators of such stations should be prepared to provide some evidence of meeting this test if called upon to do so. Global Positioning System (GPS) readings are okay for this purpose. Multiple QSL cards are not required.

For VUCC awards on 50 through 1296 MHz, all contacts must be made from a location or locations within the same grid square or locations in different grid squares no more than 50 miles apart. For SHF awards, contacts must be made from a single location, defined as within a 300-meter diameter circle.

Application Procedure (please follow carefully):

Confirmations (QSL cards) the VUCC application form (MSD-260) and VUCC field sheet (MSD-259) must be submitted to an approved VHF Awards Manager for verification. ARRL Special Service Clubs appoint VHF Awards Managers whose names are on file at HQ. If you do not know of an Awards Manager in your area, HQ will give you the name of the closest Awards Manager to you. Do not send cards to ARRL HQ.

For the convenience of the Awards Manager in checking cards, applicants may indicate in pencil (pencil only) the grid square locator on the address side of the cards that do not clearly indicate the grid locator. The applicant affirms that he/she has accurately determined the proper locator from the address information given on the card by signing the affirmation statement on the application.

Cards must be sorted alphabetically by field and numerically from 00 to 99 within that field. These cards are then listed on the VUCC field sheet (MSD-259).

Where it is necessary to mail cards for verification, the applicant must include sufficient postage for the return of their cards. In addition, a separate self-addressed stamped envelope (SASE) must be provided to the Awards Manager to send the application and appropriate fees to ARRL HQ. The use of registered mail when handling QSL cards is recommended.

There is a first-time fee of $10 for VUCC awards. Additional VUCC awards are also $10. A lapel pin is included with each new award as of January 1, 1996. VUCC lapel pins are available for $5 each.

Enclosed with the initial VUCC certificate from HQ will be a photocopy of the original list of grid squares for which the applicant has received credit (MSD-259). When applying for endorsement, the applicant will indicate in red on that photocopy those new grids for which credit is being sought, and complete a new application form (MSD-260) and submit this to the Awards Manager. A new updated photocopy listing will be returned along with appropriate endorsement stickers. Thus, a current list of grid squares worked is always in the hands of the VUCC award holder and a permanent record is on hand at HQ of the applicant's award status. For endorsement applications it is necessary to submit only those MSD-259s that indicate new grid squares worked since the previous submission, along with the application form MSD-260.

Disqualification

Altered/forged confirmations or fraudulent applications submitted may result in disqualification of the applicant from VUCC participation by action of the ARRL Awards Committee. The applicant affirms he/she has abided by all the rules of membership in the VUCC and agrees to be bound by the decisions of the ARRL Awards Committee.

Decisions of the Awards Committee regarding interpretation of the rules here printed or later amended shall be final.

Operating Ethics

Fair play and good sportsmanship in operating are required of all VUCC participants.

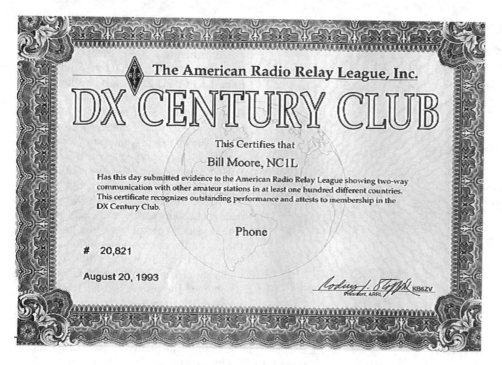

The American Radio Relay League, Inc.

DX CENTURY CLUB

This Certifies that

Bill Moore, NC1L

Has this day submitted evidence to the American Radio Relay League showing two-way communication with other amateur stations in at least one hundred different countries. This certificate recognizes outstanding performance and attests to membership in the DX Century Club.

Phone

20,821

August 20, 1993

President, ARRL   KB6ZV

DX Century Club (DXCC)

DXCC is the premier operation award in Amateur Radio. The initial DXCC certificate (a nominal fee of $2 US is charged for the DXCC lapel pin) is available to League members in Canada, the US and possessions, and Puerto Rico, and all amateurs in the rest of the world. There are 12 separate DXCC awards available:

Mixed (general type)—contacts must be made using any mode since November 14, 1945.

Phone—contacts must be made using radiotelephone since November 14, 1945. Confirmations for cross-mode contacts for this award must be dated before October 1, 1981. Confirmations need not indicate two-way (2×) to be credited.

CW—contacts must be made using CW since January 1, 1975. Confirmations for cross-mode contacts for this award must be dated before October 1, 1981. Confirmations need not indicate two-way (2×) to be credited.

RTTY—contacts must be made using radioteletype since November 15, 1945. Confirmations for cross-mode contacts for this award must be dated before October 1, 1981. Confirmations need not indicate two-way (2×) to be credited.

160 meter, 80 meter, 40 meter, 10 meter, 6 meter, 2 meter—contacts must be made since November 15, 1945.

Satellite—contacts must be made using satellites since March 1, 1965 (non-endorsable).

DXCC Fee Schedule

Effective January 1, 1994, all amateurs applying for their very first DXCC Award will be charged a one-time Registration Fee of $10. This same fee applies to both ARRL members and foreign non-members, and both will receive one DXCC certificate and a DXCC pin. Applicants must provide funds for postage charges for QSL return.

(a) A $5 shipping and handling fee will be charged for each additional DXCC certificate issued, whether new or replacement. A DXCC pin will be included with each certificate.

(b) Endorsements and new applications may be presented at ARRL HQ, and at certain ARRL conventions. When presented in this manner, such applications shall be limited to 110 cards maximum, and a $2 handling charge will apply.

(c) Each ARRL member will be allowed one submission in each calendar year at no cost (except as in [b] above, or return postage). This annual submission may include any number of QSL cards for any number of DXCC Awards, and may be a combination of new and endorsement applications. Fees as in (a) above will apply for additional new DXCC Awards.

(d) Foreign non-members will be allowed the same annual submission as ARRL members, however, they will be charged all a $10 DXCC Award fee, in addition to return postage charges. Fees in (a) and (b) may also apply.

(e) DXCC participants who wish to submit more than once per year will be charged a DXCC fee for each additional submission made during the remainder of the calendar year. These fees are dependent upon membership status: ARRL members: $10 foreign non-members: $20. Additionally, return postage must be provided by applicant, and charges from (a) and (b) above may be applied.

Confirmations (QSL cards) must be submitted directly to ARRL HQ for all countries claimed. Confirmations for a total of 100 or more countries must be included with first application. Contacts made on all amateur bands are valid for DXCC. **[Contacts made on 10/18/24 MHz are not valid for 5-Band DXCC.]**

The ARRL DXCC Countries List criteria (see below) will be used in determining what constitutes a "country."

Confirmations must be accompanied by a list of claimed DXCC countries and stations to aid in checking and for future reference (the required DXCC application materials are available from ARRL HQ for an SASE).

Endorsement stickers for affixing to certificates or pins will be awarded as additional credits are granted. For the Mixed, Phone, CW, RTTY and 10-Meter DXCC, these stickers are in exact multiples of 25, ie, 125, 150, etc, between 100 and 250 DXCC countries; in multiples of 10 between 250 and 300, and in multiples of 5 above 300 DXCC countries. For 160-meter, 80-meter, 40-meter, 6-meter, 2-meter and Satellite DXCC, the stickers are in exact multiples of 10 starting at 100 and multiples of 5 above 200.

All contacts must be made with amateur stations working in the authorized amateur bands or with other stations licensed to work amateurs. Contacts made through "repeater" devices or any other power relay method (aside from Satellite DXCC) are invalid for DXCC credit.

In countries where amateurs are licensed in the normal manner, credit may be claimed only for stations using regular government-assigned call signs or portable call signs where reciprocal agreements exist or the host government has so authorized portable operation. No credit may be claimed for contacts with stations in any country that has temporarily or permanently closed down Amateur Radio operations by special government edict where amateur licenses were formerly issued in the normal manner. Some countries, in spite of such prohibitions, issue authorizations that are acceptable.

All stations must be "land stations." Contacts with ships and boats, anchored or under way, and airborne aircraft, cannot be counted.

All stations must be contacted from the same DXCC country.

All contacts must be made by the same station licensee. However, contact may have been made under different call signs in the same country if the licensee for all was the same.

Any altered, forged or otherwise invalid confirmations submitted by an applicant for DXCC credit may result in disqualification of the applicant. Any holder of a DXCC award submitting altered, forged or otherwise invalid confirmations may forfeit the right to continued DXCC membership. The ARRL Awards Committee shall rule in these matters and may also determine the eligibility of any DXCC applicant who was ever barred from DXCC to reapply, and the conditions for such application.

(a) Fair play and good sportsmanship in operating are required of all DXCC members. In the event of specific objections relative to continued poor operating ethics, an individual may be disqualified from DXCC by action of the ARRL Awards Committee.

(b) Credit for contacts with individuals who have displayed continued poor operating ethics may be disallowed by action of the ARRL Awards Committee.

For (a) and (b) above, "operating" includes confirmation procedures and/or documentation submitted for DXCC accreditation.

Each DXCC applicant must stipulate that he/she has observed all DXCC rules as well as all pertinent governmental regulations established for Amateur Radio in the country or countries concerned, and agrees to be bound by the decisions of the ARRL Awards Committee. Decisions of the ARRL Awards Committee regarding interpretations of the DXCC rules (either currently in effect or later amended) shall be final.

All new DXCC applications must contain sufficient postage in the form of US currency, check or money order. For DXCC endorsements, sufficient funds for return postage is available on request from ARRL HQ (and is reproduced in Chapter 17). ARRL membership is not required of foreign applicants.

Official DXCC application forms are required. These may be obtained from the DXCC Desk at ARRL HQ. Please include a business-size SASE. **The complete DXCC rules appear in Chapter 17.**

5BDXCC

A Five-Band DXCC Award has been established to encourage more uniform DX activity throughout the amateur bands, encourage the development of more versatile antenna systems and equipment, provide a challenge for DXers, and enhance amateur-band occupancy. The basic DXCC rules apply, although the starting data for valid QSOs is January 1, 1969.

The 5BDXCC certificate is issued after the applicant submits a minimum of 500 QSLs representing two-way contact with 100 different DXCC countries on each of five Amateur Radio bands. Phone and CW segments of the band do not count as separate bands for this award. Confirmations made on any legal mode are acceptable, but no cross-mode or cross-band contacts are acceptable. Contacts made on 10/18/24 MHz are not valid for 5BDXCC. All QSLs must be checked by the ARRL HQ DXCC Desk. 5BDXCC is endorsable for additional bands: 160 meters, 17 meters, 12 meters, 6 meters, 2 meters. In addition to the 5BDXCC certificate, a 5BDXCC plaque is available

at an extra charge of $25.

Countries List Criteria

The ARRL DXCC Countries List is the result of progressive changes in DXing since 1945. The full list will not necessarily conform completely with the current criteria since some of the listings were recognized from pre-WW II or were accredited from earlier versions of the criteria. While the general policy has remained the same, specific mileages and additional points have been added to the criteria over the years. The specific mileages in Point 2(a) and Point 3, mentioned in the following criteria, have been used in considerations made April 1960 and after. The specific mileage in Point 2(b) has been used in considerations made April 1963 and after.

When an area in question meets *at least one* of the following three points, it is eligible as a separate country listing for the DXCC Countries List. These criteria address considerations by virtue of Government [Point 1] or geographical separation [Points 2 and 3], while Point 4 addresses ineligible areas. All distances are given in statute miles.

Point 1, GOVERNMENT

An independent country or nation-state having *sovereignty* (that is, a body politic or society united together, occupying a definite territory and having a definite population, politically organized and controlled under one exclusive regime, and engaging in foreign relations—including the capacity to carry out obligations of international law and applicable international agreements) constitutes a separate DXCC country by reason of **Government**. They *may* be indicated by membership in the United Nations (UN). However, some nations that possess the attributes of sovereignty are *not* members of the UN, although these nations may have been *recognized* by a number of UN-member nations. Recognition is the formal act of one nation committing itself to treat an entity as a sovereign state. There are some entities that have been admitted to the UN that lack the requisite attributes of sovereignty and, as a result, are *not* recognized by a number of UN-member nations.

Other entities which are not totally independent may also be considered for separate DXCC country status by reason of Government. Included are Territories, Protectorates, Dependencies, Associated States, and so on. Such an entity may delegate to another country or international organization a measure of its authority (such as the conduct of its foreign relations in whole or in part, or other functions such as customs, communications or diplomatic protection) *without* such an entity is individually considered, based on all the available facts in the particular case. In making a reasonable determination as to whether a sufficient degree of sovereignty exists for DXCC purposes, the following characteristics (list not necessarily all-inclusive) are taken into consideration:

(a) Membership in specialized agencies of the UN, such as the International Telecommunication Union (ITU).

(b) Authorized use of ITU-assigned call sign prefixes.

(c) Diplomatic relations (entering into international agreements and/or supporting embassies and consulates), and maintaining a standing army.

(d) Regulation of foreign trade and commerce, customs, immigration and licensing (including landing and operating permits), and the issuance of currency and stamps.

An entity that qualifies under Point 1, but consists of two or more separated land areas, will be considered a single DXCC country (since none of these areas alone retains an independent capacity to carry out the obligations of sovereignty), *unless* the areas can qualify under Points 2 or 3.

Point 2, SEPARATION BY WATER

An island or a group of islands which is part of a DXCC country established by reason of **Government**, Point 1, is considered as a separate DXCC country under the following conditions.

(a) The island or islands are situated off shore, geographically separated by a minimum of 225 miles of open water from a continent, another island or group of islands that make up *any part* of the "parent" DXCC country.

For any *additional* island or islands to qualify as an additional separate DXCC country or countries, such must qualify under Point 2(b).

(b) This point applies to the "second" island or island grouping geographically separated from the "first" DXCC country created under Point 2(a). For the second island or island grouping to qualify, at least a 500-mile separation of open water from the first is required, as well as meeting the 225-mile requirement of (a) from the "parent." For any subsequent island(s) to qualify, the 500-mile separation would again have to be met. This precludes, for example, using the 225-mile measurement *for each* of several islands from the parent country to make several DXCC countries.

Point 3, SEPARATION BY ANOTHER DXCC COUNTRY

(a) Contiguous land mass: Where a country, such as that covered by Point 1, is totally separated by an intervening DXCC country into two areas which are at least 75 miles apart, *two* DXCC countries result. This straight line measurement is made at the closest point, and may include inland lakes and seas (that are part of the country) in the measurement. International waters may be included in the separation but do not contribute to the 75-mile minimum requirement.

(b) Islands: Where two islands, of the government under Point 1, are totally separated by an intervening DXCC country (also under Point 1), *each* island counts as a separate DXCC country. No minimum distance is required. The test for total separation means that a straight line cannot be drawn from any point on one island to any point on the other island without

passing through another DXCC country. This intervening country may be part of either island, another island, or part of a continent.

Point 4, INELIGIBLE AREAS

(a) Any area which is unclaimed or unowned by any recognized government does not count as a separate DXCC country.

(b) Any area which is classified as a Demilitarized Zone, Neutral Zone or Buffer Zone does not count as a separate DXCC country.

(c) The following do not count as a separate DXCC country from the host country: Embassies, consulates and extraterritorial legal entities of any nature, including, but not limited to, monuments, offices of the United Nations agencies or related organizations, other intergovernmental organizations or diplomatic missions.

RULES FOR *CQ* MAGAZINE'S DX AWARDS
Worked all Zones (WAZ)

The CQ WAZ Award will be issued to any licensed amateur station presenting proof of contact with the 40 zones of the world. This proof shall consist of proper QSL cards, which may be checked by any of the authorized CQ checkpoints or sent directly to the WAZ Award Manager. Many of the major DX clubs in the United States and Canada and most national Amateur Radio societies abroad are authorized CQ check points. If in doubt, consult the WAZ Award Manager or the *CQ* Magazine DX editor. Any legal type of emission may be used, providing communication was established after November 15, 1945.

The official CQ WAZ Zone Map and the printed zone list will be used to determine the zone in which a station is located. [A DXCC Countries List, which includes CQ Zones, appears in Chapter 17 of this book.] Confirmation must be accompanied by a list of claimed zones, using CQ form 1479, showing the call letters of the station contacted within each zone. The list should also clearly show the applicant's name, call letters and complete mailing address. The applicant should indicate the type of award for which he or she is applying, such as All SSB, All CW, Mixed, All RTTY. In remote locations and in foreign countries, a handwritten list may be submitted and will be accepted for processing, provided the above information is shown.

All contacts must be made with licensed, land-based, amateur stations operating in authorized amateur bands, 160-10 meters. All contacts submitted by the applicant must be made from within the same country. It is recommended that each QSL clearly show the station's zone number. When the applicant submits cards for multiple call signs, evidence should be provided to show that he or she also held those call letters. Any altered or forged confirmations will result in permanent disqualification of the applicant.

A processing fee ($4 for subscribers—a recent *CQ* mailing label must be included; $10 for nonsubscribers) and a self-addressed envelope with (sufficient postage or IRCs to return the QSL cards by the class of mail desired and indicated) must accompany each application. IRCs equal in redemption value to the processing fee are acceptable. Checks can be made out to the WAZ Award Manager.

In addition to the conventional certificate for all bands and modes, specially endorsed and numbered certificates are available for phone (including AM), SSB and CW operation. The phone certificate requires that all contacts be two-way phone,

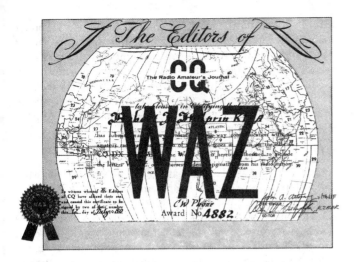

the SSB certificate requires that all contacts be two-way SSB and the CW certificate requires that all contacts be two-way CW.

If at the time of the original application, a note is made pertaining to the possibility of a subsequent application for an endorsement or special certificate, only the missing confirmations required for that endorsement need be submitted with the later application, providing a copy of the original authorization signed by the WAZ Manager is enclosed.

Decisions of the CQ DX Awards Advisory Committee on any matter pertaining to the administration of this award will be final.

All applications should be sent to the WAZ Award Manager after the QSL cards have been checked by an authorized CQ checkpoint. Zone maps, printed rules and application forms are available from the WAZ Award Manager. Send a business-size (4 × 9-inch), self-addressed envelope with two units of First-Class postage, or a self-addressed envelope and 3 IRCs.

Single Band WAZ

Effective January 1, 1973, special WAZ Awards will be issued to licensed amateur stations presenting proof of contact with the 40 zones of the world on 80, 40, 20, 15 and 10 meters. Contacts for a Single Band WAZ award must have been made after 0000 hours UTC January 1, 1973. Single-band certificates will be awarded for both two-way phone, including SSB and two-way CW.

5 Band WAZ

Effective January 1, 1979, the CQ DX Department, in cooperation with the CQ DX Awards Advisory Committee, announced a most challenging DX award—5 Band WAZ. Applicants who succeed in presenting proof of contact with the 40 zones of the world on the five HF bands—80, 40, 20, 15 and 10 meters (for a total of 200)—will receive a special certificate in recognition of this achievement.

These rules are in effect as of July 1, 1979, and supersede all other rules, Five Band WAZ will be offered for any combination of CW, SSB, phone or RTTY contacts, mixed mode only. Separate awards will not be offered for the different modes. Contacts must have been made after 0000 UTC January 1, 1979. Proof of contact shall consist of proper QSL cards checked only by the WAZ Award Manager. The first plateau

will be a total of 150 zones on a combination of the five bands. Applicants should use a separate sheet for each frequency band, using CQ Form 1479.

A regular WAZ or Single Band WAZ is a prerequisite for a 5 Band WAZ certificate. All applications should show the applicant's WAZ number. After the 150 zone certificate is earned, each 10 additional zones requires the submission of QSL cards and a $1 fee. The final objective is 200 zones for a complete 5 Band WAZ. The applicant has a choice of paying a fee for a plaque and/or applying for an endorsement sticker commemorating this achievement.

All applications should be sent to the WAZ Award Manager. The 5 Band Award is governed by the same rules as the regular WAZ Award and uses the same zone boundaries.

WARC Bands WAZ

Effective January 1, 1991, single band WAZ Awards were issued to amateurs presenting proof of contact with the 40 zones of the world on any *one* of the WARC bands: 30, 17 or 12 meters. (Each band constitutes a separate award and may be applied for separately.) This award is available for Mixed Mode, SSB, RTTY or CW. Contacts for each WARC WAZ Award must have been made after each station involved in the contact had permission from its licensing authority to operate on the band and mode.

RTTY WAZ

Special WAZ Awards are issued to Amateur Radio stations presenting proof of contact with the 40 world zones using RTTY. For the mixed band award, QSL cards must show a date of November 15, 1945 or later. The RTTY WAZ is also available with a single band endorsement for 80, 40, 20, 15 or 10 meters. QSL cards submitted for single band endorsements must show a date of January 1, 1973 or later.

WNZ

WNZ stands for "Worked Novice Zones" and is available *only* to holders of a US Novice or Technician license. Proof of contact with at least 25 of the 40 CQ zones as described by the WAZ rules is required. All contacts must be made using the 80, 40, 15 and 10-meter Novice bands. In addition, all contacts must be made while holding a Novice or Technician license, although the application may be submitted at a later date. Contacts must be made prior to receiving authorization to operate with higher class privileges. The WNZ is available as a Mixed Mode, SSB or CW award. It may also be endorsed for a single band. The WNZ award may be used to fulfill part of the application requirement for the WAZ Award when all 40 zones are confirmed.

The basic award can be obtained by submitting QSL cards for 25 zones. The processing fee is $5 for all applicants. All QSL cards must show a date of January 1, 1952 or later. Use CQ Form 1479 to apply for this award.

160 Meter WAZ

The WAZ Award for 160 meters requires that the applicant submit directly to the WAZ Manager QSL cards from at least 30 zones. All QSL cards must be dated January 1, 1975 or later and a $5 fee must accompany all applications. The 160 WAZ is a mixed mode award only. The basic 160 WAZ Award may be secured by submitting QSLs from 30 zones. Stickers for 35, 36, 37, 38, 39 and 40 zones can be obtained from the WAZ Manager upon submission of the QSL cards and $2 for each sticker.

Satellite WAZ

The Satellite WAZ Award is issued to Amateur Radio stations submitting proof of contact with all 40 CQ zones through any Amateur Radio satellite. The award is available for mixed mode only. QSL cards must show a date of January 1, 1989 or later.

The Prefix Award Program

WPX

The CQ WPX Award recognizes the accomplishments of confirmed QSOs with the many prefixes used by amateurs throughout the world. Separate distinctively marked certificates are available for two-way SSB, CW and mixed modes, as well as the VPX Award for shortwave listeners and the WPNX Award for Novices.

All applications for WPX certificates (and endorsements) must be submitted on the official application form, *CQ 1051A*. This form can be obtained by sending a self-addressed stamped, business-size (4 × 9-inch) envelope to the WPX Award Manager, Norm Koch, K6ZDL, PO Box 593, Clovis, NM 88101. All QSOs must be made from the same country. All call letters must be in strict alphabetical order and the entire call letters must be shown. All entries must be clearly legible. Certificates are issued for the following modes and number of prefixes. Cross-mode QSOs are not valid for the CW or 2× SSB certificates. Mixed (CW/phone only): 400 prefixes confirmed. CW: 300 prefixes confirmed. Separate applications are required for each mode. Cards need not be sent, but they must be in the possession of the applicant. Any and all cards may be requested by the WPX Award Manager or by the CQ DX Committee. The application fee for each certificate is $4 for subscribers (subscribers must include a recent *CQ* mailing label) and $10 for nonsubscribers, or the equivalent in IRCs. All applications and endorsements should be sent to the WPX Award Manager.

Prefix endorsements are issued for each 50 additional prefixes submitted. Band endorsements are available for working the following numbers of prefixes on the various bands: 1.8 MHZ—50, 3.5 MHz—175, 7 MHz—250, 14 MHz—300, 21 MHz—300, 28 MHz—300. Continental endorsements are given for working the following numbers of prefixes in the respective continents: North America—160, South America—95, Europe—160, Africa—90, Asia—75, Oceania—60. Endorsement applications must be submitted on CQ Form 1051A. Use separate applications for each mode and be sure to specify the mode of your endorsement application. For prefix endorsements, list only additional call letters confirmed since the last endorsement application. A self-addressed, stamped envelope or proper IRCs for surface or airmail return is required, and $1 or 5 IRCs for each endorsement sticker.

Prefixes: The two or three letter/numeral combinations which form the first part of any amateur call will be considered the prefix. Any difference in the numbering, lettering, or order of same shall constitute a separate prefix. The following would be considered different: W2, WA2, WB2, WN2, WV2, K2 and KN2. Any prefix will be considered legitimate if its use was licensed or permitted by the governing authority in that country after November 15, 1945. A suffix would designate portable operation in another country or call area and would count only if it is the normal prefix used in that area. For example, K4IIF/KP4 would count as KP4. However, KP4XX/7 would not count as KP7, since this is not a normal prefix. Suffixes such as /M, /MM, /AM, /A and /P are not counted as prefixes. An exception to this rule is granted for portable operation

within the issued call area. Thus, contact with a special prefix such as WS2JRA/2 counts for WS2; however, WS2JRA/3 would count for W3. All calls without numbers will be assigned an arbitrary Ø plus the first two letters to constitute a prefix. For example, RAEM counts as RAØ, AIR as AIØ, UPOL as UPØ. All portable suffixes that contain no numerals will be assigned an arbitrary Ø. For example, W4BPD/LX counts as LXØ, and WA6QGW/PX counts as PXØ.

WPNX

The WPNX Award can be earned by US Novices who work 100 different prefixes prior to upgrading. The application may be submitted after receiving the higher license, providing the actual contacts were made as a Novice. Prefixes worked for the WPNX Award may be used later for credit toward the WPX Award. The rules for the WPNX Award are the same as for the WPX Award except that only 100 prefixes must be confirmed. Applications are sent to the WPX Award Manager.

WPX Honor Roll

The WPX Honor Roll recognizes those operators and stations that maintain a high standing in confirmed, current prefixes. The rules therefore, reflect the belief that Honor Roll membership should be accessible to all active radio amateurs and not to be unduly advantageous to the "old-timers." With the exceptions listed below, all general rules for WPX apply toward Honor Roll credit. A minimum of 600 prefixes is required to be eligible for the Honor Roll.

Only current prefixes may be counted toward WPX Honor Roll standings, those prefixes to be listed and updated annually in *CQ* or to be available from the WPX Award Manager. Special-issue prefixes (such as OF, OS 4A) will be considered current for as long as they are assigned to a particular country and deducted as credit for Honor Roll standings after cessation of their use or assignment. Honor Roll applicants must submit their list of current prefixes (entire call required) separate from their regular WPX applications. Use regular Form 1051 and indicate "Honor Roll" at the top of the form. Forms may be obtained by sending a business-size SASE or one IRC (foreign stations send extra postage or IRC if airmail desired) to the WPX Award Manager. A separate application must be made for each mode. Endorsements for the Honor Roll may be made for 10 or more prefixes. An SASE or IRC should be included. For prefixes by countries, see Chapter 17.

WPX Award of Excellence

This is the ultimate award for the prefix DXer. The requirements are 1000 prefixes mixed mode, 600 prefixes SSB, 600 prefixes CW, all six continental endorsements, and the 5 band endorsements 80-10 meters. A special 160-meter endorsement bar is also available. The WPX Plaque fee is $60, and the 160-meter bar is $5.25.

The CQ DX Awards Program

The CQ CW DX Award and CQ SSB DX Award are issued to any amateur station submitting proof of contact with 100 or more countries on CW or SSB. The CQ DX RTTY Award is issued to any amateur station submitting proof of contact with 100 countries on RTTY. Applications should be submitted on the official CQ DX Award application (Form 1067B). All QSOs must be 2× SSB, 2× CW or 2× RTTY. Cross-mode or one-way QSOs are not valid for the CQ DX Awards. QSLs must be listed in alphabetical order by prefix, and all QSOs must be dated after November 15,

1945. QSL cards must be verified by one of the authorized checkpoints for the CQ DX Awards or must be included with the application. If the cards are sent directly to the CQ DX Awards Manager, Billy Williams, N4UF, PO Box 9673, Jacksonville, FL 32208, postage for return by First-Class mail must be included. If certified or registered mail return is desired, sufficient postage should be included. Country endorsements for 150, 200, 250, 275, 300, 310 and 320 countries will be issued.

Any altered or forged confirmations will result in permanent disqualification of the applicant. Fair play and good sportsmanship in operating are required for all amateurs working toward CQ DX Awards. Continued use of poor ethics will result in disqualification of the applicant. A fee of $4 for subscribers (subscribers must include a recent *CQ* mailing label with their application) or $10 for nonsubscribers, or the equivalent in IRCs, is required for each award to defray the cost of the certificate and handling. An SASE or 1 IRC is required for each endorsement.

The ARRL DXCC Countries List (see Chapter 17) constitutes the basis for the CQ DX Award country status. Deleted countries will not be valid for the CQ DX Awards. Once a country has lost its status as a current country, it will automatically be deleted from our records. All contacts must be with licensed land-based amateur stations working in authorized amateur bands. Contacts with ships and aircraft cannot be counted. Decisions of the CQ DX Advisory Committee on any matter pertaining to the administration of these awards shall be final.

To promote multiband usage and special operating skills, special endorsements are available for a fee of $1 each:

- A 28-MHz endorsement for 100 or more countries confirmed on the 28-MHz band.
- A 3.5/7-MHz endorsement for 100 or more countries confirmed using any combination of the 3.5 and 7-MHz bands.
- A 1.8-MHz endorsement for 50 or more countries confirmed on the 1.8-MHz band.
- A QRP endorsement for 50 or more countries confirmed using 5-watts input or less.
- A Mobile endorsement for 50 or more countries confirmed while operating mobile. The call-area requirement is waived for this endorsement.
- An SSTV endorsement for 50 or more countries confirmed using two-way slow scan TV.
- An OSCAR endorsement for 50 countries confirmed via amateur satellite.

(After the basic award is issued, only a listing of confirmed QSOs is required for these seven special endorsements. However, specific QSLs may be requested by Award Manager N4UF.)

The CQ DX Honor Roll will list all stations with a total of 275 countries or more. Separate Honor Rolls will be maintained for SSB and CW. To remain on the Honor Roll, a station's country total must be updated annually.

USA-CA Rules and Program

The United States of America Counties Award, also sponsored by *CQ*, is issued for confirmed two-way radio contacts with specified numbers of US counties under rules and conditions below. [Note: A complete list of US counties appears in Chapter 17.]

The USA-CA is issued in seven (7) different classes, each a separate achievement as endorsed on the basic certificate by

use of special seals for higher class. Also, special endorsements will be made for all one band or mode operations subject to the rules.

Class	Counties Required	States Required
USA-500	500	Any
USA-1000	1000	25
USA-1500	1500	45
USA-2000	2000	50
USA-2500	2500	50
USA-3000	3000	50

USA 3076-CA for ALL counties and Special Honors Plaque [$40]

USA-CA is available to all licensed amateurs everywhere in the world and is issued to them as individuals for all county contacts made, regardless of calls held, operating QTHs or dates. All contacts must be confirmed by QSL, and such QSLs must be in one's possession for identification by certification officials. Any QSL card found to be altered in any way disqualifies the applicant. QSOs via repeaters, satellites, moonbounce and phone patches are not valid for USA-CA. So-called "team" contacts, wherein one person acknowledges a signal report and another returns a signal report, while both amateur call signs are logged, are not valid for USA-CA. Acceptable contact can be made with only one station at a time.

The National Zip Code & Directory of Post Offices will be

helpful in some cases in determining identity of counties of contacts as ascertained by name of nearest municipality. Publication No. 65, Stock no. 039-000-00264-7, is available at your local Post Office or from the Superintendent of Documents, US Government Printing Office, Washington, DC 20402, for $12, but will be shipped only to US or Canada.

Unless otherwise indicated on QSL cards, the QTH printed on cards will determine country identity. For mobile and portable operations, the postmark shall identify the county unless information stated on QSL cards makes other positive identity. In the case of cities, parks or reservations not within counties proper, applicants may claim any one of adjoining counties for credit (once).

The USA-CA program will be administered by a *CQ* staff member acting as USA-CA Custodian, and all applications and related correspondence should be sent directly to the custodian at his or her QTH. Decisions of the Custodian in administering these rules and their interpretation, including future amendments, are final.

The scope of USA-CA makes it mandatory that special Record Books be used for application. For this purpose, *CQ* has provided a 64-page $4^1/4 \times 11$-inch Record Book which contains application and certification forms and which provides record-log space meeting the conditions of any class award and/or endorsement requested. A completed USA-CA Record Book constitutes medium of basic application and becomes the property of *CQ* for record purposes. On subsequent applications for either higher classes or for special endorsements, the applicant may use additional Record Books to list required data or may make up own alphabetical list conforming to requirements. Record Books are to be obtained directly from *CQ*, 76 North Broadway, Hicksville, NY 11801, for $1.25 each. It is recommended that two be obtained, one for application use and one for personal file copy.

Make Record Book entries necessary for county identity and enter other log data necessary to satisfy any special endorsements (band-mode) requested. Have the certification form provided signed by two licensed amateurs (General class or higher) or an official of a national-level radio organization or affiliated club verifying the QSL cards for all contacts as listed have been seen. The USA-CA custodian reserves the right to request any specific cards to satisfy any doubt whatever. In such cases, the applicant should send sufficient postage for return of cards by registered mail. Send the original completed Record Book (not a copy) and certification forms and handling fee. Fee for nonsubscribers to *CQ* is $10 US or IRCs; for subscribers, the fee is $4 or 12 IRCs. (subscribers, please include recent *CQ* mailing label.) Send applications to USA-CA Custodian, Norm Van Raay, WA3RTY, Star Route 40, Pleasant Mount, PA 18453.

For later applications for higher-class seals, send Record Book or self-prepared list per rules and $1.25 or 6 IRCs handling charge. For application for later special endorsements (band/mode) where certificates must be returned for endorsement, send certificate and $1.50 or 8 IRCs for handling charges. Note: At the time any USA-CA award certificate is being processed, there are no charges other than the basic fee, regardless of number of endorsements or seals; likewise, one may skip lower classes of USA-CA and get higher classes without losing any lower awards credits or paying any fee for them. Also note: IRCs are not accepted from US stations.

[The Mobile Emergency and County Hunters Net meets on 14,336 kHz SSB every day and on 3866 kHz evenings during the winter. The CW County Hunters Net meets on 14,066.5 MHz daily.]

GERATOL NET

During the winter of 1977, the "Unbelievable Operating Achievement Award" (otherwise known as the GERATOL Net award) was established by N4BA, K5BG, W7RQ and WØGX. This award is available to any Amateur Extra licensee who has worked all US states in the Amateur Extra portion of 75 meters, contacts being made with stations having 1×2, 2×1 or 2×2 call signs.

On March 10, 1978, an informal net went into operation to provide a meeting place and to assist others with the completion of the new award developed by N4BA and company. The net name is the acronym "GERATOL" (Greetings Extra Radio Amateurs—Tired of Operating Lately?). In many ways, this describes the essence of the GERATOL Net. More than one "GERATOLLER" will tell you that the fellowship and challenge has revitalized their amateur operations and allowed a welcome escape from the mundane "Roger 59" DX world.

The GERATOL Net is active from October through April. As many as 48 states have checked into the net during the course of an evening, including Alaska and Hawaii. Using three net control stations, it convenes around 0100 UTC each evening, and routinely runs until after 0500. Meeting around 3767 kHz, the net still pursues the original purpose developed by it founders. There are no dues or membership fees. DX stations are welcome to check in and work any W or VE, but W/VEs are not allowed to call DX stations (for obvious reasons).

For information on this award and the many endorsements that are offered, or for further information on the GERATOL Net, send your request with a business-size SASE with two units of postage to Richard Beran, WØYTZ, 300 Valley View Drive, Ord, NE 68862.

The "Directors Award" is available for working 100 GERATOL members and exchanging 100 GERATOL numbers. Additional recognition is available for each group of 100 contacts verified above the initial 100 contact requirement. The operator must have a GERATOL number to exchange numbers.

So take a shot of GERATOL, and join the group on 75 meters to experience a new operating challenge.

QRP AMATEUR RADIO CLUB INTERNATIONAL (ARCI)

The objective of the QRP ARCI, Inc awards program is to demonstrate that power is no substitute for skill, encouraging full enjoyment of Amateur Radio while running the minimum power necessary to complete a QSO and reduce QRM on crowded amateur bands. Requirements for the following awards are set forth below:

QRP-25. Issued to any amateur for working 25 members of the QRP ARCI while those members were running a power output of 5 watts or less. Endorsements are offered for 50, 100 and every 100 thereafter. To apply, send full log info with the list of the member-number in order of the member-number.

WAC-QRP. Issued to any amateur for confirmed contacts with stations in all six continents. A power output of 5 watts or less must have been used by the applicant for all QSOs. To apply, send full log info with photocopies of QSLs or a confirmed list.

WAS-QRP. Issued to any amateur for confirmed QSOs with each of the 50 states, while running a power output of 5 watts or less. Basic award issued for 20 states, endorsement seals for 30, 40 and 50. To apply, send full log data with photocopies of QSLs or a confirmed list.

1000-Mile-Per Watt (km/W). Issued to any amateur transmitting from or receiving the transmissions of a low-power station such that the great circle bearing distance between the two, divided by the power output of the low-power station, equals or exceeds 1000-miles-per-watt. Additional certificates can be earned with different modes/bands. To apply, send full log info and photocopies of QSLs or certified list including RST and power used on both sides, band, mode and exact (as nearly as possible) QTH on both sides.

QRP-Net (QNI-25). Issued to those members completing 25 check-ins into any individual QRP ARCI net. Subsequent 25 QNIs in another net will earn an endorsement seal. Net managers send a list of those qualifying for the QNI certificate to the Net Manager at the end of each month. Awards are issued FREE to those qualified by the Awards Chairman.

Notes:

1) A two-way QRP seal will be issued for all the above awards, except QRP-25, if log data indicates the power output on both sides of all QSOs was 5 watts or less.

2) Other endorsement seals available—one band, one mode, natural power, Novice. Please specify endorsements desired.

3) Certified list (GCR). QRP ARCI will accept as satisfactory proof of confirmed QSOs and that the QSLs are on hand as claimed by the applicant if the list is signed by (a) a radio club official, or (b) two Amateur Radio operators of General class or higher, or (c) a notary public, or (d) a CPA (Certified Public Accountant). If you must send QSLs, be sure to include postage for their return.

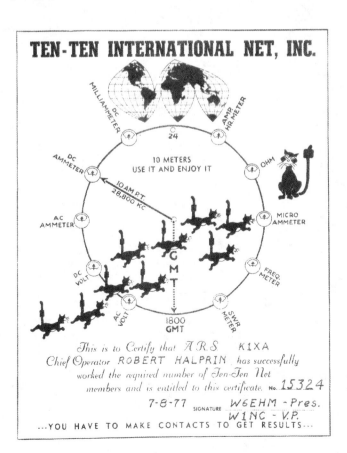

4) QRP ARCI member-numbers are not published. The Awards Manager will accept as satisfactory proof for any of the club awards a QSO with a club member giving his membership number and power output in the log data. If the QRP number and power level are not given, a QSL is required for confirmation.

5) Fee structure
   (a) original certificate and seal—$2 or 10 IRCs.
   (b) subsequent endorsement seals—$1 or five IRCs.
   (c) return postage for QSLs or other application material to be returned.

6) Send applications and the required fee to QRP ARCI Awards Chairman, Bob Gaye, K2LGJ, 25 Hampton Parkway, Buffalo, NY 14217-1217.

TEN-TEN

Amateurs operating on 10 meters are often bewildered by requests for "10-10 numbers." 10-10 numbers are assigned by the 10-10 International Net, Inc. A number is available to any amateur who works ten 10-10 members and submits the log data to the appropriate 10-10 Call Area Manager. Once the log data has been received and approved, the applicant will be issued his own 10-10 number, which can be used to exchange with others on 10 meters to obtain a "Bar" for each 100 numbers collected. 10-10 numbers are issued for life, and well over 70,000 10-10 numbers have been issued internationally.

The purpose of 10-10 is to promote interest and activity on the 10-meter band. 10-10 holds a net meeting each day, except Sunday, on 28.800 MHz at 1800Z. (An alternate net is held each Monday at 1800 UTC on 28.380 MHz.) All amateurs, 10-10 members or not, are invited to check in on the daily net. In addition to holding both CW and SSB contests twice each year, 10-10 has more than 220 Chapters which hold net meetings at least once each week. Most 10-10 Chapters issued certificates for working members of that Chapter. If chasing certificates and awards is your "thing," obtaining a 10-10 number is highly recommended. For information on how to join 10-10, including an application form and the latest list of 10-10 Chapters, meeting times and frequencies, send $1 and two First-Class stamps and an address label to Mike Elliott, KF7ZQ, 9832 W Gurdon Court, Boise, ID 83704.

Besides the collection of "Bars" for each 100 10-10 members worked, 10-10 offers some additional awards for contacts on 10 meters. 10-10 Worked All States (10-10 WAS) is awarded in recognition of confirmed contacts with members in all 50 states on 10 meters. QSL card confirmations must indicate the 10-10 numbers of stations worked. For information and application, send an SASE to Susan Brackeen, KA1CAD, Rte 6, Box 151, Boonville, MS 38829. The 10-10 Countries Award is issued for confirmed contacts with 10-10 members in 25 different countries on the 10-meter band. Endorsement may also be earned for additional countries worked. For information on this award and an application form, send an SASE to Alan Sherman, K1AS, RR 4, Box 422, Danielson, CT 06239.

WORLD ITU ZONE AWARD

This award will be issued to amateurs who work the prescribed number of ITU (International Telecommunication Union) zones of the world. There are 90 ITU zones. QSL cards are suggested, but not required. If QSLs are not submitted, a certified list of contacts (verified by a local club official or two licensed amateurs) shall be submitted.

The ITU Award is issued in the following classes:

Class D—40 zones
Class C—50 zones
Class B—65 zones
Class A—75 zones
Class AA—90 zones

The official ITU zone list [see Chapter 17 for an ITU zone list] and map will be the criteria for determining zones.

Fees are $2 US postpaid for the basic award, including endorsements requested at the same time, and 50 cents US postpaid for each future endorsement.

Applications should be sent to John Lee, K6YK, 3654 Three Oaks Rd, Stockton, CA 95205.

CHAPTER 9

HF Digital Communications

I f you've spent time exploring the Amateur Radio HF bands, you've probably heard some mysterious sounds—especially the odd noises that you'll find just above the CW portions of the bands. No doubt you've listened to the warbling, musical signals of Baudot radioteletype, otherwise known as *RTTY* (pronounced "ritty"). If you twisted the dial a bit further, you may have heard a chorus of electronic crickets. These are the chirping dialogs of *AMTOR* (*AM*ateur *T*eleprinting *O*ver *R*adio) stations. Perhaps you also picked up the sounds of *PACTOR*, *G-TOR* or *CLOVER*. And what were those raspy, high-pitched bursts? Those are the unmistakable signatures of HF *packet*.

These are the primary HF digital modes. They're called digital modes because the communication involves an exchange of data between one station and another. In the case of RTTY, for example, letters typed on a keyboard are translated into data by a computer or data terminal. Another device, usually a multimode communications processor (*MCP*), accepts this data and converts it to whatever encoded audio tones are required. The tones are sent to the transmitter and away they go! At the receiving end the same process occurs in reverse: The tones are translated back into data and displayed as text on a computer or terminal screen.

AN HF DIGITAL HISTORY LESSON

In the sense that we commonly interpret the word "digital" today, RTTY has the distinction of being the granddaddy of HF digital communications. RTTY dates back to World War II when the military began connecting mechanical teletype machines to HF radios. The idea was to send printed communications into areas where telephone lines didn't exist.

At first they tried simple on/off keying to send the text, but that didn't work very well. The receiving equipment couldn't always tell the difference between a signal and a burst of noise. After some further experimentation, the designers switched to *frequency-shift keying*

STEVE FORD, WB8IMY
ARRL HEADQUARTERS

(FSK). This approach used two specific tones to indicate the on/off (MARK/SPACE) signals. FSK was a success and RTTY as we know it today was born.

Hams adopted RTTY after the war ended and by the early '50s it was a well-established mode. Initially it found a home on VHF, but later became more popular on the HF bands. (RTTY is seldom heard on the VHF bands today.)

For several decades, hams relied on surplus teletype machines for their RTTY stations. These mechanical monsters were slow, noisy and often dirty (they had a nasty tendency to drip oil on the floor!). Operators had to read the text on paper as it was printed. The keyboard was a bit unusual, mainly due to the nature of the Baudot code.

With Baudot, all letters are capitalized (upper case). All numbers from 0 through 9 are available along with some limited punctuation. To send numbers or punctuation a special FIGS character must be sent *first*. To return to alphabetical

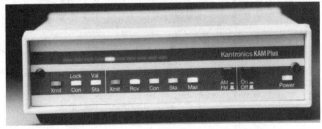

Multimode communications processors (MCPs) such as these offer RTTY, AMTOR, packet, PACTOR, CW and many other modes in a single device.

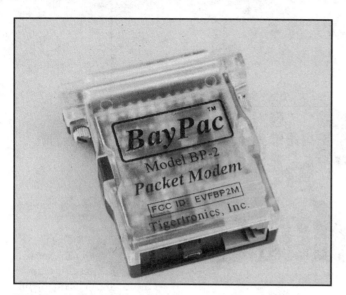

The Tigertronics BayPac BP-2M is a multimode controller for the budget conscious. Selling for under $70, the BP-2M offers packet (HF and VHF), RTTY and AMTOR using a tiny outboard modem and readily available *shareware*.

letters, a LTRS character must be sent. On the original teleprinters you had to press the FIGS key whenever you wanted to send numbers or punctuation. To return to alphabetical letters, you had to press the LTRS key (see Fig 9-1). You can imagine how difficult it must have been to master the keyboards of those old machines!

Some ancient teletype machines remain, but most RTTY enthusiasts rely on computerized systems. The Baudot code is still the same, however, and the FIGS/LTRS shift is still required. No need to worry about your typing skills. The FIGS/LTRS shift is handled automatically by most systems. All you have to do is type and your software and hardware will take care of everything else. Rather than reading the text on flowing sheets of paper, you'll see it on your monitor screen.

RTTY has fulfilled its promise of transmitting the written word throughout the world. When band conditions are good and signals are optimal, RTTY is efficient and accurate. But what happens when conditions are less than ideal? That's where RTTY shows its weak side. Interference from other transmit-ters, fading and electrical noise cause errors in RTTY communications. Mild interference will cause a few letters to be deleted here and there. Severe interference, however, can turn the entire text to gibberish!

Commercial maritime communication systems relied on RTTY for decades, but the inability to detect errors was a persistent problem. It could even have life-threatening consequences. What if a maritime weather service tried to alert ships of an approaching storm? The RTTY-transmitted warning had to be repeated over and over to give the ships a decent chance of copying the entire message.

The Advent of AMTOR

The pressure to create a more reliable teletype system lead to the development of *TOR* (*T*eleprinting *O*ver *R*adio), commonly known today as *SITOR* (*S*implex *T*eleprinting *O*ver *R*adio). Instead of sending the text in one long transmission, the TOR method sends only a few characters at a time. The receiving station checks for errors using a bit-ratio-checking scheme. If all characters are received error-free, the receiving station sends an acknowledgment or ACK signal and the next few characters are transmitted. If an error is detected, a *nonacknowledgment* or NAK signal is sent. This tells the transmitting station to repeat the characters. The result is digital communication *without* errors—a major improvement over previous RTTY systems. The rapid error-checking dialog—known as *Mode A* or *ARQ*—creates the distinctive chirping sounds associated with SITOR communications.

In the early 1980s, the Federal Communications Commission approved SITOR techniques for Amateur Radio use. Peter Martinez, G3PLX, adapted SITOR coding and developed AMTOR. AMTOR was tailor-made for the personal-computer era.

Packet, PACTOR, CLOVER and G-TOR

Packet appeared on the scene in the mid 1980s, although it was (and still is) most popular on VHF and UHF. Twenty meters is the most popular band for HF packet, but you'll encounter this mode on other bands, too. Every MCP on the market today has HF packet capability. So do a number of packet-only TNCs. The widespread availability of equipment is the engine that's driven much of the packet activity on the HF bands.

Three more HF digital modes debuted in the early 1990s: PACTOR, G-TOR and CLOVER. PACTOR is a fusion of AMTOR and packet. It features the capabilities of packet (up-

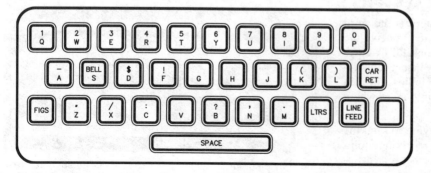

Fig 9-1—The old teleprinters featured a keyboard layout similar to the one shown here. Notice the FIGS key in the lower left corner. You had to press this key before sending punctuation or numbers. Pressing the LTRS key returned you to the "letters" mode. Although these vintage units have all but disappeared, the need to shift from FIGS to LTRS and back again remains. This is now handled automatically by software.

per/lower case characters, binary transmission and so on) with an ACK/NAK system similar to AMTOR. PACTOR I made its appearance in 1991, followed by PACTOR II in 1995. The '90s also saw the emergence of G-TOR and CLOVER.

The Nature of FSK

If you want to become a competent HF digital operator, it helps to understand frequency-shift keying, or FSK. Most HF digital modes use frequency-shift keying to pass information from one station to another.

Let's start with data. I'm sure you've heard that the fundamental language of all computers is binary *machine code*. In a binary-number system, you're only dealing with 0s and 1s. This is a natural situation for a computer since it's comprised of a multitude of solid-state logic switches that can only be *on* or *off* ("high" or "low"). So, an "on" condition represents a binary 1 while an "off" condition represents a binary 0.

If you use wires to connect two computers, the on/off voltage states are communicated from one machine to another easily. But let's make the situation more complicated and move the computers several hundred miles apart. Now what are you going to do? Radio seems like a natural choice, but you can't send high/low voltage states over the air. . . or can you?

What if you translated the changing voltages to changing tones? You could use 2,125 Hz to represent a binary 1 and 2,295 Hz to represent a binary 0. Feed those tones to the audio input of an SSB transceiver operating on lower sideband (LSB), for example, and they'll be transmitted as signals at specific points below the *suppressed carrier frequency*. The 2,125-Hz tone will create a signal 2,125 Hz below the suppressed carrier. The 2,295-Hz tone will create a signal 2,295 Hz below the suppressed carrier. Subtract the frequency of the high tone from the low tone and you get 170 Hz. In other words, the tones shift 170 Hz to represent a 1 or 0. Shifting voltages have become shifting tone frequencies! As we discussed previously, the tone that represents a binary 1 is called the MARK. The tone that represents a binary 0 is called the SPACE (see Fig 9-2).

At the receiving end of the path, you'll need to convert the tones back into binary high/low voltage states. FSK demodulators are designed with audio filters to detect the MARK and SPACE tones and produce corresponding data pulses. Feed those data pulses to a computer running terminal software and—trumpet fanfare please!—text appears on the screen.

FSK or AFSK?

In our previous example, the shifting tones were supplied to the transmitter via an audio input jack. Technically speaking, this is audio FSK or AFSK. There is another method known as *direct* FSK. What's the difference? If you're using direct FSK, the data pulses from your MCP are not converted to MARK and SPACE tones. Instead, the MCP supplies these high and low logic levels to the FSK input of an SSB transmitter where they're used to shift the frequency of the master oscillator up and down. The result at the receiver is the same: MARK and SPACE tones.

So which is better—direct FSK or AFSK? Some hams believe that FSK is a "purer" method of transmitting because it minimizes distortion and harmonics. Other hams argue that if the transmitter isn't overmodulated, AFSK transmissions are just as pure. You'll find large numbers of amateurs using one method or the other, although the majority seem to favor the AFSK approach because most transceivers do not feature an FSK input. The important thing to remember is that whether you use AFSK or direct FSK, the result at the receiving station is the same.

Baudot RTTY

Before you can enjoy a RTTY conversation, you must be capable of tuning the signals properly. Every MCP features some sort of tuning indicator, depending on the type of equipment you're using. In days gone by, RTTY operators would attach oscilloscopes to their terminal units and tune the signals until they saw the classic "crossed bananas" display (see Fig 9-3). As technology advanced, many terminal units included tiny built-in oscilloscopes that performed the same function.

MCPs often use LED indicators. Some units feature an indicator comprised of several LEDs arranged in a horizontal bar. When the LEDs at opposite ends of the bar flash in sync with the RTTY signal, you know you have it tuned properly.

What Is That Signal?

Most RTTY operators use lower sideband transmissions with a 170-Hz frequency shift between the MARK and SPACE signals. The commonly used data rate is 60 words per minute, often expressed as 45 baud. But what if you stumble upon two operators who aren't conforming to "conventional" practices? Your indicator says you're tuned in properly, but nothing coherent prints on your screen!

It looks like you'll need to do a little detective work. Check the following:

❑ Is the signal "upside down"? The RF frequency of the MARK signal is usually higher than the RF frequency of the SPACE signal, but there is no law that dictates this standard. With most MCPs, all it takes is a push of a button or keyboard key to invert the normal MARK/SPACE frequency relationship.

❑ Are the operators really using 170-Hz shift at 45 baud? For example, many RTTY operators prefer to run at 75 baud (100 WPM) when exchanging lengthy files. To complicate matters, an operator may also decide to use an 850-Hz shift.

It may be of some comfort to know that these situations are uncommon. You may encounter RTTY at 75 baud or higher, but most operators stick to 45 baud. The use of inverted MARK/SPACE signals and odd frequency shifts is relatively rare.

Listen to the Action

Let's say you've successfully tuned in a RTTY signal. Unless you happen to be monitoring during a contest, a typical RTTY conversation may look like this:

I FINALLY TOOK DOWN MY OLD VERTICAL AND RE-PLACED IT WITH A DIPOLE CUT FOR 40 METERS. YOU KNOW WHAT I FOUND OUT? I CAN FORCE FEED THAT DIPOLE WITH MY ANTENNA TUNER AND USE IT ON 20, 15 AND 10 METERS. N9GOR DE WA1MBK K

WA1MBK DE N9GOR....REALLY? THAT SOUNDS PRETTY INCREDIBLE. I HAD NO IDEA YOU COULD DO THAT. I BET THE SWR IS VERY HIGH ON THE OTHER BANDS SINCE THE DIPOLE WOULD NOT BE RESONANT ON THOSE FREQUEN-CIES. WHAT ABOUT THE RF YOU LOSE IN YOUR FEED LINE? WA1MBK DE N9GOR K

N9GOR DE WA1MBK...FEED LINE LOSS IS NOT A PROB-LEM EVEN WITH A HIGH SWR. AS LONG AS YOU USE LOW LOSS COAX OR OPEN WIRE FEED LINE, THE LOSS IS SO SMALL IT DOES NOT MATTER. A GOOD ANTENNA TUNER IS THE KEY...N9GOR DE WA1MBK K

And so it goes, back and forth at a leisurely pace. When an

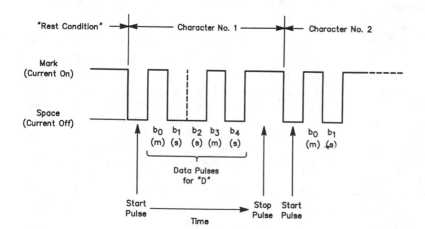

Fig 9-2—This is a diagram of RTTY MARK and SPACE signals as the letter "D" is sent. A start pulse begins the character, followed by the 5 bits that define it (b0 through b4). By looking at the bits, you can see that "D" is MARK-SPACE-SPACE-MARK-SPACE. A stop pulse signals the end of the character and another start pulse begins a new character. These MARK and SPACE pulses appear as shifting audio tones at the receiver. They are fed to a multimode communications processor (MCP) to be decoded back into data and displayed as text.

operator is not sending data, you'll hear a continuous tone or a rhythmic bee-bee-bee-bee signal. The rhythmic signal is known as the diddle. The MCP is simply switching back and forth between MARK and SPACE signals while awaiting more data from the operator.

The call signs at the beginning and end of each transmission are optional, as long as you identify at least once every 10 minutes. Still, some habits are hard to break and you'll find that RTTY operators often open and close their transmissions with an exchange of call signs. This is made easier by the fact that many modern MCPs feature automatic call-sign exchange capability. You simply enter the call sign of the other station once. After that, you can generate the entire <his call sign>-DE-<your call sign> sequence by pressing a single key.

Did you notice the K used to signify the end of each transmission? It's a RTTY custom to use CW prosigns in conversations. Depending on the operator, he or she may send a K (over to you), AR K (end of message, over to you) or KN (over to you only). When the conversation is over, it's common to use SK to signal the end of the contact. RTTY operators also adopt the CW custom of abbreviating words. In the sample conversation shown above, "frequencies" may be sent as "freqs."

It's Your Turn

What if you're eavesdropping on a QSO and you see that it's about to end?

N9GOR DE WB8IMY...I HAVE TO RUN, WAYNE. DINNER WILL BE READY IN AN HOUR AND I STILL HAVE NOT FINISHED MOWING THE LAWN. HAVE A GOOD WEEKEND AND I HOPE TO CHAT WITH YOU AGAIN....N9GOR DE WB8IMY SK

WB8IMY DE N9GOR...NO PROBLEM. GLAD I HAD THE CHANCE TO DISCUSS ANTENNAS WITH YOU. I LEARN SOMETHING EVERY DAY. SEE YOU LATER...WB8IMY DE N9GOR SK

Looks like N9GOR is still available. Why not give him a call? Since you were able to copy the transmissions, chances are good that your MCP is already set for the proper data rate, shift and MARK/SPACE frequency relationship. (A quick check never hurts, though!) Switch your MCP to the transmit mode and start typing...

N9GOR N9GOR DE N6ATQ N6ATQ N6ATQ

N9GOR N9GOR DE N6ATQ N6ATQ N6ATQ K K

[switch back to receive]

Make sure to repeat your call sign several times. After all, N9GOR knows his own call, but he doesn't know yours. Send your transmission in several short lines rather than one long line.

If the other operator is able to copy your signal, you may see a response like this:

N6ATQ N6ATQ DE N9GOR N9GOR...THANKS FOR THE CALL. NAME HERE IS WAYNE WAYNE AND I AM LOCATED IN MILWAUKEE MILWAUKEE WISCONSIN WISCONSIN. YOUR RST RST IS 579 579...BACK TO YOU...N6ATQ DE N9GOR K K

Notice how N9GOR sends the important information twice. He doesn't know how well you are receiving his signal, so he wants to make sure that you won't miss his name, city, state and so on. This is always a good technique to use when you aren't sure of the signal path between your station and another.

Wayne has sent your signal report (RST) and it's 579. The first digit from the left is your readability (R), the second is relative strength (S) and the third is tonal quality (T). In this case, a 579 RST means that he is receiving you very well, your signal strength is moderate and your RTTY tones are good. A perfect RST would be a 599, but a 579 is fine. You can be reasonably certain that he is receiving everything you're sending. Go ahead and tell him who you are and where you are. Don't forget to give him a signal report, too.

[switch to transmit]

N9GOR DE N6ATQ....HELLO WAYNE. NAME HERE IS CRAIG CRAIG AND I AM IN ESCONDIDO ESCONDIDO CALIFORNIA CALIFORNIA. YOUR RST RST IS 599 599. THIS IS MY FIRST RTTY CONTACT. I AM USING A MODEL PK232 MCP AND A KENWOOD TS-820S TRANSCEIVER. ANTENNA IS A DIPOLE UP 30 FEET. SO HOW COPY? N9GOR DE N6ATQ K

[switch to receive]

These are the preliminaries of most RTTY contacts. Some hams are very proud of their station equipment and will give you a brief rundown of their entire setup. In the earlier days of RTTY, messages such as these were created in advance and stored on reels of paper tape. When fed to a teleprinter, the holes punched in the paper tape were translated into MARK and SPACE signals for transmission.

If you own a computer, you can create and store "canned" messages of your own and save them on diskette or magnetic tape. Depending on the type of software you're using, a single keystroke will send the entire message automatically! Many RTTY operators store and send their station descriptions in this manner, although the process is still known by its old name-

(A)   (B)   (C)

Fig 9-3—Oscilloscope-type tuning indicators produce patterns like these. Pattern A is the classic "crossed bananas," showing that the RTTY signal is tuned properly. At B the receiver is slightly off frequency, while C indicates that the transmitting station is using a shift that differs from the MCP setting.

sake: the brag tape. Consult your software manual to learn how to create your own brag tapes and other stored messages.

Is There Anybody Out There?

If you can't find someone to talk to, consider calling CQ. You never know what you'll turn up . . .

[switch transmitter on]

CQ CQ CQ CQ CQ CQ CQ DE WB8IMY WB8IMY WB8IMY

CQ CQ CQ CQ CQ CQ CQ DE WB8IMY WB8IMY WB8IMY

CQ CQ CQ CQ CQ CQ CQ DE WB8IMY WB8IMY WB8IMY K K

[switch to receive]

A CQ should be long enough to attract attention, but short enough to avoid boring the other station. Repeat your call sign often so the operator on the other end has a decent chance of getting it right. You may have to send your CQ more than once before you're noticed. Storing your CQ on disk makes it easy to send it again without retyping.

If you don't receive an answer, don't lose hope. Just move to another frequency or band. Before calling CQ, it's good

Bill Price, WA4MCZ, is a picture of concentration as he operates the ARRL RTTY Roundup contest

practice to ask if the frequency is in use. Perhaps you've just tuned onto a frequency and it seems to be unoccupied. Don't let appearances fool you! You often hear only one side of a conversation—and that side may be listening at the moment!

Checking the frequency is easy. For example, I'd send: QRL? QRL? DE WB8IMY WB8IMY K K

If the frequency is occupied, someone will let me know right away. On the other hand, if no one replies, I can assume it's safe for me to go ahead and call CQ.

Working the DX Pileups

Sooner or later you're bound to encounter the fascinating phenomenon known as the DX pileup. You'll know when you've found a pileup because it will sound like pure pandemonium.

Pileups are the result of a desirable DX station coming on the air. What makes a DX station worthy of a pileup? If the operator is in a country that is not heard on the air often, that country is considered "rare DX." Any transmission from that part of the globe is a major event and it quickly attracts hordes of hams eager to make a contact.

The first CQs from a rare DX station snag the few hams who are lucky enough to be near the frequency at the time. More join the fray as they discover what's going on. DX *PacketClusters* also sound the alarm on the VHF frequencies and bring operators by the dozens. Within minutes you have a huge number of stations chasing the same goal: the hapless DX operator. As soon as he finishes a contact, everyone starts calling at once! You can imagine how this would sound in your receiver.

When you hear a pileup in progress, the first thing to do is monitor the exchanges. Determine the DX station's call sign and see if you can copy his signal. When the FR5ZU/G group began calling for contacts from Reunion Island, here is how it looked at my station.

DE XE1/JA1QXY..XE1/JA1QXY..XE1/JA1QXY..BK BK

Bad Habits

As you scan the RTTY subbands, you'll find some operators sending long streams of RYs at the beginning of their transmissions.

RYRYRYRYRYRYRYRYRYRYRYRYRYRYRYRY
CQ CQ CQ CQ CQ CQ CQ CQ DE WB8XYZ WB8XYZ
WB8XYZ K K

This is another artifact from the early days of RTTY when it was necessary to make sure that mechanical teleprinters were ready to copy a transmission. In the modern era of computers and data terminals, this is unnecessary. Many operators find these long RY streams highly irritating since they do nothing but waste time and band space. Try to avoid the RY habit. If you have something to say, such as calling CQ, go ahead and say it. You don't need to send a meaningless string of letters in advance.

You may also find operators who make frequent use of the date/time function included in their MCPs. Many units have this feature but, thankfully, most operators don't use it except in contest or public service situations. What does it do? It inserts the time and date (local or UTC) as part of the transmitted text. Usually it appears at the end of a transmission, but may pop up at the beginning. You may also see it at the end of stored messages.

Unless you're in desperate need of a clock, do you really want to know what time it is? Probably not. Now turn the tables. Do you think the other station cares to be informed of the date and time? I think you get the idea! Time and date "stamping" is another waste of RF in most cases. Veteran RTTY operators consider it a nuisance, although most will be too polite to tell you!

The well-equipped HF digital station of Frank Fallon, N2FF

XE1/JA1QXY DE FR5ZU/G...UR 599 599 BK TO U...KN

QSL UR 559-559 TKS QSO DE XE1/JA1QXY

XE1/JA1QXY sends his call sign several times and is heard by FR5ZU. Signal reports are exchanged quickly and the contact is over in a matter of seconds! Notice the heavy use of Q-signals and abbreviated words to speed the process. The DX station is fair game once again and two operators slug it out for the prize ...

DE W2JGR DE W2JGR DE W2JGR DE W2JGR K

FR5ZU/G DE NJØM NJØM NJØM PSE K

NJØM uses the traditional approach while W2JGR tries repeating his call sign preceded by DE ("from"). His tactic pays off and he wins.

W2JGR DE FR5ZU/G...GOOD MORNING...UR 569 569...BK

FR5ZU/G DE W2JGR...TNX...UR 579 579 NAME JULES...QSL??? BK

FR5ZU passes along a signal report and shoots it right back to W2JGR. W2JGR gives his report and sends his name (Jules) as well. At the end of his transmission, he asks if the DX station copied everything ("QSL?").

QSL ES 73...FR5ZU/G QRZ KK

FR5ZU send a quick "QSL" to mean, "Yes, I got it all" along with his best wishes. He immediately sends QRZ to signal that he is ready for another contact. That's NJØM's cue to try again!

DE NJØM NJØM NJØM PSE KK

What you've seen here is less than five minutes of a DX pileup that lasted over an hour! Pity the DX operator at the other end of this melee. He has to do his best to sort out readable call signs among the rampaging signals. Sometimes the interference is so severe, he sees nothing but a wild jumble of letters on his screen.

The best you can do is be patient and keep calling. The rules of DX courtesy say that you shouldn't transmit if a contact has already been established. If you keep transmitting, you may

get the attention of the DX station in a way you'd never expect—he'll refuse to answer you for the remainder of the operation! Also, if the DX operator tries to control the mayhem through techniques such as working stations by call sign areas (1s, 2s, 3s and so on), don't buck the system. All you'll manage to do is anger the person you're trying to contact!

It's a "Split Decision"

What if you discover a rare DX station, but you don't hear a pileup? No matter how long you listen, you only seem to copy one side of the conversation—his! This is the telltale sign of a DX station that is working split. In other words, he is listening on one frequency and transmitting on another.

For some DX stations, working split is the only way to manage a pileup. This is especially true when a pileup gets too large and begins to disintegrate into chaos. Without the ability to work split frequencies, the DX station may be buried under a torrent of competing signals. Even if he manages to sort out a call sign, making contact is difficult because of interference from other stations who are continuing to call.

A good DX operator will always make it clear that he is working split and will indicate where he's listening for replies. You may see something like this:

CQ CQ DE 5U7M 5U7M UP 10

or...

QRZ DE 5U7M LISTENING 21.085 TO 21.090

The DX operator, 5U7M, is telling everyone that he is listening for calls 10 kHz above this frequency ("UP 10"), or that he is listening between 21.085 and 21.090 MHz specifically

To contact a DX station using a split-frequency scheme, you'll need a transceiver that can transmit on a frequency other than the receive frequency. Many modern transceivers feature dual VFOs for this purpose. Other rigs have the capability to use a remote VFO (a second VFO in a separate enclosure).

If you have remote or dual-VFO capability, leave your receiver tuned to the DX station and move your transmit frequency to the desired spot. Check your equipment carefully and make sure you know exactly where you are transmitting and receiving. Whatever you do, don't transmit on the DX station's frequency.

The "Burst" Modes

Aside from RTTY, all other digital modes in use on the ham bands today send their information in byte-sized (please excuse the pun!) chunks. Unlike RTTY, which sends a continuous stream of data, the burst modes send pieces of information that are checked for errors on the receiving end. Some sophisticated modes are capable of "repairing" damaged data to a certain extent. If the data is damaged beyond repair, however, they request repeat transmissions. This technique is known as ARQ, *A*utomatic *R*epeat re*Q*uest

The advantage of the burst modes is that they can send information that arrives error-free at the receiving station. And some of the more advanced burst modes can perform this trick under abominable conditions, making it possible to receive clean text in the face of noise, weak signals or interference.

So why do hams still continue to use RTTY? Despite the advantages the burst modes offer, RTTY is still king when it comes to making quick contacts. With a burst mode your MCP must establish the link, get the handshaking underway and then, after the information is exchanged, properly terminate the link. That's a serious hassle if you're working a contest or a DX pileup. Because of its lack of handshaking, RTTY gets the job done much faster, albeit without error detection.

Let's take a look at the burst modes in current use, and see how they work.

AMTOR

AMTOR is a close cousin to RTTY. It uses the same character set, but adds the advantage of error detection.

In the ARQ mode, the AMTOR *information sending station* (*ISS*) sends its text in bits and pieces. It transmits a group of three characters in a single burst and then switches to the receive mode. The *information-receiving station* (*IRS*) checks the characters for the 4:3 bit ratio and then transmits a control character. The control character means "Acknowledged. Send the next three" (*ACK*) or, "Not acknowledged. Repeat the last three" (*NAK*). If the ISS doesn't receive a reply (due to fading signals or interference), it repeats the characters anyway. Each station gets its turn to be the IRS or ISS (see Fig 9-4).

AMTOR's reign as undisputed king of the burst modes lasted only 10 years. In fact, if you tune through the HF digital subbands, you'll hear relatively few AMTOR stations on the air today. The problem with AMTOR is its inability to use the complete ASCII character set. AMTOR stations are limited to the exchange of text only; they cannot send binary information. And while AMTOR provides reasonable good performance in rough conditions, it lacks the sophistication to deal with fading, high noise levels and so on. That's why AMTOR eventually yielded its crown to PACTOR.

PACTOR

PACTOR is the most popular form of amateur HF digital communication. Every MCP on the market today offers PACTOR as an operating mode. When this edition went to press, PACTOR II (its high-octane cousin) was available on only one MCP—the SCS PCT-II (see the sidebar, "Hail the New King?")

PACTOR supports the complete ASCII character set. This means that you can send upper- and lower-case letters as well as binary files (computer software and other information).

PACTOR sends error-free information by using a handshaking system. When the data is received intact, the receiving station sends an ACK signal (for *acknowledgment*). If the data contains errors, a NAK is sent (for *nonacknowledgment*). In simple terms, ACK means, "I've received the last group of characters okay. Send the next group." NAK means, "There are errors in the last group of characters, send them again." This back-and-forth data conversation sounds like crickets chirping. In the case of PACTOR, the long chirp is the data and the short chirp is the ACK or NAK. AMTOR and PACTOR sound similar when you hear them, but PACTOR is the mode with the extended chirps.

Memory ARQ

With AMTOR, a group of characters must be repeated over and over if that's what it takes to deliver the information error-free. This results in slow communications, especially when conditions are poor.

PACTOR handles the challenge of sending error-free data in a different and interesting way. As with AMTOR, each character block is sent and acknowledged if it's received intact. If signal fading or interference destroys some of the data, a NAK is sent and the block is repeated. The big difference, however, involves memory.

When a PACTOR controller receives a mangled character block, it analyzes the parts and temporarily memorizes whatever information appears to be error-free. If the block is shot full of holes on the next transmission as well, the controller quickly compares the new data fragments with what it has memorized. It fills the gaps as much as possible and then, if necessary, asks for another repeat. Eventually, the controller gathers enough fragments to construct the entire block. PACTOR's *memory ARQ* feature dramatically reduces the need to make repeat transmissions of damaged data. This translates into much higher throughput.

With memory ARQ, and the ability to change data rates automatically, PACTOR allows you to enjoy conversations under conditions that would easily disrupt AMTOR or RTTY. Any ham, even those with very modest stations, can use PACTOR to communicate effectively on the HF bands. Amplifiers, beams and other expensive toys are definitely not re-

---

What's Your SELCAL?

Before you can operate AMTOR ARQ, you have to choose your *selective call identifier*, or *SELCAL*. When AMTOR stations wish to communicate in ARQ, this is the code that must be used to establish the link. The SELCAL code uses *only* letters, and we choose letters that match at least part of our call signs. Some examples are:

	CCIR-476	CCIR-625
Call Sign	SELCAL	SELCAL
KS9I	KKSI	KSIIXXX
WA9YLB	WYLB	WAIYLBX
W1AW	WWAW	WAAWXXX
WB8IMY	WIMY	WBHIMYX

CCIR-476 is the name of the recommended technical specifications for the version of TOR that's most popular on the amateur bands today.

Most amateurs use only the CCIR-476 SELCAL configuration. The letter combinations shown for CCIR-625 are strictly my own choice—you can use others.

Most MCPs feature a special command that will allow you to store your SELCAL in memory. (In the Kantronics KAM, for example, the command is "MYSEL." In the PK-232, it's "MYSELCAL.") This is an important step. Without a SELCAL, other AMTOR stations will not be able to call and link to you in the ARQ mode.

But Wait a Minute! What is CCIR-625?

CCIR-625 is a revised AMTOR/SITOR international standard. It was devised to address two problems: (1) the four-character CCIR-476 code was too limited to provide different SELCALs to all stations, and (2) under some circumstances, a CCIR-476 station could re-link with an incorrect station if the original link failed. CCIR-625 allows *seven* characters in its SELCAL, automatically identifies both stations at link-up, and also tightens the specifications for FEC synchronization.

Newer AMTOR controllers include both modes, but CCIR-625 is compatible with both new and old equipment. In a few years, however, CCIR-625 may become the dominant Amateur Radio AMTOR format.

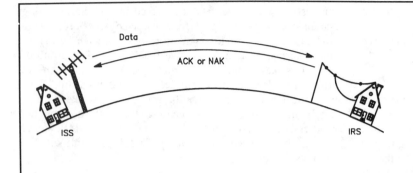

Fig 9-4—AMTOR, PACTOR and G-TOR use a system of acknowledgments (ACKs) and non acknowledgments (NAKs) to send information. The information sending station (ISS) sends a group of data bits to the information receiving station (IRS). If the data is received without errors, an ACK is sent and the next group is transmitted. A NAK means that errors occurred and the data must be sent again.

quired to enter the world of PACTOR.

Eavesdropping on PACTOR

Most MCPs include some sort of *listen* mode. The listen mode allows you to tune to an ARQ contact in progress and print the text *without* being part of the link. You need a bit of patience when using the listen mode since it doesn't include the error-correcting feature. You'll see plenty of errors and repeated characters, but at least you can follow the discussion.

Listen mode also works differently, depending on the MCP you're using. The listen mode in some controllers automatically senses and switches to ARQ or FEC. Other units monitor only ARQ transmissions in the listen mode and must be manually switched to the standby mode to monitor FEC signals.

Switch your transceiver to LSB and your MCP to the listen or monitor mode. Now hunt for a chirping PACTOR signal. When you've tuned the signal correctly—remember to watch your indicator—you should see characters within 15 to 20 seconds. If not, try another signal. On some controllers, you may have to reset the listen mode to restart the synchronizing process. With practice, you should be able to copy a PACTOR ARQ signal with ease. The listen mode may drop out of sync at times, especially when one station ends its transmission and

switches from ISS to IRS. This is normal since the listen mode can only synchronize to one station at a time.

Calling CQ on PACTOR

You send CQ on PACTOR by using its FEC mode. See the sidebar, "It Looks so Nice, I Sent it Twice." Put your MCP in the PACTOR standby mode and call CQ

[Switch your MCP to FEC transmit]
[Send a blank line]

CQ CQ CQ DE WB8IMY WB8IMY WB8IMY

CQ CQ CQ DE WB8IMY WB8IMY WB8IMY

CQ CQ CQ DE WB8IMY WB8IMY WB8IMY K K

[Switch your MCP back to receive]

Any station that's monitoring your signal will reply accordingly. When you send CQ, use several short lines of text rather than a few long lines. This helps stations synchronize to your signal more easily.

Once you make contact, the conversation flows along just like AMTOR. The station that answered your call is the *ISS* (information sending station) and you're the *IRS* (information receiving station). When he sends his "hello," he'll probably

---

Hail the New King?

PACTOR II is the newest of all HF digital modes. You won't hear many PACTOR II signals on the air at the moment, but its popularity is growing.

PACTOR II is *backward compatible* with PACTOR. That is, a PACTOR station can contact a PACTOR II station and vice versa. PACTOR II also retains many of the features that made PACTOR so popular: Support of the complete ASCII character set; memory ARQ to reduce the number of retransmissions; on-line data compression to speed transfers; automatic data-rate selection; and independence from sideband selection (it doesn't matter if you operate with your rig set to LSB or USB).

The additional treats that PACTOR II brings to the table are considerable. Using differential phase-shift keying (DPSK) and innovative convolutional coding (with a true Viterbi decoder), PACTOR II boasts data transfer rates at least *three times* faster than PACTOR. Under optimum band conditions, the rate could soar to six times faster. In terms of effective data rates, that translates to about 800 bit/s. PACTOR-II has the option of using Huffman coding for data compression (the same as PACTOR I), or Pseudo-Markov coding (PMC) for even greater performance and efficiency.

One of the most astonishing aspects of PACTOR II is that

it achieves all this wondrous performance while using only 500-Hz of spectrum (at –50 dB). You need fast hardware and digital signal processing (DSP) techniques to make this practical. That's why you won't find PACTOR II included in the current crop of multimode controllers. In fact, the only place you'll find PACTOR II at the moment is in the PTC-II manufactured by Special Communications System (SCS), Roentgenstrasse 36, D-63454 Hanau, Germany; tel 49-6181-23368; World Wide Web **http://www.scs-ptc.com**.

The PTC-II is an advanced multimode controller that offers PACTOR , PACTOR II, AMTOR, RTTY and CW in its list of operating modes. It also provides limited transceiver frequency control and the ability to function as a DSP audio filter.

If you demand high-performance HF digital communication regardless of cost, PACTOR II may be your dream come true. (When this edition went to press, the PTC-II was selling for $900.) There is no question that PACTOR II offers superb performance, and its implementation in the PTC-II is outstanding. Is PACTOR II better than the other contenders for the high-performance market—CLOVER or G-TOR? PACTOR II proponents would answer "yes," but CLOVER and G-TOR disciples would argue otherwise, depending on specific conditions.

use the over command to flip the link. (Consult your MCP manual and/or software documentation concerning the over command.) Now you're the ISS and he is the IRS. Introduce yourself and ask a question about where he lives, or what he does for a living. Use the over command to flip the link again. A conversation is underway!

Answering CQ

Answering a CQ in PACTOR is straightforward. Depending on the software you're using, it may be as simple as entering:

CALL N1BKE

or

CONNECT N1BKE

at the cmd: prompt. Some types of software streamline the process even further. There may be pop-up boxes where you simply enter a call sign.

PACTOR has a clever feature that comes in handy when you're trying to connect to a distant station. If the station is more than about 5000 miles away, use the exclamation mark (!) in your connect request. This forces the MCP to lengthen the time it waits to receive an ACK signal from the station, allowing more time for your signal to reach him and his ACK to return. For example:

CONNECT !HS1CHG

Anatomy of a PACTOR Conversation

Once a PACTOR link begins, the exchange of ACKs and NAKs goes on continuously—even if no one is sending information. If you and the other operator decided to leave your keyboards and grab a snack, your stations would chirp mindlessly back and forth to each other.

When the link is first established, you're the IRS. Just wait patiently for the other operator to send his greeting...

WB8IMY DE KU7G...

HELLO! MY NAME IS BOB AND I LIVE IN WASHINGTON STATE

IN THE SHADOW OF MOUNT ST HELENS. YOUR RST IS 589.

BACK TO YOU...K

Over to You

There is a very important point to remember about most "live" ARQ conversations: each station must turn the link over to the other station at the end of every transmission. With many MCPs the "over" is sent by tapping a single key. This automatically sends a control code that says, "Let's turn the link around. You're the ISS now." When the link switches, you'll hear a distinct change in the chirping rhythm.

Recognizing that there are times when the IRS operator (receiving station) would like to immediately break in and make a comment, most MCPs include a *forced over* command. A forced over causes an immediate link reversal, even if the ISS operator is still typing or has text in his transmit buffer. The exact command used to cause a forced over varies between units. Use the forced over sparingly; it's rarely needed, but very handy at times.

The Conversation Continues

So Bob has turned the link over to you. Now you're the ISS and he's the IRS. Why not send a short greeting and a signal report, too?

HELLO, BOB. YOU ARE SOUNDING FINE HERE IN CONNECTICUT. I LIVE IN A TOWN CALLED WALLINGFORD AND MY NAME IS STEVE. YOUR RST IS 599.

There is no need to repeat the information as you might do during a RTTY exchange. With the ARQ ACK/NAK system, the other station either receives your text or doesn't! The text is never garbled on the receiving end. If there is trouble on the frequency (due to fading or interference), the flow of incoming or outgoing text will slow down or stop altogether.

It sounds like Bob lives in an interesting place. Maybe you should ask him about it.

WERE YOU LIVING NEAR MOUNT ST HELENS WHEN IT ERUPTED BACK IN 1979? KU7G DE WB8IMY K

You remembered to turn the link over. Now Bob can respond.

YES, I WAS JUST MOVING INTO MY NEW HOME WHEN THE MOUNTAIN BLEW ITS TOP. THERE WERE ASHES EVERYWHERE! I COULDNT DRIVE TO MY NEW JOB BECAUSE OF ALL THE ASH IN THE ENGINE. I STILL HAVE SOME OF IT IN JARS IN MY BASEMENT! K

Now you have the start of a fascinating conversation! As with RTTY, you can type your comments while you're receiving his. When the link turns over, your system will begin sending the pretyped text automatically. All good things must come to an end, however ...

WELL, STEVE, I HAVE TO GET UP VERY EARLY TOMORROW AND DRIVE ALL THE WAY TO BOISE, IDAHO. I THINK I SHOULD GET TO BED SOON OR I WILL NEVER HEAR THE ALARM CLOCK. THANKS FOR THE GREAT CONVERSATION. I REALLY ENJOYED IT. GO AHEAD AND MAKE YOUR FINAL COMMENTS AND THEN YOU CAN DOWN THE LINK. 73 . . . WB8IMY DE KU7G SK

Bob says you can "down the link?" What does that mean? It simply means that he is asking you to send the end command that will terminate the ARQ link between your stations.
GOODNIGHT, BOB. HAVE A GOOD TRIP. HOPE TO LINK UP WITH YOU AGAIN ONE OF THESE DAYS. 73 . . . KU7G DE WB8IMY SK

Most MCPs feature a single keystroke that sends the end command (some label it "disconnect").

Working a PACTOR BBS

Although live conversations are common with PACTOR, there are lots of BBSs as well. You'll find that they're almost identical to packet systems. When you connect to a PACTOR BBS, you'll see something like this:

**Welcome to WB8SVN's PACTOR BBS in Placentia, CA.
Enter command: A,B,C,D,G,H,I,J,K,L,M,N,P,R,S,T,U,V,
W,X,?,* >**

After you enter a command you do not need to send the over code. The BBS will flip the link automatically after it receives the command from you.

You can use PACTOR BBSs to exchange messages with other PACTOR operators. And with PACTOR's ability to handle binary data, you can even download small programs from the BBS and run them on your computer!

Most PACTOR-capable MCPs have mailbox functions. You can use your PACTOR mailbox as a convenient way to pick up messages from friends when you're not at the keyboard.

HF Packet

Noise and interference are deadly to packet, and the HF bands have both in abundance. All it takes is the corruption of one bit of data and an entire packet data frame is ruined. The receiving station won't accept it, which means the transmitting station must try again. When you listen to packet signals, the first thing you notice is that some of the bursts are relatively long—up to 3 seconds or more. That's an eternity when you're sending data. It's also more than enough time for a pop of static or a blip of interference to wreck everything.

The HF bands are also prone to fading. Even a slight fade is enough to cause data loss. By combining the effects of noise, interference and fading, you set the stage for frustration. When conditions deteriorate, throughput suffers. The act of sending a simple greeting can take several minutes. (Oops! Lost a couple of bits! We'll have to send that frame again. And again. And again . . .)

This is why so many hams ultimately turn to other digital modes such as AMTOR, G-TOR, PACTOR and CLOVER when they need to pass information on HF efficiently. This is not to say that packet doesn't have its place on the HF bands. When signal conditions are good, it performs quite well. There are still many VHF and UHF bulletin boards throughout the world that depend on HF packet to move mail over great distances. And there are great opportunities to enjoy fascinating conversations on HF packet. The key to success is patience!

An HF Packet Tour

Before you enjoy your first HF packet conversation, spend some time prowling through the bands. HF packet can be a little confusing at first glance.

Set up your TNC or MCP so that you can monitor all packet data—even packets not intended for you. In some units the **MCOM** command is used (**MCOM ON**). In others it is part of the **MONITOR** command. Check your TNC or MCP manual.

When you find a packet signal, tune *very carefully*. If your TNC or MCP has a bargraph-type tuning indicator, you're ahead of the game. You can use it to get your radio on the correct frequency with little trouble. Packet signals can be tuned without an indicator, but it is quite difficult. With practice, however, you'll learn to do it by ear. That is, you'll know you're close when you hear the packet signal at a certain pitch.

Don't expect to see information on your screen right away.

You may have to wait for the next transmission (after you've stopped tuning) before data is displayed. And if conditions are so-so, you may have to wait even longer.

Mail Forwarding

Of all the activity you'll see on HF packet, *mail forwarding* is the most common. The global packet system is made up of VHF and UHF networks that serve local areas. The bulletin boards in these networks pass mail between each other on VHF and UHF frequencies when they can, but sometimes the routes require leaps of hundreds or thousands of miles. Unless the bulletin board operator has the ability to relay mail via amateur satellite, the next best choice is HF packet.

The act of relaying packet mail from one point to another is known as *forwarding*. Messages and bulletins travel on VHF or UHF until they reach stations with HF packet forwarding capability. A station must have more than an HF transceiver and a TNC to be a forwarder. An HF mail forwarder keeps appointments with other stations on specific frequencies. Elaborate software checks each incoming message from the local network and "decides" which HF station should receive it. The messages are stored until they can be relayed via HF packet to the appropriate station.

Let's say that you've stumbled upon some forwarding in progress. It may look like this:

W0XK>N4HOG:

ö_îywl⁺°_ÛòÊ$£¶ØæHÅ'G^›2%T´JÜ·_y7™o,ÖÓTnx for QSLW0XK>N4HOG:

**ç
‰»_áY∂lÚîf'_x"ªxö®;=»8À‡¿Î¥c_CaueJó &__ÓÛúÿ_OQçy@}/
YÔÂÂ_J›3—sé ¿ −Ÿ‹mïV=**

No, there's nothing wrong with your computer! These stations are using a type of data compression to make it easier to send large amounts of data. The result on your screen, however, looks like this. If you see stations passing traffic without compression, this is more typical of what you'll see:

XF3R>XE1M-15:

Another ham (80 years old!) told me about a mode called PACTOR II

XF3R>XE1M-15:

Another ham (80 years old!) told me about a mode called PACTOR II

XF3R>XE1M-15:

Another ham (80 years old!) told me about a mode called PACTOR II

XF3R>XE1M-15:

I never heard of this before!

XF3R>XE1M-15:

Has anybody further info?

XF3R>XE1M-15:

Thanx in advance

XF3R>XE1M-15:

73...Joss

Notice that parts of this message were repeated several times. I bet you can guess why! Noise, interference or fading must have caused errors on the receiving end. A few repeats were required to get the message through.

Bulletin Boards and Mailboxes

HF packet gives you the opportunity to connect with bulletin board systems (PBBSs) throughout the world. By connecting with a bulletin board, you can view messages and bulletins concerning every topic imaginable! Of course, you can do the same thing on VHF packet, but DX bulletin boards offer a perspective you won't find elsewhere. What are the big issues among hams in Argentina, for example? Connect to an Argentinean PBBS and find out!

PBBS activity is easy to spot. Most hams request a list of the latest bulletins when they connect, so look for any station that seems to be sending such a list. Here is an actual example copied on 20 meters:

VE4GQ>WB4TDB:

8965 BF 1967 ALL   @MAN   VE4SYG 07-Jun MRS BYLAWS

8964 B$ 1587 SALE @ALLCAN VO2APL 04-Jun LOTS OF THINGS FOR SALE

8961 B$ 1012 SALE @ALLCAN VO2APL

8960 B$ 414 WANTED@CANADA VE4YE 07-Jun

8956 B$ 2986 RSGB @CANADA VE6EQ 05-Jun RSGB MAIN News 5th June

8955 B$ 2603 RAC   @CANADA VE3LVO 05-Jun $$$ STONEWALLING? $$$

Some PBBSs are able to serve more than one station at a time, so you may see several messages going to one station, then another.

As you comb through the packet signals you may also run into data that looks like it's coming from a PBBS, but it isn't. For example:

[KPC2-5.00-HM$]

16000 BYTES AVAILABLE

Thanks for checking in. Please leave a message and I'll get back to you. 73, Bill/W1KKF

ENTER COMMAND: B,J,K,L,R,S, or Help >

This is the typical sign-on message you'll receive from a packet mailbox. These mailbox functions are common features on TNCs and MCPs. Depending on the model you've purchased, you may have one at your fingertips right now! You'll also notice that the command line

(ENTER COMMAND: B,J,K,L,R,S, or Help >)

is much shorter than those you'll see on most packet bulletin boards. That's because packet mailboxes have very limited capabilities. This one will allow you only to send mail (S), read mail (R), list messages (L), kill messages (K), see a list of stations heard recently (J) and disconnect (B).

CLOVER

CLOVER is an advanced HF digital communication system that Ray Petit, W7GHM, developed in a joint venture with HAL Communications of Urbana, Illinois. CLOVER uses a four-tone modulation scheme. Depending on signal conditions, any of 10 modulation formats can be selected manually or automatically. Six of the modulation systems employ phase-shift modulation (PSM); two use amplitude-shift modulation (ASM); and two use frequency-shift modulation (FSM). Each tone is phase- and/or amplitude-modulated as a separate, narrow-bandwidth data channel. As you might guess, the resulting CLOVER signal is very complex!

For example, when the tone pulses are modulated using quadrature phase-shift modulation (QPSM), the differential phase of each tone shifts in 90° increments. Two bits of data are carried by each tone for a total of eight bits in each 32-ms frame. The resulting block data rate is about 250 bits per second. The complex, higher-speed modulation systems are used when conditions are favorable. When the going gets rough, CLOVER automatically brings several slower (but more robust) modes into play.

Even with these ingenious adaptive modulation systems, errors are bound to occur. That's where CLOVER's Reed-Solomon coding fills the gaps. Reed-Solomon coding is used in all CLOVER modes. Errors are detected at the receiving station by comparing check bytes that are inserted in each block of transmitted text. When operating in the ARQ mode, CLOVER's damaged data can often be reconstructed without the need to request repeat transmissions! This is a major departure from the techniques used by packet, AMTOR and PACTOR. Of course, CLOVER can't always repair data; repeat transmissions—that CLOVER handles automatically— are sometimes required to get everything right. With the combination of adaptive modulation systems and Reed-Solomon coding, CLOVER boasts remarkable performance—even under the worst HF conditions.

CLOVER Handshaking

As you may recall, PACTOR and AMTOR both use an *over* command to switch the link so that one station can send while the other receives. CLOVER links must be switched as well, but the switching takes place *without* using *over* commands

When two CLOVER stations make contact, they can send limited amounts of data to each other (up to 30 characters in each block) in what is known as the *chat mode*. If the amount of data waiting for transmission at one station exceeds 30 characters, CLOVER automatically switches to the *block data mode*. The transmitted blocks immediately become larger and are sent much faster. The other station, however, remains in the chat mode. Because of precise frame timing, this takes place without the need for either operator to change settings, or send *over* commands. The CLOVER controllers at both stations "know" when to switch from transmit to receive and vice versa. And what if both stations have large amounts of data to send at the same time? Then they *both* switch to the data block mode. This high degree of efficiency is transparent to you, the operator. All you have to do is type your comments or select the file you want to send—CLOVER takes care of everything else!

CLOVER features an FEC mode similar to that used by AMTOR, PACTOR and G-TOR. You use the CLOVER FEC to call CQ, or to send transmissions that can be received by several stations at once. (In the CLOVER ARQ mode, only two stations can communicate at a time.) CLOVER shares another characteristic with AMTOR: the use of SELCALs. When attempting to contact another CLOVER station, you must send its SELCAL first.

What Do I Need to Run CLOVER?

The requirements for a CLOVER station differ substantially from those of other HF digital modes. They are:

❏ An SSB transceiver. The transceiver must be very stable (less than 30-Hz drift per hour). It should also include a frequency display with 10-Hz resolution. The audio output from the CLOVER controller is fed to the audio input of the transceiver (CLOVER uses AFSK, not FSK). Receive audio is supplied to the controller from the external speaker jack or other source.

☐ An IBM-PC computer or compatible. The computer must be at least a 286-level machine.

☐ A CLOVER controller board such as the P-38. All CLOVER controllers are available exclusively from HAL Communications. The HAL P-38 CLOVER controller is installed *inside* the computer using any available expansion slot (see Fig 9-5). The board uses a dual-microprocessor design and digital signal processing to achieve signal modulation and demodulation. It also operates on modes other than CLOVER, such as RTTY and PACTOR.

☐ HAL PC-CLOVER software. This is supplied by HAL Communications and is included with every controller. It is *not* a terminal program. The PC-CLOVER software is the instruction set of the P-38 itself. It's loaded into the P-38's memory each time you decide to operate. This approach makes it easy to update the controller in the future. You simply buy a new diskette or download the software from a BBS.

CLOVER Conversations

When it comes to on-the-air operating, CLOVER is different from any of the modes we've discussed so far. To call CQ, for example, you switch to the **MODE** menu, highlight **CQ** and press **ENTER**. The P-38 sends a CW identification followed by a raucous stream of data. Unlike other digital modes, you do not see "CQ CQ CQ" flowing across your screen. In fact, you see nothing at all.

The P-38 sends CQ in the form of data signals that appear as CQ "flags" to other CLOVER stations. When another CLOVER operator tunes in your signal, all he sees is a statement on his screen announcing that you are calling CQ. At that point he can ignore you or press a single key to establish a CLOVER connection

Once the conversation has started, it is similar to HF packet. Because there are no *over* commands, you need to let the other station know when you've completed a statement. For example:

Hello! My name is Steve

and I live in Wallingford, Connecticut. I am new to

CLOVER. What do you think of it? >>>

Without >>>, BTU, K or a similar symbol at the end of my statement, the other operator might inadvertently jump in after the end of the first line. This can be very confusing for everyone!

While you're watching the conversation, it's easy to get distracted by the receive/transmit status table in the upper right corner of the screen. The table displays the modulation format in use at the moment, the signal-to-noise ratio, tuning error, phase dispersion, error-correction capacity and transmitter output power (as a percentage of full output. The table is split into horizontal rows labeled "MY" and "HIS." Not only do you see your own parameters changing, you see the changes taking place at the other station! (CLOVER accomplishes this feat by periodically swapping station data.) Who is enjoying the best receive conditions? Which station is doing the greatest amount of error correcting at the moment? Just look at the table.

G-TOR

G-TOR is another contender for the high-performance category. It was developed by the Kantronics Corporation and is only available in the Kantronics *KAM Plus* MCPs, specifically those units sold after March 1, 1994. Owners of earlier KAM Plus MCPs can upgrade to G-TOR by simply purchasing and installing a version 7.0P EPROM chip. Hams who own older KAMs can also add G-TOR, but they must purchase the version 7.0E enhancement board. If you already own a KAM with an enhancement board, all you need is a 7.0E EPROM.

If you have an SSB transceiver that will switch from transmit to receive in less than 100 ms (most will), you should be able to use it for G-TOR with little difficulty. You can operate G-TOR in direct FSK or AFSK. Most G-TOR operators are using AFSK, however.

Who Put the "G" in G-TOR?

G-TOR is an acronym for *Golay-coded Teleprinting Over Radio*. Golay coding is the error-correction system created by M. J. E. Golay and used by the *Voyager* spacecraft. Sending billions of bytes of data across the Solar System required a scheme to ensure that the information could be recovered despite errors caused by interference, noise, and so on.

To create G-TOR, Kantronics combined the Golay coding system with full-frame data interleaving, on-demand Huffman compression, run-length encoding, a variable data rate capability (100 to 300 bit/s) and 16-bit CRC error detection. G-TOR system timing is liberal enough to permit long-distance communication. Overall, the performance is excellent, rivaling CLOVER in many instances.

G-TOR in Action

G-TOR is surprisingly easy to operate. If you're familiar with AMTOR or PACTOR operating, G-TOR is essentially the same. When you're in the G-

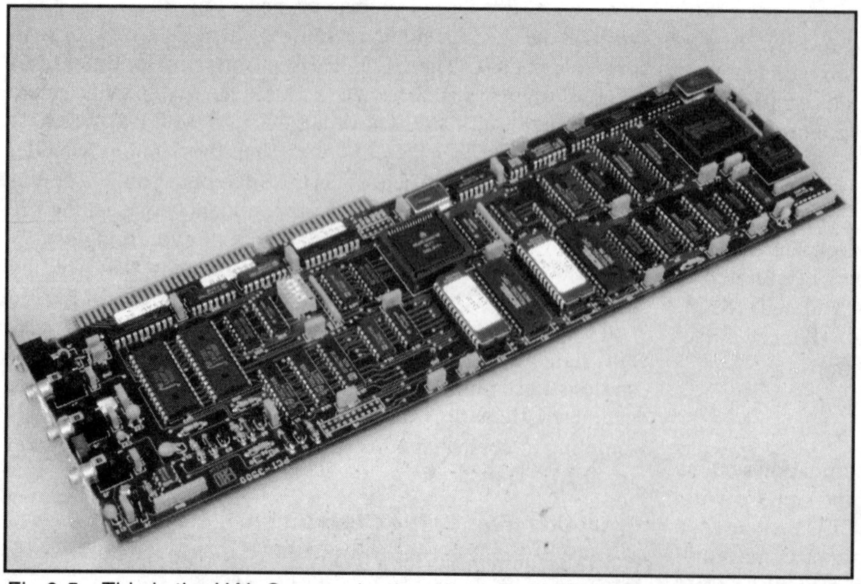

Fig 9-5—This is the HAL Communications P-38 CLOVER controller. It fits conveniently inside your IBM-PC or compatible computer. The only external wiring are audio cables to and from your SSB transceiver

TOR mode, you use its FEC mode to call CQ. This allows your transmission to be copied by as many stations as possible. If someone wants to talk to you, they respond in G-TOR using your full call sign.

When you call CQ, let the stations know that you're fishing for G-TOR contacts. Remember that AMTOR stations may be tuning in. Here's an example of a typical G-TOR CQ:

CQ CQ CQ CQ CQ—G-TOR

CQ CQ CQ CQ CQ—G-TOR

CQ CQ CQ CQ CQ—G-TOR

DE WB8IMY WB8IMY WB8IMY

Standing by for G-TOR calls. K K

Once the G-TOR link is established, the station that called you is the ISS, or *information sending station*. It's up to him to speak first. While he is transmitting, you're the *information receiving station*, or IRS. You flip these IRS/ISS roles back and forth by sending an "over" code. Like PACTOR, G-TOR can transfer ASCII or binary information.

If you copy someone calling CQ G-TOR, you simply switch to the G-TOR standby mode. Establishing a connection is as easy as entering the word G-TOR followed by the call sign of the station you wish to contact. For example:

G-TOR WB8IMY

You can monitor the G-TOR data rate by glancing at the front panel of the KAM Plus. When the **STA** LED is off, the rate is 100 bit/s. A flashing **STA** indicates 200 bit/s and a steady **STA** means that you're perking along at 300 bit/s. The data rates change automatically based on the quality of the link. Links always begin at 100 bit/s. If the number of correctly received frames exceeds a preset value, the receiving station will request a speed increase to 200 or 300 bit/s. If the link deteriorates, the data rate will automatically ratchet downward. You can also set the data rate manually.

CHAPTER 10

Packet Radio

STAN HORZEPA, WA1LOU
1 GLEN AVE
WOLCOTT, CT 06716-1442

In a very short time, packet has become one of the most popular operating modes in Amateur Radio. Since packet experiments began, in 1978, loads of packet equipment have been sold and, today, most active hams have some kind of packet installation in their ham shacks.

A big reason for packet's popularity was its timing. It came on the scene at the beginning of the home-computer revolution. Hams have a natural curiosity when it comes to new electronic gadgets, so a lot of hams built and bought computers, many intending to use them for ham radio applications.

A substantial amount of ham radio software was written for home computers. Some of the software even performed communications (CW and RTTY, for example). However, there was something lacking in the communications software. It wasn't that the software couldn't do CW and RTTY, but that CW and RTTY did not use the full potential of the computer as a communications tool. The computer hardly worked up a sweat when performing these tasks. It had an appetite for something faster.

For a number of years, a mode of computer communication called packet *switching* was being used commercially to transfer data between computers at high speeds. Although packet switching usually operated over wire lines of some sort, there had been some success using packet switching over radio waves. It was only a matter of time before someone got the idea to try it over Amateur Radio as well. The rest is history.

Besides speed, packet offers other attractive advantages.

❏ It operates without error. Whatever you send is received perfectly.

❏ It uses the radio spectrum efficiently. Multiple communications may be conducted by multiple stations on the same frequency at the same time.

❏ It provides time-shifting communication. By storing messages on packet bulletin boards (PBBS) or mailboxes, stations can communicate with other stations who are not on the air.

❏ It operates over networks when necessary. Any station can access packet networks and greatly expand their communication capabilities.

❏ With all of these features, hams now are using packet for numerous diverse applications including traffic handling, satellite QSOs, DXing, emergency communications and more.

There seems to be a niche for every ham in the packet mode. See if you can find yours.

THE PACKET STATION

There are three basic parts of a packet station: the terminal, the radio equipment and the terminal node controller (TNC).

How small can a packet station be? N1RWY put together a lightweight, portable packet station made up of a Radio Shack HTX-202 transceiver, an AEA Hot Rod 1/2-wave antenna, a Tigertronics Baypac BP-1 1200-baud TNC, and an HP200LX palmtop computer that runs *Baycom* software. The setup runs 2.5 W with the radio's built-in battery pack, or a full 8 W with a 12-V camcorder battery.

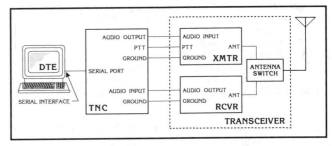

Fig 10-1—The components and interconnections of a typical packet station.

Put these parts together and you are on the air.

Terminal

A terminal interfaces the user to the TNC. By means of the terminal's keyboard, the user enters commands that control the TNC. Once communications are established, the user enters information on that same keyboard for transmission to the other station. Via its display, the terminal allows the user to read the TNC's responses to commands and to read the information that's being sent by the other station.

Computers Acting Like Terminals

In some cases, real terminals are used in packet stations. However, in the typical packet station, computers *emulating* terminals are used. Terminal emulation is achieved by running *terminal emulation software* on the computer. A variety of terminal software is available, from programs used to access mainframe computers over wire links to software that provides access to bulletin board systems (BBSs) via telephone lines. There is even software available that is specifically designed for packet applications. (For simplicity's sake, the remainder of this chapter will use the terms *DTE* and *terminal* when referring to a computer emulating a terminal, as well as a real terminal.)

If you have software for your computer that allows it to use a telephone modem, it can probably be used for packet, too. After all, a TNC is essentially an "intelligent" modem that communicates over the airwaves rather than over telephone lines. Such software will usually serve you well in the packet mode.

Software that is specifically designed for packet terminal emulation offers features that are optimized for packet communications. For example, if you communicate with more than one station simultaneously (a multiconnect situation), some packet terminal programs allow you to open separate communication windows for each multiple connection. No telephone line software package offers that feature!

If you become dissatisfied using telephone line modem software in the packet mode, consider getting a program specifically designed for packet. The sidebar titled "Packet Radio Terminal Emulation Software" lists examples of available software.

The Real Thing

Computers emulating terminals are not the only kind of DTEs you will find in a packet station. Some hams use real terminals! Surplus DTEs are inexpensive and plentiful (go to any ham/computer flea market and see what I mean). A preowned DTE can cost less than a preowned computer, and if you use a real DTE for packet you don't have to tie up your computer doing terminal emulation.

The disadvantage of using a real DTE is that it lacks features that computer terminal emulation software provides. (Some "intelligent" DTEs do include some of the features of terminal emulation software, but these DTEs are much more expensive than their "dumb" brethren.) The solution is to use a computer that runs more than one program simultaneously (this is called multitasking). Then you can run a terminal emulator for packet while your computer is doing other more important chores such as duping your Field Day log or running PBBS software!

Facing the Interface Issue

Whatever DTE you use, it must be capable of interfacing with your TNC. Almost all TNCs use the EIA/TIA-232-E ("EIA-232" for short) interface with its ubiquitous 25-pin connector to provide a serial connection to a DTE, so your DTE should have an EIA-232 interface as well.

Most real DTEs have an EIA-232 interface, so connecting them to a TNC is simple. On the other hand, some computers don't necessarily have EIA-232 interfaces. Some computers

---

Packet Radio Terminal Emulation Software

Packet radio terminal emulators are available as commercial packages (you pay for it before you use it), shareware (you use it before you pay for it and only if you like it), and freeware (you use it for free). A sampling of this software for our more popular computers follows.

Apple II

APR—An Apple II/II+/IIe/IIc/IIgs packet program that is available by sending a blank 5.25 or 3.25-inch disk and a postage-paid, self-addressed disk mailer to Larry East, W1HUE, 1355 Rimline Dr., Idaho Falls, ID 83401.

Apple Macintosh

Host Master Mac—Kantronics' packet terminal program for the Mac that is optimized for the Kantronics TNC line.

MacRATT with FAX—A full-featured packet terminal program from AEA that is optimized for AEA's TNCs, but runs well with other TNC brands, too.

Savant—Sigma Design Associates' Macintosh DTE and TNC emulator that does *not* require a TNC to do packet. All you need is an external modem, the PacketMac, and *SoftKiss*, the modem software driver. *Savant* will also run with any TNC in the KISS mode.

Atari

Packet—A packet terminal emulator for Atari 8-bit computers from Electrosoft.

IBM-PC

Host Master II+—Kantronics' packet terminal program for the PC that is optimized for the Kantronics TNC line.

PacketPet for Windows—A Windows packet terminal program that is compatible with all TNCs from the Big Four.

PaKet—A full-featured packet terminal program that runs under DOS, OS/2, and UNIX-DOS/Merge. It is available from Jim Flannery, WB0NZW, 8098 S Carr Ct., Littleton, CO 80123.

PC-PAKRATT for DOS—A full-featured packet terminal program from AEA that is optimized for AEA's TNCs.

PHSOS2.ZIP—An OS/2 multimode terminal emulator for the AEA PK-232 and PK-900 controllers may be downloaded from CompuServe's HamNet.

XPCOM—A Windows multimode terminal emulator for the AEA PK-232 and MFJ-1278 multimode controllers may be downloaded from CompuServe's HamNet.

Tandy Color Computer

COCOPACT—A Color Computer packet terminal emulator from Monty Haley, WJ5W, Rte. 1, Box 210b, Evening Shade, AR 72532.

---

have no interface at all and EIA-232 may be an option that's available through the addition of a PC board. (If you have an IBM PC or compatible, you can add a TNC-on-an-expansion-card to your PC instead of the EIA-232 interface and avoid the serial-interface middleman altogether.)

There are computers that use other interfaces. For example, TIA/EIA-422-B ("EIA-422" for short) is found on some computers like the Macintosh. EIA-422 is close enough to EIA-232 that it can be made to work by properly wiring the interfaces. (Cables that interface a Macintosh computer to a telephone line modem will interface a Mac to a TNC, too.)

Some older computers had a TTL interface, which required signal conversion before they could be interfaced to EIA-232. Since some TTL-interfaced computers were very popular (for example, the Commodore C64), a number of older TNC models offered the TTL interface as an alternative to the EIA interface. If your computer uses TTL, find a pre-owned TNC that is TTL compatible.

Although EIA-232 supports 25 signals, any TNC you use will need only eight of them (the signals on pins 2-8 and pin 20) and many TNCs can get by using even fewer signals. Check the TNC's manual and see what you can do to economize your DTE-to-TNC cabling.

The Radio

There can be a lot of equipment at the RF end of a packet station. Some of it is of little concern to us. For example, as long as the antennas and feed line are capable of putting a signal on the desired packet frequency, it satisfies our requirements. Other RF hardware needs closer inspection, however.

Our primary concern is the radio equipment's receive-to-transmit and transmit-to-receive turnaround time. A TNC can switch between the transmit and receive modes very quickly. So quickly, in fact, that it must wait for the RF equipment to switch before it can continue to communicate.

According to a study conducted by Tucson Amateur Packet (TAPR), most amateur FM radios have receive-to-transmit and transmit-to-receive turnaround times between 150 and 400 milliseconds (ms). This dramatically reduces the amount of data that can be sent and increases the chance that two or more stations will interfere with one another. Such delays slow down what is intended to be a fast mode of communication.

The physical switching of an antenna, internally in a transceiver or externally with a separate transmitter and receiver, affects the turnaround time. The older the transceiver, the more likely that switching is performed mechanically by a relay. If a separate transmitter and receiver are used with one antenna, a mechanical relay is probably performing the switching function as well. In addition, if an external power amplifier and/or receive preamplifier is used, more mechanical switching may be involved.

With newer equipment, the switching is often accomplished electronically. This speeds up the process, but the improvement may be compromised by the frequency synthesizer circuitry that is also found in the newer equipment. After switching between the transmit and receive modes, synthesizers require some time to lock on frequency before they are ready. New equipment is being designed with synthesizers that can lock more quickly. (Older RF equipment does not use frequency synthesis; therefore, it does not suffer from this delay.) If you're hunting for a new transceiver for packet applications, keep this feature in mind.

Another problem cited by TAPR is that the modem-to-radio interface of most radios used for packet depends on audio response filters and audio levels *intended for microphones and*

speakers. More often than not, this leads to incorrect deviation of the transmitted signal, noise and hum on the audio and so on. Splatter filters and deviation limiters distort frequency response and further reduce the performance of the packet-radio system. You are stuck in this environment unless you want to modify the radio. Performing surgery on your typical modern VHF/UHF FM voice transceiver may be difficult because of the use of LSIs, surface mounting and miniaturization.

Instead of using a typical amateur transceiver for packet, there are alternatives on the market today that solve many of the RF equipment problems. Check the pages of *QST* for news, reviews and advertisements for these radios that are designed with packet in mind.

The TNC

Packet is the communication mode in which the output of a DTE is assembled into bundles or "packets" according to a set of rules or a *protocol*. Each packet is transmitted to a remote radio station where it's checked for its integrity, then disassembled and its information fed to a terminal for you to read.

Packets are assembled and disassembled by a packet *assembler/disassembler* or *PAD*. Since the input and output of a PAD is digital, there is a digital-to-analog/analog-to-digital conversion between output/input of the PAD and radio transmitter/receiver. This conversion is performed by a *modulator-demodulator* or *modem*. Typically, a PAD and its associated modem are packaged into one unit called a *terminal node controller* or *TNC*, which is connected between the DTE and the radio equipment as illustrated in Fig 10-1.

You can think of a TNC as a very intelligent modem. Whereas a telephone modem permits a computer to communicate by means of the telephone network, a packet modem permits a computer to communicate by means of a radio network. Just as an intelligent telephone modem augments its functions by including a wide range of built-in commands to facilitate computer-telephone communications, the built-in intelligence of a packet-radio modem facilitates packet-radio communications.

Both a telephone modem and a packet modem are connected to a communications medium: the telephone modem to the telephone line and the packet "modem" to the radio. Telephone modems and packet modems are both connected to the serial interface of a computer.

To initiate com-

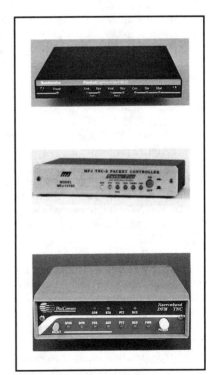

A potpourri of TNCs. From top to bottom: Kantronics KPC-9612, MFJ-1270C and the PacComm TNC/NB96. All three TNCs are capable of 1200 and 9600 bit/s operation.

puter-telephone communication, you command an intelligent modem to address (dial the telephone number of) the other computer. Similarly, to initiate computer packet-radio communications, you command a TNC to address (make a connection with the call sign of) another Amateur Radio station.

Computers Acting Like TNCs

Not only have computers been programmed to emulate terminals, but they have been programmed to emulate TNCs, too. Once upon a time, Bob Richardson, W4UCH, programmed the Radio Shack TRS-80 Model I computer to perform the functions of a TNC in software. All you had to do is connect the appropriate modem between the TRS-80's serial port and your radio and you were on the air.

The TRS-80 Model I is a computer museum piece today, but Bob's idea is still viable and has been implemented in Commodore, IBM and Macintosh computers. As in W4UCH's implementation, once you have the software, simply connect a modem to your computer's serial port and radio equipment and run the software.

See the sidebar "Packet-Radio Terminal Emulation Software" for a list of available TNC-emulation software. Modems in kit and assembled form are available from a number of sources for approximately $50. Typically, when you purchase the modem, it is bundled with terminal-emulation software.

Packet Protocols and Commands

The manner in which packet communication is conducted is called a *protocol*. The protocol consists of a standard set of rules and procedures that are programmed into each TNC so that they all will communicate in a compatible manner. In packet today, the most widely used protocol is called "AX.25." However, in the past, AX.25 had a competitor known as the "VADCG protocol."

After the Canadian Department of Communications authorized amateur packet in 1978, the VADCG protocol and a companion TNC were developed by Doug Lockhart, VE7APU, and the organization he founded, Vancouver Amateur Digital Communications Group (VADCG). Two years later, the FCC authorized amateur packet in the United States and, after a year of experimentation, the US packet pioneers agreed on a new protocol: AX.25.

The original TNC that was designed by VADCG supported only VADCG protocol (the only protocol that existed at the time). These TNCs were commonly known as "VADCG" TNCs. When AX.25 was implemented for the first time (by TAPR in their TNC 1) the VADCG protocol was also included. (VADCG TNCs outnumbered the new TAPR TNCs at that time.) The TNC 1 also had a new design and command set to support AX.25.

In 1985, TAPR introduced an AX.25-only TNC, the *TNC 2*, which used an expanded TNC 1 command set (memory limitations precluded implementation of VADCG protocol). The TNC 2 was quickly cloned by various manufacturers. After cloning, these manufacturers added new features to the clones, which resulted in "TNC 2 compatibles." Some TNC 2 compatibles not only had new bells and whistles for packet, but also permitted operation in other modes.

It was during the time that these clones and compatibles became available that the popularity of packet took off! Tens of thousands of these units were sold worldwide and the TNC 2 command set became predominant in packet. As a result, this chapter uses TNC 2 commands throughout its description of packet-radio operation.

To configure and control a TNC, you enter commands into the keyboard of a DTE. Commands may be entered only when the TNC's command prompt (cmd:) is displayed by your DTE. When this prompt is displayed, type the desired command and follow it with a carriage return (<CR>). For example, to command the TNC to disconnect, you type "DISCONNE" and a carriage return at the command prompt. In this chapter, entering commands will be represented as:

cmd: DISCONNE <CR>

To save keystrokes, many TNC commands may be used in an abbreviated version. For example, instead of typing DISCONNE for the disconnect command, you can simply type the letter D. In this chapter, each command will be printed partially in uppercase characters and partially in lowercase characters. The uppercase and lowercase characters together represent the full name of the command, whereas the uppercase characters alone represent the abbreviated version of the command. For example, the disconnect command will be represented as:

Disconne

where Disconne is the full name of the command and D is the abbreviated version.

Tailoring the TNC to Your Station

Before you transmit your first packet, you must configure the TNC to your needs. These include the requirements of your station equipment, specifically your DTE and your radio, and the requirements of the radio spectrum you will use, either HF or VHF/UHF.

DTE Parameters

Your TNC must be compatible with your DTE's operating parameters.

Data Rate—Set the data rate of your TNC's serial port to the bit/s rate of your DTE (your DTE is connected to the TNC's serial port). On the TNC 2, the data rate (300, 1200, 2400, 4800 or 9600 bit/s) is set by means of a rear panel DIP switch (positions 1 through 5) while the TNC is powered off. (All other parameters are set while the TNC is powered on.)

Note that many TNC 2 compatibles set their data rate using "autobaud" (the TNC automatically adapts to the data rate of the DTE). Typically, this is accomplished by entering carriage returns at the DTE keyboard after powering the TNC. After receiving some carriage returns, the TNC calculates the correct data rate, sets itself to that rate and sends its sign-on preamble to the DTE at the correct data rate.

Echo—The echo function causes the TNC to print on your DTE's display each character that you type into your DTE's keyboard. This TNC function should be turned on or off depending on whether your DTE does or does not provide this function. (Echo *on* is the TNC default.)

If two characters are displayed for each one that you type into the keyboard (you type HELLO and HHEELLLLOO is displayed), then both your DTE and TNC are echoing, so turn off the TNC's echo by typing:

cmd: Echo OFF <CR>

If nothing is displayed when you type, then the echo function is turned off in both your DTE and TNC, so turn on the TNC's echo function by typing:

cmd: Echo ON <CR>

Character Length—Set the TNC to the character length used by your DTE (7 bits per character is the default value.)

To select 7 bits per character, type:

cmd: AWlen 7 <CR>

To select 8 bits per character, type:

cmd: AWlen 8 <CR>

Parity—Set the TNC to the parity used by your DTE. (Even parity is the TNC default.)

To select even parity, type:

cmd: PARity 3 <CR>

To select odd parity, type:

cmd: PARity 1 <CR>

To select no parity, type:

cmd: PARity 0 <CR> or

cmd: PARity 2 <CR>

Screen Length—Set the TNC to the maximum number of characters that is displayed on each line of your DTE (80 characters per line is the TNC default), by typing:

cmd: SCREENLN n <CR>

where n is number 0 to 255, representing the number of characters per line displayed by your DTE.

Line Feeds—The automatic line-feed function causes the TNC to insert a line feed after each carriage return that is received in incoming packets. This TNC function should be turned on or off depending on whether your DTE does or does not provide this function. (Automatic line feed *on* is the TNC default.)

If a blank line follows each line of received text, then both your DTE and TNC are adding line feeds. Turn off the TNC's automatic line feed function by typing:

cmd: AUtolf OFF <CR>

If each line of received text is printed over the previously received line of text, then neither your DTE or TNC are adding line feeds, so turn on the TNC's automatic line feed function by typing:

cmd: AUtolf ON <CR>

Lowercase and Uppercase Characters—This TNC function normally sends both lowercase and uppercase characters to your DTE. If your DTE does not accept lowercase characters, this TNC function may be set to automatically translate all lowercase characters to uppercase characters.

If your DTE accepts both lowercase and uppercase characters, type:

cmd: LCok ON <CR>

If your DTE accepts only uppercase characters, type:

cmd: LCok OFF <CR>

Those are the most important parameters that you need to set to configure your TNC for compatibility with your DTE.

By setting these parameters correctly, your DTE should now display text from your TNC in a clear, legible manner. If not, you should recheck these parameters and also check the physical connection between your TNC and DTE.

(In addition to the previously described DTE parameter commands, there are other DTE-related commands that allow you to select ASCII characters that perform various functions: input editing, flow control and so on. Refer to the sidebar titled "Control Characters" for more information.

Radio Parameters

Use TNC commands to customize it to the needs of your radio equipment and the radio spectrum that you are using.

Data Rate—Set the radio-port data rate of the TNC to the data rate you will use (300 bit/s for HF, 1200 or 9600 bit/s for VHF/UHF). On the TNC 2, the radio-port data rate is set by means of a rear-panel DIP switch positions 6 through 9 while the TNC is powered off. (All other parameters are set while the TNC is powered on.) With some TNC 2 compatibles, you set the radio-port data rate by means of software, so, refer to the manual that accompanies your TNC if it does not use a rear-panel DIP switch for setting the radio-port data rate.

By default, the TNC data rate is 1200 bit/s because this is the data rate that is used on the most popular packet-radio subband: 2-meter FM. The FCC allows 1200-bit/s operation anywhere above 28 MHz, so you may operate on 10 meters or 23 centimeters without adjusting this timing parameter. If you wish to operate below 28 MHz, however, the maximum legal data rate is 300 bit/s; this means that an adjustment of the radio-port data rate is necessary.

Turnaround Time—As discussed earlier, when a transceiver switches from receive to transmit, there is a short delay after the switching begins and before intelligence can be sent over the air. Similarly, when a transceiver switches from transmit to receive, there is a short delay after the switching begins and before intelligence can be received. Although this delay is only measured in milliseconds (ms), it is critical in packet because a TNC is capable of enabling a transmitter and sending intelligence almost instantaneously. Therefore, some delay must be programmed into the TNC to be compatible with the receive-to-transmit turnaround time of your transceiver. This

---

Control Characters

Control characters, that is, characters that are typed while simultaneously pressing your DTE's control <CTRL> key, perform a number of useful functions. Some of the TNC default selections for these characters are the same as those that are commonly used by other DTEs. If your DTE is unable to generate a particular character or if you simply wish to select a different character to perform the same function, the following commands allow you to change the default character to another ASCII character of your choosing. (Note that the following list also includes control characters <CR>, <DELETE> and <I>, which are not used in conjunction with the <CTRL> key.)

Default Character	Command	Function
<CR>	SEndpac	Send previous keyboard input as a packet
<CTRL-C>	COMmand	Enter command mode (from converse mode)
<CTRL-Q>	STArt	Restart sending to DTE (from TNC)
<CTRL-Q>	XON	Restart sending to TNC (from DTE)
<CTRL-R>	REDisplay	Redisplay typed line before <CR>
<CTRL-S>	STOp	Stop sending to DTE (from TNC)
<CTRL-S>	XOFF	Stop sending to TNC (from DTE)
<CTRL-V>	PASs	Insert special character
<CTRL-X>	CANline	Cancel typed line before <CR>
<CTRL-Y>	CANPac	Cancel typed packet before <CR>
<DELETE>	DELete	Delete typed character
<I>	STReamsw	Change streams

delay may be varied from 0 to 1200 ms by typing:

cmd: TX delay n <CR>

where n is number 0 to 120 representing a delay in 10-ms increments [30 (300 ms) is the TNC default]. For example, to set the delay to 500 ms, set n to 50 (500 / 10 ms = 50 ms).

Your transceiver's turnaround time may be specified in your transceiver's manual. Typically, it is not specified, so you will have to experiment to find a delay that is suitable. In general, if your transceiver uses solid-state switching, you may set the delay relatively low. If your transceiver uses a mechanical relay for switching, you should set the delay higher. Also, synthesized transceivers require a greater delay than crystal-controlled transceivers. If you are using an external amplifier, you must take its switching delay into account, too.

Maximum Number of Unacknowledged Packets and Packet Length—The MAXframe and Paclen parameters are critical TNC parameters that you should adjust depending on the operating conditions (propagation and channel activity).

MAXframe selects the maximum number of outstanding unacknowledged packets the TNC will allow at any time. In other words, if MAXframe is set to 4, the TNC may send as many as four packets without receiving acknowledgments for any of them. Once the MAXframe limit is reached, however, the TNC will not send a new packet until one of the outstanding packets is acknowledged.

Paclen selects the maximum number of bytes of data in each packet. The TNC will never send a packet longer than the selected Paclen value. As data is entered from the DTE to the TNC, the TNC counts each byte of data. When the Paclen value is attained, the TNC makes up a packet containing the data, sends it over the air, and begins counting the number of bytes of data for the next packet. The TNC will send packets shorter than the selected Paclen value when it is specifically commanded to do so (whenever the SEndpac control character <CR> is entered).

The default values for the MAXframe and Paclen parameters (4 outstanding packets each 128 bytes long) are selected for good VHF operating conditions. When you are operating on HF or when VHF operating conditions are less than optimal (there is a high level of channel activity), the MAXframe and Paclen values should be reduced.

At 300 bit/s, it takes approximately four times as long to send the same packet as it does at 1200 bit/s. If you reduce the packet length by half, it still takes approximately twice as long to send a packet at 300 bit/s than at 1200 bit/s. This means that your packet is on the air approximately twice as long and the chances of encountering interference are twice as great! Add the fact that HF conditions are usually not optimal for packet operation, and you may end up with a lot of long unacknowledged packets hanging out there in the air. The best thing to do is to shorten the Paclen parameter and set the MAXframe parameter to 1 to force the TNC to send one short packet, wait for an acknowledgment, and then deal with the next short packet.

To change the value of the MAXframe parameter, type:

cmd: MAXframe n <CR>

where n is a number from 1 to 7 representing the maximum number of outstanding unacknowledged packets.

To change the value of the Paclen parameter, type:

cmd: Paclen n <CR>

where n is a number from 0 to 255 representing the maximum

number of bytes of data in each packet. Note that 0 actually represents 256 bytes.

Number of Packet Retries—When a TNC sends a packet, it waits a preset time for an acknowledgment that its packet was received without error. If the time limit is reached, the TNC again tries to obtain an acknowledgment from the receiving TNC by resending the packet. When the maximum number of retries is reached, the sending TNC enters the *disconnected* state. The FRack parameter sets the amount of time between retries, and the REtry parameter controls the number of allowable retransmission attempts.

The REtry and FRack parameters should be adjusted upward or downward, depending on the operating conditions. If conditions are good (good propagation and a low level of activity on the channel), the REtry and FRack parameters may be adjusted downward. If a packet cannot get through after one or two attempts under good conditions, there is probably an insurmountable problem with the link. For example, the intended receiving station may have gone off the air, so you might as well abandon the effort immediately.

If conditions are marginal (marginal propagation and/or a medium level of activity on the channel), it may only take a little longer to get the packet through to the intended receiving station, so the REtry and FRack parameters may be adjusted upward.

If conditions are poor (poor propagation and/or a high level of activity on the channel), wait until conditions improve. You already have two strikes against you: it is very difficult to get the packet through to the intended receiving station because of poor propagation, and your packets are likely to collide with other packets because of the crowded channel.

By default, the REtry parameter value is set to 10 retries, but that may be changed by typing:

cmd: REtry n <CR>

where n is a number from 0 to 15. Zero represents an infinite number of retries and 1 to 15 represents the maximum number of retries that will be attempted before the TNC stops repeating a packet. Note that the REtry parameter does not include the initial transmission of the packet; it only represents the number of retries after the initial packet transmission.

By default, the FRack parameter value is set to 3 seconds, but this may be changed by typing:

cmd: FRack n <CR>

where n is a number from 1 to 15, representing the number of seconds that the TNC will wait for a packet acknowledgment before it again tries to obtain an acknowledgment. Note that the TNC automatically adjusts this value higher depending on the number of digipeaters used in the selected path of the packet according to the formula:

$$FRack \times (2 \times dr + 1) = adjusted\ FRack$$

where dr is the number of digipeaters in the selected path. For example, if FRack is set to 4 seconds and 2 digipeaters are in the selected path, the adjusted FRack value is 20 seconds (4 sec × ((2 × 2 digipeaters) + 1) = 20 sec).

Digipeater Timing—When one station's packet collides with another station's packet, each must take steps to rectify the problem. When one station's packet collides with a digipeater's packet, the digipeater is not responsible for requesting a retransmission; the station that originated the packet must keep track of the outstanding packets. In cases where several digipeaters are used in a link, a packet may be lost by any one of them.

Digipeated packets are at a distinct disadvantage. To help counteract this, a timing parameter is included in the TNC to give digipeated packets a break. When a TNC is digipeating packets, it sends its packets as soon as the channel is clear of activity. If a TNC is originating its own packets, however, it waits a selected time period (set by the DWait command parameter value) after the channel is clear before it sends its packets. As a result, when a channel is clear, the digipeater always transmits first. This reduces the chance of digipeated packets colliding with non-digipeated TNC packets. The default setting of the DWait parameter is 16, which represents 160 ms. To change this timing parameter, type:

cmd: DWait n <CR>

where n is a number from 0 to 250 representing a delay in 10 ms increments. For example, to select a delay of 120 ms, set n to 12 (120 ms / 10 ms = 12). Note that DWait should be similarly applied in network node environments (in most areas, network nodes have replaced digipeaters).

To work effectively, all of the TNCs on a channel used by one or more digipeaters should use the same DWait value. Unless you know that the packet operators in your area are using a DWait value other than the default, you should leave your TNC set at the default value.

Prioritized Acknowledgment—TNCs that are compatible with TAPR TNC 2 software release 1.1.7 or later are capable of using the channel-sharing protocol called "prioritized acknowledgment." As its name implies, this protocol gives priority to packet acknowledgments (ACKs) on a channel.

An ACK indicates that a packet has been received correctly. By giving an ACK priority, the station that sent the packet will receive the ACK more quickly and, as a result, will be less likely to resend the packet. (On channels where ACKs are not given priority, ACKs may be delayed long enough to cause the waiting station to give up and resend the packet, thus wasting precious time on the channel.)

Configuring a TNC for prioritized acknowledgment involves setting the following parameters as indicated.

For 1200-bit/s VHF FM:

ACkprior	ON
ACKTime	14
DEAdtime	33
DWait	33
FRack	8
MAXframe	1 to 7
RESptime	0
SLots	3

For 300-bit/s HF:

ACkprior	ON
ACKTime	52
DEAdtime	8
DWait	8
FRack	16
MAXframe	1
Paclen	32 to 128
RESptime	0
SLots	3

These settings are recommended by TAPR as a starting point. Once you are on the air, you may have to adjust these settings as conditions warrant. Also, check your TNC's manual in case it recommends different settings.

Station Identification—Perhaps the most important TNC command is the one that inserts your call sign in the TNC. To program your TNC with your call sign, type:

cmd: MYcall WA1LOU <CR>

where WA1LOU is the call sign of your station.

An optional *secondary station identification* (SSID), number 1 through 15, may be appended to your call sign by typing a hyphen and the desired number immediately after the call sign in the Mycall command (for example, WA1LOU-4). SSIDs are used to differentiate digital repeaters (digipeaters), network nodes, PBBSs, and other secondary packet station operations from individual packet stations that use the same call sign. For example, the author's individual packet station is identified as WA1LOU (or WA1LOU-0 because 0 is assumed to be the SSID unless specified otherwise), whereas, his PBBS may be identified as WA1LOU-4 (to differentiate it from WA1LOU-0).

ON-THE-AIR

Once your TNC is tailored to your station and your mode of operation, you can send your first packet. The following paragraphs describe how to initiate and conduct packet communications from the keyboard of your DTE.

Connecting and Disconnecting

To contact another station you must make a connection with that station. To do so, type:

cmd: Connect WØRLI <CR>

where WØRLI is the call sign of the station to contact.

If WØRLI's packet station is on the air and receives your connect request, your stations will exchange packets to set up a connection between stations. When the connection is completed, your DTE will display:

*** CONNECTED to WØRLI

and your TNC automatically switches to the Converse Mode.

Now, everything you type into the DTE keyboard is "packeted" and sent to the other station. A packet is sent whenever you enter a carriage return <CR> (the TNC default for the Sendpac character), or whenever the byte length of your keyboard input equals the number of bytes selected with the Paclen command (the default is 128 bytes).

When you are finished conversing with the other station, return to the Command Mode by typing <CTRL-C> (which is the TNC default for the Command character). When the command prompt (cmd:) is displayed, type:

cmd: Disconne <CR>

and your station will exchange packets with the other station to break the connection. When the connection is broken, your DTE displays:

*** DISCONNECTED

If, in mid-contact, you wish to enter a command and then continue with the contact, enter the Command Mode by typing <CTRL-C>. When the command prompt (cmd:) is displayed, enter the desired command. When you are ready to return to your contact, exit the Command Mode and type:

cmd: CONVers <CR>

and you are back in the Converse Mode. Anything you type now is packeted and sent to the other station.

If, for some reason, the other station does not respond to your initial connect request, your TNC will resend the request.

It will continue to do so until the number of additional attempts equals the number selected with the Retry command (the default is 10 attempts).

When the number of additional attempts exceeds the Retry command selection, your TNC stops sending connect requests and your DTE displays:

*** retry count exceeded

*** DISCONNECTED

(Retry not only sets the maximum number of times that a connect request is resent, it also sets the number of times any packet is resent without acknowledgment.)

Monitoring Mode

A rule of thumb in Amateur Radio says that you should listen before you transmit. In support of this rule, your TNC has an array of commands for the monitoring mode.

The Monitor command causes the TNC to display on your DTE the contents of each packet received by your station while in the Command Mode. The display will include the call sign of the station originating each packet and the call sign of the station that is the intended recipient of each packet. A monitored packet will be displayed as:

WS1O>KU7G: DATA

where WS1O is the call sign of the station originating the packet, KU7G is the call sign of intended destination of the packet and DATA is the contents of the packet. (The Monitor function is on by default.) To enable (or disable) this function, type:

cmd: Monitor ON (or OFF) <CR>

The Mrpt command makes the monitoring function more revealing by displaying the call sign(s) of the station(s) that repeat a monitored packet, in addition to the call signs of the originating and destination stations. With Mrpt on, a monitored packet is displayed as:

WS1O>KU7G,KA6M-9*,WA8DED-5:DATA

---

where WS1O is the station originating the packet, KU7G is the intended destination of the packet, KA6M-9 and WA8DED-5 are the stations repeating the packet, and DATA is the packet's contents. The asterisk indicates that KA6M-9 is actually transmitting the packet you are receiving. (The Mrpt function is on by default.) To enable (or disable) this function, type:

cmd: MRpt ON (or OFF) <CR>

Other monitoring commands that may be useful are:

MAll (default: on) allows you to monitor both connected and unconnected packets;

MCOM (default: off) allows you to monitor connect requests, disconnect requests, connect/disconnect acknowledgments and non-connect/disconnect acknowledgments, as well, as packets containing data;

MCon (default: on) display packets from stations on frequency, while connected to another station;

MHeard lists the last 18 stations received on frequency by typing:

cmd: MHeard <CR>

MHClear clears the list of received stations that are logged for Mheard command recall by typing:

cmd: MHClear <CR>

MStamp (default: off) stamps the date and time of each monitored packet. To use this function, the date and time must be entered into the TNC using the *Daytime* command, by typing:

cmd: DAytime yymmddhhmm <CR>

where yy is the last two digits of the year (00-99), mm is two digits representing the month (00-12), dd is the day of the month (00-31), hh is the hour (00-23) and mm is the minute of the hour (00-59). (Note that your TNC has a 24-hour, rather than a 12-hour clock.) For example, to enter March 8, 1951, 1:30 PM, type:

cmd: 5103081330 <CR>

where 51 is the last two digits of 1951, 03 represents the third month (March), 08 represents the eighth day of the month, 13 represents the thirteenth hour of the day and 30 represents the thirtieth minute of the hour.

The *Budlist* and *Lcalls* commands may be used together to limit your monitoring. Use the Lcalls command to list a maximum of eight stations that you do or do not wish to monitor by typing:

cmd: LCAlls aaaaaa,bbbbbb...,hhhhhh <CR>

where aaaaaa, bbbbbb..., hhhhhh are the call signs of stations you do or do not wish to monitor.

Next, use the Budlist command to limit your monitoring according to the Lcalls list of stations. Your DTE will only display packets originating from stations in the Lcalls list if you type:

cmd: BUDlist ON <CR>

Your DTE will display packets of all stations except the stations in the Lcalls list (the TNC default). if you type:

cmd: BUDlist OFF <CR>

Most of your monitoring needs will be met with the wide range of TNC monitoring commands at your disposal.

VHF/UHF Operating

Today, most amateur packet activity occurs at VHF, on 2 meters to be specific. The most commonly used data rate is 1200 bit/s with frequency modulated AFSK mark and space tones of 1200 and 2200 Hz, respectively. This is referred to as the "Bell 202" telephone modem standard.

The majority of TNCs are optimized for VHF/UHF FM operation, so getting on the air is a simple matter of turning on your packet equipment and tuning to your favorite packet frequency.

Most activity hovers on or near 145.010 MHz including digipeater, network node, PBBS and direct connect contacts. (Refer to the sidebar entitled "Packet Frequencies" for information concerning other packet hot spots.) In some areas, an effort has been undertaken to establish band plans for packet. If there is a packet band plan in effect in your area, it is recommended that you comply with it. If there is no established band plan in your area, the following guidelines should be observed:

If you are conducting a direct connect contact (sans digipeater or network node), move your contact to an unused frequency. It is very inefficient to try to exchange packets on a frequency where other stations, especially digipeaters and network nodes, are also exchanging packets. The competition slows down your station's ability to exchange packets and, in turn, you are also slowing down all of the other stations. You should use a frequency occupied by a digipeater or network node only if you are using that repeater or node.

VHF/UHF packet operation is similar to VHF/UHF FM voice operation. Under normal propagation conditions, you may communicate only with stations that are in line of sight of your station. By using a repeater or node, the line-of-sight limitation is expanded. For example, if a repeater is within your line of sight, you may communicate with other stations that are also within line of sight of the same repeater.

Packet expands upon this by permitting the simultaneous use of eight digipeaters and an infinite number of nodes. Packets are relayed from one station to another station, even though the destination may be at the end of a long (and distant!) chain of nodes or digipeaters.

Besides digipeaters and nodes, some packet operations are conducted using voice repeaters. Such operations usually take place in the absence of voice activity and with the blessing of the sponsor of the voice repeater. It is imperative that you monitor the repeater before initiating packet-radio operation. One way to wear out your welcome quickly is to send a few packets in the middle of a voice conversation! To take full advantage of the voice repeater, you should set your TNC's *Axdelay* and *Axhang* parameters correctly and, when entering the Connect command, do not include Via and the call sign of the voice repeater.

Packet Networking

The packet network is a complex evolutionary system of packet stations that have been organized to transfer packets between local or distant points. Depending on the locations of those points, the packets being transferred between them may travel through a simple network consisting of one station operating on one frequency. They may also travel through a variety of networks consisting of tens of stations operating on HF, VHF and UHF.

In the past, if you wanted to connect to another packet station, you had to know the exact path through the network to make the connection. If one or more stations were in that path, you had to know (and list) each station's call sign and use them when you invoked the Connect command.

Can you imagine what would happen if you had to know the name of each intermediary post office for every letter you mailed? If you accidentally omitted a single name, your letter would not be delivered!

Luckily, the post office does not require that you be familiar with its vast network to mail a letter. Similarly, in many parts of the packet network, you do not have to be familiar with the network to make a connection with another station.

The PBBS automatic mail-forwarding system is part of the packet network. Sending a message via the PBBS automatic mail-forwarding system is similar to sending a letter through the postal system. In either case, you do not have to know who will be handling your mail before it is delivered. All you need to know is the identity of the intended recipient, and the address where the intended recipient picks up his or her mail.

For example, to send mail via the postal system, you address the envelope with the intended recipient's identification (the recipient's name) followed by the address where the recipient picks up his mail. To send mail via the PBBS automatic mail-forwarding system, you address the mail with the intended recipient's Amateur Radio identification (the recipient's call sign) followed by the at-sign (@) and the address where the recipient picks up packet mail. Packet operators usually receive their mail at their local PBBS, so the address for a packet message is simply the call sign of the destination PBBS. Once you have properly addressed your mail, the system (either postal or PBBS) does the rest, automatically forwarding your mail to its destination.

Digipeaters Are History!

Only a few years ago, packet networking was completely dependent on digipeaters. Digipeating is a function built into every AX.25-compatible TNC that permits the TNC to receive, temporarily store and retransmit (repeat) the packet transmissions of other stations. A digipeater repeats only transmissions that are specifically addressed for routing through that digipeater, as opposed to a typical voice repeater that retransmits everything it hears.

In light of the other packet networking options that are available today, digipeating is a very basic form of networking. Only eight digipeater stations can be used between any two points that are attempting to transfer packets. In order to use the digipeaters, the call sign of each must be known and specified when invoking the initial Connect command. Even though digipeating is rudimentary in comparison to what is available today, it still serves a purpose when an intermediary station or two is needed in a pinch to complete a connection.

When you are unable to make a connection with another station, the problem may be caused by terrain or propagation that prevents your signal from being received by the other station. The digipeater func-tion provides the ability to circumvent this problem. You only need to determine which on-the-air packet stations can send and receive signals between your station and the station you are trying to reach. Once the existence of an intermediary station is known, type:

cmd: Connect W4RI Via NK6K-5

where W4RI is the call sign of the station to connect and NK6K-5 is the call sign of the station that will act as the digipeater.

When NK6K-5 receives your connect request, it automatically enters the digipeater mode and stores your request in memory until the frequency is quiet. It then retransmits your request to W4RI on the same frequency. If W4RI's packet station is on the air and receives your connect request, your station and his will exchange packets through NK6K-5 to set up a connection between stations. Once the connection is established, your DTE displays:

*** CONNECTED to W4RI VIA NK6K-5

Your stations continue to use the facilities of the digipeater until the connection is broken.

Digital and voice repeaters both repeat, but the similarity ends there. A digipeater receives and transmits on the same frequency (whereas a voice repeater receives and transmits on different frequencies). As a result, a digipeater operates half-duplex; that is, it does not receive and transmit at the same time (as compared to a voice repeater, which transmits whatever it receives simultaneously).

If one station in the digipeater mode is insufficient to establish a connection, as many as eight stations can be called up on for digipeater operation in your connect request. Additional digipeaters are appended to the Connect command separated by commas. (For example, by typing "Connect W4RI Via NK6K-5,W3IWI-1" after the command prompt, your TNC will send the W4RI connect request to NK6K-5 which relays it to W3IWI-1. Then, W3IWI-1 relays it to W4RI.)

It is recommended that you not use more than one or two digipeaters at any one time because digipeater throughput decreases dramatically as the number of digipeaters increases. Each time you use a digipeater, you are competing with other stations using the same digipeater. Each competing station has the potential to generate a packet that may collide with your packet!

Any packet station can act as a digipeater. This occurs automatically without any intervention by the operator of the station being used as a digipeater. You do not need his permission, only his cooperation because he can disable his station's digipeater function by means of the Digipeat command. (In the spirit of Amateur Radio, most packet operators leave the digipeater func-

---

Packet Frequencies

On HF, most packet activity straddles the traditional RTTY subbands using 300-bit/s AFSK on LSB. The list of RTTY subbands in the United States follows.

1800-1830 kHz	14070-14099.5 kHz
3605-3645 kHz	18100-18110 kHz
7080-7100 kHz	21070-21100 kHz
10140-10150 kHz	28070-28150 kHz

Hotbeds of HF packet activity are 3606, 3630, 3642, 7093, 7097, 10145, 14101-14105, 21099-21105 and 28099-28105 kHz.

On 6 meters, 50.62 through 51.78 MHz is the packet subband with 51.70 MHz serving as the 6-meter packet calling frequency.

On 2 meters, 145.010 MHz (FM) is "the" packet frequency. Because of 145.010's popularity, packet activity has spread upward and downward to 144.910, 144.930, 144.950, 144.970, 144.990, 145.030, 145.050, 145.070, and 145.090 MHz as well as to unused simplex voice channels and on voice and RTTY repeaters.

On 222 MHz, 223.52 through 223.64 MHz is the packet subband. Another segment exists from 219 to 220 MHz, but this is restricted to network links only.

On 440 MHz, 430.05-431.025 and 440.975-441.075 MHz are the packet subbands. Above 450 MHz, there are packet subbands at 903-906, 915-918, 1248-1252, 1296 and 1297-1300 MHz.

---

tion enabled and only disable it under special circumstances.)

Similar to VHF/UHF voice repeaters, some stations are set up as dedicated digipeaters. They are usually set up in good radio locations by packet clubs. Aside from location, the other advantage of a dedicated digipeater is that it is always there (barring a calamity). Stations do not have to depend on the whims of other packet stations, which may or may not be on the air when their digipeater functions are needed most.

NET/ROM and TheNet

A network *node* is a relay station that allows you to connect to stations that are not connectable directly (a function similar to that of a digipeater). A node differs from the digipeater function because it is more "intelligent."

Say you want to connect to station X, who is five digipeater hops away (digipeaters A, B, C, D and E). To connect to X, you enter the command "Connect X via A, B, C, D, E." If A, B, C, D and E were network nodes instead of digipeaters, however, you would first connect with node A, then you would command node A to connect with node E. Once connected to node E, you would command node E to connect with station X.

If you use digipeaters, you have to know the path between you and station X (via A, B, C, D and E). If you use nodes, the nodes already "know" the path to other nodes. All you have to know is which node is local to station X. If a path changes (a node disappears or a new one appears), the node is aware of the change because nodes exchange information with each other regularly.

Another advantage of nodes is that their throughput is better than digipeaters. With digipeaters, only the last station in the circuit acknowledges received packets. With nodes, each node acknowledges each packet it receives. If a packet is lost in a chain of digipeaters, your TNC is not aware of that fact until it fails to receive an acknowledgment from the station at the opposite end of the circuit. If a packet is lost in a node sequence, the node that fails to receive an acknowledgment will retransmit the lost packet until the next node in the circuit acknowledges it. As you might guess, the recovery of lost packets is lightning fast.

Packet nodes were the brainchild of Ron Raikes, WA8DED, who released NET/ROM to the packet community in 1987. NET/ROM became very popular because it was inexpensive and easy to install (just replace one of your TNC's EPROMs with a NET/ROM EPROM). Since then, NET/ROM has begotten several children (some legitimate, some not) that now inhabit every nook and cranny of the packet world. *TheNet* is NET/ROM's best known offspring.

Many nodes use a mnemonic identifier instead of their FCC-given call signs. Typically, the identifier is a three- to six-character acronym that identifies the node's location. The reason for using identifiers is to simplify what the user has to remember. For example, instead of having to remember that XK1NG is the node located on Kong Island, all you have to remember is the node's mnemonic identifier, KONG.

To use a node, you first connect to your local node by entering a Connect command using your local node's call sign or mnemonic identifier. For example, to connect with the Kong Island node, type either:

cmd: Connect XK1NG <CR>

   or

cmd: Connect KONG <CR>

After you receive the connected message from your local

node, you may connect to another station that is also local to your node, or you may connect to another node.

To connect to another station, enter the Connect command followed by the call sign of the other station. To connect to another node, enter the Connect command followed by the call sign or mnemonic identifier of the other node. After you connect to the other node, use the Connect command again to connect to a station that is local to the other node. When you disconnect from another station, the node(s) disconnect automatically.

The Connect command used to connect to a node and the Connect command used to ask a node to connect to another station are different. When you connect to a node, you are using a TNC command. When you ask a node to connect to another station, you are using a node command. In order for a node to receive a node command, it must pass through your TNC as data (your TNC must not interpret a node command as if it were a TNC command). To ensure that this does not occur, do not send a node command *until you connect to a node* (after you receive the node connection message). If you send the node command too soon, the TNC will try to interpret it and, when it can't, will send you an error message. Note that after you connect to your local node, you will not receive another command prompt until you disconnect from the node, so don't wait for a command prompt before sending commands to a node.

Nodes will respond to other commands besides Connect. Here is a quick summary of other node commands.

CQ lets you call CQ through a local or distant node. When you send CQ through a node, you may send along a maximum of 77 characters of text to indicate who is calling CQ. (For example, "CQ—This is Bruce in downtown Gotham.")

Ident lists the node's mnemonic identifier.

Nodes lists the other known nodes by their mnemonic identifier, call sign and SSID. Nodes followed by the mnemonic identifier or call sign and SSID of a known node (for example, NODES K1WJ-1) lists routing information about that node.

Parms lists the settings of the node's parameters.

Routes lists instructions for routing to other nodes.

Users lists the call signs of who is using the node.

KA-Node

KA-Node is the Kantronics implementation of a node-to-node acknowledgment protocol. It is available in every Kantronics TNC (except the original KPC-1), and the Kantronics All Mode (KAM) controller. As explained in the description of NET/ROM and TheNet above, node-to-node acknowledgment provides improved throughput over the standard AX.25 end-to-end acknowledgment. Besides node-to-node acknowledgment, KA-Node allows you to gateway from one port to another when you are connected to a dual-port KA-Node (available in some Kantronics models).

Whereas KA-Node, NET/ROM and TheNet are similar in that they all offer node-to-node acknowledgment, they are dissimilar in other ways. The most important difference is that KA-Node does not perform automatic routing as does NET/ROM and TheNet. The user must command the KA-Node as to the desired path to another node or station. In this way, KA-Node is more like an AX.25 digipeater than a NET/ROM or TheNet node. Although a KA-Node does not achieve the same functionality of NET/ROM and TheNet, it is still an improvement over digipeating and, as a result, it is a popular packet tool.

ROSE

The *RATS Open System Environment* (ROSE) was devel-

oped over a number of years by The Radio Amateur Telecommunications Society (RATS) of New Jersey. The current elements of ROSE are:

ROSE X.25 Packet Switch—This is the X.25 Packet Networking solution for the TAPR TNC 2 and its clones. The ROSE Switch is based upon international standard protocols as defined by CCITT and ISO. It was written by Tom Moulton, W2VY.

ROSErver/PRMBS, the Packet MailBox System—ROSErver/PRMBS is a WØRLI-compatible PBBS with an advanced user interface including an Internet-compatible mailer. PRMBS provides the user and SysOp with many powerful features.

ROSErver/OCS On-line Callbook Server—OCS provides either a text-based or an efficient graphic user interface depending on the terminal capabilities of the user. Developed by Keith Sproul, WU2Z, and Mark Sproul, KB2ICI, this advanced packet server operates on a Macintosh computer.

ROSE/BBC Bulletin Broadcast Controller—BBC provides a mechanism for simultaneous reception of packet bulletins by several stations with complete data integrity and error correction. ROSE/BBC protocol was written by Gordon Beattie, N2DSY, and the software has been implemented for UNIX by Marsh Gosnell, AD2H.

ROSE/STS Station Traffic System—A complete NTS message management system, especially suited for packet, STS runs on MS-DOS and UNIX computers and is written by Frank Warren, KB4CYC.

Networks made up of ROSE X.25 Packet Switches provide amateurs with a vehicle through which they can easily establish packet connections with other stations. Connections through a ROSE Network can be thought of as a "reliable data pipe."

Using ROSE X.25 Packet Networks is similar to using the telephone network. Phones are easy to use. All one has to do is find a phone (the entry point into the network), dial the desired telephone number (the "network address" of the exit point from the network) and ask, by name, for the desired party (the person's "User ID"). A call from New York to California may go through Chicago, St. Louis or Atlanta, but all that matters to the caller is that it goes through automatically and reliably.

With ROSE Networks, users must know their entry point into the network (their local ROSE Switch), the exit point to reach the desired station (the ROSE address of the Switch used by that station) and the call sign of the desired station ("User ID"). Like the telephone network, the ROSE Network determines the path taken by the call.

ROSE Addresses follow the international standards used by commercial X.25 data networks. These addresses are based upon the telephone numbering plan used in each country. In North America, this is the telephone Area Code and Exchange. Since Area Codes and Exchanges are assigned geographically, ROSE addresses indicate a geographical area, too.

ROSE Networks are easy to use, as you will see in the following example.

In the ROSE Network, there is the KB2ICI-7 ROSErver/ OCS On-line Callbook Server at ROSE Address 609426 (central New Jersey) and there is a ROSE Switch in New York City, KB7UV-3. A user of the KB7UV-3 ROSE Switch can connect to the On-line Callbook Server by sending the following command to a TNC:

C KB2ICI-7 VIA KB7UV-3,609426

where KB7UV-3 is the call sign of the entry ROSE Switch and

609426 is the desired ROSE address. The user's TNC will first display the "connected to" message from the TNC, followed by a network message that acknowledges the call request:

*** CONNECTED to KB2ICI-7 VIA KB7UV-3,609426
  Call being Setup

An additional message will be sent by the network once the connection has been established:

Call Complete to KB2ICI-7 @ 3100609426

This message includes the call sign of the station connected to (KB2ICI-7) and the complete ROSE address of the switch that made the final link (3100609426). This "complete" address also includes country identification information ("3100" is the Amateur Radio Data Network Identification Code [DNIC] for the US).

At this point, there is a connection between the two stations through the network. Either station can end the connection by simply sending the Disconnect command to their TNC. The network will advise the other station of this by sending the message:

*** Call Clearing *** 0000 3100609426 Remote Station cleared connection

ROSE X.25 Networks also inform users of other conditions by means of similar messages (other station busy, and so on). Each of these messages, as indicated in the example above, includes a ROSE address that indicates the point in the network where the condition described in the message has occurred.

Help is always available to users of ROSE Networks. Each ROSE Switch has informative text stored in memory that may be obtained by connecting directly to the switch, then sending a carriage return or waiting 30 seconds. The switch will send you the text, which usually includes information on the network and basic instructions on switch operation.

ROSE Networks can also have directories of Services (PBBSs, ROSE Switches, etc.) and other information available to users (if implemented by the network managers). This is done through the ROSE INFO application and special ROSE addresses.

The RATS ROSE Network uses special addresses made up of the Area Code and either "555" or "411" for directories. The 555 directories are for services and 411 directories are for users. If a user of the N2DSY-3 ROSE Switch wants information on services available in Southern New Jersey (Area Code 609), the user sends the following command to a TNC:

C INFO VIA N2DSY-3, 609555

This is automatically routed to the switch containing the directory information. Once the call was completed, the directory text is sent to the user.

Optionally, ROSE Networks can have other useful applications. The most popular are HEARD and USERS. HEARD provides a list of stations heard by a given switch and USERS provides a list of switch activity. Both are used in the same way, that is, you initiate a connection to the desired application at the ROSE address of interest and, if the application is loaded at that switch, the requested information will be sent.

TCP/IP

TCP/IP for amateur packet is based on a networking scheme known as the Defense Advanced Research Project Agency's (DARPA) Transmission Control Protocol (TCP) and Internet Protocol (IP). Despite what the name "TCP/IP" implies, TCP and IP are only two parts of a collection or *suite* of protocols that comprise the complete DARPA protocol used for landline data communications networking between mainframe computers worldwide.

Phil Karn, KA9Q, wrote the amateur packet implementation of TCP/IP (for DOS-based computers) and called it *NET*. Later, he rewrote and refined the software and called it *NOS*. Phil graciously allowed his handiwork to be used freely for Amateur Radio (non-commercial) purposes.

Like the TNC-emulation software mentioned earlier, TCP/IP software emulates many of the functions of a TNC. As a result, your computer is no longer limited to the functions programmed into the TNC's read-only memory, and it may be programmed to do much more.

One of the main functions of the TCP/IP software is to allow each TCP/IP station to function as an intelligent network node. These nodes are "intelligent" because they automatically route packets to their intended destination without any operator intervention or direction. TCP/IP frees the operator of the task of figuring out which string of digipeaters, or local and remote network nodes, or alphanumeric addresses must be used to get packets delivered to their intended destinations. The operator simply commands the software to communicate with station X and the network does the work, automatically determining the route to get to the other station.

TCP/IP software improves data throughput on busy channels by using "intelligent backoff" techniques, rather than the pseudorandom method used by plain AX.25. TCP/IP senses when a frequency gets busy and waits longer between sending packets. As the channel's activity dies down, your station automatically starts sending more rapidly. This reduces collisions and retries and gets everyone's packets moving smoother and more efficiently.

Besides intelligent networking, the TCP/IP software has a packet-radio terminal function that allows you to communicate keyboard-to-keyboard with another packet operator. A file-transfer function, FTP, allows you to send and receive (upload and download) ASCII (text) and binary (executables, graphics, etc) files to and from any other TCP/IP station. The TCP/IP software includes a built-in mailbox that automatically sends and delivers mail between TCP/IP stations and "regular" PBBSs. What's more, TCP/IP software is multitasking on any machine. All of the aforementioned functions, and more, can be used simultaneously.

Here is what you need to get started with TCP/IP:

❑ An IP address, which is a unique number assigned to the computer used at your packet station for communications over the TCP/IP network. To get an IP address, contact your local IP address coordinator. An up-to-date list of IP address coordinators is available by ftp in the ham radio archives at **ftp.ucsd.edu**.

❑ Since the TCP/IP software emulates many of the functions of a TNC, you may have thought that it would allow you to get on packet without it. Well, you still need a TNC and it must be one that supports the *KISS* mode. The good news is that virtually all current TNCs support KISS and many of the older ones that don't support KISS can be made "KISSable" by updating the ROM containing the TNC software. KISS is the acronym for Keep It Simple, Stupid. When you invoke the KISS mode, it makes your TNC less intelligent. (If the software programmed in your TNC's ROM is compatible with TAPR TNC 2 software release 1.1.6 or later, then it is KISSable.)

❑ TCP/IP software. There are many versions of the software and there are many sources. Some of the sources are TAPR

(8987-309 E Tanque Verde Rd, #337, Tucson, AZ 85749-9399), ham-oriented landline BBSs, such as N8EMR's (telephone 614-895-2553) or KA1SVW's ChowdaNet (telephone 401-331-0907) and the Amateur Radio sections of commercial computer services such as CompuServe, GEnie and America Online.

Once you obtain these three items, you set up and use the software as described by its accompanying documentation. (A very useful book that details the setup and operation of NOS is *NOSintro* by Ian Wade, G3NRW. The book is available from the ARRL.)

After you have everything set, let's see if it works!

Before you run your TCP/IP program, you must place your TNC in the KISS mode. To do so, at the TNC's normal command prompt, type:

cmd: KISS ON <CR>

After your TNC is in the KISS mode, power off the TNC and then power it on. As your TNC initializes itself, its STA and CON front panel LEDs will blink three times to indicate that it is in the KISS mode. Some TNCs only require you to enter *RESTART* or *RESET*. You don't need to turn them on and off. Check your TNC manual to be sure.

Now, run your TCP/IP program and, after the system prompt appears (net>), type:

FINGER (insert your call sign) @ (insert your call sign)

The system will read and display the file that you created in your \FINGER subdirectory. (The FINGER file is a short description of you and your TCP/IP station.) Now it's time to perform a test to see if it will work on the air. ("FINGERing" yourself is an operation that occurs within the confines of your computer; nothing is sent over the air.)

Try to get a complete DOMAIN.TXT or HOSTS.NET file from another station—K1ZZ in this example. (DOMAIN.TXT or HOSTS.NET files are look-up tables of the TCP/IP stations that, at a minimum, list each station's call sign and IP address.) You can achieve this by using the File Transfer Protocol (or FTP). At the system prompt, type:

ftp k1zz

If all goes well, your computer indicates that a session has been "Established" and that "k1zz.ampr.org FTP" is "ready" for you to log on. After entering "anonymous," K1ZZ asks that you "Enter PASS command," so you enter pass <your call sign> in response. ("Pass" is short for password.)

The "Logged in" message should appear shortly indicating that you are in K1ZZ's *public* subdirectory. Send "dir" to get a list of the contents (a "directory") of the public subdirectory. If the directory contains a DOMAIN.TXT or HOSTS.NET file, you can get a copy by entering "get domain.txt" or "get hosts.net."

After entering the get command, your computer indicates that a data connection for retrieval of domain.txt or hosts.net is opening, then displays nothing while the file is actually being transferred. The file transfer can go on for quite a while. The keying of your transmitter is the only indication that anything is happening. When the transfer is completed, your computer displays "Get complete" followed by the number of bytes transferred and the "File sent ok" message. At the system prompt, enter close to end the FTP session and log off K1ZZ.

There are misconceptions about TCP/IP that have caused some potential TCP/IP users to steer clear of it. Let me assure you that the following *aren't* true:

- You must keep your radio equipment and computer on 24 hours a day.

  For the viability of the TCP/IP network (not to mention your local electric company), around-the-clock operation is preferable, but it's not a necessity as far as receiving mail is concerned. If your TCP/IP station is on the air all the time, it is always ready to receive traffic heading its way. If it is not active around the clock, however, other TCP/IP stations can intercept and hold your station's traffic for automatic delivery to you when your station is on.

- You must use a DOS-based computer.

  Although most computers using TCP/IP are DOS-based, versions of TCP/IP for other computers do exist. I use an Apple Macintosh, and there are versions for Commodore Amiga and Atari ST computers.

- You must use a fast computer with a lot of RAM and hard disk storage.

  Relatively slow computers, such as IBM PCs and Macintosh Pluses are used successfully in the amateur TCP/IP world. Fast clocks and oodles of memory and disk storage are nice, but you can TCP/IP without them.

- TCP/IP cuts you off from the rest of the packet world.

Many of the bulletins distributed in the AX.25 packet world also get distributed in the TCP/IP world. Moreover, TCP/IP bulletin distribution is better because, instead of wading through all the bulletins on your local PBBS trying to find something of interest to read, TCP/IP automatically routes only bulletins of interest to your TCP/IP mailbox. For instance, I'm

---

Organizations

Organizational periodicals, if any, follow the organization address.

The Amateur Radio Research and Development Corp. (AMRAD), PO Drawer 6148, McLean, VA 22106-6148, *The AMRAD Newsletter*

American Radio Relay League (ARRL), 225 Main St., Newington, CT 06111, *QST*

AMSAT-NA, The Radio Amateur Satellite Corp., 850 Sligo Ave., Silver Spring, MD 20910-4702, *The AMSAT Journal*

Chicago Area Packet Radio Association (CAPRA), PO Box 8251, Rolling Meadows, IL 60008, *The CAPRA Beacon*

Georgia Radio Amateur Packet Enthusiast Society (GRAPES), PO Box 871, Alpharetta, GA 30239-0871, *Grapevine*

Hamilton and Area Packet Network (HAPN), 5193 Whitechurch Rd., Mt. Hope, ON L0R 1W0, Canada

Northern California Packet Association (NCPA), PO Box 61716, Sunnyvale, CA 94088-1761, *NCPA Downlink*

Northwest Amateur Packet Radio Association (NAPRA), PO Box 70405, Bellevue, WA 98007, *Dedicated Link*

Ottawa Amateur Radio Club, Packet Working Group, PO Box 8873, Ottawa, ON K1G 3J2, Canada

Radio Amateur Telecommunications Society (RATS), 206 North Vivyen St., Bergenfield, NJ 07621

Southern California Digital Communications Council (SCDCC), PO Box 2744-1307, Huntington Beach, CA 92647, *I-Frame*

Tucson Amateur Packet Radio Corporation (TAPR), 8987-309 E. Tanque Verde Rd, Tucson, AZ 85749-9399, *Packet Status Register*

Texas Packet Radio Society, Inc., PO Box 50238, Denton, TX 76206-0238, *The TPRS Quarterly Report*

on the mailing list for Macintosh computer-related bulletins and messages and, as a result, all Mac-related bulletins and messages addressed to the Macintosh mailing list are automatically delivered to my mailbox. You can be on as many mailing lists as you desire and your mailbox will runneth over.

You aren't cut off from AX.25 mail forwarding, because many AX.25-to-TCP/IP gateway stations and PBBSs will forward your mail between the networks.

My station (IP address 44.88.4.8) has been TCP/IPing 24 hours a day for years with few problems. The S-meter panel light on my 2-meter rig is the only casualty I can recall. I admit to pulling the plug when thunderstorms pass through downtown Wolcott, or when I'm away for extended periods of time (like during my treks to Dayton). Other than that, my station is on all the time.

TexNet

The Texas Packet Society (TPRS) has developed a packet network called *TexNet*. It's composed of dual-port network-control processors (NCP) that provide AX.25-compatible, 1200-bit/s user-access on 2 meters and node-to-node linking at 9600 bit/s on 70-cm. This high-speed network is transparent to users, but results in a higher user data throughput than can be achieved over a 1200-bit/s network.

TexNet firmware can support 256 nodes within one network. Each node can provide several services including access to the TexNet network, a bulletin board, digipeaters, weather information and a conference bridge that provides a connected-mode round table service.

To use TexNet, you first connect with a local network node. You accomplish this by using the Connect command followed by the call sign of the local node and the appropriate SSID. Different SSIDs provide different services as delineated in the following list.

SSID	Service
0	Digipeating
2 and 3	Conference bridging
4	Network access
5	Local node terminal interfacing, also known as Local Node Console
6	Packet Message System or PMS (TexNet's PBBS service)
8 and 9	Cross-band digipeating

If you wish to use TexNet to communicate with another station, you access the network by using the Connect command followed by the call sign of the local node and the SSID of 4 (for example, Connect W5ABC-4). After you connect with a local node, you will receive the Network Cmd prompt. At the prompt, you can ask the network to set up communications with the other station by invoking the Connect command (with your TNC in the Converse Mode) followed by the call sign of the other station and the alias of that station's local node (Connect WA1LOU @ WOLCOTT). Local node alias' are usually based on city names. Your local node commences communications with the WOLCOTT node at 9600 bit/s to set up a connection with WA1LOU. At this speed, it will not be long before communication is established (assuming that WA1LOU is currently active on the WOLCOTT node's 2-meter frequency).

In addition to high-speed network communications, TexNet provides users with a variety of other useful services. You access these services by invoking the Connect command (again with your TNC in the Converse Mode) followed by the call sign of the node and the appropriate SSID. The following services

are accessible via TexNet.

Conference Bridging—Each TexNet node supports two independent local conference bridges (SSIDs 2 and 3). These bridges allow three or more stations at a local node to communicate with each other in roundtable fashion. (They do not provide conferencing between stations located at different nodes in the network.)

Packet Message System (PMS)—Typically, each TexNet network is supported by one PMS. One PMS per network eliminates message forwarding, which is one of the prime culprits that bogs down throughput in an AX.25 multi-PBBS network. Because there is only one PMS in the network, you can also access it from any node on the network by simply invoking the Message command at the Network Cmd prompt.

Local Node Console—This function allows you to access the node "directly," that is, by means of the node's controller: the TexNet Network Control Processor (NCP) board.

Digipeating—When all else fails, TexNet provides the old standby: standard AX.25 digipeating. Normally, you should avoid this function in favor of the network and conference bridging functions.

Cross-band Digipeating—If other channels are supported by the local node, this function allows you to access them.

The following summarizes the commands that allow you to use TexNet from its Network Cmd prompt.

Bye—disconnects you from a TexNet node.

Locations Served—lists all the TexNet nodes on the network that you can access.

Connect W5ABC @ Location—connects you to station W5ABC at the TexNet node whose alias is Location.

Connect % W5ABC @ Location—makes a connection across TexNet, where Location is the alias of the far end node. To access a secondary port, append a comma and the port number to Location (for example, Connect NØCCW @ ALAMO, 2). The percent sign (%) is used for automated stations that require a disconnect upon failure. (This command permits users to connect to anything that is not part of the network. Typically a node will permit connections with as many as two digipeaters on either side of a connection. When a user requests a connection, the call sign of the local node is used with "*** LINKED TO" followed by the call sign of the far end station.)

Connect CQ @ Location—sends CQ at the remote TexNet node whose alias is Location.

Statistics @ Location—lists information about activity over a 24-hour period at the TexNet node whose alias is Location.

Statistics Yesterday @ Location—lists statistics from the previous day for the TexNet node whose alias is Location.

Message—connects you to the PMS.

Message @ Location—connects you to a PMS system at the TexNet node whose alias is Location rather than to the network's default PMS.

Weather—connects you to the PMS designated as the weather server.

Route @ Location—lists the routing table of the TexNet node whose alias is Location.

VHF/UHF Operating Procedures

The packet network consists of an assortment of systems. TexNet has haciendas and the ROSE blooms in some garden states. TCP/IP has its defenders and plain vanilla digipeaters still perk along, while NET/ROM and its brood are everywhere. Meanwhile, a variety of bulletin boards continue to push megabytes of mail through this pipeline of dissimilar plumbing. It may seem hard to believe that it all works!

But, it could work better, especially those parts of the network that are burdened by user applications operating at slower data rates. For example, user-to-user and user-to-PBBS communications operate at 1200 bit/s when the user is typing on a keyboard at a typical rate of 25 words per minute, or approximately 23 bits/s. If everyone conformed to a few rules for operation on such a user-intensive network, the network could be utilized more efficiently.

Say No to Nodes

The first rule states that you shouldn't use the network unless it is absolutely necessary. If fewer people used the network, there would be less demand for its use, less contention between its users, and, as a result, it would operate more efficiently.

If you are connected to another station and you do not need a digipeater or network node to maintain the connection, move off the network frequency. Similarly, if you are a user who is connected to a PBBS and you do not need a digipeater or node to maintain that connection, then do not use a digipeater or node to conduct your communication. Freeing up a digipeater or node will promote increased network efficiency, especially if you can maintain the connection using reduced output power.

The converse of the first rule is the second rule, which states that a user should utilize the network when it provides the only means of maintaining a reliable connection with a "local" or "remote" user or a "local" PBBS.

Users and PBBSs are considered to be local if they can all access the same digipeater or node directly without using any intervening digipeater(s) or node(s). On the other hand, users or PBBSs are considered remote if their nearest digipeater or node is beyond the normal operating range of your station, or your local PBBS. In other words, to reach the remote station (or vice versa), intervening distant digipeaters or nodes would have to be employed to make the connection.

If you must use a digipeater or node to maintain a connection with a local user or PBBS, then, by all means, use the network facilities to maintain the connection. Similarly, it is permissible to use the network facilities to maintain a connection with a remote user. However, using the network to communicate with a remote PBBS leads us to the third rule: Do not attempt communication with a remote PBBS. To enforce this rule, many PBBSs are already configured to reject access by remote users; that is, the PBBS will not accept connect requests from users beyond a SysOp-selectable maximum number of intervening digipeaters or nodes. For example, if the maximum number is set to two, users who require three or more digipeaters or nodes to connect to that PBBS will have their connect requests rejected.

The purpose of the third rule is simply a matter of supply and demand. While accessing PBBSs is a very popular pursuit among packet operators, the supply of PBBSs is low relative to the demand. The number of PBBSs that can exist in any one locality is limited by the number of frequencies that are available and the amount of co-channel interference. (Co-channel interference results in a decrease in the network's overall efficiency.)

To solve this supply and demand problem, preferential treatment is given to local users of the PBBS especially when the PBBS is located in a well-populated area. On the other hand, if a PBBS is located in the "boonies," then there is no great supply and demand problem and remote users need not be excluded. In such a case, a remote PBBS is considered to be a "local" PBBS for a remote user.

You can send traffic to a remote PBBS without making a connection to the remote system itself. All you have to do is leave a message on your local PBBS addressed to the station at (or "@") the remote PBBS. The network will automatically forward the message to the remote PBBS as quickly as possible (see "Packet BBS Operating Procedures" for more information).

Incidentally, don't worry about missing something of interest that may be posted on a remote PBBS. General interest bulletins are usually sent to all of the PBBSs in the particular geographic area of interest, whether that area is a county, state, region, nation or planet.

HF Operating Procedures

HF packet is very different than VHF/UHF packet. A data rate of 300 bit/s is used rather than 1200 bit/s and the AFSK tones are not frequency modulated; LSB is used instead (1600 and 1800 Hz are commonly used for mark and space, respectively.)

The majority of TNCs are optimized for VHF/UHF FM operation, so getting on HF requires more effort than getting on VHF/UHF. In some TNCs, there are two modems, so switching between VHF/UHF and HF is painless. Other TNCs are not so versatile. The modems in these TNCs must be modified and recalibrated for HF operation or, TNC permitting, an external modem must be used that is compatible with HF packet operation.

While some TNC parameters may be set to whatever suits your particular VHF/UHF operation, the setting of these same parameters is much more critical for HF operation. HF packet is affected by interference, both atmospheric and man-made, and by the fickleness of propagation. Low signal-to-noise ratios also makes HF packet more difficult. Luckily, certain TNC parameters may be set to compensate for these conditions.

To increase throughput and decrease congestion on the frequency, your packets should be short. In addition, no more than one unacknowledged packet should be outstanding at any time. Use the Paclen and MAXframe commands, as described earlier in this chapter, to shorten the length of your packets and minimize the number of unacknowledged packets.

HF packet is adversely affected by spurious noise, which is considered by the TNC as a valid signal. This causes the TNC timing parameters to be reset each time such a "signal" is detected, resulting in a packet transmission delay for no legitimate reason. To avoid this, simply disable all of the timing parameters by setting DWait and TXdelay to zero, as described earlier in this chapter.

Changing the frequency of a VHF/UHF receiver is simply a matter of twisting the dial, but HF packet requires a steady hand at the controls. Tune your receiver very slowly (in 10 Hz increments, if possible) until your DTE begins displaying packets. When tuning, do not change frequency until a whole packet is received. If you shift frequency in mid-packet, that packet will not be received properly and will not be displayed on your DTE even if you were on the correct frequency before or after the frequency shift. Some TNCs have special indicators that make tuning easier.

The same rule of thumb that applies to VHF/UHF packet operation also applies to HF operation. Move to an unused frequency if there is other packet activity on the frequency you are presently using. Often, one frequency may be used as a calling frequency where stations transmit packets to attract the attention of other stations. Once a connection is established, the stations often move to another frequency where there is less activity. This clears the calling frequency and increases their own stations' throughput.

Space Communications

Packet is a space-age mode of communication and, befitting such a mode, you can find packet in space as well as on land. Packet in space has taken place aboard the American space shuttles, the Russian space station and Amateur Radio satellites.

Space Shuttles and Stations

Shuttle Amateur Radio EXperiments (SAREX) have been conducted by astronauts holding Amateur Radio licenses on a number of American space shuttle missions. Typical shuttle operations feature unique robot software that allow the maximum number of stations to contact the shuttle via packet. This software recognizes a terrestrial connect request, sends a sequential QSO number to the station, summarily disconnects the station and then logs the contact. Beacons listing successful connections are transmitted intermittently.

Russian cosmonauts often conducted live operator-to-operator packet contacts using an Amateur Radio station aboard the space station *Mir*. When a cosmonaut was not on hand to operate, the station is switched into the PBBS mode to receive messages and news from Earth.

No special packet equipment is required to contact the American shuttle packet stations or the *Mir* space station. Both use standard TNCs operating at 1200 bit/s. Shuttle operations usually occur on split 2-meter frequencies and a simple Connect command with the appropriate call sign on the right frequency is all that is needed to initiate a connection with the manned space vehicle. The frequencies are published prior to launch. The *Mir* space station can be found on 145.55 MHz simplex.

Amateur Satellites

After years of packet experimentation in space via manmade satellites, the groundwork has been laid for a packet network that relies on TNCs in orbit around the Earth. There are many packet satellites in operation (with more to come) and some of these are being configured to be integral parts of the worldwide packet network. The typical operating mode of a packet satellite is the digital store-and-forward mode (a ground station uploads data to the satellite and the satellite stores the data until it can be relayed to another ground station at a later time).

The satellites use a variety of modulation schemes and data rates with uplinks and downlinks on different frequency bands. As a result, standard TNCs must be outfitted with different modems in order to use some of these satellites. Radio equipment for various VHF/UHF bands must be available, too. With a properly equipped station, using a satellite's store-and-forward mode is similar to using a land-based PBBS.

Refer to Chapter 13 ("Satellites") for more information.

APPLICATIONS

Now that we have this high-speed, error-free mode of communications up and running, what can we do with it? The answer is "plenty." The following paragraphs describe some of the more popular pursuits that have been applied to the packet mode.

Packet BBS Operating Procedures

In 1982 there was a warehouse in Texas full of surplus Xerox 820-I computers. The "820" was a single board computer using the Zilog Z80 central processing unit (CPU). It ran under the CP/M operating system and featured 64-kBytes of RAM, one parallel port, two serial ports, a disk controller, and an 80 × 24 video display format. As surplus, the computer could be purchased for as little as $50!

Hank Oredson, WØRLI, became the father of PBBS when he decided to put this bargain computer to good use by writing software that permitted it to function as a BBS; not a telephone BBS, but a packet BBS. In no time, WØRLI PBBSs began appearing throughout packet country, not solely because the 820 was inexpensive, but also because Hank's software worked well.

PBBSs became so popular that other hams wrote PBBS software for their favorite computers. Jeff Jacobsen, WA7MBL, wrote the first PBBS software for the most popular computer in Amateur Radio, the IBM PC, and when WØRLI's code began outgrowing the Xerox 820-I, Hank rewrote his software for the IBM PC as well. Today, the IBM PC has a wide variety of PBBS software available for it.

So, what does a PBBS do? Its mailbox function allows you to post mail on the system for later retrieval by the addressee. You can also retrieve mail that's addressed to you. The mailbox does not limit your message posting to stations that frequent the local PBBS. If you know a station that checks into a *distant* PBBS, you can post a message to that station and it will be routed automatically to its destination.

Besides storing individual messages, files of interest to the general packet population may be stored as well. Here you will find packet network maps, programs, ARRL bulletins, newsletters and so on.

To use a PBBS, you must find one. On HF, a number of PBBSs are active in the packet subbands at various times of the day (refer to the sidebar "Packet Frequencies"). On VHF/UHF, there is probably at least one active PBBS on each packet channel.

The accompanying sidebar "WØRLI Mailbox Command Set" describes the commands of WØRLI's fabled software, and the PBBS procedures described below are based on this command set as well. If your local PBBS is not a WØRLI Mailbox, the sidebar and following description will still be applicable because most PBBSs use similar, if not identical, commands for their basic operations.

Logging On

Once you locate a PBBS, you must "log on" in order to use it. Logging onto a PBBS is as simple as initiating a contact with any other packet station. Just use the Connect command. For example:

cmd: Connect K8KA-4 <CR>

where K8KA-4 is the call sign and SSID of the PBBS. After you are connected to the PBBS, the PBBS preamble is displayed on your DTE followed by a request for commands from you.

Logging Off

Whenever you are finished using a PBBS, you log off the system by using the B (for "bye") command. When the PBBS receives the B command, it logs you off the system and disconnects.

Reading the Mail

In order to read messages that are posted on the PBBS, you must know what messages are available. If a message stored on the PBBS is addressed to you, you are informed of that fact when you log onto the system. By using the RM (for "read mine") command, you can read whatever mail is waiting for you. When the PBBS receives the RM command, it retrieves each message and sends it to your DTE. After you have read your messages, use the KM (for "kill mine") command to delete all of the messages you have read. (Note that you can only "kill" messages originated by you or addressed to you.)

WØRLI Mailbox Command Set

The following commands are available with version 10.11 of the WØRLI Mailbox public-domain software.

General commands:

B	Log off PBBS.
J*x*	Display call signs of stations recently heard or connected on TNC port *x*.
N *x*	Enter your name (x) in system (12 characters maximum).
NE	Toggle between short and extended command menu.
NH *x*	Enter the call sign (*x*) of the PBBS where you normally send and receive mail.
NQ *x*	Enter your location (*x*).
NZ *n*	Enter your ZIP Code (*n*).
P *x*	Display information concerning station whose call sign is *x*.
S	Display PBBS status.
T	Ring bell at the SYSOP's DTE for one minute.

Information commands:

? *	Display description of all PBBS commands.
?	Display summary of all PBBS commands.
? *x*	Display summary of command *x*.
H *	Display description of all PBBS commands.
H	Display summary of all PBBS commands.
H *x*	Display description of command *x*.
I	Display information about PBBS.
I *x*	Display information about station whose call sign is *x*.
IL	Display list of local users of the PBBS.
IZ *n*	List users at ZIP Code *n*.
V	Display PBBS software version.

Message commands:

K *n*	Kill message numbered *n*.
KM	Kill all messages addressed to you that you have read.
KT *n*	Kill NTS traffic numbered *n*.
L	List all messages entered since you last logged on PBBS.
L *n*	List message numbered *n* and messages numbered higher than *n*.
L< *x*	List messages from station whose call sign is *x*.
L> *x*	List messages addressed to station whose call sign is *x*.
L@ *x*	List messages addressed for forwarding to PBBS whose call sign is *x*.
L *n1 n2*	List messages numbered *n1* through *n2*.
LA *n*	List the first *n* messages stored on PBBS.
LB	List all bulletin messages.
LF	List all messages that have been forwarded.
LL *n*	List the last *n* messages stored on PBBS.
LM	List all messages addressed to you.
LT	List all NTS traffic.
R *n*	Read message numbered *n*.

RH *n*	Read message numbered *n* with full message header displayed.
RM	Read all messages addressed to you that you have not read.
S *x* @ *y*	Send a message to station whose call sign is *x* at PBBS whose call sign is *y*.
S *x*	Send message to station whose call sign is *x* at this PBBS.
SB *x*	Send a bulletin message to *x* at this PBBS.
SB *x* @ *y*	Send a bulletin message to *x* at PBBS whose call sign is *y*.
SP *x* @ *y*	Send a private message to station whose call sign is *x* at PBBS whose call sign is *y*.
SP *x*	Send a private message to station whose call sign is *x* at this PBBS.
SR	Send a message in response to a message you have just read.
ST *x* @ *y*	Send an NTS message to station whose call sign is *x* at PBBS whose call sign is *y*.
ST *x*	Send an NTS message to station whose call sign is *x* at this PBBS.

File transfer commands:

D*x* *y*	From directory named *x*, download file named *y*.
U *x*	Upload file named *x*.
W	List what directories are available.
W*x*	List what files are available in directory named *x*.
W*x* *y*	List files in directory named *x* whose file name matches *y*.

Port commands:

C *x y*	Via port *x*, send connect request to station whose call sign is *y*.
C *x*	Send data via port *x*.
CM *x y*	Send message numbered *x* to station whose call sign is *y*.
CM *x y* @ *z*	Send message numbered *x* to station whose call sign is *y* at PBBS whose call sign is *z*.
M*x*	Monitor port *x*.

Roundtable commands:

RT		Initiate roundtable function.
<ESC>	D *x*	Allows roundtable control station to disconnect station from roundtable whose call sign is *x*.
<ESC>	H	Obtain assistance.
<ESC>	P	Display ports available to roundtable.
<ESC>	N *x*	Enter your name (*x*).
<ESC>	Q *x*	Enter your location (*x*).
<ESC>	U	Display list of stations in roundtable.

To read other messages that are on the PBBS, use the L (for "list") command to obtain a list of all of the messages that have been stored on the PBBS since the last time you logged onto the system. The PBBS lists each message by its message number. If you wish to read a particular message, use the R (for "read") command by typing:

R n <CR>

where n is the number of the message to be read.

Sending Mail

To send a message by means of a PBBS, use the SP (for "send personal") command. When the PBBS receives the command, it asks you for the call sign of the addressee. To post the message for retrieval on the local PBBS, simply type the call sign of the station to receive the message followed by <CR>. If you want to have your message forwarded to a station at another PBBS, type the call sign of the station to receive the message, followed by the at-sign (@), the call sign of the des-

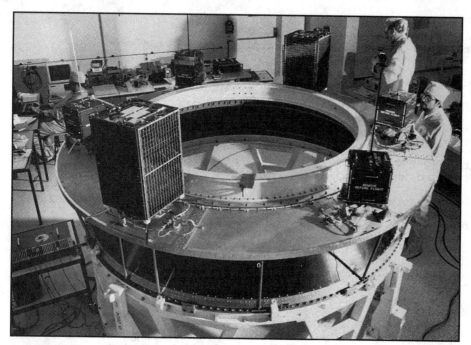

Four MicroSATs and two UoSATs mounted on the launch platform of the Ariane rocket that propelled them into space. *(photo courtesy of Joe Kasser)*

Musa Manarov, UV3AM, was active on packet from the Soviet *Mir* space station (U2MIR) on 145.55 MHz. He is shown here preparing for a mission at Moscow club station UK3R. *(UW3AX photo via WA2LQQ)*

@ KV7D).

To assist the mail-forwarding network, you can append geographic information to the message's address. For example, to give the message addressed to K9NG @ KV7D a little push, you can append .AZ.US.NA to KV7D, which results in an address of K9NG @ KV7D.AZ.US.NA. (AZ, US and NA are the abbreviations for the state, country and continent, respectively, in which KV7D is located.) This is known as *hierarchical addressing.*

Next, the PBBS asks you for the title of the message. Type a short one- or two-word title that represents the contents of the message followed by <CR>. Next, type the actual contents of the message. After you have typed the last line of the message, type a <CTRL-Z> and <CR> to indicate the end of the message. When the PBBS receives the <CTRL-Z>, it stores the message for later retrieval or forwarding. To learn more about sending mail and bulletins via the packet network, pick up a copy of *Your Packet*

tination PBBS and <CR>. For example:

K9NG @ KV7D <CR>

where K9NG is the call sign of the station that is to receive the message and KV7D is the call sign of the destination PBBS where the message will be sent for retrieval). To save a step, you can address the message at the same time you invoke the SP command (for example, SP K9NG

Companion by Steve Ford, WB8IMY, at your favorite dealer. This book is also available from ARRL Headquarters directly.

PBBS Procedures

Those are the basic commands for using a PBBS. A few basic operating rules need to be mentioned also.

If the PBBS does not respond to a command immediately, be patient and do not resend it. One of the unique features of packet is that whatever you send is received perfectly at the other end. To achieve this result sometimes takes a number of attempts, especially if there is a lot of activity on frequency. If you send the same command twice, the PBBS will eventually receive it twice and will respond twice! If the response is a long one, your repeated command will waste valuable time.

Do not perform lengthy operations during prime operating hours. Use good judgment before deciding to perform a time-consuming task. To save time, you can send more than one command at a time by preceding the <CR> that follows each command with the pass character. <CTRL-V> is the default pass character which, in this case, prevents each <CR> from causing the commands to be sent individually. For example, to send the RM, KM and L commands at one time, you would type:

RM <CTRL-V> <CR> KM <CTRL-V> <CR> L <CR>

The last <CR> is not preceded by a <CTRL-V> because you want the last <CR> to force the transmission of this complete packet. The <CTRL-V>s preceding the other <CR>s prevent those <CR>s from forcing transmission of each command individually. Without the <CTRL-V>s, RM would be sent in one packet, KM would be sent as a second packet and L would be sent as a third packet.

Public Service Communications

Packet has become integral in serving the public by providing communications during emergencies and disasters. With portable computers, TNCs and radio equipment, packet stations have been rapidly set up at disaster sites to get all-important messages on the air and to their destinations quickly and accurately. Accuracy is very important in emergency communications. When a packet station sends NEED A SURGEON, you can be sure that the station receiving the message will not have a fish delivered to the originating station!

Since each packet station is a self-contained digipeater, packet networks can be set up quickly in times of emergencies. The value of packet for emergency communications has been recognized by various public service agencies and some have allocated funds towards the purchase of packet equipment to be available to hams in times of emergencies.

Hams have worked hard to establish a means of efficiently handling traffic through the packet network. As more and more packet stations appear on the air each day, the packet network expands. Each new station has the potential of being an outlet for emergency traffic. The mail-forwarding function of the PBBS network is a key part of the packet traffic system. This system is constantly being refined to handle greater volumes of emergency traffic whenever the need arises.

DX *PacketCluster*

The DX community has a tendency to share information about the latest and greatest DX. That is the reason why there are so many successful Amateur Radio newsletters devoted to spreading the word about DX. The problem with newsletters is that their news is not always current. DXpedition plans may change at the last minute or a new country may pop on the air without warning.

DX spotting helps alleviate this problem. When someone spots new DX, he or she shares that fact by making an announcement on the local DX spotting frequency. During the 1970s and 1980s, the DX spotting frequency was typically a 2-meter FM voice repeater. Packet and DX *PacketCluster* software have changed all that forever and DX spotting has become one of the most popular applications of packet communication.

A DX *PacketCluster* consists of specialized software running on a computer attached to a TNC and a transceiver. The software allows multiple stations to connect to the *PacketCluster* station or "node." It also allows each *PacketCluster* node to connect to other nodes to form a *PacketCluster* network. When one station connects to a node and makes an announcement, all stations connected to that node and all the nodes in the network receive the same announcement.

For example, let's say that K3NA announces that 9H3JR is operating on 7005.0 kHz. If K3NA connects to a *PacketCluster* node that is part of your local *PacketCluster* network, you will be able to copy his announcement even if you cannot copy K3NA directly.

If you wanted to talk with K3NA, the *PacketCluster* permits you to do that, too. What if you wanted to know K3NA's location? You can use the *PacketCluster* to ask K3NA exactly where he is located. Besides conversing with any station on the *PacketCluster* network, users may conduct conferences (discussions involving more than two people).

Another function of the *PacketCluster* is its bulletin board system that operates in a fashion similar to a standard PBBS. Like other PBBSs, the *PacketCluster* system tells you when you have new mail. To read mail, or any message or bulletin, you use the Read command. To post a message or bulletin of your own, you use the Send command. To obtain a list of the messages and bulletins that are on the board, you use the Directory command instead of the PBBS List command, and to remove a message or bulletin from the board, you use the Delete command instead of the PBBS Kill command. The other major difference between the *PacketCluster* and WØRLI-compatible systems is that the *PacketCluster* BBS does not support mail forwarding. (The sidebar "*DX PacketCluster* Command Set" describes all of the *PacketCluster* commands.)

Perhaps, the most powerful command in the *PacketCluster* command set is the Show command (SH for short). Invoke the Show DX (SH/DX) command and the *PacketCluster* will list the last five DX announcements including the DX station's operating frequency, call sign, the time and date of the announcement, other pertinent information concerning the DX station (long path, LOUD, weak and so on). and the call sign of the station that made the announcement. If you wish to limit the Show DX command to one band, you may do so. For example, "SH/DX 10" causes the *PacketCluster* to recount the last five DX announcements for 10 meters only. You also may limit the Show DX command to one country. For example, "SH/DX SP*" will cause the *PacketCluster* to list the last five announcements for Polish stations only.

When V51NAM is at the bottom of a pileup on 20 meters, you can use the *PacketCluster* to get some ammunition before you try to work him. Assuming you have already entered your longitude and latitude in the *PacketCluster*'s database (using the SET/L command), invoke the Show Heading command for Namibia (SH/HEADING V51). The *PacketCluster* will perform some calculations and send results that will look something like this:

> V51 Namibia: 107 degs
> Q dist: 7238 mi, 11649 km
> Reciprocal heading: 309 degs

Now you can use the Show Sun command (SH/SUN V51) and the *PacketCluster* will respond with:

V51 Namibia Sunrise: 0455Z Sunset: 1703Z

The Show Sun command is also handy for finding out your local sunrise and sunset. Simply invoke the command without appending a prefix (SH/SUN) and the *PacketCluster* will send:

<your call> QTH Q Sunrise: 1055Z Sunset: 2303Z

Finally, try using the Show MUF command (SH/MUF V51) and the *PacketCluster* will respond:

Namibia propagation: MUF: 22.7 MHz LUF: 2.3 MHz

If you work V51NAM, you could invoke the Show QSL (SH/QSL) command to obtain QSL information for that station. After invoking the command, the *PacketCluster* would search its database for the QSL route for V51NAM.

Other variations of the Show command include the Show WWV command to obtain WWV propagation information and the Show Users command to obtain a list of all the other stations connected to the *PacketCluster* node or network. The Show command also may access any database installed on your local *PacketCluster*. Such databases may include the contest or DX club membership rosters, contest information, DX news, FCC rules and regulations, IRC data, QSL bureau addresses and so on.

As you can see, the *PacketCluster* is a powerful tool. Although similar tools existed in the past, only packet could make it as powerful as it is today. Faster packets will only make it more powerful in the future.

Automatic Packet Reporting System

Long-time packeteer Bob Bruninga, WB4APR, developed the Automatic Packet Reporting System (APRS), which allows packet radio to track real-time events. It deviates markedly from message- and text-transfer applications and concentrates instead on the *graphic* display of station and object locations and movements. By running APRS software, you see the locations of other APRS stations on a computer-generated map.

APRS is the result of Bob's experience using packet for real-time communications and support for public-service events. Packet has great potential, but so far, has been used mostly for passing large volumes of message traffic from point-to-point or into the national traffic-distribution system. It has been difficult to apply packet to real-time events where information has a short life span. This is because several steps are involved in preparing and passing messages, including decisions about routing and connectivity.

Bob developed APRS based on his observation that operators at most events, emergencies, exercises, weather nets and general communications spend more time concerned with where things (people, stations and so on) are—and where they are going—than any other category. Regardless of whether APRS is the best solution, it demonstrates the potential for

DX Packet Cluster Command Set

The following commands are available with Pavillion Software's DX PacketCluster software:

ANNOUNCE	Make an announcement.
A *x*	Send message *x* to all stations connected to the local node.
A/F *x*	Send message *x* to all stations connected to the cluster.
A/*x y*	Send message *y* to stations connected to node *x*.
A/*x y*	Send message *y* to stations on distribution list *x*.
BYE	Disconnect from cluster.
B	Disconnect from cluster.
CONFERENCE	Enter the conference mode on the local node.
CONFER	Enter the conference mode on the local node. Send <CTRL-Z> or /EXIT to terminate conference mode.
CONFER/ F	Enter the conference mode on the cluster. Send <CTRL-Z> or /EXIT to terminate conference mode.
DELETE	Delete a message.
DE	Delete last message you read.
DE *n*	Delete message numbered *n*.
DIRECTORY	List active messages on local node.
DIR/ALL	List all active messages on local node.
DIR/BULLETIN	List active messages addressed to ''all.''
DIR/*n*	List the *n* most recent active messages.
DIR/NEW	List active messages added since you last invoked the DIR command.
DIR/OWN	List active messages addressed from or to you.
DX	Announce DX station.
DX *x y z*	Announce DX station whose call sign is *x* on frequency *y* followed by comment *z*, e.g., DX SP1N 14.205 up 2.
DX/*a x y z*	Announce DX station whose call sign is *x* on frequency *y* followed by comment *z* with credit given to station whose call sign is *a*, e.g., DX/K1CC SP1N 14.205 up 2.
FINDFILE	Find file.
FI *x*	Ask the node to find file named *x*.
HELP or ?	Display a summary of all commands.
HELP *x*	Display help for command *x*.
READ	Read message.
R	Read oldest message not read by you.
R *n*	Read message numbered *n*.
R/*x y*	Read file named *y* stored in file area named *x*.
REPLY	Reply to the last message read by you.
REP	Reply to the last message read by you.
REP/D	Reply to and delete the last message read by you.
SEND	Send a message.
S/P	Send a private message.
S/NOP	Send a public message.
SET	Set user-specific parameters.
SE/A	Indicate that your DTE is ANSI-compatible.
SE/A/ALT	Indicate that your DTE is reverse video ANSI-compatible.
SE/H	Indicate that you are in your radio shack.
SE/L *a b c d e f*	Set your station's latitude as *a* degrees *b* minutes *c* north or south and longitude *d* degrees *e* minutes *f* east or west, e.g., SE/L 41 33 N 73 0 W.
SE/N *x*	Set your name as *x*.
SE/NEED *x*	Store in database that you need country(s) whose prefix(s) is *x* on CW and SSB, e.g., SE/NEED XX9.
SE/NEED/BAND = (*x*) *y*	Store in database that on frequency band(s) *x*, you need country(s) whose prefix(s) is *y*, e.g., SE/NEED/BAND = (10) YA.
SE/NEED/*x y*	Store in database that in mode *x* (where *x* equals CW, SSB or RTTY), you need country(s) whose prefix(s) is *y*, e.g., SE/NEED/RTTY YA.
SE/NEED/*x*/BAND = (*y*)*z*	Store in database that in mode *x* (where *x* equals CW, SSB or RTTY) on frequency band(s) *y*, you need country(s) whose prefix(s) is *z*, e.g., SE/NEED/RTTY/BAND = (10) ZS9.
SE/NOA	Indicate that your DTE is not ANSI-compatible.
SE/NOH	Indicate that you are not in your shack.
SE/Q *x*	Set your QTH as location *x*.
SHOW	Display requested information.
SH/A	Display names of files in archive file area.
SH/B	Display names of files in bulletin file area.
SH/C	Display physical configuration of cluster.
SH/C *x*	Display station connected to node whose call sign is *x*.
SH/CL	Display names of nodes in clusters, number of local users, number of total users and highest number of connected stations.
SH/COM	Display available Show commands.
SH/DX	Display the last five DX announcements.
SH/DX *x*	Display the last five DX announcements for frequency band *x*.
SH/DX/*n*	Display the last *n* DX announcements.
SH/DX/*n x*	Display the last *n* DX announcements for frequency band *x*.
SH/FI	Display names of files in general files area.
SH/FO	Display mail-forwarding database.
SH/H *x*	Display heading and distance to country whose prefix is *x*.
SH/I	Display status of inactivity function and inactivity timer value.
SH/LOC	Display your station's longitude and latitude.
SH/LOC *x*	Display the longitude and latitude of station whose call sign is *x*.
SH/LOG	Display last five entries in cluster's log.
SH/LOG *n*	Display last *n* entries in cluster's log.
SH/M *x*	Display MUF for country whose prefix is *x*.
SH/NE *x*	Display needed countries for station whose call sign is *x*.
SH/NE *x*	Display stations needing country whose prefix is *x*.
SH/NE/*x*	Display needed countries for mode *x* where *x* equals CW, SSB or RTTY.
SH/NO	Display system notice.
SH/P *x*	Display prefix(s) starting with letter(s) *x*.
SH/S *x*	Display sunrise and sunset times for country whose prefix is *x*.
SH/U	Display call signs of stations connected to the cluster.
SH/V	Display version of the cluster software.
SH/W	Display last five WWV propagation announcements.
TALK	Talk to another station.
T *x*	Talk to station whose call sign is *x*. Send <CTRL-Z> to terminate talk mode.
T *x y*	Send one-line message *y* to station whose call sign is *x*.
TYPE	Display a file.
TY/*x y*	Display file named *y* stored in file area named *x*.
TY/*x*/*n y*	Display *n* lines of file named *y* stored in file area named *x*.
UPDATE	Update a custom database.
UPDATE/*x*	Update the database named *x*.
UPDATE/*x*/APPEND	Add text to your entry in the database named *x*.
UPLOAD	Upload a file.
UP *x*	Upload a file named *x*.
UP/B *x*	Upload a bulletin named *x*.
UP/F *x*	Upload a file named *x*.
WWV	Announce and log WWV propagation information
W SF = *xxx*, A = *yy*, K = *zz*,*a*	Announce and log WWV propagation information where *xxx* is the solar flux, *yy* is the A-index, *zz* is the K-index and *a* is the forecast.

tactical communications via packet. Furthermore, it attempts to define a standard format using unconnected information (UI) frames for object positioning and reporting, which could be universally applicable in Amateur Radio. (UI frames can be thought of as packet "broadcasts." The information is sent to no one in particular and can be received by anyone.)

APRS avoids the complexity and limitations of trying to maintain a connected network. It permits any number of stations to participate and exchange data, just as voice users do on a phone net. Any station that has information to contribute simply transmits it and all stations receive it and log it.

Although APRS' mapping capability was developed to display the movement of hand-held global-positioning satellite (GPS) navigation devices interfaced to TNCs, most features evolved from earlier efforts to support real-time packet communications at special events. Any person in the network, upon determining where an object is located, can move his cursor and mark the object on his map screen. This action is then transmitted to all screens in the network, so everyone gains, at a glance, the combined knowledge of all network participants. Furthermore, the map screen retains this information for future reference. This means that moving objects can be dead-reckoned to their current locations with one keystroke—based on their previous positions—and this can be accomplished without using a single GPS device.

The availability of GPS receiver *cards* is icing on the cake. With a GPS card, TNC and hand-held transceiver stuffed in a cigar box, almost any object can be tracked by packet stations running APRS software. For example, such cigar-box technology was installed in a football helmet for the Army-Navy football game run. You can also place these boxes on bicycles for a marathon event, and, of course, in automobiles. The GPS receiver picks up precise position information from military satellites. This data is fed to the TNC which, in turn, incorporates the data into a UI frame. The radio then transmits the data to all receiving stations within range.

Bob's system is an excellent tool for triangulating the location of a hidden transmitter. A jammer command displays the intersection of bearing lines from a number of reporting stations. To use APRS in this manner, each station having a bearing on a jammer enters it into his APRS system. His station then reports its location and its bearing relative to the jammer. All stations running APRS can simply hit the J key to display the intersection of these bearing lines. Furthermore, if a direction-finding vehicle has a GPS or LORAN-C device on board, it can be tracked and directed to the location of the jammer.

APRS may also be a solution to the effective use of orbiting packet-radio digipeaters, such as on space shuttles, *Mir*, AO-21

and other satellites. The problem with space digipeaters is saturation on the uplink channel, which makes the use of a normal connected protocol impractical. For a connected contact, five successive and successful packet transmissions are required. Not only does APRS reduce this to one packet, it also capitalizes on its most fascinating aspects—its ability to display the locations of stations on a map. If all stations simply insert their latitude and longitude as the first 19 characters of their packet beacon texts, everyone monitoring the satellite with APRS software will see the location of every successful uplink. Because satellites are rapidly moving objects, the locations of successful uplink stations will move progressively along the ground track.

Bob's software is powerful and can perform a variety of tasks, yet it is simple to use. As a result, it has the potential to become an important application in amateur packet.

The APRS software is distributed as shareware and may be copied for any amateur application. Registered copies for IBM PCs are available for $19 from Bob Bruninga, WB4APR, 115 Old Farm Ct, Glen Burnie, MD 21060. A Macintosh version of APRS, MacAPRS, was developed by Keith Sproul, WU2Z, and Mark Sproul, KB2ICI. APRS is described in detail in the ARRL publication *Getting on Track With APRS*.

Other Applications

Packet has begotten a host of other useful Amateur Radio applications. Database applications are particularly popular. One well-known database application is called *White Pages*. White Pages provided a solution to a problem that faced packet operators at the onset of PBBS mail forwarding. That is, how do you know who uses which PBBS? Now, by sending a short message to a White Pages database, you can find out which PBBS your fellow packet operator calls home.

Another popular database application came about with the advent of call sign directories on CD-ROMs. It wasn't long before packet call sign directories came on the air.

With enough patience and perseverance, almost any program that can run on a computer can be made to operate with packet. In this regard, the ultimate packet application may be NM1D's DOSgate, which allows you to run software over the air on a remote IBM-class computer.

THE FUTURE

As it exists today, packet is only beginning to realize its full potential. When higher data rates, improved networks, better radio equipment and more flexible protocols are in place, more complex and fascinating applications will be accessible. Interconnections with the Internet adds still another dimension to packet radio.

CHAPTER 11

FM and Repeaters

BRIAN BATTLES, WS1O
746 EAST STREET
NEW BRITAIN, CT 06051

Pizza and beer, Abbott and Costello, FM and repeaters: Some things make natural combinations. In Amateur Radio, the combination of frequency modulation (FM) and HF/UHF repeaters has made it in a big way. FM and repeaters have been the most popular communication modes in ham radio since the early 1970s and are in the forefront of the hobby today. Nearly every amateur has used an FM repeater at some point.

The reason for the popularity of FM and repeaters is their versatility. They offer something for everyone. HF DXers swap information with fellow DXers, traffic handlers pass traffic between the local and section levels, and packet and radioteletype enthusiasts communicate digitally using repeaters dedicated to their specific modes. If you just like to talk, strike up a conversation on any repeater. Adding to the versatility of repeater operation is the ability to communicate while walking, driving or simply relaxing in the comfort of your home. You can even boost your range on HF through the use of a 10-meter FM repeater.

What's a Repeater?

The term "repeater" can be confusing to newcomers. A standard repeater is simply a relay station. It consists of a separate "input" receiver and an "output" transmitter connected to each other and tuned to two separate frequencies within the same band. When the receiver picks up a signal on the input frequency, it simultaneously retransmits the same signal on the output frequency. In this way a repeater forms a link between two stations that may not be able to communicate with each other directly.

A repeater system may include connections to receiver/transmitter combinations on other bands. For example, a 2-meter repeater linked to the 70-cm band may receive on 147.69 MHz and transmit on 147.09, while it also receives on 449.625 and transmits on 444.625. If a signal is present on 147.69 or on 449.625, it's retransmitted on both 147.09 and 449.625. These are called *crosslinked* repeaters, and they may include coverage for several bands. Such systems may operate on all available bands at all times, or have remote-control features that allow the licensee or selected users to control which crossband links are on or off at a particular time.

There are more complex repeaters that form integrated wide-coverage systems. These consist of machines miles apart that are connected by two-way VHF or UHF links. Such

---

A New Way to Start

One of the most exciting developments in the history of Amateur Radio came about in December 1990 when the FCC announced it had created the codeless Technician class license. The ruling took effect on February 14, 1991.

For the first time since ham radio licenses were instituted, US amateurs could get on the air without passing a Morse code test. A candidate who passes written exam Elements 2 and 3A receives a Tech ticket and may operate with all amateur privileges above 30 MHz. Having an alternative to the traditional Novice class of entry-level license appeals to people who are not as interested in CW, HF and DX activities and prefer to explore VHF, UHF and microwave communications.

In most parts of the US, it's increasingly difficult to locate space on the VHF and UHF bands to establish new repeaters. Even so, the existing 144, 222, 440 and 1270 MHz repeaters are not too busy to allow more amateurs to use them. Although 2-meter packet has saturated many regions, more hams are discovering and exploiting the vast potential of this mode on higher frequencies.

The codeless Technician license is a great opportunity for FM and repeater enthusiasts. Not only has the door been opened for technically minded hams to experiment on VHF, UHF and microwaves, thousands of people who are initially only interested in local communications are also able to gain access to FM repeaters. This includes friends and family members of hams who enjoy being able to get in on the excitement and rewards of Amateur Radio public service, emergency communications support, traffic handling and keeping in touch with their ham friends. In addition, repeater-oriented clubs can welcome greater numbers of new members to join their activities, enhance projects and support repeaters.

The Ins and Outs of Repeater Operating

Repeater users should know how repeaters work. Users who are not familiar with the workings of a repeater may unintentionally misuse the equipment and interfere with other users.

A repeater is like any other Amateur Radio station. It has a receiver, a transmitter and an antenna. The difference is that a repeater's receiver and transmitter are tuned to different frequencies and the output of the receiver is connected (through a carrier-operated relay or *COR*) to the input of the transmitter; thus, anything the receiver hears is simultaneously retransmitted, or *repeated*, by the transmitter.

Your transceiver's transmitter and receiver are also tuned to different frequencies. Your transmitter is tuned to the repeater's receive or *input* frequency, so whatever you transmit is received by the repeater. Meanwhile, your receiver is tuned to the repeater's transmit or *output* frequency, so whatever the repeater transmits (repeats) will be received by your transceiver.

To tune an FM transceiver, you turn its frequency control to the transmit/output frequency of the desired repeater. Your transmitter's frequency automatically follows the tuning of your receiver and is adjusted to the repeater's receive/input frequency. (The frequency display on your transceiver will indicate its *receive* frequency.) Usually, you must also select a switch position that chooses between the "simplex" and "duplex" mode. In the simplex mode, your transmitter and receiver are tuned to the same frequency. In the duplex mode, your transmitter and receiver are tuned to different frequencies, as required for repeater operation.

The frequency separation between the transmitter and receiver differs with each repeater band. On 144 MHz, the standard separation between the transmitter and receiver frequencies is 600 kHz; on 222 MHz, the separation is 1.6 MHz; and on 450 MHz, the separation is 5 MHz. A 2-meter repeater with an input frequency of 146.34 MHz has an output of 146.94 MHz (146.34 MHz + 600 kHz = 146.94 MHz [146.34 + 0.6 = 146.94]). Hams refer to a repeater by its output frequency. For example, the repeater mentioned above would be called "the 94 machine."

Transmitter-receiver frequency separation is usually built into a modern transceiver's memory chips. You *do* have to know whether the output frequency is higher or lower than the input frequency, however. In some segments of the 2-meter band, the input frequency is 600 kHz lower than the output frequency (as in our 146.34/146.94 repeater example above), while in other parts of the band, the input frequency is 600 kHz higher than the output

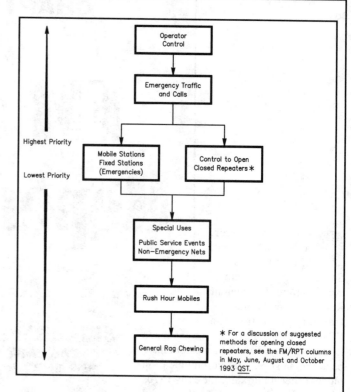

Fig A—Block diagrams of a typical FM repeater and FM transceivers show the relationship of frequency X, the repeater's input frequency, and frequency Y, the repeater's output frequency. Transceiver A transmits on frequency X and is received by the repeater's receiver, which is also tuned to frequency X. The repeater's transmitter retransmits (repeats) the received signal on frequency Y and is received by transceiver B's receiver, which is also tuned to frequency Y.

frequency. For example, a repeater with an output frequency of 147.09 MHz has an input frequency of 147.69 MHz.

Usually, when you select the duplex mode on your transceiver, you also have to select whether your transmitter frequency will be higher or lower than your receiver frequency. The simplex/duplex switch designated ± performs this function; the + position sets the transmitter frequency higher than the receiver frequency, while the − position sets the transmitter lower than the receiver frequency. The position between the + and − often selects the simplex mode.

---

systems allow operators to use a local repeater to make contacts with hams in distant cities or states. For example, the Evergreen Intertie includes more than 23 repeater stations throughout California, Oregon, Washington, Idaho, Montana, British Columbia and Alberta. An amateur on Mt Shasta can chat with a ham in Edmonton—even though both hams are using low-power hand-held transceivers! Integrated wide-coverage repeater systems allow users to turn links on and off as needed and activate telephone autopatching and other features by using the dual-tone multifrequency (DTMF) keypads on their radios.

Licensing

Except for the Novice VHF/UHF subbands, you need at least a Technician class license (or higher) to operate in the VHF and UHF Amateur Radio spectrum (above 30 MHz). A General

class license (or higher) is required to operate 10-meter FM and repeaters. Technician class hams may also use VHF/UHF repeaters that transmit on frequencies not normally permitted for use by Technicians. For example, a 450-MHz repeater may have an output on 29.640 MHz. Novices have considerable VHF/UHF privileges: All authorized modes from 222-225 MHz (25 watts maximum) and 1270-1295 MHz (5 watts). Novices can operate through all existing repeaters in those two subbands. For details, see Chapter 2.

OPERATING

FM repeaters provide the means to communicate efficiently. Before making your first transmission in the world of FM and repeater communications, however, you should be aware of some basic operating techniques.

PRESENT ARMS! The Northern Indiana VHF drill team, consisting of W9XD, WZ9M, KB9ATR, KC9XT, N9LVL and N9LBJ in the rear row. Up front are N9IOX, Chris Kratzer and KB9GRP. (*Photo courtesy of KB9GNU*)

Finding a Repeater

To use a repeater, you must know one exists. There are various ways to find a repeater. Local hams can provide information about repeater activity or you can consult a repeater listing. Various clubs, sometimes those associated with the local frequency coordinator, publish statewide and regional directories of repeaters. Each spring, the ARRL publishes *The ARRL Repeater Directory*, a comprehensive listing of repeaters throughout the US, Canada and other parts of the world. Besides identifying local repeater activity, the *Directory* is handy for finding repeaters during vacations and business trips. Once you find a repeater to use, listen and familiarize yourself with its operating procedures.

Your First Transmission

If the repeater is quiet, pick up your microphone, press the switch, and transmit your call sign as "N6ATQ listening" or "N6ATQ monitoring" to attract someone's attention. After you stop transmitting, the repeater sends an unmodulated carrier for a couple of seconds to let you know it's working. If anyone is interested in talking to you, they'll call after your initial transmission. Some repeaters have specific rules for making yourself heard, but usually your call sign is all you need.

Don't call CQ to initiate a conversation on a repeater. It takes a lot longer to complete a long CQ than to simply transmit your call sign. (In some areas, a solitary "CQ" is permissible.) Efficient communication is the goal. You're not trying to attract the attention of someone who is casually tuning his receiver across the band. Except for scanner operation, there isn't much tuning through the repeater bands.

If you want to join a conversation already in progress, transmit your call sign during a break between transmissions. The station that transmits next should acknowledge you. Don't use the word "break" to join a conversation (unless it's the operating practice in your area). "Break" usually suggests an emergency and indicates that all stations should stand by for the station with emergency traffic.

If you want to call another station and the repeater is inactive, simply call the other station. (For example, "WB8SVN, this is WB8OFR.") If the repeater is active, but the conversation in progress sounds as though it's about to end, be patient and wait until it's over before calling another station. If the conversation sounds like it's going to continue

FM and Repeater-Speak

Here are some definitions of terms used in the world of Amateur Radio FM and repeaters:

access code: one or more numbers and/or symbols that are keyed into the repeater with a telephone tone pad to activate a repeater function, such as an autopatch.

autopatch: a device that interfaces a repeater to the telephone system to permit repeater users to make telephone calls. Often just called a "patch."

break: the word used to interrupt a conversation on a repeater only to indicate that there is an emergency.

carrier-operated relay (COR): a device that causes the repeater to transmit in response to a received signal.

channel: the pair of frequencies (input and output) used by a repeater.

closed repeater: a repeater whose access is limited to a select group. (see *open repeater*)

control operator: the Amateur Radio operator who is designated to "control" the operation of the repeater, as required by FCC regulations.

courtesy beeper: an audible indication that a repeater user may go ahead and transmit.

coverage: the geographic area within which the repeater provides communications.

digipeater, digital repeater: a packet radio repeater.

duplex: a mode of communication in which you transmit on one frequency and receive on another frequency.

frequency coordinator: an individual or group responsible for assigning channels to new repeaters without interference to existing repeaters.

full duplex: a mode of communication in which you transmit and receive simultaneously.

full quieting: a received signal that contains no noise.

half duplex: a mode of communication in which you transmit at one time and receive at another time.

hand-held: a portable transceiver small enough to fit in the palm of your hand, clipped to your belt or even in a shirt pocket.

input frequency: the frequency of the repeater's receiver (and your transceiver's transmitter).

key up: to turn on a repeater by transmitting on its input frequency.

machine: a repeater system (slang).

magnetic mount, mag-mount: an antenna with a magnetic base that permits quick installation and removal from a motor vehicle or other metal surface.

NiCd: a nickel-cadmium battery that may be recharged many times; often used to power portable transceivers. Pronounced "NYE-cad."

open repeater: a repeater whose access is not limited.

output frequency: the frequency of the repeater's transmitter (and your transceiver's receiver).

over: a word used to indicate the end of a voice transmission.

radio direction finding (RDF): the art and science of locating a hidden transmitter.

Repeater Directory: an annual ARRL publication that lists repeaters in the US, Canada and other areas.

separation, split: the difference (in kHz) between a repeater's transmitter and receiver frequencies. Repeaters that use unusual separations, such as 1 MHz on 2 meters, are sometimes said to have "odd splits."

simplex: a mode of communication in which you transmit and receive on the same frequency.

time-out: to cause the repeater or a repeater function to turn off because you have transmitted for too long.

timer: a device that measures the length of each transmission and causes the repeater or a repeater function to turn off after a transmission has exceeded a certain length.

tone pad: an array of 12 or 16 numbered keys that generate the standard telephone dual-tone multifrequency (DTMF) dialing signals. Resembles a standard telephone keypad.

"Canine Wheelchair Mobile"

KE3MB

Erik Johnson, KE3MB, with a K9 on his lap, frequents the local repeater.

for a while, transmit your call sign between transmissions. After one of the other hams acknowledges you, politely ask to make a quick call. Usually, the other stations will acquiesce. Make your call short. If your friend responds to your call, ask her to move to a simplex frequency or another repeater, or to stand by until the present conversation is over. Thank the other users for letting you interrupt them to place your call.

Acknowledging Stations

If you're in the midst of a conversation and a station transmits its call sign between transmissions, the next station in queue to transmit should acknowledge that station and permit the newcomer to make a call or join the conversation. It's discourteous not to acknowledge him and it's impolite to acknowledge him but not let him speak. You never know; the calling station may need to use the repeater immediately. He may have an emergency on his hands, so let him make a transmission promptly.

The Pause That Refreshes

A brief pause before you begin each transmission allows other stations to participate in the conversation. Don't key your microphone as soon as someone else releases his. If your exchanges are too quick, you'll block other stations from getting in.

The "courtesy beepers" on some repeaters compel users to leave spaces between transmissions. The beep sounds a second or two after each transmission to permit new stations to transmit their call signs in the intervening time period. The conversation may continue only after the beep sounds. If a station is too quick and begins transmitting before the beep, the repeater may respond to the violation by shutting down!

Brevity

Keep each transmission as short as possible. Short transmissions permit more people to use the repeater. All repeaters promote this practice by having timers that "time-out," temporarily shutting down the repeater whenever the length of a transmission exceeds the preset time limit. With this in the

back of their minds, most users keep their transmissions brief.

Learn the length of the repeater's timer and stay well within its limits. The length may vary with each repeater; some are as short as 15 seconds and others are as long as three minutes. Some repeaters vary their timer length depending on the amount of traffic on frequency; the more traffic, the shorter the timer. The other purpose of a repeater timer is to prevent extraneous signals from holding the repeater on the air continuously, potentially causing damage to the repeater's transmitter.

Because of the nature of FM radio, if more than one signal is on the same frequency at one time, it creates a muffled buzz or an unnerving squawk. If two hams try to talk on a repeater at once, the resulting noise is known as a "double." If you're in a roundtable conversation, it is easy to lose track of which station is next in line to talk. There's one simple solution to eradicate this problem forever: *Always pass off to another ham by name or call sign.* Saying, "What do you think, Jennifer?" or "Go ahead, WS1O" eliminates confusion and avoids doubling. Try to hand off to whoever is next in the queue, although picking out anyone in the roundtable is better than just tossing the repeater up for grabs and inviting chaos.

The key to professional-sounding FM repeater operation is to be brisk and to the point, and to leave plenty of room for others. Keep it moving. Don't drone; dart in and out. Don't hem and haw or be reluctant to "yield the floor." Your turn will come again in a moment. Turn it over, pause for others, get things rolling. Snappy, clearheaded exchanges sound sharp and are more enjoyable for your QSO partners.

Identification

You must give your call sign at the end of each transmission or series of transmissions and at least every 10 minutes during the course of a contact. You don't have to transmit the call sign of any other station, including the one you're contacting. (Exception: You must transmit the other station's call sign when passing third-party traffic to a foreign country.)

It's illegal to transmit without identification. Aside from breaking FCC rules, it's poor operating practice to key your microphone to turn on a repeater without identifying your station. This is called "kerchunking" the repeater. If you don't want to have a conversation, but simply want to check whether your radio works or if you're able to access a particular repeater, just say "KB4ZKD testing." This way you accomplish what you want to do legally.

Go Simplex

Simplex is a fancy-sounding word for a direct contact on a single frequency. After you've made a contact on a repeater, move the conversation to a simplex frequency, if possible.

The function of a repeater is to provide communications between stations not able to communicate directly because of terrain or equipment limitations; see Fig 11-1. If stations are able to communicate without a repeater, they shouldn't use a repeater. Always use simplex whenever possible so that the repeater will be available for stations that need its facilities.

Simplex communications offer a degree of privacy impossible to achieve on a repeater. There's also no timer to worry about or courtesy beep to wait for. When selecting a frequency, make sure it's designated for FM simplex operation. Each band is subdivided for specific modes of operation, such as satellite communications and weak-signal CW and SSB. If you select a simplex frequency indiscriminately, you may interfere with stations operating in other modes (and you may not be aware of it).

VHF and UHF FM are also acceptable modes for contesting in sweepstakes, sprints, Field Day and other events. The only stipulation is that contacts must be made without repeaters and generally conducted off the national simplex or calling frequencies. Watch for announcements in *QST* and bring that 2- or 6-meter FM rig out to a mountaintop and call "CQ contest" sometime!

Fixed Stations and Prime Time

Repeaters are intended to enhance mobile and portable communications. During mobile operating prime time, fixed stations should yield to mobile stations. (Some repeaters have an explicit policy to this effect.) But when you're operating as a fixed station, don't abandon the repeater completely; monitor mobile activity. Your help may be needed in an emergency.

No Free Lunch

It takes time, money, knowledge and energy to operate a reliable repeater system. How often do you stop to think of what goes into the machine you conveniently key up any time, 24 hours a day, 365 days a year? Nobody should feel compelled to join any group, and you can use thousands of repeaters across the US without joining any clubs. On the other hand, if you frequent a system—or just want to contribute to the cost of its upkeep so it can be counted on in an emergency—support your local repeater.

Many repeaters are run by clubs whose members pool their resources to operate the repeater. Some members provide the equipment that makes up the repeater, other members provide funds to pay the electric, phone, maintenance and site-rental bills, and others provide manpower at the repeater site whenever their help is needed.

Limited Access

Most Amateur Radio repeaters are open to all users. There

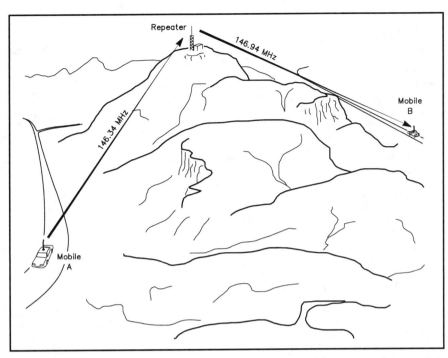

Fig 11-1—The mountain blocks direct (simplex) communication between the two mobile stations. When mobile A transmits on 146.34 MHz the repeater receives this signal and retransmits it on 146.94 MHz. Mobile B's receiver is tuned to 146.94 to capture this signal from the repeater. When it's mobile B's turn to talk it also transmits on 146.34 MHz and mobile A listens for the retransmitted signal on 146.94 MHz.

are no restrictions on the use of the repeater's facilities. Limited-access repeaters do exist, however. Although such operations go against the spirit of our hobby, "closed" repeaters are legal according to FCC regulations.

There are "open" repeaters that require the use of special codes or subaudible tones to gain access. The reason for limiting the access to some "open" repeaters is to prevent interference. In cases where extraneous transmissions often

---

Putting a New Repeater on the Air

Amateurs contemplating putting a repeater on the air may wonder how to obtain the frequency pairs for their repeaters. *Frequency coordination* is the recommended method. Across the US and Canada, volunteer individuals and groups have taken on the task of recommending frequencies for proposed repeaters within their "jurisdiction." These frequency coordinators keep extensive records on repeater operation in their area and adjacent areas. With this information, they're usually able to recommend the best pair of frequencies for proposed repeaters.

In cases of repeater-to-repeater interference, the two repeater stations are equally and fully responsible for resolving the interference unless one repeater is coordinated and the other is not. In that case, the noncoordinated repeater has primary responsibility for resolving the interference (according to the FCC).

Because of the amount of work involved in finding the best frequency pair, there is a time lag between the request for a frequency pair and receipt of the coordinated channel. If you plan to request a pair of frequencies, do it early. Consult the latest edition of *The ARRL Repeater Directory* for the listing of frequency coordinators.

Band Plans

Band plan refers to an agreement among concerned VHF and UHF operators about how each Amateur Radio band should be arranged. The goal of a band plan is to reduce interference between the modes sharing each band. Besides FM repeater and simplex activity, CW, SSB, AM, satellite, amateur television (ATV) and radio control operations also use these bands. (For example, a powerful FM signal at 144.08 MHz could spoil someone else's long-distance CW contact.) The VHF and UHF bands offer a wide variety of amateur activities, so hams have agreed to set aside space for each type.

The most heavily used FM activity is on 2 meters, so a description of the 2-meter band plan will be useful. The band is divided according to the following proposal by the ARRL VHF-UHF Advisory Committee:

Frequency (MHz)	Operation
144.00-144.05	EME (moonbounce) CW
144.05-144.10	General CW and weak signals
144.10-144.20	EME and weak-signal SSB
144.20	National calling frequency
144.20-144.275	General SSB operation
144.275-144.30	Propagation beacons
144.30-144.50	OSCAR subband
144.50-144.60	Linear translator inputs
144.60-144.90	FM repeater inputs
144.90-145.10*	Weak signal and FM simplex
145.10-145.20	Linear translator outputs
145.20-145.50	FM repeater outputs
145.50-145.80	Miscellaneous and experimental modes
145.80-146.00	OSCAR subband
146.01 -146.40	Repeater inputs
146.415-146.595	Simplex
146.61 -147.39	Repeater outputs
147.42-147.57	Simplex
147.60-147.99	Repeater inputs

*The following packet radio frequency recommendations were adopted by the ARRL Board of Directors in July 1987:

1) Automatic/unattended operations should be conducted on 145.01, 145.03, 145.05, 145.07 and 145.09 MHz.

   a) 145.01 should be reserved for interLAN use.

   b) Use of the remaining frequencies (above) should be determined by local user groups.

2) Additional frequencies in the 2-meter band may be designated for packet radio use by local coordinators.

Specific channels recommended above may not be applicable in all areas of the US. For example, amateurs in many areas conduct TCP/IP operations on 144.91 MHz and many *PacketCluster* nodes congregate on 144.95 MHz.

Prior to establishing regular packet radio use on any VHF/UHF channel, it's advisable to check with the local frequency coordinator. The decision on how the available channels are used should be based on coordination between local packet users.

Note: Please contact your regional frequency coordinator for information on channel availability.

Repeater Frequency Pairs (input/output):

144.51/145.11	144.83/145.43	146.37/146.97
144.53/145.13	144.85/145.45	146.40 or 147.60/147.00*
144.55/145.15	144.87/145.47	146.43 or 147.63/147.03*
144.57/145.17	144.89/145.49	146.46 or 147.66/147.06*
144.59/145.19	146.01/146.61	147.69/147.09
144.61/145.21	146.04/146.64	147.72/147.12
144.63/145.23	146.07/146.67	147.75/147.15
144.65/145.25	146.10/146.70	147.78/147.18
144.67/145.27	146.13/146.73	147.81/147.21
144.69/145.29	146.16/146.76	147.84/147.24
144.71/145.31	146.19/146.79	147.87/147.27
144.73/145.33	146.22/146.82	147.90/147.30
144.75/145.35	146.25/146.85	147.93/147.33
144.77/145.37	146.28/146.88	147.96/147.36
144.79/145.39	146.31/146.91	147.99/147.39
144.81/145.41	146.34/146.94	

*local option

Some areas use 146.40-146.60 and 147.40-147.60 MHz for simplex or for repeater inputs and outputs. Frequency pairs in those areas are:

147.415/146.415	147.475/146.475
147.43/146.43	147.49/146.49
147.445/146.445	147.505/146.505
147.46/146.46	147.595/146.595

Suggested 15-kHz Splinter Channels (input/output):

146.025/146.625	146.265/146.865	147.735/147.135
146.055/146.655	146.295/146.895	147.765/147.165
146.085/146.685	146.325/146.925	147.825/147.225
146.115/146.715	146.355/146.955	147.855/147.255
146.145/146.745	146.385/146.985	147.885/147.285
146.175/146.775	147.615/147.015	147.915/147.315
146.205/146.805	147.645/147.045	147.945/147.345
146.235/146.835	147.675/147.075	147.975/147.375
	147.705/147.105	

Simplex Frequencies:

*146.415	146.535	147.48
*146.43	146.55	147.495
*146.445	146.565	147.51
*146.46	146.58	147.525
*146.475	146.595	147.54
*146.49	147.42	147.555
*146.505	147.435	147.57
**146.52	147.45	147.585
	147.465	

*May also be a repeater (input/output). See repeater pairs listing.

**National simplex frequency

Several states have chosen to realign the 146-148 MHz subband, using 20-kHz spacing between channels. This choice was made to gain additional repeater pairs.

The transition from 30- to 20-kHz spacing is taking place on a case-by-case basis as the need for additional pairs occurs. Typically, the repeater on an odd-numbered pair will shift 10 kHz up or down, creating a new set on an even-numbered channel. For example, the pair of 146.13/.73 MHz would change to 146.12/.72 or 146.14/.74, while the pairs of 146.10/.70 and 146.16/.76 would be left unchanged.

activate the repeater, limiting access is the only way to resolve the problem.

How is access to these repeaters controlled? Most often, via a technique called *continuous tone-controlled squelch system (CTCSS)*. Many hams refer to CTCSS as *PL*, Motorola's trademark term that stands for Private Line, used on commercial radio gear, with a series of alphanumeric names that designate each of the tones. There are 42 standard CTCSS tone frequencies, as developed by the Electronic Industries Association (EIA). When a transmitter is configured for CTCSS, it sends a subaudible tone along with the transmitted voice or other signals. The frequency of the CTCSS tone is below the lowest audio frequency other stations will pass to their speakers, but it is sensed by a suitably equipped repeater. The repeater is programmed to respond only to carriers that send the proper tone. This effectively locks out signals that don't carry the correct CTCSS tone. Most modern VHF and UHF transceivers include the necessary circuitry to generate CTCSS tones, so if you know the one you need, you can simply program it on your rig.

There's a complete table of CTCSS frequencies in *The ARRL Repeater Directory*, and many repeaters listed in that book publish their CTCSS tones.

Minimum Power

The VHF and UHF repeater bands are paradise for QRP fans. Make it a habit to run your transceiver on a low-power setting. There's usually no need to pump out heavy watts on VHF or UHF FM if you're within a reasonable range of a repeater or other station operating simplex.

Inspect your station regularly for loose connections, broken wires, antenna problems, intermittent grounds and other potential weak spots. Mobile installations are most prone to wear and damage. A well-designed antenna, quality feed line and properly installed connectors will improve your transmitted and received signals. Never use a high-power external amplifier for a local contact. If a few watts won't bring up the 2-meter repeater five miles away, 160 watts probably won't cut it, either.

When you're on a repeater frequency, use the minimum power necessary to maintain communications to avoid the possibility of accessing distant repeaters on the same frequency. Using minimum power is not only a courtesy to the distant repeaters, but an FCC requirement.

Autopatch

An autopatch is a device that allows hams to make telephone calls through a repeater. In most repeater autopatch systems, the user simply has to generate the standard telephone company tones to access and dial through the system. This is usually done by interfacing a telephone tone pad with the user's transceiver. Dual-tone multifrequency (DTMF) tone keypads are often mounted on the front of hand-held portable transceivers or on the front of fixed or mobile transceiver microphones.

There are a number of important rules that govern autopatch operation. See the *ARRL Autopatch Guidelines* sidebar in this chapter for a thorough treatment of the subject.

If you have a legitimate reason to use the autopatch, how do you use it? First, you must access (turn on) the autopatch, usually by pressing a designated key or combination of keys on the tone pad. Check to make sure the frequency is clear before you begin and be sure to identify your station. The # (pound) and * (star) signs are common access keys. When you hear a dial tone, you've successfully accessed the autopatch. Now, punch in the telephone number you wish to call.

When you establish a call, tell the person you're calling what you're doing so that they'll know they're going over the ham bands. If you just say, "Hi, I'm calling you from the car," they might think you have a mobile cellular telephone. If the person isn't familiar with ham radio, you may want to explain that the conversation is not private and may be overheard. They should also be told that they have to wait their turn to talk. Use the word *over* to inform the person at the other end of the line that you've finished talking and that he or she can speak. Many autopatches have timers that terminate the connection after a certain period of time, so keep your telephone conversation short and end the autopatch as quickly as possible.

The procedure for turning off the autopatch is similar to the procedure for accessing it. A key or combination of keys must be punched to return the repeater to normal operation. Don't forget to identify when you finish. Autopatch users should consult the repeater group that sponsors the autopatch for specific information about access codes and timer specifics.

Traffic

Traffic handling is a natural for repeater operation. Where else can you find as many local outlets for the traffic coming down from the Section-level nets? Everyone who has a transceiver capable of repeater operation is a potential traffic handler. As a result, local traffic nets have flourished on repeaters.

The procedures for handling traffic on a repeater are about the same as handling traffic anywhere else (consult Chapter 15 for full details). However, there are two things unique to repeater traffic handling: time-out timers and only occasional use of phonetics.

Repeaters have timers that shut down the repeater if a transmission is too long. Therefore, when you relay a message by repeater, release your push-to-talk (PTT) switch during natural breaks in the message to reset the timer. If you read the message in one long breath without resetting the timer, the repeater may shut down in the middle of your message. Then you'll hold up the net—and earn a red face—while you repeat the message.

Under optimal signal conditions, the audio quality of the FM mode of communications is excellent. It lacks the noise, static and interference common to other modes. Therefore, the use of phonetics and repetition is not necessary when relaying a message over a repeater. The only time phonetics are necessary is to spell out an unfamiliar word or words with similar-sounding letters. (An exception: The word "emergency" is always spelled out when it appears in the preamble of a formal radiogram.) On the rare occasion that a receiving station misses something, she can ask you to fill in the missing information.

The efficiency provided by repeaters has made repeater traffic nets popular. There are probably one or more active repeater traffic nets in your area (consult the latest edition of *The ARRL Net Directory* to find a local repeater traffic net). If there's no repeater traffic net in your vicinity, why not fill the need by starting one yourself? Contact your ARRL Section Manager (SM) or Section Traffic Manager (STM) for details. *QST* lists your local officials; check page 8 or Section News in the current issue.

RTTY

Digital operation is not as common as voice, but it's popular on VHF and UHF FM repeaters. With FM, there's minimal noise

ARRL Autopatch Guidelines

Autopatch operation involves using a repeater as an interface to a local telephone exchange. Hams operating mobile or portable stations are able to use the autopatch to access the telephone system and place a call. Hams use autopatches to report traffic accidents, fires and other emergencies. There's no way to calculate the value of the lives and property saved by the intelligent use of autopatch facilities in emergencies. The public interest has been well served by amateurs with interconnect capabilities. As with any privilege, this one can be abused and the penalty for abuse could be the loss of the privilege for all amateurs. The suggested guidelines here are based on conventions that have been in use for years on a local or regional basis throughout the country. The ideas they represent have widespread support in the amateur community. Amateurs are urged to observe these standards carefully so our traditional freedom from government regulation may be preserved as much as possible.

1) Although it's not the intent of the FCC Rules to let Amateur Radio operation be used to conduct an individual's or an organization's commercial affairs, autopatching involving business affairs may be conducted on Amateur Radio. (The FCC has stated that it considers nonprofit and noncommercial organizations "businesses.") On the other hand, amateurs are strictly prohibited from accepting any form of payment for operating their ham transmitters, they may not use Amateur Radio to conduct any form of business in which they have a financial interest and they may not use Amateur Radio in a way that economically benefits their employers.

Amateurs should generally avoid using Amateur Radio for any purpose that may be perceived as abuse of the privilege. The point of allowing hams to involve themselves in "business" communication is to make it more convenient and to remove obstacles from ham operations in support of public service activities. Before this rule was revised in 1993, it was often technically illegal for amateurs to participate in many charitable and community service events because the FCC regarded any organization, commercial or noncommercial, as a business with respect to the rules, and prohibited hams from making any communications to in any way facilitate the business affairs of any party. That meant that operating a talk-in station for a local nonprofit radio club's hamfest constituted a violation!

So now it's legal to use ham frequencies, including autopatch facilities, to communicate in such a way as to facilitate a business transaction. The distinction is essentially whether the amateur operator or his employer has a financial stake in the communication. This means that a ham may use a patch to call someone about a club event or activity, to make a dentist appointment, to order a pizza or to see if a load of dry cleaning is ready to be picked up. In such situations, the ham isn't in it for the money. However, no one may use the ham bands to dispatch taxicabs or delivery vans, to send paid messages, to place a sales call to a customer, or to cover news stories for the local media (except in emergencies if no other means of communication is available). If the ham is paid for or will profit from the communication, it may not be conducted on an amateur frequency. That's why there are telephones and commercial business radio services available.

Use care in calling a business telephone via an amateur autopatch. Calls may be legally made to one's office to receive or to leave personal messages, although using Amateur Radio to avoid the cost of public telephones, commercial cellular telephones or two-way business-band radio isn't considered appropriate to the purpose of the amateur service. Calls made in the interests of highway safety, such as for the removal of injured persons from the scene of an accident or for the removal of a disabled vehicle from a hazardous location, are clearly permitted.

A final word on business communications: Just because the FCC says that a ham can place a call involving business matters on a repeater or autopatch doesn't mean that a repeater licensee or control operator must allow you to do so! If he or she prefers to restrict all such contacts, he or she has the right to terminate your access to the system. A club, for instance, may decide that it would rather not have members order commercial goods over the repeater autopatch and may vote to forbid members from doing so. The radio station's licensee and control operator are responsible for what goes over the air and have the right to refuse anyone access to the station for any reason.

2) All interconnections must be made in accordance with telephone company rules and fee schedules (tariffs). If you have trouble obtaining information about them from telephone company representatives, the tariffs are available for public inspection at your telephone company office. Although some local telephone companies consider Amateur Radio organizations to be commercial entities and subject to business telephone rates, many repeater organizations, as noncommercial volunteer public service groups, have successfully arranged for telephone lines at repeater sites to be charged at the lower residential rate.

3) Autopatches should not be made solely to avoid telephone toll charges. Autopatches should never be made when normal telephone service could be just as easily used. The primary purpose of an autopatch is to provide vital, convenient access to authorities during emergencies. Operators should exercise care, judgment and restraint in placing routine calls.

4) Third parties (nonhams) should not be put on the air until the responsible control operator has explained to them the nature of Amateur Radio. Control of the station must never be relinquished to an unlicensed person. Permitting a person you don't know well to conduct a patch in a language you don't understand amounts to relinquishing control because you don't know whether what they are discussing is permitted by FCC rules.

5) Autopatches must be terminated immediately in the event of any illegality or impropriety.

6) Station identification must be strictly observed.

7) Phone patches should be kept as brief as possible, as a courtesy to other amateurs; the amateur bands are intended primarily for communication among radio amateurs, not to permit hams to communicate with nonhams who can only be reached by telephone.

8) If you have any doubt as to the legality or advisability of a patch, don't make it. Compliance with these guidelines will help ensure that amateur autopatch privileges will continue to be available in the future, which helps the Amateur Radio service contribute to the public interest.

and interference, making it possible to send and receive error-free RTTY messages. Above 50 MHz, any digital code is permissible, whereas HF RTTY operators are limited to three digital codes: Baudot/Murray, ASCII and AMTOR. (Note: Although any digital code is permissible above 50 MHz, station identification at the end of each communication and every 10 minutes during any communication must be made using voice [where permitted], CW, Baudot/Murray, ASCII or AMTOR.)

The operating procedures for RTTY repeater operation are similar to HF RTTY procedures (consult Chapter 9 for details). One difference, though, is the length of transmissions. RTTY operators must keep their transmissions short enough so they don't time-out the repeater. On many RTTY repeaters, circuits are configured to disable the timer when a RTTY signal is detected.

Some repeaters used for RTTY are also used for voice communications. Under these circumstances, RTTY operators must take care not to disrupt voice communications by indiscriminately transmitting a RTTY message.

The operating procedures for radioteletype repeater operation are simple, so the transition from HF to VHF/UHF with RTTY should be easy.

Packet Radio

Packet radio on VHF and UHF FM is a digital communications mode that has become tremendously popular in the past few years. Hams all over the world are using terminal node controllers (TNCs) and their personal computers to create networks through which they can exchange messages and program files, handle traffic and enjoy direct keyboard-to-keyboard QSOs. Just as there are dedicated voice repeaters, there are dedicated packet *nodes* that perform the same function as their voice counterparts. That is, they act as intermediaries between stations that otherwise could not communicate. Packet nodes use the same frequency for transmission and reception, yet more than one—and usually several—data exchanges can be conducted on the same frequency (using one repeater). How's that for efficient repeater communications?

The operating procedures for packet are different from other forms of Amateur Radio communications because you command your TNC to do various tasks. The commands

HF Repeaters—Your DX Connection!

Imagine having a QSO with a station in Ecuador while you're on your way to work. Instead of a large HF rig under your dashboard, there is only a small transceiver that might easily be mistaken for a 2-meter FM unit at first glance. You're only running about 10 or 20 watts, yet the Ecuadorian station is giving you a "59" report. How is this possible?

Well, all the FM repeater action isn't confined to the VHF and UHF bands. Narrowband voice FM is permitted from 28.3-29.7 MHz and wider-band FM voice is allowed from 29.0- 29.7 MHz. The upper end of the 29.0-29.7 MHz segment is populated with a number of FM repeaters. When propagation conditions are good, these repeaters can provide worldwide coverage!

It can be confusing to use a 10-meter FM repeater during good conditions. Your input signal may activate distant repeaters as easily as those close by. You may also hear multiple QSOs as several different machines battle for the repeaters' (and your receiver's) front end! Sometimes band openings are as much a burden as they are a blessing. In 1980, the ARRL Board of Directors adopted a set of CTCSS access tones to be used voluntarily to provide a uniform nationwide system for 10-meter repeaters (see the table in *The ARRL Repeater Directory*).

In 1987, the FCC adopted the Novice Enhancement plan that gave Novices and Technicians their first taste of voice privileges on HF. Using SSB on 28.3-28.5 MHz, Novice and Technician Plus licensees have worked hundreds of DX countries from before dawn to past local sunset.

Even at the nadir of sunspot-cycle-related propagation, band conditions will make frequent dramatic, unpredictable shifts that produce remarkable openings. How will we know 10 is open if nobody's there to listen—or transmit? That's where 10-meter FM repeaters linked to 2 meters, 222 MHz and 440 MHz are an asset. When 10 pops open to Europe, Asia, Africa or Australia, you'll hear it first if you're monitoring your local HF-to-VHF repeater.

Where are They?

Most 10-meter FM repeaters operate between 29.520-29.680, with the FM simplex calling frequency at 29.600 MHz. (Avoid using 29.3-29.510 MHz for terrestrial contacts—the ARRL recommends this segment be reserved exclusively for amateur satellite downlinks. Because these downlinks use CW or SSB, a powerful local FM signal would easily wipe out a weak-signal QSO, so respect this portion.)

There are four standard repeater frequency pairs, as follows (input/output MHz):

29.520/29.620
29.540/29.640
29.560/29.660
29.580/29.680

Not all 10-meter repeater activity is confined to hams with General or higher-class licenses. Technicians can use these machines via VHF or UHF links. (Novices can access 10-meter repeaters from 222-225 or 1270-1295 MHz links.)

An HT at 10,000 feet comes in handy for Dan Cui, N1CJD. He uses it as he pilots a "Paramotor," which is a powered paraglider. Not all hams will want to repeat Dan's operation.

The family that talks together cruises together. Here Harriet Raynor, KB8JVY, has a dockside QSO with her spouse George, KB8JVZ.

Rick Applegate, N7WKS, combined the use of a 440 MHz H-T with a paddle mounted satellite antenna on a canoe trip. He worked in the FM mode through OSCARs 13 and 21. A separate 2-meter rig was used for the downlink. A different paddle was used for the canoe.

differ slightly, depending on which type of TNC you use (and there are a lot of commands available to do a lot of things besides communications). They all work roughly the same way, however.

To call another packet radio station, you enter a one or two-letter command followed by the call sign of the station you wish to contact; for example, C WR1B. Once you have entered the command and call sign, your TNC will attempt to contact the other station automatically.

To use a network node to contact another station, you must first connect to the node (for example, C KA1JES-1 where KA1JES-1 is the call sign of the node). Once you are connected to the node, you need only enter C WR1B and the node will complete the connection and relay the data between you and WR1B. As with voice FM and repeater operation, it's advisable in packet radio operation to use a node only if necessary.

You can call CQ on packet radio by entering a CQ message in your TNC and commanding the TNC to transmit the message over the air. Don't send a CQ message through several nodes, though. Trying to maintain a QSO through more than two nodes is a difficult proposition at best. When the frequency is very active, it is often impossible.

If your TNC establishes contact with another station (or if someone else's TNC establishes contact with yours), your terminal will display a message to that effect (***CONNECTED TO WS1O). Once you make contact, what you type at your keyboard is transmitted to the other station after you enter each control character or carriage return (depending on your TNC). During prime-time operating hours, it's good procedure to keep your messages short because you're sharing the channel with other stations. Short transmissions are more likely to make it through to the other station than long transmissions. (Short transmissions are less likely to collide with ones emanating from another conversation on the same frequency; a collision causes your TNC to repeat the transmission.)

To turn the conversation over to the other station, end your transmission with >>, K, KN, ? or something else that lets the other operator know that you're done. To end a conversation, command your TNC to do the dirty work (the simple command

Direction Finding

Radio direction finding (RDF), also known as foxhunting, bunny hunting and hidden-transmitter hunting, has become a popular radiosport in VHF and UHF FM operation. Someone hides a transmitter in the woods, a park or other inconspicuous place, and the troops, with direction-finding equipment in hand, go out to try to be the first to find the hidden transmitter. Many hamfests and club outings include a hidden-transmitter hunt as part of their agenda, and the competition is vigorous.

RDF also has a serious side. Repeater operators use these skills to track down repeater jammers and unlicensed intruders. It's also used to track down stolen transceivers that suddenly pop up on the air operated by people who appear unfamiliar with Amateur Radio operating procedures.

Locating a hidden transmitter is an art and, like any other art, practice makes perfect. So, if you have the opportunity to participate in a foxhunt, take advantage of it to hone your direction-finding skills for the real thing.

Further reading:

J. D. Moell and T. N. Curlee, *Transmitter Hunting* (Blue Ridge Summit, PA: Tab Books, 1987).

P. Danzer, Ed., *The 1997 ARRL Handbook* (Newington, CT: ARRL, 1996), Chapters 2 and 23.

D for disconnect, does the job). When that task is completed, the TNC will indicate it by displaying ***DISCONNECTED on your screen.

Another form of packet radio communications is transmission control protocol/internet protocol (TCP/IP). This is based on a more sophisticated computer networking model and permits some fancier functions for the user. To use TCP/IP, configure your TNC to turn control over to your computer by issuing a command to put the TNC in *KISS* mode (KISS stands for Keep It Simple, Stupid). The necessary software, called *NET* or *NOS*, is freely available from telephone and packet bulletin boards or from other sources by mail. The TCP/IP world interfaces with "regular" packet through gateway setups in many areas. For more information, see Chapter 10.

Satellite Gateway

A satellite gateway is a specially equipped station that links your local FM repeater to an Amateur Radio satellite. This makes it easier to try out a ham satellite without having to set up your own new equipment. It's your "gateway" to OSCAR operation because it allows you to use the simplest ham radio station, a hand-held transceiver, to work stations around the world.

Operating through a gateway is easy. The Operations Controller (OC) is in charge of what goes into the repeater and keeps things in order. A "list" operation works well to ensure minimum confusion by those trying to use the gateway through the repeater. The OC instructs users to keep each contact within a maximum length, to speak clearly, and preferably, to say "over" to avoid doubling. Good discipline in gateway operation is essential. The signals from the gateway are heard over a large portion of the Earth, so efforts to minimize confusion are desirable.

It's permissible to call CQ through a repeater to work someone by satellite. More stations will respond to "CQ from WS1O via gateway WB2NOM" than "QRZ OSCAR?" Beyond these considerations, operating through a satellite gateway is straightforward. Common sense dictates the rest.

FM—The Fun Mode

Monitor your local repeater and listen to how these procedures (and variations thereof) work in actual practice; you may discover that the local repeater community has its own special operating quirks. A little reading (of this chapter) and a little listening (to the local repeater) will prepare you to communicate efficiently and have fun with FM and repeaters.

Standard Frequency Offsets for Repeaters

Band	Offset	Band	Offset
29 MHz	100 kHz	440 MHz	5 MHz
52 MHz	1 MHz	902 MHz	12 MHz
144 MHz	600 kHz	1240 MHz	12 MHz
222 MHz	1.6 MHz		

10-M CTCSS Frequencies

In 1980 the ARRL Board of Directors adopted the 10-m CTCSS (PL) tone-controlled squelch frequencies listed below for voluntary incorporation into 10-m repeater systems to provide a uniform national system.

Call Area	Tone 1		Tone 2	
W1	131.8 Hz	-3B	91.5 Hz	-ZZ
W2	136.5	-4Z	94.8	-ZA
W3	141.3	-4A	97.4	-ZB
W4	146.2	-4B	100.0	-1Z
W5	151.4	-5Z	103.5	-1A
W6	156.7	-5A	107.2	-1B
W7	162.2	-5B	110.9	-2Z
W8	167.9	-6Z	114.8	-2A
W9	173.8	-6A	118.8	-2B
W0	179.9	-6B	123.0	-3Z
VE	127.3	-3A	88.5	-YB

CTCSS (PL) Tone Frequencies

The purpose of CTCSS (PL) is to reduce cochannel interference during band openings. CTCSS (PL) equipped repeaters would respond only to signals having the CTCSS tone required for that repeater. These repeaters would not respond to weak distant signals on their inputs and correspondingly not transmit and repeat to add to the congestion.

The standard Electronic Industries Association (EIA) frequency codes, in hertz, with their Motorola alphanumeric designators, are as follows:

67.0—XZ	107.2—1B	173.8—6A
69.3—WZ	110.9—2Z	179.9—6B
71.9—XA	114.8—2A	186.2—7Z
74.4—WA	118.8—2B	192.8—7A
77.0—XB	123.0—3Z	203.5—M1
79.7—WB	127.3—3A	206.5—8Z
82.5—YZ	131.8—3B	210.7—M2
85.4—YA	136.5—4Z	218.1—M3
88.5—YB	141.3—4A	225.7—M4
91.5—ZZ	146.2—4B	229.1—9Z
94.8—ZA	151.4—5Z	233.6—M5
97.4—ZB	156.7—5A	241.8—M6
100.0—1Z	162.2—5B	250.3—M7
103.5—1A	167.9—6Z	254.1—0Z

CHAPTER 12

VHF/UHF Operating

MICHAEL OWEN, W9IP
STAR ROUTE BOX 60
CANTON, NY 13617

The radio spectrum between 30 and 3,000 MHz is one of the greatest resources available to the radio amateur. The VHF and UHF amateur bands are a haven for ragchewers and experimenters alike; new modes of emission, new antennas and state-of-the-art equipment are all developed in this territory. Plenty of commercial equipment is available for the more popular bands, and "rolling your own" is very popular as well. Propagation conditions may change rapidly and seemingly unpredictably, but the keen observer can take advantage of subtle clues to make the most of the bands. Most North American hams are already well acquainted with 2-meter or 440-MHz FM. For many, channelized repeater operation is their first exposure to VHF or UHF. However, FM is only part of the story! This chapter will discuss the "other end" of the VHF/UHF bands, the low end, where almost all weak-signal SSB and CW activity takes place. It will discuss the bands from 50 MHz up through microwaves, but will emphasize the lower VHF frequency bands because the vast majority of amateur activity occurs there

HOW ARE THE BANDS ORGANIZED?

One of the keys to using this immense resource properly is knowing how the bands are organized. For example, if you are unaware of the calling frequencies on most of the VHF bands, you might spend weeks tuning around before you find someone to talk to! But, knowing the best frequencies and times, you will have little trouble making plenty of contacts, working DX, and enjoying the "world above 50 MHz." See Fig 12-1 for suggested band use in the VHF/UHF range.

Currently, activity on VHF/UHF is concentrated in the two lower VHF bands, 6 and 2 meters (50 and 144 MHz). The number of active stations on these bands is about equal. Above the 2-meter band, there are considerably more active stations on 432 MHz than any other. Because it is not available worldwide, 222 MHz is not inhabited as much as it could be. The 902-928 MHz band is gaining in popularity and 1296 MHz has quite a few occupants near major cities. For US amateurs, the 2300-MHz band is split into two parts (2300-2310 and 2390-2450 MHz). Higher frequencies, above 3300 MHz, are the territory of the true experimenters because little commercial equipment is available for those bands. See accompanying sidebar on the microwave bands.

The VHF/UHF bands are roomy. That is, there is lots of frequency spectrum available for ragchewing, experimenting or working DX. Although some of the more popular frequencies occasionally get very crowded, most VHF/UHF allocations are available all the time. This means that whatever you like to do—chat with your friends across town, test amateur television or bounce signals off the moon—there is usually plenty of spectrum available. VHF/UHF is a great resource!

The key to enjoyable use of this resource is to know how everyone else is using it, and to follow their lead. Basically, this means to listen first. Pay attention to the segments of the band already in use, and follow the operating practices that experienced operators are using. This way, you won't interfere with ongoing use of the band by others, and you'll "fit in" right away.

All of the bands between 50 and 1296 MHz have widely accepted calling frequencies (see Table 12-1). These are the frequencies most operators monitor most of the time, where you just park the receiver when you're doing something else around the shack. If someone wants to call you, or someone calls CQ, he or she will use the calling frequency and you will already be "on frequency" to hear them call.

The most important thing to remember about the calling frequencies is that they are not for ragchewing. After all, if a dozen other stations want to have a place to monitor for calls, it's really impolite to carry on a long-winded conversation on that frequency, isn't it?

In most areas of the country, everyone uses the calling frequency to establish a contact, and then the two stations move up or down a few tens of kHz to chat. This way, everyone can share the calling frequency without having to listen to each other's QSOs. You can easily tell if the band is open by monitoring the call signs of the stations making contact on the calling frequency—you sure don't need to hear their whole QSO!

On 6 meters, a *DX window* has been established in order to reduce interference to DX stations. This window, which extends from 50.100-50.125 MHz, is intended for DX QSOs only. The DX calling frequency is 50.110 MHz. If you make a DX contact and expect to ragchew, you should move up a few kHz to clear the DX calling frequency. US and Canadian 6-meter operators should use the domestic calling frequency of 50.125 MHz for non-DX work. Once again, when contact is established, you should move off the calling frequency as quickly as possible.

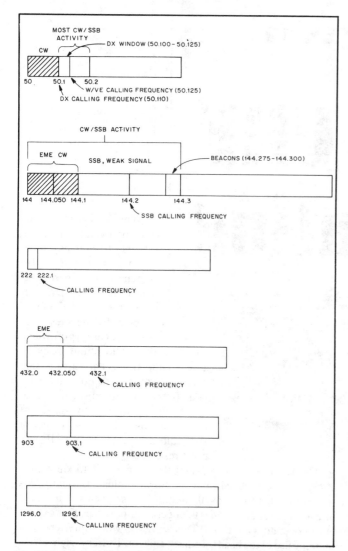

Fig 12-1—Suggested US VHF/UHF band usage (in MHz). On 222 MHz and above, activity is usually centered around the calling frequency. On the 50-MHz band, a DX window exists between 50.100 and 50.125 MHz. US and Canadian operators should leave this window clear except when working or calling DX. Domestic and local SSB QSOs should take place above 50.125 MHz.

ACTIVITY NIGHTS

Although you can scare up a QSO on 50 or 144 MHz almost any evening (especially during the summer), in some areas of the country there is not always enough activity to make it easy to find someone to talk with. Therefore, informal "activity nights" have been established in many parts of the country so you will know when to expect some activity. Each band has its own night.

There is a lot of variation in activity nights from place to place, so don't trust the following list to be entirely accurate. Check with someone in your area to find out about local activity nights.

Activity nights are particularly important for 222 MHz and above. On these bands, there is often little activity during most "normal" nights. If you have just finished a new transverter or antenna for one of these bands, you will have a much better chance to try them out during the band's weekly activity night. That doesn't mean there is no activity on other nights, especially if the band is open. It may just take longer to get someone's attention during other times

Local VHF/UHF nets often meet during activity nights. Two

Microwaves

The region above 1000 MHz is known as the microwave spectrum. Although European amateurs have made tremendous use of this rich resource, few Americans have done so. Little commercial equipment is available and entirely different construction techniques are required if amateurs are to build their own equipment. Nevertheless, the microwave bands are receiving a lot of attention from the experimentally minded among us. We can expect a considerable amount of progress as more amateurs try out the microwave bands.

10 GHz is the most popular microwave band, and VHF enthusiasts—such as Cheryl, KA1IXI—find 10-GHz mountaintopping particularly effective. (*NA1L photo*)

The most popular microwave band in the USA is the 10-GHz (3 cm) band. Using narrowband techniques (SSB/CW) and high-gain dish antennas, hams have made QSOs of over 1000 miles.

Most microwave activity is by prearranged schedules (at least in the US). This is partly because activity is so sparse and partly because antenna bandwidths are so narrow. With a 5° beamwidth, you would have to call CQ 72 times to cover a complete circle!

national organizations, SMIRK (Six Meter International Radio Klub) and SWOT (Sidewinders on Two), run nets in many parts of the country. These nets provide a meeting place for active users of the 50- or 144-MHz bands. In addition, the Southeastern VHF Society runs a net on 432 MHz. For those whose location is far away from the net control's location, the nets may provide a means of telling if your station is operating up to snuff, or if propagation is enhanced. Furthermore, you can sometimes catch a rare state or grid square checking into the net. For information on the meeting times and frequencies of the nets which are run by SMIRK and SWOT, ask other local

Common Activity Nights

Band (MHz)	Day	Local Time
50	Sunday	6:00 PM
144	Monday	7:00 PM
222	Tuesday	8:00 PM
432	Wednesday	9:00 PM
902	Friday	9:00 PM
1296	Thursday	10:00 PM

Table 12-1
North American Calling Frequencies

Band (MHz)	Calling Frequency
50	50.110 DX calling
	50.125 US, local
144	144.010 EME calling
	144.100, 144.110 CW
	144.200 SSB calling
222	222.100 CW/SSB
432	432.010 EME calling
	432.100 CW/SSB
902	903.1 (902.1 on West Coast)
1296	1296.100 CW/SSB
2304	2304.1
10,000	10,368.1 (w/SSB)
	10,280 WBFM

occupants of the bands in your area or write to these organizations directly: SMIRK c/o Ray Clark, K5ZMS, 7158 Stone Fence Dr, San Antonio, TX 78227; SWOT c/o Harry Arsenault, K1PLR, 48 Crane Rd, North Stamford, CT 06902.

WHERE AM I?

One of the first things you will notice when you tune the low end of any VHF band is that most QSOs include an exchange of "grid squares." What are grid squares? Well, first of all, grid squares aren't squares. They're more like rectangles, and they're just a way of dividing up the surface of the Earth. Grid squares are a shorthand means of describing your general location anywhere on the Earth. (For example, instead of trying to tell a distant station that I'm in Canton, New York, I tell them I'm in "grid square FN24." It sounds strange, but FN24 is a lot easier to locate on a map than my small town!) Grid squares are sometimes called "Maidenhead" grid squares.

Grid squares are based on latitude and longitude. The world is first divided into 324 very large areas called "fields," which cover 10∞ of latitude by 20° of longitude. Each field is divided into 100 "squares" from which grid squares get their name. Grid squares are 1° × 2° in size. When the full grid square designator is used, even smaller "sub-square" divisions are made, down to areas only a few kilometers wide. For most of us, knowing about "squares" is sufficient.

Grid squares are coded with a 2-letter/2-number/2-letter code (such as FN24KP). Most people just use the first four characters (such as FN24). This handy designator uniquely identifies the grid square; no two have the same identifying code. This is the biggest advantage of the Maidenhead system over the older QRA-locator system used for decades in Europe. The European system is not applicable worldwide because the same QRA designator might be used for more than one place.

There are several ways to find out your own grid square identifier. You can start by consulting Tables 12-2 and 12-3. By following the instructions shown in the tables, you will be able to locate your own grid square identifier. The hardest part is finding your QTH on a good map that has latitude and longitude on it; the rest is easy. Most high-quality road maps have this on the margins. Or, you could go a step further and purchase the topographic map of your immediate area from the US Geological Survey. For maps, write to US Geological Survey Information Services, Box 25286, Denver, CO 80225, tel 1-800-USA-MAPS. Once you have your latitude and longitude, the rest is a snap!

The ARRL publishes a map (see Fig 12-2) of the continental United States and most populated areas of Canada, on which grid squares are marked. This map is available from ARRL for $1. It really helps to keep a copy of the map handy to help locate stations as you work them. If you are keeping track of grid squares for the purpose of VUCC (the VHF/UHF Century Club award—see below), you can color in each square as you work it. Some operators use a light color when they work the grid for the first time, then color it more darkly when they receive a

QSL. Others use color or pattern schemes to indicate different propagation modes. [Note: The ARRL also publishes a *World Grid Locator Atlas*, available for $5.]

If you own a computer, a program known as *GRIDLOC* will allow you to determine your grid square if you know your latitude and longitude. The program also works in reverse, providing latitude and longitude for any grid square identifier. *GRIDLOC* is just one of 26 handy programs that are available on the *UHF/Microwave Experimenter's Software* diskette (3½-inch, for IBM-PCs and compatibles). The diskette is available from the ARRL for $10.

PROPAGATION
Normal Conditions

If you are new to the world above 50 MHz, you might wonder what sort of range is considered "normal." To a large extent, your range on VHF is determined by your location and the quality of your station. After all, you can't expect the same

Pete, N8YEL, enjoys portable VHF operating along the shores of Lake Michigan.

Table 12-2
How to Determine Your Grid Location†

†For those geographical areas not encompassed here, a complete explanation appears in the April 1982, issue of *The Lunar Letter*, entitled "Worldwide QTH Locator System Proposed by Region 1," by Lance Collister, WA1JXN.

1st and 2nd characters: Read directly from the map.

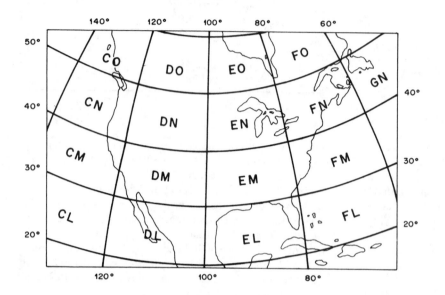

3rd character: Take the number of whole degrees west longitude, and consult the following chart.

Degrees West Longitude	Third Character	Degrees West Longitude	Third Character	Degrees West Longitude	Third Character
60-61	9	88- 89	5	114-115	2
62-63	8	90- 91	4	116-117	1
64-65	7	92- 93	3	118-119	0
66-67	6	94- 95	2	120-121	9
68-69	5	96- 97	1	122-123	8
70-71	4	98- 99	0	124-125	7
72-73	3	100-101	9	126-127	6
74-75	2	102-103	8	128-129	5
76-77	1	104-105	7	130-131	4
78-79	0	106-107	6	132-133	3
80-81	9	108-109	5	134-135	2
82-83	8	110-111	4	136-137	1
84-85	7	112-113	3	138-139	0
86-87	6				

4th character: This number is the same as the *2nd single digit* of your latitude. For example, if your latitude is 41° N, the 4th character is 1; for 29° N, it's 9, etc.

This four-character (2-letter, 2-number) designator indicates your 2° × 1° square for VUCC award purposes.

Table 12-3
More Precise Locator

To indicate location more precisely, the addition of 5th and 6th characters will define the *sub-square*, measuring about 4 × 3 miles. Longitude-latitude coordinates on maps, such as U.S. Department of the Interior Surveys, can be extrapolated to the nearest tenth of a minute, necessary for this level of locator precision. *This is not necessary in the VUCC awards program.*

5th character: If your number of degrees longitude is an *odd* number, see Fig. A. If your number of degrees longitude is an *even* number, see Fig. B.

Odd Longitude° (Fig. A)	
Minutes W. Longitude	5th Character
0- 5	L
5-10	K
10-15	J
15-20	I
20-25	H
25-30	G
30-35	F
35-40	E
40-45	D
45-50	C
50-55	B
55-60	A

Even Longitude° (Fig. B)	
Minutes W. Longitude	5th Character
0- 5	X
5-10	W
10-15	V
15-20	U
20-25	T
25-30	S
30-35	R
35-40	Q
40-45	P
45-50	O
50-55	N
55-60	M

6th character: Take the number of *minutes of latitude* (following the number of degrees) and consult the following chart.

Minutes N. Latitude	6th Character
0- 2.5	A
2.5- 5.0	B
5.0- 7.5	C
7.5-10.0	D
10.0-12.5	E
12.5-15.0	F
15.0-17.5	G
17.5-20.0	H
20.0-22.5	I
22.5-25.0	J
25.0-27.5	K
27.5-30.0	L
30.0-32.5	M
32.5-35.0	N
35.0-37.5	O
37.5-40.0	P
40.0-42.5	Q
42.5-45.0	R
45.0-47.5	S
47.5-50.0	T
50.0-52.5	U
52.5-55.0	V
55.0-57.5	W
57.5-60.0	X

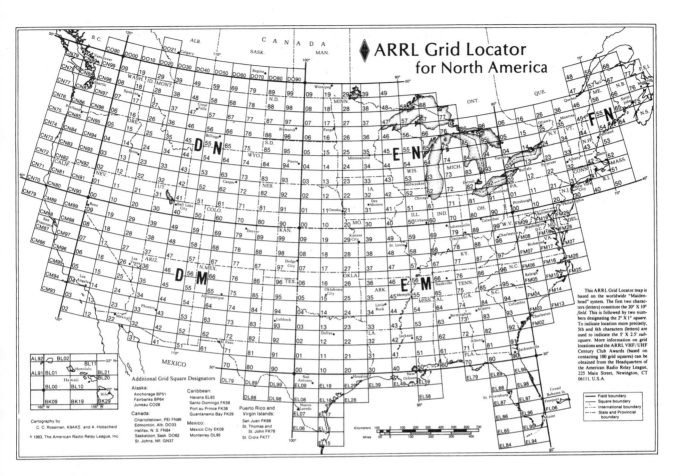

Fig 12-2—ARRL Grid Locator Map

WA6YBT's elegant station layout: a 2-meter multimode transceiver, antenna tuner and HF rig, 2-meter and 70-cm transverter with a solid-state 70-cm amplifier on top of it, antenna rotor controls and a wattmeter (*WA6YBT photo*)

performance from a 10-watt rig and a small antenna on the roof as you might from a kilowatt and stacked beams at 100 feet.

For the sake of discussion, consider a more-or-less "typical" station. On 2-meter SSB, a hypothetical typical station would probably consist of a low-powered rig, perhaps a multimode rig (SSB/CW/FM), followed by a 100-watt amplifier. The antenna of our typical station might be a single long Yagi at around 50 feet, fed with low-loss coax.

Using SSB or CW, how far could this station cover on an average night? Location plays a big role, but it's probably safe to estimate that you could talk to a similarly equipped station about 200 miles away almost 100% of the time. Naturally, higher-power stations with tall antennas and low-noise receive preamps will have a greater range than this, up to a practical maximum of about 350-400 miles in the Midwest (less in the hilly West and East).

On 222 MHz, a similar station might expect to cover about the same distance, and somewhat less (150 miles) on 432 MHz. This assumes normal propagation conditions and a reasonably unobstructed horizon. This range is a lot greater than you would get for noise-free communication on FM, and it represents the sort of capability the typical station should seek. Increase the height of the antenna to 80 feet and the range might increase to 250 miles, probably more, depending on your location. That's not bad for reliable communication!

Band Openings and DX

The main thrill of the VHF and UHF bands for most of us is the occasional "band opening," when signals from far away are received as if they are next door. DX of well over 1000 miles on 6 meters is commonplace during the summer, and occurs at least once or twice a year on 144, 222 and 432 MHz.

DX propagation on the VHF/UHF bands is strongly influenced by the seasons. Summer and fall are definitely the most active times in the spectrum above 50 MHz, although band openings occur at other times as well.

The following is a review of the different types of "band openings," what bands they affect and how they may be predicted. Later on, how to operate these modes will be discussed. Remember that there is a lot of variation, and that no two band openings are alike. This uncertainty is part of what makes VHF/UHF interesting! Here is a list of the main types:

1) *Tropospheric—or simply "tropo"—openings.* Tropo is the most common form of DX-producing propagation on the bands above 144 MHz. It comes in several forms, depending on

local and regional weather patterns. This is because it is caused by the weather. Tropo may cover only a few hundred miles, or it may include huge areas of the country at once.

a) *"Radiation Inversion"* (mostly summer). This common type of tropo is caused by the Earth cooling off in the evening. The air just above the Earth's surface also cools, while the air a few hundred meters above remains warm. This creates an inversion that refracts VHF/UHF radio signals. If there is little or no wind, you will notice a gradual improvement in the strength of signals out to a range of 100-200 miles (less in hilly areas) as evening passes into night. This form of tropo mostly affects the bands above 144 MHz, and is seldom noticeable at 50 MHz.

b) *Broad, regional tropo openings* (late summer and early fall). These are the DXer's dream! In one of these openings, stations as far away as 1200 miles (maybe more!) are brought into range on VHF/UHF. The broad, regional type of tropo opening is caused by stagnation of a large, slow-moving high-pressure system. The stations that benefit the most are often on the south or western sides (the so-called "back side") of the system. This sort of sluggish weather system may often be forecast just by looking at a weather map.

c) *Wave-cyclone tropo* (spring). These openings don't usually last long. They are brought about by an advancing cold front that interacts with the warm sector ahead of it. The resulting contrast in air temperatures may cause thunderstorms, but sometimes don't. If conditions are just right, a band opening may result. These openings usually involve stations in the warm sector ahead of the cold front. You may feel that there's a pipeline between you and a DX station 1000 miles or more away. You may not be able to contact anybody else!

There are several other types of tropo openings, but space doesn't allow more than a brief discussion here. Coastal breezes sometimes cause long, narrow tropo openings along the East and Gulf Coasts. Rarely, cold fronts may cause brief openings as they slide under warmer air. US West Coast VHFers are always on the alert for the California-to-Hawaii duct that permits 2500-mile DX. For a very complete discussion of all the major forms of weather-related VHF propagation, see "The Weather Which Brings VHF DX" by Emil Pocock, W3EP, May 1983 *QST*.

2) *The Scatter modes.* Long-distance communication on the VHF and lower UHF bands is possible using the ionosphere. Some modes are within the reach of modest stations, whereas others require very large antennas and high power. They all have one thing in common: They take advantage of the scattering of radio waves whenever there is an irregularity in the uppermost atmosphere or ionosphere. Very briefly, the main types are:

a) *Meteor scatter.* When meteors enter the Earth's atmosphere, they ionize a small trail through the E layer. This ionization doesn't last long, often only a second. However, before it dissipates, the ionization can scatter, or sometimes reflect, VHF radio waves. This sort of propagation may be identified by the fact that it's very brief, but signals may be rather strong. It is quite common on 6 meters, less so on 2 meters, and rare indeed on 222 and 432 MHz. No successful QSOs have been completed at higher frequencies. Operating techniques for this mode are discussed later.

b) *Tropo scatter.* This takes place in the troposphere, and is the result of wave scattering from bundles of turbulent air. Signals are always weak, with a large amount of fading. Depending on frequency, tropo scatter may be assisted by scatter in the D and E layers of the ionosphere. Contacts out to about 1000 miles are possible, but difficult; definitely the territory of well-equipped stations.

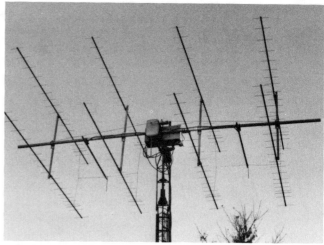

W9IP/2 in Western New York: an 8 x 19-element RIW 432-MHz EME array fed with open-wire line.

3) *Sporadic E (often called "E_s " for "E sporadic").* This type of propagation is the most spectacular DX-producer on the 50-MHz band, where it may occur almost every day during late June and July and early August. A short E_s season also occurs during December-January. Sporadic E is most common in mid-morning and again around sunset, during the summer months, but it can occur any time, any date. E_s occurs once or twice a year on the 2-meter band in most areas.

E_s results from small patches of ionization in the E layer of the ionosphere. These patches move about, intensify or disappear altogether. The fast-moving patches of ionization appear to follow a general SE-NW track as they fly across the mid-latitude Northern Hemisphere.

Reflections from E_s patches make coast-to-coast DX possible on 6 meters. Single-hop distances of 1200-1400 miles are most common, but multi-hop openings are common on 6 meters. Multi-hop has also been observed on 2 meters. It is exceedingly rare, but it can provide the most spectacular 144-MHz DX available without moonbounce. QSOs above 2 meters are possible, but they are exceedingly rare.

E_s signals are usually quite strong, but they may fade away altogether within a short time. On 6 meters, some days are seemingly filled with E skip. Strong signals from all parts of the country can pour in for hours and hours, or sometimes only a small area is covered. However, it is also common to go through "dry periods" when there may be no E_s for a week or more, even in the middle of the summer. That's one reason they call it "sporadic" E!

Sporadic E openings usually last only about half an hour on 2 meters, but some have lasted much longer. They are much more rare on 2 meters than on 6. Probably fewer than one in 20 strong E_s openings on 6 meters ever materialize on 2 meters.

4) *Aurora ("Au").* The aurora borealis, or "northern lights," is a beautiful spectacle which is seen occasionally by those who live in Canada, the northern part of the USA and Europe. Similar "southern lights" are sometimes visible in the southernmost parts of South America and Africa. The aurora is caused by the Earth intercepting a massive number of charged particles thrown from the Sun during a solar "storm." These particles are funnelled into the polar regions of the Earth by its magnetic field. As the charged particles interact with the upper atmosphere, the air glows, which we see as the aurora. And,

important to VHFers, these particles also provide an irregular, moving curtain of ionization which can propagate signals for many hundreds of miles.

Like sporadic E, aurora is much more common on 6 meters than on 2 meters. Nevertheless, 2-meter Au is far more common than 2-meter sporadic E, at least above 40° N latitude. Au is also possible on 222 and 432 MHz, and many tremendous DX contacts have been made on these bands. Current record distances are over 1000 miles on 144, 222 and 432 MHz.

Aurora can be predicted by listening to radio station WWV at the National Institute of Science and Technology. At 18 minutes past each hour, WWV transmits a summary of the condition of the Earth's geomagnetic field. If the K index is 4 or above, you should watch for Au. However, as many VHFers have learned, a high K index is no guarantee of an aurora. Similarly, K indices of only 3 have occasionally produced spectacular radio auroras at middle latitudes. When in doubt, point the antenna north and listen!

Auroral DX signals are highly distorted. CW is the most practical mode, although SSB is sometimes used. Stations point their antennas generally northwards and listen for the telltale hissing note that is characteristic of auroral signals. More about operating Au is presented later.

5) *Transequatorial propagation (TE)*. This strange mode is responsible for almost unbelievable DX on 50 through 222 MHz. It is a strictly north-south path, crossing the equator. Stations which are equally spaced north and south of the geomagnetic equator have communicated over 4800 miles on 2 meters. Apparently some form of ionospheric ducting is responsible, although there is little agreement on the exact mechanism. TE is most common during and after periods of geomagnetic unrest. Signals have a weak, fluttery quality. This form of propagation is quite worthy of continued experimentation by amateurs.

6) *EME, or Earth-Moon-Earth, often called "moonbounce."* This is the ultimate VHF/UHF DX medium! Moonbouncers use the Moon as a passive reflector for their signals, and QSO distance is limited only by the diameter of the Earth. Any two stations who can simultaneously see the Moon may be able to work each other via EME. QSOs between the USA and Europe or Japan are commonplace on VHF and UHF by using this mode. That's DX!

Previously the territory of only the biggest and most serious VHFers, moonbounce has now become more widely popular. Thanks to the efforts of pioneer moonbouncers such as Bob Sutherland, W6PO, and Al Katz, K2UYH, hundreds of stations are active, mainly on 144 and 432 MHz. This huge increase in activity is partly the result of the "snowball" effect: More stations are EME-capable, so there's more incentive for others to "get on the Moon." Improvements in technology—low-noise amplifiers and better designs—have made it easier to get started. Also, several individuals have assembled gigantic antenna arrays, which make up for the inadequacy of smaller antennas. The result is that even modestly equipped VHF stations (150 watts and one or two Yagis) are capable of making a few moonbounce contacts. Activity is constantly increasing. There is even an EME contest in which moonbouncers compete on an international scale.

Moonbounce requires larger antennas than most terrestrial VHF/UHF work. In addition, you must have a high-power transmitting amplifier and a low-noise receiving preamplifier if you wish to be able to work more than the biggest guns. A "typical" EME station on 144 MHz probably consists of four long-boom Yagi antennas on an azimuth-elevation mount (for pointing at the Moon), a kilowatt amplifier and a GaAsFET preamplifier. On 432 MHz, the "average" antenna is eight long Yagis. You can make contacts with a smaller antenna, but they will be with only larger stations on the other end. Yagi antennas aren't the only type available; collinears and quagis are also widely used. Several UHFers have also built large parabolic dish antennas.

HOW DO I OPERATE ON VHF/UHF?
Normal Conditions

The most important rule to follow, on VHF/UHF like all other amateur bands, is to listen first. Even on the relatively uncrowded VHF bands, interference is common near the calling frequencies. The first thing to do when you switch on the radio is to tune around, listening for activity. Of course, the calling frequencies are the best place to start listening. If you listen for a few minutes, you'll probably hear someone make a call, even if the band isn't open. If you don't hear anyone, then it's time to make some noise yourself!

Stirring up activity on the lower VHF bands is usually just a matter of pointing the antenna and calling CQ. Because most VHF beams are rather narrow, you might have to call CQ in several directions before you find someone. Several short CQs are always more productive than one long-winded CQ. But don't make CQs too short; you have to give the other station time to turn the antenna toward you.

Give your rotator lots of exercise; don't point the antenna at the same place all the time. You never know if a new station or some DX might be available at some odd beam heading. VHFers in out-of-the-way locations, far from major cities, monitor the bands in the hope of hearing you.

Band Openings

How about DX? What is the best way to work DX when the band is open? There's no simple answer. Each main type of band opening or propagation mode requires its own techniques. This is natural, because the strength and duration of openings vary considerably. For example, you wouldn't expect to oper-

WA8NLC's 17-foot dish for 1296-MHz EME.

We thought they only did this at airports! Jack, N3DQZ, fixes his VHF and UHF antennas after a nasty winter storm.

ate the same way during a 10-second meteor-burst QSO as during a three-day tropo opening.

The following is a review of the different main types of propagation and descriptions of the ways that most VHFers take advantage of them.

Tropo

Tropospheric propagation—usually called "tropo"—is usually most noticeable on the 144- through 1296-MHz bands. It often lasts at least several hours, and sometimes several days, so there's no big hurry to make contacts. You have time to listen carefully and determine how the opening is developing, how it is drifting and who is active. For stations on the West Coast, the openings usually extend up the coast and inland a couple of hundred miles or so. Therefore, the biggest question is how far north or south, and how far east, the opening will extend. Hawaii is usually at the other end of the path from the West Coast. In the middle part of the country and the East Coast, openings sometimes extend for well over a thousand miles, and in several directions.

[Recalling the ARRL VHF QSO Party in September 1979, in which the Midwest, South and East Coast experienced one of the greatest tropo openings in VHF history, my multioperator contest log shows literally dozens of QSOs with stations more than 750 miles away, and several over 1000 miles. The VHF bands were open over a huge area; within one hour, for example, stations in southern Texas and Vermont were worked from my QTH in Illinois on 2 meters. The 432-MHz band was almost as good.]

East of the Rockies, big tropo openings occur at least once a year, although they aren't often as massive or long-lived as the September '79 opening. Unfortunately, amateurs in the mountainous West seldom experience tropo of this scale. In California, most openings are either along the coast or between the mainland and Hawaii. The latter, which accounts for the world's record tropo distances (greater than 2500 miles!) on VHF and several UHF bands, is rare.

To take best advantage of tropo openings, you need to know how they develop with time. Here are some tips: First, remember that most weather systems drift to the east with time. Large high-pressure systems tend to drift east and spread out as they mature. The long-haul DX may be available just before the opening terminates in your area. In other words, it may appear that the opening gets better and better and then all of a sudden

you're out of it. If you're in the middle of the opening, sometimes stations will be "talking over your head." Be patient. Your time will come as the opening drifts east. Second, remember that most tropo-type openings improve as the evening progresses. If the opening is good and the weather is stable at sunset, then start the coffee pot, because you'll be DXing all night! In fact, the best tropo DX seems to come just around sunrise. But don't forget that if nobody is on the air, it may seem like the opening has ended. In the small hours of the morning, you might have trouble finding anyone on the air.

Tropo/Ionospheric Scatter

Tropo scatter usually requires prearranged schedules, a lot of patience and good equipment at each end of the path. It is often assisted by D-layer scattering, airplanes or even meteor scatter. Some form of tropo scatter is usable on all the VHF and UHF bands. The best means to take advantage of tropo scatter is to use medium-speed CW and follow a sequence in transmit-receive between the two stations; 30-second sequences seem to work best. One station transmits for the first 30 seconds of each minute, and the other station listens, then they trade-off. If you're lucky, a passing airplane or some other anomaly will briefly enhance signals enough to make the QSO.

Meteor Scatter

Meteor scatter is very widely used on 50 and 144 MHz, and it has been used on 222 and 432 as well. Operation with this exciting mode of DX comes under two main headings: prearranged schedules and random contacts. Either SSB or CW may be used, although SSB is more popular in North America. European meteor-scatter enthusiasts use very high speed CW (50-100 WPM) which is recorded on tape.

Most meteor-scatter work is done during major meteor showers, and many stations "run skeds" (prearranged schedules) with others in states or grid squares that they need. In a "sked," 15-second transmit-receive sequences are the norm for North America (Europeans use longer sequences). One station, almost always the westernmost, will "take the first and third." This means that he/she transmits during the first and third 15-second interval of each minute, and the other station transmits during the other two segments. This is a very simple procedure that ensures that only one station is transmitting when a meteor falls. See accompanying sidebar.

A specific frequency, far removed from local activity centers, is chosen when the sked is set up (on the HF bands, the Internet or on the telephone). It is important that both stations have accurate frequency readout and synchronized clocks, but with today's technology this is not the big problem that it once was. Schedules normally run for 1/2 hour or 1 hour. On 222 and 432 MHz, however, two-hour skeds are not uncommon because many fewer meteor trails are available at those frequencies. Meteor-scatter QSOs on 222 and 432 are well earned!

The best way to get the feel for the meteor-scatter QSO format is to listen to a couple of skeds between experienced operators. Then, look in *QST* or ask around for the name/call sign of veteran meteor-scatter operators in the 800-1000 mile range from you (this is the easiest distance for meteor scatter). Call them on the telephone or catch them on one of the national VHF nets and arrange a sked. After you cut your teeth on "easy" skeds, you'll be ready for more difficult DX.

A lot of stations make plenty of meteor-scatter QSOs without the help of skeds. Especially during major meteor showers, VHFers congregate near the calling frequency of each band. There they wait for meteor bursts like hunters waiting for ducks.

Energetic operators call CQ while everyone else listens; when the big "Blue Whizzer" meteors blast in, the band comes alive with dozens of quick QSOs. For a brief time, normally five seconds to perhaps 30 seconds, 2 meters may sound like 20 meters! Then the band is quiet again. . . until the next meteor burst!

The quality of shower-related meteor-scatter DX depends on three factors. These three factors are well known or can be predicted. The most important is the "radiant effect." This refers to where the shower's radiant is located in the sky. The radiant is the spot in the sky from which the meteors appear to fall. If the radiant is below the horizon, or too high in the sky, you will hear very few meteors. The most productive spot for the radiant is at an elevation of about 45° and an azimuth of 90° from the path you're trying to work. The second important factor is the velocity of the meteors. Slow meteors cannot ionize sufficiently to propagate 144- or 222-MHz signals, no matter how many meteors there are. For 144 MHz, meteors slower than 50 km/s are usually inadequate (see accompanying sidebar detailing major meteor showers). Third, the shower will have a peak in the number of meteors that the Earth intercepts. However, because the peak of many meteor showers is more than a day in length, the exact time of the peak is not as important as most people think. Two interesting references are "Improving Meteor Scatter Communications" by Joe Reisert, W1JR, in June 1984 *Ham Radio*, and "VHF Meteor Scatter: an Astronomical Perspective," by M. R. Owen, W9IP, in June 1986 *QST*.

It takes a lot of persistence and a good station to be successful with "random" meteor scatter. This is mainly because you must overcome tremendous QRM in addition to the fluctuations of meteor propagation. At least 100 watts is necessary for much success in meteor scatter, and a full kilowatt will help a lot. One or two Yagis, stacked vertically, is a good antenna system.

In populated areas, it can be difficult to hear incoming meteor-scatter DX if many local stations are calling CQ. Therefore, many areas observe 15-second sequencing for random meteor-scatter QSOs, just as for skeds. Those who want to call CQ do so at the same time so everyone can listen for responses between transmissions. Sometimes a bit of peer pressure is necessary to keep everyone together, but it pays off in more QSOs for all. The same QSO format is used for scheduled and "random" meteor-scatter QSOs.

Sporadic E

Sporadic E (E$_s$) propagation is very common on 6 meters during the summer and very early fall, and rare at other times. It is always rare on 2 meters, but the greatest chance of it happening is during July and early August. On 6 meters, the band may open for hours on end, constantly shifting, disappearing briefly, then reappearing. Signals are often very strong, and 10-watt stations can work 1000 miles with ease. On 2 meters, the openings are equally strong but often very short; a half-hour sporadic E opening on 2 meters is rare but wonderful.

Predicting these openings in advance is very difficult, if not impossible. A very weak correlation may exist between sporadic E and large thunderstorms, but this might be because big storms and E skip are both more common during the summer. The best way to alert yourself to the possibility of sporadic E is to monitor lower frequencies, because this type of propagation usually rises in frequency. An opening may begin as "short skip" on 10 meters, then rise to include 6 meters. Then, distant FM broadcast and TV stations will become audible as the

---

Meteor-Scatter Procedure

In a meteor-scatter QSO, neither station can hear the other except when a meteor trail exists to scatter or reflect their signals. The two stations take turns transmitting so that they can be sure of hearing the other if a meteor happens to fall. They agree beforehand on the sequence of transmission. One station agrees to transmit the 1st and 3rd 15 seconds of each minute, and the other station takes the 2nd and 4th. It is standard procedure for the western-most station to transmit during the 1st and 3rd.

It's important to have a format for transmissions so you know what the other station has heard. This format is used by most US stations;

Transmitting	Means you have copied;
Call signs	nothing, or only partial calls
Call plus signal report (or grid square or state)	full calls—both sets
ROGER plus signal report (or grid square or state)	full calls, plus signal report (or grid square or state)
ROGER	ROGER from other station

Remember, for a valid QSO to take place, you must exchange full call signs, some piece of information, and acknowledgment. Too many meteor QSOs have not been completed for lack of ROGERs. Don't quit too soon; be sure the other station has received your acknowledgment. Often, stations will add "73" when they want to indicate that they have heard the other station's ROGER.

Until a few years ago, it was universal practice to give a signal report which indicated the length of the meteor burst. "S1" meant that you were just hearing pings, "S2" meant 1-5 second bursts, and so on. Unfortunately, virtu- ally everyone was sending "S2," so there was no mystery at all, and no significant information was being exchanged.

Grid squares have become popular as the "piece of information" in meteor-scatter QSOs. More and more stations are sending their grid-square designator instead of "S2". This is especially true on "random" meteor-scatter QSOs, where you might not know in advance where the station is located. Other stations prefer to give their state instead. For an excellent summary of meteor-scatter procedure, see "Meteor Scatter Communications" by Clarke Greene, K1JX, in January 1986 *QST* (pp 14-17). Also see "Hooked on Meteors!" by Tom Hammond, WD8BKM, in May 1995 *QST* (p 74).

---

maximum frequency of the E cloud reflections goes higher. Then, if you're very lucky indeed, the E cloud will create an opening on 2 meters.

You'll need to be quick on your toes to take advantage of sporadic E. You might not even notice an E$_s$ band opening in progress because signals are so strong ("DX is never that loud!"). Quick, brief calls and short exchanges are needed if you are to make the most of the opening. Don't spend a half hour calling a long-winded CQ. The opening will be gone by the time you're through. Once the E$_s$ opening is in progress, most stations don't waste a lot of time hunting around. Just work everything you can hear, and quickly. Almost all stations will be DX, so you can't lose!

Don't worry if you aren't running a lot of power and a huge antenna; sporadic E is the greatest equalizer in VHF. The size of your station hardly matters at all. It's mostly location that counts! You may be having the greatest opening of all time, and your friend a few miles away might hear little or nothing. That's the way E$_s$ is. Next week, your buddy may be the lucky one and you will be left out, so make the most of it!

Aurora

Aurora (Au) is a propagation mode which favors stations at

Meteor Showers

Every day, the Earth is bombarded by billions of tiny grains of interplanetary debris, called meteors. They create short-lived trails of E-layer ionization which can be used as reflectors for VHF radio waves. On a normal morning, careful listeners can hear about 3-5 meteor pings (short bursts of meteor-reflected signal) per hour on 2 meters.

At several times during the year, the Earth passes through huge clouds of concentrated meteoric debris, and VHFers enjoy a "meteor shower". During meteor showers, 2-meter operators may hear 50 or more pings and bursts per hour. Here are some data on the major meteor showers of the year. Other showers also occur, but they are very minor.

Major Meteor Showers

Shower	Date range	Peak date	Time above quarter max	Approximate visual rate	Speed km/s	Best Path	Time (local)
Quadrantids	Jan 1-6	Jan 3/4	14 hours	40-150	41	NE-SW;SE-NW	13-15;05-07
Eta Aquarids	Apr 21-May 12	May 4/5	3 days	10-40	65	NE-SW;E-W;SE-NW	05-07;06-09;09-11
Arietids	May 29-Jun 19	Jun 7	?	60	37	N-S	06-07 & 13-14
Perseids	Jul 23-Aug 20	Aug 12	4.6 days	50-100	59	NE-SW;SE-NW	09-11;01-03
Orionids	Oct 2-Nov 7	Oct 22	2 days	10-70	66	NE-SW;N-S;NW-SE	01-03;01-02 & 07-09;07-08
Geminids	Dec 4-16	Dec 13/14	2.6 days	50-80	34	N-S	22-24 & 05-07

Meteor-Shower Peaks

These calculated predictions of meteor-shower peak times/dates are based on the latest astronomical data. Nevertheless, meteor showers are notoriously fickle about meeting their predicted peaks. Therefore, don't put too much faith in these estimates. The time at which the shower's radiant is at a 45° elevation, perpendicular to the path you want to work, is much more important than the actual peak of the shower. Best times and paths are listed above. These values are calculated for Kansas City and are valid for the entire USA because the time indicated is local standard time. One hour before and after the given time ranges should generally be good as well.

Meteor Shower Peak Times (UTC): 1997-1998

Note: The first line indicates month-day (eg, 1-3 is January 3) and the second line (eg, 1900) indicates UTC.

Shower	1997	1998	Shower	1997	1998
Quadrantids	1-3 1300	1-3 1900	Perseids	8-12 1300	8-12 1900
Eta Aquarids	5-6 0400	5-6 1000	Orionids	10-22 0100	10-22 0700
Arietids	6-7 0900	6-7 1500	Geminids	12-13 1930	12-14 0130

high latitudes. It is a wonderful blessing for those who must suffer through long, cold winters because other forms of propagation are rare during the winter. Aurora can come at almost any time of the year. New England stations get Au on 2 meters about five to 10 times a year, whereas stations in Tennessee get it once a year if they're lucky. Central and Southern California almost never see aurora.

The MUF of the aurora seems to rise quickly, so don't wait for the lower-frequency VHF bands to get exhausted before moving up in frequency. Check the higher bands right away.

You'll notice Au by its characteristic hiss. Signals are distorted by reflection and scattering off the rapidly moving curtain of ionization. They sound like they are being transmitted by a leaking high-pressure steam vent rather than radio. SSB voice signals are so badly distorted that often you cannot understand them unless the speaker talks very slowly and clearly. The amount of distortion increases with frequency. Most 50-MHz Au contacts are made on SSB, where distortion is the least. On 144, 222 and 432 MHz, CW is the only really useful means of communicating via AU.

If you suspect Au, tune to the CW calling frequency (144.100) or the SSB calling frequency (144.200) and listen with the antenna to the north. Maybe you'll hear some signals. Try swinging the antenna as much as 45° either side of due north to peak signals. In general, the longest-distance DX sta-

tions peak the farthest away from due north. Also, it is possible to work stations far south of you by using the aurora; in that case your antenna is often pointed north.

High power isn't necessary for aurora, but it helps. Ten-watt stations have made Au QSOs but it takes a lot of perseverance. Increasing your power to 100 watts will greatly improve your chances of making Au QSOs. As with most short-lived DX openings, it pays to keep transmissions brief.

Aurora openings may last only a few minutes or they may last many hours, and the opening may return the next night, too. If WWV indicates a geomagnetic storm, begin listening on 2 meters in the late afternoon. Many spectacular Au openings begin before sunset and continue all evening. If you get the feeling that the Au has faded away, don't give up too soon. Aurora has a habit of "dying" and then returning several times, often around midnight. And remember: If you experience a terrific Au opening, look for an encore performance about 27-28 days later, because of the rotation of the sun.

Propagation Indicators

Active VHFers keep a careful eye on various propagation indicators to tell if the VHF bands will be open. The kind of indicator you monitor is related to the kind of propagation you are expecting. For example, during the summer it pays to watch closely for sporadic E because openings may be very brief. If your area only gets one or two per year on 2 meters, you sure don't want to miss them!

Many forms of VHF/UHF propagation develop first at low frequency and then move upward to include the higher bands. Aurora is a good example. Usually it is heard first on 10 meters, then 6 meters, then 2 meters. Depending on your location and the intensity of the aurora, the time delay between hearing it on 6 meters and its appearance on 2 meters may be only a few minutes, to as much as an hour, later. Still, it shows up first at low frequency. The same is true of sporadic E; it will be noticed first on 10 meters, then 6, then 2.

Tropospheric propagation, particularly tropo ducting, acts in just the reverse manner. Inversions and ducts form at higher frequencies first. As the inversion layer grows in thickness, it refracts lower and lower frequencies. However, most of us don't notice this because amateur activity drops off with higher frequencies, and atmospheric losses increase. So, it may be true that the band opens high in frequency first, but it may appear to open first on 2 meters because that is where much of the country's VHF activity is. Six-meter tropo openings are very rare because few inversion layers ever develop sufficient thickness to enhance such long-wavelength signals.

How can you monitor for band openings? The best way is to take advantage of the "beacons" which are inadvertently provided by commercial TV and FM broadcast stations. Television Channels 2 through 6 (54-88 MHz) are great for catching sporadic E. As the E opening develops, you will see one (or more) stations appear on each channel. First, you may see Channel 2 get cluttered with stations 1000 miles or more away, then the higher channels may follow. If you see strong DX stations on Channel 6, better get the 2-meter rig warmed up. Channel 7 is at 175 MHz, so you rarely see any sporadic E there or on any higher channels. If you do, however, it means that a major 2-meter opening is in progress!

The gap between TV Channel 6 and 7 is occupied partly by the FM broadcast band (88-108 MHz). Monitoring that spectrum will give you a similar "feel" for propagation conditions.

Several amateurs have built converters to monitor TV video carrier frequencies. A system like this can be used to keep track of meteor showers. It's a way of checking to tell if meteors are plentiful or not, even if there appears to be little activity on the amateur VHF bands. It also can alert you to aurora and sporadic E. A variety of systems for monitoring propagation have been discussed by Joe Reisert, W1JR, in the June 1984 issue of *Ham Radio*.

EME: Earth-Moon-Earth

EME, or moonbounce, is available any time the Moon is in a favorable position. Fortunately, the Moon's position may be easily calculated in advance, so you always know when this form of DX will be ready. It's not actually that simple, of course, because the Earth's geomagnetic field can play havoc with EME signals as they are leaving the Earth's atmosphere and as they return. Not only can absorption (path loss) vary, but the polarization of the radio waves can rotate, causing abnormally high path losses at some times. Still, most EME activity is predictable.

You will find many more signals "off the Moon" during times when the Moon is nearest the Earth (perigee), when it is overhead at relatively high northern latitudes ("positive declination"), and when the Moon is nearly at Full phase. The one weekend per month which has the best combination of these three factors is informally designated the "activity" or "skeds" weekend, and most EMEers will be on the air then. This is particularly true for 432 EME; 144-MHz EME is active during the week and on non-skeds weekends as well. See accompanying sidebar on EME operating practices.

Hilltopping and Portable Operation

One of the nice things about the VHF/UHF bands is that antennas are relatively small, and station equipment can be packed up and easily transported. Portable operation, commonly called "hilltopping" or "mountaintopping," is a favorite activity for many amateurs. This is especially true during VHF and UHF contests, where a station can be very popular by being located in a rare grid square. If you are on a hilltop or mountaintop as well, you will have a very competitive signal. See accompanying sidebar for further information.

Hilltopping is fun and exciting because hills elevate your antenna far above surrounding terrain and therefore your VHF/UHF range is greatly extended. If you live in a low-lying area such as a valley, a drive up to the top of a nearby hill or mountain will have the same effect as buying a new tower and antenna, a high-power amplifier and a preamplifier, all in one!

The popularity of hilltopping has grown as more stations acquire solid-state rigs and amplifiers. The "box and brick" (compact multimode VHF rig and solid-state 100-200 watt amplifier) combination is ideally suited for mobile and portable operation. You need no other power source than a car battery, and even with a simple antenna your signals will be outstanding.

Many VHF/UHFers drive to the top of hills or mountains and set up their station. A hilltop park, rest area or farmer's field are equally good sites, so long as they are clear of trees and obstructions. You should watch out for high-power FM or TV broadcasters who may also be taking advantage of the hill's good location; their powerful signals may cause inter-modulation problems in your receiver.

Antennas, on a couple of 10-foot mast sections, may be turned by hand as the operator sits in the passenger seat of the car. A few hours of operating like this can be wonderfully enjoyable and can net you a lot of good VHF/UHF DX. In fact, some VHF enthusiasts have very modest home stations but rather elaborate hilltopping stations. When they notice that

Charles Suckling, G3WDG, (left) operating 10-GHz EME with Kent Britain, WA5VJB, using Kent's 3-meter dish.

Moonbounce Operating Practices

After traveling 400,000 km, bouncing off a poorly reflective Moon, and returning 400,000 km, EME signals are quite weak. A large antenna, high transmitter power, low-noise preamplifier and very careful listening are all essential for EME. Nevertheless, hundreds of amateurs have made EME contacts, and their numbers are growing.

The most popular band for EME is 144 MHz, followed by 432 MHz. Other bands with regular EME activity are 1296 and 2304/2320 MHz.

Moonbounce QSO procedure is different on 144 and 432 MHz. On 144 MHz, schedule transmissions are 2 minutes long, whereas on 432 they are 2 ½ minutes long. In addition, the meaning of signal reports is different on the two bands. The difference in procedure is somewhat confusing to newcomers. Most EMEers operate on only one band, so they grow accustomed to the procedure on that band pretty quickly.

Signal

Report		Meaning	
	144 MHz		432 MHz
T	Signal just detectable		Portions of calls copyable
M	Portions of calls copyable		Complete calls copied
O	Both calls fully copied		Good signal, easily copied
R	Both calls, and "O" signal report copied		Calls and report copied

What does this difference in reporting mean? Well, the biggest difference is that "M" reports aren't good enough for a valid QSO on 2 meters but they are good enough on 432. So long as everyone understands the system, then there is no confusion. On both bands, if signals are really good, then normal RST reports are exchanged.

The majority of EME QSOs are made without any pre-arranged schedules. Almost always, there is no rigid transmitting time-slot sequencing. Just as in CW QSOs on HF, you transmit when the other station turns it over to you. This is particularly true during EME contests, where time-slot transmissions slow down the exchange of information. Why take 10-15 minutes when two or three will do?

On the other hand, many QSOs on EME are made with the assistance of "skeds". This is especially the case for newcomers and rare DX stations.

On 144 MHz, each station transmits for 2 minutes, then listens as the other station transmits. Which 2-minute sequence you transmit in is agreed to in advance. The common terminology is "even" and "odd". The even station transmits the 2nd, 4th, 6th, 8th. . . 2-minute segment of the hour, while the odd station transmits the others (1st, 3rd, 5th. . .). Therefore, 0030-0032 is an "even" slot, even if the sked begins on the half-hour. In most cases, the westernmost station takes the odd sequence. An operating chart, similar to many which have been published, will help you kep track.

Transmitting Sequence

Odd (eastern)	Even (western)
00-02	02-04
04-06	06-08
08-10	10-12
12-14	14-16
16-18	18-20
20-22	22-24
24-26	26-28
28-30	30-32
32-34	34-36
36-38	38-40
40-42	42-44
44-46	46-48
48-50	50-52
52-54	54-56
56-58	58-00

During the schedule, at the point when you've copied portions of both calls, the last 30 seconds of your 2-minute sequence is reserved for signal reports; otherwise, call sets are transmitted for the full 2 minutes.

On 432 MHz, sequences are longer. Each transmitting slot is 2½ minutes. You either transmit "first" or "second". Naturally, "first" means that you transmit for the first 2½ minutes of each 5 minutes.

144-MHz Procedure—2-Min Sequence

Period	1 ½ minutes	30 seconds
1	Calls (W6XXX DE W1XXX)	
2	W1XXXDE W6XXX	T T T T
3	W6XXX DE W1XXX	O O O O
4	RO RO RO RO	DE W1XXX K
5	R R R R R	DE W6XXX K
6	QRZ? EME	DE W1XXX K

432-MHz Procedure—2½-Min Sequence

Period	2 minutes	30 seconds
1	VE7BBG DE K2UYH	
2	K2UYH DE VE7BBG	
3	VE7BBG DE K2UYH	T T T
4	K2UYH DE VE7BBG	M M M
5	RM RM RM RM	DE K2UYH K
6	R R R R R	DE VE7BBG SK

On 222 MHz, some stations use the 144-MHz procedure and some use the 432-MHz procedure. Which one is used is determined in advance. On 1296 and above, the 432 procedure is always used.

EME operation generally takes place in the lowest parts of the VHF bands: 144.000-144.070; 222.000-222.025; 432.000-432.070; 1296.000-1296.050. Terrestrial QSOs are strongly discouraged in these portions of the bands. Recently, 10 GHz is being used by a number of stations for EME.

band conditions are improving, they hop in the car and "head for the hills." There, they have a really excellent site and can make many more QSOs.

In some places, there are no roads to the tops of hills or mountains where you might wish to operate. In this case, it is a simple (but sometimes strenuous) affair to hike to the top, carrying the car battery, rig and antenna. Many hilltoppers have had great fun by setting up on the top of a firetower on a hilltop, relying on a battery for power.

Some VHF/UHFers, especially contesters, like to take the entire station, high power amplifiers and all, to hilltops for extended operation. They may stay there for several days, camping out and DXing. Probably the most outstanding ex-

ample of this kind of operation is put on regularly by the group at W2SZ/1. Dozens of operators and helpers assemble this multiband station, often with moonbounce capability, and operate major VHF and UHF contests from Mount Greylock in Western Massachusetts. Their winning scores in virtually every VHF contest they have entered testifies to their skill and the effectiveness of hilltopping!

Contests

The greatest amount of activity on the VHF/UHF bands occurs during contests. VHF/UHF contests are scheduled for some of the best propagation dates during the year. Not only are propagation conditions generally good, but activity is always very high. Many stations "come out of the woodwork" just for the contest, and many individuals and groups go hilltopping to rare states or grid squares.

A VHF/UHF contest is a challenging but friendly battle between you, your station, other contesters, propagation and Murphy's Law. Your score is determined by a combination of skill and luck. It is not always the biggest or loudest station that scores well. The ability to listen, switch bands quickly and to take advantage of rapidly changing propagation conditions is more important than brute strength.

There are quite a few VHF/UHF contests. Some are for all the VHF and UHF bands, while others are for one band only. Some run for entire weekends and others for only a few hours. Despite their differences, all contests share a basic similarity. In all North American contests, your score is determined by the number of contacts (or more precisely, the number of QSO points) you make, multiplied by some "multiplier," which may be ARRL DXCC countries, grid squares, or some other factor. In most of the current major contests, you keep track of QSOs and multipliers by band. In other words, you can work the same station on each band for separate QSO and multiplier credit.

Listed below are the major contests of the VHF/UHF realm in North America. Other contests are popular in Europe and Asia, but these are not listed here because information about them is not usually available to most of us. All the listed contests are open to all licensed amateurs, regardless of their affiliation with any club or organization. Detailed rules for each contest are published in *QST*, *CQ* and the newsletters of the sponsoring organizations. For more information about contesting, see Chapter 7.

1) *ARRL VHF Sweepstakes*: This one-weekend contest occurs in January. It is favored by clubs because it permits club members to pool their individual scores for the club's total. In addition, individuals and multioperator groups compete. Scores are determined by total QSO points per band multiplied by the number of grid squares worked per band.

2) *ARRL Spring Sprints*: These contests, as their names imply, are short: Only 4 hours. There is a separate Sprint for each of the major VHF and UHF bands. Each one takes place on a different evening. The dates vary, but are usually in the spring. The evenings for the contests are usually chosen to coincide with each band's activity night. So, the 2-meter Sprint takes place on a Monday, the 222-MHz Sprint on Tuesday and so on. The 6-meter Sprint usually is held on a weekend at the end of the sequence, to place it closer to the sporadic-E season. Your score is the number of QSO points multiplied by the number of grid squares worked in the Sprint's 4-hour time period. Sprint scores are reported in *NCJ*, the *National Contest Journal* published by the ARRL.

3) *ARRL June VHF QSO Party*: This contest is the highlight of the contest season for most VHFers. Scheduled for the second full weekend of June, the "June Contest" sees the most activity of any North American VHF contest. Conditions on all the VHF bands are usually good, with 6 meters leading the way. This contest covers all the VHF and UHF bands, and QSOs and multipliers are accumulated for each band. Your score is determined by multiplying QSO points per band by grid squares per band. Singleband and multiband awards are given to high-scoring individuals in each ARRL section. In addition, multioperator groups compete against each other (and the competition can be fierce!).

4) *CQ Magazine Worldwide VHF WPX Contest*: This contest is international in scope. Your score is the number of different grid locators worked per band times the number of QSO

CT1WW is one of the most widely heard 6-meter European stations in the US. Tiago Frederico's shack holds antennas for several VHF bands and is also the location of the CT0WW 6-meter beacon. At 800 meters above sea level, Tiago has a clear shot in all directions.

points per band. All bands above 50 MHz are permitted. A multitude of awards are available.

5) *ARRL UHF Contest*: As its name suggests, this contest is restricted to the UHF bands (plus 222 MHz). It takes place over a full weekend in August. All UHF bands are permitted, and grid squares are the multiplier. Less equipment is involved for this contest than for contests which cover all VHF/UHF bands, so many groups go hilltopping for the UHF Contest.

6) *ARRL September VHF QSO Party*: The rules for this contest are identical to those for the June VHF QSO Party. This contest is also very popular, and many multioperator groups travel to rare states and grid squares for it. By the second weekend of September, most 6-meter sporadic E has disappeared, but tropo conditions are often extremely good. Therefore, the bands above 144 MHz are the scene of tremendous activity during the September contest. Some of the best tropo openings of recent decades have taken place during this contest, and they are made even better by the high level of activity.

7) *ARRL EME Contest*: This contest is devoted to moonbounce. It takes place over two full weekends spaced almost a month apart, usually in the fall. The date of the contest is different each year because of the variable phase of the Moon. The dates are usually chosen by active moon-bouncers to coincide with the best combination of high lunar declination, perigee and the full phase of the Moon. This contest is international in scope. Your score is the number of QSO points made via EME per band, multiplied by the number of US call districts and ARRL countries per band. Hundreds of EMEers participate in this challenging test of moonbounce capability.

Contests are lots of fun, whether you're actively competing or not. You don't have to be a full-time competitor to participate or to enjoy yourself! In fact, most of the participants in contests aren't really "in" the contest. Lots of operators get on the air to pass out points, to have fun for a while, and to listen for rare DX. Others try their hardest for the entire contest, keep track of their score and send their logs in for awards. Either way, the contest is a fun challenge.

If you don't plan on being a serious competitor, then you just need to know what the "exchange" is. The exchange is the minimum information that must be passed between each station to validly count the QSO in the contest. In all of the terrestrial VHF contests sponsored by the ARRL, you need to pass only your grid square designator to the other station, and re-

VHF Mountaintopping for the '90s

Recently, VHF has seen a resurgence of an activity as old as VHF itself—mountaintopping. Few are blessed with a home operating location that facilitates a total command of the frequency. Consequently, ardent VHFers construct bigger and bigger arrays and amplifiers for the home station in order to produce that booming signal. Some, however, utilize the "great equalizer" to compete on an equal or superior footing with the big home stations—namely, a mountaintop location. Here, perched high above all the home stations, a small portable rig with a single Yagi antenna only a few feet off the ground suddenly sounds like a kilowatt feeding a killer antenna installed at home. Simple equipment performs amazingly well from a mountaintop QTH on VHF.

A mountaintop expedition can vary from a spur-of-the-moment Sunday afternoon picnic to a full-fledged weekend contest. Quick trips can also be conducted during band openings. Since a contest optimizes the opportunity to work a lot of stations on VHF, this sidebar is mostly a "how-to" for weekend contest operation conducted by one or two people. But this can be scaled down to a mountaintop stay of shorter duration.

The Times Are a Changin'

Old-timers will remember the drudgery of lugging "boat anchors" up rocky crevices. Dragging equipment and generators weighing hundreds of pounds up steep mountainsides was no picnic. Those who suffered sprained backs soon gave it up. The advent of solid-state equipment, however, has made mountaintopping a far less strenuous activity. Even some of the highly competitive HF types have found new worlds to conquer above 50 MHz. A key factor in this revival has been the introduction of a worldwide grid locator system—now much in vogue of VHF. The use of grid squares in the major VHF contests has tickled the innermost secret desire of every radio amateur—to be on the receiving end of a DX pileup. Now instead of going on safari to some distant DX land, you can head for the mountains—some nearby mountaintop located in a rare grid square.

Choosing a Site

Choosing a mountaintop site involves considering how far you want to travel to get there, accessibility to the top of the mountain and its all-important grid square location. Ideally, your mountain is only a short driving distance away, towers into the cirrusphere, has a six-lane interstate to the top and rises within a grid square that has never been on the air before!

Obviously, some of these considerations may have to be compromised. Your first step to finding VHF heaven involves extensive study of a road atlas. How far do you want to travel? Where are the mountains? How high are they? Can you drive to the top? Draw in the grid line boundaries so you can tell which square it is in. Ask some active VHFers which are the difficult squares to work. When you start zeroing in on a potential site, you may want to get a topographic survey map of the area to determine access roads and direction of "drop-off" from the summit.

I've never operated from a mountaintop without first scoping it out in person. Access is most important. Thus far, I've operated from sites which I have reached by car, ferry, gondola, 4-wheel drive, motor home and hiking. Unless you are going with mini-radios and gel-cells, you want to get there without backpacking it. A passable road to the top is ideal. When checking out a 2-meter site, bring a compass and 2-meter FM hand-held. A call on 146.52-MHz simplex should tell you how good the location is. Are you blocked in any direction? Is it already "RF-city" with commercial installations—a potential source of interference? Will you be able to clear any trees with a lightweight mast? Then the prime requisite: Is there a picnic table permanently at the site? If not, plan on bringing an operating table and chair—which adds considerable bulk and weight to transport.

Once you've selected an operating site, be sure you have secured the necessary permission to use the site. This may simply require verbal permission from some authority or the owner, or, it could involve a lengthy exchange of correspondence with a state environmental agency of forests and parks and the signing of a liability release. But be sure you have permission. The last thing you want is the local sheriff shining a flashlight in your eyes at 3 AM, rousing you out of a fantastic tropo opening on 2 meters. You'll find rangers on fire watch most helpful in pointing out how to obtain necessary permission.

Power Source

Unless you are awfully lucky to find a location that will permit you to just "plug in," make plans for providing your own power. With a single-band operation from a car (with antenna mast mounted just outside the car window), the car battery will probably suffice. Run a set of heavy-duty jumper cables directly to the car battery. Even a solid-state "brick" amplifier can be run off the car battery without ill effects. Just in case, park the car facing downhill!

For a more serious effort utilizing several VHF and UHF bands, a small generator is recommended. If the word generator conjures up an image of an ugly engine block from a 1947 LaSalle, then tune in to the modern world. Small, even attractive, generators in the 500- 1000 W category, that look more like American Tourister luggage, are now available. Mine is a 650-W beauty that weighs in at 43 pounds, and runs for four hours on a half gallon of petrol. And quiet? You can hear the wings of a Monarch butterfly flutter at 50 paces. A 5-gallon gerry-can provides more than enough flammable juice for a contest weekend.

Equipment

A mountaintop location effectively places your antenna atop a natural tower of hundreds, or perhaps, thousands of feet. With this height advantage, compact, lightweight, low-powered radios that can be boosted up to the 50- to 100-W range with solid-state amplifiers will perform nicely. Low-powered portable transceivers are manufactured for just this purpose. The popular 10- and 25-W multimode rigs are also quite adequate. Many discontinued models can be obtained at a substantial savings through the Ham Ads section of *QST*. Use of transverters operating with mobile-type HF radios should also be considered.

If you don't have any sizable trees to get over, you can use simple mast sections that fit together. I use 5-foot sections available at the popular shack of radios. They are easily transportable. Important too is the method of antenna rotation. If you can install the antenna mast right next to the operating position, do it. It will save all the hassle of installing motorized rotators. Nothing beats the "Armstrong" method for speed and simplicity. I use a cross-piece of aluminum tubing mounted with U-bolts to the mast at arm level. See Fig 12-3. This provides instan-taneous antenna-peaking capability—a necessity on V/UHF. While home stations are twirling their antennas in every direction trying to peak a weak signal, I've already worked him!

Installing antennas for several bands on the same mast is recommended. They should be oriented in the same direction. Many contacts on UHF are the result of moving stations over from other bands. For example, in a contest if you move a multiplier to 432 MHz after first working on 2 meters, and both antennas are on the same mast, you will first want to peak the signal on 144 MHz. Then, when you move to 432 MHz where the antennas are probably a bit more sharp and propagation perhaps marginal, both an- tennas will be pointing at each other for maximum signal. This can make the difference in whether the contact is made.

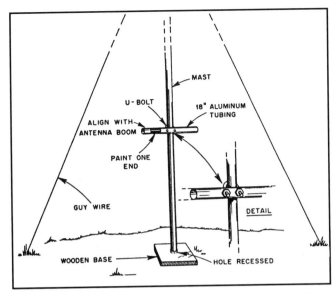

Fig 12-3—A closeup view of the "Armstrong" rotator, used for quick peaking of signals.

If your mountaintop operation involves staying overnight, additional attention must be paid to having the proper survival equipment. The most luxurious way to go is a van or RV. Otherwise, a tent will be required. For the rugged outdoors type, this can be as appealing as the radio part. I find that cooking steaks over a campfire with a canopy of stars overhead (while a programmable keyer is calling CQ) is half the fun. But keep this aspect of the operation also simple as possible. Champagne and caviar can be held for another time. I've also found out the hard way that one can expect heavy winds on mountaintops. Large tents blow down easily in such weather.

Further on the subject of weather, just because you're topping it in July or August, don't expect it will always be T-shirt and shorts weather. No matter what the season, expect to need a heavy jacket after dark. I always bring a heavy flannel shirt and ski jacket for night, and shorts in the daytime. And bring lightweight raingear, just in case. And depending on the habitat, don't be surprised to be introduced to a critter or two, especially after dark!

Getting Started

Okay, you've read this far and are beginning to say to yourself: "Self, I think I'd like to try that." But there is a little voice of caution in you that says: "Don't go bonkers until you've sampled a little first." Good advice!

Start out by setting up on an easily accessible mountain for an afternoon during a contest period on a single band. For the first effort, I recommend 2 meters. With so many 10-W multimode rigs out there in radioland, 2 meters is your "bread and butter" band. Using a multi-element Yagi a few feet off the ground of a strategically located mountain or hill can whet your appetite. I first got hooked by operating from the side of a highway on Hogback Mountain, Vermont, with a 3-W portable 2-meter radio to a 30-W brick and 11-element Yagi. I was astounded by the results, with contacts hundreds of miles away. This launched my interest in acquiring more equipment for portable mountaintop use, each operation adding a new band or better antenna. The basic formula of keeping it lightweight and simple has prevailed, however.

Now what's keeping you from operating from Mount Everest?—*John F. Lindholm, W1XX*

A Checklist of Typical Items Needed for a Weekend Portable Mountaintop Operation

Radio Equipment
- ☐ Transceivers for each band
- ☐ Solid-state amplifiers
- ☐ Antennas
- ☐ Coax (Belden 9913 or equiv)
- ☐ Keyer
- ☐ Paddle
- ☐ Antenna Masts
- ☐ Earphones
- ☐ Coax connector cables
- ☐ Coax adapters
- ☐ DC cables with plugs
- ☐ Power supply(ies)
- ☐ Fuses
- ☐ Antenna rotator crosspiece
- ☐ Microphones
- ☐ Key line with plug
- ☐ Multi-ac plug outlet
- ☐ SWR meter
- ☐ DVM
- ☐ Multi-dc plug box
- ☐ Clip leads
- ☐ Cell-phone

Tools
- ☐ Wrenches
- ☐ Pliers
- ☐ Screwdrivers
- ☐ Hammer

Power Source
- ☐ Generator
- ☐ Power cable
- ☐ Gasoline
- ☐ Oil
- ☐ Jumper cables (if on car battery)
- ☐ Gas funnel

Personal
- ☐ Toothbrush/toothpaste
- ☐ Soap
- ☐ Towel
- ☐ Suntan lotion
- ☐ Change of clothes
- ☐ Rain gear
- ☐ Hat
- ☐ Warm jacket
- ☐ Alarm clock
- ☐ Bug spray
- ☐ Insect repellent
- ☐ Electric shaver
- ☐ Toilet paper

Camping Equipment
- ☐ Matches
- ☐ Tent stakes (enough for tent and masts)
- ☐ Tarpaulin canopy
- ☐ Tent
- ☐ Ground cloth
- ☐ Extra rain cover
- ☐ Cot
- ☐ Lantern
- ☐ Flashlight and extra batteries
- ☐ Pot and pan set
- ☐ Pot holder
- ☐ Spatula
- ☐ Can opener
- ☐ Water bottle (5 gal)
- ☐ Rope
- ☐ Charcoal briquettes
- ☐ Table and chair (if needed)
- ☐ Sleeping bag
- ☐ Food and drink
- ☐ Knife and fork set
- ☐ Paper plates
- ☐ Paper towels
- ☐ Paper cups
- ☐ Cooler with ice
- ☐ Cook stove
- ☐ Stove fuel
- ☐ Funnel for fuel
- ☐ Fluorescent-type bat. lantern
- ☐ Old newspaper
- ☐ Coffee cup
- ☐ Aluminum foil
- ☐ Trash bag

Miscellaneous
- ☐ Compass
- ☐ 24-hour clock
- ☐ Black plastic tape
- ☐ Masking tape
- ☐ Logs
- ☐ Grid square maps
- ☐ Pencils (many)
- ☐ Clipboard
- ☐ Highlighter
- ☐ First-aid kit

One of the many multiop contest efforts by N2SB, this during the VHF Sweepstakes.

ceive theirs, plus acknowledgment, to count the QSO. Other contests require some different information, such as serial numbers or signal reports, so it pays to check the rules to make sure.

A serious contest effort requires dedication and effort, as well as a station that can withstand a real workout. Contests are a challenge to operators and equipment alike. A good contest score is the result of hard work, a good station and favorable propagation. A good score is something to be proud of, especially if there is lots of stiff competition. And with the popularity of VHF and UHF contests these days, competition is always stiff!

If you are a competitor in the contest, you will probably need to keep a "dupe sheet" which helps prevent duplicate QSOs. A dupe sheet is a large piece of paper on which you record each call sign that you work in a sort of a matrix so you can check it quickly. If the other station's call sign is already in the dupe sheet, working him/her again won't count for contest credit. Don't rely on your memory. After 24 hours of contesting, most of us have trouble remembering our own names! Also, if you make more than 200 QSOs in an ARRL VHF contest, you must submit your dupe sheet along with your logs. See the sidebar titled "Where Do I Find Rules and Entry Forms?" in Chapter 7 for this information on ARRL contests.

Many operators don't feel that their stations are competitive on all the VHF and UHF bands. Not to worry; in most contests, it is possible to compete on only one band if you want to. This has the advantage of concentrating your efforts where your station is the strongest, allowing you to devote full time to just one band. You don't need to be high powered to compete as a single-band entry. Location makes a lot of difference, and hilltopping single-banders have had tremendous success, particularly if they go to a rare grid square.

Other operators like to try for all-band competition. In this case, it's a real advantage to be able to hop from one band to another. You can quickly check 6 meters for activity while also tuning 2 meters. Or, if you work a rare grid on one band, you can take advantage of the opportunity by asking the other station to switch bands right then. Some contesters work one station on all possible bands within two minutes by band hopping! This is a speedy way to increase your grids and QSOs.

Multioperator stations, in which more than one person performs operating tasks such as logging, dupe checking or making contacts, are very popular in VHF/UHF contests. Multiop contest entries come in two varieties: *Limited* and *Unlimited*.

Limited multiop entries can compete on as many bands as they want during the contest, but submit scores for their best four bands. This category is a good environment for small groups of contesters who have the equipment necessary to operate on only a few UHF or microwave bands. On the other hand, Unlimited multiop stations compete on *all* VHF, UHF and microwave bands. Another class of competition is the *Rover Class*. A Rover is a station that operates from more than one grid square during the contest. Usually, Rovers operate from a vehicle that takes advantage of hilltops in rare grid squares, moving from one to the other throughout the contest period. This class is becoming very popular.

In ARRL contests, all multiop stations are also multiband. For multiop stations, the ability to operate several bands simultaneously is an obvious advantage. Being able to pass messages from one operating position to another is also important. It allows one band-station to alert another to changing propagation conditions. For example, the 6-meter position may notice the beginning of an aurora several minutes before it will appear on 2 meters. If the 6-meter operator can alert the other operators, valuable time can be saved.

Similarly, it is advantageous to be able to pass "referrals" from one station to another in a multiop contest effort. A referral is an on-the-spot schedule to call another station at a particular time and frequency. It's the same as a quick band change for single op stations. For example, if you are working the multioperator contest station W8VP on 2 meters, you might request an immediate sked on 222. If the 2-meter operator at W8VP can pass a quick message to the 222 station, they'll be looking for you, and you will probably make the QSO promptly.

During VHF/UHF contests, the above-listed times for activity nights are often observed for special activity, particularly on 222 MHz and up. In other words, look for increased activity on 222 MHz at 8 PM, 432 MHz at 9 PM, and 1296 MHz at 10 PM local during major contests. In some areas, this scheme also applies to morning (AM) as well. Bigger stations sometimes turn their antennas in accordance with the minute hand of the clock, looking north at the beginning of the hour, east at 15 minutes past, and so forth.

SOURCES OF INFORMATION

Many VHF/UHF operators like to keep abreast of the latest happenings on the bands such as new DX records, band openings or new designs for equipment. There are many excellent sources for current information about VHF/UHF. In addition, they provide a way to share ideas and ask questions, and for newcomers to become familiar with operation above 50 MHz.

The main sources are nets, newsletters and published columns. Each has its own use and appeal; active VHFers usually seek out at least one of them.

Nets

Several nets meet regularly on the HF and VHF bands so that VHF/UHFers can chat with each other. These are listed below. It's a good idea to listen first, before checking in the first time, so you'll know the format of the net. Some nets like to get urgent news, scheduling information and other "hot" topics out of the way early, and save questions and discussion until later. Others are more free-form. You can learn quite a bit just by tuning in to these nets for a few weeks. Regular participants in the nets are often very knowledgeable and experienced. The technical discussions which sometimes take place can be very informative.

1) *Central States VHF Net.* Meets each Sunday at 0230 UTC, 3.818 MHz. Open to all interested stations. A terrific source of information, as well as skeds for meteor scatter and EME. Operates informally during major meteor showers.

2) *144 and 432 MHz EME Nets.* These meet each Saturday and Sunday at 1700 and 1600 UTC, respectively, on 14.345 MHz. These are international nets which serve those interested in moonbounce. Each net meeting is usually occupied by a review of recent EME conditions, news from participants and setting up schedules for future EME work. The 432 EME net is one hour long, whereas the 2-meter net often continues for several hours.

3) *Sidewinders On Two (SWOT) Net.* This net meets on 2-meter SSB, at regular intervals which vary by region. It is a good place to find active 2-meter operators. For information about the SWOT net in your area, consult the SWOT newsletter (c/o Howard Hallman, WD5DJT, 3230 Springfield, Lancaster, TX 75134).

4) *East Coast 432 Net.* This net is called each Wednesday at 9-11 PM local on 432.090 MHz. It is particularly popular in the Southeast, although they have had checkins from as many as 30 states in the past few years.

5) *Local Nets.* Lots of these nets, both formal and informal, are scattered across the country. It's impossible to tabulate their meeting times and frequencies. The best thing to do is to get on the air and ask other VHF operators about nets in your area.

Awards

Several awards have helped to spur activity on the VHF/UHF bands. The most popular are WAS, WAC and VUCC.

It was long thought that WAS, Worked All States, was an impossible dream for VHFers. After all, it was "common knowledge" that no VHFer could work more than a thousand miles or so. To think of working coast-to-coast was silly, and to consider Alaska and Hawaii was crazy. To work all 50 states was just impossible. Fortunately, not everyone listens to "common knowledge," and WAS has been attained by several hundred amateurs on the 50, 144, 222 and 432-MHz bands.

WAS requires that you work each state in the United States and confirm the QSOs with QSL cards. The quest for this award has probably been responsible for much of the technical advancement of VHF/UHFers during the past three decades. Moonbounce activity has benefited most directly because WAS on any band above 144 MHz requires EME capability. (It

should be noted, however, that a handful of diehards have worked 48 states on 2 meters *without* moonbounce!)

The 6-meter band has been host to more than a hundred WAS-achievers. During the peak of the most recent sunspot cycle, F_2 openings made transcontinental contacts common for many months. Amateurs in the continental US worked Alaska, Hawaii and tons of DX during that time. As of late 1991, there is little hope of widespread F_2 for the next few years as we gradually head into the sunspot minimum. Nevertheless, multi-hop E_s openings occur each summer to provide the slim chance of WAS on 6 meters.

Using portable moonbounce stations, several enterprising groups have mounted EME-DXpeditions to rare states. This has allowed quite a few hard-working VHF/UHFers to complete WAS, even when there was no resident EME activity in some states. At last count, over 100 stations had received WAS on 2 meters. At least one station, W1JR, has achieved WAS on 50, 144 and 432 MHz.

WAC, Worked All Continents, has also been achieved on 6 meters with the assistance of F_2 propagation. The chances of sporadic E of sufficient distance for catching all continents are almost nil, so we'll have to wait a while for more WAC awards on this band.

WAC is available to EMEers on 144, 432 and perhaps 1296 MHz. Stations from all continents have been active on these bands in recent years, although 1296 is not fully represented.

Jim Mead, WB2BYW, (left) and Tom Williams, WA1MBA, explore the frontiers of microwave communication. In 1994 they made the first amateur contact on 120 GHz over a distance of 1.15 km.

This impressive station belongs to Dan Gautschi, HB9CRQ. With a 2-meter DXCC award on his wall, Dan is a member of a select group!

In fact, several top-scoring stations in the annual EME contests have made QSOs on all continents during a single weekend. The rarest continent is probably South America, where only a small handful of EMEers are active. WAC is not possible on 222 or 902 MHz because these bands are not authorized for amateur use outside of ITU Region 2 (North and South America).

ARRL breathed new life into the VHF and UHF bands in January 1983 when the League began sponsoring the VUCC (VHF/UHF Century Club) awards. These awards, based upon grid squares, rejuvenated the VHF and UHF bands tremendously. Somewhat later, the required exchange in all terrestrial ARRL-sponsored VHF and UHF contests was changed to include grid squares, and activity in those contests increased as well. You can do a lot of grid-hunting during a good contest!

The VUCC award is based on working and confirming a certain number of grid squares on the VHF/UHF bands. For 50 and 144 MHz, the number is 100. On 222 and 432 MHz, the number is 50, and on 902 and 1296 MHz the minimum number is 25. 2.3-GHz operators need to work 10 grids, and five grids are required on the higher microwave bands. This is quite a challenge, and qualifying for the VUCC award is a real accomplishment!

VUCC endorsement stickers are available for those who work specified numbers of additional grids above the minimum required for the basic award. Several stations have exceeded 300 grids on 6 meters and 200 on 2 meters, and a few have worked over 100 grids on 432. Impressive!

Another award should be mentioned although only a handful of VHFers can hope to attain it: *DXCC*. The DX Century Club award, based on working 100 ARRL DX countries, has been a popular goal among low-band (HF) DXers for decades. Many people have considered it preposterous to even think that DXCC could ever be achieved by VHFers. However, DXCC has been conquered by a number of 50-MHz stations. Excellent F_2 propagation during 1989-1990 and the avalanche of 6-meter operating permits in Europe made many of these DXCC awards possible. Even more impressive has been the achievement of a few hams who have earned DXCC on the 144-MHz band. Moonbounce, plus a lot of patience and hard work, did the trick for W5UN. For specific details concerning the ARRL awards program, see Chapter 8.

Problems

With the ever-increasing sharing of the VHF and UHF spectrum by commercial, industrial and private radio services, a certain amount of interference with amateur operation is almost inevitable. Amateurs are well acquainted with interference, and so we normally solve interference problems by ourselves.

Several main types of interference are common. The first is our old friend, television interference (TVI). Fundamental overload, particularly of TV Channels 2 and 3 from 6-meter transmitters, is still common. Some Channel 12 and 13 viewing is bothered by 222-MHz transmissions. Fortunately, as modern TV manufacturers have slowly improved the quality of their sets, the amount of TVI is beginning to decline.

A more common form of TVI is called CATVI, Cable Television Interference. It results from many cable systems distributing their signals, inside shielded cables, on frequencies which are allocated to amateurs. As long as the cable remains a "closed system," in which all of their signal stays inside the cable and our signals stay out, then everything is usually okay. However, it is a sad fact that many cable systems are deterio-

rating because of age, and have begun to leak. When a cable system leaks, your perfectly clean and proper VHF signal can get into the cable and cause enormous amounts of mischief. By the same token, the cable company's signals can leak out and interfere with legitimate amateur reception.

RFI—radio frequency interference—is a common complaint of owners of unshielded or poorly designed electronic entertainment equipment. Amateur transmissions, especially high power, may be picked up and rectified, causing very annoying problems. RFI may include stereos, video-cassette recorders and telephones. The symptoms of RFI usually include muffled noises which coincide with keying or SSB voice peaks, or partial to total disruption of VCR pictures.

This chapter cannot cover all the complex methods which are used to track down and correct TVI/RFI problems. However, a few general principles may help in beginning your search for a solution:

1) With ordinary TVI, be sure your transmitter is clean, all coax connectors are tightened, and a good dc and RF ground is provided in your shack—before you look elsewhere for the cause of the problem. Then, find out if TVI affects all televisions or just one. If it's just one, then the problem is probably in the set and not your station.

2) With CATVI, remember that it is the cable company's responsibility to keep its system "closed" to the limits of FCC rules. Unfortunately, the FCC's limits are loose enough that in some cases there will be interference-causing leakage from the system which is still within FCC limits. In that case, there is no easy solution to the problem. However, you may be pleasantly surprised by the cooperative attitude of some cable TV operators. (Cable company technicians often have worked overtime trying to solve CATVI complaints, but not everyone is so fortunate.)

3) In dealing with RFI, the main goal is to keep your RF out of the entertainment system. This is often solved rather simply by bypassing the speaker, microphone and power leads with disc ceramic capacitors. In other cases, particularly some telephones, you must also employ RF chokes. For further information on RFI and CATVI, consult *Radio Frequency Interference: How to Find It and Fix It,* published by ARRL.

Several other types of interference plague VHF/UHF amateurs. These are the receive-only kinds of interference, which just affect your receiving capability. One form has already been mentioned: CATV leakage. For example, cable channel "E" is distributed on a frequency in the middle of the amateur 2-meter band. If the cable system leaks, you may experience very disruptive interference. Reducing leakage or perhaps eliminating the use of channel "E" may be satisfactory. This problem has vexed many stalwart VHFers already, and no end is in sight.

Scanner "birdies" may be a problem in your area. All scanners are superheterodyne-type receivers which generate local oscillator signals, just as your receiving system does. Unfortunately, many scanners have inadequate shielding between the local oscillator and the scanner's antenna. The result is radiation of the oscillator's signal each time its channel is scanned—a very annoying chirp, swoosh or buzz sound every second or so. If the scanner's local oscillator frequency happens to fall near a frequency you're listening to, you'll hear the scanner instead. Amateurs can easily pick up 10-15 scanner birdies within 5 kHz of the 2-meter calling frequency (for example, a New York State Police frequency is about 10.7 MHz above 144.200!). Very little can be done about this problem aside

Ralph, N5RZ, poses beside his "rover machine."

from installing tuned traps in everyone's scanners—an unattractive prospect to most scanner owners.

Many amateurs who operate 432 MHz have lived with radar interference for years. Radar interference is identified by a very rapid burst of noise which sounds vaguely like ignition noise, repeated on a regular basis. Although some radars are being phased out (no pun intended), many remain. Amateurs are secondary users of the 420-450 MHz band, so we must accept this interference. The only solution may be directional antennas which may null out the interference, not a very satisfying alternative in some cases. 432-MHz EMEers sometimes hear radar interference off the Moon!

NEW FRONTIERS

The VHF/UHF spectrum offers amateurs a real opportunity to contribute to their own knowledge as well as the advancement of science in general. We as amateurs are capable of several aspects of personal research which are not possible for the limited resources of most scientific research organizations. Although these organizations are able to conduct costly experiments, they are limited in the time available, geographic coverage and, sometimes, by the "it's impossible" syndrome. Hams, on the air 24 hours a day, scattered across the globe, don't know that some aspects of propagation are "impossible," so they occasionally happen for us.

For example, it is commonly "known" that VHF meteor scatter is limited to frequencies below about 150 MHz, yet hams have made dozens of contacts at 222 and 432 MHz. A spirit of curiosity and a willingness to learn is all that is required to turn VHF/UHF operating into an interesting scientific investigation.

What are some of the topics amateurs can contribute to? The following list includes just a few of the possibilities:

1) *Aurora on 902 or 1296*. Many aurora contacts have been made on 432 MHz; is Au "impossible" on 902? 1296?

2) *Unusual forms of F-layer ionospheric propagation*—FAI, TE and what else? Little is known about their characteristics, how they relate to overall geomagnetic activity and frequency. Amateurs can discover a lot here.

3) *Multiple-hop sporadic E on 50 MHz*. Single-hop propagation on this band is a daily occurrence during the summer. In recent years, multiple-hop openings have not been uncommon. Now that many European countries are beginning to allow 6-meter amateur operation, we can find out how many hops are actually possible!

4) *Meteor scatter*. Astronomers are very interested in the orbits of comets and their swarms of debris which give rise to meteor showers. However, it is often difficult for astronomers to observe meteors (because of clouds, moonlight and so on). Amateurs can make very substantial contributions to the study of meteors by keeping accurate records of meteor-scatter contacts.

5) *Polarization of E-layer signals*. Do VHF signals rotate in polarity when reflected by sporadic-E or meteor-trail ionization? Are vertical antennas better than horizontal antennas for E-layer signals?

6) *How effective is diversity reception* in different types of VHF/UHF reception? Stations with two or more antennas could investigate.

7) *New modes of data transfer*. Packet, AMTOR and the like—what are some of the limits to their use? Who knows how many amateurs may attempt to send packet information via EME.

Chapter 13

Satellites

JON BLOOM, KE3Z
and
STEVE FORD, WB8IMY

ARRL HEADQUARTERS

When asked about satellites, many amateurs will respond by saying, "Satellites are really interesting, but they're too complicated for me." Those who have experienced the fascination of amateur satellite operation know differently! Like any specialized activity, satellite operation has its unique requirements, and like many facets of Amateur Radio, going whole-hog with an elaborate station can be costly. In truth, you don't need to invest a lot of time or money to get involved in amateur satellite operation—there are a number of reasonably painless ways to get started.

SATELLITES: ORBITING REPEATERS

Most amateurs are familiar with repeater stations that retransmit signals to provide wider coverage. This is essentially the function of an amateur satellite as well. Of course, while a repeater antenna may be as much as a few thousand meters above the surrounding terrain, the satellite is hundreds or thousands of kilometers above the surface of the Earth. The area of the Earth that the satellite's signals can reach is therefore much larger than the coverage area of even the best Earth-bound repeaters. It is this characteristic of satellites that makes them attractive for communications. Most amateur satellites act either as analog repeaters, retransmitting signals exactly as they are received, or as packet store-and-forward systems that receive whole messages from ground stations for later relay. We'll concentrate on the analog satellites first.

Analog Satellites

Analog satellites contain *linear transponders*. These are devices that retransmit a band of frequencies, usually 50 to 100 kHz wide. Since the linear transponder retransmits the entire band, a number of signals may be retransmitted simultaneously. For example, if four SSB signals (each separated by 20 kHz) were transmitted to the satellite, the satellite would retransmit all four signals—still separated by 20 kHz each. Just like a terrestrial repeater, the retransmission takes place on a frequency that is different from the one on which the signals were originally received.

In the case of amateur satellites the difference between the transmit and receive frequencies is similar to what you might encounter on a cross-band terrestrial repeater. In other words, retransmission occurs on a different *band* from the original signal. For example, a transmission received by the satellite on 2 meters might be retransmitted on 10 meters. This cross-band operation allows the use of simple filters in the satellite to keep its transmitter from interfering with its receiver. Cross-band operation also has the happy effect of allowing ground stations to use simple filters in the same way. Because it's relatively easy to do, most satellite stations operate *full duplex*, meaning

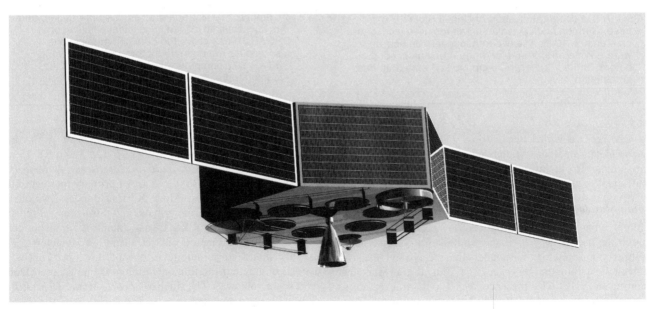

AMSAT's new Phase 3D spacecraft in orbit. Image courtesy Dick Jansson, WD4FAB; computer enhancement by David Pingree, N1NAS.

Amateur Satellites: A History

OSCAR 1, the first of the "Phase 1" satellites, was launched on December 12, 1961. It sent information concerning its internal temperature back to Earth on the 2-meter band. The now traditional CW identification HI was varied in code speed as the temperature changed. The 0.10-watt transmitter onboard discharged its batteries after only three weeks, but the mission was so successful that construction of a second satellite was begun.

OSCAR 2 was launched on June 2, 1962. Virtually identical to OSCAR 1, Amateur Radio's second venture into space lasted 18 days.

OSCAR 3, launched March 9, 1965, was the world's first free access communications satellite. During its two-week life, over 100 pioneering amateurs in 16 countries communicated through the 1-watt linear transponder on 2 meters.

OSCAR 4 was launched on December 21, 1965. More advanced in design than OSCAR 3, this satellite had a 2-meter-to-70-cm linear transponder with an output of 3 watts. Unfortunately, a launch-vehicle defect placed OSCAR 4 into a high elliptical orbit, which prevented widespread amateur use. A handful of hams did communicate through it, however, including the first US to USSR satellite contact of any kind.

OSCAR 5 was built by students at Melbourne University, in Australia. A new amateur organization, AMSAT, prepared the satellite for launch and coordinated ground activities. OSCAR 5 transmitted telemetry about its operating parameters on both 2 meters and 10 meters. Its batteries lasted for over a month.

OSCAR 6 was the first of the Phase 2 satellites. Launched on October 15, 1972, it carried a 1-watt-output 2-meter-to-10-meter Mode A linear transponder. Not only was the satellite magnetically stabilized, but the internal battery package was continuously recharged by solar cells. Tens of thousands of contacts were made during the nearly five-year lifespan of this satellite.

OSCAR 7 was built by hams from West Germany, Canada, Australia and the United States. Two elaborate linear transponders each ran 2-watts output; one was a 2-meter-to-10-meter Mode A like the one on OSCAR 6, the other a 70-cm-to-2-meter Mode B unit. Internal circuitry, as well as ground commands, controlled this sophisticated satellite's functions. OSCAR 7 was launched on November 15, 1974.

OSCAR 8, a cooperative effort of United States, Japanese, German and Canadian amateurs, was launched on March 5, 1978. The two transponders on board were a 2-meter-to-10-meter Mode A unit as used before, and a 2-meter-to-70-cm Mode J unit. Both had an output of 2 watts.

Radio Sputniks 1 and 2 were launched from the Soviet Union on October 26, 1978. Each satellite carried a 2-meter-to-10-meter transponder with extremely high sensitivity. Like their OSCAR counterparts, the RS satellites transmitted telemetry data relating to their well-being, and were commandable from the ground.

OSCAR Phase 3-A, the first of a planned series of satellites that would finally realize the dream of reliable high-altitude long-distance satellite communications over extended periods, was launched on May 23, 1980, a date that unfortunately came to be known as "Black Friday." Phase 3-A piggybacked aboard the European Space Agency Ariane LO2 rocket. A few minutes after liftoff from the ESA launch facility in Kourou, French Guiana, the Ariane rocket failed, dumping Phase 3-A and the dream that rode with it into the Atlantic.

OSCAR 9, built by a group of radio amateurs and educators at the University of Surrey in England, went aloft in October 1981 as part of a secondary payload aboard a NASA Delta rocket. This was a scientific/educational low-orbit satellite containing many experiments but no amateur transponders. These included HF beacons at 7.050, 14.002, 21.002 and 29.510 MHz for propagation studies, general and engineering data beacons on 2 meters and 70 cm, and two additional beacons on 13 cm and 3 cm. An Earth-imaging camera, magnetometer, synthesized voice telemetry capability and onboard computer rounded out the experimental hardware. UoSAT-OSCAR 9 became a "Silent Key" on October 13, 1989

Radio Sputniks 3-8 were launched simultaneously aboard a single vehicle in December 1981. Several of these RS satellites carried 2-meter-to-10-meter transponders that permitted communications over distances greater than 5000 miles. Also aboard two of them was a unique device nicknamed "Robot" that could automatically handle a CW QSO with terrestrial stations.

Iskra 2, was launched manually by two Soviet cosmonauts from the *Salyut 7* space station in May 1982. With the call sign RK02, Iskra 2 sported a telemetry beacon on 10 meters and a 15-meter-to-10-meter HF transponder. Iskra 2 was destroyed on reentering the atmosphere a few weeks after launch.

Iskra 3, launched in November 1982 from *Salyut 7*, was even shorter lived than its predecessor.

OSCAR 10, the second Phase 3 satellite, was launched on June 16, 1983, aboard an ESA Ariane rocket, and was successfully placed in its initial elliptical orbit. OSCAR 10 has proved to be a communications resource for radio amateurs throughout the world. OSCAR 10's computer (IHU) began showing the accumulated effects of near-Earth space particle

---

they can receive while transmitting. (The phrase *satellite stations* means amateur stations that use the satellite to relay their signals.) In most instances the built-in receiver filtering is sufficient to allow full-duplex operation.

Getting Started with the RS Satellites

Many amateurs were introduced to satellite operating through the use of the RS (Radio Sputnik) satellites. These satellites contain analog transponders that receive and retransmit signals within the HF bands, as well as transponders operating at VHF. The presence of HF-only transponders means that many amateurs already have the equipment they need to operate through an RS satellite.

The RS satellites are built and launched in the Soviet Union.

There have been a number of RS satellites in orbit. At the time of this writing, satellites RS-10/11, RS-12/13 and RS-15 are operating (see sidebar: *Amateur Satellites: A History*). RS-10/11, RS-15 and RS-12/13 are of primary interest to the beginner. RS-12 carries "Mode K" transponders, which receive signals on the 15-meter band and retransmit on the 10-meter band. RS-10 and RS-15 use "mode A" transponders, which receive on 2 meters and retransmit on 10 meters. (The use of "mode" terminology is one of the most confusing aspects of amateur satellite operation to the newcomer. Don't let it scare you away. The sidebar *Satellite Modes Demystified* should clear up any confusion.)

A Matter of Perspective: When you transmit a signal to a satellite, the satellite retransmits it, just like a repeater does. If

radiation in mid-1986. Faulty operation of the IHU means that the satellite has become uncommandable and uncontrollable.

OSCAR 11, another scientific/educational low-orbit satellite like OSCAR 9, was built at the University of Surrey in England and launched on March 1, 1984. UoSAT-OSCAR 11 has also demonstrated the feasibility of store-and-forward packet digital communications and is fully operational as this is written.

OSCAR 12 or *Fuji-OSCAR 12* (FO-12) was the first Japanese OSCAR, a joint effort between the Japan Amateur Radio League (JARL) and the Japan AMSAT (JAMSAT) organization. It carried a Mode J transponder that served two functions: linear SSB service (Mode JA) and an advanced packet store-and-forward global message service with a 1.5-megabyte RAM storage bank called Mode JD (for Mode J digital).

Radio Sputnik (RS) 10/11 carries a dual designation because it consists of two electronics packages integrated into a single spacecraft. It was launched on June 23, 1987 and, like the Iskra satellites, carries 15-meter-to-10-meter transponders. These work in conjunction with a Mode A transponder to provide a variety of HF/VHF operating modes.

OSCAR 13, the third Phase 3 satellite and the second operational one, enjoyed a flawless launch and deployment on June 15, 1988. Its subsequent "burn" to final orbit also was flawless. OSCAR 13 has since provided worldwide communications using Modes B, J, L (23 cm to 70 cm) and S (70 cm to 13 cm). An experimental packet transponder called RUDAK, contributed by AMSAT-DL, failed after launch. OSCAR 13 re-entered the atmosphere in December 1996.

OSCAR 14 and 15, built by the University of Surrey were launched on January 22, 1990. OSCAR 14 carries a 2-meter-to-70-cm store-and-forward packet-radio transponder called the Packet Communications Experiment (PCE) along with scientific experiments. OSCAR 15, which carried only experiments and a solid-state TV camera failed catastrophically shortly after launch.

OSCARs 16 through 19 are the Microsats. Designed by a cooperative effort of amateurs in the Western Hemisphere, these tiny satellites (approximately 9-inch cubes) were launched, along with OSCARs 14 and 15, aboard a single Ariane rocket. At the time, this single launch more than doubled the number of operational amateur satellites! PACSAT-OSCAR 16, sponsored by AMSAT-NA, carries a 2-meter-to-70-cm store-and-forward packet transponder, as does Webersat-OSCAR 18 and Lusat-OSCAR 19. Webersat, sponsored by Weber State University, carries a small TV camera for Earth imaging. LUSAT, sponsored by

AMSAT-LU (Argentina), includes a 70-cm CW beacon transmitter. DOVE-OSCAR-17 was sponsored by AMSAT-Brazil. It consists of a 2-meter transmitter that can transmit packet or digitized voice messages.

OSCAR-20, launched on February 7, 1990 is almost a clone of OSCAR 12. It has a somewhat improved power subsystem for more consistent operation.

RS-12/13, like RS-10/11, carries multiple HF/VHF transponders. It was launched on February 5, 1991.

OSCAR 21, also known as RS-14, was the first Soviet amateur satellite to carry an OSCAR designation. It was deactivated in November 1994.

UoSAT-OSCAR 22 was designed by the University of Surrey in the United Kingdom. It provided packet store and forward capability at 9600 bit/s as well as a high resolution camera.

KITSAT-OSCAR 23 was the first Korean amateur satellite to carry an OSCAR designation. Launched in 1992, it offered packet capability at 9600 bit/s. KITSAT also provided views from *two* onboard cameras.

ARSENE, launched May 12, 1993, failed several weeks later. It was intended to function as a packet digipeater.

KITSAT-OSCAR 25 is a virtual clone of OSCAR 23. Launched in September 1993, it also has imaging cameras as well as 9600 bit/s packet capability.

ITAMSAT-OSCAR 26 was launched in September 1993 along with OSCAR 25. It was the first Italian amateur satellite. It operates primarily as a 1200 bit/s store-and-forward packet system.

AMRAD-OSCAR 27 was also a launch companion to OSCARs 25 and 26. It is an experimental module that operates as part of *Eyesat*, a commercial satellite. AMRAD-OSCAR 27 is intended to be used as a test bed for new Amateur Radio technology. When this book went to press, it was functioning as an FM repeater.

PoSAT-OSCAR 28 is the first Portuguese amateur satellite. It was launched along with OSCARs 25, 26 and 27. Like OSCAR 27, it is part of a commercial satellite. The satellite is functioning in its commercial mode, but amateur activity has been minimal. At the time of this writing, its future as a ham satellite is in question.

RS-15 was launched in December 1994. The satellite carries a Mode A transponder.

Fuji-OSCAR 29 was launched to orbit from Japan in 1996. The satellite is essentially a clone of OSCAR 20, with the exception that its packet BBS has 9600-baud capability.

Mexico-OSCAR 30 was intended to be a 1200-baud packet satellite similar to OSCAR 16. Unfortunately, the bird failed several weeks after achieving orbit.

you can receive while you transmit (highly recommended), you'll hear your own signal coming back from the satellite a fraction of a second after you transmit. Even dyed-in-the-wool satellite operators get a thrill from hearing their own signal being repeated by the satellite as it comes into view!

This brings up an important point. Unlike commercial television and telephone relay satellites, an amateur satellite is not always immediately accessible. Commercial satellites are *geostationary*, which means that they appear to be motionless from our perspective. On the other hand, amateur satellites, including the RS series, utilize *low-Earth orbits* (LEOs). The LEO orbit describes a circle, with the satellites traveling about 1000 km above the surface at a speed that causes the satellite to complete the circle about every hour and a half. As the

satellite orbits, the Earth rotates on its axis, bringing different portions of the Earth "in view" of the satellite at different times of the day.

From our perspective a LEO satellite rises above the horizon, travels across the sky in an arc and then sets again. It may do so six to eight times a day. For "passes" in which the satellite goes nearly overhead, this rise and set cycle takes 15 or 20 minutes. On some orbits the satellite's path is such that it rises only a short distance above the horizon, much like the winter sun near the Arctic Circle. As you might expect, the time the satellite is in view is much shorter. The total amount of time that any particular LEO satellite is available for use at a given location is perhaps an hour or so each day. It doesn't seem like a long time, but it is more than enough to provide outstanding

Satellite Modes Demystified

Since there are a number of amateur satellites operating, and since they operate on a variety of bands, some way of distinguishing the various combinations of bands became necessary. The first amateur transponder in space was the one carried by OSCAR 6. It received uplink signals at 2 meters and retransmitted them on the downlink at 10 meters. When OSCAR 7 was launched it included a transponder with a 70-cm uplink and a 2-meter downlink. To distinguish between the two, the 2-meter-to-10-meter transponder operation was called "Mode A" and the 70-cm-to-2-meter operation "Mode B."

The launch of OSCAR 8 complicated matters by using a transponder with a 2-meter uplink and a 70-cm downlink, just the reverse of the Mode-B transponder of OSCAR 7. This new transponder was built by Japanese amateurs, and in their honor the new setup was dubbed "Mode J."

OSCAR 10 presented yet another possibility with its 23-cm-to-70-cm transponder. The 23-cm band is in that portion of the spectrum called "L band" by microwave engineers, so this uplink/downlink setup acquired the name "Mode L." A similar situation occurred when OSCAR 13 deployed a 70-cm-to-13-cm transponder. Since the 13-cm band is in the S-band part of the microwave spectrum, this mode is called—you guessed it—"Mode S."

The digital satellites, the first of which was the Japanese OSCAR 12, threw a monkey wrench into this (relatively) simple naming scheme. For the first time, amateurs had to contend with transponders that were different in kind rather than just frequency. FO-12 (and its successor, FO-20) carried both an analog transponder and a digital packet-radio transponder. Both transponders received uplink signals on 2 meters and transmitted on 70 cm, conforming to the Mode J band selection. To indicate the band selection while distinguishing between the analog and digital transponders, the analog operation is called Mode JA and the digital operation Mode JD.

Finally, the RS-12 satellite uses a unique uplink/downlink band pair: 15 meters to 10 meters (Mode K).

It's Very Easy to Work RS-10!

(Reprinted from *OSCAR Satellite Report*)

Think you need much to work RS-10? Wrong! I called CQ this morning using a Kenwood TH-26AT hand-held and its 4-inch "rubber duck" antenna, sending CW with the mike button. Back came VE3DJ in Ancaster, Ontario, totally without prearrangement, and we completed a QSO giving me a 569 report! Power output at this end was about 2 watts for the QSO, but I subsequently heard my downlink Q5 with half a watt output. I used my station receiver for the [10-meter] downlink with a long-wire antenna. The rubber duck was indoors!

This morning I called CQ again on RS-10 with my TH-26AT's push-to-talk button and was answered by W1SJM in New Hampshire, who gave me a 449 report. No rubber duck this time, though. RS-10 was only 17 degrees high to the west, so I needed a really large antenna: an AEA half-wave whip!—*by Ray Soifer, W2RS*

---

operating enjoyment on a regular basis.

Working the Satellite: When using the RS-12 mode K transponders you don't need to know precisely where the satellite is positioned in the sky. After all, you aren't likely to be using narrow beamwidth antennas. (Often heard on RS-12/13: "Antenna here is a dipole.") Mainly, you want to know when the satellite will be in view. Of course, 15 and 10-meter signals are subject to ionospheric bending, so it pays to listen for the satellite before and after the predicted visibility period.

Once you've determined when the satellite is due to rise above the horizon at your location, listen for the satellite's telemetry beacon (see Table 13-1). This signal is transmitted constantly by the satellite and carries information about the state of the satellite's systems, such as its battery voltage, solar-panel currents, temperatures and so on. You should hear it just as the satellite rises above the horizon. As soon as you can hear the beacon, start tuning across the downlink passband.

On an active day you should pick up several signals. They will sound like normal amateur voice and CW contacts. Nothing unusual about them at all, except... it sounds like the signals are slowly drifting downward in frequency! That's the effect of Doppler shift. It's not too serious on the 10-meter downlink, but it can be a challenge when the downlink is at 70 cm because the shift is proportional to the transmitted frequency. (More about the Doppler factor later.) Now tune your transmitter's frequency to the satellite's uplink passband. Send a series of dits while you tune the transmitter frequency (yes, the *transmitter* frequency). With any luck you'll soon

hear your own signal coming back down from the satellite. A tentative CQ might bring a call from another station that also has the satellite in view. Give it a try. Did you hear an answer? Congratulations! You've just become a satellite operator.

Movin' On Up—Mode A: While Mode K is fun and exciting, there is much to be said for VHF operation. The noise levels at both the satellite and ground station tend to be lower, allowing better reception. If you have an all-mode (CW and SSB) 2-meter transceiver and an HF transceiver, you're set to operate Mode A on RS-10 or RS-15. (While FM signals will pass through the satellites' transponders, use of FM is strongly discouraged. Because FM transmissions use full power at all times, they require a disproportionate amount of the satellite's available downlink power. A satellite can handle many more simultaneous SSB signals than FM signals, and even one FM signal can degrade the usability of the satellite. But there *is* a way you can use your FM transmitter: as a CW transmitter. See the sidebar: *It's Very Easy to Work RS-10!*)

The only significant difference when using a 2-meter uplink or downlink is that the Doppler shift becomes more pronounced. Doppler shift is the same effect you hear from a high-speed passing truck: the truck's sound is higher than normal as it approaches, lower as it retreats from you. The same thing happens to radio signals traveling to and from the satellite. See the sidebar: *Doppler Shift*.

Exotica: Phase 3 Satellites

The limitations of LEO satellites, especially their brief periods of availability, are overcome by a class of satellites called "Phase 3." The name comes from the various phases in the development of amateur satellites. The earliest ones, during Phase 1, contained beacon and telemetry transmitters, but not transponders. These early satellites were all in circular, low-Earth orbits—as were the Phase 2 satellites, which carried communications transponders.

Phase 3 satellites are not in low-Earth orbits. Rather, their orbits describe an ellipse. A Phase 3 satellite swings within a few hundred kilometers of the Earth's surface at one end of the ellipse (the *perigee*) and streaks out to 30,000 km or so at the other end (the *apogee*). The physics of an orbiting body dictates that the satellite spends much more of its time near apogee than perigee. Therefore, the Phase 3 satellites spend most of their time at very high altitudes. From a typical point in the Northern Hemisphere, a particular Phase 3 satellite is available for more than 10 hours per day. This is a remarkable improvement over

Table 13-1
Analog Transponder Frequencies

RS Satellites

	RS-10	RS-11	RS-12	RS-13	RS-15
Mode A					
Uplink	145.860-145.900	145.910-145.950	145.910-145.950	145.960-146.000	145.858-145.898
Downlink	29.360-29.400	29.410-29.450	29.410-29.450	29.460-29.500	29.354-29.394
Beacons	29.357/29.403	29.407/29.453	29.408/29.454	29.458/29.504	29.353/29.399
Mode A Robot					
Uplink	145.820	145.830	145.830	145.840	
Downlink	29.357/29.403	29.407/29.453	29.454	29.504	
Mode K					
Uplink	21.160-21.200	21.210-21.250	21.210-21.250	21.260-21.300	
Downlink	29.360-29.400	29.410-29.450	29.410-29.450	29.460-29.500	
Beacons	29.357/29.403	29.403/29.453	29.408/29.454	29.458/29.504	
Mode K Robot					
Uplink	21.120	21.130	21.129	21.138	
Downlink	29.357/29.403	29.403/29.453	29.454	29.504	
Mode T					
Uplink	21.160-21.200	21.210-21.250	21.210-21.250	21.260-21.300	
Downlink	145.860-145.900	145.910-145.950	145.910-145.950	145.960-146.000	
Beacons	145.857/145.903	145.907/145.953	145.912/145.958	145.862/145.908	
Mode T Robot					
Uplink	21.120	21.130	21.129	21.138	
Downlink	145.857/145.903	145.907/145.953	145.958	145.908	

Phase 3 Satellites

Satellite	Mode	Uplink (MHz)	Downlink (MHz)
AO-10	B	435.030-435.180	145.825-145.975
	Beacon		145.810

Other Satellites

Satellite	Mode	Uplink (MHz)	Downlink (MHz)
FO-20	J(A)	145.900-146.000	435.800-435.900
	Beacon		435.795
FO-29	J(A)	145.900-146.000	435.800-435.900
	Beacon		435.795

the LEO satellites! And because the Phase 3 satellite is so much higher, it is visible from a greater fraction of the Earth's surface, too. The result is a vast improvement in the communications capability of the satellite.

There is a downside, however. The greater distance to the Phase 3 satellite means that more transmitted power is needed to access it, and a weaker signal is received from the satellite at the ground station. (This problem is alleviated somewhat by the use of gain antennas on the satellite.) The signal levels are such that operation using the first generation of Phase 3 satellites is feasible only with ground-station antennas that exhibit significant gain (10 dBi or more). Such directional antennas must be pointed directly at the satellite. The satellite's position in the sky changes over time, however, so the antenna's position must change as well. In fact, the satellite changes both its bearing from the ground station (azimuth) and its height above the horizon (elevation). Most ground stations that access Phase 3 satellites use antenna systems that can be rotated in both the azimuth and elevation planes.

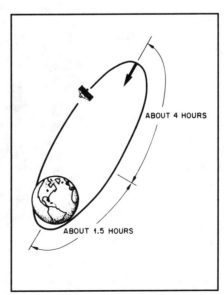

Fig 13-1—The orbit of AMSAT OSCAR 10 provides hours of coverage on each orbit to stations in the Northern Hemisphere.

Doppler Shift

Doppler shifting of signals is caused by the relative motion between you and the satellite. As the satellite moves toward you, the frequency of the downlink signals will increase by a small amount as the velocity of the satellite adds to the velocity of the transmitted signal. As the satellite passes overhead and starts to move away from you, there will be a rapid drop in frequency of a few kilohertz, much the same way as the tone of a car horn or a train whistle drops as the vehicle moves past the observer. Martin Davidoff, K2UBC, provides a very complete and understandable discussion of Doppler shift in *The Satellite Experimenter's*

Handbook. This brief outline on Doppler shifts provides highlights of the effect.

The Doppler effect is different for stations located at different distances from the satellite because the relative velocity of the satellite with respect to the observer is dependent on the observer's distance from the satellite. The result of all this is that signals passing through the satellite transponder shift slowly around the calculated downlink frequency. Locating your downlink signal is more than a simple computation, since tuning is needed to compensate for the Doppler shift.

Doppler shift through a transponder becomes the sum of the Doppler shifts of both the uplink and downlink signals. In the case of an inverting type transponder (as in OSCAR 10 Mode B), a Doppler-shifted increase in the uplink frequency causes a corresponding decrease in downlink frequency, so the resultant Doppler shift is the *difference* of the Doppler shifts, rather than the *sum*. The shifts tend to cancel.

While Doppler shifts can be observed on OSCAR 10, it's not nearly so apparent as it was for OSCARs 7 and 8. The time required to observe a given change in Doppler shifts is much longer for Phase 3 satellites than for the low-altitude satellites, and is therefore much less of a problem. For instance, the total OSCAR 10 Doppler shift is approximately 4.5 kHz over the period of time that normal operations are conducted, approximately four hours on each side of apogee. This rate of change amounts to approximately 9 Hz per minute. For OSCAR 8, operators experienced a total shift of up to 12 kHz over a period of 16 minutes, a rate of approximately 750 Hz per minute!

It must be pointed out that the complete whole-orbit Doppler-shift curve is not a straight line. However, over the portion of the orbit that encompasses the majority of operations, a straight-line approximation will suffice quite well. These ranges cover approximately MA = 35 to MA = 220, and the total shift averages to be 4529 Hz, or 8.96 Hz/minute.

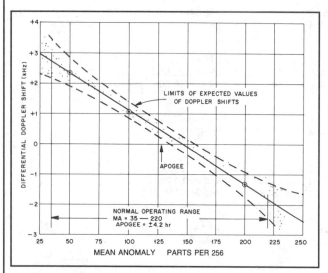

OSCAR 10 Mode B differential Doppler shift, normal for MA operating range.

At the time of this writing, the only Phase 3 satellite in orbit is OSCAR 10. It is only intermittently available; its control computer suffered accumulated radiation damage that rendered the satellite uncontrollable. It occasionally operates when it gets sufficient sunlight.

Phase 3D

If all goes as planned, an Ariane 5 rocket will carry the Phase 3D satellite into orbit in 1997. It will be the largest, most complex Amateur Radio satellite ever created.

But to fully understand Phase 3D, it helps to know the story of OSCAR 13. Launched in 1988, OSCAR 13 quickly became the most popular satellite in Amateur Radio history. It traveled in an elliptical orbit that sent it almost 37,000 kilometers into space at its greatest distance from Earth. From that vantage point, OSCAR 13 could "see" huge portions of the globe. Any station within its *footprint* could use the satellite as a relay to work anyone else. Its transponders functioned like repeaters in space, except that they could relay many signals at once. When the satellite was positioned over the Atlantic, for ex-

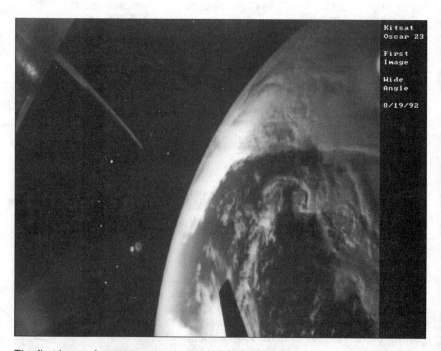

The first image from cameras aboard KITSAT-OSCAR 23. This shot was taken only hours after it achieved orbit. The long, narrow objects in the image are the satellite's antennas. Similar images are also available from cameras on OSCARs 22 and 25. *(Courtesy Harold Price, NK6K)*

ample, hams east of the Mississippi could enjoy SSB and CW conversations with hams in Europe, the Middle East and much of Africa.

Any tracking program could be used to determine when OSCAR 13 was in range, but there was an extra complication: Every 90 days the satellite shifted its orientation in space so that its solar cells could receive sufficient sunlight. When it did, its antennas would point away from Earth, making it difficult to use the bird. So, not only did you have to know when you were within OSCAR 13's footprint, you had to determine where its antennas were pointing relative to your station. This was the so-called "squint angle" or "off-pointing angle."

OSCAR 13 was equipped with several transmitters, but they were not very powerful. This meant that most OSCAR 13 users needed sizable antennas.

But despite these difficulties, OSCAR 13 never had a shortage of activity. On weekends in particular, the downlink sounded like 20 meters. DXpeditions would bring satellite equipment and treat OSCAR 13 like another "band," generating huge pileups!

You can blame the Sun and the Moon for OSCAR 13's premature demise. Their gravitational influences disturbed its orbit, causing it to gradually deteriorate. Each time it passed near the Earth (its *perigee*), OSCAR 13 sank deeper into the atmosphere. On December 6, 1997, the satellite finally lost its battle with gravity and plunged to fiery destruction.

More than a Replacement

To say that Phase 3D is a mere replacement for OSCAR 13 is like saying that a new Ferrari is a "mere replacement" for a Ford Taurus. Consider the output power of Phase 3D's transmitters compared to OSCAR 13's. On 2 meters OSCAR 13 managed to generate about 50 W PEP. On the same downlink Phase 3D will pump more than *100* W PEP to an 11-dB gain antenna. On its 2.4 GHz downlink, OSCAR 13 produced a single watt of output. On 2.4 GHz, Phase 3D will furnish *50* W to a 19-dB gain antenna!

With Phase 3D's substantial output power and high-gain antennas, you won't need large antennas on many bands, or great amounts of RF power on *any* bands. That translates into smaller, more affordable stations—particularly if you make the leap to microwaves. Apartment dwellers and hams suffering under antenna restrictions will appreciate this!

When it comes to uplink and downlink frequencies, Phase 3D has phenomenal flexibility (see Table 13-2). The transponders can be mixed and matched as the need arises. For example, the satellite could be configured to listen on 70 cm and relay on 10 GHz. Or listen on 1.2 GHz and relay on 24 GHz. And if the power budget is as great as expected, Phase 3D will operate in two uplink/downlink modes *simultaneously*. (We'll know if this will be possible once the bird is in final orbit and all tests are complete.)

Phase 3D's orbit will be higher at apogee and more stable than OSCAR 13's. It may also allow the satellite to perform a neat trick. That is, Phase 3D may appear in the same positions in your local sky every 48 hours. If the satellite is 10° above your neighbor's apple tree at 8 o'clock on Monday morning, it would be there again at 8 o'clock on Wednesday morning. This doesn't mean that you can throw away your satellite tracking software, but it certainly won't be as critical to your station.

Phase 3D may also dispose of the squint-angle problem that bedeviled the OSCAR 13 community. If the large solar panels deploy properly, Phase 3D *won't* have to reorient itself periodically for proper Sun exposure. Instead, it will keep its antennas

pointed directly at the Earth at all times. Translation: If Phase 3D is above your horizon, you can probably use it.

Phase 3D will open the door to long-distance communication regardless of solar conditions. Whenever the satellite is above your horizon, the band will be "open" for you. And I'm not talking about chats over a few hundred miles. Each time the satellite is at apogee, half of the Earth's disc will be "visible" to its antennas and wide-bandwidth transponders. That's a sizable footprint with opportunities for lots of DX.

Technician-class amateurs will be among the big beneficiaries of Phase 3D. The satellite will open the world to them, allowing many to work DX on a *regular basis*. After Phase 3D reaches orbit, we'll have to make room for many more Technicians on the DXCC rolls!

Phase 3D will also function as an ideal platform for those who want to extend their horizons into the microwave bands. Imagine being able to receive the satellite's 24-GHz downlink using a tiny, 2-foot dish. How about a portable microwave satellite station that fits into a standard suitcase? We desperately need more activity on our microwave frequencies and Phase 3D will give hams a powerful incentive.— *Steve Ford, WB8IMY*

What About Phase 4?

It has long been the dream of amateur satellite designers to launch a geostationary amateur satellite. Fixed, high-gain antennas such as parabolic dishes could be pointed at the satellite, greatly reducing the complexity of the ground station. Unfortunately, the costs of producing and maintaining such satellites puts them out of the reach of the amateur community—for now.

You may wonder who sponsors these amateur satellites. The answer is that amateurs themselves provide most of the sponsorship of satellites built in the US and Europe. The organization that has performed most of this work in the last 20 years or so is AMSAT, the Radio Amateur Satellite Corporation, a nonprofit amateur group. If you are interested in satellite operation, join AMSAT. Aside from helping to sponsor future satellites, you'll receive *The AMSAT Journal* and discounts on AMSAT software and other membership benefits. For more information send a self-addressed, stamped envelope to AMSAT, PO Box 27, Washington, DC 20044.

PACSATs: Digital Wonders

Technology marches on, and anyone who has observed Amateur Radio in the past decade will recognize that the most fundamental change has been the advent of packet radio. Even in packet's infancy, visionaries saw the potential marriage of packet and satellite technology. The combination of the two has resulted in the PACSAT: a satellite carrying a packet radio transponder and a computer. PACSATs operate in a fundamentally different way from satellites with analog transponders.

A ground station transmits a message (digitally, of course) to the satellite. The satellite stores the entire message in its onboard memory, which typically can hold several million characters of message text. Later, when the satellite is over the ground station for which the message is intended, it transmits the message to that station. This kind of *store-and-forward* operation provides true worldwide communications using low-Earth-orbit satellites. Because PACSATs can hold a lot of data, and because they are optimized for transmitting data rather than voice or CW, they provide an unsurpassed bulletin transmission system.

The software in the on-board computer works in conjunction with personal computers at ground stations to deliver

Table 13-2
Phase 3D Frequencies

Note: These are inverting transponders. For example, if you transmit upper sideband in the lower portion of the uplink passband, the satellite repeats in lower sideband in the upper portion of the downlink passband.

Uplinks

Band	Digital (MHz)	Analog (MHz)
15 meters	——	21.210—21.250
2 meters	145.800—145.840	145.840—145.990
70 cm	435.300—435.550	435.550—435.800
23 cm(1)	1269.000—1269.250	1269.250—1269.500
23 cm(2)	1268.075—1268.325	1268.325—1268.575
13 cm(1)	2400.100—2400.350	2400.350—2400.600
13 cm(2)	2446.200—2446.450	2446.450—2446.700
6 cm	5668.300—5668.550	5668.550—5668.800

Downlinks

Band	Digital (MHz)	Analog (MHz)
10 meters	29.330 MHz (To be used for AM voice bulletins)	
2 meters	145.955—145.990	145.805—145.955
70 cm	435.900—436.200	435.475—435.725
13 cm	2400.650—2400.950	2400.225—2400.475
3 cm	10451.450—10451.750	10451.025—10451.275
1.5 cm	24048.450—24048.750	24048.025—24048.275

Beacons

Band	Beacon-1 (MHz)	Beacon-2 (MHz)
70 cm	435.450	435.850
13 cm	2400.200	2400.600
3 cm	10451.000	10451.400
1.5 cm	24048.000	24048.400

error-free copies of bulletins from around the world. As the unique capabilities of PACSATs become better understood, they are being intergrated into terrestrial packet networks. For the first time, hundreds of thousands of amateurs can have their messages relayed via amateur satellites.

In general, packet satellite operation is similar to terrestrial packet. The same AX.25 protocol is used to communicate with the satellite, but the radio equipment is somewhat different. The combination of weaker signals and Doppler shifting burdens a satellite-transmitted packet with more critical modem requirements. To solve this problem, phase-shift keying (PSK) or direct FSK modulation techniques are used. For AO-16, WO-18, LO-19, IO-26, and FO-29, a PSK modem is used in conjunction with a 2-meter FM transmitter and a 70-cm SSB receiver. OSCARs 22, 23 and 25 require 9600 bit/s FSK modems with 2-meter and 70-cm FM rigs. The modems typically control the frequency of the receiver via the UP/DOWN pins on the microphone connector. Although this automatic tuning isn't strictly required, one quickly tires of adjusting the frequency to compensate for the large Doppler shift of the downlink signal! Most modern receivers that feature computer control capability can be used to directly compensate for Doppler shift.

To date, most of the store-and-forward PACSATs use a common scheme for communicating with ground stations. This scheme requires special-purpose software at the ground station. (Software packages are available from AMSAT.) For those interested in creating their own software to access these satellites, the specifics have been published in the *Proceedings of the 9th Computer Networking Conference.*

FINDING A SATELLITE

In order to use satellites, you'll need to have some idea when each one is "visible" at your location. Unfortunately, the time at which a particular satellite is above your horizon varies from day to day. So does its location in the sky from your point of view. Therefore, you need some means of determining where and when the satellite is going to appear. If you are using steerable directional antennas, you also need to know the satellite's exact position in the sky.

Computer-based calculation of the satellite position is the most common technique used by amateurs today. The calculations can be performed by most popular home computers using one of a variety of programs. Commercial programs that calculate satellite positions are available from AMSAT at a modest cost. For computers that have graphic display capability, there are programs that will display the satellite orbit on a map of the Earth.

The techniques used to mathematically calculate satellite position are well outside the scope of this book but are thoroughly explained in *The Satellite Experimenter's Handbook*. For our purposes, it is sufficient to know that in order to perform these calculations, the computer programs require data, called *Keplerian elements*, that mathematically describe the orbit of the satellite.

Since a satellite's orbit is affected by the inconstant gravity of the Earth, moon and sun, and by atmospheric drag in the case of the lower satellites, it tends to gradually change. This

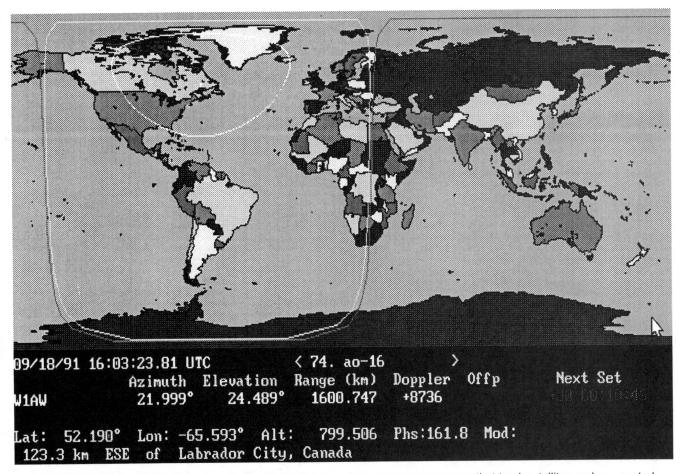

09/18/91 16:03:23.81 UTC   < 74. ao-16   >
                Azimuth Elevation Range (km) Doppler Offp   Next Set
W1AW   21.999°   24.489°   1600.747   +8736

Lat: 52.190° Lon: -65.593° Alt: 799.506 Phs:161.8 Mod:
123.3 km ESE of Labrador City, Canada

Fig 13-2—AMSAT's InstantTrack is an example of one of the modern computer programs that track satellites and can control the antenna system.

means that a new set of Keplerian elements are required from time to time. AMSAT issues Keplerian elements for a variety of satellites on a weekly basis. These are distributed by packet, the Internet and landline bulletin-board systems, and in the AMSAT newsletter. They are also transmitted by W1AW using RTTY modes.

SETTING UP A SATELLITE STATION

A basic station for use with an analog transponder is shown at the beginning of this chapter. If you want to try AO-10, you'll probably need to add receiving preamplifiers and appropriate switching circuitry. (The preamps must be located on the tower near the antennas to minimize transmission line losses.) Since CW and SSB are the required modes, you may want to start on either Mode K or Mode A. Both modes require a 10-meter receiving system at the ground station. While some ground stations use specialized 10-meter satellite antennas such as a dipole above a reflecting screen or a "turnstile" design, these aren't really necessary to get started. An existing dipole, vertical or small Yagi antenna will work just fine.

Modern amateur receivers have sufficient sensitivity, selectivity and stability for 10-meter satellite operation, but be careful of older equipment or inexpensive shortwave receivers. They may be lacking in one or more of these characteristics. Almost any 15-meter CW/SSB transmitter can be used for Mode K and, again, existing antennas should prove adequate. For Mode A you can still use existing antennas, but best performance will result from using a circularly polarized antenna such as a turnstile. (See the sidebar: *Circular Polariza-*

tion.) For LEO satellites you need only a few watts of 2-meter transmitter power, so most CW/SSB transmitters should do the job "barefoot."

Mode B requires 2-meter reception and 70-cm transmission capability. There are numerous ways of setting up Mode-B stations, ranging from the use of transverters along with HF transmitters and receivers to the purchase of multiband satellite transceivers. But these are the easiest of the VHF-and-above modes for the beginner. Equipment for the 2-meter and 70-cm bands is easier to find, cheaper to buy and easier to build than equipment for higher frequency bands.

One important factor to consider at VHF and above is transmission line loss. Losses tend to increase with frequency. A 100-foot run of coax that works perfectly well at 10 meters, for example, may be all but useless at 70 cm. Setting up a satellite station for Mode B or the higher-frequency modes requires a bit of engineering. It isn't difficult, but you have to consider the antennas, transmission line(s) and radios as a whole. See *The Satellite Experimenter's Handbook* for complete details.

To operate via the digital satellites you will also need a packet terminal node controller (TNC). Several of the satellites use a 1200-baud PSK downlink and 1200-baud Manchester FSK uplink scheme. At the ground station, this system requires a 70-cm SSB receiver and a 2-meter FM transmitter. A specialized modem must be connected between the TNC and the radio. 1200-baud PSK/FSK modems are available from the following sources:

Circular Polarization

As VHF-and-above operators know, antenna polarization is critical to effective communications between stations. This presents a problem when using a satellite because the satellite changes its orientation with respect to the ground station as the satellite moves. Using horizontal or vertical polarization would result in deep fades when the satellite's antenna was cross-polarized to that of the ground station.

Fortunately, there's an answer: circular polarization (CP). A CP signal rotates in polarization once per cycle. That is, the polarization is constantly changing. This kind of polarization is immune to orientation changes of the satellite (assuming the satellite antenna is pointed directly at the ground station while the satellite spins). Circular polarization can be "right hand" (RHCP) or "left hand" (LHCP). Amateur satellites that use CP antennas typically use RHCP. While you can receive CP signals on a linearly polarized antenna, you lose about half the signal as compared to using a CP antenna.

Although RHCP is the norm, you may experience a change to LHCP signals when the spacecraft antenna is pointed away from you; often the antenna sidelobes have odd polarizations. Being able to switch between RHCP and LHCP is a useful antenna feature, and many satellite antennas have this option.

Tucson Amateur Packet Radio
8987-309 E Tanque Verde Rd No. 337
Tucson, AZ 85749-9399
tel 817-383-0000

Pac-Comm, Inc
4413 N Hesperides St
Tampa, FL 33614-7618
tel 813-874-2980

These same sources can supply modems for use with the 9600-baud packet satellites.

A few of the digital satellites are actually intended for monitoring purposes only. For example, DOVE (or DO-17) simply transmits. It is capable of transmitting digitized voice messages or packet telemetry. DOVE's primary mission is education. Its target audience is schoolchildren and others interested in learning about satellites and Amateur Radio. DOVE also transmits telemetry and short bulletin messages uploaded by the command stations. These transmissions use 1200-baud *AFSK* packet—the same kind of transmission used by terrestrial VHF packet stations. The strong signals of DOVE, coupled with the compatibility of its packet transmissions, means that literally tens of thousands of stations are already equipped to receive DOVE. If you operate 2-meter packet, your station is DOVE-capable right now! AMSAT has software that will help decode the telemetry from DOVE. Monitoring DOVE is a great way to get your feet wet in digital satellite operation.

Other transmit-only satellites include UO-11 and WO-18. UO-11 transmits on 145.825 MHz sending mostly telemetry, although it also can send digitized speech. While UO-11 does contain a digital store-and-forward system, its use is restricted because of the relatively small amount (128 kBytes) of on-board memory. UO-11 also includes several scientific/educational experiments.

WEBERSAT (WO-18) transmits using the 1200-baud PSK standard. It contains an on-board camera for Earth imaging.

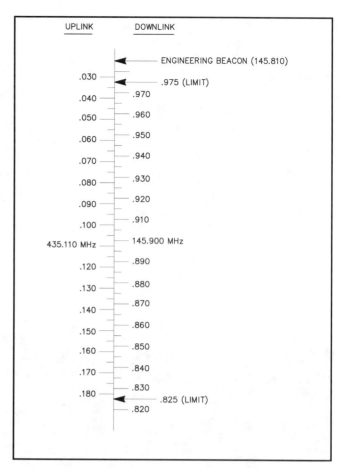

Fig 13-3—OSCAR 10 Mode B Frequency Translation Chart.

Again, software for decoding and viewing WEBERSAT images is available from AMSAT.

SATELLITE OPERATING

Linear Transponders

Operating via an analog transponder is much like operating on the HF bands. Then again, it's also similar to repeater operation since the transponder does essentially the same job as a repeater. But satellite operation is unique in many ways. The most dramatic difference is that in satellite operation you can hear the satellite's transponder repeating your own signal. While this is marvelous in itself, it leads to some different operating techniques. This is particularly true when you couple this characteristic with the need to compensate the operating frequency for the effects of Doppler shift.

To compensate for Doppler, the two stations in a QSO must occasionally adjust frequency. But there is yet another problem: Since the amount of frequency shift you experience is dependent on the satellite's speed *relative to your location*, and since the satellite has different relative speeds at different stations, two QSOs that were originally separated in frequency can "walk" into one another as the relative speeds change.

To minimize this problem, satellite operators use a technique that requires the least amount of retuning. One of the two stations in the QSO is chosen as a reference. This station holds the *lower* of its two frequencies (transmit or receive) constant. Since this station can hear its own signals (full duplex), it adjusts the *higher* frequency to match. That is, if the lower frequency is the receive (as in Mode B, for example), that remains constant while the

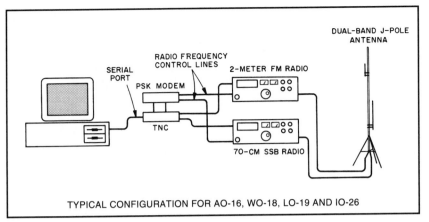

Fig 13-4—A typical Microsat ground station. For reception of DO-17 (DOVE) telemetry, however, only a 2-meter-FM transceiver and a standard 1200-baud TNC are required.

transmit frequency is "swooped" onto the receive frequency. The transmit frequency is subsequently readjusted to keep the downlink signal stable. If the *transmit* frequency is the lower one (as in Mode J), it is held constant while the receive frequency is adjusted to match. Because different stations experience different Doppler shifts, only one of the stations can use this procedure. The other station adjusts its receive frequency to receive the first station, then adjusts its transmit frequency to match.

The technique described above works best when the reference station experiences the least amount of Doppler drift. Of the two stations, the one furthest removed from the ground track experiences the least drift. If you are using a tracking method that tells you the maximum elevation of the pass, you can compare that to the maximum elevation of the other station. The station with the lesser number is further from the ground track of the satellite. It's worth noting that the Phase 3 satellites don't suffer from severe Doppler effects except when near perigee. We generally use the same procedure, though, since it still minimizes what Doppler effects are present.

Finding the Downlink Signal

Since a satellite's linear transponders aren't channelized, you'll need to be able to figure out where your signal will appear in the downlink passband when you're transmitting on a particular uplink frequency. While Doppler shift complicates this question, you can easily determine the frequency within a few kilohertz. This is usually sufficient since you can hear the downlink signal yourself and tune it exactly.

Two important results occur when inverted transponders are used. First, the total Doppler shift is minimized since the shift experienced on the downlink partially compensates for the (inverted) shift experienced on the uplink. (The uplink shifts more than the downlink if it's on a higher frequency.) Second, SSB signals are inverted. Although there isn't an established rule, most stations transmit on USB, resulting in LSB downlink signals.

An informal band plan has evolved for linear transponder operation. Generally, the lower third of the passband is used for CW operation, the upper third for SSB, and the middle third for mixed CW/SSB and special modes. AO-10 had a more complete band plan that included reserved channels for special functions. The difficulties now being encountered by AO-10 have rendered this band plan moot.

Don't Be a Transponder Hog: A linear transponder retransmits a faithful reproduction of the signals received in the uplink passband. This means that the loudest signal received will be the loudest signal retransmitted. If a received signal is so strong that retransmitting it would cause an overload, automatic gain control (AGC) within the transponder reduces the transponder's amplification. However, this reduces all of the signals passing through it!

Strong signals don't receive any benefit from being loud since the downlink is "maxed out" anyway. Instead, all of the other users are disrupted since their signals are reduced. Avoiding this condition is just common sense—and good operating practice. It's easy, too: just note the signal level of the satellite's beacon transmission and adjust your transmit power to a level just sufficient to make your downlink appear at the same level as the beacon. Any additional power is too much!

Digital Transponders

The subject of digital transponders is a bit more complex than that of analog transponders. The first amateur digital satellite communications were performed through analog transponders. While it was successful, it wasn't particularly efficient. Analog transponders are optimized for low-duty-cycle modulations such as CW and SSB. Digital transmissions are usually constant-power signals that take a disproportionate fraction of the transponder power. This isn't a problem with digital transponders since they demodulate received signals and transmit using an independent digital transmitter. More importantly, digital transponders receive transmissions from the ground, process them using an onboard computer and respond appropriately. This process lets the digital transponder screen the incoming data for errors and enables one of the most important and useful results of this type of system: *store-and-forward* message relay.

Describing digital satellite operation is a little like shooting at a moving target. The modulation schemes are fixed by hardware. What the satellite does with the received data is wholly dependent on the programming of the on-board computer—programming that can be changed from the ground command stations. As you might guess, the operational specifics of a given digital satellite may change over time.

The "file server" satellites can be accessed using the AMSAT PACSAT software. The IBM PC software package (software is also available for the Apple Macintosh) consists of several programs: *PB* receives broadcasts from the satellites, *PG* lets you transfer files to and from the satellite, and *PHS* and *PFHADD* process files for satellite transfer. *WISP* is a complete easy-to-use software package designed to work in Microsoft *Windows*. These programs or their equivalent must be used to access the PACSATs. Both software packages are available from AMSAT.

Manned Missions

One of the more exciting operations available to today's amateur is communicating with men and women in space. At present, a semi-permanent 2-meter FM voice and packet station is operational from the Russian Mir space station. While operation is generally unscheduled, numerous voice and packet contacts with Mir have occurred. Mir uses a split-frequency scheme for

Table 13-3

Digital Transponder Frequencies and Modes

Satellite	Uplink(s) (MHz)	Downlink(s) (MHz)	Data Format
AMSAT-OSCAR-16	145.90 145.92 145.94 145.96	437.025 437.05 2401.10	1200 bit/s PSK AX.25
DOVE-OSCAR-17	None	145.825 2401.22	Digitized voice with 1200 bit/s AFSK AX.25 (FM) telemetry
WEBERSAT-OSCAR-18	None	437.075 437.10	1200 bit/s PSK AX .25
LUSAT-OSCAR-19	145.84 145.86 145.88 145.90	437.125 437.15	1200 bit/s PSK AX .25
UoSAT-OSCAR 22	145.90 145.975	435.120	9600 bit/s FSK (FM)
KITSAT-OSCAR 23	145.85 145.90	435.175	9600 bit/s FSK (FM)
KITSAT-OSCAR 25	145.87 145.98	436.50	9600 bit/s FM
ITAMSAT-OSCAR 26	145.875 145.900 145.925 145.950	435.870	1200 bit/s PSK
FUJI-OSCAR 29	145.85 145.89 145.91	435.910	1200 bit/s PSK

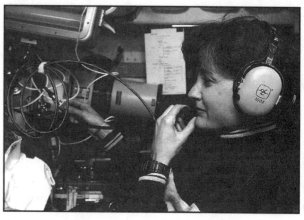

Astronaut Linda Godwin, N5RAX, Pauses during her mission duties to operate the SAREX station aboard the space shuttle *Endeavour. (Photo courtesy of NASA)*

1200-baud AFSK packet), and a voice/packet/slow-scan television system with fast-scan uplink capability. Which configuration can be flown on a shuttle mission depends mostly on the available stowage space, which in turn depends on the duration of the flight (because longer flights need to devote more space to storage of food and other consumables).

Like Mir, SAREX uses split-frequency operation. A 600-kHz split is used to accommodate fixed-offset 2-meter FM radios. It is very important *not* to transmit on the SAREX downlink frequency. Aside from the futility of doing so—the astronaut isn't listening to that frequency—you'll likely interfere with other stations that are trying to receive the SAREX transmissions.

Because SAREX is of secondary importance to the shuttle mission as a whole, SAREX operations are at the mercy of the needs of the primary payload. Open operations are also scheduled around space-to-classroom links, which complicates the picture somewhat. If you plan to operate via SAREX, it's wise to monitor W1AW and WA3NAN (the Goddard Space Center Amateur Radio Club station) for late-breaking schedule changes.

voice and packet. You transmit to Mir on 145.200 MHz and listen on 145.800 MHz. Mir also has a 70-cm FM repeater: Input 437.750 MHz, Output 437.950, 141.3-Hz CTCSS. The packet system aboard Mir uses standard 1200-baud AFSK packet. It often runs in the "mailbox" mode, allowing ground stations to exchange messages with the cosmonauts aboard Mir.

SAREX, the Shuttle Amateur Radio EXper- iment, is a continuing series of Amateur Radio operations from US space shuttle missions. While the focus of these missions has evolved toward space-to-classroom use of Amateur Radio, there are still many opportunities for the average ham to communicate with shuttle astronauts. The current SAREX suite comprises several equipment configurations: a simple battery-operated hand-held transceiver for FM voice-only QSOs, a voice-plus-packet setup (using

Try It Yourself

As you can see, it's easier than ever to add satellite communications to your list of operating modes. With new technologies on the horizon, it promises to become even easier in the future. Just like CW, RTTY, DXing, contesting and so on, satellites can become an integral part of your enjoyment of Amateur Radio. Don't be fooled into thinking that satellite operating is only for wealthy hams with lots of equipment and real estate. It's for you too!

For more information...

Publications

The Satellite Experimenter's Handbook, 2nd edition, ARRL (Order #3185, $20). This is *the* book to have. It's a complete guide to amateur satellites.

The *AMSAT Journal* comes with your membership in AMSAT ($30 in the United States, $36 Mexico and Canada and $45 in all other areas). Contains news, technical articles and commentary. Available from AMSAT, PO Box 27, Washington, DC 20044.

OSCAR Satellite Report, a semimonthly newsletter, contains late-breaking satellite news, current Keplerian elements and commentary. Available from R. Myers Communications, PO Box 17108, Fountain Hills, AZ 85269-7108.

The ARRL Satellite Anthology (order #5595, $12) is a collection of articles on a wide variety of satellite topics.

Computer Bulletin Board

ARRL BBS   860-594-0306

World Wide Web

http://www.amsat.org

CHAPTER 14

Emergency Communications

People in desperate need of help can depend on Amateur Radio to serve them. Since 1913, ham communicators have been dedicated volunteers for the public interest, convenience and necessity by handling free and reliable communications for people in disaster-stricken areas, until normal communications are restored.

RICHARD R. REGENT, K9GDF
5003 SOUTH 26TH ST
MILWAUKEE, WI 53221

Emergency communication is an Amateur Radio communication directly relating to the immediate safety of human life or the immediate protection of property, and usually concerns disasters, severe weather or vehicular accidents. The ability of amateurs to respond effectively to these situations with emergency communications depends on practical plans, formalized procedures and trained operators.

a particular response and careful planning. Large cities usually have capable relief efforts handled by paid professionals, and there always seems to be some equipment and facilities that remain operable. Even though damage may be more concentrated outside a city, it can be remote from fire-fighting or public-works equipment and law enforcement authorities. The rural public then, with

PLANS

More than any other facet of Amateur Radio, emergency communication requires a plan—an orderly arrangement of time, talent and activities that ensures that performance is smooth and objectives are met. Basically, a plan is a method of achieving a goal. Lack of an emergency-communications plan could hamper urgent operations, defer crucial decisions or delay critical supplies. Be sure to analyze what emergencies are likely to occur in your area, develop guidelines for providing communications after a disaster, know the proper contact people and inform local authorities of your group's capabilities.

Start with a small plan, such as developing a community awareness program for severe-weather emergencies. Next, test the plan a piece at a time, but redefine the plan if it is unsatisfactory. Testing with simulated-emergency drills teaches communicators what to do in a real emergency, without a great deal of risk. Finally, prepare a few contingency plans just in case the original plan fails.

One of the best ways to learn how to plan for emergencies is to join your local Amateur Radio Emergency Service (ARES) group, where members train and prepare constantly in an organized way. Before an emergency occurs, register with your ARRL Emergency Coordinator (EC). The EC will explain to civic and relief agencies in your community what the Amateur Service can offer in time of disaster. ARES and the EC are explained later in this chapter.

Communications for city or rural emergencies each require

Jim Funk, N9JF, and his daughter Melanie, N9IQV, operate from the Adams County American Red Cross Chapter House in Quincy, Illinois, during the Great Flood of 1993. (*photo by Sandy Martin*)

few volunteers spread over a wide area, may be isolated, unable to call for help or incapable of reporting all of the damage.

It's futile to look back regretfully at past emergencies and wish you had been better prepared. Prepare yourself now for emergency communications by maintaining a dependable transmitter-receiver setup and an emergency-power source. Have a plan ready and learn proper procedures.

PROCEDURES

Besides having plans, it is also necessary to have procedures—the best methods or ways to do a job. Procedures become habits, independent of a plan, when everyone knows what happens next and can tell others what to do. Actually, the size of a disaster affects the size of the response, but not the procedures.

Before disasters occur, there are many existing procedures; for example, how to correctly coordinate or deploy people, equipment and supplies. There are procedures to use a repeater and an autopatch, to check into a net and to format or handle traffic. Because it takes time to learn activities that are not normally used every day, excessively detailed procedures will confuse people and should be avoided.

Specific Procedures

Your EC will have developed a procedure to activate the ARES group, but will need your help to make it work. A telephone alerting "tree" call-up, even if based on a current list of phone numbers, might fail if there are gaps in the calling sequence, members are not near a phone or there is no phone service.

Consider alternative procedures, use alerting tones and frequent announcements on a well-monitored repeater to round up many operators at once. An unused 2-meter simplex frequency can function for alerting; instead of turning radios off, your group would monitor this frequency for alerts without the need of any equipment modifications. Since this channel is normally quiet, any activity on it would probably be an alert announcement.

During an emergency, report to the EC so that up-to-the-minute data on operators will be available. Don't rely on one leader; everyone should keep an emergency reference list of relief-agency officials, police, sheriff and fire departments, ambulance service and NTS nets. Be ready to help, but stay off the air unless there is a specific job to be done that you can efficiently handle. Always listen before you transmit. Work and cooperate with the local civic and relief agencies as the EC suggests; offer these agencies your services directly in the absence of an EC. During a major emergency, copy special W1AW bulletins for the latest developments.

Afterward, let your EC know about your activities, so that a timely report can be submitted to League HQ. Amateur Radio has won glowing public tribute in emergencies. Help maintain this record.

For a comprehensive study of emergency communications procedures, see the ARRL *Emergency Coordinator's Manual* (FSD-9). Contact the ARRL HQ Field Services Department for details on how to obtain a copy. "Experience is the worst teacher: it gives the test before presenting the lesson" (Vernon's Law). Train and drill now so you can be prepared for an emergency.

TRAINING

Amateurs need training in operating procedures and communications skills. In an emergency, radios don't communicate, but people do. Because amateurs with all sorts of

---

Southeastern Floods

Macon (Georgia) Amateur Radio Club members, on July 4, 1994, begin communicating reports of damage caused by spring flooding and Tropical Storm Alberto. Torrential rains pour down more than 12 inches of water on ground already saturated by several previous days' precipitation. Amateurs report bridges down, roads out, homes under water, and youth camps cut off from roads. Red Cross officials open shelters where hams relay instructions and messages. Meanwhile mobile and home-based operators assist the Georgia Emergency Management Association to keep in touch with officials in Atlanta.

The ARES works with Red Cross and National Weather Service staff. The Red Cross deploys damage assessment teams who need communications in order to decide the structural safety of bridges and other public facilities.

As the Ocmulgee River nears flood stage, the situation becomes alarming. But on the scene are hams, often the first to see the need for evacuation or support. A raging river breaks over its banks and levees to inundate a water treatment plant. Fresh water is dispersed in trucks to the most needy sites, coordinated with Amateur Radio communications. This frees law enforcement officers to handle a backlog of other duties.

After the weather abates on July 8, net operators relay Health-and-Welfare messages. In the southeastern floods disaster Amateur Radio communication is found to be flexible and adaptable by providing trained operators and hardware to others. It is effective at providing a total communications network.

---

varied interests participate, many of those who offer to help may not have experience in public-service activities. Rarely are there enough trained operators, especially if a crisis persists for a long time.

Proper disaster training replaces chaotic pleas with smooth organized communications. The ARRL recognizes the need for emergency preparedness and emergency-communications training through sponsorship of the ARES. Well-trained communicators respond during drills or actual emergencies with quick, effective and efficient communications. Each understands his or her role in the plan and sets a good example by knowing the proper procedures to use.

Whether training takes place with programs at club meetings, on-the-air or with a personal approach, the basic subjects should cover emergency communications, traffic handling, net or repeater operation and technical knowledge. Get as many people involved as possible to learn how emergency communications should be handled. Explain what's going on and assign each participant a useful role.

Practical on-the-air activities, such as the ARRL's Field Day and Simulated Emergency Test offer training opportunities on a nationwide basis for individuals and groups. Participation in such events reveals weak areas where discussions and more training are needed. In addition, drills and tests can be designed specifically to check dependability of emergency equipment, or to rate training in the local area.

Field Day

The League's Field Day (FD) gets more amateurs out of their cozy shacks and into tents on hilltops than any other event. You may not be operating from a tent after a disaster but the training you will get from FD is invaluable.

In the League Field Day event, a premium is placed on sharp operating skills, adapting equipment that can meet challenges of emergency preparedness and flexible logistics. Amateurs assemble portable stations capable of long-range communications at almost any place and under varying conditions. Alternatives to commercial power in the form of generators, car batteries, windmills or solar power are used to power equipment to make as many contacts as possible. FD is held on the fourth full weekend of June, but enthusiasts get the most out of their training by keeping preparedness programs alive during the rest of the year.

Simulated Emergency Test

The ARRL Simulated Emergency Test (SET) builds emergency-communications character.

The purposes of SET are to:
- Help amateurs gain experience in communicating using standard procedures under simulated emergency conditions, and to experiment with some new concepts.
- Determine strong points, capabilities and limitations in providing emergency communications to improve the response to a real emergency.
- Provide a demonstration, to served agencies and the public through the news media, of the value of Amateur Radio, particularly in time of need.

The goals of SET are to:
- Strengthen VHF-to-HF links at the local level, ensuring that ARES and NTS work in concert.
- Encourage greater use of digital modes for handling high-volume traffic and point-to-point Welfare messages of the affected simulated-disaster area.
- Implement the Memoranda of Understanding between the League, the users and cooperative agencies.
- Focus energies on ARES communications at the local level. Increase use and recognition of tactical communication on behalf of served agencies; using less amateur-to-amateur formal radiogram traffic.

Help promote the SET on nets and repeaters with announcements or bulletins, or at club meetings and publicize it in club newsletters. SET is conducted on the third full weekend of October. However, some groups have their SETs any time during the period of September 1 through November 30, especially if an alternate date coincides more favorably with a planned communications activity and provides greater publicity. Specific SET rules are published in *QST*.

Drills and Tests

A drill or test that includes interest and practical value makes a group glad to participate because it seems worthy of their efforts. Formulate training around a simulated disaster such as weather-caused disasters or vehicle accidents. Elaborate on the situation to develop a realistic scenario or have the drill in conjunction with a local event. Many ARRL Section Emergency Coordinators (SECs) have developed training activities that are specifically designed for your state, section or local area.

During a drill:

1) Announce the simulated emergency situation, activate the emergency net and dispatch mobiles to served agencies.

2) Originate messages and requests for supplies on behalf of served agencies by using tactical communications.

3) Use emergency-powered repeaters and employ digital modes.

4) As warranted by traffic loads, assign liaison stations to receive traffic on the local net and relay to your section net. Be

Field Day provides annual practice before an emergency happens. KA5WVE and AA5BC are seated operating packet under the watchful eye of KA5SKO.

sure there is a representative on each session of the section nets to receive traffic coming to your area.

After a drill:

1) Determine the results of the emergency communications.

2) Critique the drill.

3) Report your efforts, including any photos, clippings and other items of interest, to your SEC or ARRL HQ.

ARRL Emergency Coordinator and Certification Course

The League has a certification program to provide training and formal recognition of amateur achievement in the field of emergency communications. This course is administered to ECs (or potential ECs) by the Section Emergency Coordinator for each section. The ARRL also offers the public-service-oriented awards program of certificates for Public Service Honor Roll, Emergency Communications Commendation and Public Service Commendation.

Net Operator Training

Network discipline and message-handling procedures are fundamental concepts of Amateur Radio operation. Training should involve as many different operators as possible in Net Control Station and liaison functions; don't have the same operator performing the same functions repeatedly. There should be plenty of work for everyone. Good liaison and cooperation at all levels of NTS requires versatile operators who can operate either phone or CW. Even though phone operators may not feel comfortable on CW and vice versa, encourage net operators to gain familiarity on both modes by giving them proper training.

The liaison duties to serve between different NTS region net cycles as well as between section nets are examples of the need for versatile operators.

Ask your ARRL Section Traffic Manager (STM) to visit your club to conduct a training seminar and to provide any operating tips pertaining to traffic nets in your section. If no local traffic net exists, your club should consider initiating a net on an available 2-meter repeater. Coordinate these efforts with the STM and the trustee(s) of the repeater you'll use. Encourage club members to participate in traffic-handling activities, either from their home stations or as a group activity from a message center. For more information on NTS and traffic handling, see Chapter 15.

Los Angeles Area Earthquake

The earthquake tremor jolts Los Angeles area residents from their sleep on January 17, 1994, at 4:31 AM. Without warning, 300 schools are seriously damaged, thousands of small business crumble in ruins and power lines tear apart. The quake severed three aqueducts, cutting off water service to thousands of Angelenos. Older, unreinforced masonry buildings and modern wood-frame apartments built on top of ground floor garages sustain some of the heaviest damage. The tremor destroys thousands of homes. Eleven major roads are blocked from debris or damaged. The Santa Monica Freeway, hardest hit, was scheduled to be steel retrofit in just one month. Gas lines and water mains burst sending geysers of flames and water into the air.

SEC K6TMJ activates the ARES Emergency Communications Van at the San Fernando Valley area hospital to handle emergency traffic. He moves to recovery efforts the next day near the Devonshire Police Station in Northridge. Officials use fiber-optics cameras and dogs to find victims. Medical facilities need more trained personnel to cope with massive casualties. A Red Cross shelter is set up in the Birmingham High School.

AA6RK starts his generator running since there is no electricity in the San Fernando Valley where he lives. He relays more than 300 messages on 20 meters both in and out of the epicenter, the 818 telephone area code, in two days.

Amateurs comply with rules not to release names of survivors for the first 48 hours after the disaster and the dusk to dawn curfew. Other hams send traffic into the Valley using repeaters and to the Filmore Memorial building shelter operations. Some packet radio bulletin boards handle over 4000 messages for the San Fernando and Santa Clarita Valleys. In the Los Angeles Section alone, ARES had more than 100 hams volunteer for hospital and agency emergency communications services.

It is LA's worst earthquake since the 1971 Sylmar quake. Although 6.6 on the Richter scale is violent, it is not The Big One. Yet 55 people died due to this quake. It will be remembered that Amateur Radio definitely provided significant reliable communications to many relief agencies in a time of crisis.

METHODS OF HANDLING INFORMATION

Emergency Operations Center

Amateur Radio emergency communications frequently use the combined concepts of a Command Post (CP) and an Emergency Operations Center (EOC). See Fig 14-1.

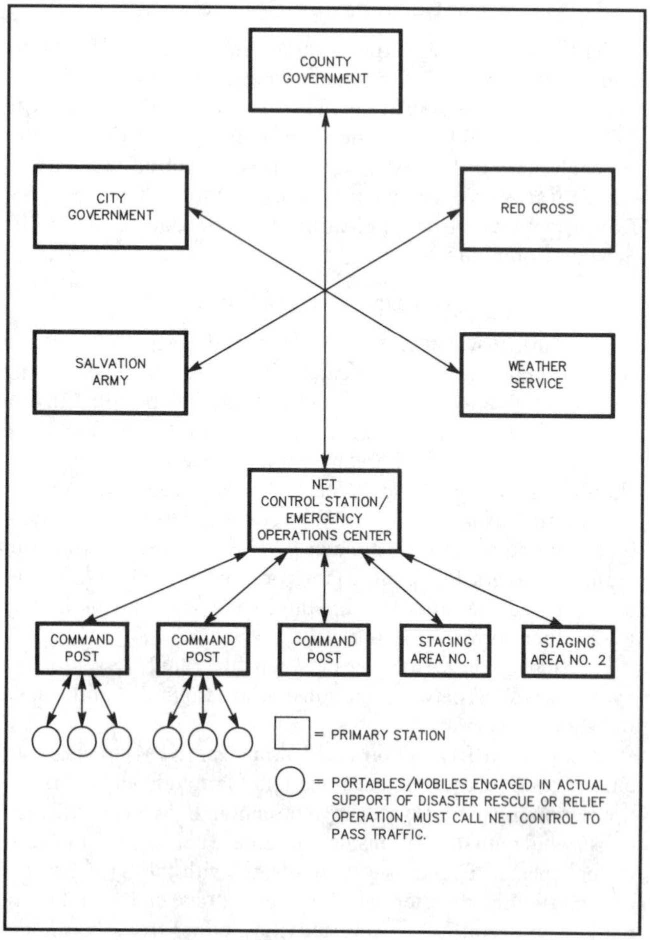

Although a Command Post controls initial activities in emergency and disaster situations, the CP may be unapparent because it is self-starting and automatic. The CP's general procedure is to assess the situation, report to a dispatcher and ask for equipment and people.

Consider an automobile accident where a citizen or an amateur, first on the scene, becomes a temporary Command Post to call or radio for help. A law-enforcement officer is dispatched to the accident scene in a squad car and, upon arriving, takes over the CP tasks. Incidentally, a Command Post may expand into multiple CPs or move to accommodate the situation. Relief efforts, like those in this simple example of an automobile accident, begin when someone takes charge, makes a decision and directs the efforts of others.

The Emergency Operations Center responds to a Command Post by dispatching equipment and helpers, anticipating needs to supply support and assistance, and may send more equipment and people to a staging area to be stored where they can be available almost instantly.

If the status of an accident changes (a car hits a utility pole, which later causes a fire), the CP gives the EOC an updated report then keeps control until the support agencies arrive and take over their specific responsibilities: Injuries—medical; fires—fire department; disabled vehicles—law enforcement or tow truck; and utility poles—utility company. By being outside the perimeter of dangerous activities, the EOC can use the proper type of radio communications, concentrate on gathering data from other agencies and then provide the right response.

As an analogy, think of the CPs, who request action and provide information, being similar to net participants checking into an amateur net with emergency or priority traffic. The EOC, who coordinates relief efforts, then functions as a Net Control Station.

Whether there is a minor vehicle accident or a major disaster operation, the effectiveness of the amateur effort in an emergency depends mainly on handling information.

Fig 14-1— The interaction between the EOC/NCS and the command post(s) in a local emergency.

Incident Command System

The Incident Command System (ICS) is a management tool

that provides a coordinated system of command, communications, organization and accountability in managing emergency events and is rapidly being adopted by professional emergency responders throughout the country. Amateurs should become familiar with ICS to work with agencies in a variety of multiple jurisdictions and political boundaries situations.

Incident Command Systems use:

- Clear text and common terms. Participants are expected to be familiar with ICS terminology. When the Incident Commander orders "a strike team of Type 2 trucks," everyone affiliated with filling the order knows exactly what is being requested.
- Unified Command. The Incident Commander is the only boss and is responsible for the overall operation.
- Flexibility. Functions such as planning, logistics, operations, finance and working with the press are described in detail so the organization size can change to match the particular incident's requirements. The IC can contract to a single individual for a small incident or expand to a Command Staff for a large incident.
- Concise Span of Control. Since management works well with a small number of people, the ICS typically uses a "strike team": a group of about five people having similar resources, plus a leader, who use common communications.

In some areas the ICS evaluates and determines what resources will be needed to start recovery. Amateur communicators are within the Logistics Section, Service Branch and Communications Unit of an ICS (Fig 14-2).

Tactical Traffic

Whether traffic is tactical, by formal message, packet radio or amateur television, success depends on knowing which to use.

Tactical traffic is first-response communications in an emergency situation involving a few operators in a small area. It may be urgent instructions or inquiries such as "send an ambulance" or "who has the medical supplies?" Tactical traffic, even though unformatted and seldom written, is particularly important in localized communications when working with government and law-enforcement agencies.

The 146.52 MHz FM calling frequency—or VHF and UHF

Stress Management

Emergency responders should understand and practice stress management. A little stress helps you to perform your job with more enthusiasm and focus, but too much stress can drive you to exhaustion or death.

Watch for these physiological symptoms:
- Increased pulse, respiration or blood pressure
- Trouble breathing, increase in allergies, skin conditions or asthma
- Nausea, upset stomach or diarrhea
- Muffled hearing
- Headaches
- Increased perspiration, chills, cold hands or feet or clammy skin
- Feeling weakness, numbness or tingling in part of body
- Feeling uncoordinated
- Lump in throat
- Chest pains

Cognitive reactions may next occur in acute stress situations; many of these signs are difficult to self-diagnose.
- Short term memory loss
- Disorientation or mental confusion
- Difficulty naming objects or calculating
- Poor judgment or making decisions
- Lack of concentration and attention span
- Loss of logic or objectivity to solve problems

Perhaps the best thing to do as you start a shift is to find someone that you trust and ask them to let you know if you are acting a bit off. If at some time they tell you they've noticed you're having difficulty, then perhaps it's time to ask for some relief. Another idea is to have some sort of stress management training of your group before a disaster occurs.

---

repeaters and net frequencies (see Fig 14-3)—are typically used for tactical communications. This is a natural choice because FM mobile, portable and fixed-station equipment is so plentiful and popular.

One way to make tactical net operation clear is to use tactical call signs—words that describe a function, location or agency. Their use prevents confusing listeners or agencies who are monitoring. When operators change shifts or locations, the set of tactical calls remains the same. Amateurs may use tactical call signs like "parade headquarters," "finish line," "Red Cross," "Net Control" or "Weather Center" to promote efficiency and coordination in public-service communication activities. However, amateurs must identify their station operation with its FCC-assigned call sign at the end of a transmission or series of transmissions and at intervals not to exceed 10 minutes.

Another tip is to use the 12-hour local-time system for time and dates when working with relief agencies, unless they understand the 24-hour or UTC systems.

Taking part in a tactical net as an ARES team member requires some discipline and following a few rules:

1) Report to the Net Control Station (NCS) promptly as soon as you arrive at your station.

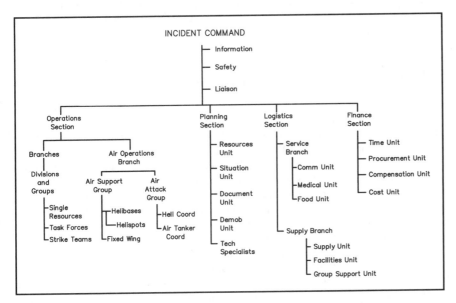

Fig 14-2—The Incident Command System structure.

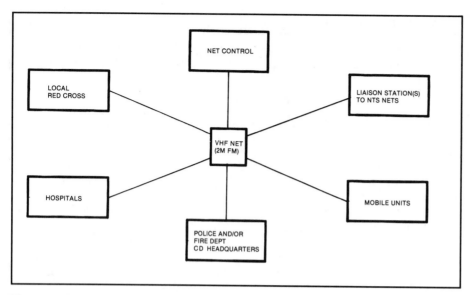

Fig 14-3—Typical station deployment for local ARES net coverage in an emergency.

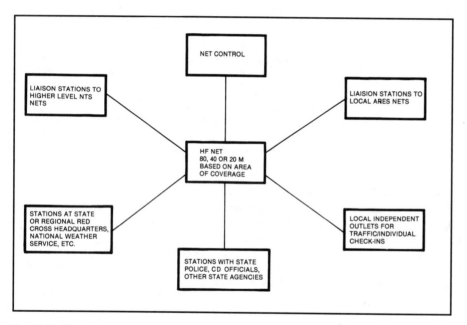

Fig 14-4—Typical structure of an HF network for emergency communications.

2) Ask the NCS for permission before you use the frequency.

3) Use the frequency for traffic, not chit-chat.

4) Answer promptly when called by the NCS.

5) Use tactical call signs.

6) Follow the net protocol established by the NCS.

In some relief activities, tactical nets become resource or command nets. A resource net is used for an event which goes beyond the boundaries of a single jurisdiction and when mutual aid is needed. A command net is used for communications between EOCs and ARES leaders. Yet with all the variety of nets, sometimes the act of simply putting the parties directly on the radio—instead of trying to interpret their words—is the best approach.

Formal Message Traffic

Formal message traffic is long-term communications that involve many people over a large area. It's generally cast in standard ARRL-message format and handled on well-established National Traffic System (NTS) nets, primarily on 75-meter SSB, 80-meter CW or 2-meter FM (see Fig 14-4). [In addition, there is a regular liaison to the International Assistance and Traffic Net, IATN, now officially designated the NTS Atlantic Region Net (ARN). The net meets on 14.303 MHz daily at 1130 UTC (1100 during the summer), to provide international traffic outlets, as suggested in Fig 14-5.]

Formal messages can be used for severe weather and disaster reports. These radiograms, already familiar to many agency officials and to the public, avoid message duplication while ensuring accuracy. Messages should be read to the originators before sending them, since the originators are responsible for their content. When accuracy is more important than speed, getting the message on paper before it is transmitted is an inherent advantage of formal traffic.

Packet Radio

Packet radio is a powerful tool for traffic handling, especially with detailed or lengthy text (see Chapter 10). Prepare and edit messages off line as text files. These can then be sent error free in just seconds, an important time-saver for busy traffic channels. Public service agencies are impressed by fast and accurate printed messages. Packet radio stations can even be mobile or portable. Relaying might be supplemented by AMTOR-Packet Link (APLink), a system equipped to handle messages between AMTOR HF and packet-radio VHF stations.

Image Communications

Image communication offers live pictures of an area to allow, for example, damage assessment by authorities. Amateur Television (ATV) in its public-service role usually employs portable Fast Scan Television (FSTV) which displays full motion, has excellent detail, can be in color and has a simultaneous sound channel. Although a picture is worth a thousand words, an ATV system requires more equipment, operating skill and preparation than using a simple hand-held radio.

Video cameras and 420-430 or 1240-1294 MHz radios can transmit public-service images from a helicopter to a ground base station equipped with video monitors and a VCR for taping. Image communication works well on the ground, too. Video coverage of parades, severe weather, and even operation Santa Claus in hospitals adds another dimension to your information.

AMATEUR RADIO GROUPS

The Amateur Radio Emergency Service

In 1935, the League developed what is now called the Amateur Radio Emergency Service (ARES), an organization of radio amateurs who have voluntarily registered their capabilities and equipment for emergency communications (see Fig 14-6). They are groups of trained operators ready to serve the public when disaster strikes and regular communications fail. ARES often recruits members from existing clubs, and includes amateurs outside the club area since emergencies do not recognize boundary lines.

Are you interested in public-service activities, or preparing for emergency communications? Join ARES in your area and exchange ideas, ask questions and help with message centers, sports events or weather spotting. Any licensed amateur with a sincere desire to serve is eligible for ARES membership. The possession of emergency-powered equipment is recommended, but it is not a requirement. An ARES group needs to refresh its training with meetings, scheduled nets, drills or real emergencies. An effective ARES group is to our benefit, as well as to the benefit of the entire community. Information about ARES may be obtained from your ARRL Section Manager (listed on page 12 of *QST*) or League Headquarters.

The Amateur Radio Emergency Service has responded countless times to communications emergencies. ARES also introduces Amateur Radio to the ever changing stream of agency officials. Experience has proven that radio amateurs react and work together more capably in time of emergency when practice has been conducted in an organized group. There is no substitute for experience gained—before the need arises.

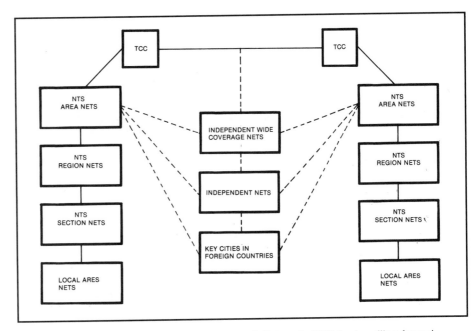

Fig 14-5—Emergency communications through liaison to NTS for handling formal message traffic.

Many Amateur Radio emergency groups use ham fax software to receive early severe weather warnings—direct from NOAA satellites or HF broadcasts. Hurricane Henrietta set off a few alerts in 1995, but moved out to sea.

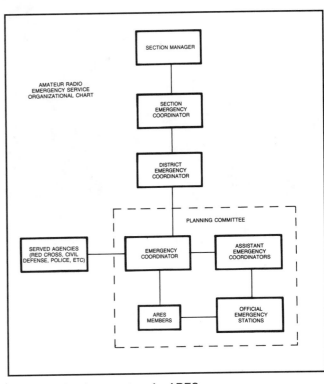

Fig 14-6—Section structure for ARES.

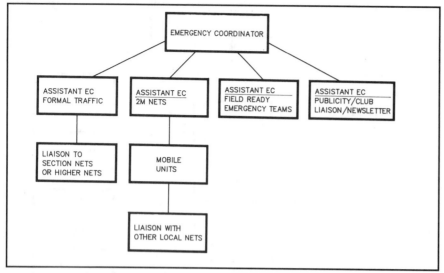

Fig 14-7—Local ARES structure for a county, city or other area of coverage.

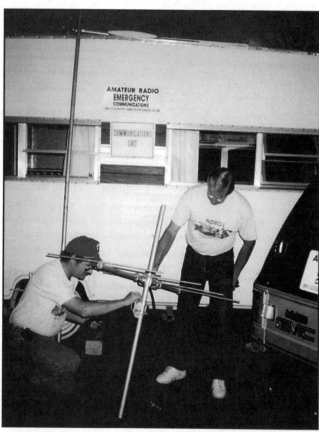

ARES EC Dick Doyal, KBØAHR, and Dale Morrie, NØROJ, prepare a 2-meter antenna for the communication van in Glenwood Springs, Colorado, where fires ravaged a large area. (*photo by John Radloff, NØMOR*)

Official Emergency Station

After you get some ARES training and practice, you might want to refine your skills in emergency communications. If you possess a full ARRL membership there are several opportunities available. The first is the OES appointment, which requires regular participation in ARES including drills, emer-

gency nets and, possibly, real emergency situations. An OES aims for high standards of activity, emergency-preparedness and operating skills.

Emergency Coordinator

Next, when you feel qualified enough to become a team leader of your local ARES group, consider the EC appointment, if that position is vacant in your area. An EC, usually responsible for a county, is the person who can plan, organize, maintain response-readiness and coordinate for emergency communications (see Fig 14-7).

Much of the work involves promoting a working relationship with local government and agencies. The busy EC can hold meetings, train members, keep records, encourage newcomers, determine equipment availability, lead others in drills or be first on the scene in an actual disaster. Some highly populated or emergency-prone areas may also need one or more Assistant Emergency Coordinators (AEC) to help the EC. The AEC is an unofficial appointment made by an EC and, as a specific exception, can hold any class license and need not be an ARRL member.

District Emergency Coordinator

If there are many ARES groups in an area, a DEC may be appointed. Usually responsible for several counties, the DEC coordinates emergency plans between local ARES groups, encourages activity on ARES nets, directs the overall communication needs of a large area or can be a backup for an EC. As a model emergency communicator, the DEC trains clubs in tactical traffic, formal traffic, disaster communications and operating skills.

Section Emergency Coordinator

Finally, there is one rare individual who can qualify as top leader of each ARRL Section emergency structure, the SEC. Only the Section Manager can appoint a candidate to become the SEC. The SEC does some fairly hefty work on a section-wide level: making policies and plans and establishing goals, selecting the DECs and ECs, promoting ARES membership and keeping tabs on emergency preparedness. During an actual emergency, the SEC follows activities from behind the scenes, making sure that plans work and section communications are effective.

Section Manager

The overall leader of the ARRL Field Organization in each section is the SM, who is elected by the League membership in that section. The SM not only appoints the SEC to handle details of ARES and emergency-preparedness activities, but also appoints a Section Traffic Manager (STM) to handle details of the NTS and formal message traffic operation. Response to emergency and public service needs combines the ARES and the NTS. [For further details on the ARRL Field Organization, see Chapter 17.]

RADIO AMATEUR CIVIL EMERGENCY SERVICE

The Radio Amateur Civil Emergency Service (RACES) was set up in 1952 as a special phase of the Amateur Radio service conducted by volunteer licensed amateurs. It is designed to provide emergency communications to local or state civil-preparedness agencies.

RACES operation is authorized during periods of local, regional or national civil emergencies by the FCC upon request of a state or federal official. While RACES was originally based on potential use during wartime, it has evolved over the years (as has the meaning of "civil defense" which is now called "civil preparedness"), to encompass all types of emergencies and natural disasters. RACES is sponsored by Federal Emergency Management Agency.

Amateurs operating in a local RACES organization must be officially enrolled in that local civil-preparedness group. RACES operation is conducted by amateurs using their own primary station licenses, and by existing RACES stations. The FCC no longer issues new RACES (WC prefix) station call signs. Operator privileges in RACES are dependent upon, and identical to, those for the class of license held in the Amateur Radio service. All of the authorized frequencies and emissions allocated to the Amateur Radio service are also available to RACES on a shared basis. But in the event that the President invokes his War Emergency Powers, amateurs involved with RACES would be limited to the certain frequencies, while all other amateur operation would be silenced.

When operating in a RACES capacity, RACES stations and amateurs registered in the local RACES organization may not communicate with amateurs not operating in a similar capacity. See FCC regulations for further information.

Although RACES and ARES are separate entities, the League advocates dual membership and cooperative efforts between both groups whenever possible. The RACES regulations make it simple for an ARES group whose members are all enrolled in and certified by RACES to operate in an emergency with great flexibility. Using the same operators and the same frequencies, an ARES group also enrolled as RACES can "switch hats" from ARES to RACES or RACES to ARES to meet the requirements of the situation as it develops. For example, during a "non-declared emergency," ARES can operate under ARES, but when an emergency or disaster is officially declared by a state or federal authority, the operation can become RACES with no change in personnel or frequencies.

Where there currently is no RACES, it would be a simple matter for an ARES group to enroll in that capacity, after a presentation to the civil-preparedness authorities. For more information on RACES, contact your State Emergency Management, Civil-Preparedness Office, or FEMA.

THE NATIONAL TRAFFIC SYSTEM

In 1949 the League created the National Traffic System to handle medium- and long-haul formal message traffic through networks whose operations can be expedited to meet the needs of an emergency situation.

The main function of NTS in an emergency is to link various local activities and to allow traffic destined outside of a local area to be systematically relayed to the addressee. In a few rare cases, a message can be handled by taking it directly to a net in the state where the addressee lives for rapid delivery by an amateur there within toll-free calling distance. However, NTS is set up on the basis of being able to relay large amounts of traffic systematically, efficiently and according to an established flow pattern. This proven and dependable scheme is what makes NTS so vital to emergency communications. See Chapter 15 for more information on NTS.

Additional details on ARES and NTS can be found in the *Public Service Communications Manual*, the *Emergency Communicator's Manual* and *The ARRL Net Directory*, all published by the ARRL (send an SASE for ordering information). Information on emergency communications and traffic handling also appears regularly in *QST*.

GOVERNMENT AND RELIEF AGENCIES

Government and relief agencies provide effective emergency management to help communities in disasters. They must rely on normal communications, but even reliable communications systems may fail, be unavailable or become overloaded in an emergency. In disaster situations, agency-to-agency radio systems may be incompatible.

Fortunately, Amateur Radio communicators can serve and support in these situations. We can bridge communications gaps with mobile, portable and fixed stations. We also supply trained volunteers needed by the agencies for collection and exchange of critical emergency information.

As government and relief agencies are restrained by budget cuts, the general public is becoming more dependent on volunteer programs. They quickly recognize the value of efforts of radio amateurs who serve the public interest. This public recognition is important support for the continued existence and justification of Amateur Radio.

By using Amateur Radio operators in the amateur frequency bands, the ARRL has been and continues to be in the forefront of supplying emergency communications, either directly to the general public or through various agencies. In fact, there are several organizations that have signed official Memorandas of Understanding with the League:

- The American National Red Cross
- The Salvation Army
- The Federal Emergency Management Agency
- The National Communications System
- The Associated Public Safety Communications Officers, Inc.
- National Weather Service

The American National Red Cross and the Salvation Army extend assistance to individuals and families in times of disaster. Red Cross chapters, for example, establish, coordinate and maintain continuity of communications during disaster-relief operations (both national and international). These agencies have long recognized that the Amateur Radio service, because of its excellent geographical station coverage, can render valuable aid.

The Federal Emergency Management Agency (FEMA) is a

federal agency that supports state and local civil-preparedness and emergency-management agencies. With headquarters in Washington and 10 regional offices throughout the country, it can provide technical assistance, guidance and financial aid to state and local governments wishing to upgrade their emergency communications and warning systems. Since FEMA is in charge of the RACES program, its recognition of ARRL-sponsored emergency preparedness programs can be a powerful tool in selling ARES capability to local emergency-management officials.

The National Communication System (NCS) is a confederation of government agencies and departments established by a Presidential Memorandum to ensure that the critical telecommunication needs of the Federal Government can be met in an emergency. The ARRL Field Organization continues to participate in communications tests sponsored by NCS to study telecommunications readiness in any conceivable national emergency. Through this participation, radio amateurs have received recognition at the highest levels of government.

The Association of Public Safety Communications Officials—International (APCO) and amateurs share common bonds of communications in the public interest. APCO is made up of law-enforcement, fire and public-safety communications personnel. These officials have primary responsibility for the management, maintenance and operation of communications facilities in the public domain. They also establish international standards for public-safety communications, professionalism and continuity of communications through education, standardization and the exchange of information.

The National Weather Service (NWS) consists of a national headquarters in Washington, DC, six regional offices and over 200 local offices throughout the US. The ARRL Field Organization cooperates with NWS in establishing SKYWARN networks for weather spotting and communications. SKYWARN is a plan sponsored by NWS to report and track destructive storms or other severe weather conditions.

An increased awareness of radio amateur capabilities has also been fostered by ARRL's active participation as a member of the National Volunteer Organizations Active in Disaster (NVOAD). NVOAD coordinates the volunteer efforts of its 23 member-agencies (the Red Cross and Salvation Army are among its members).

ARRL and radio amateurs continue to accelerate their presence with these agencies This enhanced image and recognition of Amateur Radio attracts more "customers" for amateurs and our communications skills.

PUBLIC SERVICE EVENTS

Amateur Radio has a unique responsibility to render emergency communications in times of disaster when normal communications are not available. Progressive experience can be gained by helping with public-service events such as message centers, parades and sports events. Besides serving the community, participation teaches skills that can be applied during an emergency.

Amateurs, according to FCC regulations, are responsible to aid during emergencies, but cannot receive compensation for this aid. This requirement does not remove public events sponsored by for-profit organizations from the list of events at which you can help protect the public welfare.

Message Centers

Message centers are Amateur Radio showcase stations set up and operated in conspicuous places to send free radiogram messages for the public. Many clubs provide such radiogram message services at shopping malls, fairs, information booths, festivals, exhibits, conferences and at special local events to demonstrate Amateur Radio while handling traffic. Message centers also afford opportunities to train operators, to practice handling messages, which are similar to disaster Welfare traffic, and to show the public that amateurs are capable, serious and responsible communicators. A plan for a message center should include a display, equipment, public relations and operators.

Successful centers include eye-catching organized displays and attractive stations that promote Amateur Radio and lure newcomers to the hobby. Cover a table with bright-colored cloth. Put out a few radio magazines and hang up some QSLs, maps, posters and operating aids. ARRL Headquarters can supply free literature, pamphlets, press handouts and various informative videotapes to help make your club display a success.

Equipment should be simple, safe and uncomplicated. Try only a few of the following: HF stations with CW, SSB, or digital capability and VHF stations with FM, packet radio, satellite or ATV activities. Keep expensive radios secure from possible theft, out of reach from visitors or stored under a protected table. Decide on whether to use power from batteries, commercial power or generators. Check prior to the event for receiver hash or public-address system interference. To instill interest, have an easy-to-operate shortwave receiver accessible for spectators to tune and listen. To promote a good public-relations image, consider the visitors; use an extension loudspeaker, desk microphone or computer readout of code. Clear, local contacts impress people more than long-distance, scratchy QSOs.

It may be necessary to form a barrier to separate visitors and busy operators. Even so, have friendly and knowledgeable club members available. They should clearly describe to the passerby what Amateur Radio can do, answer questions, explain the message process, assist in preparing messages and so on. Avoid ham jargon. Keep explanations simple and informative; because of the varying degrees of interest the public has, only a few will be interested in specific details. Discuss the variety of Amateur Radio, even if it is not demonstrated. Most visitors will want to know about the role of Amateur Radio in the community, about FCC exams and the license process.

Well-planned message centers can handle thousands of messages at one event. Some visitors bring their address books and excitedly write out bunches of messages. Others inquisitively watch or wait for their turns. Handling messages for the local community is a valuable way amateurs can enhance public recognition. It also keeps you in practice and your equipment ready for emergencies. Before getting involved in message centers, you should become familiar with the traffic-handling procedures as described in Chapter 15 (which also contains additional hints on exhibit-station operation).

Parades

From a communications viewpoint, parades simply relocate people and equipment, with a few requests for supplies or medical aid along the way. Therefore, parades simulate evacuations.

The first step in handling parade communications is to select a group representative who will contact the parade organizers and other officials during the planning stages. This representative can then instruct the ham group and describe what must be done.

As for most public-service events, arrive early. Wear proper identification, whether it is a special jacket, cap or badge. The basic order for the day is to provide communications. Leave parade decisions to officials and first-aid or medical care to Red Cross personnel.

Certain operators should be assigned to find parade officials and help them exchange information. Others work in pairs or teams with the parade marshall and parade specialists like the band, float, vehicle and marcher officials. These officials will be busy once the bus loads of uniformed marchers arrive at the staging area. Set up net controls at parade headquarters where a telephone is available, at an announcement or information booth or at the staging area near the starting line. Fill vacant positions with standby operators. Establish a procedure for helping lost children.

Next, after initial parade staging is underway, many other operators have specific on-route assignments. Position hand-held operators at intervals along the parade route where assignments can be carried out immediately. Operators can walk, march or ride along with the parade to report any specific problems. If possible, it's a good idea to place one amateur on top of a high building to spot general problems. Communicators near loud marching bands will need earphones. Hot weather can keep first-aid volunteers busy with victims of dehydration, sunstroke and heat exhaustion. In some cases, a request for an ambulance, paramedic or water will be needed. Parade floats may need repairs, fuel or towing.

Then, at the de-staging area, more officials rely on communications to prevent clogging the area and stopping the parade. Large parades can require over 75 amateurs to handle communications, and they must plan and practice many months before the event.

As a result, amateurs provide reliable communications to resolve unanticipated problems and handle last-minute changes with immediate action. For more information on public service events, see the *ARRL Special Events Communications Manual*. Contact the ARRL Headquarters Field Services Department for details on how to obtain a copy.

Sports Events

Sports events include all types of outdoor athletic activities where there is competition or movement, or when participants are timed. Again, it is important to assess the communication needs and plan with event officials, law-enforcement officers, medical and first-aid personnel. Know the route and preview the course. Everyone should understand their functions.

Like parades, there are strategic locations where sport-event

amateur communicators could be positioned to be most useful:

- Stationed with or shadowing event officials and organizers or law-enforcement officers.
- At check-in points, staging areas or starting lines.
- At water stops, aid stations and checkpoints around the course or route.
- Near road intersections or sharp turns where safety of crowds or participants is crucial.
- In the course preview car, pace car, follow-up vehicle or roving sheriff's squad.
- At the net control station.
- Alongside first aid or Red Cross stations, medical personnel or ambulance.
- At message centers.
- Near a telephone or at a home station with a telephone.
- At the de-staging area or finish line.

There are many types of sports events in which amateurs serve. The list is nearly endless: air shows, balloon races, soap-box derbies, road rallies, solar car races, Boy Scout hikes, bicycle tours, regattas, golf tournaments, kayak and river-raft races, football games, mini-olympics, Special Olympics, World Olympics, ski-lift operations, slalom courses, cross-country meets, and so on.

A popular event is the *A-Thon*: the walk-athon, marathon, sport-athon, bike-athon or even a triathlon (usually running, bicycling and swimming or canoeing). Long-distance marathons require a large, disciplined force of communicators. This is the case with the New York City Marathon, which is served by well over 200 ham radio operators. Good coverage allows operators to radio the identifying numbers and locations of marathon participants who drop out of the course. This information is posted at a base station to help friends and relatives locate their favorite runners, or to help account for all participants.

Physical endurance and fitness is not only for the participants; hams use roller skates, bicycles, golf-carts, motorcycles or cars around the routes of sport events. Some communicators cover a race at a particular point until all participants have passed, then they become mobile and leapfrog up the route to repeat the procedure. Roving mobiles, sometimes called *roamers*, can keep track of everyone's location, determine race miles remaining, warn of potential traffic hazards, find lost people or items, report harassing spectators or post signs for riders or runners. Yet, even with this variety, keep in mind that an amateur is not to be used as a parking-lot attendant or crowd-control police. Take only those tasks you're capable of handling, like being a communicator.

Portable operators, those not on wheels, can report medical emergencies, give weather reports and handle messages for participants or spectators. Separate UHF channels for administrative purposes are useful for getting supplies like drinking cups, water or bandages, and to report malfunctioning facilities or request more volunteers. Telephones may be crowded in the area, so first-aid stations or checkpoints with bulletin boards can double as NTS message centers. It is then easy for the NCS to give assignments and tell the officials that the route is ready and participants are set to go.

During major sports events, the Red Cross can dispatch their own first-aid vehicles and keep them near crowds or hazardous areas for fast accident response. A specific medical-radio frequency is often assigned. Be prepared to report anything from minor medical situations to serious problems where an ambulance might be called to take the victim to a hospital for examination. If you do call for the ambulance, assign someone to flag it down and give the driver final directions to the victim's location.

Public-service events are great teachers. Once you gain skill in working with the public in hectic and near-emergency conditions, you're well prepared to tackle communications for more difficult natural-disaster assignments.

NATURAL DISASTERS AND CALAMITIES

Nature relentlessly concocts severe weather and natural calamities that can cause human suffering and create needs which the victims cannot alleviate without assistance. Despite the spectrum of requirements desperately needed to help people in a disaster, it is generally understood and agreed that amateurs will neither seek nor accept any duties other than Amateur Radio communications. Volunteer communicators do not, for example, enforce local laws, make major decisions, work as common laborers, rent generators, tents or lights to the public and so on. Instead, amateurs simply handle *radio communications*.

Here are several typical categories of amateur disaster and calamity communications:

- Severe-weather spotting and reporting
- Supporting evacuation of people to safe areas
- Shelter operations
- Assisting government groups and agencies
- Victim rescue operations
- Medical help requests
- Critical supplies requests
- Health-and-Welfare traffic
- Property damage surveys and cleanup

Severe Weather Spotting and Reporting

Nasty weather hits somewhere every day. Long ago, amateurs exchanged simple information among themselves about the approach and progress of storms. Next, concerned hams phoned the weather offices to share a few reports they thought might be of particular interest to the public. National Weather Service forecasters were relying on spotters: police, sheriff, highway patrol, emergency government and trained individuals who reported weather information by telephone. But when severe weather strikes, professional spotters may be burdened with law-enforcement tasks, phone lines may become overloaded, special communication circuits might go out of service or, worse yet, there can be a loss of electrical power. Because of these uncertainties, weather-center officials welcomed amateur operators and encouraged them to install their battery-powered radio equipment on-site so forecasters could monitor the weather nets, request specific area observations and maintain communications in a serious emergency.

Those first, informal weather nets had great potential to access perhaps hundreds of observers in a wide area. Many Meteorologists-in-Charge eagerly began to instruct hams in the types of information needed during severe-weather emergencies, including radar interpreting. Eventually, Amateur Radio SKYWARN operations developed as an important part of community disaster preparedness programs. Accurate observations and rapid communications during extreme weather situations now proves to be fundamental to the NWS. Amateur Radio operators nationwide are a first-response group invaluable to the success of an early storm-warning effort. Weather spotting is popular because the procedures are easy to learn and reports can be given from the relative safety and convenience of a home or an auto.

For example, during a severe-storm episode in Wisconsin in 1994, there was danger of local flooding. A quick check of conditions at the homes of hams operating in an ARES net rapidly revealed trouble spots over a 40 square mile area. This information proved invaluable to emergency government and weather officials, yet no ham had to leave home to participate.

Weather reports on a severe-weather net are limited to drastic weather data, unless specifically requested by the net-control operator. So, most amateurs monitor net operations and transmit only when they can help.

Weather forecasters, depending on their geographical location, need certain information.

During the summer or thunderstorm season report:
- Tornadoes, funnels or wall clouds.
- Hail.
- Damaging winds, usually 50 miles per hour or greater.
- Flash flooding.
- Heavy rains, rate of 2 inches per hour or more.

During the winter or snow season report:
- High winds.
- Heavy, drifting snow.
- Freezing precipitation.
- Sleet.
- New snow accumulation of 2 or more inches.

Here's a four-step method to describe the weather you spot:
1) *What*: Tornadoes, funnels, heavy rain and so on.
2) *Where*: Direction and distance from a known location; for example, 3 miles south of Newington.
3) *When:* Time of observation.
4) *How:* Storm's direction, speed of travel, size, intensity and destructiveness. Include uncertainty as needed. ("Funnel cloud, but too far away to be certain it is on the ground.")

Alerting the Weather Net

The Net Control Station, using a VHF repeater, directs and maintains control over traffic being passed on the Weather Net. The station also collates reports, relates pertinent material to the Weather Service and organizes liaison with other area repeaters. Priority Stations, those that are assigned tactical call signs, may call any other station without going through net control. The NCS might start the net upon hearing a National Oceanic and Atmospheric Administration (NOAA) radio alert, or upon request by NWS or the EC. The NCS should keep in mind that the general public or government officials might be listening to net operations with scanners.

Here are some guidelines an NCS might use to initiate and handle a severe-weather net on a repeater:
1) Activate alert tone on repeater.
2) Read weather-net activation format.
3) Appoint a backup NCS to copy and log all traffic—and to take over in the event the NCS goes off-the-air or needs relief.
4) Ask NWS for the current weather status.
5) Check in all available operators.
6) Assign operators to priority stations and liaisons.
7) Give severe-weather report outline and updates.
8) Be apprised of situations and assignments by EC.
9) Periodically read instructions on net procedures and types of severe weather to report.
10) Acknowledge and respond to all calls immediately.
11) Require that net stations request permission to leave the net.

ARES Personal Checklist

The following represents recommendations of equipment and supplies ARES members should consider having available for use during an emergency or public-service activity.

Forms of Identification
- ARES Identification Card
- FCC Amateur Radio license
- driver's license

Radio Gear
- rig (VHF)
- microphone
- headphones
- power supply/extra batteries
- antennas with mounts
- spare fuses
- patch cords/adapters (BNC to PL259/RCA phono to PL259)
- SWR meter
- extra coax

Writing Gear
- pen/pencil/eraser
- clipboard
- message forms
- logbook
- note paper

Personal Gear (short duration)
- snacks
- liquid refreshments
- throat lozenges
- personal medicine
- aspirin
- extra pair of prescription glasses
- sweater/jacket

Personal Gear (72-hour duration)
- foul-weather gear
- three-day supply of drinking water
- cooler with three-day supply of food
- mess kit with cleaning supplies
- first-aid kit
- personal medicine
- aspirin
- throat lozenges
- sleeping bag
- toilet articles
- mechanical or battery powered alarm clock
- flashlight with batteries/lantern
- candles
- waterproof matches
- extra pair of prescription glasses

Tool Box (72-hour duration)
- screwdrivers
- pliers
- socket wrenches
- electrical tape
- 12/120-V soldering iron
- solder
- volt-ohm meter

Other (72-hour duration)
- rig (HF)
- hatchet/ax
- saw
- pick
- shovel
- siphon
- jumper cables
- generator, spare plugs and oil
- kerosene lights, camping lantern or candles
- $^3/_8$-inch hemp rope
- highway flares
- extra gasoline and oil

12) During periods of inactivity and to keep the frequency open, make periodic announcements that a net is in progress.

13) Close the net after operations conclude.

At the Weather Service

The NCS position at the Weather Service, when practical, can be handled by the EC and other ARES personnel. Operators assigned there must have 2-meter hand-helds with fully charged batteries. The station located at the NWS office may also be connected to other positions with an off-the-air intercom system. This allows some traffic handling without loading up the repeater. Designate a supplementary radio channel in anticipation of an overload or loss of primary communications circuits.

If traffic is flowing faster than you can easily copy and relay, NWS personnel may request that a hand-held radio be placed at the severe-weather desk. This arrangement allows them to monitor incoming traffic directly. Nevertheless, all traffic should be written on report forms. If a disaster should occur during a severe-weather net, shift to disaster-relief operations.

Repeater Liaisons

Assign properly equipped and located stations to act as liaisons with other repeaters. Two stations should be appointed to each liaison assignment. One monitors the weather repeater at all times and switches to the assigned repeater just long enough to pass traffic. The second monitors the assigned repeater and switches to the weather repeater just long enough to pass traffic. If there aren't enough qualified liaison stations,

one station can be given both assignments.

Weather Warnings

NWS policy is to issue warnings only when there is absolute certainty, for fear of the "cry wolf" syndrome (premature warnings cause the public to ignore later warnings). Public confidence increases with reliable weather warnings. When NWS calls a weather alert, it will contact the local EC by phone or voice-message pager, or the EC may call NWS to check on a weather situation.

Hurricanes

A hurricane is declared when a storm's winds reach 75 miles per hour or more. These strong winds may cause storm-surge waves along shores and flooding inland.

A Hurricane *Watch* means a hurricane may threaten coastal and inland areas. Storm landfall is a possibility, but it is not necessarily imminent. Listen for further advisories and be prepared to act promptly if a warning is issued.

A Hurricane *Warning* is issued when a hurricane is expected to strike within 24 hours. It may include an assessment of flood danger, small-craft or gale warnings, estimated storm effects and recommended emergency procedures.

Amateurs, in the 4th and 5th call areas in particular, can spot and report the approach of hurricanes well ahead of any news service. In fact, their information is sometimes edited and then broadcast on the local radio or TV to keep citizens informed.

The Hurricane Watch Net on 14.325 MHz, for example, serves either the Atlantic or Pacific during a watch or warning period and keeps in touch with the National Hurricane Center.

Frequent, detailed information is issued on nets when storms pose a threat to the US mainland. In addition to hurricane spotting, local communicators may announce that residents have evacuated from low-lying flood areas and coastal shore. Other amateurs across the country can help by relaying information, keeping the net frequency clear and by listening.

Tornadoes

A tornado is an intensely destructive whirlwind formed from strongly rising air currents. With winds of up to 300 miles per hour, tornadoes appear as rotating, funnel-shaped clouds from gray to black in color. They extend toward the ground from the base of a thundercloud. Tornadoes may sound like the roaring of an airplane or locomotive. Even though they are short lived over a small area, tornadoes are the most violent of all atmospheric phenomena. Tornadoes that don't touch the ground are called *funnels*.

A Tornado *Watch* is issued when a tornado may occur near your area. Carefully watch the sky.

A Tornado *Warning* means take shelter immediately, a tornado has actually been sighted or indicated by radar. Protect yourself from being blown away, struck by falling objects or injured by flying debris.

"Tornado alley" runs in the 5th, 9th and 10th US call areas. Amateurs in these areas often receive Tornado Spotter's Training and refresher courses presented by NWS personnel.

Amateur Radio's quick-response capability has reduced injuries and fatalities by giving early warnings to residents. Veteran operators know exactly how serious a tornado can be. They've seen tornadoes knock out telephone and electrical services just as quickly as they flip over trucks or destroy homes. Traffic lights and gas pumps won't work without electricity, creating problems for motorists and fuel shortages for electric generators.

After a tornado strikes, amateurs provide communications in cooperation with local government and relief agencies. Welfare messages are sent from shelters where survivors receive assistance. Teams of amateurs and officials also survey and report property damage.

Floods, Mud Slides and Tidal Waves

Floods occur when excessive rainfall causes rivers to overflow their banks, when heavy rains and warmer-than-usual temperatures melt excessive quantities of snow, or when dams break. Floods can be minor, moderate or major. Floods or volcanic eruptions may melt mountain snow, causing mud slides. A tidal wave or tsunami is actually a series of waves caused by a disturbance which may be associated with earthquakes, volcanoes or sometimes hurricanes.

Don't wait for the water level to rise or for officials to ask for help; sound the alarm and activate a weather net immediately. Besides handling weather data for NWS, enact the response plans necessary to relay tactical flood information to local officials. Assist their decision making by answering the following questions:

1) Which rivers and streams are affected and what are their conditions?

2) When will flooding probably begin, and where are the flood plain areas?

3) What are river-level or depth-gauge readings for comparison to flood levels?

Mobile operators may find roads flooded and bridges washed out. Flood-rescue operations then may be handled by marine police boats with an amateur aboard. If power and telephones are out, portable radio operators can help with relief operations to evacuate families to care facilities where a fixed station should be set up. The officials will need to know the number and location of evacuees. As the river recedes, the water level drops in some areas, but it may rise elsewhere to threaten residents. Liaisons to repeaters downstream can warn others, possibly through the Emergency Broadcast System, of impending flooding. Property-damage reports and welfare traffic will usually be followed by disaster relief and clean-up operations.

Winter Storms

A Winter Storm *Watch* indicates there is a threat of severe winter weather in a particular area. A Winter Storm *Warning* is issued when heavy snow (6 inches or more in a 12-hour period, 8 inches or more in a 24-hour period), freezing rain, sleet or a substantial layer of ice is expected to accumulate.

Freezing rain or freezing drizzle is forecast when expected rain is likely to freeze when it strikes the ground. Freezing rain or ice storms can bring down wires, causing telephone and power outages.

Sleet consists of small particles of ice, usually mixed with rain. If enough sleet accumulates on the ground, it will make roads slippery.

Blizzards are the most dangerous of all winter storms. They are a combination of cold air, heavy snow and strong winds. Blizzards can isolate communities. A Blizzard *Warning* is issued when there is considerable snow and winds of 35 miles per hour or more. *A Severe Blizzard Warning* means that a heavy snow is expected, with winds of at least 45 miles per hour and temperatures 10° F or lower.

Travelers advisories are issued when ice and snow are expected to hinder travel, but not seriously enough to require warnings. Blizzards and snowstorms can create vehicle-traffic problems by making roads impassable, stranding motorists or drastically delaying their progress.

In some areas, local snowmobile club members or 4-wheel-drive-vehicle enthusiasts cooperate with hams to coordinate and transport key medical personnel when snowdrifts render roads almost impassable. They may also assist search teams looking for motorists.

Brush and Forest Fires

A prolonged period of hot, dry weather parches shrubs, brush and trees. This dry vegetation, ignited by lightning, arson or even a helicopter crash, can start a forest fire. The fires quickly become worse when winds spread the burning material.

An amateur who spots a blaze that has the potential of growing into a forest fire can radio the Park Service and ask them to dispatch a district ranger and fire-fighting equipment. When fires are out of control, hams help with communications to evacuate people, report the fires' movement or radio requests for supplies and volunteers. ARES groups over a wide geographical area can set up portable repeaters, digital links or ATV stations.

Amateurs, in the 6th and 7th US call areas in particular, should get adequate safety training, including fire line safety and fire shelter deployment. Even then, only travel to a fire operation upon receiving clear dispatch instructions from a competent authority. If you don't have proper training, inform whoever is in charge of your lack of training to prevent being given an inappropriate or dangerous assignment.

Fire safety rules are of special importance in an emergency, but also should be observed every day to prevent disaster. Since most fires occur in the home, even an alert urban amateur can spot and report a building on fire. Fires are extinguished by taking away the fuel or air (smother it), or by cooling it with water and fire-extinguishing chemicals. A radio call to the fire department will bring the needed control. The Red Cross generally helps find shelter for the homeless after fires make large buildings, such as apartments, uninhabitable. So even in cities, communications for routing people during and after fires may be needed.

Earthquakes and Volcanic Eruptions

Earthquakes are caused when underground forces break and shift rock beneath the surface, causing the Earth's crust to shake or tremble. The actual movement of the earth is seldom a direct cause of death or injury, but can cause buildings and other structures to collapse and may annihilate communications, telephone service and electrical power. Most casualties result from falling objects and debris, splintering glass and fires.

Earthquakes strike without warning, but everyone knows that the quake happened. Amateurs should first ensure that the immediate surroundings and loved ones are safe, and then begin monitoring the ARES frequencies. Amateurs are often the first to alert communities immediately after earthquakes and volcanic eruptions occur. Their warnings are definitely credited with lessening personal injuries in the area.

Amateurs may assist with rescue operations, getting medical help or critical supplies and helping with damage appraisals. There will be communications, usually on VHF or UHF, for Red Cross logistics and government agencies. After vital communications is handled, the rest of the world will be trying to get information through Health-and-Welfare traffic.

Shelter Operations

A shelter or relief center is a temporary place of protection where rescuers can bring disaster victims and where supplies can be dispensed. Many displaced people can stay at the homes of friends or relatives, but those searching for family members or in need are housed in shelters.

Whether a shelter is for a few stranded motorists during a snowstorm, or a whole community of homeless residents after a disaster, it is an ideal location to set up an Amateur Radio communication base station. An alternate station location would be an ARES mobile communications van, if available, near the shelter.

Once officials determine the locations for shelters, radio operators can be assigned to set up equipment at the sites. In fact, the amateur station operators could share a table with shelter registration workers. Make sure you obtain permission for access to the shelter to assist and do not upset the evacuees. Use of repeaters and autopatches allows Welfare phone calls for those inside the shelters to inform and reassure friends, families and relatives. Use of a computer may help to control and distribute information. *ARESdata* is a packet-radio database program that works well with emergency and public-service communications (see December 1990 *QST*, page 75).

Health-and-Welfare Traffic

There can be a tremendous amount of radio traffic to handle during a disaster. This is due in part because phone lines that remain in working order should be reserved for emergency use by those people in peril.

Shortly after a major disaster, Emergency messages within the disaster area often have life-and-death urgency. Of course, they receive primary emphasis. Much of their local traffic will be on VHF or UHF. Next, Priority traffic, messages of an emergency-related nature but not of the utmost urgency, are handled. Then, Welfare traffic is originated by evacuees at shelters or by the injured at hospitals and relayed by Amateur Radio. It flows one way and results in timely advisories to those waiting outside the disaster area.

Incoming Health-and-Welfare traffic should be handled only after all emergency and priority traffic is cleared. Don't solicit traffic going to an emergency because it can severely overload an already busy system. Welfare inquiries can take time to discover hard-to-find answers. An advisory to the inquirer uses even more time. Meanwhile, some questions might have already been answered through restored circuits.

Shelter stations, acting as net control stations, can exchange information on the HF bands directly with destination areas as propagation permits. Or, they can handle formal traffic through a few outside operators on VHF who, in turn, can link to NTS stations. By having many NTS-trained amateurs, it's easy to adapt to whatever communications are required.

Property Damage Surveys

Damage caused by natural disasters can be sudden and extensive. Responsible officials near the disaster area, paralyzed without communications, will need help to contact appropriate officials outside to give damage reports. Such data will be used to initiate and coordinate disaster relief. Amateur Radio operators offer to help but often are unable to cross roadblocks established to limit access by sightseers and potential looters. Proper emergency responder identification will be required to gain access into these areas. In some instances, call-letter license plates on the front of the car or placards inside windshields may help. It's important for amateurs to keep complete and accurate logs for use by officials to survey damage, or to use as a guide for replacement operators.

Accidents and Hazards

The most difficult scenarios to prepare for are accidents and hazardous situations. They are unpredictable and can happen anywhere. Generally, an emergency *autopatch* is used only to report incidents that pose threats to life or personal safety, such as vehicle accidents, disabled vehicles or debris in traffic, injured persons, criminal activities and fires.

Using the keypad featured on most modern VHF hand-held and mobile radios, the operator activates a repeater autopatch by sending a particular code. The repeater connects to a telephone line and routes the incoming and outgoing audio accordingly. By dialing 9-1-1 (or another emergency number), the operator has direct access to law-enforcement agencies.

Vehicle Accidents

Vehicle-accident reports, by far the most common public-service activity on repeaters, can involve anything from bikes, motorcycles and automobiles, to buses, trucks, trains and airplanes. Law-enforcement offices usually accept reports of such incidents anywhere in their county and will relay information to the proper agency when it pertains to adjacent areas.

Here's a typical autopatch procedure:

1) Give your call and say "emergency patch."
2) Drop your carrier momentarily.
3) Key in the access code.
4) Wait for police or fire operator to answer.

5) Answer the questions that the operator asks.

6) After operator acknowledges, dump the patch by keying in the dump code.

7) Give your call and say "patch clear."

8) Don't stay keydown! Talk in short sentences, releasing your push-to-talk switch after each one, so the operator can ask you questions.

When you report a vehicle accident, remain calm and get as much information as you can. This is one time you certainly have the right to break into a conversation on a repeater. Use plain language, say exactly what you mean, and be brief and to the point. Do not guess about injuries; if you don't know, say so. Some accidents may look worse than they really are; requesting an ambulance to be sent needlessly could divert it away from a bona-fide accident injury occurring at the same time elsewhere. And besides, police cruisers are generally only minutes away in an urban area.

Here's what you should report for a vehicle accident:

1) Highway number (eg, I-43, I-94, US-45).

2) Direction of travel (north, south, east, west).

3) Address or street intersection, if on city streets, closest exit on highway.

4) Traffic blocked, or if accident is out of traffic.

5) Apparent injuries, number and extent.

6) Vehicles on fire, smoking or a fuel spill.

Example: "This is WB8IMY, reporting a two-car accident, I-94 at Edgerton, northbound, blocking lane number two, property damage only."

The first activities handled by experts at a vehicle-accident scene are keyed to rescue, stabilize and transport the victims. Then they ensure security, develop a perimeter, handle vehicle traffic and control or prevent fires from gasoline spills. Finally, routine operations restore the area with towing, wrecking and salvage.

The ability to call the police or for an ambulance, without depending on another amateur to monitor the frequency, saves precious minutes. Quick reaction and minimum delay is what makes an autopatch useful in emergencies. The autopatch, when used responsibly, is a valuable asset to the community.

Freeway Warning

Many public-safety agencies recommend that you do not stop on freeways or expressways to render assistance at an accident scene unless you are involved, are a witness or have sufficient medical training. Freeways are extremely dangerous because of the heavy flow of high-speed traffic. Even under ideal conditions, driver fatigue or inattentiveness, high speeds and short distances between vehicles often make it impossible to stop a vehicle from striking stationary objects. If you must stop on a freeway, pull out of traffic and onto distress lanes. Exercise extreme caution to protect yourself. Don't add to the traffic problem. Instead, radio for help.

WORKING WITH PUBLIC-SAFETY AGENCIES

Amateur Radio affords public-safety agencies, such as local police and fire officials, with an extremely valuable resource in times of emergency. Once initial acceptance by the authorities is achieved, an ongoing working relationship between amateurs and safety agencies is based on the efficiency of our performance. Officials tend to be very cautious and skeptical about those who are not members of the public-safety professions. At times, officials may have trouble separating problem solvers from problem makers, but being understaffed and on a limited budget, they often accept communications help if it is offered in the proper spirit.

Here are several image-building rules for working with safety agencies:

- Maintain group unity. Work within ARES, RACES or local club groups. Position your EC as the direct link with the agencies.
- Be honest. If your group cannot handle a request, say so and explain why. Safety personnel often risk their lives based on a fellow disaster-worker's promise to perform.
- Equip conservatively. Do not have more flashing lights, signs, decals and antennas than used by the average police or fire vehicle. Safety professionals are trained against overkill and use the minimum resources necessary to get the job done.
- Look professional. In the field, wear a simple jump-suit or jacket with an ARES patch to give a professional image and to help officials identify radio operators.
- Respect authority. Only assume the level of authority and responsibility that has been given to you.
- Publicize Amateur Radio. If contacted by members of the press, restrict comments solely to the amateurs' role in the situation. Emergency status and names of victims should only come from a press information officer, or the government agency concerned.

The public-service lifeline provided by Amateur Radio must be understood by public-safety agencies before the next disaster occurs. Have an amateur representative meet with public-safety officials in advance of major emergencies so each group will know the capabilities of the other. The representative should appear professional with a calm, businesslike manner and wear conservative attire. It is up to you to invite the local agencies to observe or cooperatively participate with your group.

Assisting the Police

Act as a communicator and radio for the police when you spot criminal activities. Memorize a description of the suspects for apprehension, or information about a vehicle for later recovery. Use caution—uniformed personnel arriving on the scene do not know who you are. Be an observer; let the officers act. For instance, if you see a vehicle traveling at high speed at night without lights, report the location and direction of travel to the law enforcement agency. Don't follow the vehicle.

Some Amateur Radio groups provide police communication assistance involving free taxi service to citizens who do not wish to drive home on New Year's Eve. They also act as lookouts for vandals on city/town streets and on freeway overpasses on Halloween. Hams have even reported sighting a person who is about to make a life-threatening jump into a river. Assisting the police in small ways can make a big difference in the community.

Search and Rescue

Amateurs helping search for an injured climber use repeaters to coordinate the rescue. A small airplane crashes, and amateurs direct the search by tracking from its Emergency Locator Transmitter. No matter what the situation, it's reassuring to team up with local search-and-rescue organizations who have familiarity with the area. Once a victim is found, the hams can radio the status, autopatch for medical information, guide further help to the area and plan for a return transportation. If the victim is found in good condition, Amateur Radio can bolster the hopes of base-camp personnel and the family of the victim with direct communications.

Even in cities, searches are occasionally necessary. An elderly person out for a walk gets lost and doesn't return home. After a reasonable time, a local search team plans and coordinates a search. Amateurs take part by providing communications, a valuable part of any search. When the missing person is discovered, there may be a need to radio for an ambulance for transportation to a nearby hospital.

Hospital Communications

Hospital phones can fail. For example, when a construction crew using a backhoe may accidentally cut though the main trunk line supplying telephone service to several hundred users, including the hospital. Such major hospital telephone outages can block incoming emergency phone calls. In addition, the hospital staff cannot telephone to discuss medical treatment with outside specialists.

Several hand-held equipped amateurs can first handle emergency calls from nursing homes, fire departments and police stations. They can also provide communications to temporarily replace a defective hospital paging system. Next, they can help restore critical interdepartmental hospital communications and, finally, communications with nearby hospitals.

Preparation for hospital communications begins by cooperating with administrators and public-relations personnel. You'll need their permission to perform inside signal checks, install outside antennas or set up net control stations in the hospital.

Working with local hospitals doesn't always involve extreme situations. You may simply be asked to relay information from the poison control center to a campsite victim. Or you may participate in an emergency exercise where reports of the "victims'" conditions are sent to the hospital from a disaster site. One typical drill involved a simulated crop-duster plane crash on an elementary school playground during recess. The doctors and officials depended on communications to find casualties who were contaminated by crop-dusting chemicals. Again, never give the names of victims or fatalities over the air since this information must be handled only by the proper agency.

Toxic-Chemical Spills and Hazardous Materials

A toxic-chemical spill suddenly appears when gasoline pours from a ruptured bulk-storage tank. A water supply is unexpectedly contaminated, or a fire causes chlorine gas to escape at an apartment swimming pool. On a highway, a faulty shutoff valve lets chemicals leak from a truck, or drums of chemicals fall onto the highway and rupture. Amateur communications have helped in all these situations.

Caution: don't rush into a Hazardous Materials (HAZMAT) incident area without knowing what's involved. Vehicles carrying 1000 pounds or more of a HAZMAT are required by Federal regulations to display a placard bearing a four-digit identification number. Radio the placard number to the authorities, and they will decide whether to send HAZMAT experts to contain the spills.

Follow directions from those in command. Provide communications to help them evacuate residents in the immediate area and coordinate between the spill site and the shelter buildings. Hams also assist public-service agencies by setting flares for traffic control, helping reroute motorists and so on. We're occasionally asked to make autopatch calls for police or fire-department workers on the scene as they try to determine the nature of the chemicals.

The National Transportation Safety Board, the Environmental Protection Agency and many local police, fire, and emergency government departments continue to praise ARES volunteers in their assistance with toxic spills.

CHAPTER 15

Traffic Handling

MARIA L. EVANS, KT5Y
8461 BROWNS STATION DR
COLUMBIA, MO 65202

For just pennies a day, you can protect you and your family from all sorts of catastrophic illnesses with the Mutual of Podunk health-care policy....

Yes, at the amazing low price of $9.95, you can turn boring old potatoes, carrots and okra into culinary masterpieces with the Super Veggie-Whatchamadoodler—a modest investment for your family's mealtime happiness....

Tired of all that ugly fat on your otherwise-beautiful body? Our Exer-Torture home gym will trim those thunder thighs in 30 days or your money back....

Ah, those pitches. Everybody is trying to sell something—even the participants in the specialty modes of Amateur Radio. Our eyes light up thinking about all those wondrous gizmos that digitize, packetize and equalize. Those kinds of specialty modes are easily remembered and can quickly gain popularity. Most people, though, forget about the oldest specialty mode in Amateur Radio—traffic handling.

Admittedly, it's hard for traffic handlers to compete with other specialty modes because it just doesn't look exciting. After all, tossing messages from Great-Uncle Levi and Grandma Strauss sure doesn't put you on the leading edge of our high-tech hobby, does it? Yet, over 500 nets are members of the ARRL National Traffic System (NTS), probably one of the most highly organized special interests of Amateur Radio. Today's traffic handlers are at this very moment setting new standards for traffic handling via AMTOR and packet—two of the hottest specialty modes going—as well as using traditional modes. If you enjoy emergency-communications preparation, traffic handling is for you. Sure, it's true that over 90% of all messages handled via Amateur Radio are of the "at the state fair, wish you were here" or the "happy holidays" variety—certainly not life-and-death stuff. But consider this: Your local fire department often conducts drills without ever putting out a real fire, your civil defense goes out regularly and looks for "pink tornadoes," and many department stores hire people to come in and pretend to be shoplifters to check the alertness of their employees. Similarly, when real emergencies rear their ugly heads, traffic handlers just take it all in stride and churn messages out like they always do.

Traffic handling is also an excellent way to paint a friendly picture of Amateur Radio to the nonamateur public. What sometimes seems to be unimportant to us is usually never unimportant to that person in the address block of a message. Almost every traffic handler can relate stories of delivering a Christmas message from some long-lost friend or relative that touched the heart of the recipient such that they could hear the tears welling up at the other end of the phone. Those happy recipients will always mentally connect Amateur Radio traffic handling with good and happy things, and often this is the most satisfying part of this hobby within a hobby.

Next time you go to a hamfest, see if you can spot the traffic handlers. They are almost always reveling in a big social cluster, sharing stories and enjoying a unique camaraderie. Traffickers always enthusiastically look forward to the next hamfest, because it means another chance to spend time with their special friends of the airwaves. These friendships often last a lifetime, transcending barriers of age, geographical distance, background, gender or physical ability.

Young or old, rural or urban, there's a place reserved for you on the traffic nets. Young people often can gain respect among a much older peer group and obtain high levels of responsibility through the traffic nets, good preparation for job opportunities and scholarships. "Nine to fivers" on a tight schedule can still manage to get a regular dose of Amateur Radio in just 15 to 30 minutes of net operation—a lot of hamming in a little time. Retired people can stay active in an important activity and provide a service to the general public.

Even if you live in the sticks, where you rarely get a delivery, you can still perform a vital function in NTS as a net control station (NCS) or as a representative to the upper echelons of NTS. You don't have to check in every night (a popular myth about traffic handling—if you can donate time just once a week you are certainly welcome on NTS). You don't need fancy antennas or huge amplifiers, and you don't even need those ARRL message pads. For the cost of a pad or paper and a pencil, you can interface with a system that covers thousands of miles and consists of tens of thousands of users. A world of fun and friendship is waiting for you in traffic handling. You only have to check into a net to become part of it.

MAKING THE BIG STEP: CHOOSING A NET AND CHECKING INTO IT

Checking into a net for the first time is a lot like making your first dive off the high board. It's not usually very pretty, but it's a start. Once you've gotten over the initial shock of hitting the water that hard, the next one comes a lot easier. But, like a beginning diver, you can do a little advance preparation that will get you emotionally prepared for your first plunge.

Before you attempt to interface with the world of NTS, you need the right "software." Get a copy of *The ARRL Net Directory*, available from ARRL ($3) and two free ARRL operating aids—FSD-3, the list of ARRL numbered radiograms, and FSD-218, the pink "Q-signals for nets and message form'' card. Read all the articles in the *Net Directory*, and keep the two operating aids between the covers of your logbook for quick reference. After you have a basic understanding of the materials, you're ready to pick out a net in your ARRL section/state that suits you.

Go through the *Net Directory* and match up your time schedule with the nets in your section/state. If you're a Novice, your section/state slow-speed net is a good choice (or your neighboring section, if yours doesn't have one). But if you aren't a Novice, and still don't have the confidence to try the ''big'' CW section net, don't feel embarrassed about checking into a slow-speed net. Slow-speed nets are chock-full of veterans that help with NCS and NTS duties, and they're willing to help you, too. Perhaps you'd like to try the section phone net or weather net or a local 2-meter net. Some sections have also established RTTY nets. At any rate, you are the sole judge of what you want to try first.

Once you've chosen a net, it's a good idea to listen to it for a few days before you check in. Although this chapter will deal with a generalized format for net operation, each net is ''fingerprinted'' with its own special style of operation, and it's best to become acquainted with it before you jump in. If you have a friend who checks into the net, let your friend ''take you by the hand'' and tell you about the ins and outs of the net.

When the big day arrives, keep in mind that everybody on that net had to check into the net for the first time once. You aren't doing anything different than the rest of them, and this, like your first QSO as a Novice, is just another ''rite of passage'' in the ham world. You will discover that it doesn't hurt, and that many other folks will be pleased, even happy, that you checked in with them. (See accompanying sidebar for general recommendations on how to make your NCS love you!)

On CW

First, we'll pay a visit to a session of the Mowegia CW Net, not so long ago, on an 80-meter frequency not so far away. KØSI is calling tonight's session, using that peculiar CW net shorthand that we aren't too used to yet.

KØSI: MCWN MCWN MCWN DE KØSI KØSI KØSI GE AND PSE QNZ V V V MCWN QNI QTC?

(Translation: Calling the Mowegia CW Net, calling the Mowegia CW Net, this is KØSI. Good evening, this is a directed net, please zero beat with me. [Then, he sent some Vs to aid stations in zero beating him.] Calling the Mowegia CW Net...any checkins, any traffic?)

In the meantime, NDØN, K2ONP and WØOUD are waiting in the wings to check in. Since each is an experienced traffic handler, each listens carefully before jumping in, so as to not step on anyone.

NDØN: N

KØSI: N

(Notice that NDØN just sent the first letter of his suffix, and the NCS acknowledged it. This is common practice, but if you have the letter E or T or K as the first letter of your suffix, or if it is the same as another net operator's first letter, you might use another letter. It's not a hard-and-fast rule.)

NDØN: DE NDØN GE PETE QTC1 SOUTH CORNER 1 AIØO K

(NDØN has one piece of traffic for South Corner, and one for AIØO.)

KØSI: NDØN DE KØSI GE JOHN QSL TU $\overline{AS}$

(Good evening, John, I acknowledge your traffic. Thank you, please stand by.)

How to be the Kind of Net Operator the Net Control Station (NCS) Loves

As a net operator, you have a duty to be self-disciplined. A net is only as good as its worst operator. You can be an exemplary net operator by following a few easy guidelines.

1) *Zero beat the NCS.* The NCS doesn't have time to chase all over the band for you. Make sure you're on frequency, and you will never be known at the annual net picnic as ''old so-and-so who's always off frequency.''

2) *Don't be late.* There's no such thing as ''fashionably late'' on a net. Liaison stations are on a tight timetable. Don't hold them up by checking in 10 minutes late with three pieces of traffic.

3) *Speak only when spoken to by the NCS.* Unless it is a bona fide emergency situation, you don't need to ''help'' the NCS unless asked. If you need to contact the NCS, make it brief. Resist the urge to help clear the frequency for the NCS or to ''advise'' the NCS. The NCS, not you, is boss.

4) Unless otherwise instructed by the NCS, *transmit only to the NCS.* Side comments to another station in the net are out of order.

5) *Stay until you are excused.* If the NCS calls you and you don't respond because you're getting a ''cold one'' from the fridge, the NCS may assume you've left the net, and net business may be stymied. If you need to leave the net prematurely, contact the NCS and simply ask to be excused (QNX PSE on CW).

6) *Be brief when transmitting to the NCS.* A simple ''yes'' (C) or ''no'' (N) will usually suffice. Shaggy dog tales only waste valuable net time.

7) *Know how the net runs.* The NCS doesn't have time to explain procedure to you. After you have been on the net for a while, you should already know these things.

K2ONP: M

KØSI: M

K2ONP: DE K2ONP GE PETE TEN REP QRU K

(K2ONP is representative to the NTS Tenth Region Net tonight, and he has no traffic.)

KØSI: K2ONP DE KØSI HI GEO TU $\overline{AS}$

WØOUD: BK

KØSI: BK

WØOUD: DE WØOUD GE PETE QRU K

KØSI: WØOUD DE KØSI GE LETHA QNU TU $\overline{AS}$

(Good evening, Letha, the net has traffic for you. Please stand by.)

(Since Letha lives in South Corner, the NCS is going to move WØOUD and NDØN off frequency to pass the South Corner traffic.)

KØSI: ØN?

NDØN: HR

(HR [here], or C [yes] are both acceptable ways to answer the NCS, who will usually use only your suffix from here on out to address questions to you.)

KØSI: OUD?

WØOUD: C

KØSI: NDØN ES WØOUD PSE UP 4 1 SOUTH CORNER K

(Please go up 4 kHz and pass the South Corner traffic.)

NDØN: GG (going)

WØOUD: GG

(When two stations go off frequency, the receiving station always calls the transmitting station. If the NCS had said WØOUD PSE QNV NDØN UP 4 GET 1 SOUTH CORNER, WØOUD would have called NDØN first on frequency to see if she copied him. This is done often when conditions are bad. If they don't make connection, they will return to net frequency. If they

ARRL QN Signals for CW Net Use

QNA*	Answer in prearranged order.
QNB*	Act as relay *Between* _____ and _____.
QNC	All net stations *Copy*.
	I have a message for all net stations.
QND*	Net is *Directed* (controlled by net control station.)
QNE*	*Entire* net stand by.
QNF	Net is *Free* (not controlled.)
QNG	Take over as net control station.
QNH	Your net frequency is *High*.
QNI	Net stations report *In*.
	I am reporting into the net. (Follow with a list of traffic or QRU.)
QNJ	Can you copy me?
QNK*	Transmit messages for _____ to _____.
QNL	Your net frequency is *Low*.
QNM*	You are QRMing the net. Stand by.
QNN	*Net* control station is _____.
	What station has net control?
QNO	Station is leaving the net.
QNP	Unable to copy you.
	Unable to copy _____.
QNQ*	Move frequency to _____ and wait for _____ to finish handling traffic. Then send him traffic for _____.
QNR*	Answer _____ and *Receive* traffic.
QNS	Following *Stations* are in the net.*
	(Follow with list.)
	Request list of stations in the net.
QNT	I request permission to leave the net for _____ minutes.
QNU*	The net has traffic for *you*. Stand by.
QNV*	Establish contact with _____ on this frequency. If successful, move to _____ and send him traffic for _____.
QNW	How do I route messages for _____?
QNX	You are excused from the net.*
	Request to be excused from the net.
QNY*	Shift to another frequency (or to _____ kHz) to clear traffic with _____.
QNZ	Zero beat your signal with mine.

*For use only by the Net Control Station.

Notes on Use of QN Signals

The QN signals listed above are special ARRL signals for use in amateur CW nets *only*. They are not for use in casual amateur conversation. Other meanings that may be used in other services do not apply. Do not use QN signals on phone nets. *Say it with words.* QN signals need not be followed by a question mark, even though the meaning may be interrogatory.

These "Special QN Signals for New Use" originated in the late 1940s in the Michigan QMN Net, and were first known to Headquarters through the then head traffic honcho W8FX. Ev Battey, W1UE, then ARRL assistant communications manager, thought enough of them to print them in *QST* and later to make them standard for ARRL nets, with a few modifications. The original list was designed to make them easy to remember by association. For example, QNA meant "Answer in *A*lphabetical order," QNB meant "Act as relay *B*etween...," QNC meant "All *Net* Copy," QND meant "*Net* is *Directed*," etc. Subsequent modifications have tended away from this very principle, however, in order that some of the less-used signals could be changed to another, more needed, use.

Since the QN signals started being used by amateurs, international QN signals having entirely different meanings have been adopted. Concerned that this might make our use of QN signals with our own meanings at best obsolete, at worst illegal, ARRL informally queried FCC's legal branch. The opinion then was that no difficulty was foreseen as long as we continued to use them only in amateur nets.

Occasionally, a purist will insist that our use of QN signals for net purposes is illegal. Should anyone feel so strongly about it as to make a legal test, we might find ourselves deprived of their use. Until or unless we reach such an eventuality, however, let's continue to use them, and use them right. After all, we were using them first.

---

do make connection, and pass the traffic, they will return as soon as they are done.)

AIØO: O

KØSI: O

AIØO: DE AIØO GE PETE QRU K

KØSI: AIØO DE KØSI GE ROB QNU PSE UP 4 1 AIØO WID NDØN AFTER WØOUD THEN BOTH QNX 73 K

(Good evening, Rob, the net has traffic for you. Please go up 4 and get one for you from NDØN [WID means "with"] after he finishes with WØOUD. Then, when you are both finished, you and NDØN are both excused from the net. 73!)

AIØO: 73 GG

As you can see, it doesn't take much to say a lot on a CW net. Now that all the net business is taken care of, The NCS will start excusing other stations. Since K2ONP has a schedule to make with the Tenth Region Net, he will be excused first.

KØSI: K2ONP DE KØSI TU GEO FER QNI NW QRU QNX TU 73 K

K2ONP: TU PETE CUL 73 DE K2ONP $\overline{SK}$

WØOUD: OUD

(Letha is back from receiving her traffic.)

KØSI: OUD TU LETHA NW QRU QNX 88 K

WØOUD: GN PETE CUL 88 DE WØOUD $\overline{SK}$

Now, the NCS will close the net.

KØSI: MCWN NW QNF [the net is free] GN DE KØSI CL

When you check into a CW net for the first time, don't worry about speed. The NCS will answer you at about the speed you check into the net. You will discover that everyone

on a CW net checks in with a different speed, just as everyone has a different voice on SSB. It's nothing to be self-conscious about. As the saying goes, "We're all in this together." The goal is to pass the traffic correctly, with 100 percent accuracy, not burn up the ether with our spiffy fists. Likewise, don't hesitate to slow down for someone else. Remember, when handling traffic, 100 percent is the minimum acceptable performance level!

On SSB

Now, let's journey through a world with plenty of shadow and substance, the Mowegia Single Sideband Net.

As we look in on KØPCK calling tonight's session, keep in mind these few pointers:

1) The net preamble, given at the beginning of each session, will usually give you all the information you need to survive on the net. Method of checking in varies greatly from net to net. For instance, some section nets have a prearranged net roll, some take checkins by alphabetical order, or some even take checkins by city, county or geographical area. Don't feel intimidated by a prearranged net roll. Those nets will always stand by near the end of the session to take stations not on the net roster.

2) As you listen to a net, you will find that on phone, formality also varies. Some SSB nets are strictly business, while others are "chattier." However, don't always confuse lack of formality with looseness on a net. There is still a

definite net procedure to adhere to.

3) Once, on a close play, the catcher asked umpire Bill Klem, "Well, what is it?"

"It ain't nothin' till I call it," he growled.

By the same token, you need to keep in mind that the NCS is the absolute boss when the net is in session. On CW nets, this doesn't seem to be much of a problem to the tightness of operation, but on phone nets, sometimes a group of "well-meaners" can really slow down the net. So don't "help" unless NCS tells you to.

Since net time is upon us, let's get back to the beginning of the Mowegia Single Sideband Net.

"Calling the Mowegia Single Sideband Net, calling the Mowegia Single Sideband Net. This is KØPCK, net control. The Mowegia Single Sideband Net meets on 3988 kHz nightly at 6 PM for the purpose of handling traffic in Mowegia and to provide a link for out-of-section traffic through the ARRL National Traffic System. My name is Ben, Bravo Echo November, located in Toad Lick, Mowegia. When I call for the letter corresponding to the first letter of the suffix of your call, please give your call sign only."

"Is there any emergency, priority or time value traffic for the net?"

(wait 5 seconds)

"Any relays?"

(wait 5 seconds)

"Any low-power, mobile or portable stations wishing to check in?"

(wait 5 seconds)

"Any relays?"

(wait 5 seconds)

"This is KØPCK for the Mowegia Sideband Net. Do we have any formal written traffic?"

"KØORB."

"KØORB, Good evening Bill. List your traffic, please."

"Good evening, Ben. I have two pieces to go out-of-state."

"Very good. Who is our Tenth Region Rep tonight?"

"Good evening, Ben. This is NIØR, Ten Rep."

"NIØR, this is KØPCK. Hi, Roger. Please call KØORB, move him to 3973 and get Bill's out-of-state traffic."

"KØORB, this NIØR. See you on 73."

"Going. KØORB."

Now, let's sit back and analyze this. As you can see, the format is pretty much identical except that it takes more words. Oh, yes, one other difference... you may have noticed that not one single Q-signal was used! Q-signals should not be used on phone nets; using them detracts from the "professionalism" of a net. Work hard at avoiding them, and you will reduce the "lingo barrier," making it a little less intimidating for a potential new checkin.

Two Special Cases

RTTY nets are almost identical to CW nets, except for the method of transmission, but you need to remember two extra tips:

1) It is always important to zero-beat the NCS, but on RTTY it is crucial! It doesn't take much to turn QTC1 Corn Shuck Holler to QTHGCRRX ULUSK HXLBELZZ4.

2) Stay far away from the key that sends RYRYRYRYRYRYRY. If you follow rule number one, RYing isn't necessary.

The other "exception to the rule" is the local 2-meter FM net. Many 2-meter nets are designed for ragchewing or weather spotting, so often if you bring traffic to the net, be prepared to coach someone in the nuances of traffic handling.

Who Owns the Frequency?

Traffic nets sometimes have difficulties when it comes time for the call-up, and a ragchew is taking place on the published net frequency. What to do? Well, you could break in on the ragchew and ask politely if the participants would mind relinquishing the frequency. This usually works, but what if it doesn't? The net has no more right to the frequency than the stations occupying it at net time, and the ragchew stations would be perfectly within their rights to decline to relinquish it.

The best thing to do in such a case is to call the net near, but not directly on, the normal frequency—far enough (hopefully) to avoid causing QRM, but not so far that net stations can't find the net. Usually, the ragchewers will hear the net and move a bit farther away—or even if they don't, the net can usually live with the situation until the ragchew is over.

It is possible to conceive of a situation, especially on 75-meter phone, in which the net frequency is occupied and the entire segment from 3850 to 4000 kHz is wall-to-wall stations. In such a case, the net will just have to do the best it can and take its chances on QRMing and being QRMed. In any case, it is *not* productive to argue about who has the most right to the frequency. Common courtesy says that the first occupants have, but there are many extenuating circumstances. Avoid such controversies, especially on the air.

Accordingly, net frequencies should be considered "approximate," inasmuch as it may be necessary for nets to vary their frequencies according to band conditions at the time. Further, no amateur or organization has any preemptory right to any specific amateur frequency.

A few other things to remember:

1) Unlike a "double" on CW or SSB, where the NCS might even still get both calls, a "double" on an FM repeater either captures only one station, or makes an ear-splitting squeally heterodyne. Drag your feet a little before you check in, so you are less likely to double.

2) Be especially aware to wait for the squelch tail or courtesy beep. More people seem to time out the repeater on net night than any other!

3) Remember that a lot of people have scanners, many with the local repeater programmed in on one of the channels. Design your behavior in such a way that it attracts nonhams to Amateur Radio. In other words, don't do anything you wouldn't do in front of the whole town! [For more information on RTTY operating, see Chapter 9; see Chapter 11 for additional details about FM.]

Oh, yes. One other thing will insure your success and build your reputation as a good traffic handler. It's a simple thing, really, but you'd be surprised how much mileage you can get out of saying "please" and "thank you."

MAKING IT, TAKING IT AND GIVING IT AWAY: MESSAGE HANDLING AND MESSAGE FORM

By this stage, you have probably been checking into the net for a while, and things have started to move along quite smoothly on your journey as a rookie traffic handler. But in the life of any new traffic op, the fateful day comes along when the NCS points RF at you and says those words that strike fear in almost every newcomer: "Pse up 4 get one QTC."

Now what? You could suddenly have "rig trouble" or a

"power outage" or a "telephone call," or just bump your dial and disappear. After all, it has been done before, and everyone that didn't do it sure thought of it the first time they were asked to take traffic! Of course, there is a more honorable route—go ahead and take it! Chances are you will be no worse than anyone else your first time out. The late Bob Peavler, WØBV, a well-known and respected trafficker, used to say, "Don't be afraid of making a mistake. Chances are we have already made all of them before you."

To ease the shock of your first piece of traffic, maybe it would be a good idea to go over message form "by the book"—ARRL message form, that is.

ARRL Message Form—the Right Way, and the Right Way

A common line of non-traffickers is, "Aw, why do they have to go through that ARRL message-form stuff? It just confuses people and besides, my message is just a few words or so. It's silly to go through all that rigamarole."

Well, then, let's imagine that you are going to write a letter to your best friend. What do you think would happen to your letter if you decided that the standardized method the post office used was "silly," so you signed the front, put the addressee's address where the return address is supposed to go, and stamped the inside of the letter? It would probably end up in the Dead Letter Office.

Amateur message form is standardized so it will reach its destination speedily and correctly. It is very important for every amateur to understand correct message form, because you never know when you will be called upon in a emergency. Most nonhams think all hams know how to handle messages, and it's troublesome to discover how few do. You can completely change the meaning of a piece of traffic by accident if you don't know the ARRL message form, and as you will see later, this can be a real "disaster." Learn it the right way, and this will never happen.

If you will examine the sample message in this section, you will notice that the message is essentially broken into four parts: the preamble, the address block, the text and the signature. The preamble is analogous to the return address in a letter and contains the following:

1) The number denotes the message number of the originating station. Most traffic handlers begin with number 1 on January 1, but some stations with heavy volumes of traffic begin the numbering sequence every quarter or every month.

2) The precedence indicates the relative importance of the message. Most messages are Routine (R) precedence—in fact, about 99 out of 100 are in this category. You might ask, then why use any precedence on routine messages? The reason is that operators should get used to having a precedence on messages so they will be accustomed to it and be alerted in case a message shows up with a different precedence. A Routine message is one that has no urgency aspect of any kind, such as a greeting. And that's what most amateur messages are—just greetings.

The Welfare (W) precedence refers to either an inquiry as to the health and welfare of an individual in the disaster area or an advisory from the disaster area that indicates all is well. Welfare traffic is handled only after all emergency and priority traffic is cleared. The Red Cross equivalent to an incoming Welfare message is DWI (Disaster Welfare Inquiry).

The Priority (P) precedence is getting into the category of high importance and is applicable in a number of circumstances: (1) important messages having a specific time limit, (2) official messages not covered in the emergency category, (3) press dispatches and emergency-related traffic not of the utmost urgency, and (4) notice of death or injury in a disaster area, personal or official.

The highest order of precedence is EMERGENCY (always spelled out, regardless of mode). This indicates any message having life-and-death urgency to any person or group of persons, which is transmitted by Amateur Radio in the absence of regular commercial facilities. This includes official messages of welfare agencies during emergencies requesting supplies, materials or instructions vital to relief of stricken populace in emergency areas. During normal times, it will be very rare.

3) Handling Instructions are optional cues to handle a message in a specific way. For instance, HXG tells us to cancel delivery if it requires a toll call or mail delivery, and to service it back instead. Most messages will not contain handling instructions.

4) Although the station of origin block seems self-explanatory, many new traffic handlers make the common

Handling Instructions

HXA—(Followed by number.) Collect landline delivery authorized by addressee within _____ miles. (If no number, authorization is unlimited.)

HXB—(Followed by number.) Cancel message if not delivered within _____ hours of filing time; service originating station.

HXC—Report date and time of delivery (TOD) to originating station.

HXD—Report to originating station the identity of station from which received, plus date, time and method of delivery.

HXE—Delivering station get reply from addressee, originate message back.

HXF—(Followed by number.) Hold delivery until _____ (date).

HXG—Delivery by mail or landline toll call not required. If toll or other expense involved, cancel message and service originating station.

An HX prosign (when used) will be inserted in the message preamble before the station of origin, thus: NR 207 R HXA50 W1AW 12. . . (etc). If more than one HX prosign is used, they can be combined if no numbers are to be inserted; otherwise the HX should be repeated, thus: NR 207 R HXAC W1AW. . . (etc), but: NR 207 R HXA50 HXC W1AW . . . (etc). On phone, use phonetics for the letter or letters following the HX, to insure accuracy.

mistake of exchanging their call sign for the station of origin after handling it. The station of origin never changes. That call serves as the return route should the message encounter trouble, and replacing it with your call will eliminate that route. A good rule of thumb is *never* to change any part of a message.

5) The check is merely the word count of the text of the message. The signature is not counted in the check. If you discover that the check is wrong, you may not change it, but you may amend it by putting a slash bar and the amended count after the original count. [See below for additional information on the message check.]

Another common mistake of new traffickers involves "ARL" checks. A check of ARL 8 merely means the text has an ARL numbered radiogram message text in it, and a word count of 8. It does not mean ARRL numbered message no. 8. This confusion has happened before, with unpleasant results. For instance, an amateur with limited traffic experience once received a message with a check of ARL 13. The message itself was an innocuous little greeting from some sort of fair, but the amateur receiving it thought the message was ARRL numbered message 13—"Medical emergency exists here." Consequently, he unknowingly put a family through a great deal of unnecessary stress. When the smoke cleared, the family was on the verge of bringing legal action against the ham, who himself developed an intense hatred for traffic of any sort and refused to ever handle another message. These kinds of episodes certainly don't help the "white hat" image of Amateur Radio!

When counting messages, don't forget that each "X-ray" (instead of periods), "Query" and initial group counts as a word. Ten-digit telephone numbers count as three words; the ARRL-recommended procedure for counting the telephone number in the text of a radiogram message is to separate the telephone number into groups, with the area code (if any) counting as one word, the three-digit exchange counting as one word, and the last four digits counting as one word. Separating the telephone number into separate groups also helps to minimize garbling. Also remember that closings such as "love" or "sincerely" (that would be in the signature of a letter) are considered part of the text in a piece of amateur traffic.

6) The place of origin can either be the location (City/State or City/Province) of the originating station or the location of the third party wishing to initiate a message through the originating station. Use standard abbreviations for state or province. ZIP or postal codes are not necessary. For messages from outside the US and Canada, city and country is usually used. If a message came from the MARS (Military Affiliate Radio System) system, place of origin may read something like "Korea via Mars."

7) The filing time is another option, usually used if speed of delivery is of significant importance. Filing times should be in UTC or "Zulu" time.

8) The final part of the preamble, the date, is the month and day the message was filed—year isn't necessary.

Next in the message is the address. Although things like ZIP code and phone number aren't entirely necessary, the more items included in the address, the better its chances of reaching its destination. To experienced traffic handlers, ZIP codes and telephone area codes can be tip-offs to what area of the state the traffic goes, and can serve as a method of verification in case of garbling. For example, all ZIP codes in Minnesota start with a 5. Therefore, if a piece of traffic

Checking Your Message

Traffic handlers don't have to dine out to fight over the check! Even good ops find much confusion when counting up the text of a message. You can eliminate some of this confusion by remembering these basic rules:

1) Punctuation ("X-rays," "Querys") count separately as a word.

2) Mixed letter-number groups (1700Z, for instance) count as one word.

3) Initial or number groups count as one word if sent together, two if sent separately.

4) The signature does not count as part of the text, but any closing lines, such as "Love" or "Best wishes" do.

Here are some examples:
• Charles J McClain—3 words
• W B Stewart—3 words
• St Louis—2 words
• 3 PM—2 words
• SASE—1 word
• ARL FORTY SIX—3 words
• 2N1601—1 word
• Seventy three—2 words
• 73—1 word

Telephone numbers count as 3 words (area code, prefix, number), and ZIP codes count as one. ZIP + 4 codes count as two words. Canadian postal codes count as two words (first three characters, last three characters.)

Although it is improper to change the text of a message, you may change the check. Always do this by following the original check with a slash bar, then the corrected check. On phone, use the words "corrected to."

Book Messages

When sending book traffic, always send the common parts first, followed by the "uncommon" parts. For example:

R N0FQW ARL 8 BETHEL MO SEP 7 B̅T̅

ARL FIFTY ONE B̲E̲T̲H̲E̲L̲ SHEEP FESTIVAL X LOVE B̅T̅
PHIL AND JANE B̅T̅

NR 107 TONY AND LYN C̲A̲L̲H̲O̲U̲N̲ A̅A̅
160 NORTH DOUGLAS A̅A̅ SPRINGFIELD IL 62702 B̅T̅

NR 108 JOE WOOD AJ0X A̅A̅
84 MAIN STREET A̅A̲
LAUREL MS 39440 B̅T̅

NR 109 JEAN WILCOX A̅A̲
1243 EDGEWOOD DRIVE A̅A̅
LODI CA 95240 A̅R̅ N

Before sending the book traffic to another operator, announce beforehand that it is book traffic. Say "Follows book traffic." Then use the above format. On CW, a simple HR BK TFC will do.

sent as St Joseph, MO, with a ZIP of 56374 has been garbled along the way, it conceivably can be rerouted. So, when it comes to addresses, the adage "the more, the better" applies.

The text, of course, is the message itself. You can expedite the counting of the check by following this simple rule—when copying by hand, write five words to a line. When copying with a typewriter, or when sending a message via RTTY, type the message 10 words to a line. You will discover that this is a quick way to see if your message count agrees with the check. If you don't agree, nine times out of 10 you have dropped or added an "X-ray," so copy carefully. Another important thing to remember is that you never end a text with a "X-ray"—it just wastes space and makes the word count longer.

Finally, the signature. Remember, complimentary closing words like "sincerely" belong in the text, not the signature. In addition, signatures like "Dody, Vanessa, Jeremy,

MARS

Most modern-day traffic handlers don't realize that our standard message preamble is largely fashioned after the form used in the Army Amateur Radio System, which had its heyday in the '30s. The ARRL standard preamble prior to that time was quite different, but the AARS form was adopted because it had advantages and was so widely used. AARS nets were numerous in the '30s, using WL calls on two frequencies (3497.5 and 6990 kc.) outside the amateur bands and many frequencies, using amateur calls and amateur participants, inside the bands.

MARS, the post-World War II successor to AARS, encompasses all three of the US armed services. Although it operates numerous nets, all of which are outside the amateur bands on military frequencies, thousands of amateurs participate in MARS. This service performs some traffic coverage that we amateurs are not permitted to perform. MARS, which stands for "Military Affiliate Radio System," is conducted in three different organizations under the direction of the Army, Navy and Air Force. Therefore, MARS is not in the strictest sense Amateur Radio.

Nevertheless, many amateur messages find their way into MARS circuits, and MARS messages find their way into amateur nets. In fact, NTS has a semiformal liaison with MARS to handle the many messages from families in the States to their sons and daughters serving overseas.

Traffic for some points overseas at which US military personnel are stationed can be handled via MARS, provided a complete military address is given, even though some of these points cannot be covered by Amateur Radio or NTS. The traffic is originated in standard ARRL form and refiled into MARS form (now quite a bit different from ours) when it is introduced into a MARS circuit for transmission overseas. In this manner, traffic may be exchanged with military personnel in West Germany, Japan and a few other countries that do not otherwise permit the handling of international third-party traffic.

Traffic coming from MARS circuits into amateur nets for delivery by Amateur Radio are converted from MARS to amateur form and handled as any other amateur message. The *exception* occurs when traffic originates overseas. In this case, the name of the country in which it originates, followed by "via MARS," should appear as the place of origin, so it does not appear that such messages were handled illegally by Amateur Radio. There are places where US military personnel are stationed that even MARS cannot handle traffic with, presumably because of objections by the host country. This information is in the hands of MARS "gateway" stations and changes from time to time.

The amount of MARS traffic appearing on amateur nets is not great, since MARS has a pretty good system of handling it on MARS frequencies, but it is important that we maintain liaison as closely as possible since all civilian MARS members are US licensed amateurs.

Information concerning MARS may be obtained directly from the individual branches at these addresses:

Air Force MARS
Chief, Air Force MARS
HQ, AFCC/DOOCC
Scott AFB, IL 62225-6001

Navy-Marine Corps MARS
Director, Navy-Marine Corps MARS
Naval Communications Unit
Washington, DC 20390-5161

HQ Army MARS
US Army Information Systems Command
AS-OPS-OA
Ft Huachuca, AZ 85613

Ashleigh, and Uncle Porter," no matter how long, go entirely on the signature line.

At the bottom of our sample message you will see call signs next to the blanks marked "sent" and "received." These are not sent as the message, but are just bookkeeping notes for your own files. If necessary, you could help the originating station trace the path of the message.

Keeping It Legal

In the FCC rules under the "Prohibited transmissions" heading (§97.113), it states that "no amateur station shall transmit any communication the purpose of which is to facilitate the business or commercial affairs of any party...except as necessary to providing emergency communications." Under the same heading, it further states that "no station shall transmit messages for hire or for material compensation, direct or indirect, paid or promised."

The FCC rules also have a section directly addressing traffic handling—§97.115, "Third party communications," which reads as follows:

(a) an amateur station may transmit messages for a third party to:

(1) Any station within the jurisdiction of the United States.

(2) Any station within the jurisdiction of any foreign government whose administration has made arrangements with the United States to allow amateur stations to be used for transmitting international communications on behalf of third parties. No station shall transmit messages for a third party to any station within the jurisdiction of any foreign government whose administration has not made such an arrangement. This prohibition does not apply to a message for any third party who is eligible to be a control operator of the station.

(b) The third party may participate in stating the message where:

(1) The control operator is present at the control point and is continuously monitoring and supervising the third party's participation; and

(2) The third party is not a prior amateur service licensee whose license was revoked; suspended for less than the balance of the license term and the suspension is still in effect; suspended for the balance of the license term and relicensing has not taken place; or surrendered for cancellation following notice of revocation suspension or monetary forfeiture proceedings. The third party may not be the subject of a cease and desist order which relates to amateur service operation and which is still in effect.

(c) At the end of an exchange of international third party communications, the station must also transmit in the station identification procedure the call sign of the station with which a third party message was exchanged.

Note that emergency communications is defined in §97.403 as providing "essential communication needs in connection with the immediate safety of human life and immediate protection of property when normal communications systems are not available."

It's self-explanatory. Every amateur should be familiar with these rules. Also, while third-party traffic is permitted in the US and Canada, this is not so for most other nations. A special legal agreement is required in each country to make such traffic permissible, both internally and externally (except if the message is addressed to another amateur). A list of third-party agreements in effect when this book was published appears in Chapter 5. Check this list before agreeing to handle *any kind* of international traffic.

Such agreements specify that only unimportant, personal, non-business communications be handled—things that ordinarily would not utilize commercial facilities. (In an emergency situation, amateurs generally handle traffic *first* and face the possible consequences *later*. It is not unusual for a special limited-duration third-party agreement to be instituted by the affected country during an overseas disaster.) The key point here is, particularly under routine day-to-day nonemergency conditions, if we value our privileges, we must take care not to abuse any regulations, whether it be on the national or international level.

Some Helpful Hints about Receiving Traffic

1) Once you have committed the format of ARRL message form to memory, there's no need to use the "official" message pads from ARRL except for deliveries. Traffic handlers have many varied materials on hand for message handling. Some just use scrap paper. Many buy inexpensive 200-sheet 6- × 9-inch plain tablets available at stationery stores. Those who like to copy with a typewriter often use roll paper or fan-fold computer paper to provide a continuous stream of paper, separating each message as needed. As long as you can keep track of it, anything goes in the way of writing material.

2) Don't say "QSL" or "I roger number..." unless you mean it! It's not "Roger" unless you've received the contents of the message 100 percent. It's no shame to ask for fills (ie, repeats of parts of the message). Make sure you have received the traffic correctly before going on to the next one.

3) Full- (QSK) or semi-breakin can be very useful in handling traffic on CW (or VOX on SSB). If you get behind, saying "break" or sending a string of dits will alert the other op that you need a fill.

4) You can get a fill by asking for "word before" (WB), "word after" (WA), "all before" (AB), "all after" (AA) or "between" (BN).

Sending the Traffic

Just because you've taken a few messages, don't get the notion that being good at receiving traffic makes you a good sender, too. In the "dark ages" of hamdom, when one had to journey to the FCC office to take exams, one not only had to receive code in the code test, but send it as well. It used to be amazing how many folks could copy perfect 20-WPM code only to be inadequate on the sending portion. A lot of good ops became quite amazed that they could be so ham-fisted on a straight key!

Good traffic operators know they have to learn the nuances of sending messages as well as getting them. Your ability to send can "make or break" the other operator's ability to receive traffic in poor conditions. Imagine yourself and the other operator as a pair of computers interfaced over the telephone. Your computer (your brain) must successfully transfer data through your modem (your rig) over the lines of communication (the amateur frequencies) to the other modem, and ultimately, to the other computer.

Of course, for this transfer to be successful, two major items must be just right. The modems must operate at the same baud rate, and they must operate on matching protocol. Likewise, you must be careful to send your traffic at a comfortable speed for the receiving op, and use standardized protocol (standard ARRL message form). As you will see, these are slightly different for phone and CW, with even a couple of other deviations for RTTY or FM.

Sending the Traffic by CW

Someone once remarked, "The nice thing about CW traffic handling is that you have to spell it as you go along, so you don't usually have to spell words over." Also, the other main difference in CW traffic handling is that you tell someone when to go to the next line or section by use of the prosigns $\overline{AA}$ or $\overline{BT}$. Keeping this in mind, let's show how our sample message would be sent:

NR 133 R HXG WØMME ARL8 MOUNT PLEASANT IA 1700Z SEP 1
MR MRS JEFF HOLTZCLAW $\overline{AA}$
ROUTE 1 BOX 127 $\overline{AA}$
TONGANOXIE KS 66086 $\overline{AA}$
913 555 1212 $\overline{BT}$
ARL FIFTY ONE OLD THRESHERS REUNION X LOVE $\overline{BT}$
UNCLE CHUCKIE $\overline{AR}$ N (if you have no more messages) or $\overline{AR}$ B (if you have further messages)

Now, let's examine a few points of interest:

1) You don't need to send "preamble words" such as precedence and check. The other operator is probably as familiar with standard ARRL form as you are (maybe more!).

2) The first three letters of the month are sufficient when sending the date.

3) In the address, always spell out words like "route" and "street."

4) Do not send dashes in the body of telephone numbers; it just wastes time.

5) Always, always, always spell out each word in the text! For example, "ur" for "your" could be misconstrued as the first two letters of the next word. Abbreviations are great for ragchewing, but not for the text of a radiogram.

6) Sometimes, if you have sent a number of messages, when you get to the next-to-last message, it's a good idea to send $\overline{AR}$ 1 instead of $\overline{AR}$ B to alert the other station that you have just one more.

7) If the other operator "breaks" you with a string of dits, stop sending and wait for the last word received by the other side. Then, when you resume sending, start up with that word, and continue through the message.

Becoming a proficient CW traffic sender is tough at first, but once you've mastered the basics, it will become second nature—no kidding!

Sending the Traffic by Phone

Phone traffic handling is a lot like the infield fly rule in baseball—everyone thinks they know the rule, but in truth few really do. Correct message handling via phone can be just as efficient as via CW if and only if the two operators follow these basic rules:

1) If it's not an actual part of the message, don't say it.

2) Unless it's a very weird spelling, don't spell it.

3) Don't spell it phonetically unless it's a letter group or mixed group, or the receiving station didn't get it when you spelled it alphabetically.

Keeping these key points in mind, let's waltz through our sample message. This is how an efficient phone traffic handler would send the message:

"Number one hundred thirty three, routine, Hotel X-ray Golf, WØ Mike Mike Echo, ARL EIGHT, Mount Pleasant, Iowa, seventeen hundred Zulu, September one."

"Mr and Mrs Jeff Holtzclaw, route one, box twenty seven, Tonganoxie, T-O-N, G-A-N, O-X, I-E, Kansas, six six zero eight six. Nine one three, five five five, one two, one two. Break." You would then let up on the PTT switch (or pause if using VOX) and give the operator any fills needed in the first half of the message.

"ARL FIFTY ONE Old Threshers Reunion X-ray Love

Break Uncle Chuckie. End, no more'' (if you have no more messages), ''more'' (if you have more messages).

Notice that in phone traffic handling, a pause is the counterpart for $\overline{AA}$. Also notice the lack of extraneous words. You don't need to say ''check,'' or ''signature,'' or ''Jones, common spelling.'' (If it's common spelling, why tell someone?) You only spell the uncommon. Most importantly, you speak at about half reading speed to give the other person time to write. If the receiving operator types, or if you have worked with the other op a long time and know his capabilities, you can speak faster. Always remember that any fill slows down the message more than if you had sent the message slowly to begin with!

Oh, Yes... Those Exceptions!

Once again, RTTY and VHF FM provide the exceptions to the rules. RTTY traffic is very much like CW traffic. You include the prosigns $\overline{AA}$, $\overline{BT}$, and $\overline{AR}$, as a typed AA, BT or AR. Single-space your message (you don't need to waste paper) and send ''BBBB'' or ''NNNN'' as the last line of your message depending on whether you have more traffic or not. Use three or four lines between messages. This allows you to get four or five average messages on a standard 8½ × 11-inch sheet of paper. For further details, see Chapter 9.

When sending a message over your local repeater, remember that you often will be working with someone who isn't a traffic handler. It may be necessary to break more often (between the preamble and the address, for instance). Also, always make sure they understand about ARL numbered radiogram texts, and if they don't have a list, tell them what the message means. [The complete list of ARRL numbered radiogram texts appears in Chapter 17.] FM is a quiet mode, so you can get away with less spelling than you do on SSB.

If you yourself are already into traffic, don't try to force-feed correct traffic procedure in the case of someone just starting out; ease the person into it a little at a time. You will give a nontrafficker a more positive impression of traffic handling and may even make someone more receptive to joining a net. After all, our goal as traffic handlers is to make it look fun!

Delivering a Message

Up to now, all our traffic work has been carried out in the safe, secure world of Amateur Radio. All of this changes, though, when we get a piece of traffic for delivery. Suddenly, we're thrust out among the general public in the real world with this little message and expected to give it to someone. Unfortunately, many hams don't realize the importance of this little action and miss an opportunity to engrave a favorable impression of Amateur Radio on nonhams. It's ironic that many hams can chat for hours on CW or SSB, but can't pick up the telephone and deliver a 15-word message without mumbling, stuttering or acting embarrassed. Delivering messages should be a treat, not a chore.

Let's go through a few guidelines for delivery, and if you keep these tips in mind, you and the party on the other end of the phone will enjoy the delivery.

1) Ask for the person named in the message. If he or she is not home, ask the person on the phone if they would take a message for that person.

2) Introduce yourself. Don't you hate phone calls from people you don't know and don't bother to give a name? Chances are they're trying to sell you something, and you brush them off. Most people have no idea what Amateur Radio is about, and it's up to you to make a good first impression.

3) Tell who the message is from before you give the message. Since the signature appears at the end of the message, most hams give it last, but you will hold the deliveree's attention longer if you give it first. When you get letters in the mail, you check out the return addresses first, don't you? Then you open them in some sort of order of importance. Likewise, the party on the phone will want to know the sender of the message first.

A good way to start off a delivery is to say something like, ''Hi, Mr/Mrs/Miss so-and-so, my name is whatever, and I received a greeting message via Amateur Radio for you from wherever from such-and-such person.'' This usually gives you some credibility with your listener, because you mentioned someone they know. They will usually respond by telling such-and-such is their relative, college friend, etc. At that point, you have become less of a stranger in their eyes, and now they don't have to worry about you trying to sell them some aluminum siding or a lake lot at Casa Burrito Estates. Make sure you say it's a *greeting* message, too, to allay any fears of the addressee that some bad news is imminent.

4) When delivering the message, skip the preamble and just give the text, avoiding ARL text abbreviations. Chances are, Grandma Ollie doesn't give two hoots about the check of a message, and thinks ARL FORTY SIX is an all-purpose cleaner. Always give the ''translation'' of an ARL numbered text, even if the message is going to another ham.

5) Unless it is a ''mass origination'' message from a fair or other special event, ask the party if they would like to send a return message. Explain that it's absolutely free, and that you would be happy to send a reply if they wish. Experienced traffickers can vouch that it's easy to get a lot of return and repeat business once you've opened the door to someone. It's not uncommon for strangers to ask for your name or phone number once they discover Amateur Radio is a handy way to tell relatives ''Arriving 3 PM on Flight 202 next Wednesday.''

To Mail or Not to Mail

Suppose you get a message that doesn't have a phone number, or the message would require a toll call. Then what? If you don't know anyone on 2 meters that could deliver it, or Directory Assistance is of no help, you are faced with the decision whether to mail it or not. There is no hard and fast rule on this (unless, of course, the message has an HXG attached). Always remember that since this is a free service, you are under no obligation to shell out a stamp just because you accepted the message.

Many factors may influence your decision. If you live in a large urban area, you probably have more deliveries than most folks, and mail delivery could be a big out-of-pocket expense that you're not willing to accept. If you live out in the wide open spaces, you may be the only ham for miles around, and probably consider mail delivery more often than most. Are you a big softie on Christmas or Mother's Day? If so, you may be willing to brunt the expense of a few stamps during those times of the year when you wouldn't otherwise. At any rate, the decision is entirely up to you.

Although you may be absolved from the responsibility of mailing a message, you don't just chuck the message in the trash. You do have a duty to inform the originating station that the message could not be delivered. A simple ARL SIXTY SEVEN followed by a brief reason (no listing, no one home for three days, mail delivery returned by post office, and so forth) will suffice. This message always goes to the station of origin, not the person in the signature. The originating station will appreciate your courtesy.

Now That You're Moving Up in the World

By now, you are starting to get a grasp of the traffic world. You've been checking in to a net on a regular basis, and you're pretty good at message form. Maybe you've even delivered a few messages. Now you are ready to graduate from Basic Traffic 101 and enroll in Intermediate Traffic 102. Good for you! You have now surpassed 80 percent of your peers in a skillful specialty area of Amateur Radio. However, there's still a lot to learn, so let's move on.

Book Messages

Over the years, book messages have caused a lot of needless headaches and consternation among even the best traffic handlers. Many hams avoid booking anything just because they think it's too confusing. Truthfully, book messages are fairly simple to understand, but folks tend to make them harder than they actually are.

So, just what are book messages? Book messages are merely messages with the same text and different addresses. They come in two categories—ones with different signatures, and ones with the same signatures. Elsewhere in the chapter are examples. Often you will see book messages around holiday times and during fairs or other public events.

Oh, yes . . . one other thing about book messages. When you check into a net with a bunch of book messages, give the regular message count only. Don't say, "I have a book of seven for Outer Baldonia." (The NCS has enough to keep track of without having to break his train of thought to divide by three.) Say instead, "Seven Outer Baldonia." Then, when you and the station from Outer Baldonia go off frequency to pass the traffic, tell him that it is book traffic. When he tells you to begin sending, give common parts first, then the "uncommon" parts (addresses and possibly signatures.) By following this procedure, you will avoid a lot of confusion.

Suppose you get a book of traffic on the NTS Region net bound for your state, but to different towns. When you take them to your section net, you will not be able to send them as a book, since they must be sent to different stations. Now what? Simply "unbook" them, and send them as individual messages. For instance, let's say you get a book of three messages for the Mowegia section from the region net. Two are for Mowegia City, and one is for Swan Valley. Simply list your traffic as Mowegia City 2 and Swan Valley 1, for a total count of 3. Books aren't ironclad chunks of traffic, but a step-saver that can be used to your advantage. They can be unbooked at any time. Use them whenever you can, and don't be afraid of them.

BECOMING NCS AND LIVING TO TELL ABOUT IT

Some momentous evening in your traffic career, you may be called upon to take the net. Perhaps the NCS had a power failure, or is on vacation, or perhaps a vacancy occurred in the daily NCS rotation on your favorite net. Should this be the case, consider yourself lucky. Net Managers entrust few members with net-control duties.

Of course, you probably won't be thinking how lucky you are when the Net Manager says "QNG" and sticks your call after it. Once again, just like your first checkin or your first piece of traffic, you will just have to grit your teeth and live through it. However, you can make the jump easier by following these hints *long before* you are asked to be a net control:

1) Become familiar with the other stations on the net. Even if you never become NCS, it pays to know who you work with and where they live.

2) Pay close attention to the stations that go off frequency to pass traffic. What frequencies does the net use to move traffic? Which stations are off frequency at the moment? You will gain a feel for the net control job just by keeping track of the action.

3) Try to guess what the NCS will do next. You will discover many dilemmas when you try to second-guess the NCS. Often different amounts of traffic with equal precedence appear on the net, and a skillful NCS must rank them in order of importance. For instance, if you follow the NCS closely, you will discover that traffic for the NTS rep, such as out-of-state traffic, gets higher priority than one for the NTS rep's city. Situations like these are fun to second-guess when you are standing by on the net and will better prepare you for the day *you* might get to run the net.

Should that day arrive, just keep your cool and try to implement the techniques used by your favorite net-control stations. After a few rounds of NCS duties, you will develop your own style, and who knows? Perhaps some new hopeful for NCS will try to emulate *you* some day! See accompanying sidebars for further hints on developing a "type-NCS" personality and on proper net-control methods.

HANDLING TRAFFIC AT FAIRS OR OTHER PUBLIC EVENTS

A very special and important aspect of message handling is that of how to handle traffic at public events. If the event is of any size, like a state fair, it doesn't take long to swamp a group of operators with traffic. Only by efficient, tight organization can a handful of amateurs keep a lid on the backlog.

No matter what size your public event, the following points need to be considered for any traffic station accessible to the public:

1) Often, more nontraffic handlers than traffic handlers will be working in the booth. This means your group will have to lay out a standard operating procedure to help the non-traffickers assist the experienced ops.

Jobs such as meeting the public, filling out the message blanks, sorting the "in" and "out" piles, and keeping the booth tidied up, can all be performed by people with little or no traffic skill, and is a good way to introduce those people into the world of traffic.

Are You a Type-NCS Personality?

As net control station (NCS), it pays to remember that the net regulars are the net. Your function is to preside over the net in the most efficient, businesslike way possible so that the net participants can promptly finish their duties and go on to other ones. You must be tolerant and calm, yet confident and quick in your decisions. An ability to "take things as they come" is a must. Remember that you were appointed NCS because your Net Manager believes in you and your abilities.

1) *Be the boss, but don't be bossy.* It's your job to teach net discipline and train new net operators (and retrain some old ones!). You are the absolute boss when the net is in session, even over your Net Manager. However, you must be a "benevolent monarch" rather than a tyrant. Nets lose participation quickly one night a week when it's Captain Bligh's turn to call the net. If the net has a good turnout every night but one, that tells something about its NCS.

2) *Be punctual.* Many of the net participants have other commitments or nets to attend to; liaison stations are often on a tight schedule to make the NTS region or area net. If you, as NCS, don't care when the net starts, others will think it's okay for them to be late, too. Then traffic doesn't get passed in time, and someone may miss his NTS liaison. In short, the system is close to breaking down.

3) *Know your territory.* Your members have names—use them. They also live somewhere—by knowing their locations, you can quickly ascertain who needs to get the traffic. As NCS, it's your responsibility to know the geography of your net. You also need to understand where your net fits into the scheme of NTS.

4) *Take extra care to keep your antennas in good shape,* because an NCS can't run a net with a "wimpy" signal. Although you don't have to be the loudest one on the net, you do have to be heard. You will discover that the best way to do this is to have a good antenna system. A linear amplifier alone won't help you hear those weak checkins!

5) *The NCS establishes the net frequency.* Just because the *Net Directory* lists a certain frequency doesn't give you squatters rights to it if a QSO is already in progress there. Move to a nearby clear frequency, close enough for the net to find you. QRM is a fact of life on HF, especially on 75/80 meters, so live with it.

6) *Keep a log of every net session.* Just because the FCC dropped the logging requirements doesn't mean that you have to drop it. It's a personal decision. The Net Manager may need information about a checkin or a piece of traffic, and your log details can be helpful to him in determining what happened on a particular night.

7) *Don't hamstring the net by waiting to move the traffic.* Your duty is to get traffic moving as quickly as possible. As soon as you can get two stations moving, send them off to clear the traffic. If you have more than one station holding traffic for the same city, let the "singles" (stations with only one piece for that city) go before the ones with more than one piece for that city. The quicker the net gets the traffic moved, the sooner the net can be finished and the net operators can be free to do whatever they want.

---

2) If you plan to handle fairly large amounts of traffic, the incoming traffic needs to be sorted. A good system is to have an "in-state" pile, an "in-region" pile, and three piles for the three areas of NTS. After the traffic has been sent, it needs to be stacked in numerical order in the "out" pile. Keeping it in numerical order makes it easier to find should it need to be referred to.

Since your station will be on a number of hours, plan to check into your region and area, as well as your section, net.

Another good idea is to have "helpers" on 2 meters who can also take some of your traffic to the region and/or area net. These arrangements need to be worked out in advance.

3) Make up your radiogram blanks so that most of the preamble is already on them, and all you need to fill in is the number, check and date. In the message portion, put only about 20 word lines to discourage lengthy messages. Try to convince your "customer" to use a standard ARL text so you can book your messages.

The Net Control Sheet

Net controlling is no easy task, requiring much talent on the part of the operator. A useful prop is a set system for keeping track of net operation so that you don't get mixed up and start to lose control. This can best be effected by a sheet of paper on which you record who has what traffic, who covers what and who is on what side frequency. Trying to keep this information in your head is a losing battle unless you have a remarkable memory.

There are many methods for doing this, depending to a great extent on what net is being controlled and the exact procedure used. In general, however, the best method is to list the calls of stations reporting in vertically down the page, followed by their coverage, if known. The coverage may be unnecessary if the NCS knows his stations, but it is a good idea to leave space for it in case an unfamiliar station reports in and you have to ask his location.

Next, list horizontally across the page the traffic reported by each station coming into the net, using destination (abbreviated) followed by the number of messages. From this you can see at a glance when traffic flow can start. In most nets, it is best to start it right away, and not wait until all stations have reported in. As traffic is passed, it can be crossed off. Whenever you get a station who has no traffic and for whom there is none, that station can be excused (QNX) and crossed off the list. As stations clear traffic and there is none for them, they also can be excused.

If side frequency (QNY) procedure is used, net controlling is a bit more complicated, but the use of such frequencies vastly speeds up the process. In this case, you will need to keep track of who is on which side frequency clearing which traffic, and you will be kept busy dispatching them, sending stations up or down to meet stations already on side frequencies, checking stations back into the net as they return, etc. Both your fingers and your key or mike button will be kept going, and it can be a nightmare if not handled properly.

Probably the best method is to keep the side frequencies on a separate sheet, each side frequency utilizing a separate column, labeled up 5, up 10, down 5, down 10, or whatever spacing intervals you find practical. As two stations are dispatched to a side frequency, enter the suffixes of their calls in the appropriate column, at the same time crossing out the traffic they are sent there to handle. When they return, cross them out of the side-frequency column; this side frequency can then accommodate two other stations. Of course, if you dispatch a third station to the side frequency to wait to clear traffic to one of the stations already there, enter him in the column also, and then only one of the two originally dispatched will return, so just cross off that one. When all your listed traffic is crossed off, the net is ready to secure (QNX QNF).

There are a number of refinements to this method, but the above is basic and a good way to start. Experience will soon indicate better ways to do it. For instance, K2KIR's novel approach follows.

Net Controlling EAN

At the NTS area and region levels, the complexity of the net control task often suggests the desirability of using a matrix form of log sheet. Many of the Eastern Area Net (EAN) Net Control Stations use the form here. A similar form for region nets can be made by replacing the RN column headings with the section net designations and the CAN/PAN columns with single column titled THRU (or EAN, CAN or PAN). Although this form can be used with no other accessories than a pencil, the use of moveable objects such as 6-32 hex nuts, buttons or push pins materially aids the visualization of which stations are on the net frequency and hence which stations are available for pairing off and passing traffic.

As stations are sent off frequency, their hex nuts are moved from the net column to the appropriate side frequency and a single diagonal line is drawn through the traffic being cleared. If, for any reason, the traffic is not cleared, a circle is drawn around the traffic total as a reminder to the NCS that it still must be cleared. (In other words, the circle overrides the slash.) Assuming the traffic is cleared on the second attempt, an opposite diagonal is drawn, thus totally crossing out the traffic quantity.

For the example shown, the following notes should explain how the various features of this NCS sheet are utilized:

Halfway through the Eastern Area Net session:

• W1EFW and VE3GOL are both clear and have been QNXed at 0055 and 0057, respectively.

• W8PMJ has cleared his CAN and PAN traffic, as well as 8RN traffic from W1TN. He has yet to clear his 3RN, plus receive 8RN from W3YQ. He is presently DOWN TEN with W2CS.

• W2CS was previously sent off to clear his 4RN with W4NTO, but they were unable to complete the pairing. The NCS will try again later, perhaps by using a relay station.

• WB4PNY has a net report for K2KIR, to be sent direct to him if he QNIs.

• N2AKZ and KW1U are UP TEN, clearing 2RN traffic.

• KQ3T, W4NTO and K4ZK are standing by on the net frequency, waiting for new assignments.

There are many techniques for determining in what order to clear the traffic listed, but a couple of fairly common ones are worth mentioning.

1) Assign highest priorities to stations having the largest totals. Thus, W1NJM (PAN RX), W8PMJ (8RN TX and RX) and W3YQ (3RN TX) should be kept waiting as little as possible. If conditions are good, QNQing other stations to these three will speed things up immensely.

2) Tackle the smaller individual destination totals first. This allows early excusing of stations having small traffic totals to clear, and avoids a last-minute panic near the end of the net.

3) Clear short-haul pairings first. In the winter near a sunspot minimum, short-haul communications are most

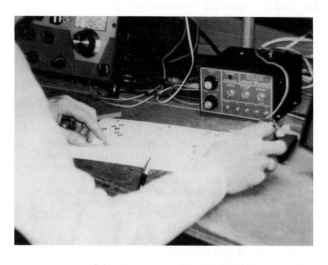

+30	+25	+20	+15	+10	+5	IN	STATION	OUT	1RN	2RN	3RN	4RN	8RN	ECN	CAN	PAN	OTR	−5	−10	−15	−20	−25
						35	W1EFW	55	▨													
					⊙	31	N2AKZ			▨												
						33	KQ3T				▨											
						33	W4NTO					▨										
						33	W8PMJ			2	▨		X	7				⊙				
						36	VE3GOL	57	X					▨	X							
					⊙	34	WB4PNY							▨			KIR					
						30	W1NJM									▨		⊙				
					⊙	30	W3YQ		X	2		X	4	X	X	2X						
						31	W1TN			X	X		3			X		⊙				
						32	W2CS					⊗	3			3			⊙			
						33	K4ZK				X					7						
					⊙	34	KW1U		X	X	1											

FIG. 1

likely to be successful early in the net session before the skip has gone out. At all other times, especially in the summer with nets on Daylight Savings Time, it delays long-haul pairings until band conditions have improved for those paths.

Obviously, there are times when these (or any other algorithms) will conflict with each other. In the final analysis, a good NCS bases his decisions on the specifics of a given net session: the band conditions, the operators, the traffic distribution and the time remaining.—*Bud Hippisley, K2KIR*

A real time- and headache-saver in this department is to fill out the message blank for the sender. This way, you can write in the X-rays and other jargon that the sender is unaware of.

4) Most importantly, realize that you sometimes have to work at getting "customers" as much as if you were selling something! Most people have no concept of Amateur Radio at all, and don't understand how message handling works. ("How can they get it? They don't have a radio like that" is a very common question!)

Use posters to make your booth appealing to the eye. Make sure one of the posters is of the "How your message gets to its destination" variety, such as the one shown in this section. Don't go over someone's head when answering a question—explain it simply and succinctly.

Finally, don't be afraid to "solicit" business. Get up in front of the booth and say "hi" to folks. If they say "hi" back, ask them if they would like to send a free greeting to a friend or relative anywhere in the US (grandparents and grandchildren are the easiest to convince!). Even if they decide not to send a message, your friendliness will help keep our image of "good guys in white hats" viable among the general public, which is every bit as great a service as message handling. For further hints, see accompanying sidebar.

THE ARRL NATIONAL TRAFFIC SYSTEM— MESSAGE HANDLING'S "ROAD MAP"

Although you probably never think about it, when you check into your local net or section net, you are participating in one of the most cleverly designed game plans ever written—the National Traffic System (NTS). Even though the ARRL conceived NTS way back in 1949, and it has grown from one

Handy Hints for Handling Traffic at Fairs or Other Public Events

1) Although you may only be there a day or two, don't compromise your station too much. Try to put up the most you can for an antenna system because band conditions on traffic nets in the summer can really be the pits! Usually, you will be surrounded by electrical lines at fairs, so a line filter is a must. An inboard SSB or CW filter in your rig is a definite plus, too, and may save you many headaches.

2) Don't huddle around the rigs or seat yourself in the back of the booth. Get up front and meet the people. After all, your purpose is to "show off" Amateur Radio to perk the interest of nonhams.

3) Most people will not volunteer to send a piece of traffic, nor will they believe a message is really "free." It's up to you to solicit "business." Be cheerful.

4) Always use "layman's language" when explaining about Amateur Radio to nonhams. Say "message," not "traffic." Don't ramble about the workings of NTS or repeaters; your listener just wants to know how Aunt Patty will get the message. "We take the message and send it via Amateur Radio to Aunt Patty's town, and the ham there will call her on the phone and deliver it to her," will do.

5) Make sure your pencils or pens are attached to the booth with a long string, or you will be out of writing utensils in the first hour!

6) Make sure there are plenty of instructions around for hams not familiar with traffic handling to help them "get the hang of" the situation.

7) Make sure your booth is colorful and attractive. You will catch the public's eye better if you give them something to notice, such as this suggested poster idea.

regular cycle to two to three, NTS hasn't outgrown itself and remains the most streamlined method of traffic handling in the world. (During this discussion, please refer to the accompanying Section/Region/Area map, the NTS Routing Guide and the NTS Flow Chart.)

Actually, the National Traffic System can trace its roots to the railroad's adoption of Standard Time back in 1883, when radio was still only a wild dream. Three of the Standard Time Zones are the basis for the three NTS areas—Eastern Area (Eastern Time Zone), Central Area (Central Time Zone) and Pacific area (Mountain and Pacific Time Zones). Within these areas are a total of 12 regions. You may wonder why NTS has 12 regions—why not just break it up into 10 regions, one for each area? Ah, but check the map. You will discover that NTS not only covers the US, but our Canadian neighbors as well. Then, of course, the region nets are linked to section/local nets.

The interconnecting lines between the boxes on the flow chart represent liaison stations to and from each level of NTS. The liaisons from area net to area net have a special name, the Transcontinental Corps (TCC). In addition to the functions shown, TCC stations also link the various cycles of NTS to each other.

The clever part about the NTS setup, though, is that in any given cycle of NTS, all nets in the same level commence at approximately the same local time. This allows time for liaisons to the next level to pick up any outgoing traffic and meet the next net. In addition, this gives the TCC stations at least an hour before their duties commence on another area net or their schedule begins with another TCC station.

The original NTS plan calls for four cycles of traffic nets, but usually two cycles are sufficient to handle a normal load of traffic on the system. However, during the holiday season, or in times of emergency, many more messages are dumped into the system, forcing NTS to expand to four cycles temporarily. The cycles of normal operation are Cycle Two, the daytime cycle, which consists primarily of phone nets, and

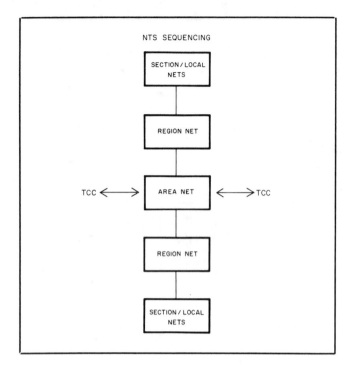

NTS SEQUENCING

Cycle Four, the nighttime cycle, made up mostly of CW nets. In addition, Cycle One has been implemented in the Pacific and Eastern areas. Now that the rudiments of NTS have been covered, let's see where you fit in.

NTS and You

Before the adoption of NTS, upper-level traffic handlers worked a system called the "trunk line" system, where a handful of stations carried the burden of cross-country traffic, day in day out. Nowadays, no one has to be an "iron man" or "iron woman" within NTS if they choose not to. If each liaison slot and TCC slot were filled by one person, one day a week, this would allow over 1000 hams to participate in NTS! Unfortunately, many hams have to double- and triple-up duties, so there is plenty of room for any interested amateur.

An NTS liaison spot one day (or night) a week is a great way to stay active in the traffic circuit. Many hams who don't have time to make the section nets get satisfaction in the traffic world by holding a TCC slot or area liasion once a week. If you would enjoy such a post, drop a note to your STM, Net Manager or TCC Director. They will be happy to add another to their fold.

However, remember that the area and region net are very different from your section or local net in one aspect. The function of the section or local net is to "saturate" its jurisdiction, so the more checkins, the better. On the upper-level nets, though, the name of the game is to move the traffic as quickly and efficiently as possible. Therefore, additional checkins—other than specified liaisons and stations holding traffic—only slow down the net. [However, if you are a station holding traffic to be moved, you can enter NTS at any level to pass your traffic, even if you've never been on an upper-level net before. Entering the system at the section or local level is preferred.]

If you are interested in finding out more about the "brass tacks" of the workings to NTS, get a copy of the *Public Service Communications Manual* (FSD-235), available free from ARRL HQ. Every aspect of NTS is explained, as well

as information about local net operating procedure, RACES and ARES operation. The *PSCM* will also orient you with net procedure of region and area NTS nets. It takes a little more skill and savvy to become a regular part of NTS, but the rewards are worth the effort. If you have the chance, go for it!

So There You Have It

Although this chapter is by no means a complete guide to

Table 1
National Traffic System Routing Guide

State/Province	Abbrev.	Region	Area
Alaska	AK	7	PAN
Alabama	AL	5	CAN
Alberta	AB	7	PAN
Arizona	AZ	12	PAN
Arkansas	AR	5	CAN
British Columbia	BC	7	PAN
California	CA	6	PAN
Colorado	CO	12	PAN
Connecticut	CT	1	EAN
Delaware	DE	3	EAN
Dist. of Columbia	DC	3	EAN
Florida	FL	4	EAN
Georgia	GA	4	EAN
Guam	GU	6	PAN
Hawaii	HI	6	PAN
Idaho	ID	7	PAN
Illinois	IL	9	CAN
Indiana	IN	9	CAN
Iowa	IA	10	CAN
Kansas	KS	10	CAN
Kentucky	KY	9	CAN
Labrador	LB	11	EAN
Louisiana	LA	5	CAN
Maine	ME	1	EAN
Manitoba	MB	10	CAN
Maryland	MD	3	EAN
Massachusetts	MA	1	EAN
Michigan	MI	8	EAN
Minnesota	MN	10	CAN
Mississippi	MS	5	CAN
Missouri	MO	10	CAN
Montana	MT	7	PAN
Nebraska	NE	10	CAN
Nevada	NV	6	PAN
New Brunswick	NB	11	EAN
New Hampshire	NH	1	EAN
New Jersey	NJ	2	EAN
New Mexico	NM	12	PAN
New York	NY	2	EAN
Newfoundland	NF	11	EAN
North Carolina	NC	4	EAN
North Dakota	ND	10	CAN
Nova Scotia	NS	11	EAN
Ohio	OH	8	EAN
Oklahoma	OK	5	CAN
Ontario	ON	11	EAN
Oregon	OR	7	PAN
Pennsylvania	PA	3	EAN
Prince Edward Is.	PEI	11	EAN
Puerto Rico	PR	4	EAN
Quebec	PQ	11	EAN
Rhode Island	RI	1	EAN
Saskatchewan	SK	10	CAN
South Carolina	SC	4	EAN
South Dakota	SD	10	CAN
Tennessee	TN	5	CAN
Texas	TX	5	CAN
Utah	UT	12	PAN
Vermont	VT	1	EAN
Virginia	VA	4	EAN
Washington	WA	7	PAN
West Virginia	WV	8	EAN
Wisconsin	WI	9	CAN
Wyoming	WY	12	PAN
Virgin Islands	VI	4	EAN
APO/FPO	AE	2	EAN
APO/FPO	AA	4	EAN
APO/FPO	AP	6	PAN

Fame and Glory: Your Traffic Total, PSHR Report, and Traffic Awards and Appointments

Even if you handle only one message in a month's time, you should send a message to your ARRL Section Traffic Manager (STM) reporting your activity. Your report should include your total originations, messages received, messages sent and deliveries.

An *origination* is any message obtained from a third party for sending from your station. If you send a message to Uncle Filbert on his birthday, you don't get an origination. However, if your mom or your neighbor wants you to send him one with her signature, it qualifies (it counts as one originated and one sent). The origination category is essentially an "extra" credit for an off-the-air function. This is because of the critical value of contact with the general public and to motivate traffickers to be somewhat more aggressive in making their message-handling services known to the general public.

Any formal piece of traffic you get via Amateur Radio counts as a message *received.* Any message you send via Amateur Radio, even if you originated it, counts as a message *sent.* Therefore, any time you relay a message, you get two points: one received and one sent.

Any time you take a message and give it to the party it's addressed to, on a mode other than Amateur Radio, you are credited with *a delivery.* (It's okay if the addressee is a ham.) As long as you deliver it off the air (eg, telephone, mail, in person), you get a delivery point.

Your monthly report to your STM, if sent in radiogram *format,* should look something like this:

```
NR 111 R NIØR 15 ST JOSEPH MO NOV 2
BLAIR CARMICHAEL WBØPLY
MISSOURI STM
FULTON MO 65251

OCTOBER TRAFFIC ORIG 2 RCVD 5 SENT 6 DLVD 1
TOTAL 14 X 73
ROGER NIØR
```

If you have a traffic total of 500 or more in any month, or have over 100 originations-plus-deliveries in a month, you are eligible for the Brass Pounders League (BPL) certificate, even if you did all that traffic on SSB! If you make BPL three times, you receive a handsome medallion for your shack.

Another mark of distinction is the Public Service Honor Roll (PSHR). You don't have to handle a single message to get PSHR, so it's a favorite among traffickers in rural areas. The categories have traditionally been listed in the Public Service column of *QST,* and they include checking into public service nets, acting as net control, handling an emergency message, and so on. See the Public Service column for particulars.

Almost any station, with regular participation, can get an ARRL Net Certificate through the Net Manager. Once you have participated regularly in a net for three months, you are eligible. If you are a League member, you can also become eligible for an Official Relay Station appointment in the ARRL Field Organization (for details concerning the ARRL Field Organization, see Chapter 17). Either certificate makes a handsome addition to your shack.

---

traffic handling, it should serve as a good reference for verterans and newcomers alike. If you've never been involved in message handling, perhaps your interest has been piqued. Should that be the case, don't put it off. Find a net that's custom-made for you and check into it! You'll find plenty of fine folks that will soon become close friends as you begin to work with them, learn from them, and yes, even chat with them when the net is over.

The roots of traffic handling run deep into the history of Amateur Radio, yet its branches reach out toward many tomorrows. Our future lies in proving our worth to the nonham public, and what better way to ensure the continuance of our hobby than by uniting family and friends via Amateur Radio?

Sure, it takes some effort, but any trafficker will tell you he stays with it because of the satisfaction he gets from hearing those voices on the other end of the phone say, "Oh, isn't that nice!" We've plenty of room for you—come grow with us!

Nov tfc: AB4E 425, W4EAT 242, KI4YV 174, K4IWW 130, K4CWZ 107, N4WZH 99, N9CGD 71, KM4DY 54, AB4W 52, WD8DIN 49, KR4LS 46, W4IRE 46, AE4EC 45, KE4WCW 45, N4UE 45, KE4AHC 42, AA4YW 40, N9MN 40, WA4SRD 34, KE4JHJ 33, WA4MJF 28, KD4RYE 26, N4SHE 24, NT4K 24, K4AIF 21, W4DYW 18, @W4RAL 16, WD4MRD 15, KE4YMA 14, KC4PGN 13, KR4ZJ 13, K4ROK 10, W2CS 10, KE4TES 7, N2JLE 7, KA4KTU 6, KF4EML 6, KT4CD 6, KE4TYJ 5, KE4KFZ 4, K4MPJ 3.

Every month, the SM (Section Manager) of each ARRL section sends *QST* a report of net activities. This November, 1996 report of W. Reed Whitten, AB4W (North Carolina section) has stations handling from 425 to 3 messages each.

Tips on Handling NTS Traffic By Packet Radio

Listing Messages

* After logging on to your local NTS-supported bulletin board, type the command LT, meaning List Traffic. The BBS will sort and display an index of all NTSXX traffic awaiting delivery. The index will contain information that looks something like this:

MSG#	TR	SIZE	TO	FROM	@BBS	DATE	Title
200	TN	282	NTSCT	KY1T		870225	QTC 1 Hartford
198	TN	302	NTSRI	K1CE		870224	QTC 1 Cranston
192	TN	215	NTSIN	WF4R		870224	QTC 1 Indianapolis
190	TN	200	NTSCT	AJ6F		870224	QTC 1 Waterbury
188	TN	315	NTSCT	KH6WZ		870224	QTC 1 Newington
187	TN	320	NTSCT	K6TP		870224	QTC 1 New Haven
186	TN	300	NTSAL	WA4STO		870224	QTC 1 Birmingham
184	TN	295	NTSCA	WB4FDT		870224	QTC 1 Fresno

Receiving Messages

* To take a message off the Bulletin Board for telephone delivery to the third party, or for relay to a NTS Local or Section Net, type the R or RT command, meaning Read Traffic, and the message number. R 188 will cause the BBS to find and send the message text file containing the RADIOGRAM for the third party in Newington. The RADIOGRAM will look like any other, with preamble, address, text and signature, only some additional packet-related message header information is added. This information includes the routing path of the message for auditing purposes; e.g., to discern any excessive delays in the system.

* After the message is saved to the printer or disk, the message should be KILLED by using the KT command, meaning Kill Traffic, and the message number. In the above case, at the BBS prompt, type KT 188. This prevents the message from being delivered twice.

* At the time the message is killed, many BBS's will automatically send a message back to the station in the FROM field with information on who took the traffic, and when it was taken!

Delivering or Relaying A Message

* A downloaded RADIOGRAM should, of course, be handled expeditiously in the traditional way: telephone delivery, or relay to a phone or cw net.

Sending Messages

* To send a RADIOGRAM, use the ST command, meaning Send Traffic. The BBS will prompt you for the NTSXX address (NTSOH, for example), the message title which should contain the city in the address of the RADIOGRAM (QTC 1 Dayton), and the text of the message in RADIOGRAM format. The BBS, usually within the hour, will check its outgoing mailpouch, find the NTSOH message and automatically forward it to the next packet station in line to the NTSOH node. Note: Some states have more than one ARRL Section. If you do not know the destination ARRL Section ("Is San Angelo in the ARRL Northern, Southern or West Texas Section?"), then simply use the state designator NTSTX.

*Note: While NTS/packet radio message forwarding is evolving rapidly, there are still some gaps. When uploading an NTS message destined for a distant state, use handling instruction "HXC" to ask the delivering station to report back to you the date and time of delivery.

* Unbundle your messages please: one NTS message per BBS message. Please remember that traffic eventually will have to be broken down to the individual addressee somewhere down the line for ultimate delivery. When you place two or more NTS messages destined for different addressees within one packet message, eventually the routing will require the messages to be broken up by either the BBS SYSOP or the relay station, placing an additional, unreasonable burden on them. Therefore it is good practice for the originator to expend the extra word processing in the first place and create individual messages per city regardless if there are common parts of other messages. This means that book messages are not suitable in packet at this time unless they are going to the same city. Bottom line: Messages should be sent unbundled. (Tnx NI6A for this tip)

We Want You!

Local and Section BBS's need to be checked daily for NTS traffic. SYSOPs and STMs can't do it alone. They need your help to clear NTS RADIOGRAMS every day, seven days a week, for delivery and relay. If you are a traffic handler/packeteer, contact your Section Traffic Manager or Section Manager for information on existing NTS/packet procedures in your Section.

If you are a packeteer, and know nothing of NTS traffic handling, contact ARRL HQ, your Section Manager or Section Traffic Manager for information on how you can put your packet radio gear to use in serving the public in routine times, but especially in times of emergency!

And, if you enjoy phone/cw traffic handling, but aren't on packet yet, discover the incredible speed, and accuracy of packet radio traffic handling. You probably already have a small computer and 2-meter rig; all you need is a packet radio "black box" to connect between your 2-meter rig and computer. Cost: about $130. For more information on packet radio, contact ARRL HQ.

Chapter 16

Image Communications

BRUCE BROWN, WA9GVK
3422 SILVER MAPLE PLACE
FALLS CHURCH, VA 22042

and

RALPH TAGGART, WB8DQT
602 SOUTH JEFFERSON
MASON, MI 48854

Amateur Radio operators have always had a strong propensity and well-deserved reputation for resourcefulness, whether it be fashioning an envelope detector out of a safety pin, razor blade and pencil lead, or making antennas out of coat hangers and toy springs. On a larger scale, hams have successfully transformed piles of surplus parts and metal scraps into radio transmitters capable of worldwide communications. Amateur image communicators continue to uphold this venerable tradition, but now with a slightly different leaning.

This time, hams have been successfully incorporating elements of the home video and personal computer revolutions into their image systems. Components such as camcorders, low-cost color video cameras and monitors, videotape recorders, computer processors, printers and large capacity solid-state memories are having a profound beneficial effect upon the operation, performance and costs related to the transmission and reception of both still and moving pictures over the airwaves. Most hams are surprised to learn that some of their home video and computer equipment can indeed serve as the foundation for building a reasonably priced, yet highly effective, image communication system. Once the virtual private domain of the experimenter, image communication is now being enjoyed by an ever-expanding segment of the amateur population.

In this chapter, the goal is to improve your image—quite literally—and you won't even have to attend charm school! The focus will be on the three main image communications systems: fast-scan amateur television (FSTV), also referred to as ATV; slow-scan television (SSTV); and facsimile (fax). Each mode has something unique and fascinating to offer. (The technical aspects of these systems are fully covered in the periodicals listed at the end of this chapter.) Regulatory requirements along with beneficial operating suggestions and procedures for achieving more effective communications will be discussed. Armed with this information and aided by some old-fashioned common sense and on-the-air courtesy, you'll be all set to see and be seen. After all, hams should be seen as well as heard!

FAST-SCAN AMATEUR TELEVISION

"If it looks like a duck, sounds like a duck, walks like a duck, then it must be a...," or so the saying goes. Well, fast-scan amateur television (FSTV), also referred to simply as amateur television (ATV), certainly looks and sounds like commercial-broadcast-TV quality, and the pictures move around on the screen like commercial-broadcast TV. But alas, ATV is still just "amateur." The similarity between the systems is by no means accidental since ATVers use the same basic transmission standards as does the commercial world. Amateurs are prohibited, however, from engaging in any overt forms of broadcasting or entertainment. Also, amateurs operate at considerably lower power levels that must be compensated through the use of high-gain antennas,

Fig 16-1—Richard Logan, WB3EPX, is an active ATV operator. His station is designed with space limitations in mind.

sensitive preamps and low-loss cable at the receiver. Nevertheless, there is no greater compliment that an ATVer can be paid than hearing from a receiving station that your picture is coming across "just like commercial broadcast quality." That's like getting a 5-by-9 report on the phone bands! Most folks viewing ATV for the first time are surprised by the performance levels attainable by the relatively simple and low-cost amateur TV equipment.

Because of its large bandwidth, ATV may not be operated below the 70-cm band. This, in turn, makes ATV largely a line-of-sight or local-coverage system. Exceptions do occur, though, especially when inversion layers in the atmosphere are present and ducting occurs. Under these conditions, DX of several hundred miles is possible.

Television is not exactly new to Amateur Radio; enterprising amateurs have been involved in this branch of the electronics art for more than 60 years. *QST* began carrying articles on TV as early as 1925. Although the early experiments used crude mechanical methods, the advent of the moderately priced iconoscope in 1940 made electronic television a practical reality for the amateur. As reported in the November 1940 issue of *Radio* magazine, "Two-way television was demonstrated for the first time by Amateur Radio men of the W2USA Radio Club, between their station in New York City and their glass-enclosed booth in the Communications Building at the World's Fair, New York." By 1960, amateurs were beginning to experiment with color TV and in the '70s hams learned how to modify inexpensive surplus UHF taxicab radios for ATV application.

In recent years, ATV has undergone some dramatic changes, most notably in performance improvements and expanded portable applications. The most evident is in the widespread use of color because of the availability of low-cost color cameras and color monitors originally aimed at the home-video entertainment market. Color transmission has also been facilitated by the expanded use of broadband solid-state transmitters and stripline amplifiers, which provide the necessary frequency response and phase characteristics to faithfully transmit color. The effectiveness of the low-power solid-state equipment has been further enhanced through the expanded use of ATV repeaters throughout the country, and low-noise GaAsFET receiver preamps. Finally, the small-size ATV transmitters are making their way out of the shack and into the field for use in scientific and public service applications.

License Requirements and Operating Frequencies

ATV can be used by any ham holding a Technician or higher-class license. Most operations can be found in the 420 to 440-MHz segment of the 70-cm band, with operation in the 910-928-MHz portion of the 33-cm band and the 1240-1294-MHz portion of the 23-cm band gaining in popularity. Novices are permitted to use the 1270 to 1295 MHz segment of the 23-cm band with a 5-watt power restriction. See the ARRL band plans in Chapter 2 for more information. Some amateurs are also experimenting with video transmission in the 2300-2450 band using modified equipment originally intended for commercial microwave TV distribution systems, and some testing can even be found at 10 GHz using Gunnplexers and surplus microwave burglar-alarm systems. It is important that you first check with your ATV club, VHF/UHF local frequency coordination organization, or the *ARRL Repeater Directory* to determine the specific ATV frequencies for your area.

Most FSTV activity occurs in the 70-cm band due to lower equipment costs and generally favorable propagation conditions. Picture carrier frequencies of 439.25, 426.25 and

421.25 MHz are the most popular. Increased levels of interference from other services sharing this band have motivated interest in the relatively clear higher frequencies. In the 33-cm band, 910.25 and 923.25 MHz are fairly standard, while for 23 cm suggested channels are 1241.25, 1253.25 and 1277.25 MHz. Many population centers have FSTV repeaters, which may be either in-band (426.25 MHz input/439.25 MHz output, for example) or cross-band (426.25 MHz input/923.25 or 1253.25 MHz output). ATVers always try to avoid interfering with weak signal work (DX/EME) on 432 MHz and 1296 MHz. Amateurs in certain areas are prohibited by the FCC from operating in the 421-430 MHz spectrum. Additionally, 50-watt PEP output power limitations apply to certain amateurs operating adjacent to designated military installations in the US. See *The FCC Rule Book*, published by the ARRL, for the details of these restrictions.

Antenna Polarization

Antenna polarization also varies throughout the country. Most nonrepeater areas tend to use horizontal, as do the DXers. Most repeaters still use vertical, although the trend is toward horizontal to increase propagation range and reduce interference from other services. It's highly important that you are using the same polarization as those with whom you intend to communicate. Using a different polarization can mean the difference between receiving a good picture and none at all. Again, check with your local ATV club or frequency coordinators.

Sound Format

Another interoperability requirement that will affect your equipment selection and operation, is the method of transmitting and receiving sound. As with frequencies and antenna polarization, voice format varies in different regions. There are three possibilities:

1) *4.5-MHz FM subcarrier audio;* audio can be heard on a standard TV set along with the picture. This is also the format used by commercial-broadcast TV.

2) *On-carrier audio;* requires a separate audio receiver to demodulate. This mode was an outgrowth of modifying amateur or surplus business transmitters having an existing "on carrier" FM voice capability and modifying the unit for AM video using collector, plate or grid modulation of the final. Some ATV repeaters will convert the on-carrier audio at its input to the subcarrier format on output so that standard TV receivers can be used in the shack to receive the audio.

3) *2 meters;* although this mode requires a separate rig, it has the advantage of letting other local hams know what is happening on ATV to publicize your activities. It allows nonATVers to participate in ATV nets and can be instrumental in picking up many new video converts. In sum, before setting up your station and attempting to operate, be sure to check:

1) simplex and repeater frequencies
2) antenna polarization
3) sound format

Equipment

Primary components of an ATV station are the camera, transmitter, receive-downconverter, standard commercial UHF/VHF television receiver, low-loss transmission line and high-gain antenna. The camcorder you purchased to chronicle the "wonder years" of your children is an ideal starting point. If you don't already have one, used black-and-white cameras can be found at hamfests in the $25 range, while color units with built-in electronic viewfinders—less recorders—are typically priced under $100. Some of the fancy, and more expensive, cameras have built-in character generators to compose on-screen messages and IDs. A video recorder (VCR) or color graphics generated on a personal computer can be used in place of the camera to show a variety of preplanned, prerecorded material. This video is fed into an ATV transmitter or transceiver connected to a high-gain antenna via lowloss cable. Transceivers, which are commercially available from several sources, are especially popular due to their ease of use and compact size. A linear power amplifier also may be used to increase transmission range.

To receive, a downconverter (or transceiver) translates the incoming signal to an unused VHF channel, typically Channel 2 or 3, for reception on an unmodified home TV. Even general purpose commercial UHF/VHF receivers such as the ICOM R7000, equipped with a special adapter, can be used to receive ATV. Add a mike and provide good lighting, and you'll be able to send and receive color television pictures that rival the quality seen on commercial broadcast stations. A typical ATV setup is shown in Fig 16-1.

Lighting

Good lighting is one of the most important and underestimated aspects of ATV. If it's a choice between good picture contrast generated by good lighting or higher transmitted power, contrast wins most every time. In other words, a high-contrast picture from a low-powered transmitter will look much better through the "snow" than a low-contrast picture from a high-powered transmitter. Fig 16-2 shows one possible lighting configuration to eliminate harsh shadows. For black-and-white work, incandescent 100-watt bulbs with reflectors will do a good job. For color, tungsten halogen lights, available at your local camera store, will give you excellent contrast while maintaining the proper color hue for home cameras. Fluorescent lights also can be effective, particularly from an overhead position, to soften background shadows. Though these different light sources appear to be "white," they all contain a degree of color as perceived by the color television camera. This amount of color, referred to as color temperature, is measured in kelvins. Typical values are:

100 W incandescent	2900
Tungsten halogen	3200
Photoflood	3400
Fluorescent	4500
Daylight	6300

The lower temperatures "redden" the picture; the higher temperatures tend to produce "bluer" pictures. Color television cameras typically have built-in filters or make automatic electronic adjustments to compensate for the different color temperatures. When using multiple light sources, try to use lamps that are nearly equivalent in color temperature to maintain consistent color accuracy throughout the scene.

Illumination is not only helpful to improve contrast but also allows cameras to operate with a smaller lens aperture (higher f stop), improving the depth of field. This means that objects both near and far can be viewed in sharp focus. You can experiment by monitoring your camera output on a closed-circuit local video monitor prior to transmitting. Once the lighting looks good, set the video gain control as high as possible before the picture smears or "whites out" as seen on a video monitor driven by the demodulated RF output of your transmitter, or by reports from remote receivers.

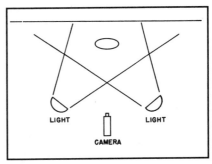

Fig 16-2—A popular lighting arrangement for ATV or SSTV pictures. Using two lights helps eliminate harsh shadows that result from a single light source. Frequently one light is set brighter or closer than the other to provide a little shading.

DXing

Besides using mast-mounted preamplifiers and low-loss Hardline, there are a few tips to increase your chances for ATV DX. The best times for optimum tropospheric conditions tend to be around sunrise and right after sunset. A clue that 70-cm propagation may be especially good is to monitor commercial UHF TV. If you begin to see distant commercial stations, especially on Channels 14 through 30, there may be a real opportunity for some good DX at 70 cm. Horizontal is the preferred polarization of DXers due to better demonstrated propagation characteristics, especially through vegetation. To punch through noise, it is helpful to use pictures with very high contrast, especially big, bold, black call letters against a white background. These pictures will stay in sync despite very weak signal conditions.

Identifying

Fast-scan TV transmissions may be identified in video, or by CW or voice on the audio on-carrier or subcarrier portion. A picture of your QSL card or vanity ham-license plate makes an ideal video identification.

Signal Reporting

Signal reporting on ATV is markedly different than in all other aspects of Amateur Radio. Instead of the RST system used for voice and CW, image communicators use the "P" system which describes the amount of "snow" or noise observed on the screen (see Fig 16-3) and described below:

P5—Excellent; no visible noise. "Closed-circuit" or commercial-broadcast quality.

P4—Very good; some noise visible.

P3—Fairly good; noticeable or moderate noise present but picture is very usable.

P2—Passable; high noise; picture can still be seen but small details are lost.

Pl—Weak picture, limited use; very high, objectionable noise.

P0—Not usable; picture is totally lost in noise; can detect sync bars or barely detect picture presence but is totally unusable.

In some cases, you can split hairs with the system. If you receive an excellent picture with very slight noise, you may want to say it is a "four point five" or P4.5.

What to Show

There are many uses for ATV around the shack to show off your equipment, prize QSLs, family, pets, friends, photographs, home videotapes and vacation movies. How about showing a videotape on how you erected your 100-foot antenna tower? (You'll be surprised how few hams really know how it's done!) How about an informative presentation showing the

AMATEUR RADIO FAST SCAN TELEVISION
VIDEO PICTURE STANDARDS

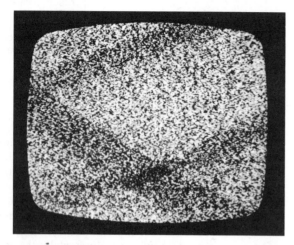

P0—Total noise visible. No picture at all or detectable Video Sync Bars.

P1—High noise visible. Weak picture.

P2—High noise visible. Fair picture. Fair detail.

P3—Noise visible. Strong picture. Recognizable detail.

P4—Slight noise visible. Very strong picture. Good detail.

P5—No noise visible. Closed circuit picture. Excellent detail.

Fig 16-3—ATV picture quality reporting system: P5—Excellent, P4—Very Good, P3—Fairly Good, P2—Passable, P1—Weak Picture, P0—No picture at all. *(Photos courtesy of Dave Williams, WBØZJP)*

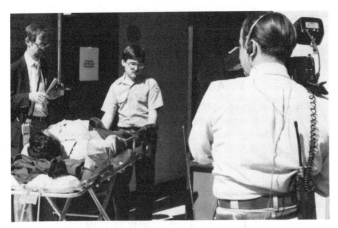

Fig 16-4—During the National Disaster Medical System drill, ATV was used to send pictures of simulated injury victims arriving at a Virginia hospital to drill coordinators at a USAF base command post in Maryland. *(WA9GVK photo)*

progress you're making on a new electronics project? You can use multiple cameras and even special-effects generators to superimpose lettering over your pictures or perform split-screen effects. If you also operate SSTV, you can retransmit the SSTV images you've received from overseas to all the local amateurs via ATV. Amateur television has been successfully transmitted from private aircraft, model rockets, and even radio-controlled airplanes and robots. The FCC has given permission to amateurs to retransmit the Space Shuttle video provided over NASA's public information channel on Satcom 1 R. As part of the Shuttle Amateur Radio Experiment (SAREX) program, ATVers successfully uplinked with 70-cm fast-scan TV directly to the STS-37 and STS-50 manned spacecraft in 1991 and 1992 respectively. Videotape exchanges with amateurs of other countries also makes interesting viewing. You can also show off the video output of your home computer to display animated color graphics.

Thanks to the availability of compact and portable gear, ATV has been making tremendous inroads in the public service arena nationwide. Hams have used ATV during national disaster drills to send video from hospitals back to emergency command posts; see Fig 16-4. In Oklahoma, a camera and ATV transmitter mounted atop a tower searches the skies for tornadoes and relays the pictures directly to the National Weather Service, providing a valuable early warning capability. Amateur TV is used annually to help monitor and coordinate the Tournament of Roses Parade, marathons and yacht races; see Fig 16-5. During the Christmas season, amateurs have taken their TV systems into hospitals to enable children to see and talk to Santa Claus over live TV. Hams have used their equipment to send Voyager 2-received pictures of Jupiter and Saturn from the Jet Propulsion Laboratory in Pasadena, California, to remote locations such as an observatory and convention center to enable more widespread viewing. Amateur TV is also always an attention-getter for Amateur Radio demonstrations held at shopping malls and Boy Scout Jamborees.

Of all the specialized applications of amateur video transmission, none have captured the excitement of the ATV community in recent years more than high-altitude ballooning. Groups throughout the US launch 5 to 8 foot diameter helium balloons equipped with 1-watt ATV transmitters, cameras,

(A)

(B)

Fig 16-5—Doug McKinney, KC3RL, checks out his portable 8-watt ATV station powered by a 10 ampere-hour motorcycle battery. This arrangement is ideal for public service applications, such as emergency drills, parades and marathon coordination. *(WA9GVK photo)*

2-meter beacons and environmental sensors. Spectacular pictures from altitudes of 100,000 feet have been received at ranges of 300 miles.

Maintaining Strong ATV Activity

Based upon the experiences of ATV organizations throughout the country, several recommendations have emerged as key factors for maintaining enthusiasm and steady growth of on-the-air activity and club support:

1) *Weekly on-the-air meeting*. Set aside a specific weekly time for an on-the-air get-together. This is the time when newcomers can be guaranteed that there will be a signal available for receiver testing and someone available to check the quality of their transmitted picture. It's a time to compare notes and help each other out by exchanging suggestions. (By the way, it's also the best time to show off the videotapes of your recent vacation trip to ensure it will be watched by a large audience!) The meeting is run by a control station who asks all "video rangers" to check in and then queries each one in a disciplined roundtable fashion so each person has a chance to transmit. The weekly get-together usually turns out to be one of the most busy, lively and enlightening aspects of ATV operation.

2) *Use 2 meters for audio*. Even though your ATV operations may use an on-carrier or subcarrier audio signal, 2 meters is recommended for voice to publicize your operations to the rest of the amateur world who may not be fully aware of all the ATV activity in the 70-cm and higher regions of the spectrum. ATV repeaters can be adapted to receive and transmit the audio portion on 2 meters.

3) *Promote a club project*. Public service efforts are ideal because they not only perform a valuable community service but they also generate a sense of purpose and enthusiasm among club members. It's a good feeling to work as a team and use your gear productively.

4) *Vary the picture content*. It's great to see your pretty face all the time on ATV but try to be creative in what you show. Variety is the spice of ATV life!

Self-Monitoring

Although you can usually depend on the received station to give you on-the-air signal reports, the quickest and most accurate method is to use an RF demodulator driving a video monitor to check the quality of your transmitted signal. You'll be able to determine contrast, focusing, framing, color level and color hue, while making sure you are not bending and rolling by excessive sync compression. Commercial ATV transceivers generally have this monitoring function as a built in capability. Video samplers, which convert RF into a video signal that can be fed directly into a TV monitor, are available from a variety of sources including PC Electronics, 2522 Paxson Ln, Arcadia, CA 91007-8537; Wyman Research, 8339 South, Waldron, IN 46182-9608; and Pauldon Associates, 210 Utica St, Tonawanda, NY 14150.

An inherent advantage of working through a crossband repeater is the ability to operate in full-duplex to monitor your signal from the repeater while simultaneously transmitting. This is possible since receiver desense—a normal problem with in-band transmit/receive operation—is virtually eliminated.

Why not adjust the transmitter for best picture on the receiver in the shack? The problem is that receiver overloading from the transmitter and downconverter will usually give misleading results. The preferable methods are to employ the sampler or crossband repeater. Both approaches allow an operator to make picture adjustments without relying on descriptions from other stations. Making effective use of these techniques will save you much on-the-air time in getting your station properly adjusted.

TV Audio Noise Squelch

One of the more annoying, but easily solvable, aspects of ATV operation is the audio noise generated by your TV when no signal is being received. When using your commercial TV set to receive video and audio subcarrier, you may notice that when you switch from receive to transmit, the TV set will emit an undesirable hiss because of the absence of a receive signal. That's the same sound that may have awakened you after you dozed off during the late, late show after the station went off the air! Hams have resorted to laboriously turning down the TV audio each time they transmit. However, there are two simple automated and effective alternatives for solving the problem:

1) *Noise squelch* . Communications Concepts, 508 Millstone Dr, Beavercreek, OH 45434-5840, markets a very inexpensive noise squelch that connects to the earphone jack of your TV. An external speaker is then connected to the unit's output. Without touching any knobs or controls, the squelch will automatically turn off the TV audio to the speaker if only noise is present on the channel. When the TV receives a signal, the squelch will sense the quieting of noise and will allow the audio to pass through to the speaker.

2) *VHF TV relay*. Another technique is to use an RFD relay connected to the TV's VHF antenna terminals along with an RF modulator capable of upconverting camera video to the same VHF channel utilized by the receive converter. The relay is activated using your ATV transmitter PTT line. When you are in the transmit mode, the up-converted video from your camera is fed via a relay into the VHF terminals of the TV. On receive, the relay connects the ATV receiver downconverter output to the TV's VHF antenna input. This not only quiets the TV on transmit but allows the TV to perform a dual function of both ATV receiver display and local camera monitor. This technique is fully described in a construction article in August 1985 *QST*, pp 32-37.

SLOW-SCAN TELEVISION

Nestled among the CW, phone and RTTY operators in the Amateur Radio bands is a sizable following of hams who regularly exchange still pictures in a matter of seconds virtually anywhere on Earth. They are using a system called slow-scan television (SSTV), which was originally designed by an amateur in the early 1960s. Over the years, the amateur community has been continually refining and improving the quality of SSTV. Amateur success with SSTV for almost three decades has led to its application by the military and commercial users as a reliable long-range, narrow-bandwidth transmission system. The worldwide appeal of SSTV is evident by the many DX stations that are now equipped for this type of picture transmission. Several amateurs have even worked over 100 DXCC countries on SSTV!

Just as the name implies, SSTV is the transmission of a picture by very slowly transmitting the picture elements, while a television monitor at the receiving end reproduces it in step. The most basic SSTV signal for black and white transmission consists of a variable frequency audio tone from 1500 Hz for black to 2300 Hz for white, with 1200 Hz used for synchronization pulses. Unlike fast-scan television, which uses 30 frames per second, a single SSTV frame takes at least eight to fill the screen. Additionally, the vertical resolution of SSTV is only 120 lines (or 128 for some digital systems) compared with 525

Table 16-1
SSTV Formats

Mode	Designator	Color Type	Scan Time (sec)	Scan Lines	Notes
AVT	24	RGB	24	120	D
	90	RGB	90	240	D
	94	RGB	94	200	D
	188	RGB	188	400	D
	125	BW	125	400	D
Martin	M1	RGB	114	240	B
	M2	RGB	58	240	B
	M3	RGB	57	120	C
	M4	RGB	29	120	C
Pasokon TV	P3	RGB	203	16+480	
	P5	RGB	305	16+480	
	P7	RGB	406	16+480	
Robot	8	BW	8	120	A,E
	12	BW	12	120	E
	24	BW	24	240	E
	36	BW	36	240	E
	12	YC	12	120	
	24	YC	24	120	
	36	YC	36	240	
	72	YC	72	240	
Scottie	S1	RGB	110	240	B
	S2	RGB	71	240	B
	S3	RGB	55	120	C
	S4	RGB	36	120	C
	DX	RGB	269	240	B
Wraase SC-1	24	RGB	24	120	C
	48	RGB	48	240	B
	96	RGB	96	240	B
Wraase SC-2	30	RGB	30	128	
	60	RGB	60	256	
	120	RGB	120	256	
	180	RGB	180	256	
Pro-Skan	J120	RGB	120	240	
WinPixPro	GVA 125	BW	125	480	
	GVA 125	RGB	125	240	
	GVA 250	RGB	250	480	
JV Fax	JV Fax Color	RGB	variable	variable	F

Notes

RGB—Red, green and blue components sent separately.

YC—Sent as Luminance (Y) and Chrominance (R-Y and B-Y).

BW—Black and white.

A—Similar to original 8-second black & white standard.

B—Top 16 lines are gray scale. 240 usable lines.

C—Top 8 lines are gray scale. 120 usable lines.

D—AVT modes have a 5-second digital header and no horizontal sync.

E—Robot 1200C doesn't really have B&W mode but it can send red, green or blue memory separately. Traditionally, just the green component is sent for a rough approximation of a b&w image.

F—JV Fax Color mode allows the user to set the number of lines sent, the maximum horizontal resolution is slightly less than 640 pixels. This produces a slow but very high resolution picture. SVGA graphics are required.

Fig 16-6—SSTV picture as seen on a standard TV set using a digitial scan converter.

lines for fast-scan. These disadvantages are offset by the fact that SSTV requires less than 1/2000 of a fast-scan TV's bandwidth. Thus, the FCC permits it in any amateur phone band.

The basic SSTV format represents a trade-off among bandwidth, picture rate and resolution. To achieve practical HF long-distance communications, the SSTV spectrum was designed to fit into a standard 3-kHz voice bandwidth through a reduction in picture resolution and frame rate. Thus, SSTV resolution is lower than FSTV and is displayed in the form of still pictures. A sample SSTV picture is shown in Fig 16-6.

Extraordinary new developments are emerging from the amalgamation of consumer color cameras, digital techniques and personal computers. The greatest advancements currently being made in the realm of color SSTV are increased resolution and improved noise immunity. That's the good news. The bad news is that these efforts have spawned over 40 SSTV modes that are not interoperable. Table 16-1 lists many of the formats. Until firm standards can be agreed upon in the SSTV community, interface and software developers are designing products with a multiple mode capability. For example, the Amiga AVT is capable of supporting 56 modes.

License Requirements and Operating Frequencies

Since SSTV operation is restricted to the phone bands, license requirements are identical to those for voice operations as listed in Chapter 2. While most activity occurs on HF using SSB, it is also well suited to FM at VHF or UHF frequencies. In the US, slow-scan TV using double-sideband AM or FM on the HF bands is not permitted.

The common accepted SSTV calling frequencies are 3.845 MHz (Advanced), 7.171 MHz (Advanced), 14.230 MHz,

14.233 MHz (General) 21.340 MHz (General), 28.680 MHz (General) and 145.5 MHz (All). Traditionally, 20 meters has been the most popular band for SSTV operations, especially on Saturday and Sunday mornings. A weekly international SSTV net is held each Saturday at 1500 UTC on one of the active 14-MHz frequencies. Many years ago, when SSTV was first authorized, the FCC recommended that SSTVers not spread out across the band even though it was legal to do so. A "gentlemen's agreement" has remained to this day that SSTVers operate as close as possible to the above calling frequencies to maintain problem-free operation.

Identifying

On SSTV, the legal identification must be made by voice or CW. Sending "This is W9NTP" on the screen is not sufficient. Most stations intersperse the picture with comments anyway, so voice ID is not much of a problem. Otherwise SSTV operating procedures are quite similar to those used on SSB.

Equipment

All you need to get started is an SSB station (or FM station for VHF/UHF), a monitor, scan converter and a video source. Like RTTY, SSTV is a 100%-duty-cycle transmission. Most sideband rigs will have to run considerably below their voice power ratings to avoid ruining the final amplifier or power supply. Early SSTV monitors used long-persistence CRTs much like classical radar displays. In a darkened environment, the image remained visible for a few seconds while the frame was completed. This type of reception is unusual today and has been replaced with digital scan converters which convert SSTV to FSTV to place a bright image on a conventional television monitor.

Observing SSTV pictures on scan converters is a result of recent advances in digital techniques. In receive mode, the device converts the incoming audio to a fast-scan video signal that is usable by a conventional fast-scan TV monitor. Similarly, on transmit the converter changes the fast-scan camera output to a standard slow-scan signal. Examples of commercial scan converters include the Robot 1200C and the Wraase SC-I. A personal computer equipped with the proper software and interface also makes a highly versatile slow-scan converter. For more information on SSTV modes and software, see *Personal Computers In The Ham Shack*, published by the ARRL.

The video source can be a black-and-white or color camera, video recorder or audio tape recorder, or the video output from a personal color-graphics computer. Since the signal emanating from the transmit-side of the scan converter is audio, it can be recorded on a cassette machine. The frequency response is not critical, but some consideration should be given for the wow and fluter specifications to minimize picture skew. You can have a friend pre-record some pictures from his/her camera onto your recorder and use them for on-the-air transmissions. The combinations of computer, camera and tapes (both audio and video) are endless. The superposition of graphics and images both generated by the operator and those received over the air can be recorded on either audio or video tapes for later use.

To send SSTV over the air, just couple the audio output of the scan converter directly into the microphone input of any single-sideband transmitter or FM transmitter on VHF or UHF. To receive, you tune in the signals on an SSB receiver and feed the tones into the scan converter which is connected to the scan monitor.

Setting Up Your Studio

Standard photographic lighting arrangements work well for TV. The set-up in Fig 16-2 is good because it reduces background shadows and harsh shadows on the subject. A pair of 100-watt bulbs with reflectors would be a minimum requirement for the typical camera. Outdoors, of course, the light level is pretty much predetermined. Care must be taken, however, to prevent direct sunlight from shining into the lens, since most TV camera pickup tubes can be damaged. Adjust the brightness and contrast controls on the camera so that the darkest screen area comes out black while the brightest areas reach maximum brightness without "blooming" on the monitor screen. This assumes that the monitor brightness and contrast controls are set correctly. (Many of the newer cameras are fully automatic and do not need any adjustment.) It is also highly useful to run a test audio tape consisting of a test pattern with various gray shades through your system to properly set up your equipment. Such a tape may be obtained by an on-the-air recording from another amateur or obtained commercially. Many computer programs and special modes of commercial scan converters also provide the test signals to adjust your signals properly.

Operating

Prior to sending a picture, it is important to voice the specific SSTV mode that will be used. It is customary to send two or three frames of each subject to ensure the receiving station gets at least one good-quality picture. Many operators then follow with descriptive verbal comments about the picture.

An SSTV signal must be tuned in properly so the picture will come out with the proper brightness and synchronization. If the signal is not "in sync," the picture will appear wildly skewed. The easiest way to tune SSTV is to wait for the transmitting operator to say something on voice and then fine tune for proper pitch. With experience, you may find you are able to zero in on an SSTV signal by listening to the sync pulses and by watching for proper synchronization on the screen. Many SSTV monitors are equipped with automatic or manual tuning aids.

If you want to record slow-scan pictures off the air, there are several ways of doing it. One is to tape record the audio signal for later playback. The other is to take a photographic picture of the image right from the SSTV screen. An instant print camera equipped with a closeup lens enables you to see the results

Fig 16-7—SSTV picture using the ROBOT 1200C format transmitted from the space shuttle during the STS-51F mission.

shortly after the picture is taken. If you want to do this without darkening the room lights, you'll have to fabricate a light-tight hood to fit between the camera and monitor screen. It is also possible to feed the converter's fast-scan output to a videotape recorder for later viewing.

Picture Subjects

The selection of things to show is endless, but you will find that high contrast black-and-white or high-saturation color pictures, that are not too cluttered, work best. A "live" close-in shot of you in front of the operating position is probably the most desirable picture. Don't forget a couple of frames with your call, name and QTH. A closeup of your QSL card is ideal, and how about showing off your prized QSLs? Illustrating a technical conversation with a simple schematic diagram can often make clear a point that would be very difficult to get across on voice alone. There is a wealth of good SSTV material in newspapers. Home-brew cartoons are also very popular subjects.

Some SSTVers find it convenient to mount the camera in front of an easel onto which various prepared cards can be inserted. These cards can be photographs, drawings or lettered signs. A kitchen-type noteboard or menu board with press-on removable letters is handy for making headings. Today it is more convenient to use the keyboard on a computer or other specialized captioning devices to produce these letters.

It is fun to make up taped "programs" of related pictures. Some operators have dozens of programs on different cassette tapes that can be selected and put on the air in seconds. People might enjoy seeing a pictorial tour of your shack and QTH. Do you have another hobby? Maybe that person you are working would like to learn about your model train collection. A closeup of your 1862 Cannonball Express would be a lot more interesting than a few bland comments about the weather.

Spectacular pictures of Saturn and Jupiter taken by the Voyager 2 spacecraft and received at the Jet Propulsion Laboratory in California have been relayed via SSTV instantaneously to amateurs around the world. Color SSTV using the ROBOT 1200C scan-converter format came into its own when Space Shuttle *Challenger* carried aloft SSTV equipment on the STS-51F mission and exchanged pictures with Amateur Radio clubs throughout the world (Fig 16-7).

Of course, there are more down-to-earth applications of SSTV. Slow-scan television has good potential for rendering service to the public by means of third-party traffic. For example, the scientists working on the Antarctic ice pack often don't see their families for months—except by amateur SSTV. This was one of the demonstrations that was used by slow-scan pioneers in 1968 to convince the FCC that slow-scan television should be permanently authorized in the ham bands. Anyone can write or call almost anywhere in the world these days, but grandma and grandpa would certainly enjoy seeing pictures of the kids from 3000 miles away by SSTV.

In whatever you show, use your imagination. But remember—never to pass up the opportunity to transmit your own image occasionally to let everyone on the frequency know with whom they are communicating!

FACSIMILE

Facsimile (fax) is a method for transmitting very high resolution still pictures using voice-bandwidth radio circuits. The narrow bandwidth of the fax signal, equivalent to SSTV, provides the potential for worldwide communications on the HF bands. Fax is the oldest of the image-transmitting technologies and has been the primary method of transmitting newspaper photos and weather charts. Facsimile is also used to transmit high-resolution cloud images from both polar-orbit and geo-stationary satellites. Many of these images are retransmitted using fax on the HF bands.

The resolution of typical fax images greatly exceeds what can be obtained using SSTV or even conventional television (typical images will be made up of 800 to 1600 scanning lines). This high resolution is achieved by slowing down the rate at which the lines are transmitted, resulting in image transmission times in the 4- to 10-minute range. Prior to the advent of digital technology, the only practical way to display such images was to print each line directly to paper as it arrived. The mechanical systems for accomplishing this are known as *facsimile recorders* and are based on either photographic media (a modulated light source exposing film or paper) or various types of direct-printing technologies including electrostatic and electrolytic papers.

Modern desktop computers have virtually eliminated bulky fax recorders from most amateur installations. Now the incoming image can be stored in computer memory and viewed on a standard TV monitor or a high-resolution computer graphics display. The use of a color display system makes it entirely practical to transmit color fax images when band conditions permit.

The same computer-based system that handles fax images is often capable of SSTV operation as well, blurring what was once a clear distinction between the two modes. The advent of the personal computer has provided amateurs with a wide range of options within a single imaging installation. SSTV images of low or moderate resolution can be transmitted when crowded band conditions favor short frame transmission times. When band conditions are stable and interference levels are low, the ability to transmit very-high-resolution fax images is just a few keystrokes away!

Hardware and Software

In the past few years, electro-mechanical fax equipment has been replaced by personal computer hardware and software. This replacement allows reception and transmission of various line per minute rates and indices of cooperation by simply pressing a key or clicking a mouse. Many fax programs are available as either commercial software or shareware. Usually, the shareware packages (and often trial versions of the commercial packages) are available by downloading from the Internet.

A good starting point is the ARRL software repositories. To get to them, set your browser to **http://www.arrl.org** and look for links to software. You can use any commercial search site to look for "fax" AND "software."

One very popular fax program is *JVFAX*. It is a DOS-based program, with a large number of options for installation. It can receive and transmit several fax formats, black and white and color. Your computer's serial port, connected to a very simple interface, provides the connection to your transceiver. For more information, take a look at *Personal Computers In the Ham Shack*, published by the ARRL.

Weatherman is a DOS-based program, using a *Sound-Blaster* (or compatible) card as the interface. The program is shareware, receive only, and a single, shielded wire from your receiver audio output to the computer audio input is the only connection needed.

WXSat was written to operate under *Windows 3.X*. While specifically set up to decode and store weather satellite APT pictures, it can also be used for HF fax reception.

Both *Weatherman* and *WXSat* are samples of what you can

find during a search on the Internet. Often, programs are offered and then either withdrawn or improved over the versions previously distributed—to get the latest and greatest you have to periodically search and see what comes up. If you use an on-line service such as CompuServe or AOL, they are another source of fax software. Check their ham forums or sections for listings.

Many commercial multi-mode controllers either contain software to receive and transmit fax, or are compatible with PC-hosted software. Available controller suppliers include MFJ (MFJ-1278B), Kantronics (ask your distributor about any needed software for the Kam Plus) and AEA. Check the advertising pages of *QST* for the latest units available.

One well-known fax page on the internet, complete with downloadable software, is **http://ourworld.compuserve.com/homepages/HFFAX**. Posted and maintained by Marius Rensen, it contains listings of commerical fax transmissions for you to test your software or just SWL for interest. Before using a program taken from any Internet source, check other sources for newer versions. It is not uncommon to have older versions posted on one place and newer versions in another. It is a good idea to virus check the software before and after unzipping.

If you would like to hear what a HF fax station sounds like, without turning on a receiver, go to:

http://www.chilton.com/scripts/radio/tuning-page/Freq=8078/Mode=USB/Filter=1.8/AGC=ON/Station=USN Norfolk

This is one long address, entered in your Internet browser without any spaces. It turns on a receiver, tuned to the US Navy fax station in Norfolk, VA on 8080 kHz.

To test your fax receive setup, you can usually find fax transmitted, 24 hours a day (weekdays) on 3357, 8080, 10865, 15959 and 20015 kHz from NMN in Norfolk, VA; 8682 and 12730 kHz from NMC in Point Reyes, CA. You may have to set your receiver approximately 1.9 kHz higher or lower than the listed frequencies, depending on the station and your selection of USB or LSB.

Formats and Standards

All facsimile transmissions using voice transmitters are accomplished by feeding a video modulated tone into the transmitter's audio input. On frequencies below 30 MHz this tone or subcarrier is frequency modulated between 1500 (black) and 2300 Hz (white). This permits the use of audio limiters at the receiving end, making reception relatively immune to fading. Accurate tuning of the SSB receiver is critical to obtaining faithful image reproduction. Many fax adapters incorporate circuits to indicate when the signal is properly tuned.

On the VHF and microwave frequencies used by weather satellites, the subcarrier is amplitude modulated (maximum amplitude = white, minimum amplitude = black). This is practical because the transmissions are made using FM modulation of the carrier, avoiding the effect of Doppler frequency shifts caused by the rapid movement of the spacecraft in orbit.

Unlike conventional television and SSTV, fax transmissions do not employ line synchronization pulses. Proper synchronization of both the transmitter and receiver in a mechanical system is maintained by operating the drums using precision synchronous motors locked to a crystal oscillator frequency standard. Computer-based systems derive their timing from crystal-locked counter chains.

One of the primary standards in facsimile is the rate at which lines are transmitted, expressed as the number of lines per minute (LPM). Successful reception of signals requires that the same line rate be used at both ends of the circuit. Commonly encountered speeds and the associated services include 90 LPM (wirephotos), 120 LPM (HF weather charts and satellite images and polar orbit satellite transmissions), 180 LPM (wirephotos) and 240 LPM (polar orbit and geostationary WEFAX satellites). Some old wirephoto equipment can still be found operating at 60 LPM, but this is too slow for practical amateur work where transmissions need to be kept shorter than 10 minutes. Some 360- and 480-LPM systems are also in use. At those speeds the video bandwidth becomes excessive for use on amateur HF bands, although they would be usable on VHF and UHF links. In general, 120 to 240 LPM systems are probably optimum for amateur use, providing the best trade-off between resolution and image transmission time.

Selection of a suitable surplus fax recorder or conversion of such a system requires dealing with the transmission rate (LPM), the scanning density (lines per inch) relative to the size of the image and the modulation format. These subjects are discussed at length, complete with conversion data, in the "Modulation Sources" chapter of any recent editions of the *ARRL Handbook*.

Color Fax

Amateur fax systems based on the use of high-resolution computer displays are capable of displaying color images using techniques similar to those in use for color SSTV. In such cases, the original color image must be scanned as separate red, green and blue images and then stored in memory for transmission.

Transmission is accomplished in one of two ways. Each red, green and blue version of the complete image can be transmitted **in its entirety** in sequence. This is known as *frame sequential* transmission. Assuming that the images are formatted correctly at the receiving end, a reproduction of the original color image will be displayed on the color graphics monitor. Alternatively, the red, green and blue versions of each **line** of the original image can be transmitted in sequence (*line sequential* transmission) such that the image will appear, line by line, in full color as it is received.

At present, two factors limit the extensive use of color on amateur facsimile. First, high-resolution fax images use a great deal of memory. (A color image requires three times the memory capacity of a grayscale image of the same resolution.) The second factor is time. A grayscale fax image requires from 4 to 10 minutes for transmission while an equivalent color image would require three times as long. HF band conditions and interference levels are not suited for such extended transmissions. Experimentation on VHF and UHF is quite practical, however, and the use of digital image compression techniques can be expected to make color more practical in the years to come.

Weather-Satellite Fax Reception

At present, the area of greatest amateur fax activity involves the reception of images transmitted by weather satellites. All of these spacecraft use AM video subcarrier modulation with line rates of 120 and 240 LPM. There are two major categories of weather satellites: those in near-polar orbits at relatively low altitudes (600-800 miles) and geostationary spacecraft located 22,000 miles above the equator.

Most polar-orbiting spacecraft are capable of transmitting both visible light and infrared (IR) images. The US NOAA

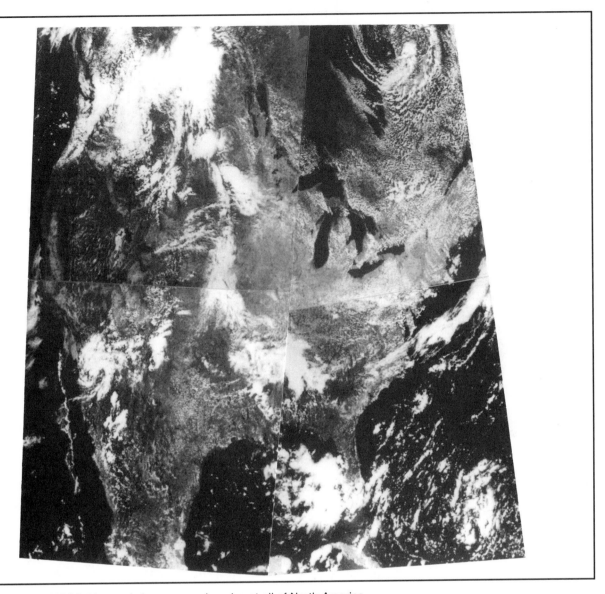

Fig 16-8—A two-pass NOAA-11 mosaic image covering almost all of North America.

spacecraft transmit simultaneous visible and IR views during daylight passes with two channels of IR data available at night. Transmissions from these spacecraft occur in the 137-138 MHz range (137.50 and 137.62 MHz in the case of the NOAA satellites) using frequency modulation of the carrier.

Reception requires the use of a receiver with an IF bandwidth of 30-40 kHz and receivers designed for this service, such as the Vanguard Labs WEPIX 2000, are among the most popular options. Modern low-noise RF preamplifiers make it possible to use omnidirectional turnstile of helix antennas without the need to track the spacecraft. Optimum passes, 10-15 minutes in duration, occur twice a day and an orbital prediction program can be a great help in determining when to listen for a specific spacecraft. AMSAT (the Radio Amateur Satellite Corporation) provides such programs on disk for a number of different personal computers. Contact AMSAT at P.O. Box 27, Washington, DC 20044.

Image data are transmitted continuously from polar orbit spacecraft. If the entire pass is recorded, you can expect images to cover an area about 1500 miles north to south and 800 miles east to west in the case of an optimum pass. Not many years ago, home-built or surplus mechanical fax recorders were the primary display systems, but most stations today are using dedicated scan converters or systems using computers with high-resolution graphics capabilities.

Geostationary spacecraft obtain images of the entire globe and transit them to earth in a series of quadrants. In addition, many of these spacecraft also relay weather charts and mosaics prepared from polar-orbit spacecraft data. The US GOES satellite system ideally consists of two primary spacecraft, one over South America and the other over the eastern Pacific, often with a third spacecraft devoted entirely to image relay functions. (Premature failure of spacecraft may reduce the number of available satellites.) A consortium of European nations operates the METEOSAT spacecraft stationed over Africa while the Japanese operate their GMS satellite over the western Pacific.

All of the geostationary satellites have FM transmission

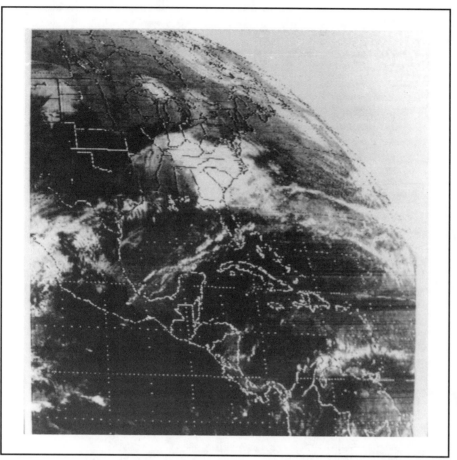

Fig 16-9—A northeast quadrant infrared image obtained from the GOES satellite. High, cold clouds are visible in white in the infrared format.

formats similar to those of the polar-orbiters, but transmissions are made at 1691 MHz. Reception typically involves the use of a small dish antenna (4-foot models will provide a reasonable gain margin) operating into a low-noise RF preamplifier/downconverter designed to convert the signal to the 137-138 MHz range. This permits the VHF receiver to serve as an IF for the geostationary system as well as provide reception of polar-orbit transmissions.

The *Weather Satellite Handbook*, published by the ARRL, provides comprehensive coverage of the various aspects of amateur weather satellite activity, including chapters on antennas, receivers, digital display systems and tracking programs.

In addition to the *ARRL Handbook* and *QST*, *ATQ* regularly contains information on image communications:

Amateur Television Quarterly
3 North Court Street
Crown Point, IN 46307
(219) 662-6396

Acknowledgments: Frame and resolution data for SSTV mode chart furnished by Advanced Electronic Applications. Dennis Bodson, W4PWF, also contributed to this chapter.

CHAPTER 17

References

	Page
ARRL DXCC Rules	17-2
ARRL DXCC Countries List	17-7
ARRL DXCC Field Representatives	17-16
DXCC Application	17-21
Russian Oblasts	17-23
Morse Code for Other Languages	17-24
Phonetic Alphabets	17-24
DX Operating Code	17-24
The Origin of "73"	17-25
The RST System	17-25
Q Signals	17-26
KØOST Ham Outline Maps	
ARRL Field Organization	17-27
California Counties	17-28
South American Countries	17-29
Brazilian States	17-30
Argentine Provinces	17-31
Antarctica and Offshore Islands	17-32
Mexican States	17-33
European Countries	17-34
French Departments	17-35
Swiss Cantons	17-36
Swedish Laens	17-37
Italian Provinces	17-38
Estonia	17-39
Slovakia	17-40
British Counties	17-41
Spanish Provinces	17-42
Belgium/Dutch Provinces	17-43
Polish Provinces	17-44
Western Russian Oblasts	17-45
Eastern Russian Oblasts	17-46
African Countries	17-47
Asian Countries	17-48

	Page
Japanese Prefectures	17-49
Chinese Provinces	17-50
Return Postage Chart for DXCC and WAS	
QSL Cards	17-51
WAS Application	17-52
5BWAS Application	17-54
VUCC Application	17-56
WAC Application	17-58
US Counties	17-59
ARRL Field Organization	17-70
N5KR Azimuthal Equidistant Maps	
W1AW	17-78
Eastern USA	17-79
Central USA	17-80
Western USA	17-81
Alaska	17-82
Hawaii	17-83
Caribbean	17-84
Eastern South America	17-85
Southern South America	17-86
Antarctica	17-87
Western Europe	17-88
Eastern Europe	17-89
Western Africa	17-90
Eastern Africa	17-91
Southern Africa	17-92
Near East	17-93
Southern Asia	17-94
Southeast Asia	17-95
Far East	17-96
Australia	17-97
South Pacific	17-98
N7BH World Time Finder	17-99

INTRODUCTION

"...the number of countries worked is increasingly becoming the criterion of excellence among outstanding DX stations."

Clinton B. DeSoto, W1CBD, October 1935 *QST*

From its simple beginnings, culminating in the announcement of the new DX award, **The DX Century Club**, in September 1937 *QST* (which was itself based on the "ARRL List of Countries" published in January 1937 *QST*), membership in the ARRL DX Century Club (DXCC) has been *the* mark of distinction among radio amateurs the world over. That it is regarded with such prestige by DXers is a testament to its integrity and level of achievement. The high standards of DXCC are intensely defended and supported by its membership. The rules established by the founders of DXCC were consistent with the art of Amateur Radio as it existed at the time. As technology improved the ability to communicate, the rules were progressively changed to maintain a competitive environment and complement the gaining popularity of DXCC.

Because of the vast changes in the international scene brought about by World War II, it logically followed that DXCC needed to be recast, as indicated in December 1945 *QST*. Ultimately, after a great deal of study, the first postwar DXCC Countries List emerged as published in February 1947 *QST*. The new DXCC Rules appeared in March 1947 *QST*. Contacts were valid from November 15, 1945, the date US amateurs were authorized by the FCC to return to the air.

The DXCC rules today represent the aggregate of experience gained from administering postwar DXCC. Some countries on the DXCC Countries List do not, of course, meet the present criteria. This includes countries "grandfathered" from the WWII era or those that met the criteria as it existed at the time and are not subject to deletion (see Section III for the appropriate grounds for deletion). Changes are announced under DXCC Notes in *QST*.

SECTION I. BASIC RULES

1) The DX Century Club Award, with certificate and lapel pin (there is a nominal fee of $5 for the DXCC lapel pin) is available to Amateur Radio operators throughout the world (see #15 below for the DXCC Award Fee Schedule). ARRL membership is required for DXCC applicants in the US and possessions, and Puerto Rico. ARRL membership is not required for foreign applicants. All DXCCs are endorsable (see Rule 5). There are 12 separate DXCC awards available, plus the DXCC Honor Roll:

(a) **Mixed** (general type): Contacts may be made using any mode since November 15, 1945.

(b) **Phone:** Contacts must be made using radiotelephone since November 15, 1945. Confirmations for cross-mode contacts for this award must be dated September 30, 1981, or earlier.

(c) **CW:** Contacts must be made using CW since January 1, 1975. Confirmations for cross-mode contacts for this award must

be dated September 30, 1981, or earlier.

(d) **RTTY:** Contacts must be made using radiotele▪ since November 15, 1945. (Baudot, ASCII, AMTOR and pa▪ count as RTTY.) Confirmations for cross-mode contacts for ▪ award must be dated September 30, 1981, or earlier.

(e) **160 Meter:** Contacts must be made on 160 me▪ since November 15, 1945.

(f) **80 Meter:** Contacts must be made on 80 meters s▪ November 15, 1945.

(g) **40 Meter:** Contacts must be made on 40 meters s▪ November 15, 1945.

(h) **10 Meter:** Contacts must be made on 10 meters s▪ November 15, 1945.

(i) **6 Meter:** Contacts must be made on 6 meters s▪ November 15, 1945

(j) **2 Meter:** Contacts must be made on 2 meters s▪ November 15, 1945.

(k) **Satellite:** Contacts must be made using satellites s▪ March 1, 1965. Confirmations must indicate satellite QSO.

(l) **Five-Band DXCC (5BDXCC):** The 5BDXCC cer▪ cate is available for working and confirming 100 current D▪ countries (deleted countries don't count for this award) on ▪ of the following five bands: 80, 40, 20, 15, and 10 Meters. ▪ tacts are valid from November 15, 1945. The 5BDXC▪ endorsable for these additional bands: 160, 17, 12, 6, a▪ meters. 5BDXCC qualifiers are eligible for an individu▪ engraved plaque (at a charge of $25.00 US plus shipping).

(m) **Honor Roll:** Attaining the DXCC Honor Roll re▪ sents the pinnacle of DX achievement:

*Mixed—To qualify, you must have a total confir▪ country count that places you among the numerical top ten D▪ countries total on the current DXCC Countries List (exampl▪ there are 326 current DXCC countries, you must have at ▪ 317 countries confirmed).

*Phone—same as Mixed.

*CW—same as Mixed.

To establish the number of DXCC country credits neede▪ qualify for the Honor Roll, the maximum possible number of ▪ rent countries available for credit is published monthly in ▪ First-time Honor Roll members are recognized monthly in ▪ Complete Honor Roll standings are published annually in ▪ usually in the July issue. See DXCC Notes in *QST* for spe▪ information on qualifying for this Honor Roll standings list. ▪ recognized on this list or in a subsequent monthly update of ▪ members, you retain your Honor Roll standing until the next st▪ ings list is published. In addition, Honor Roll members are re▪ nized in the DXCC Annual List for those who have been liste▪ the previous Honor Roll listings or have gained Honor Roll s▪ in a subsequent monthly listing. Honor Roll qualifiers receiv▪ Honor Roll endorsement sticker for their DXCC certificate an▪ eligible for an Honor Roll lapel pin ($5) and an Honor Roll pl▪ ($25 plus shipping). Write the DXCC Desk for details.

#1 Honor Roll: To qualify for a Mixed, Phone or CW Number One plaque, you must have worked every country on the current DXCC Countries List. Write the DXCC Desk for details.

2) Written proof (confirmations, ie, QSL cards) of having made two-way communication must be submitted directly to ARRL Headquarters for all DXCC countries claimed. Photocopies and electronically transmitted confirmations (including, but not limited to FAX, telex and telegram) are not acceptable for DXCC purposes. Applicants for their first DXCC award may have the cards checked by ARRL DXCC Field Representatives—see Section V for details. The use of the official DXCC application forms or an approved facsimile (eg, produced by a computer program) is required. Complete application materials are available from ARRL Headquarters. Confirmations for a total of 100 or more countries must be included with your first application. By ARRL Board of Directors action, 10-MHz confirmations are creditable to the Mixed, CW and RTTY awards only.

3) The ARRL DXCC Countries List criteria will be used in determining what constitutes a DXCC country.

4) Confirmation data for two-way communications (ie, contacts) must include the call signs of both stations, the country, mode, and date, time and frequency band.

5) Endorsement stickers for affixing to certificates or pins will be awarded as additional DXCC credits are granted. For the Mixed, Phone, CW, RTTY and 10-Meter DXCC, these stickers are in exact multiples of 25, ie 125, 150, etc, between 100 and 350 DXCC countries; in multiples of 10 between 250 and 300, and in multiples of 5 above 300 DXCC countries. For 160-Meter, 80-Meter, 40-Meter, 6-Meter, 2-Meter and Satellite DXCC, the stickers are in exact multiples of 10 starting at 100 and multiples of 5 above 200. Confirmations for DXCC countries may be submitted for credits in any increment. (See #15 for applicable fees, if any.)

6) All contacts must be made with amateur stations working in the authorized amateur bands or with other stations licensed or authorized to work amateurs. Contacts made through "repeater" devices or any other power relay method (aside from Satellite DXCC) are invalid for DXCC credit.

7) Any Amateur Radio operation should take place only with the complete approval and understanding of appropriate administration officials. In countries where amateurs are licensed in the normal manner, credit may be claimed only for stations using regular government-assigned call signs or portable call signs where reciprocal agreements exist or the host government has so authorized portable operation. No credit may be claimed for contacts with stations in any country that has temporarily or permanently closed down Amateur Radio operations by special government edict where amateur licenses were formerly issued in the normal manner. Some countries, in spite of such prohibitions, issue authorizations which are acceptable.

8) All stations contacted must be "land stations." Contacts with ships and boats, anchored or under way, and airborne aircraft, cannot be counted.

9) All stations must be contacted from the same DXCC country.

10) All contacts must be made by the same station licensee. However, contacts may have been made under different call signs in the same country if the licensee for all was the same. That is, you may simultaneously feed one DXCC from several call signs held, as long as the provisions of Rule 9 are met.

11) Any altered, forged, or otherwise invalid confirmations submitted by an applicant for DXCC credit may result in disqualification of the applicant. Any holder of a DXCC award submitting altered, forged or otherwise invalid confirmations may forfeit the right to continued DXCC membership. The ARRL Awards Committee shall rule in these matters and may also determine the eligibility of any DXCC applicant who was ever barred from DXCC to reapply and the conditions of such application.

12) Operations Ethics:

(a) Fair play and good sportsmanship in operating are required of all DXCC members. In the event of specific objections relative to continued poor operating ethics, an individual may be disqualified from DXCC by action on the ARRL Awards Committee.

(b) Credit for contacts with individuals who have displayed continued poor operating ethics may be disallowed by action of the ARRL Awards Committee.

(c) For (a) and (b) above, "operating" includes confirmation procedures and/or documentation submitted for DXCC accreditation.

13) Each DXCC applicant must stipulate that he/she has observed all DXCC rules as well as all pertinent governmental regulations established for Amateur Radio in the country or countries concerned, and agrees to be bound by the decisions of the ARRL Awards Committee. Decisions of the ARRL Awards Committee regarding interpretations of the rules here printed or later amended shall be final.

14) All DXCC applications (both new and endorsements) must include sufficient funds to cover the cost of returning all confirmations (QSL cards) via the method chosen. Funds must be in US dollars, utilizing US currency, check or money order made payable to the ARRL, or International Reply Coupons (IRCs). Address all correspondence and inquiries relating to the various DXCC awards and all applications to: ARRL Headquarters, DXCC Desk, 225 Main St, Newington, CT 06111, USA.

15) Effective January 1, 1994, all amateurs applying for their very first DXCC Award will be charged a one-time Registration Fee of $10.00. This same fee applies to both ARRL members and foreign non-members, and both will receive one DXCC certificate and a DXCC pin. Applicants must provide funds for Postage charges for QSL return.

(a) A $5.00 Shipping and handling fee ($10.00 effective January 1, 1997) will be charged for each additional DXCC certificate issued, whether new or replacement. A DXCC pin will be included with each certificate.

(b) Endorsements and new applications may be presented at ARRL HQ, and at certain ARRL conventions. When presented in this manner, such applications shall be limited to 120 cards maximum, and a $2.00 handling charge will apply.

(c) Each ARRL member will be allowed one submission in each calendar year at no cost (except as in (b) above, or return postage). This annual submission may include up to 120 QSL cards for any number of DXCC Awards, and may be a combination of new and endorsement applications. Fees as in (a) above will apply for additional new DXCC Awards.

(d) A $0.10 fee will be charged for each QSO credited beyond the limits described in (b) and (f).

(e) Foreign non-ARRL members will be allowed the same annual submission as ARRL members, however, they will be charged a $10.00 DXCC Award fee, in addition to return postage charges. Fees in (a), (b) and (d) may also apply.

(f) DXCC participants who wish to submit more than once per year will be charged a DXCC fee for each additional submission made during the remainder of the calendar year. These fees are dependent upon membership status: ARRL Members: $10.00 (for the first 100 cards) Foreign non-members: $20.00 (for the first 100 cards). Additionally, return postage must be provided by applicant, and charges from (a), (b) and (d) above may be applied.

16) The ARRL DX Advisory Committee (DXAC) requests your comments and suggestions for improving DXCC. Address correspondence, including petitions for new country consideration, to ARRL Headquarters, DXAC, 225 Main St, Newington, CT 06111, USA. You may send e-mail to: dxac@arrl.org.

SECTION II. COUNTRIES LIST CRITERIA

The ARRL DXCC Countries List is the result of progressive changes in DXing since 1945. The full list will not necessarily conform completely with current criteria since some of the listings were recognized from pre-WWII or were accredited from earlier versions of the criteria. While the general policy has remained the same, specific mileages in Point 2(a) and Point 3, mentioned in the following criteria, have been used in considerations made April 1960 and after. The specific mileage in Point 2(b) has been used in considerations made April 1963 and after.

When an area in question meets *at least one* of the following three points, it is eligible as a separate country listing for the DXCC Countries List. These criteria address considerations by virtue of Government [Point 1] or geographical separation [Points 2 and 3], while Point 4 addresses ineligible areas. All distances are given in statute miles.

Point 1, GOVERNMENT

An independent country or nation-state having *sovereignty* (that is, a body politic or society united together, occupying a definite territory and having a definite population, politically organized and controlled under one exclusive regime, and engaging in foreign relations—including the capacity to carry out obligations of international law and applicable international agreements) constitutes a separate DXCC country by reason of **Government.** This *may* be indicated by membership in the United Nations (UN). However, some nations that possess the attributes of sovereignty are *not* members of the UN, although these nations may have been *recognized* by a number of UN-member nations. Recognition is the formal act of one nation committing itself to treat an entity as a sovereign state. There are some entities that have been admitted to the UN that lack the requisite attributes of sovereignty and, as a result, are *not* recognized by a number of UN-member nations.

Other entities which are not totally independent may also be considered for separate DXCC country status by reason of Government. Included are Territories, Protectorates, Dependencies, Associated States, and so on. Such an entity may delegate to another country or international organization a measure of its authority (such as the conduct of its foreign relations in whole or in part, or other functions such as customs, communications or diplomatic protection) *without* surrendering its sovereign status. DXCC country status for such an entity is individually considered, based on all the available facts in the particular case. In making a reasonable

determination as to whether a sufficient degree of sovere[ignty] exists for DXCC purposes, the following characteristics (lis[t] necessarily all-inclusive) are taken into consideration:

(a) Membership in specialized agencies of the UN, su[ch as] the International Telecommunications Union (ITU).

(b) Authorized use of ITU-assigned call sign prefixes

(c) Diplomatic relations (entering into international a[gree]ments and/or supporting embassies and consulates), and m[ain]taining a standing army.

(d) Regulation of foreign trade and commerce, cust[oms] immigration and licensing (including landing and operating [per]mits), and the issuance of currency and stamps.

An entity that qualifies under Point 1, but consists of tw[o or] more separate land areas, will be considered a single DX[CC] country (since none of these areas alone retains an indepen[dent] capacity to carry out the obligations of sovereignty), *unles[s]* areas can qualify under Points 2 or 3.

Point 2, SEPARATION BY WATER

An island or a group of islands which is part of a DX[CC] country established by reason of **Government,** Point 1, is [con]sidered as a separate DXCC country under the following co[ndi]tions:

(a) The island or islands are situated off shore, geogra[phi]cally separated by a minimum of 225 miles of open water fr[om a] continent, another island or group of islands that make up [a] part of the "parent" DXCC country.

For any *additional* island or islands to qualify as an a[ddi]tional separate DXCC country or countries, such must qu[alify] under Point 2(b).

(b) This point applies to the "second" island or is[land] grouping geographically separated from the "first" DXCC c[oun]try created under Point 2(a). For the second island or is[land] grouping to qualify, at least a 500-mile separation of open w[ater] from the first is required, as well as meeting the 225-mile req[uire]ment of (a) from the "parent". For any subsequent island([s) to] qualify, the 500-mile separation would again have to be met. [This] precludes, for example, using the 225-mile measurement *for e[ach]* of several islands from the parent country to make several DX[CC] countries.

(c) An island is defined as a naturally formed area of [land] surrounded by water, the surface of which is above water at [high] tide. Rocks which cannot sustain human habitation shall n[ot be] considered for DXCC country status.

(d) An island must meet or exceed size standards. T[o be] eligible for consideration, the island must be visible, and na[med] on a chart with a resolution no greater than 1:1,000,000. Ch[arts] used must be from recognized national mapping agencies. [The] island must consist of a single unbroken piece of land not [less] than 10,000 square feet in area, which is above water at high [tide.] The area requirements should be demonstrated by a chart [of] appropriate resolution.

Point 3, SEPARATION BY ANOTHER DXCC COUNTRY

(a) Where a Point 1 DXCC country, composed of on[e or] more continental land areas or of continental land areas [and] islands, is totally separated by an intervening DXCC country [into] two land areas which are at least 75 miles apart, two DXCC c[ountries] es result. This distance is measured along the great circle tween the two closest points of the two areas divided. The mea- red distance may include inland lakes and seas which are part the intervening DXCC country. The test for total separation o two areas requires that a great circle cannot be drawn from y point on the continental land and/or islands of one area to y point on the continental land and/or islands of the other area thout intersecting any land of the intervening DXCC country.

(b) Where a Point 1 DXCC country, composed entirely of ands, is totally separated by an intervening DXCC country into o areas, then two DXCC countries result. No minimal distance is quired for the separation. The test for total separation into two eas requires that a great circle cannot be drawn from any point on y island of one area to any point on any island of the other area thout intersecting any land of the intervening DXCC country.

int 4, INELIGIBLE AREAS

(a) Any area which is unclaimed or unowned by any rec- nized government does not count as a separate DXCC country.

(b) Any area which is classified as a Demilitarized Zone, Neutral ne or Buffer Zone does not count as a separate DXCC country.

(c) The following do not count as a separate DXCC coun- from the host country: Embassies, consulates and extra-terri- rial legal entities of any nature, including, but not limited to, onuments, offices of the United Nations agencies or related ganizations, other inter-governmental organizations or diplo- atic missions.

CTION III. DELETION CRITERIA

A DXCC country is subject to deletion from the ARRL XCC Countries List if political change causes it to cease to eet Point 1 of the Countries List Criteria (a derivative of such ange may cause it to cease to meet Points 2 or 3) or if it falls to Point 4 of the criteria. Additions to and deletions from the XCC Countries List come about as a result of a myriad of such litical changes. Reviewing the nature of the changes which ve occurred since 1945 as they affect DXCC, these changes n be grouped into categories as follows:

(a) **Annexation.** When an area that has been recognized as separate country under Point 1 is annexed or absorbed by an jacent Point 1 country, the annexed area becomes a deleted untry. Examples: India annexed Sikkim (AC3); China annexed bet (AC4); Indonesia annexed Portuguese Timor (CR8).

(b) **Unification.** When two or more entities that have been parate DXCC countries under Point 1 unite or combine into a ngle entity under a common administration, one new DXCC untry is created and two or more DXCC countries have been leted. Example: Italian Somaliland (I5) plus British Somaliland Q6) became Somalia (6O/T5).

(c) **Partition.** When one country is divided or partitioned to two or more countries, one DXCC country is deleted and o or more DXCC countries are created. Example: French Equa- rial Africa (FQ) was deleted and replaced by Central Africa L), Congo (TN), Gabon (TR) and Chad (TT). The partition cat- ory is *not* employed when the original political entity contin- s in some form. That is, if part of country A splits off to form untry B, the original DXCC country (A) is retained and one w DXCC country (B) is added. Examples: the British vereign Bases on Cyprus (ZC4); Aruba (P4).

(d) **Independence**. Mere independence does not result in Countries List deletion. Examples: the Tonga Islands, then a ritish protectorate (VR5), is the same country as the present

listing of the Kingdom of Tonga (A3). Further, an entity already recognized as a separate DXCC country is *not* deleted because of a change in its independent status. Bangladesh (S2) is the same listing as East Pakistan (AP), which was already separate from West Pakistan by virtue of Point 3. Also, a country that merely changes its name (such as when Upper Volta became Burkina Faso) does not change its basic status as a DXCC country on the DXCC Countries List.

SECTION IV. ACCREDITATION CRITERIA

1) The many vagaries of how each nation manages its telecommunications matters does not lend itself to a hard set of rules that can be applied across the board in accrediting all Ama- teur Radio DX operations. However, during the course of more than 40 years of DXCC administration, basic standards have evolved in determining whether a DX operation meets the test of legitimate operation. The intent is to assure that DXCC credit is given only for contacts with operations that are conducted appro- priately in two respects: (1) proper licensing; and (2) physical presence in the country to be credited.

2) The following points should be of particular interest to those seeking accreditation for a DX operation:

(a) The vast majority of operations are accredited routine- ly without any requirement for submitting authenticating docu- mentation.

(b) In countries where Amateur Radio operation has not been permitted or has been suspended or where some reluctance to license amateur stations has been evidenced, authenticating documents *may* be required prior to accrediting an operation.

(c) Some DXCC countries, even though part of a country with no Amateur Radio restrictions, nevertheless require the permission of a governmental agency or private party prior to conducting Ama- teur Radio operations on territory within their jurisdiction. Exam- ples: Desecheo I. (KP5); Palmyra I. (KH5); Kingman Reef (KH5K).

3) In those cases where supporting documentation is required, the following should be used as a guide as to what infor- mation may be necessary to make a reasonable determination of the validity of the operation:

(a) Photocopy of license or operating authorization.

(b) Photocopy of passport entry and exit stamps.

(c) For islands, a landing permit and/or signed statement of the transporting ship's, boat's, or aircraft's captain, showing all pertinent data, such as date, place of landing, etc.

(d) For some locations where special permission is known to be required to gain access, evidence of this permission having been given is required.

4) These accreditation requirements are intended to preserve the DXCC program's integrity and to ensure that the program does not encourage amateurs to "bend the rules" in their enthusi- asm, thus jeopardizing the future development of Amateur Radio. Every effort will be made to apply these criteria in a uniform manner in conformity with these objectives.

SECTION V. FIELD CHECKING OF QSL CARDS

QSL cards for new DXCC awards and endorsements may be checked by two DXCC Field Representatives. This program applies to any DXCC award for an individual or station. Specifi- cally excluded from this program are 5BDXCC, 6-meter, 2-meter and Satellite DXCC, and all 160-meter QSLs.

1) Countries Eligible for Field Checking:

(a) Eligible countries will be indicated in the ARRL DXCC Countries List, and are subject to change. Only cards from these eligible countries may be checked by DXCC Field Representatives. QSLs for other DXCC countries must be submitted directly to ARRL Headquarters.

(b) The ARRL Awards Committee determines which countries are eligible for Field Checking.

2) DXCC Field Representatives:

(a) DXCC Field Representatives must be ARRL members who have a DXCC award endorsed for at least 300 countries.

(b) To become a DXCC Field Representative, a person must be nominated by a DX club. (A DX club is an ARRL-affiliated club with at least 25 members who are DXCC members and which has, as its primary interest, DX. If there are any questions regarding the validity of a DX club, the issue shall be determined by the Division Director where the DX club is located.) A person does not have to be a member to be nominated by a DX club.

(c) DXCC Field Representatives are approved by the Director of the ARRL Division in which they reside and appointed by the President of the ARRL.

(d) DXCC Field Representative appointments must be renewed annually.

3) Card Checking Process:

(a) Only cards from the list of eligible countries can be checked by DXCC Field Representatives. An application for a new award shall contain a minimum of 100 QSL confirmations from the list and shall not contain any QSLs from countries that are not on the list of eligible countries. Additional cards should not be sent to HQ with field checked applications. The application may contain the maximum number of countries that appear on the list of eligible countries. That is, if there are 245 countries on the list, field-checked applications could contain 245 countries.

(b) It is the applicant's responsibility to get cards to and from the DXCC Field Representatives.

(c) Field Representatives may, at their own discretion, handle members' cards by mail.

(d) The ARRL is not responsible for cards handled by DXCC Field Representatives and will not honor any claims.

(e) The QSL cards must be checked by two DXCC Field Representatives.

(f) The applicant and both DXCC Field Representatives must sign the application form. (See SECTION I no. 11 regarding altered, forged or otherwise invalid confirmations.)

(g) The applicant shall provide a stamped no. 10 envelope (business size) addressed to ARRL HQ to the DXCC Field Representatives. The applicant shall also provide the applicable fees (check or money order payable to ARRL—no cash).

(h) The DXCC Field Representatives will forward completed applications and appropriate fee(s) to ARRL HQ.

4) ARRL HQ involvement in the card checking process:

a) ARRL HQ staff will receive field-checked applications, enter application data into DXCC records and issue DXCC credits and awards as appropriate.

(b) ARRL HQ staff will perform random audits of applications. Applicants or members may be requested to forward cards to HQ for checking before or after credit is issued.

c) The applicant and both DXCC Field Representatives will be advised of any errors or discrepancies encountered by ARRL staff.

(d) ARRL HQ staff provides instructions and guidelines to DXCC Field Representatives.

5) Applicants and DXCC members may send cards to

Recent changes to the DXCC List

Prefix	Country	Published DXC List Changed
Added to the DXCC Country List:		
7O	Yemen	Mar 91
ZSØ, 1	Penguin Is.	Sep 91
9A, YU2	Croatia	Jan 93
S5, YU3	Slovenia	Jan 93
T9, 4N4, 4O4, YU4	Bosnia-Herzegovina	Jan 93
Z3, 4N5, YU5	Macedonia (former Yugoslav Rep.)	Jun 93
OK-OL	Czech Rep.	Jun 93
OM	Slovak Rep.	Jun 93
P5	North Korea	Oct 95
BV9P	Pratas I.	Apr 96
BS7	Scarborough Reef	Apr 96
Moved from deleted list to active status:		
E3	Eritrea	Feb 94
Moved to deleted list:		
DM, Y2-9	German Democratic Rep	Mar 91
4W	Yemen Arab Rep (North)	Mar 91
7O	PDR of Yemen (South)	Mar 91
—	Abu Ail	Jun 93
OK-OM	Czechoslovakia	Jun 93
ZSØ, 1	Penguin Is.	Jul 94
ZS9	Walvis Bay	Jul 94

* DXCC List edition date when the change first appear See notes in the countries list for effective dates.

Rules changes:	
SECTION II, Point 3	Sep 91
SECTION V (added)	Sep 91
SECTION V, 2), (d) new wording	Jan 92
SECTION I, 1), (l) 5BDXCC Start Date	Jun 93
SECTION I, 15) Fees change	Feb 94
SECTION I, 15), (c) "Walk in" card limit	Feb 94
SECTION V, field checking for any new DXCCs expanded	Feb 94
SECTION V, 160-m QSLs ineligible for field checking	Jul 94
SECTION I, 15), new fees	Apr 96
SECTION V, endorsements may be field checked	Apr 96

Note: There were several changes to countries eligible field checking made effective January 1993.

In the February 1994 list, several "U" block prefi changed. See Prefix Cross References section t follows the Countries List.

ARRL Headquarters at any time for review or recheck if the vidual feels that an incorrect determination has been made.

ARRL DXCC COUNTRIES LIST

NOTE: • INDICATES COUNTRIES ELIGIBLE FOR FIELD CHECKING.

NOTE: * INDICATES CURRENT LIST OF COUNTRIES FOR WHICH QSLs MAY BE FORWARDED BY THE ARRL MEMBERSHIP OUTGOING QSL SERVICE.

NOTE: † INDICATES COUNTRIES WITH WHICH U.S. AMATEURS MAY LEGALLY HANDLE THIRD-PARTY MESSAGE TRAFFIC.

Prefix	Country	CONTINENT	ZONE ITU	ZONE CQ	MIXED	PHONE	CW	RTTY	SAT	160	80	40	20	17	15	12	10
1AØ[1]	Sov. Mil. Order of Malta	EU	28	15													
1S[1]	Spratly Is.	AS	50	26													
3A*	Monaco	EU	27	14													
3B6, 7*	Agalega & St. Brandon	AF	53	39													
3B8*	Mauritius	AF	53	39													
3B9*	Rodriguez I.	AF	53	39													
3C	Equatorial Guinea	AF	47	36													
3CØ	Annobon I.	AF	52	36													
3D2*	Fiji	OC	56	32													
3D2*	Conway Reef	OC	56	32													
3D2*	Rotuma I.	OC	56	32													
3DA†*	Swaziland	AF	57	38													
3V	Tunisia	AF	37	33													
3W, XV	Vietnam	AS	49	26													
3X	Guinea	AF	46	35													
3Y*	Bouvet	AF	67	38													
3Y*	Peter I I.	AN	72	12													
4J, 4K*	Azerbaijan	AS	29	21													
4L*	Georgia	AS	29	21													
4P-4S*	Sri Lanka	AS	41	22													
4U_ITU†*	ITU HQ	EU	28	14													
4U_UN*	United Nations HQ	NA	08	05													
4X, 4Z†*	Israel	AS	39	20													
5A	Libya	AF	38	34													
5B*	Cyprus	AS	39	20													
5H-5I*	Tanzania	AF	53	37													
5N-5O*	Nigeria	AF	46	35													
5R-5S	Madagascar	AF	53	39													
5T[2]	Mauritania	AF	46	35													
5U[3]	Niger	AF	46	35													
5V*	Togo	AF	46	35													
5W*	Western Samoa	OC	62	32													

	Prefix	Country	CONTINENT	ITU	CQ	MIXED	PHONE	CW	RTTY	SAT	160	80	40	20	17	15	12	10				
	5X*	Uganda	AF	48	37																	
●	5Y-5Z*	Kenya	AF	48	37																	
●	6V-6W4*	Senegal	AF	46	35																	
●	6Y†*	Jamaica	NA	11	08																	
	7O5	Yemen	AS	39	21, 37																	
●	7P*	Lesotho	AF	57	38																	
●	7Q	Malawi	AF	53	37																	
●	7T-7Y*	Algeria	AF	37	33																	
●	8P*	Barbados	NA	11	08																	
●	8Q	Maldives	AS/AF	41	22																	
●	8R†*	Guyana	SA	12	09																	
	9A6*	Croatia	EU	28	15																	
	9G7†*	Ghana	AF	46	35																	
●	9H*	Malta	EU	28	15																	
●	9I-9J*	Zambia	AF	53	36																	
●	9K*	Kuwait	AS	39	21																	
●	9L†*	Sierra Leone	AF	46	35																	
●	9M2, 48*	West Malaysia	AS	54	28																	
●	9M6, 88*	East Malaysia	OC	54	28																	
●	9N	Nepal	AS	42	22																	
●	9Q-9T	Zaire	AF	52	36																	
	9U9	Burundi	AF	52	36																	
●	9V10*	Singapore	AS	54	28																	
●	9X9	Rwanda	AF	52	36																	
●	9Y-9Z†*	Trinidad & Tobago	SA	11	09																	
●	A2*	Botswana	AF	57	38																	
●	A3*	Tonga	OC	62	32																	
●	A4*	Oman	AS	39	21																	
	A5	Bhutan	AS	41	22																	
●	A6	United Arab Emirates	AS	39	21																	
●	A7*	Qatar	AS	39	21																	
●	A9*	Bahrain	AS	39	21																	
●	AP-AS*	Pakistan	AS	41	21																	
	BS711	Scarborough Reef	AS	50	27																	
●	BV*	Taiwan	AS	44	24																	
	BV9P12	Pratas I.	AS	44	24																	
●	BY,BT*	China	AS	(A)	23,24																	
●	C2*	Nauru	OC	65	31																	
●	C3*	Andorra	EU	27	14																	
●	C5†*	The Gambia	AF	46	35																	

Prefix	Country	Continent	ITU	CQ	MIXED	PHONE	CW	RTTY	SAT	160	80	40	20	17	15	12	10		
			ZONE																
C6*	Bahamas	NA	11	08															
C8-9*	Mozambique	AF	53	37															
CA-CE†*	Chile	SA	14,16	12															
CE0†*	Easter I.	SA	63	12															
CE0†*	Juan Fernandez Is.	SA	14	12															
CE0†*	San Felix & San Ambrosio	SA	14	12															
CE9/KC4▲*	Antarctica	AN	(B)	(C)															
CM, CO†*	Cuba	NA	11	08															
CN*	Morocco	AF	37	33															
CP†*	Bolivia	SA	12,14	10															
CT*	Portugal	EU	37	14															
CT3*	Madeira Is.	AF	36	33															
CU*	Azores	EU	36	14															
CV-CX†*	Uruguay	SA	14	13															
CY0*	Sable I.	NA	09	05															
CY9*	St. Paul I.	NA	09	05															
D2-3	Angola	AF	52	36															
D4*	Cape Verde	AF	46	35															
D6†13*	Comoros	AF	53	39															
DA-DL14*	Fed. Rep. of Germany	EU	28	14															
DU-DZ†*	Philippines	OC	50	27															
E315	Eritrea	AF	48	37															
EA-EH*	Spain	EU	37	14															
EA6-EH6*	Balearic Is.	EU	37	14															
EA8-EH8*	Canary Is.	AF	36	33															
EA9-EH9*	Ceuta & Melilla	AF	37	33															
EI-EJ*	Ireland	EU	27	14															
EK*	Armenia	AS	29	21															
EL†*	Liberia	AF	46	35															
EP-EQ	Iran	AS	40	21															
ER*	Moldova	EU	29	16															
ES*	Estonia	EU	29	15															
ET*	Ethiopia	AF	48	37															
EU, EV, EW*	Belarus	EU	29	16															
EX*	Kyrgyzstan	AS	30,31	17															
EY*	Tajikistan	AS	30	17															
EZ*	Turkmenistan	AS	30	17															
F*	France	EU	27	14															
FG*	Guadeloupe	NA	11	08															
FJ, FS1*	Saint Martin	NA	11	08															

Prefix	Country	CONTINENT	ITU	CQ	MIXED	PHONE	CW	RTTY	SAT	160	80	40	20	17	15	12	10
FH[11]*	Mayotte	AF	53	39													
FK*	New Caledonia	OC	56	32													
FM*	Martinique	NA	11	08													
FO*	Clipperton I.	NA	10	07													
FO*	French Polynesia	OC	63	32													
FP*	St. Pierre & Miquelon	NA	09	05													
FR/G[16]*	Glorioso Is.	AF	53	39													
FR/J, E[16]*	Juan de Nova, Europa	AF	53	39													
FR*	Reunion	AF	53	39													
FR/T*	Tromelin I.	AF	53	39													
FT8W*	Crozet I.	AF	68	39													
FT8X*	Kerguelen Is.	AF	68	39													
FT8Z*	Amsterdam & St. Paul Is.	AF	68	39													
FW*	Wallis & Futuna Is.	OC	62	32													
FY*	French Guiana	SA	12	09													
G, GX*#	England	EU	27	14													
GD, GT*	Isle of Man	EU	27	14													
GI, GN*	Northern Ireland	EU	27	14													
GJ, GH*	Jersey	EU	27	14													
GM, GS*	Scotland	EU	27	14													
GU, GP*	Guernsey	EU	27	14													
GW, GC*	Wales	EU	27	14													
H4*	Solomon Is.	OC	51	28													
HA, HG*	Hungary	EU	28	15													
HB*	Switzerland	EU	28	14													
HBØ*	Liechtenstein	EU	28	14													
HC-HD†*	Ecuador	SA	12	10													
HC8-HD8†*	Galapagos Is.	SA	12	10													
HH†*	Haiti	NA	11	08													
HI†*	Dominican Republic	NA	11	08													
HJ-HK†*	Colombia	SA	12	09													
HKØ†*	Malpelo I.	SA	12	09													
HKØ†*	San Andres & Providencia	NA	11	07													
HL, DS-DT, 6K-6N†	South Korea	AS	44	25													
HO-HP†*	Panama	NA	11	07													
HQ-HR†*	Honduras	NA	11	07													
HS, E2*	Thailand	AS	49	26													
HV*	Vatican	EU	28	15													
HZ*	Saudi Arabia	AS	39	21													
I*	Italy	EU	28	15, 33													

#Third-party traffic permitted with special-events stations in the United Kingdom having the prefix GB *only*, with the exception that GB3 stations are not included in this agreement.

| Prefix | Country | CONTINENT | ZONE | | MIXED | PHONE | CW | RTTY | SAT | 160 | 80 | 40 | 20 | 17 | 15 | 12 | 10 | |
			ITU	CQ														
IS0, IM0 *	Sardinia	EU	28	15														
J2*	Djibouti	AF	48	37														
J3†*	Grenada	NA	11	08														
J5	Guinea-Bissau	AF	46	35														
J6†*	St. Lucia	NA	11	08														
J7†*	Dominica	NA	11	08														
J8†*	St. Vincent	NA	11	08														
JA-JS*	Japan	AS	45	25														
JD1[17]*	Minami Torishima	OC	90	27														
JD1[18]*	Ogasawara	AS	45	27														
JT-JV*	Mongolia	AS	32,33	23														
JW*	Svalbard	EU	18	40														
JX*	Jan Mayen	EU	18	40														
JY†*	Jordan	AS	39	20														
K,W, N, AA-AK†	United States of America	NA	6,7,8	3,4,5														
KC6, T8[19]	Pelau (W. Caroline Is.)	OC	64	27														
KG4†*	Guantanamo Bay	NA	11	08														
KH0†	Mariana Is.	OC	64	27														
KH1†	Baker & Howland Is.	OC	61	31														
KH2†*	Guam	OC	64	27														
KH3†*	Johnston I.	OC	61	31														
KH4†	Midway I.	OC	61	31														
KH5†	Palmyra & Jarvis Is.	OC	61,62	31														
KH5K†	Kingman Reef	OC	61	31														
KH6, 7†*	Hawaii	OC	61	31														
KH7K†	Kure I.	OC	61	31														
KH8†	American Samoa	OC	62	32														
KH9†	Wake I.	OC	65	31														
KL7†*	Alaska	NA	1, 2	1														
KP1†	Navassa I.	NA	11	08														
KP2†*	Virgin Is.	NA	11	08														
KP3, 4†*	Puerto Rico	NA	11	08														
KP5[20]†	Desecheo I.	NA	11	08														
LA-LN*	Norway	EU	18	14														
LO-LW†*	Argentina	SA	14,16	13														
LX*	Luxembourg	EU	27	14														
LY*	Lithuania	EU	29	15														
LZ*	Bulgaria	EU	28	20														
OA-OC†*	Peru	SA	12	10														
OD*	Lebanon	AS	39	20														

Prefix	Country	CONTINENT	ZONE ITU	ZONE CQ	MIXED	PHONE	CW	RTTY	SAT	160	80	40	20	17	15	12	10
● OE*	Austria	EU	28	15													
● OF-OI*	Finland	EU	18	15													
● OHØ*	Aland Is.	EU	18	15													
● OJØ, OHØM*	Market Reef	EU	18	15													
OK-OL21*	Czech Rep.	EU	28	15													
OM21*	Slovak Rep.	EU	28	15													
● ON-OT*	Belgium	EU	27	14													
● OX*	Greenland	NA	5, 75	40													
● OY*	Faroe Is.	EU	18	14													
● OZ*	Denmark	EU	18	14													
● P222*	Papua New Guinea	OC	51	28													
● P423*	Aruba	SA	11	09													
P524	North Korea	AS	44	25													
● PA-PI*	Netherlands	EU	27	14													
● PJ2, 4, 9*	Bonaire,Curacao(Neth. Antilles)	SA	11	09													
● PJ5-8*	St.Maarten,Saba,St.Eustatius	NA	11	08													
● PP-PY†*	Brazil	SA	(D)	11													
● PPØ-PYØF†*	Fernando de Noronha	SA	13	11													
PPØ-PYØS†*	St. Peter & St. Paul Rocks	SA	13	11													
● PPØ-PYØT†*	Trindade & Martim Vaz Is.	SA	15	11													
● PZ*	Suriname	SA	12	09													
● R1FJ*	Franz Josef Land	EU	75	40													
R1MV	Malyj Vysotskij I	EU	29	16													
SØ1,25*	Western Sahara	AF	46	33													
S2*	Bangladesh	AS	41	22													
S56*	Slovenia	EU	28	15													
● S7	Seychelles	AF	53	39													
● S9*	Sao Tome & Principe	AF	47	36													
● SA-SM*	Sweden	EU	18	14													
● SN-SR*	Poland	EU	28	15													
● ST*	Sudan	AF	47, 48	34													
● STØ*	Southern Sudan	AF	47, 48	34													
● SU*	Egypt	AF	38	34													
● SV-SZ*	Greece	EU	28	20													
SV/A*	Mount Athos	EU	28	20													
● SV5*	Dodecanese	EU	28	20													
● SV9*	Crete	EU	28	20													
● T226	Tuvalu	OC	65	31													
● T3Ø	W. Kiribati (Gilbert Is.)	OC	65	31													
● T31	C. Kiribati (Brit. Phoenix Is.)	OC	62	31													

Prefix	Country	Continent	ZONE ITU	ZONE CQ	MIXED	PHONE	CW	RTTY	SAT	160	80	40	20	17	15	12	10		
T32	E. Kiribati (Line Is.)	OC	61,63	31															
T33	Banaba I. (Ocean I.)	OC	65	31															
T5	Somalia	AF	48	37															
T7*	San Marino	EU	28	15															
T9²⁷*	Bosnia-Herzegovina	EU	28	15															
TA-TC*	Turkey	EU/AS	39	20															
TF*	Iceland	EU	17	40															
TG, TD†*	Guatemala	NA	11	07															
TI, TE†*	Costa Rica	NA	11	07															
TI9†*	Cocos I.	NA	11	07															
TJ	Cameroon	AF	47	36															
TK*	Corsica	EU	28	15															
TL²⁸	Central Africa	AF	47	36															
TN²⁹	Congo	AF	52	36															
TR³⁰*	Gabon	AF	52	36															
TT³¹	Chad	AF	47	36															
TU³²*	Côte d'Ivoire	AF	46	35															
TY³³	Benin	AF	46	35															
TZ³⁴*	Mali	AF	46	35															
UA-UI1,3,4,6 RA-RZ*	European Russia	EU	(E)	16															
UA2*	Kaliningrad	EU	29	15															
UA-UI8, 9, Ø RA-RZ*	Asiatic Russia	AS	(F)	(G)															
UJ-UM*	Uzbekistan	AS	30	17															
UN-UQ*	Kazakhstan	AS	29-31	17															
UR-UZ, EM-EO*	Ukraine	EU	29	16															
V2†*	Antigua & Barbuda	NA	11	08															
V3†*	Belize	NA	11	07															
V4³⁵†*	St. Kitts & Nevis	NA	11	08															
V5*	Namibia	AF	57	38															
V6³⁶†	Micronesia (E. Caroline Is.)	OC	65	27															
V7†*	Marshall Is.	OC	65	31															
V8*	Brunei	OC	54	28															
VE, VO, VY†*	Canada	NA	(H)	1-5															
VK†*	Australia	OC	(I)	29,30															
VKØ†*	Heard I.	AF	68	39															
VKØ†*	Macquarie I.	OC	60	30															
VK9C†*	Cocos-Keeling Is.	OC	54	29															
VK9L†*	Lord Howe I.	OC	60	30															
VK9M†*	Mellish Reef	OC	56	30															
VK9N†*	Norfolk I.	OC	60	32															

Prefix	Country	Continent	ITU	CQ	MIXED	PHONE	CW	RTTY	SAT	160	80	40	20	17	15	12	10
● VK9W†*	Willis I.	OC	55	30													
● VK9X†*	Christmas I.	OC	54	29													
● VP2E35*	Anguilla	NA	11	08													
● VP2M35	Montserrat	NA	11	08													
● VP2V35*	British Virgin Is.	NA	11	08													
● VP5*	Turks & Caicos Is.	NA	11	08													
● VP8*	Falkland Is.	SA	16	13													
● VP8, LU*	South Georgia I.	SA	73	13													
● VP8, LU*	South Orkney Is.	SA	73	13													
● VP8, LU*	South Sandwich Is.	SA	73	13													
VP8, LU, CE9, HF0, 4K1*	South Shetland Is.	SA	73	13													
● VP9*	Bermuda	NA	11	05													
● VQ9*	Chagos Is.	AF	41	39													
● VR6†*	Pitcairn I.	OC	63	32													
● VS6, VR2*	Hong Kong	AS	44	24													
● VU*	India	AS	41	22													
● VU*	Andaman & Nicobar Is.	AS	49	26													
● VU*	Lakshadweep Is.	AS	41	22													
● XA-XI†*	Mexico	NA	10	06													
XA4-XI4†*	Revilla Gigedo	NA	10	06													
● XT37*	Burkina Faso	AF	46	35													
XU	Cambodia	AS	49	26													
XW	Laos	AS	49	26													
● XX9	Macao	AS	44	24													
XY-XZ	Myanmar	AS	49	26													
YA	Afghanistan	AS	40	21													
● YB-YH38*	Indonesia	OC	51,54	28													
YI*	Iraq	AS	39	21													
● YJ*	Vanuatu	OC	56	32													
YK*	Syria	AS	39	20													
● YL*	Latvia	EU	29	15													
● YN†*	Nicaragua	NA	11	07													
● YO-YR*	Romania	EU	28	20													
● YS†*	El Salvador	NA	11	07													
● YT-YU, YZ*	Yugoslavia	EU	28	15													
● YV-YY†*	Venezuela	SA	12	09													
YV0†*	Aves I.	NA	11	08													
● Z2*	Zimbabwe	AF	53	38													
Z339*	Macedonia (former Yugoslav Rep.)	EU	28	15													
ZA*	Albania	EU	28	15													

| Prefix | Country | Continent | ZONE | | MIXED | PHONE | CW | RTTY | SAT | 160 | 80 | 40 | 20 | 17 | 15 | 12 | 10 |
			ITU	CQ													
ZB2*	Gibraltar	EU	37	14													
ZC4 [40] *	UK Sov. Base Areas on Cyprus	AS	39	20													
ZD7*	St. Helena	AF	66	36													
ZD8*	Ascension I.	AF	66	36													
ZD9	Tristan da Cunha & Gough I.	AF	66	38													
ZF*	Cayman Is.	NA	11	08													
ZK1*	N. Cook Is.	OC	62	32													
ZK1*	S. Cook Is.	OC	62	32													
ZK2*	Niue	OC	62	32													
ZK3*	Tokelau Is.	OC	62	31													
ZL-ZM*	New Zealand	OC	60	32													
ZL7*	Chatham Is.	OC	60	32													
ZL8*	Kermadec Is.	OC	60	32													
ZL9*	Auckland & Campbell Is.	OC	60	32													
ZP†*	Paraguay	SA	14	11													
ZR-ZU*	South Africa	AF	57	38													
ZS8*	Prince Edward & Marion Is.	AF	57	38													

[1] Unofficial prefix.

[2] (5T) Only contacts made June 20, 1960, and after, count for this country.

[3] (5U) Only contacts made August 3, 1960, and after, count for this country.

[4] (6W) Only contacts made June 20, 1960, and after, count for this country.

[5] (7O) Only contacts made May 22, 1990, and after, count for this country.

[6] (9A, S5) Only contacts made June 26, 1991, and after, count for this country.

[7] (9G) Only contacts made March 5, 1957, and after, count for this country.

[8] (9M2, 4, 6, 8) Only contacts made September 16, 1963, and after, count for this country.

[9] (9U, 9X) Only contacts made July 1, 1962, and after, count for this country.

[10] (9V) Contacts made from September 16, 1963 to August 8, 1965, count for West Malaysia.

[11] (BS7) Only contacts made January 1, 1995, and after, count for this country.

[12] (BV9P) Only contacts made January 1, 1994, and after, count for this country.

[13] (D6, FH8) Only contacts made July 5, 1975, and after, count for this country.

[14] (DA-DL) Only contacts made with DA-DL stations September 17, 1973, and after, and contacts made with Y2-Y9 stations October 3, 1990 and after, count for this country.

[15] (E3) Only contacts made November 14, 1962, and before, or May 24, 1991, and after, count for this country.

[16] (FR) Only contacts made June 25, 1960, and after, count for this country.

[17] (JD) Formerly Marcus Island.

[18] (JD) Formerly Bonin and Volcano Islands.

[19] (KC6) Includes Yap Islands December 31, 1980, and before.

[20] (KP5, KP4) Only contacts made March 1, 1979, and after, count for this country.

[21] (OK-OL, OM) Only contacts made January 1, 1993, and after, count for this country.

[22] (P2) Only contacts made September 16, 1975, and after, count for this country.

[23] (P4) Only contacts made January 1, 1986, and after, count for this country.

[24] (P5) Only contacts made May 14, 1995, and after count for this country.

[25] (SØ) Contacts with Rio de Oro (Spanish Sahara), EA9, also count for this country.

[26] (T2) Only contacts made January 1, 1976, and after, count for this country.

[27] (T9) Only contacts made October 15, 1991 and after count for this counrty.

[28] (TL) Only contacts made August 13, 1960, and after, count for this country.

[29] (TN) Only contacts made August 15, 1960, and after, count for this country.

[30] (TR) Only contacts made August 17, 1960, and after, count for this country.

[31] (TT) Only contacts made August 11, 1960, and after count for this country.

[32] (TU) Only contacts made August 7, 1960, and after, count for this country.

[33] (TY) Only contacts made August 1, 1960, and after, count for this country.

[34] (TZ) Only contacts made June 20, 1960, and after, count for this country.

[35] (V4, VP2) For DXCC credit for contacts made May 31, 1958 and before, see page 97, June 1958 QST.

[36] (V6) Includes Yap Islands January 1, 1981, and after.

[37] (XT) Only contacts made August 5, 1960, and after, count for this country.

[38] (YB) Only contacts made May 1, 1963, and after, count for this country.

[39] (Z3) Only contacts made September 8, 1991, and after, count for this country.

[40] (ZC4) Only contacts made August 16, 1960, and after, count for this country.

▲Also 3Y, 8J1, ATØ, DPØ, FT8Y, LU, OR4, R1AN, VKØ, VP8, ZL5, ZS1, ZXØ, etc. QSL via country under whose auspices the particular station is operating. The availability of a third-party traffic agreement and a QSL Bureau applies to the country under whose auspices the particular station is operating.

Zone Notes can be found with Prefix Cross References.

Atlantic Division

EPA

M.Stokley Benson, W3SB
1192 Divot Drive
Wescosville, PA 18106

Norman Zoltack, K3NZ
4333 Locust Drive
Schnecksville, PA 18078

MDC

Murray Green, K3BEQ
5730 Lockwood Road
Cheverly, MD 20785

Everett C. Bollin, WA3DVO
8000 Ray Leonard Court
Palmer Park, MD 20785

Michael C. Cizek, KO7V
Box 227
Oxon Hill, MD 20750

SNJ

John M. Fisher, K2JF
538 Wesley Avenue
Pitman, NJ 08071-1924

James M. Mollica, Sr. K2OWE
22 Wright Loop
Williamstown, NJ 08094

Steve Branca, K2SB
202 Minnetonka Road
Somerdale, NJ 08083

John D. Imhof, N2VW
3 Seeley Drive
Mt. Holly, NJ 08060

WNY

John C. Yodis, K2VV
PO Box 460
Hagaman, NY 12086

Eugene W. Nadolny, W2FXA
21 Hidden Valley Drive
Elma, NY 14059

Bob Rossi, NA2X
75 Olde Erie Trail
Rochester, NY 14626

Elmer Wagner, WB2BNJ
23 Red Barn Circle
Pittsford, NY 14534

Robert Dow, WB2CJL
115 Puritan Road
Tonawanda, NY 14150

James R. Ciurczak, WB2IVO
10404 Cayuga Drive
Niagara Falls, NY 14304

Robert E. Nadolny, WB2YQH
135 Wetherstone Drive
West Seneca, NY 14224

Stephen A. Licht, WF2S
3268 West Main St Rd
Batavia, 14020-9109

Central Division

IL

Tim Wright, KØBFR
1131 Warren
Alton, IL 62002

Joe Pontek, Sr. K8JP
PO Box 59573
Schaumburg, IL 60159-0573

Jerry Brunning, K9BG
15307 Shamrock Lane
Woodstock, IL 60098

Andrew White, K9CW
1725 Counrty Rd, 2500 N
Thomasboro, IL 61878

Merv Schweigert, K9FD
Rt 2, Box 138A
Red Bud, IL 62278

Robert E. Nielsen, K9RN
1372 Skyridge Drive
Crystal Lake, IL 60014

James J. Coleman, KA6A
250 County Road 700N
Ivesdale, IL 61851

Douglas E. Williams, KD9Q
1717 S. 28th Street
Quincy, IL 62301-6305

Greg Wilson, N4CC
3865 Staunton Road
Edwardsville, IL 62025

Michael L. Nowack, NA9Q
2011 N. Sheridan Dr
Quincy, IL 62301-8929

Ed Doubek, N9RF
25W063 Wood Court
Naperville, IL 60563

Jim Dawson, K9DD
257 S. 9th Street
East Alton, IL 62024

Gary H. Hilker, K9LJN
804 Otto Road
Machesney Park, IL 61115

IN

David Bunte, K9FN
129 Ivy Hill Drive
West Lafayette, IN 47906

Victor Keller, N9GK
4011 Daner Drive
Fort Wayne, IN 46815

John C. Goller, K9UWA
4836 Ranch Road
Leo, IN 46765

James R. Weigand, N9BW
1422 Buckskin Drive
Fort Wayne, IN 46804

Mark Reese, NB9F
3511 Windlass Court
Fort Wayne, IN 46815

Vernon Seitz, W9HLY
45 Homestead
Decatur, IN 46733

Bill Gibbons, W9KBV
535 Grapevine Lane
Fort Wayne, IN 46825

Robert C. Webb, W9OCL
8208 Valley Estates Drive
Indianapolis, IN 46227

William D. DeGeer, W9TY
3601 Tyler Street
Gary, IN 46408

Mark E. Musick, WB9CIF
P.O. Box 575
Plainfield, IN 46168

David L Zeph, W9ZRX
16310 Springmill Rd
Westfield, IN 46074

Michael Goode, N9NS
13040 Broadway St
Indianapolis, IN 46280

WI

James Hugo, KE9ET
713 Vernon Avenue
Madison, WI 53714

Peter Byfield, K9VAL
4534 S. Hill Ct.
Deforest, WI 53532

Ron Gorski, N9AU
3241 S. Clement Avenue
Milwaukee, WI 53207

Gerald Scherkenbach, N9AW
5952 So. Elaine Avenue
Cudahy, WI 53110

Herb Jordan, W9LA
6318 Putnam Road
Madison, WI 53711

Richard F. Roll, W9TA
14880 W. Maple Ridge Court
New Berlin, WI 53151

Frank Holliday, WB9NOV
23 Walworth Ct.
Madison, WI 53705

Ed Toal, N9MW
5141 Sunrise Ridge Trail
Middleton, WI 53562

Jim Promis, WY9E
3745 Rolling Heights
Oneida, WI 54155

Robert Gorsiski, WA9BDX
1123 David Street
Racine, WI 53404

Ken Tate, WB9OBX
728 South Twenty Second St
Manitowoc, WI 54220

David Jaeger, WN9Q
11913 Sunnyslope Road
Whitelaw, WI 54247

George R. Croy, W9MDP
2113 Twin Willows Drive
Appleton, WI 54915

John Meyer, NZ9Z
PO Box 146
Kellnersville, WI 54215

Gary Banks, N9ER
1692 Shangra La
Neenah, WI 54956

Skip Caswell, KE9LK
HC2 Box 1038
Florence, WI 54121

Gary Hoehne, KB9AIT
W8244 Winnegamie Drive
Neenah, WI 54956

Marian McCone, KB9EKO
6899 K
Rhinelander, WI 54501

Myron McCone, K9JJR
6899 Cty Hwy K
Rhinelander, WI 54501

Dakota Division

MN

Curt Swenson, KØCVD
4821 Westminster Road
Minnetonka, MN 55345

Dave Wester, KØIEA
10205 217th Street North
Forest Lake, MN 55205

Gary Reichow, KNØV
15615 Harmony Way
Apple Valley, MN 55124

Keith H. Gilbertson, KØKG
HC10, Box 44
Detroit Lakes, MN 56501

John Hill, NJØM
3353-34th Avenue South
Minneapolis, MN 55406

Jack Falker, W8KR
5716 View Lane
Edina, MN 55436

ND

Ron L. Roche, KØALL
PO Box 5301
Fargo, ND 58105

Delta Division

AR

Paul E. Wynne, AF5M
5306 Marion Street
N. Little Rock, AR 72118

Stanley W. Krueger, K5AS
3169 Skillern Road
Fayetteville, AR 72703

J. Sherwood Charlton, K5GOE
306 N. Willow Avenue
Fayetteville, AR 72701

Leonard Mendel, K5OVC
309 Rolling Acres Drive
Pearcy, AR 71964

Oliver A. Gade, W5GO
156 Wildcat Estates Lane
Hot Springs, AR 71913

Earl F. Smith, KD5ZM
818 Green Oak Lane
White Hall, AR 71602

Stewart W. Long, KE5PO
28 Nob View Circle
Little Rock, AR 72205

F. Martin Hankins, W5HTY
406 East Cedar Street
Warren, AR 71671

Ben Lockerd, W5NF
3231 S. Cliff Drive
Fort Smith, AR 72903

Mark Manes, WC5I
123 Maple Shade Road
Alma, AR 72921

LA

John Wondergem, K5KR
600 Smith Drive
Metairie, LA 70005

Jim O'Brien, K5NV
PO Box 403
Harvey, LA 70059

Silvano Amenta, KB5GL
5028 Hearst Street
Metairie, LA 70001

Paul Azar, N5AN
412 Live Oak
Lafayette, LA 70503

Shirley Wondergem, N5GGO
600 Smith Drive
Metairie, LA 70005

Wes Attaway, N5WA
2048 Pepper Ridge
Shreveport, LA 71115

Michael Mayer III, W5ZPA
5836 Marcia Avenue
New Orleans, LA 70124

Roy Bonvillian, WD5DBV
PO Box 350
Cade, LA 70519

Cornell C. Bodensteiner, WD5GJB
129 Patricia Ann Place
Lafayette, LA 70508

Randy Hollier, WX5L
1421 North Atlantic
Metaire, LA 70003

MS

Louis J. Raymond, N4CSF
1720 Thomas Jefferson Drive
Biloxi, MS 39531

Walter Hopper, K5VV
8960 Countyline Rd
Hernando, MS 38632

C. T. Green, KB5WQ
128 Gandy Circle
Long Beach, MS 39560

Steven Alexander, KX5V
22 Yorkshire Pkwy
Gulfport, MS 39501

Floyd Gerald, N5FG
17 Green Hollow Rd
Wiggins, MS 39577

Gary E. Jones, W5FI
23 Pirate Drive,
Hattiesburg, MS 39402-9557

W. Paul Howard, WA5TUD
Rt 1, Box 59E
Water Valley, MS 38965

Vic West, WA5SUE
113 Ben Drive
Gulfport, MS 39503

Michael J. Parisey, WDØGML
251 Eisenhower Drive, Apt 337
Biloxi, MS 39531

George G. Bethea, K5JZ
609 Woodward Avenue
Gulfport, MS 39501

TN

Gary S. Pitts, AA4DO
2023 Kimberly Drive
Mt. Juliet, TN 37122

John L. Anderson, AG4M
1041 West Outer Drive
Oak Ridge, TN 37830

Robert M. May II, K4SE
PO Box 453
Jonesboro, TN 37659

Don Moore, K4CN
309 Brookhaven Trail
Smyrna, TN 37167

Vernon T. Underwood, WA4NIB
PO Box 3771
Knoxville, TN 37927

Lawrence D. Strader, K4LDS
11915 Turpin Lane
Knoxville, TN 37932

Don Payne, K4ID
117 Sam Davis Drive
Springfield, TN 37172

William B. Jones, WB4PUD
5081-HY 49 West
Springfield, TN 37172

Great Lakes Division

KY

James C. Vaughan, K4TXJ
5504 Datura Lane
Louisville, KY 40258

Robert K. Ledford, KC4MK
4533 Highway 155
Fisherville, KY 40023

Timothy B. Totten, N4GN
8309 Dawson Hill Rd
Louisville, KY 40299

Ralph E. Wettle, N4NTQ
884 Clarks Lane
Louisville, KY 40217

Roy M. Dobbs, W4KHL
516 Dorsey Way
Louisville, KY 40223

Gary Hext, K4UU
4953 Westgate Drive
Bowling Green, KY 42101

John W. Reasoner, WA4QMQ
125 Clearview Ave.
Bowling Green, KY 42101

Larry R. Smith, K4CMS
623 Shady Ln.
Alvaton, KY 42122

William E. Gann, N4HID
445 Elrod Rd.
Bowling Green, KY 42104

Paul D Schrader ,N4XM
7001 Briscoe Kn
Louisville, KY 40228

MI

Wayne W. Wiltse, K8BTH
14468 Bassett Avenue
Livonia, MI 48154

Merlin D. Anderson, K8EFS
4300 South Cochran
Charlotte, MI 48813-9109

Donald L. Defeyter, KC8CY
5061 East Clark Road
Bath, MI 48808

Edmund Gurney, Jr., W8EG
18712 Westbrook Drive
Livonia, MI 48152

John C. Kroll, K8LJG
3528 Craig Drive
Flint, MI 48506

OH

Pete Michaelis, N8TR
12224 East River Road
Columbia Station, OH 44028

Ron Moorefield, W8ILC
8847 Cajun Ct
Huber Heights, OH 45424-7014

Harry T. Flasher, W8KKF
7425 Barr Circle
Dayton, OH 45459

Alfred V. Altomari, W8QWI
2009 W. 36 Street
Lorain, OH 44053

Robert A. Frey, WA6EZV
8729 Sarah's Bend Drive
Cincinnati, OH 45251

Jerome Kurucz, WB8LFO
5338 Edgewood Drive
Lorain, OH 44053

John Udvari, N8RF
2896 Pine Lake Road
Uniontown, OH 44685

Reno M. Tonsi, WT8C
9248 Woodvale Court
Mentor, OH 44060

Hudson Division

ENY

Jay Musikar, AF2C
6 Cherry Lane
Putnam Valley, NJ 10579

Saul M. Abrams, K2XA
RD #1, Maple Road
Slingerlands, NY 12159

G. William Hellman, W2UD
3713 Valleyview Street
Mohegan Lake, NY 10547

Steven L. Weinstein, K2WE
45 Esterwood Avenue
Dobbs Ferry, NY 10522

Walter J. Bieber KB2RA
374 N. Little Tor Rd.
New City, NY 10956

David Fisher, KA2CYN
124 Bellows Lane
New City, NY 10956

NLI

Richard J. Tygar, AC2P
5 Clelmsford Drive
Wheatley Heights, NY
11798-1504

Arthur M. Albert, K2ENT
2476 Fortesque Avenue
Oceanside, NY 11572

Larry Strasser, K2LS
1404 Amend Drive
North Merrick, NY 11566-1301

Frank Kiefer, K2PWG
1-Sherrill Lane
Port Jefferson Station, NY 11776

Leonard Zuckerman, KB2HK
2444 Seebode Court
Bellmore, NY 11710-4522

Edward J. Manheimer, KD2OD
14 Seneca Trail
Ridge, NY 11961

Martin P. Miller, NN2C
24 Earl Road
Melville, NY 11747

L. E. Dietrich, N2TU
32 Beach Road
Massapequa, NY 11758

Jim Stiles, W2NJN
21 Branch Drive
Smithtown, NY 11787

Jean Chittenden, WA2BGE
RR2, Box 2105
Syosset, NY 11791

NNJ

Stefan Kurylko, N2TN
23 Ackerson Avenue
Pequannock, NJ 07440

William Inkrote, K2NJ
911 Route 579
Flemington, NJ 08822

Robert B. Larson, K2TK
PO Box 53
Stockholm, NJ 07460

Benjamin J Friedland, K2BF
9 Knollwood Trl W
Mendham, NJ 07945

Warren Hager, K2UFM
31 Forest Drive
Hillsdale, NJ 07642

Joyce A. Birmingham,
KA2ANF
235 Van Emburgh Ave
Ridgewood, NJ 07450

Ronald Loneker, KA2BZS
2092 Nicholl Avenue
Scotch Plains, NJ 07076

Derry Galbreath, W2XT
207 Gordon Place
Neshanic Station, NJ 08853

Guy Glaser, KE2CG
240 Grant Avenue
Piscataway, NJ 08854

Ronald Hauser, W2RD
59 Deltart Drive
Belle Mead, NJ 08502

John Hults, K2WJ
186 Mountain View Road
Asbury, NJ 08802-1026

Harry Johnson, W2FT
424 Jefferson Avenue
Hasbrouck Heights, NJ 07604

Eugene Ingraham, N2BIM
79 Stillwater Road
Newton, NJ 07860

Michael Pagan, N2GBH
1 Woodview Avenue
Long Beach, NJ 07740

Harry Westervelt, NA2K
72 Kuhlthau Avenue
Milltown, NJ 08850

John Sawina, NA2R
61 Gallmeier Road
Frenchtown, NJ 08825-3719

Richard Petermann, NF2K
79 Pompton Avenue
West Paterson, NJ 07424

Peter Pellack, NO2R
77 Money Street
Lodi, NJ 07644

Joseph H. Painter, W2BHM
38 Beech Avenue
Berkeley Heights, NJ 07922

Ted Marks, W2FG
81 Oakey Drive
Kendall Park, NJ 08824

Orion Arnold, W2HN
2 Sleepy Hollow Drive
Ho-Ho-Kus, NJ 07423

T. Edward Berzin, W2MIG
47 Palisade Road
Elizabeth, NJ 07208

Stan Dicks, W4AG
9 Settlers Court
Neshanic Station, NJ 08855

William W. Hudzik, W2UDT
111 Preston Drive
Gillette, NJ 07933

Angel M. Garcia, WA2VUY
7 Markham Drive
Long Valley, NJ 07853

Mario Karcich, K2ZD
311 Liberty Avenue
Hillsdale, NJ 07642

Andrew Birmingham, WB2RQX
235 Van Emburgh Avenue
Ridgewood, NJ 07450

Robert C. Shelton, Jr., N2EDF
11 North Clarke Street
Ogdensburg, NJ 07439

Midwest Division

IA

Thomas White, KØVZR
2027 Carter Avenue
Jessup, IA 50648

James R. Pitts, NØAMI
1221 East 35th Street
Des Moines, IA 50317

Robert W. Walstrom, WØEJ
7431 Macon Drive
Cedar Rapids, IA 52411

Thomas Noel Vinson, NYØV
10211 Hall Road
Cedar Rapids, IA 52400

Dale E. Repp, WØIZ
1618 Texas Avenue, NE
Cedar Rapids, IA 52402

Donald L. Schmidt, WØANZ
2161 NW 80th Place
Des Moines, IA 50325-5626

James L. Spencer, WØSR
3712 Tanager Drive, NE
Cedar Rapids, IA 52402

Thomas G. Vavra, WB8ZRL
682 Palisades Access Road
Ely, IA 52227

KS

Richard G. Tucker, WØRT
PO Box 875
Parsons, KS 67357

Dean Lewis, K9ZV
609 Otto Avenue
Salina, KS 67401-7261

John Shoultys, WDØBNC
2157 Edward
Salina, KS 67401

MO

Joseph F. Nemecek, KØJN
1208 N.E. 77th Street
Kansas City, MO 64118

Thomas L. Bishop, KØTLM
4936 N. Kansas Avenue
Kansas City, MO 64119

Clifford H. Ahrens, KIØW
65 Pioneer Trail
Hannibal, MO 63401-2744

Ken Kreski, KWØA
6645 St. Johns Lane
House Springs, MO 63051

Jim Higgins, NDØF
219 Birchwood Drive
St. Peters, MO 63376

Udo Heinze, NIØG
#2 Wildwood Circle
Crestwood, MO 63126

Jim Glasscock, WØFF
3416 Manhattan Avenue
St. Louis, MO 63143

Bill Wiese, WØHBH
9520 Old Bonhomme
Olivette, MO 63132

John L. Chass, WØJLC
12167 N.W. Hwy. 45
Parkville, MO 64152

Don Gaikins, WØVM
344 Hillside Avenue
Webster Groves, MO 63119

NE

Larry L Lehmann, KCØDA
528 N North Ave
Minden, NE 68959

Tim L Andersen NOØC
15 LaPlatte Rd
Kearney, NE 68847-4802

Robert W. Staub, Jr., WBØYWO
PO Box 113
Hoskins, NE 68740

New England Division

CT

Joel Wilks, AK1N
27 Champion Hill Road
East Hampton, CT 06424

Al Rousseau, W1FJ
180 Den Quarry Road
Lynn, MA 01904

William C. Stapleford, K1DII
Bird Road, PO Box 1093
Bristol, CT 06010

Ralph M. Hirsch, K1RH
172 Newton Road
Woodbridge, CT 06525

Peter Budnik, KB1HY
17 Atwood Street
Plainville, CT 06062

John P. Larson, NQ1K
34 Donna Drive
Burlington, CT 06013

EMA

William G. Bithell, Jr., N1BB
96 Greenwood Avenue
Swampscott, MA 01907

Gary Young, K2AJY
1 Sutton Place
Swampscott, MA 01907

Clifford O. Thomson, N1EOA
5 Lakeshore Drive
Beverly, MA 01915

Glenn Scanlon, N1EUO
52 Sunset Dr
Beverly, MA 01915

Ray Sylvester, Jr., NR1R
20 Gardner Road
Reading, MA 01867

Melrose R. Cole, WZ1Q
PO Box 8
Prides Crossing, MA 01965

NH

W. David Gerns, KA1CB
3 Rolling Hill Avenue
Plaistow, NH 03865

Carl Huether, KM1H
169 Jeremy Hill Road
Pelham, NH 03076

Paul I. Cleveland, W1OHA
PO Box 89, Craighill
N. Sutton, NH 03260-0089

William A. Dodge, K1BD
78 Littleworth Road
Dover, NH 03820

Ann M. Santos, WA1S
245 Colburn Road
Milford, NH 03055

VT

David A. Wilson, N4DW
Rt 1, Box 11A
East Burke, VT 05832

WMA

Gareth B. Gaudette, W1GG
21 Westview Road
Lanesboro, MA 01237

Edward Landry, WA1ZAM
140 Cliff Street
North Adams, MA 01247

Robert Tublitz WT2Q
P.O. Box 847
Stockbridge, MA 01262

Northwestern Division

EWA

Earl M. Ringle, N7ER
N. 11310 Parksmith
Mead, WA 99021

William Martinek, K7EFB
N 522 Moore Road
Veradale, WA 99037

Michael J. Klem, K7ZBV
804 Cedar Avenue
Richland, WA 99352

Herb Rode, KE7PB
P.O. Box 4173
West Richland, WA 99352

Lester S. Morgan, NR7B
N 6811 Monroe Street
Spokane, WA 99208-4129

Donald W. E. Calbick, W7GB
447 Knolls Vista Drive
Moses Lake, WA 98837

Marion Lundrigan, W7OIH
2801 Road 92
Pasco, WA 99301

Warren L. Triebwasser, W7YEM
4624 NW Blvd
Spokane, WA 99205

Jay W. Townsend, WS7I
PO Box 644
Spokane, WA 99210

MT

George E. Martin, K7ABV
3608 5th Avenue S.
Great Falls, MT 59405

Robert E. Leo, W7LR
6790 S. 3rd Road
Bozeman, MT 59715

Doug Laubach, WQ7B
Box 19
Carter, MT 59420

OR

Gerald D. Branson, AA6BB
93787 Dorsey Lane
Junction City, OR 97448

Larry R. Johnson, K7LJ
10036 SW 50th Avenue
Portland, OR 97219

Vincent J. Varnas, K8REG
3229 SW Luradel Street
Portland, OR 97219

Sanford T. Weinstein, NK7Y
2945 SW 4th Avenue
Portland, OR 97201

Robert A. Moore, W7JNC
PO Box 33197
Portland, OR 97233

Michael J. Huey, W7ZI
12385 SW Stillwell Lane
Beaverton, OR 97005

Nels H. Nelson, W7EYE
39105 SW Hartley Road
Gaston, OR 97119

Richard A. Zalewski, W7ZR
10735 SW 175th Avenue
Beaverton, OR 97007

Ronald E. Vincent, WJ7R
3715 University St
Eugene, OR, 97405-4347

WWA

Hillar Raamat, N6HR
PO Box 213
Greenbank, WA 98253

Jim Fenstermaker, K9JF
10312 NE 161st Avenue
Vancouver, WA 98662

Herbert Anderson, K7GEX
20148 6th NE
Seattle, WA 98155

Dick Moen, N7RO
2935 Plymouth Drive
Bellingham, WA 98225

John Gohndrone, N7TT
PO Box 863
Bothell, WA 98041

Brenda G. Murphy, N7SB
2612 NW 18th Avenue
Camas, WA 98607

Kermit W. Raaen, W7BG
15509 E. Mill Plain Blvd 42
Vancouver, WA 98664

Al Johnson, W7EKM
8186 Stein Road
Custer, WA 98240

Randy Stegemeyer, W7HR
PO Box 1590
Port Orchard, WA 98366

Leonard A. Westbo, Jr., W7MCU
10528 S.E. 323nd Street
Auburn, WA 98002

Alan N. Rovner, K7AR
18809 N.E. 21st Street
Vancouver, WA 98684

ID

Don M. Clower, KA7T
5103 W. Cherry Lane
Meridian, ID 83642

Pacific Division

EB

Thomas F. Jones, K6TS
1239 Hillview Drive
Livermore, CA 94550

James W. Sheley, KA6DXY
5849 Hansen Drive
Pleasanton, CA 94566

Charles E. McHenry, W6BSY
1612 Via Escondido
San Lorenzo, CA 94580

Rubin Hughes, WA6AHF
17494 Via Alamitos
San Lorenzo, CA 94580

Randy Wright, WB6CUA
18432 Milmar Road
Castro Valley, CA 94546

NV

Frank Dziurda, K7SFN
225 W. Coyote Drive
Carson City, NV 89704

JB Coats, KB7YX
1839 Deep Creek Drive
Sparks, NV 89434

James R. Frye, NW7O
4120 Oakhill Avenue
Las Vegas, NV 89121

William L. Dawson, W7TVF
HCR 65 Box 71623
Pahrump, NV 89041-9678

PAC

Lee Roger Wical KH6BZF
45-601 Luluku Road
Kane'Ohe, HI 96744-1854-25

Harry K. Nishiyama, KH6FKG
1990 Hale Hooko Street
Hilo, HI 96720

Richard I. Senones, KH6JEB
95-161 Kauopae Place
Mililani Town, HI 96789

Charles S. Y. Yee, KH6WU
27 Moe Moe Place
Wahiawa, HI 96786

John D Peters K1ER
98-1547 Akaaka St
Aiea, HI 96701

SCV

John N Knight K6CXT
322 Coleridge Dr
Salinas, CA 93901

Stan Goldstein, N6XU
791 Calabasas
Watsonville, CA 95076

Edward Muns, WØYK
PO Box 1877
Los Gatos, CA 95031-1877

James A. Maxwell, W6CF
PO Box 473
Redwood Estates, CA 95044

John G. Troster, W6ISQ
82 Belbrook Way
Atherton, CA 94025

Gerald D. Griffin, K6MD
123 Forest Avenue
Pacific Grove, CA 93950

Richard G. Whisler, WA6SLO
716 Hill Avenue
S. San Francisco, CA 94080

Richard M. Letrich, W6KM
3686 Kirk Road
San Jose, CA 95124-3816

Walt Del Conte WD6EKR
328 Sequoia St
Salinas, CA 93906

SF

Chuck Ternes, N6OJ
535 Cherry Street
Petaluma, CA 94952

Dave Stocham, W6NPY
106 Locust Avenue
Mill Valley, CA 94941

Al Lotze, W6RQ
46 Cragmont Avenue
San Francisco, CA 94116-1308

Jerry Foster, WA6BXV
16 San Ramon
Novato, CA 94947

Leonard R Geraldi K6ANP
9705 Old Redwood Hwy
Penngrove, CA 94951

SJV

Perry Foster, K6XJ
10575 E. Bullard
Clovis, CA 93612

Robert W Selbrede K6ZZ
6200 Natoma Ave
Mojave, CA 93501

Charles McConnell, W6DPD
1658 W. Mesa
Fresno, CA 93711-1944

PAC (continued)

Robert Craft, W6FAH
8136 Grenoble Way
Stockton, CA 95210

Charles Allessi, W6IEG
PO Box 1244
Oakhurst, CA 93644

Dennis J. DuGal, WG6P
2008 Sharilyn Drive
Modesto, CA 95355

Robert W. Smith, W6GR
13447 Road 23
Madera, CA 93637

Stanley R. Ostrom, W6XP
4921 E. Townsend
Fresno, CA 93727

Jules Wenglare, W6YO
1416 Seventh Avenue
Delano, CA 93215

Carl Boone, WB6VIN
1642 Seventh Place
Delano, CA 93215

Frederick L Moore W6KUS
2929 Jewetta Ave.
Bakersfield, CA 93312

Fred K. Stenger N6AWD
6000 Hesketh Dr.
Bakersfield, CA 93309

Kenneth H. Day W6YK
5211 Crystal Fall Ln.
Bakersfield, CA 93313

SV

Larry Murdoch, K6AAW
14370 Brian Road
Red Bluff, CA 96080

Ken Anderson, K6PU
PO Box 853
Pine Grove, CA 95665

John D. Brand, K6WC
9655 Tanglewood Circle
Orangevale, CA 95662

Phil Sanders, K6YS
8580 Krogh Ct.
Orangevale, CA 95662

Ted Davis, W6BJH
PO Box 494243
Redding, CA 96049

Jettie B. Hill, W6RFF
306 St. Charles Ct.
Roseville, CA 95661

Jerry Fuller, W6JRY
PO Box 363
Forest Ranch, CA 95942

Roanoke Division

NC

Bill Parris, AA4R
16741 100 Norman Place
Huntersville, NC 28078

Ron Oates, AA4VK
9908 Waterview Road
Raleigh, NC 27615

Bruce M. Gragg, K4ZO
Route 2, Box 329
Conover, NC 28613

Frank Dowd, Jr., K4BVQ
PO Box 35430
Charlotte, NC 28235

Larry Sossoman, K4CEB
4383 Ponderosa Lane
Concord, NC 28025

NC (continued)

Bill McDowell, K4CIA
2317 Norwood Road
Raleigh, NC 27614

Les Murphy, K4DY
PO Box 626
Hickory, NC 28603-0626

Lynn L Pendleton, K4NYV
3617 Country Cove Lane
Raleigh, NC 27606

Mike Jackson, N4YS
2568 Devon Drive
Dallas, NC 28034

Joe Young, N4DAZ
1912 Miles Chapel Road
Mebane, NC 27302

Roger Burt, N4ZC
RFD 1 Box 246
Mt. Holly, NC 28120

Les Murphy, K4DY
PO Box 626
Hickory, NC 28603-0626

Robert H. McNeill, II, W4MBD
PO Box 843
Morehead City, NC 28557

Mark McIntyre, WA4FFW
2903 Maple Avenue
Burlington, NC 27215

Ewell M. Brown, W4CZU
PO Box 447
Taylorsville, NC 28681

Robert C. Cranford, K4SI
772 Fernwood Road
Lincolnton, NC 28092

Joe Simpkins, K4MD
2400 Flintwood Lane
Charlotte, NC 28226

Gene Turner, WJ4T
159 Windsor Drive
Graham, NC 27253

Stephen D Budensiek KØSD
225 Beulah Ln
Salisbury, NC 28146

SC

Rick Porter, AA4SC
Box 2731
Rock Hill, SC 29731

Gary Dixon, K4MQG
1606 Crescent Ridge
Ford Mill, SC 29715

Jack Jackson, N4JJ
PO Box 12612
Florence, SC 29504

Charles Johnson, N4TJ
109 Kimberly Lane
Florence, SC 29505

Bill Jennings, W4UNP
630 Whitepine Drive
Catawba, SC 29704

Murphy Ratterree, W4WMQ
264 Wayland Drive
Rock Hill, SC 29732

C. W. Eldridge, WA4PLR
220 Pinewood Lane
Rock Hill, SC 29730

Ted F. Goldthorpe, W4VHF
209 Swamp Fox Drive
Fort Mill, SC 29715

VA

M. David Gaskins, K4AV
228 Unser Drive
Chesapeake, VA 23320

Ralph King, K1KOB
3409 Mornington Drive
Chesapeake, VA 23321

John N. Kirkham, KC4B
10920 Byrd Drive
Fairfax, VA 22030

Gordon Garrett, K1GG
4528 North Fork Road
Elliston, VA 24087

Emmett Stine, KD4OS
1269 Parkside Place
Virginia Beach, VA 23454

Fran Sledge, N4CRU
3004 Oakley Hall Road
Portsmouth, VA 23702

Clay Partin, W7CP
6642 Shingle Ridge Road, SW
Roanoke, VA 24018

Karl H. Oyster, Jr., K1KO
524 Vicksdell Cres.
Chesapeake, VA 23320-3551

Nap Perry, W4DHZ
9317 Marlow Avenue
Norfolk, VA 23503

Art Westmont, W4EEU
4717 Little John Road
Virginia Beach, VA 23455

Grover Brewer, W4FPW
1359 Eagle Avenue
Norfolk, VA 23518

Howell G. Gwaltney, W4IF
209 Pennington
Portsworkt, VA 23701

Pradyumna Rana, WB4NFO
29 East Chapman Street
Alexandria, VA 22301

Richard T. Williams, WE6H
7902 Viola Street
Springfield, VA 22152

Vance B. Williams K4TQ
824 Moffat Ln.
Virginia Beach, VA 23464

Harold E. Sager KO4OG
308 Plaza Trail Ct.
Virginia Beach,VA 23452

James L. Young K4JDJ
556 Babbtown Rd.
Suffolk, VA 23434

Donald F. Lynch Jr. W4ZYT
1517 W. Little Neck Rd.
Virginia Beach, VA 23452

David R. Klimaj W4JVN
5637 Heming Ave.
Springfield, VA 22151

Rocky Mountain Division
CO

Steve P. Gecewicz, KØCS
PO Box 1048
Elizabeth, CO 80107

Karen Schultz, KAØCDN
15643 East 35th Place
Aurora, CO 80011

Glenn Schultz, WØIJR
15643 East 35th Place
Aurora, CO 80011

Jim Spaulding, WØUO
6277 S. Niagara Court
Englewood, CO 80111

NM

Robert L. Norton, N5EPA
116 Pinon Heights Road
Sandia Park, NM 87047

Arne Gjerning, N7KA
PO Box 1485
Corrales, NM 87048

Fred F. Seifert, W5FS
3106 Florida Street NE
Albuquerque, NM 87110-2617

Eugene F. Carter, W5OLN
320 Camino Tres SW
Albuquerque, NM 87105

Paul Rubinfield, WF5T
PO Box 4909
Santa Fe, NM 87502

UT

Steven C. Salmon, K7OXB
2184 East 6070 South
Salt Lake City, UT 84121

WY

Wayne Mills, N7NG
PO Box 1945
Jackson, WY 83001

John F. Hall, W7CA
Box 975
Mills, WY 82644

Southeastern Division
AL

Bill Christian, K4IKR
2800 Cave Avenue
Huntsville, AL 35810

Tim Pearson, KU4J
6214 Rime Village Drive #205
Huntsville, AL 35806

GA

Frank G. Deak, AF4Y
2566 Glen Circle
Lawrenceville, GA 30244

Randy Tudor, K4ODL
1032 Rockcrest Drive
Marietta, GA 30062

Samuel I. Silverman, KB4NJ
3960 Green Forest Parkway
Smyrna, GA 30082

Neil R. Foster, N4FN
3185 Friar Tuck Way
Atlanta, GA 30340

John Smith, KI4XO
450 Chaffin Road
Roswell, GA 30075

Roy E. Conley, N4ONI
1429 Dallas Circle
Marietta, GA 30064-2913

Verne Fowler, W8BLA
113215 Stroup Road
Roswell, GA 30075-2225

Robert Varone, WA4ETN
5177 Holly Springs Dr
Douglasville, GA 30135

R.L."Gad"Williamson,WA4CUG
5877 Brookside Drive
Mapleton, GA 30059

NFL

Homer J. Cumm, AB4NS
3599 Peoria Rd
Orange Park, FL 32065

William H. Bosley, K3NN
201 Highland Street
Valparaiso, FL 32580-1222

William R. Hicks, K4UTE
7002 Deauville Road
Jacksonville, FL 33205

Jim Buth, K9CJK
1 Sommer Road
Crawford, FL 32327

H. J. Huddleston, K4BU
925 Forest Avenue
Fort Walton Beach, FL 32547

Leslie E. Smallwood, KW4V
24 Apple Hill Hollow
Casselberry, FL 32707

Clarence J. Kerous, KX8N
1104 Buggywhip Train
Middleburg, FL 32608

Ronald E. Blake, N4KE
258 Wesley Road
Green Cove Springs, FL 32043

Richard A. Knox, WR4K
5172 Pine Avenue
Orange Park, FL 32072

Gary L Letchford KÿLUZ
214 Tollgate Trail
Longwood, FL 32750

PR

Telesforo Figueroa, KP4P
PO Box 1651
Yabucoa, PR 00767-1651

Eduardo Negron, KP4EQF
Urb. Valle Real AG-78
Ponce, PR 00731

SFL

Harry D. Belock, AA2X
8108 N. W. 102 Terr.
Tamarac, FL 33321

Leon A. Katz, K2EWB
4136 Lakespur Circle North
Palm Beach Gardens, FL 33410

Edwin L. Chinnock, W2FZY
5641 Bayview Way
Fort Lauderdale, FL 33308

David Beckwith, W2QM
3115 NW 13th Street
Delray Beach, FL 33445

Stephen Sarasohn, W2ZR
3288 NW 60 Street
Boca Raton, FL 33496

Michael M. Raskin, K4KUZ
561 West Tropical Way
Plantation, FL 33317

Gary Fowks, K4MF
635 W. 64 Drive
Hialeah, FL 33012

Robert R. Wilson, KI4LP
2323 Treetop Ct
Melbourne, FL 32934

Ernie Greeson, K4RD
1500 E. Terra Mar Drive
Pompano Beach, FL 33062

David Novoa, KP4AM
1994 SW 142nd Avenue
Miami, FL 33175

Peter Rimmel, K8UNP
4209 Madison Street
Hollywood, FL 33021

Marv Westerdahl, KC2KU
1934 Hawaii Avenue N.E.
St. Petersburg, FL 33703

Lefty Boggess, KE4VU
11636 Grove Street North
Seminole, FL 34642

Ray Riker, NY2E
360 Ponte Vedra Road
Palm Springs, FL 33461

Donald B. Search, W3AZD
10550 State Road 84
#147 Park City West
Ft. Lauderdale, FL 33324

Gene Sykes, W4OO
6510 Carambola Circle
West Palm Beach, FL 33406

Mark Horowitz, WA2YMX
6831 SW 16th Street
Plantation, FL 33317

Joseph L. Picior, WB4OSN
61 Eden Lane
Lake Placid, FL 33852

Harry T. Henley, WK9Z
2013 Massachusetts Avenue NE
St. Petersburg, FL 33703

Edward Grogan, WW1N
902 Bay Point Drive
Madeira Beach, FL 33708

Bruce Paul Phegley, W4OVU
3940 N.W. 4 Ct.
Coconut Creek, FL 33066

Ralph E. Small, K1ZLA
969 Waialae Cir. Ne.
Palm Bay, FL 32905

John A. Mann Sr. ,N4OLE
4400 Sw 3rd St.
Miami, FL 33134

Douglas A McDuff, W4OX
10380 SW 112th St
Miami, FL 33176

Southwestern Division
AZ

Ned Stearns, AA7A
7038 E. Aster Drive
Scattsdale, AZ 85254

Mike Fulcher, KC7V
6545 E. Montgomery Road
Cave Creek, AZ 85331

Lee Finkel, KY7M
6928 E. Ludlow
Scottsdale, AZ 85254

Hardy Landskov, N7RT
15814 N. 44th Street
Phoenix, AZ 85032

Jim McDonald, N7US
7233 N. 16th Ave
Phoenix, AZ 85021

Steve Towne, NN7X
12002 N. Oakhurst Way
Scottsdale, AZ 85254

Frank Schottke, W2UE
5049 E. Laurel Lane
Scottsdale, AZ 85254

Stuart Greene, WA2MOE
7537 N. 28th Avenue
Phoenix, AZ 85051

William G. Snyder, KF8N
8930 N. Camino De La Tierra
Tuscon, AZ 85742

Bernie J. Sasek, W7KQ
8925 N. Morning View Dr.
Tuscon, AZ 85737

William J. Stange, W9DDP
P.O. Box 927
Patagonia, AZ 85624

Ronald G. Swann ,WA7BOD
5656 W. Utah St.
Tuscon, AZ 85746

Douglas C. Schaber KG7IZ
2611 N. Camino De Oeste
Tuscon, AZ 85745

LAX

William A. Angenent, KN6DV
43150 6th St East
Lancaster, CA 93535

Leonard Svidor, W6AUG
17760 Alonzo Place
Encino, CA 91316

Sheldon C. Shallon, W6EL
11058 Queensland Street
Los Angeles, CA 90034

ORG

Rick Samoian, W6SR
5302 Cedarlawn Drive
Placentia, CA 92670

SB

Milt Bramer, N6MB
4161 Shadyglade Drive
Santa Maria, CA 93455

John O. Norback, W6KFV
133 Pino Solo Court
Nipomo, CA 93444

Neil B. Taylor, W6MUS
PO Box 392
Atascadero, CA 93423

Jim Robb, W6OUL
501 N. Poppy
Lompoc, CA 93436

Steve Nolan, K6MX
904 N. Western Avenue
Santa Maria, CA 93454

SDG

Bruce Clark, K6JYO
508 Washingtonia Drive
San Marcos, CA 92069

Cliff Smith, K6TWU
10565 Rancho Road
La Mesa, CA 91941

Art Charette, K6XT
16604 Adrienne Way
Ramona, CA 92065

Rick Craig, N6ND
PO Box 741
Ramona, CA 92065

Ed Andress, W6KUT
12821 Corte Dorotea
Poway, CA 92064

George Pugsley, W6ZZ
1362 Via Rancho Prky
Escondido, CA 92029

Pat Bunsold, WA6MHZ
14291 Rios Canyon Road #33
El Cajon, CA 92021

Charles L. Roy, WS6F
6231 Lake Shore Drive
San Diego, CA 92110

West Gulf Division
NTX

Phil Clements, K5PC
1313 Applegate Lane
Lewisville, TX 75069

Mike Krzystyniak, K9MK
249 Bayne Road
Haslet, TX 76052

Joel Rubenstein, KA5W
3601 Larkin Lane
Rowlett, TX 75088

Vernon Brunson, KD5HO
2717 Whispering Trail
Arlington, TX 76013

Mickey Heimlich, N5AJW
9207 Mill Hollow
Dallas, TX 75243

John Clifford, NK5K
125 Alta Mesa Drive
Fort Worth, TX 76108

David Jaksa, WÿVX
626 Torrey Pines Lane
Garland, TX 75044

Thomas Little, W5GVP
2109 Miriam Lane
Arlington, TX 76010

Ken Knudson, W5PLN
5725 Wales Avenue
Fort Worth, TX 76133

Richard Pruitt, WB5TED
2232 Jensen
Fort Worth, TX 76112

OK

George W. Adkins, AD1S
PO Box 88
Wellston, OK 74881

Ed Murta, K5LIL
4408 NW 47th Street
Oklahoma City, OK 73112

Ross Hunt, K5RH
17910 E. 15 Street
Tulsa, OK 74108-5130

Jeff Martin, K5WE
5337 S. 71 E. Avenue
Tulsa, OK 74145

Elton Abbott, KE5JE
216 N. Pecan Street
Pauls Valley, OK 73075

Kenneth L. Adams, K5KA
5201 Philson Farm Road
Bartlesville, OK 74006

Mark Byard, N5OGP
504 Foster Avenue
Ponca City, OK 74601

Coy C. Day, N5OK
Rt 1, Skyline View Estates
Claremore, OK 74017-9801

Bob See, N5PC
1708 E. Woodland Road
Ponca City, OK 74604

Charles E. Calhoun, WÿRRY
13914 N. 90 E. Avenue
Collinsville, OK 74021

W. Clegg Hahn, W5CKT
2816 SE Hampden Road
Bartlesville, OK 74006

Gil Wood, W5NUT
36806 Old Hwy 270
Shawnee, OK 74801-8701

Larry Watson, W5VHP
4212 Beacon Court
Bartlesville, OK 74006

Gary N. Gompf, W7FG
3300 Wayside Drive
Bartlesville, OK 74006

Jim Hood, WV5S
11623 Smoking Oaks Drive
Oklahoma City, OK 73150

STX

Robert C. Walworth, N5ET
3210 Chaparral Way
Spring, TX 77380

Don Hall, K5AQ
4870 Enchanted Oaks Drive
College Station, TX 77845

William E. Barnes, K5GE
434 Windcrest Drive
San Antonio, TX 78239

Charles R. Crutchfield, K5BC
3031 Rose Lane
La Marque, TX 77568

Linda G. Walworth, KE5TF
3210 Chaparral Way
Spring, TX 77380

Evonne M. Lane, KF5MY
1602 Chestnut Ridge
Kingwood, TX 77339

G.B. Jim Lane II, N5DC
1602 Chestnut Ridge
Kingwood, TX 77339

Edward L. Linde, KÿGEX
3900 Sorrell Cove
Austin, TX 78730

Joseph A. Castorina, N5FJY
4437 F. M. 646 N
Sante Fe, TX 77510

Harley L. Dillon, N5HB
4902 Lambeth
San Antonio, TX 78228

John P. Warren, NT5C
3517 Gattis School Road
Round Rock, TX 78664

Robert A. Wood, W5AJ
1013 Lewis Drive
Kemah, TX 77565

Wayne Wyatt, N5BV
7114 Quail Landing
San Antonio, TX 78250

Randy Light, WC5Q
2100 Bent Oak Street
College Station, TX 77840

WTX

David Zulawski, KA5TQF
2808 Catnip Street
El Paso, TX 79925

Joe T. Milam, W5UA
432 W. Dengar
Midland, TX 79707

Leslie A. Bannon, WF5E
3400 Bedford
Midland, TX 79703

DXCC AWARD APPLICATION

DIRECTIONS: PLEASE PRINT. Use this form for both new and endorsement applications. Complete both sides of this form. (The reverse side is optional if you are sending QSL cards to ARRL HQ.) **Do NOT include other correspondence with this application.**

- I am applying for the following DXCC awards (check ALL as appropriate)

Award Type	New app	Endorsement	*Required Last submission date	*Required Cert No.
Mixed				
Phone				
CW				
RTTY				
SAT.				
160				
80				
40				
10				
6				
2				
* 5BDXCC		160-17-12-6-2 ◄— Circle Band(s)		

* 5BDXCC Award Number _____
Needed for (Endorsements)

- FEES: Effective 1 January, 1994:All Amateurs applying for their first DXCC Award - $10 (includes certificate and DXCC pin) Each additional DXCC certificate (new or replacement) - $5 shipping and handling fee (includes certificate and DXCC pin)

  Endorsements and new application presented at ARRL HQ or conventions. (These applications will be limited to 110 QSOs) - $2 handling fee.

  ARRL members may make one free submission each year, subject to charges above, and return postage. These may be a combination of new awards and endorsements.

  Additional member submissions within a calendar year will be $10, subject to charges above.

  Foreign non-ARRL members will be charged $10 for the first submission in a calendar year, subject to charges above.

  Foreign non-ARRL members will be charged $20 for additional submissions in a calendar year, subject to charges above.

  Return postage or SASE must be provided by the applicant for any cards or requests for information.

- ALL Applicants in the US and possessions, Puerto Rico, and foreign members MUST complete the following statement:

"I currently hold ARRL membership expiring _____." OR attach label from *QST* wrapper;
(month, year)

- ALL applicants must complete the following statement:
"I affirm that I have observed all DXCC rules (see attached) as well as all pertinent governmental regulations established for Amateur Radio in my country. I agree to be bound by the decisions of the ARRL Awards Committee. (Decisions of the ARRL Awards Committee shall be final.)"

Signature _____ Call _____ Date _____

- Send application forms, QSL cards, fees (if applicable) and return postage to : DXCC Desk, ARRL HQ, 225 Main St, Newington, CT 06111, USA. **To confirm receipt include an SASE or post card with your application.**

- For any questions or clarifications, please write to the DXCC Desk separately at the above address. The DXCC Desk can also be contacted as follows:

Telephone 203-666-1541
Fax 203-665-7531 (24-hour direct line to ARRL HQ)

Telex 640215-5052 MCI
MCI Mail 215-5052

- Thank you for your cooperation, and **Good DX!**

- Call Sign _____

  Ex Calls _____

- Name _____
  Last (*Spanish* Apellido),   First

  Address _____
  (Number and street)

  _____
  (City, State/Country, ZIP)

  ☐ Check here if new address

- Fill this line out ONLY if you will receive a certificate(s) from this submission. Otherwise, leave blank.

  Name _____
  (Print exactly as you want it to appear on certificate)

- Number of QSL cards enclosed _____

- Number of (QSOs) on QSL cards _____
  (Confirmations)

- ☐ Postcard enclosed for confirmation of receipt of cards.

- Postage = $ _____
  Awards Fee = $ _____   TOTAL = $ _____
  ☐ US Currency _____   ☐ Stamps
  ☐ Check or money order_____   ☐ IRCs

- Return my QSL cards via:
  ☐ Registered mail (**Recommended**)   ☐ First class mail
  ☐ Certified (US only)   ☐ Airmail
  ☐ United Parcel Service (US only)   ☐ Other _____

Note: IRCs valued at $0.50 US. Check with your local Post Office for current mailing rates.

ARRL DXCC FIELD REPRESENTATIVE VERIFICATION

I hereby verify that I have personally inspected the confirmations and verify that this application is correct and true.

Signature _____ Call _____ Date _____

Signature _____ Call _____ Date _____

MSD-505(495)
Printed in USA

DIRECTIONS: (1) Sort cards and list below first by band, e.g. all the 80-meter cards together, then the 40-meter cards, etc. (2) Within each band, sort and list below by mode, e.g. all the 80-meter phone cards, then the 80-meter CW, 40-meter phone, etc. (3) Make one entry below for each QSO credit. (4) Cards indicating multiple contacts should be placed at the end, listing each contact on a separate line below. (5) QSO Date = Day, Month, Year. (6) Bands =160, 80, 40, 30, 20, 17, 15, 12, 10, 6, 2. (7) Modes = PHONE, CW, RTTY, SAT.

	CALL	QSO DATE (DD/MM/YY)	BAND	MODE	COUNTRY
1		/ /			
2		/ /			
3		/ /			
4		/ /			
5		/ /			
6		/ /			
7		/ /			
8		/ /			
9		/ /			
10		/ /			
11		/ /			
12		/ /			
13		/ /			
14		/ /			
15		/ /			
16		/ /			
17		/ /			
18		/ /			
19		/ /			
20		/ /			
21		/ /			
22		/ /			
23		/ /			
24		/ /			
25		/ /			
26		/ /			
27		/ /			
28		/ /			
29		/ /			
30		/ /			

This side of form may be photocopied if more pages are needed.

Russian Oblasts

Updated January 1996

Call sign number and first letter of suffix	Russian	English	Designator
1A, 1B	Sankt-Peterburg[1]	St. Petersburg	SP
Also 1D, 1F, 1G, 1H, 1I, 1J, 1L, 1M			
1C,	Leningradskaya obl.[1]	Leningradskaya Oblast	LO
1N	Karel'skaya ASSR	Karel'skaya Autonomous Soviet Socialist Republic	KL
1O	Arkhangel'skaya obl.	Arkhangel'skaya Oblast	AR
1P	Nenetskiy AO	Nenetskiy Autonomous District (Okrug)	NO
1Q,1R 1S	Vologodskaya obl.	Vologodskaya Oblast	VO
1T,1U	Novgorodskaya obl.	Novgorodskaya Oblast	NV
1W, 1X	Pskovskaya obl.	Pskovskaya Oblast	PS
1Y, 1Z	Murmanskaya obl.	Murmanskaya Oblast	MU
2F	Kaliningradskaya obl.[2]	Kaliningradskaya Oblast	KA
3A, 3B	g. Moskva[3]	Moscow City	MA
Also 3C, 3F, 3H			
3D	Moskovskaya obl.[3]	Moskovskaya Oblast	MO
3E	Orlovskaya obl.	Orlovskaya Oblast	OR
3G	Lipetskaya obl.	Lipetskaya Oblast	LP
3I, 3J	Tverskaya obl.	Tverskaya Oblast	TV
3L	Smolenskaya obl.	Smolenskaya Oblast	SM
3M	Yaroslavskaya obl.	Yaroslavskaya Oblast	JA
3N, 3O	Kostromskaya obl.	Kostromskaya Oblast	KS
3P	Tul'skaya obl.	Tul'skaya Oblast	TL
3Q	Voronezhskaya obl.	Voronezhskaya Oblast	VH
3R	Tambovskaya obl.	Tambovskaya Oblast	TB
3S	Ryazanskaya obl.	Ryazanskaya Oblast	RA
3T	Nizhegorodskaya obl.	Nizhegorodskaya Oblast	NN
3U	Ivanovskay aobl.	Ivanovskaya Oblast	IV
3V	Vladimirskaya obl.	Vladimirskaya Oblast	VL
3W	Kurskaya obl.	Kurskaya Oblast	KU
3X	Kaluzhskaya obl.	Kaluzhskaya Oblast	KG
3Y	Bryanskaya obl.	Bryanskaya Oblast	BR
3Z	Belgorodskaya obl.	Belgorodskaya Oblast	BO
4A, 4B	Volgogradskaya obl.	Volgogradskaya Oblast	VG
4C, 4D	Saratovskaya obl.	Saratovskaya Oblast	SA
4F	Penzenskaya obl.	Penzenskaya Oblast	PE
4H. 4I	Samarskaya obl.	Samarskaya Oblast	SR
4L, 4M	Ul'yanovskaya obl.	Ul'yanovskaya Oblast	UL
4N, 4O	Kirovskaya obl.	Kirovskaya Oblast	KI
4P, 4Q	Respublika Tatarstan	Republic of Tatarstan	TA
Also 4R			
4S, 4T	Respublika Mariy-El	Republic of Mariy-El	MR
4U	Mordovskaya SSR	Mordovskaya Soviet Socialist Republic	MD
4W	Respublika Udmurtiya	Republic of Udmurtiya	UD
4Y, 4Z	Respublika Chuvashiya	Republic of Chuvashiya	CU
6A, 6B	Krasnodarskiy kray	Krasnodarskiy Kray	KR
Also 6C, 6D			
6E	Karachayevo-Cherkasskaya	Karachayevo-Cherkasskaya Respublika Republic	KC
6H, 6F	Stavropol'skiy kray	Stavropol'skiy Kray	ST
Also 6H			
6I	Respublika Kalmykiya	Republic of Kalmykiya	KM
6J	Severo-Osetinskaya SSR	North-Osetinskaya Soviet Socialist Republic	SO
6L, 6M	Rostovskaya obl.	Rostovskaya Oblast	RO
Also 6N, 6O			
6P	Chechenskaya Respublika	Chechenskaya Republic	CN
Also 6R			
6U, 6V	Astrakhanskaya obl.	Astrakhanskaya Oblast	AO
6W	Respublika Dagestan	Republic of Dagestan	DA
6X	Respublik Kabardino-Balkariya	Republic of Kabardino-Balkariya	KB
6Y	Respublika Adygeya	Republic of Adygeya	AD
8T	Ust'-Ordynskiy Buryatskiy AO	Ust'-Ordynskiy Buryatskiy Autonomous District (Okrug)	UO
8V	Aginskiy Buryatskiy AO	Aginskiy Buryatskiy Autonomous District (Okrug)	AB
9A, 9B	Chelyabinskaya obl.	Chelyanbinskaya Oblast	CB
9C, 9D	Sverdlovskaya obl.	Sverdlovskaya Oblast	SV
Also 9E			
9F PM	Permskaya obl.	Permskaya Oblast	
9G	Komi-Permyatskiy AO	Komi-Permyatskiy Autonomous District (Okrug)	KP
9H, 9I	Tomskaya obl.	Tomskaya Oblast	TO
9J	Khanty-Mansiyskiy AO	Khanty-Mansiyskiy Autonomous District (Okrug)	HM
9K	Yamalo-Nenetskiy AO	Yamalo-Nenetskiy Autonomous District (Okrug)	JN
9L	Tyumenskaya obl.	Tyumenskaya Oblast	TN
9M, 9N	Omskaya obl.	Omskaya Oblast	OM
9O, 9P	Novosibirskaya obl.	Novosibirskaya Oblast	NS
9Q, 9R	Kurganskaya obl.	Kurganskaya Oblast	KN
9S, 9T	Orenburgskaya obl.	Orenburgskaya Oblast	OB
9U, 9V	Kemerovskaya obl.	Kemerovskaya Oblast	KE
9W	Respublika Bashkortostan	Republic of Bashkortostan	BA
9X	Respublika Komi	Republic of Komi	KO
9Y	Altayskiy kray	Altayskiy Kray	AL
9Z	Gorno-Altayskiy AO	Gorno-Altayskiy Autonomous District (Okrug)	GA
0A	Krasnoyarskiy kray	Krasnoyarskiy Kray	KK
0B	Taymyrskiy AO	Taymyrskiy Autonomous District (Okrug)	TM
0C	Khabarovskiy kray	Khabarovskiy Kray	HK
0D	Yevreyskaya AO	Jewish Autonomous Oblast[4]	EA
0E, 0F	Sakhalinskaya obl.	Sakhalinskaya Oblast	SL
Also 0G			
0H	Evenkiyskiy AO	Evenkiyskiy Autonomous District (Okrug)	EW
0I	Magadanskaya obl	Magadanskaya Oblast	MG
0J	Amurskaya obl.	Amurskaya Oblast	AM
0K	Chukotskiy AO	Chukotskiy Autonomous District (Okrug)	CK
0L, 0M	Primorskiy kray	Primorskiy Kray	PK
Also 0N			
0O, 0P	Respublika Buryatiya	Republic of Buryatiya	BU
0Q, 0R	Respublika Sakha	Republic of Sakha	YA
0S, 0T	Irkutskaya obl.	Irkutskaya Oblast	IR
0U, 0V	Chitinskaya obl.	Chitinskaya Oblast	CT
0W	Respublika Khakassiya	Republic of Khakassiya	HA
0X	Koryakskiy AO	Koryakskiy Autonomous District (Okrug)	KJ
0Y	Respublika Tuva	Republic of Tuva[5]	TU
0Z	Kamchatskaya obl.	Kamchatskaya Oblast	KT

[1]St. Petersburg is within, and separate from, Leningradskaya Oblast.

[2]Kaliningradskaya Oblast covers the northern part of the former East Prussia. Annexed by the USSR at the end of World War II and made part of the RSFSR, the area is separated from the rest of Russia by Lithuanian and Belarussian territory. The administrative center of the Oblast is Kaliningrad, formerly Konigsberg.

[3]Moscow city is within, and separate from, Moskovskaya Oblast.

[4]According to a U.S. reference dating back to 1961, the Jewish Autonomous Oblast (Yevreyskaya avtonomnaya oblast or YeAO) was established in 1934 in the Soviet Far East but attracted few Jews; its population was 165,000, including no more than 30,000 Jews.

[5]This is the former Tannu Tuva, which was part of the Chinese Empire until 1911, became a Russian Protectorate in 1914, and proclaimed its independence in 1921. It was annexed by the USSR in 1944 and given the status of an autonomous oblast within the RSFSR. In 1961, it was elevated to the status of an autonomous SSR, still within the RSFSR.

Morse Code for Other Languages

Code	Japanese		Korean	Arabic	Hebrew	Russian	Greek	
•	ヘ he	├ a			ˈ vav	E,Э E	E epsilon
—	ム mu	┤ ŏ	ت ta	תת tav		Т T	Т tau
••	· nigori	├ ya	ڥ ya	· yod	И I	I iota
•—	イ i	⊥ o	ا alif	א aleph	А A	А alpha
—•	タ ta	ㅛ yo	ن noon	נ nun	Н N	N nu
——	ヨ yo	ㅁ m	ر meem	מם mem	М M	M mu
•••	ラ ra	ㅓ yŏ	س seen	שׁ shin	С S	Σ sigma
••—	ウ u		ط ta	ט tet	У U	ОТ omicron ypsilon
•—•	ナ na	ㅠ yu	ر ra	ר reish	Р R	Р rho
•——	ヤ ya	ㅂ p(b)	و waw	צ tzadi	В V	Ω omega
—••	ホ ho	┥ dal		ד dalet	Д D	Δ delta
—•—	ワ wa	ㅇ -ng	ك kaf	כך chaf	Н K	К kappa
——•	リ ri	ㅅ s	غ ghain	ג gimmel	Г G	Г gamma
———	レ re	ㅍ p'	خ kha	ה heh	О O	O omicron
••••	ヌ nu	┬ u	ح ha	ח chet	Х H	Н eta
•••—	ク ku	ㄹ r-(l)	ض dad		Ж J	НТ eta ypsilon
••—•	チ ti	ㄴ n	ف fa	מ feh	Ф F	Φ phi
••——	ノ no				Ю yu	АТ alpha ypsilon
•—••	カ ka	ㄱ k(g)	ل lam	ל lamed	Л L	Λ lambda
•—•—	ロ ro	ㅊ ch(j)	ع ain		Я ya	АI alpha iota
•——•	ツ tu	ㅈ ch(j)		ב peh	П P	П pi
•———	ヲ wo	ㅌ h	ج jeem	ע ayen	Й Y	ТI ypsilon iota
—•••	ハ ha	ㄷ t(d)	ب ba	בב bet	Б B	В beta
—••—	マ ma	ㅋ k'	ص sad		Ц TS	Ξ xi
—•—•	ニ ni	ㅊ ch'	ث tha	ס samech	Ы Z	Θ theta
—•——	ケ ke	ㅖ e	ظ za		Ь Z	Т ypsilon
——••	フ hu	ㄸ t'	ذ dhal	ז zain	З Z	Ζ zeta
——•—	ネ ne	ㅐ ae	ق qaf	פ kof	Ш SHCH	Ψ psi
———•	ソ so		ز zay		Ч CH	ЕТ epsilon ypsilon
————	コ ko		ش sheen		Ш SH	Х khi
•• ••	ト to		غ he			
••—•—	ミ mi					
••— ••	● han-nigori					
•—••—	オ o					
•—•—•	ヰ (w)i					
•—•——	ン n					
•—— •—	テ te					
•—— ••	ヱ (w)e					
•—•—•	‐ hyphen					
•—••—	セ se					
——•——	メ me					
——•—•	モ mo					
—••—•	ユ yu					
—••—•	キ ki					
—•—••	サ sa					
—•—•—	ル ru					
—•——•	エ e					
——••—	ヒ hi					
——•—•	シ si					
———•—	ア a					
———•—	ス su					
•—•••—				ی lam-alif			

Spanish Phonetics

America	ah-MAIR-ika
Brasil	brah-SIL
Canada	cana-DAH
Dinamarca	dina-MAR-ka
Espana	es-PAHN-yah
Francia	FRAHN-seeah
Grenada	gre-NAH-dah
Holanda	oh-LONN-dah
Italia	i-TAL-eeah
Japon	hop-OWN
Kilowatio	kilo-WAT-eeoh
Lima	LIMA
Mejico	MEH-heeco
Norvega	nor-WAY-gah
Ontario	on-TAR-eeoh
Portugal	portu-GAL
Quito	KEY-toe
Roma	ROW-mah
Santiago	santee-AH-go
Toronto	tor-ON-toe
Uniforme	oonee-FORM-eh
Victoria	vic-TOR-eeah
Washington, Wisky	washingtone, wisky
Xilofono	see-LOW-phono
Yucatan	yuca-TAN
Zelandia	see-LAND-eeah

W		DOE-bleh-vay
Ø	cero	SEH-roe
1	uno	OO-no
2	dos	DOS
3	tres	TRAYCE
4	cuatro	KWAT-roe
5	cinco	SINK-oh
6	seis	SAYCE
7	siete	see-AY-teh
8	ocho	OCH-oh
9	nueve	new-AY-veh

—John Mason Jr., EA4AXW

A large selection of phonetic alphabets is at
http://www.cl.cam.ac.uk/users/bck1/menu.html
and
http://www.columbia.edu/~fuat/cuarc/phonetic. html

DX Operating Code

For W/VE Amateurs

Some amateur DXers have caused considerable confusion and interference in their efforts to work DX stations. The points below, if observed by all W/VE amateurs, will help make DX more enjoyable for all.

1) *Call* DX only after he calls CQ, QRZ? or signs $\overline{SK}$, or voice equivalents thereof. Make your calls short.

2) Do not call a DX station:
   a) On the frequency of the station he is calling until you are sure the QSO is over ($\overline{SK}$).
   b) Because you hear someone else calling him.
   c) When he signs $\overline{KN}$, $\overline{AR}$ or $\overline{CL}$.
   d) Exactly on his frequency.
   e) After he calls a directional CQ, unless of course you are in the right direction or area.

3) Keep within frequency band limits. Some DX stations can get away with working outside, but you cannot.

4) Observe calling instructions given by DX stations. Example: 15U means "call 15 kHz *up* from my frequency." 15D means *down*, etc.

5) Give honest reports. Many DX stations *depend* on W/VE reports for adjustment of station and equipment.

6) Keep your signal clean. Key clicks, ripple, feedback or splatter gives you a bad reputation and may get you a citation from FCC.

7) *Listen* and call the station you want. Calling CQ DX is not the best assurance that the rare DX will reply.

8) When there are several W or VE stations waiting, avoid asking DX to "listen for a friend." Also avoid engaging him in a ragchew against his wishes.

For Overseas Amateurs

To all overseas amateur stations:

In their eagerness to work you, many W and VE amateurs resort to practices that cause confusion and QRM. Most of this is good-intentional but ill-advised; some of it is intentional and selfish. The key to the cessation of unethical DX operating practices is in your hands. We believe that your adoption of certain operating habits will increase your enjoyment of Amateur Radio and that of amateurs on this side who are eager to work you. We recommend your adoption of the following principles:

1) Do not answer calls on your own frequency.

2) Answer calls from W/VE stations only when their signals are of good quality.

3) Refuse to answer calls from other stations when you are already in contact with someone, and do not acknowledge calls from amateurs who indicate they wish to be "next."

4) Give *everybody* a break. When many W/VE amateurs are patiently and quietly waiting to work you, avoid complying with requests to "listen for a friend."

5) Tell listeners where to call you by indicating how many kilohertz up (U) or down (D) from your frequency you are listening.

6) Use the ARRL-recommended ending signals, especially $\overline{KN}$ to indicate to impatient listeners the status of the QSO. $\overline{KN}$ means "Go ahead (specific station); all others keep out."

7) Let it be known that you avoid working amateurs who are constant violators of these principles.

The Origin of "73"

The traditional expression "73" goes right back to the beginning of the landline telegraph days. It is found in some of the earliest editions of the numerical codes, each with a different definition, but each with the same idea in mind—it indicated that the end, or signature, was coming up. But there are no data to prove that any of these were used.

The first authentic use of 73 is in the publication *The National Telegraphic Review and Operators' Guide*, first published in April 1857. At that time, 73 meant "My love to you"! Succeeding issues of this publication continued to use this definition of the term. Curiously enough, some of the other numerals used then had the same definition as they have now, but within a short time, the use of 73 began to change.

In the National Telegraph Convention, the numeral was changed from the Valentine-type sentiment to a vague sign of fraternalism. Here, 73 was a greeting, a friendly "word" between operators and it was so used on all wires.

In 1859, the Western Union Company set up the standard "92 Code." A list of numerals from one to 92 was compiled to indicate a series of prepared phrases for use by the operators on the wires. Here, in the 92 Code, 73 changes from a fraternal sign to a very flowery "accept my compliments," which was in keeping with the florid language of that era.

Over the years from 1859 to 1900, the many manuals of telegraphy show variations of this meaning. Dodge's *The Telegraph Instructor* shows it merely as "compliments." The *Twentieth Century Manual of Railway and Commercial Telegraphy* defines it two ways, one listing as "my compliments to you"; but in the glossary of abbreviations it is merely "compliments." Theodore A. Edison's *Telegraphy Self-Taught* shows a return to "accept my compliments." By 1908, however, a later edition of the Dodge Manual gives us today's definition of "best regards" with a backward look at the older meaning in another part of the work where it also lists it as "compliments."

"Best regards" has remained ever since as the "put-it-down-in-black-and-white" meaning of 73 but it has acquired overtones of much warmer meaning. Today, amateurs use it more in the manner that James Reid had intended that it be used—a "friendly word between operators."—*Louise Ramsey Moreau, W3WRE*

The RST System

READABILITY
1—Unreadable.
2—Barely readable, occasional words distinguishable.
3—Readable with considerable difficulty.
4—Readable with practically no difficulty.
5—Perfectly readable.
SIGNAL STRENGTH
1—Faint signals barely perceptible.
2—Very weak signals.
3—Weak signals.
4—Fair signals.
5—Fairly good signals.
6—Good signals.
7—Moderately strong signals.
8—Strong signals.
9—Extremely strong signals.
TONE
1—Sixty-cycle ac or less, very rough and broad.
2—Very rough ac, very harsh and broad.
3—Rough ac tone, rectified but not filtered.
4—Rough note, some trace of filtering.
5—Filtered rectified ac but strongly ripple-modulated.
6—Filtered tone, definite trace of ripple modulation.
7—Near pure tone, trace of ripple modulation.
8—Near perfect tone, slight trace of modulation.
9—Perfect tone, no trace of ripple or modulation of any kind.

The "tone" report refers only to the purity of the signal, and has no connection with its stability or freedom from clicks or chirps. If the signal has the characteristic steadiness of crystal control, add X to the report (e.g., RST 469X). If it has a chirp or "tail" (either on "make" or "break") add C (e.g., 469C). If it has clicks or noticeable other keying transients, add K (e.g., 469K). Of course a signal could have both chirps and clicks, in which case both C and K could be used (e.g., RST 469CK).

Q Signals

Given below are a number of Q signals whose meanings most often need to be expressed with brevity and clarity in amateur work. (Q abbreviations take the form of questions only when each is sent followed by a question mark.)

QRG Will you tell me my exact frequency (or that of ____)? Your exact frequency (or that of ____) is ____ kHz.
QRH Does my frequency vary? Your frequency varies.
QRI How is the tone of my transmission? The tone of your transmission is ____ (1. Good; 2. Variable; 3. Bad).
QRJ Are you receiving me badly? I cannot receive you. Your signals are too weak.
QRK What is the intelligibility of my signals (or those of ____)? The intelligibility of your signals (or those of ____) is ____ (1. Bad; 2. Poor; 3. Fair; 4. Good; 5. Excellent).
QRL Are you busy? I am busy (or I am busy with ____). Please do not interfere.
QRM Is my transmission being interfered with? Your transmission is being interfered with ____ (1. Nil; 2. Slightly; 3. Moderately; 4. Severely; 5. Extremely.)
QRN Are you troubled by static? I am troubled by static ____ (1-5 as under QRM).
QRO Shall I increase power? Increase power.
QRP Shall I decrease power? Decrease power.
QRQ Shall I send faster? Send faster (____ WPM).
QRS Shall I send more slowly? Send more slowly (____ WPM).
QRT Shall I stop sending? Stop sending.
QRU Have you anything for me? I have nothing for you.
QRV Are you ready? I am ready.
QRW Shall I inform ____ that you are calling on ____ kHz? Please inform ____ that I am calling on ____ kHz.
QRX When will you call me again? I will call you again at ____ hours (on ____ kHz).
QRY What is my turn? Your turn is numbered ____
QRZ Who is calling me? You are being called by ____ (on ____ kHz).
QSA What is the strength of my signals (or those of ____)? The strength of your signals (or those of ____) is ____ (1. Scarcely perceptible; 2. Weak; 3. Fairly good; 4. Good; 5. Very good).
QSB Are my signals fading? Your signals are fading.
QSD Is my keying defective? Your keying is defective.

QSG Shall I send ____ messages at a time? Send ____ messages at a time.
QSK Can you hear me between your signals and if so can I break in on your transmission? I can hear you between my signals; break in on my transmission.
QSL Can you acknowledge receipt? I am acknowledging receipt
QSM Shall I repeat the last message which I sent you, or some previous message? Repeat the last message which you sent me [or message(s) number(s) ____].
QSN Did you hear me (or ____) on ____ kHz? I did hear you (or ____) on ____ kHz.
QSO Can you communicate with ____ direct or by relay? I can communicate with ____ direct (or by relay through ____).
QSP Will you relay to ____? I will relay to ____
QST General call preceding a message addressed to all amateurs and ARRL members. This is in effect "CQ ARRL."
QSU Shall I send or reply on this frequency (or on ____ kHz)? Send a series of Vs on this frequency (or ____ kHz).
QSW Will you send on this frequency (or on ____ kHz)? I am going to send on this frequency (or on ____ kHz).
QSX Will you listen to ____ on ____ kHz? I am listening to ____ on ____ kHz.
QSY Shall I change to transmission on another frequency? Change to transmission on another frequency (or on ____ kHz).
QSZ Shall I send each word or group more than once? Send each word or group twice (or ____ times).
QTA Shall I cancel message number ____? Cancel message number ____
QTB Do you agree with my counting of words? I do not agree with your counting of words. I will repeat the first letter or digit of each word or group.
QTC How many messages have you to send? I have ____ messages for you (or for ____).
QTH What is your location? My location is ____
QTR What is the correct time? The time is ____

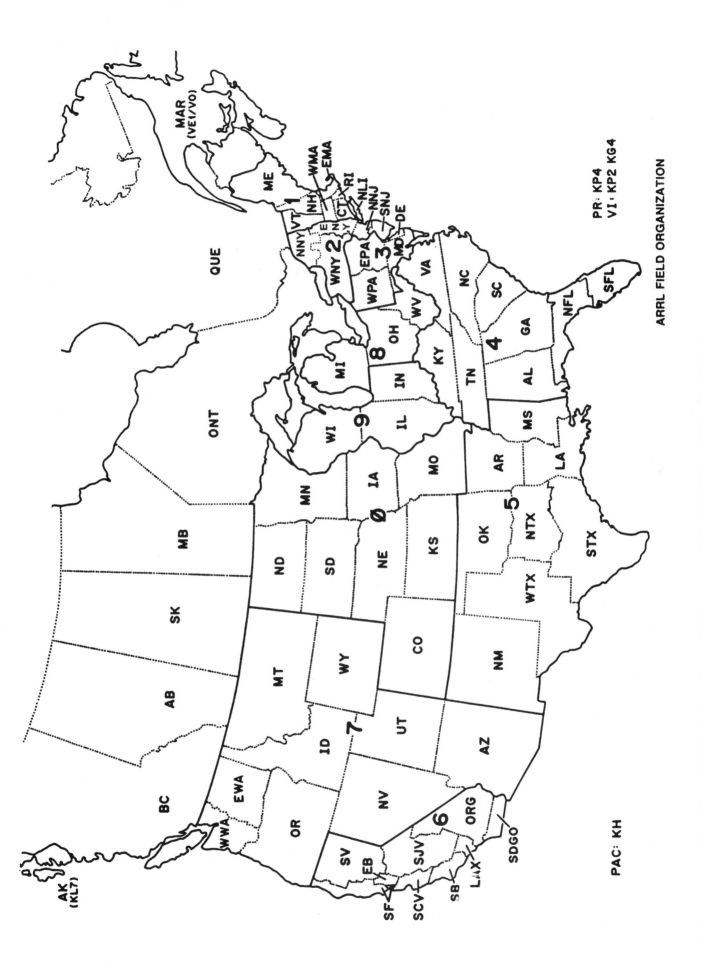

ARRL FIELD ORGANIZATION

PR: KP4
VI: KP2 KG4

PAC: KH

**CALIFORNIA COUNTIES/ARRL SECTIONS
HAM OUTLINE MAP No 27**

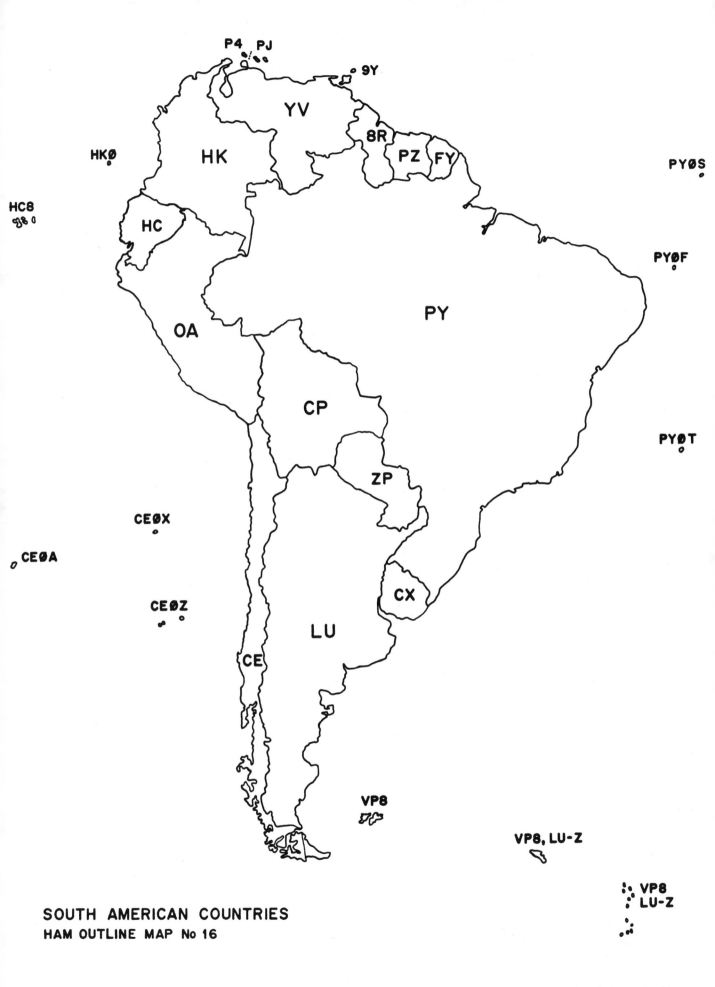

P4 PJ

9Y

YV

HKØ

HK

8R

PZ FY

PYØS

HC8

HC

PYØF

OA

PY

CP

PYØT

ZP

CEØX

CEØA

CX

CEØZ

LU

CE

VP8

VP8, LU-Z

VP8
LU-Z

SOUTH AMERICAN COUNTRIES
HAM OUTLINE MAP No 16

PP1—Espirito Santo ES
PP2—Goias GO
PP5—Santa Catarina SC
PP6—Sergipe SE
PP7—Alagoas AL
PP8—Amazonas AM
PR7—Paraiba PB
PR8—Maranhao MA
PS7—Rio Grande do Norte
PS8—Piaui PI
PT2—Distrite Federal DF (Brasilia)
PT7—Ceara CE
PT8—Acre AC
PT9—Mato Grosso do Sul MS
PU8—Amapa AP
PV8—Roraima RR
PW8—Rondonia RO
PY1—Rio de Janeiro RJ
PY2—Sao Paulo SP
PY3—Rio Gran de do Sul RS
PY4—Minas Gerais MG
PY5—Parana PR
PY6—Bahia BA
PY7—Pernambuco PE
PY8—Para PA
PY9—Mato Grosso MT
PY0T—Trinidade Island
PY0F—Fernando de Noronha Is.
PY0S—Sao Pedro/S. Paulo Rocks

BRAZILIAN STATES
PREFIX: PP-PY ZV-ZZ
HAM OUTLINE MAP No 17

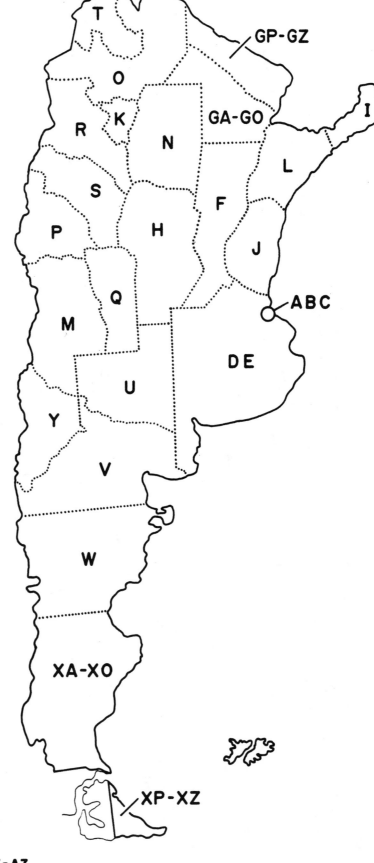

Province is indicated by letter(s)
after number in call sign

A B C	Buenos Aires City
D E	Buenos Aires Province
F	Santa Fe
GA-GO	Chaco
GP-GZ	Formosa
H	Cordoba
I	Misiones
J	Entre Rios
K	Tucuman
L	Corrientes
M	Mendoza
N	Santiago Del Estero
O	Salta
P	San Juan
Q	San Luis
R	Catamarga
S	La Rioja
T	Jujuy
U	La Pampas
V	Rio Negro
W	Chubut
XA-XO	Santa Cruz
XP-XZ	Tierra Del Fuego
Y	Neuquen
Z	Antarctica

ARGENTINE PROVINCES
  PREFIX: LO-LW L2-L9 AY-AZ
HAM OUTLINE MAP No 18

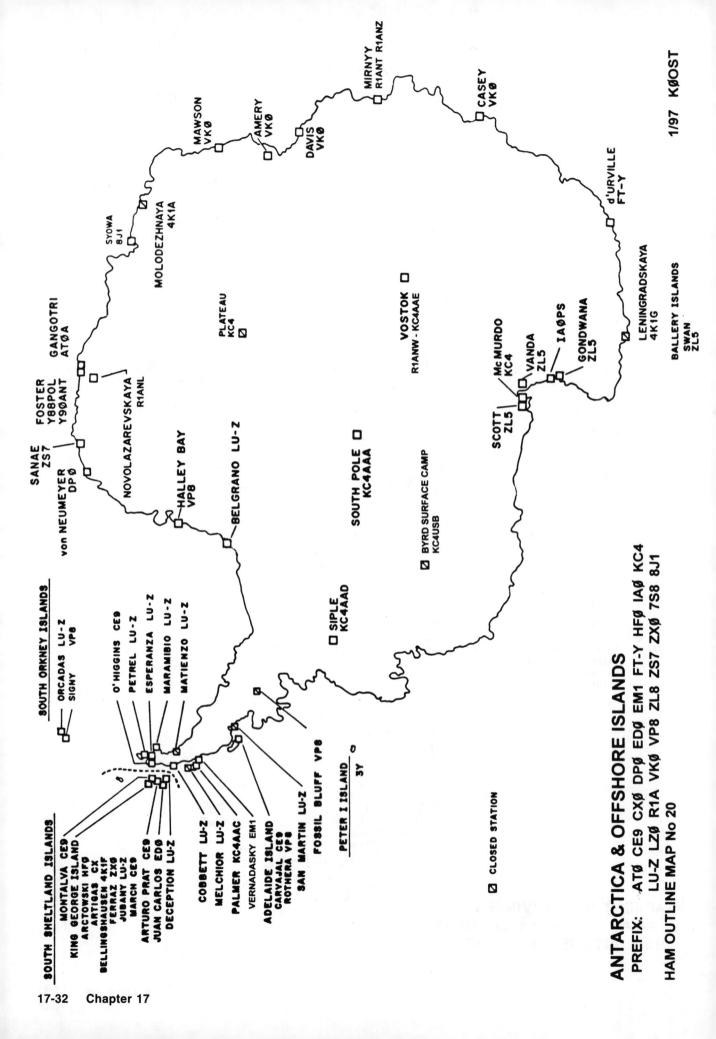

ANTARCTICA & OFFSHORE ISLANDS

PREFIX: ATØ CE9 CXØ DPØ EDØ EM1 FT-Y HFØ IAØ KC4
LU-Z LZØ R1A VKØ VP8 ZL8 ZS7 ZXØ 7S8 8J1

HAM OUTLINE MAP No 20

1/97 KØOST

SYOWA
8J1

MAWSON
VKØ

AMERY
VKØ

DAVIS
VKØ

MIRNYY
R1ANT R1ANZ

CASEY
VKØ

d'URVILLE
FT-Y

MOLODEZHNAYA
4K1A

PLATEAU
KC4

VOSTOK
R1ANW - KC4AAE

LENINGRADSKAYA
4K1G

BALLERY ISLANDS
SWAN
ZL5

McMURDO
KC4

VANDA
ZL5

IAØPS

GONDWANA
ZL5

SCOTT
ZL5

SANAE
ZS7

FOSTER
Y88POL
Y9ØANT

GANGOTRI
ATØA

von NEUMEYER
DPØ

NOVOLAZAREVSKAYA
R1ANL

HALLEY BAY
VP8

BELGRANO LU-Z

SOUTH POLE
KC4AAA

BYRD SURFACE CAMP
KC4USB

SOUTH ORKNEY ISLANDS

— ORCADAS   LU-Z
SIGNY   VP8

SOUTH SHETLAND ISLANDS

MONTALVA CE9
KING GEORGE ISLAND
ARCTOWSKI HFØ
ARTIGAS CX
BELLINGSHAUSEN 4K1F
FERRAZ ZXØ
JUBANY LU-Z
MARCH CE9

ARTURO PRAT CE9
JUAN CARLOS EDØ
DECEPTION LU-Z

COBBETT LU-Z
MELCHIOR LU-Z
PALMER KC4AAC
VERNADASKY EM1
ADELAIDE ISLAND
CARVAJAL CE9
ROTHERA VP8
SAN MARTIN LU-Z
FOSSIL BLUFF VP8

O'HIGGINS CE9
PETREL LU-Z
ESPERANZA LU-Z
MARAMIBIO LU-Z
MATIENZO LU-Z

SIPLE
KC4AAD

PETER I ISLAND

3Y

⊠ CLOSED STATION

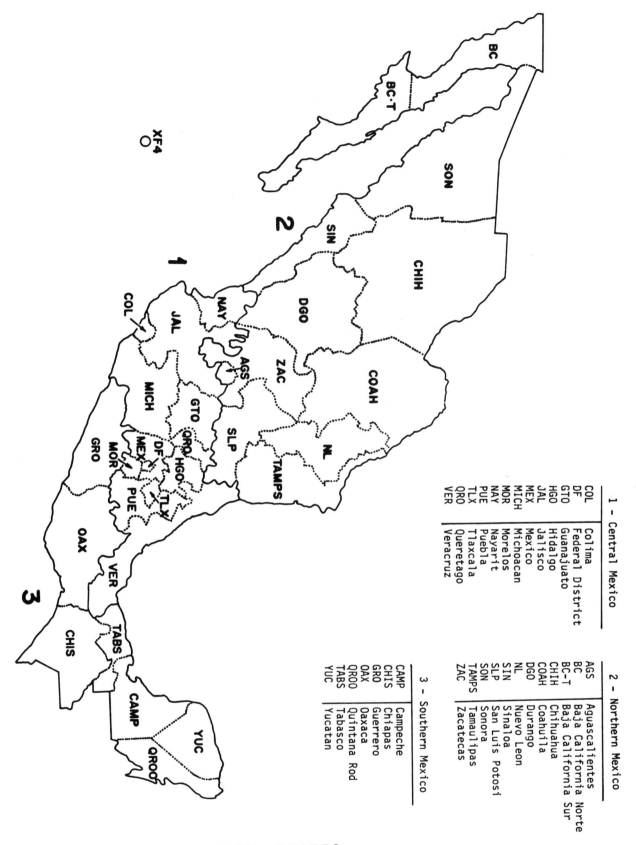

MEXICAN STATES

PREFIX: XA-XI 4A-4C 6D-6J

HAM OUTLINE MAP No 4

1 - Central Mexico

COL	Colima
DF	Federal District
GTO	Guanajuato
HGO	Hidalgo
JAL	Jalisco
MEX	Mexico
MICH	Michoacan
MOR	Morelos
NAY	Nayarit
PUE	Puebla
TLX	Tlaxcala
QRO	Queretago
VER	Veracruz

2 - Northern Mexico

AGS	Aguascalientes
BC	Baja California Norte
BC-T	Baja California Sur
CHIH	Chihuahua
COAH	Coahuila
DGO	Durango
NL	Nuevo Leon
SIN	Sinaloa
SLP	San Luis Potosi
SON	Sonora
TAMPS	Tamaulipas
ZAC	Zacatecas

3 - Southern Mexico

CAMP	Campeche
CHIS	Chiapas
GRO	Guerrero
OAX	Oaxaca
QROO	Quintana Rod
TABS	Tabasco
YUC	Yucatan

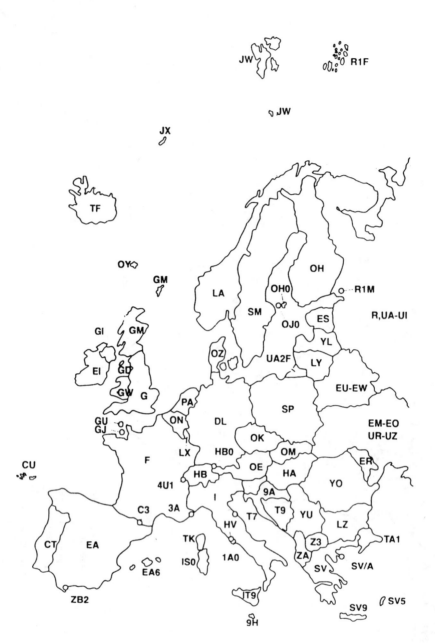

EUROPEAN COUNTRIES
HAM OUTLINE MAP No 6

12/95   KØOST

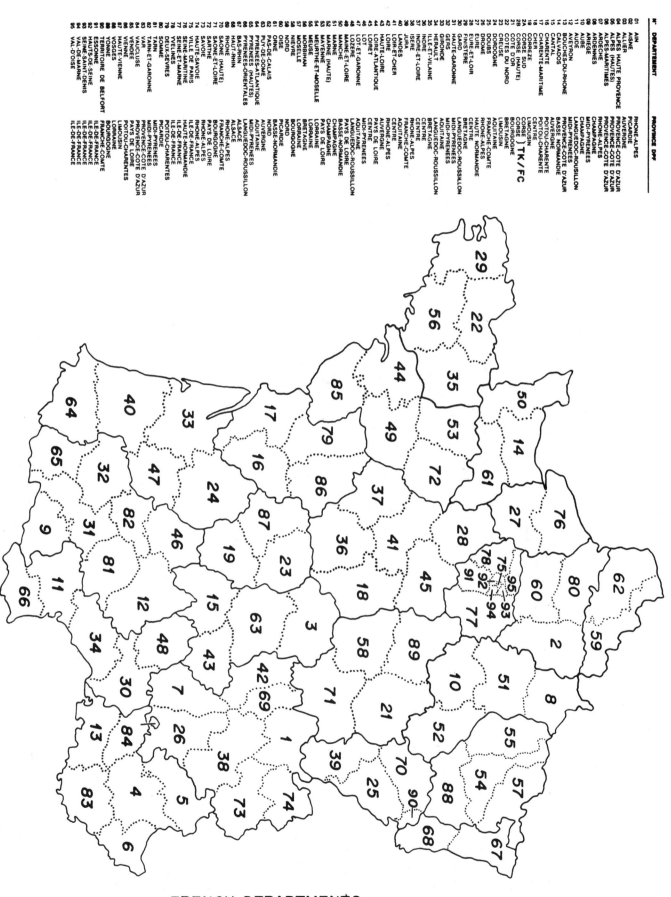

FRENCH DEPARTMENTS
PREFIX: F HW-HY TH TK TM
HAM OUTLINE MAP No 7

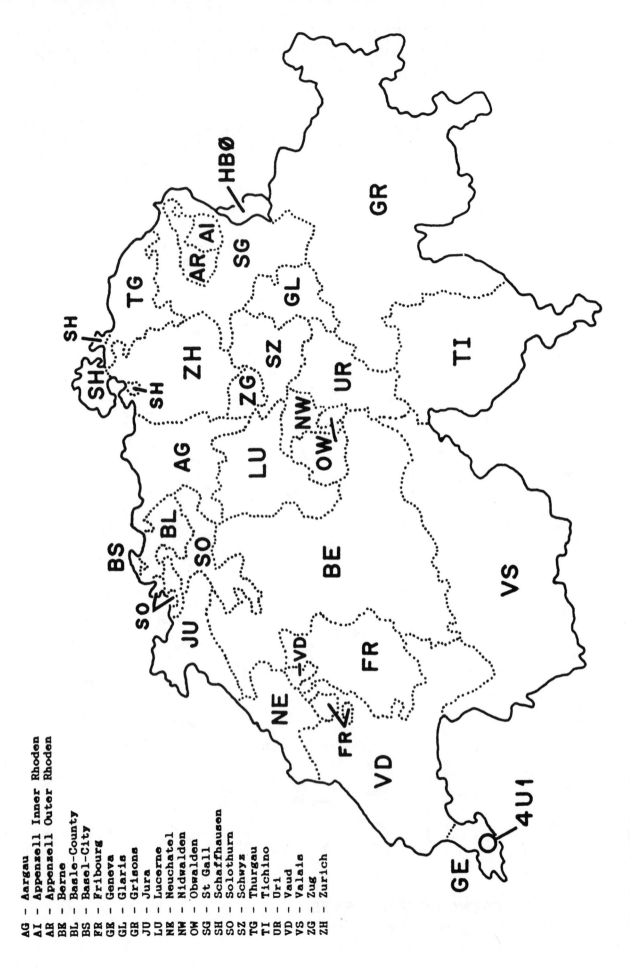

SWISS CANTONS

PREFIX: HB HE

HAM OUTLINE MAP No 44

AG – Aargau
AI – Appenzell Inner Rhoden
AR – Appenzell Outer Rhoden
BE – Berne
BL – Basle-County
BS – Basel-City
FR – Fribourg
GE – Geneva
GL – Glaris
GR – Grisons
JU – Jura
LU – Lucerne
NE – Neuchatel
NW – Nidwalden
OW – Obwalden
SG – St Gall
SH – Schaffhausen
SO – Solothurn
SZ – Schwyz
TG – Thurgau
TI – Tichino
UR – Uri
VD – Vaud
VS – Valais
ZG – Zug
ZH – Zurich

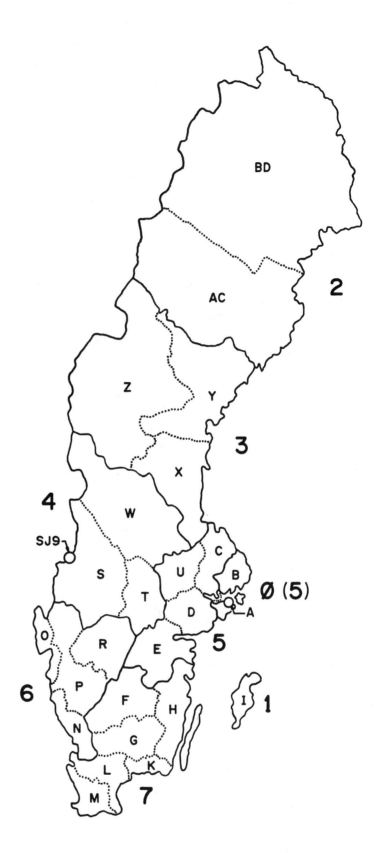

1 - Gotland

| I | Gotland |

2 - Northern

| AC | Vasterbotten |
| BD | Norrbotten |

3 - Northern

X	Gavleborg
Y	Vasternorrland
Z	Jamtland

4 - West Central

S	Varmland
T	Orebro
W	Kopparberg

5 - East Central

A	Stockholm City
B	Stockholm County
C	Uppsala
D	Sodermanland
E	Ostergotland
U	Vastmanland

6 - Southwestern

N	Halland
O	Goteborg och Bohus
P	Alvsborg
R	Skaraborg

7 - Southern

F	Jonkoping
G	Kronoberg
H	Kalmar
K	Blekinge
L	Kristianstad
M	Malmohus

Ø - Stockholm Area

| A | Stockholm City |
| B | Stockholm County |

| SM8 | Maritime Mobile |
| SJ9 | Morokulien |

SK	Club
SL	Military
SM	Individual

SWEDISH COUNTIES (LAENS)
PREFIX: SA-SM 7S 8S
HAM OUTLINE MAP No 8

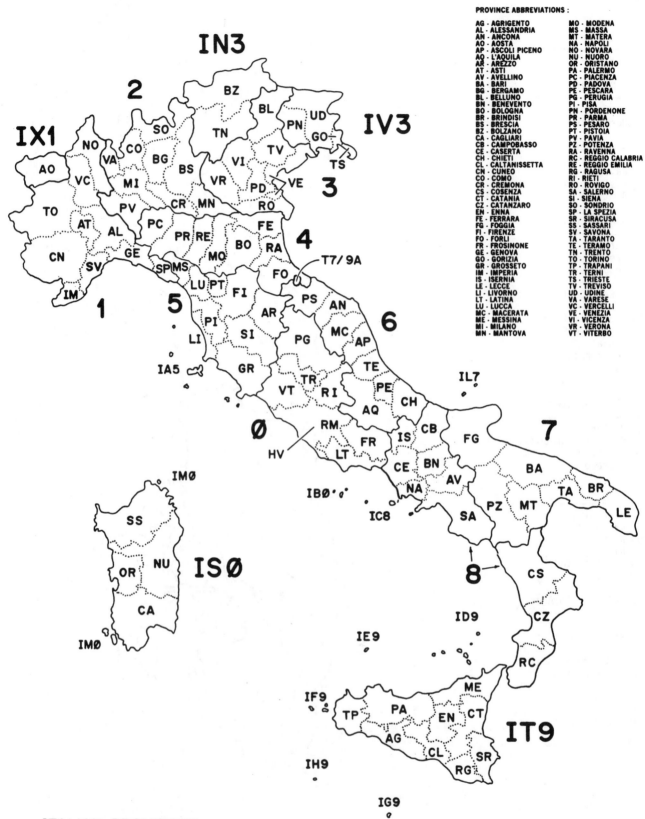

PROVINCE ABBREVIATIONS :

AG - AGRIGENTO	MO - MODENA
AL - ALESSANDRIA	MS - MASSA
AN - ANCONA	MT - MATERA
AO - AOSTA	NA - NAPOLI
AP - ASCOLI PICENO	NO - NOVARA
AQ - L'AQUILA	NU - NUORO
AR - AREZZO	OR - ORISTANO
AT - ASTI	PA - PALERMO
AV - AVELLINO	PC - PIACENZA
BA - BARI	PD - PADOVA
BG - BERGAMO	PE - PESCARA
BL - BELLUNO	PG - PERUGIA
BN - BENEVENTO	PI - PISA
BO - BOLOGNA	PN - PORDENONE
BR - BRINDISI	PR - PARMA
BS - BRESCIA	PS - PESARO
BZ - BOLZANO	PT - PISTOIA
CA - CAGLIARI	PV - PAVIA
CB - CAMPOBASSO	PZ - POTENZA
CE - CASERTA	RA - RAVENNA
CH - CHIETI	RC - REGGIO CALABRIA
CL - CALTANISSETTA	RE - REGGIO EMILIA
CN - CUNEO	RG - RAGUSA
CO - COMO	RI - RIETI
CR - CREMONA	RO - ROVIGO
CS - COSENZA	SA - SALERNO
CT - CATANIA	SI - SIENA
CZ - CATANZARO	SO - SONDRIO
EN - ENNA	SP - LA SPEZIA
FE - FERRARA	SR - SIRACUSA
FG - FOGGIA	SS - SASSARI
FI - FIRENZE	SV - SAVONA
FO - FORLI	TA - TARANTO
FR - FROSINONE	TE - TERAMO
GE - GENOVA	TN - TRENTO
GO - GORIZIA	TO - TORINO
GR - GROSSETO	TP - TRAPANI
IM - IMPERIA	TR - TERNI
IS - ISERNIA	TS - TRIESTE
LE - LECCE	TV - TREVISO
LI - LIVORNO	UD - UDINE
LT - LATINA	VA - VARESE
LU - LUCCA	VC - VERCELLI
MC - MACERATA	VE - VENEZIA
ME - MESSINA	VI - VICENZA
MI - MILANO	VR - VERONA
MN - MANTOVA	VT - VITERBO

ITALIAN PROVINCES
PREFIX: I IA-IZ
HAM OUTLINE MAP No 9

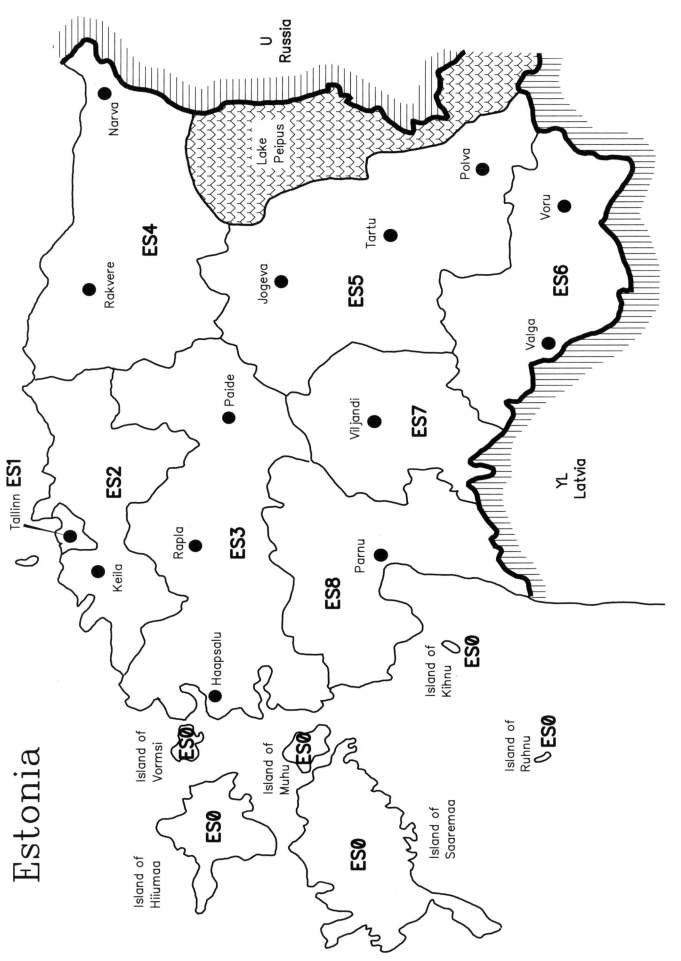

Estonia

Island of Hiiumaa

ES0

Island of Vormsi

ES0

Island of Muhu

ES0

Island of Saaremaa

ES0

Island of Kihnu

ES0

Island of Ruhnu

ES0

Tallinn ES1

Keila

ES2

Rapla

ES3

Haapsalu

Paide

ES8

Parnu

Viljandi

ES7

Rakvere

ES4

Narva

Jogeva

ES5

Tartu

Lake Peipus

Polva

Voru

ES6

Valga

YL Latvia

U Russia

SLOVAKIA – OM

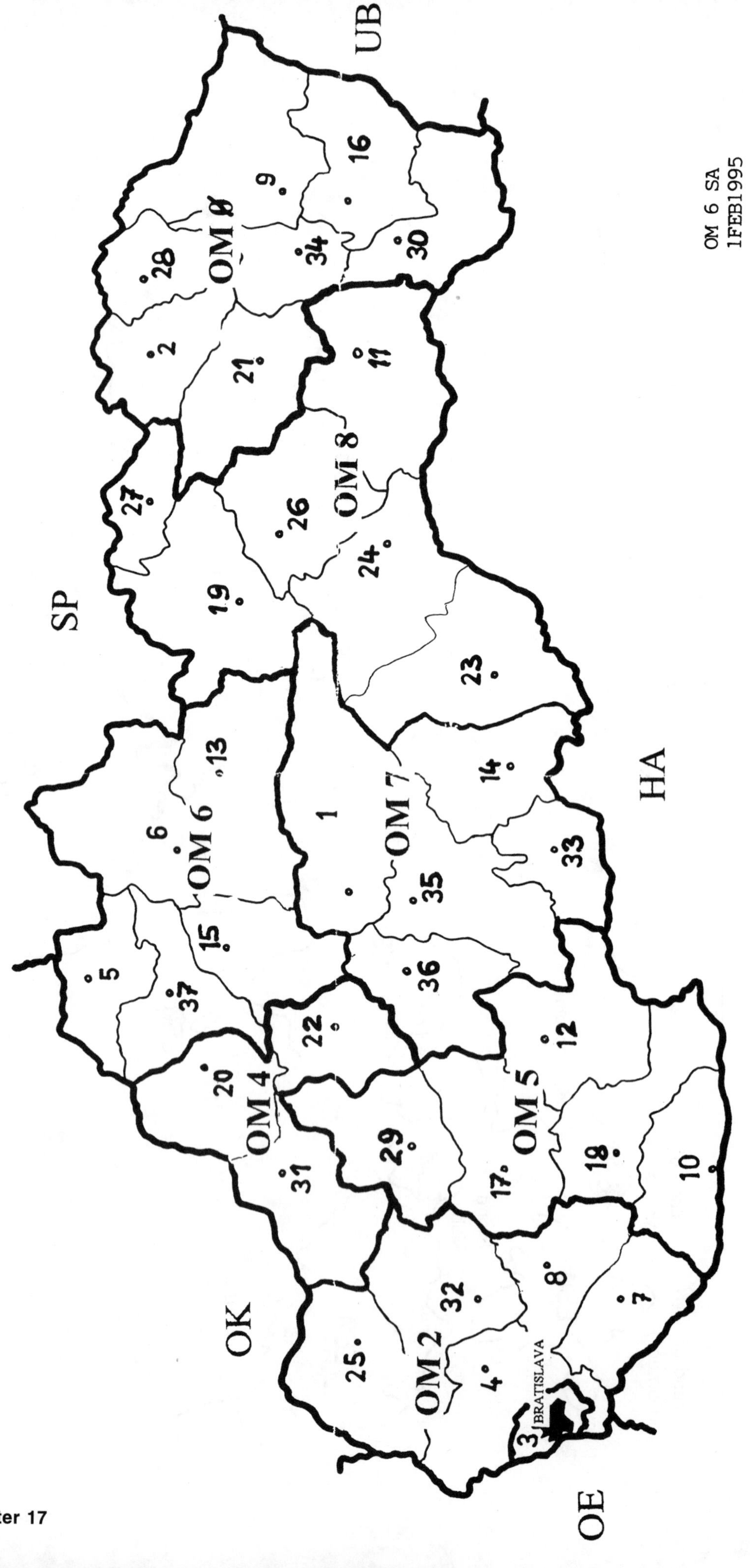

UB

SP

OK

OE

HA

OM 6 SA
1FEB1995

OM 1

28
9
16
34
30
2
21
11
OM 8
27
26
24
19
23
13
OM 6
6
14
1
OM 7
35
33
15
5
37
36
12
22
OM 5
20
OM 4
29
17
18
31
OM 2
32
10
25
8
4
7
BRATISLAVA
3

SHETLAND

ORKNEY

WESTERN ISLES

HIGHLANDS

GRAMPIAN

TAYSIDE

FIFE

CENTRAL

LOTHIAN

STRATH-CLYDE

BORDERS

DUMFRIES & GALLOWAY

NORTH-UMBERLAND

TYNE & WEAR

DURHAM

CLEVELAND

CUMBRIA

NORTH YORKSHIRE

ISLE MAN

HUMBER SIDE

LANCA-SHIRE

WEST YORKSHIRE

MERSEYSIDE

GM

SOUTH YORKSHIRE

CLWYD

CHESHIRE

DERBY-SHIRE

NT

LINCOLN-SHIRE

GWYNEDD

STAFFORD-SHIRE

NORFOLK

SHROP-SHIRE

LEICESTER SHIRE

WM

CAM-BRIDGE

POWYS

HEREFORD WORCESTER

WARWICK-SHIRE

NORTHAMPTON-SHIRE

SUFFOLK

BD

DYFED

GLOUCESTER-SHIRE

OXFORD-SHIRE

BK

HT

ESSEX

WG

MG

GWENT

GREATER LONDON

SG

AVON

BERKSHIRE

SURREY

KENT

WILT-SHIRE

HAMP-SHIRE

SOMERSET

WEST SUSSEX

EAST SUSSEX

DORSET

ISLE WRIGHT

DEVON

CORNWALL

BD BEDFORDSHIRE
BK BUCKINGHAMSHIRE
GM GREATER MANCHESTER
HT HERTFORDSHIRE
MG MID GLAMORGAN
NT NOTTINGHAMSHIRE
SG SOUTH GLAMORGAN
WG WEST GLAMORGAN
WM WEST MIDLANDS

BRITISH COUNTIES
HAM OUTLINE MAP No 10

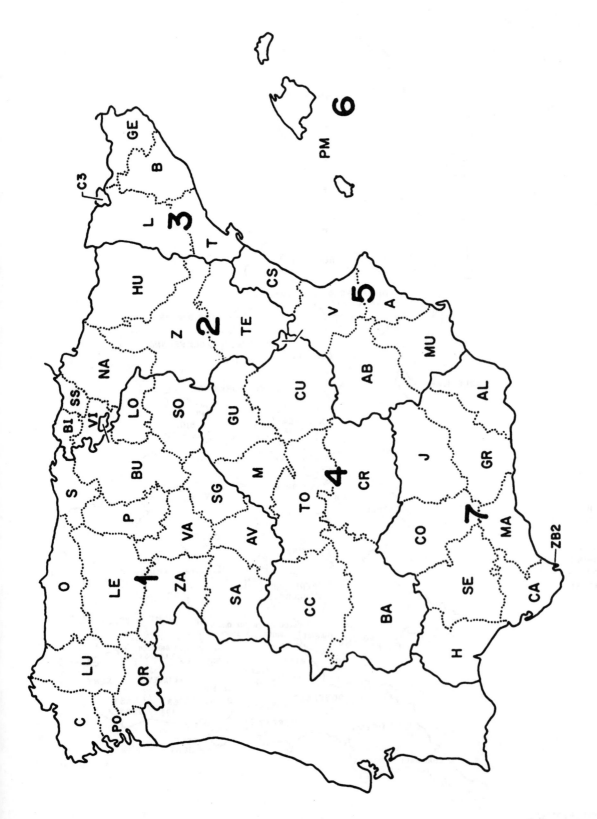

SPANISH PROVINCES
PREFIX: AM-AO EA-EH
HAM OUTLINE MAP No 41

A	5	ALICANTE
AB	5	ALBACETE
AL	7	ALMERIA
AV	1	AVILA
B	3	BARCELONA
BA	4	BARDAJOZ
BI	1	VIZCAYA
BU	1	BURGOS
C	1	LA CORUNA
CA	7	CADIZ
CC	4	CACERES
CE	9	CEUTA
CO	7	CORDOBA
CR	4	CIUDAD READ
CS	5	CASTELLON
CU	5	CUENCA
GC	8	LAS PALMAS
GE	3	GERONA
GR	7	GRANADA
GU	4	GUADALAJARA
H	7	HUELVA
HU	2	HUESCA
J	7	JAEN
L	3	LERIDA
LE	1	LEON
LO	1	LA RIOJA
LU	1	LUGO
M	4	MADRID
MA	7	MALAGA
ML	9	MELILLA
MU	5	MURCIA
NA	2	NAVARRA
O	1	ASTURIAS
OR	1	ORENSE
P	1	PALENCIA
PM	6	BALEARES
PO	1	PONTEVEDRA
S	1	CANTABRIA
SA	1	SALAMANCA
SE	7	SEVILLA
SG	1	SEGOVIA
SO	1	SORIA
SS	2	GUIPUZCOA
T	3	TARRAGONA
TE	2	TERUEL
TF	8	TENERIFE
TO	4	TOLEDO
V	5	VALENCIA
VA	1	VALLADOLID
VI	1	ALAVA
Z	2	ZARAGOZA
ZA	1	ZAMORIA

4/91 KØOST

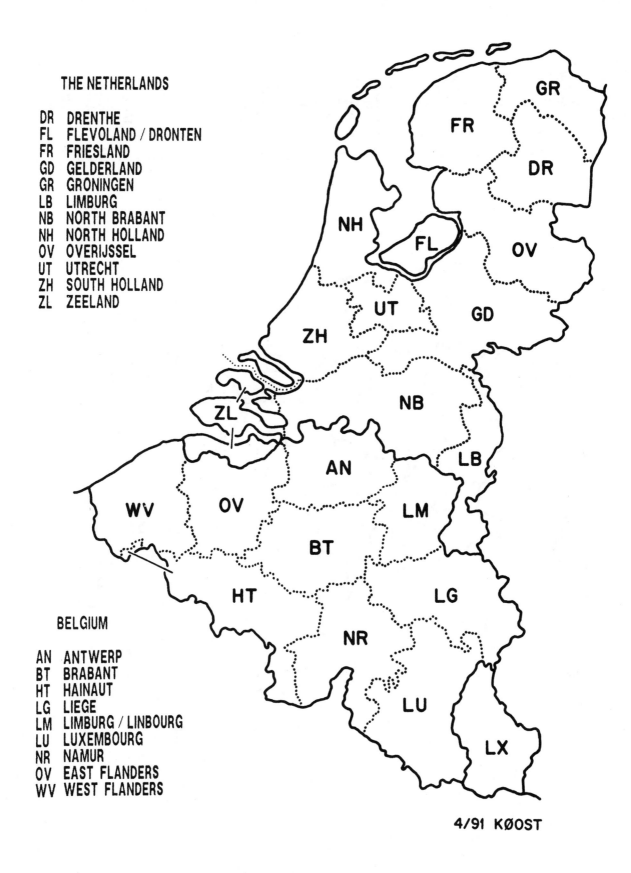

THE NETHERLANDS

DR   DRENTHE
FL   FLEVOLAND / DRONTEN
FR   FRIESLAND
GD   GELDERLAND
GR   GRONINGEN
LB   LIMBURG
NB   NORTH BRABANT
NH   NORTH HOLLAND
OV   OVERIJSSEL
UT   UTRECHT
ZH   SOUTH HOLLAND
ZL   ZEELAND

BELGIUM

AN   ANTWERP
BT   BRABANT
HT   HAINAUT
LG   LIEGE
LM   LIMBURG / LINBOURG
LU   LUXEMBOURG
NR   NAMUR
OV   EAST FLANDERS
WV   WEST FLANDERS

4/91 KØOST

BELGIAN / DUTCH PROVINCES
PREFIX: ON-OT / PA-PI
HAM OUTLINE MAP No 42

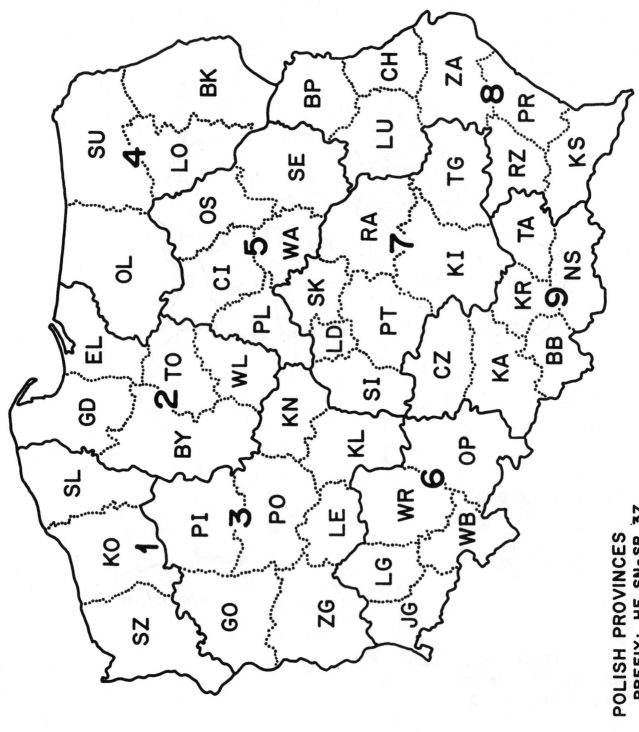

BB 9 BIELSKO BLALA
BK 4 BIALYSTOK
BP 8 BIATA PODLASKA
BY 2 BYDGOSZCZ
CH 8 CHELM
CI 5 CIECHANOW
CZ 9 CZESTOCHOWA
EL 2 ELBLAG
GD 2 GDANSK
GO 3 GORZOW
JG 6 JELENIA GORA
KA 9 KATOWICE
KI 7 KIELCE
KL 3 KALISZ
KN 3 KONIN
KO 1 KOSZALIN
KR 9 KRAKOW
KS 8 KROSNO
LD 7 LODZ
LE 3 LESZNO
LG 6 LEGNICA
LO 4 LOMZA
LU 8 LUBIN
NS 9 NOWY SACZ
OL 4 OLSZTYN
OP 6 OPOLE
OS 5 OSTROLEKA
PI 5 PILA
PL 5 PLOCK
PO 3 POZNAN
PR 8 PRZEMYSL
PT 7 PIOTRKOW
RA 7 RADOM
RZ 8 RZESZOW
SE 5 SIEDLCE
SI 7 SIERADZ
SK 7 SKIERNIEWICE
SL 1 SLUPSK
SU 4 SUWALKI
SZ 1 SZCZECIN
TA 9 TARNOW
TG 7 TARNOBRZEG
TO 2 TORUN
WA 5 WARSAWA
WB 6 WALBRZYCH
WL 2 WLOCLAWEK
WR 6 WROCLAW
ZA 6 ZAMOSC
ZG 3 ZIELONA GORA

4/91 KØOST

POLISH PROVINCES
PREFIX: HF SN-SR 3Z
HAM OUTLINE MAP No 43

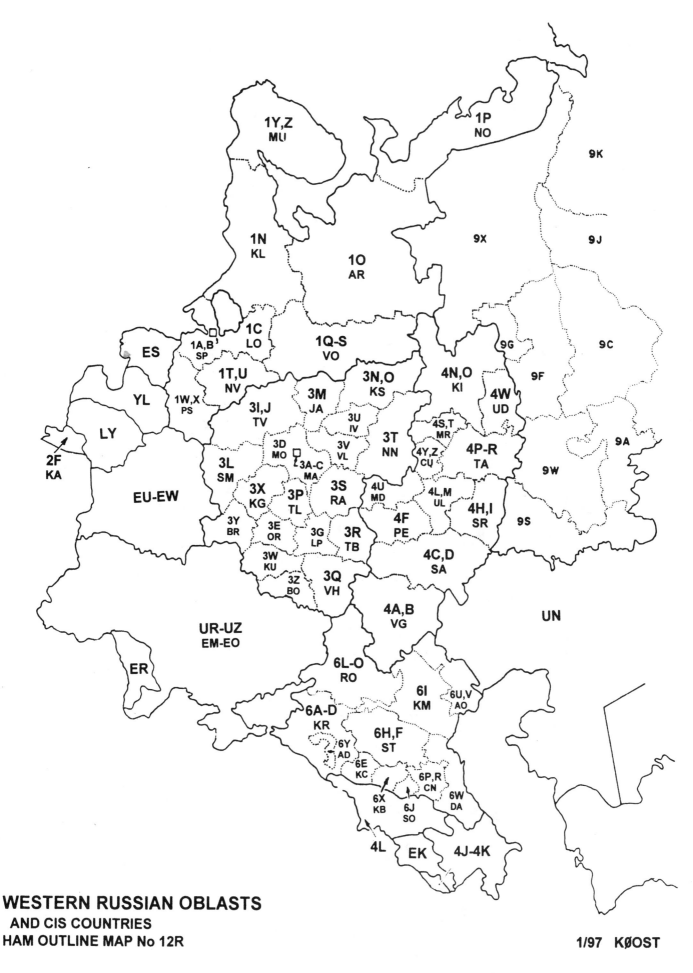

WESTERN RUSSIAN OBLASTS
AND CIS COUNTRIES
HAM OUTLINE MAP No 12R

1/97   KØOST

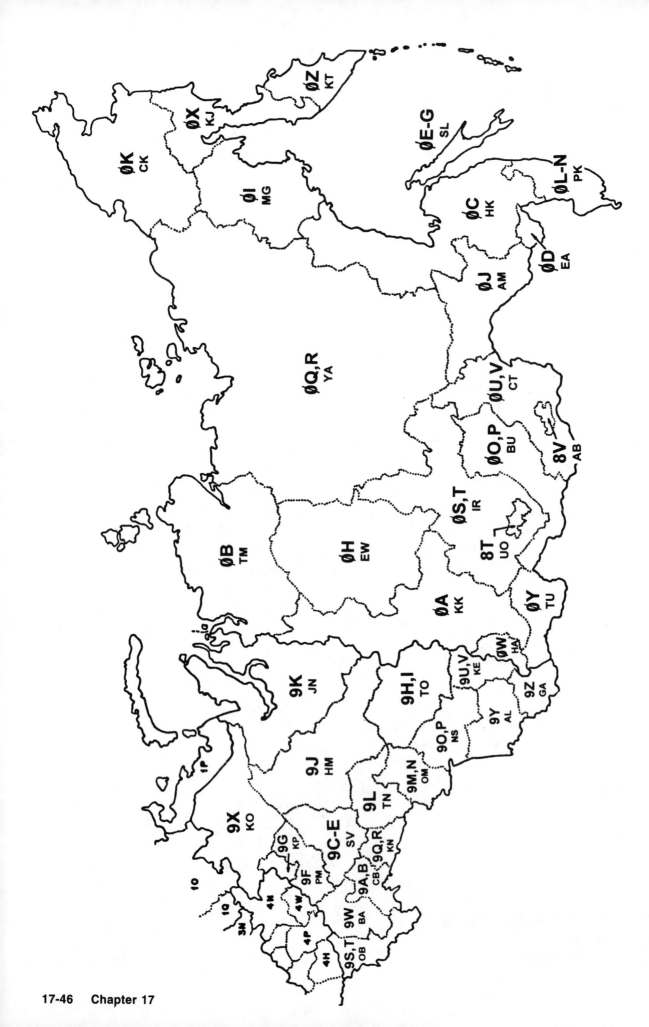

EASTERN RUSSIAN OBLASTS

FORMER UA9, Ø CALL AREAS

HAM OUTLINE MAP No 13

1/97 KØOST

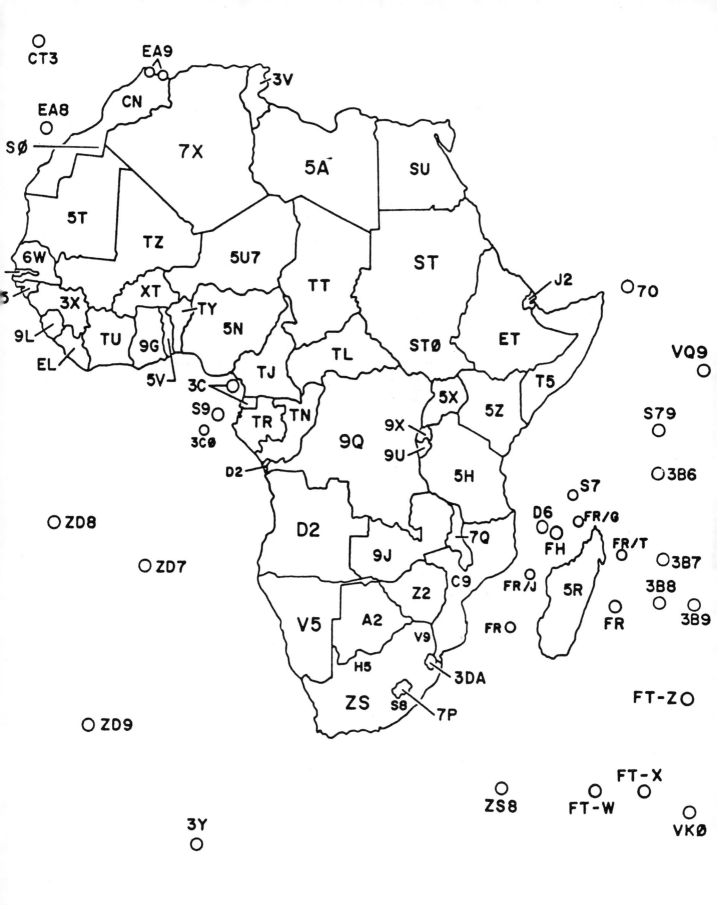

AFRICAN COUNTRIES
HAM OUTLINE MAP No 22

1/97   KØOST

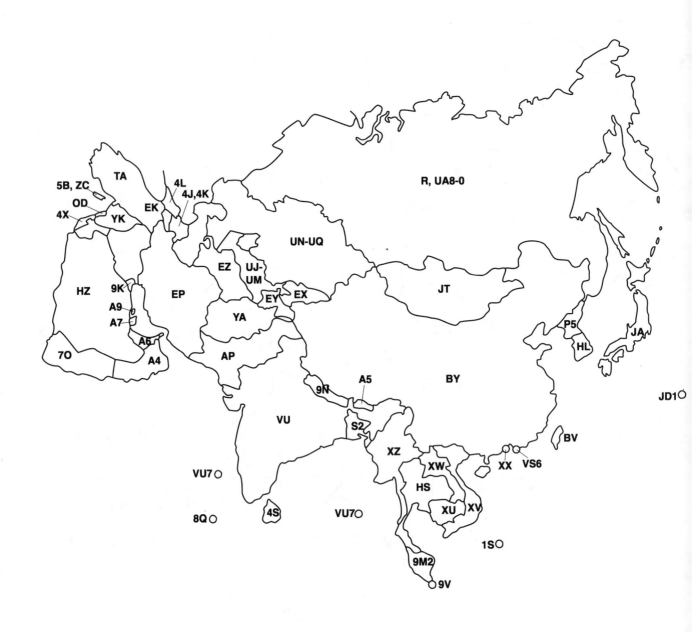

**ASIAN COUNTRIES
HAM OUTLINE MAP No 1**

12/95   KØOST

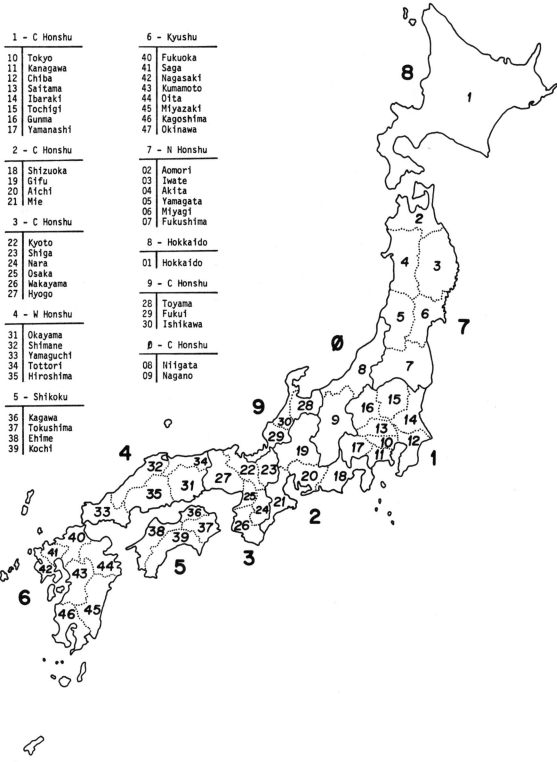

1 - C Honshu

10	Tokyo
11	Kanagawa
12	Chiba
13	Saitama
14	Ibaraki
15	Tochigi
16	Gunma
17	Yamanashi

2 - C Honshu

18	Shizuoka
19	Gifu
20	Aichi
21	Mie

3 - C Honshu

22	Kyoto
23	Shiga
24	Nara
25	Osaka
26	Wakayama
27	Hyogo

4 - W Honshu

31	Okayama
32	Shimane
33	Yamaguchi
34	Tottori
35	Hiroshima

5 - Shikoku

36	Kagawa
37	Tokushima
38	Ehime
39	Kochi

6 - Kyushu

40	Fukuoka
41	Saga
42	Nagasaki
43	Kumamoto
44	Oita
45	Miyazaki
46	Kagoshima
47	Okinawa

7 - N Honshu

02	Aomori
03	Iwate
04	Akita
05	Yamagata
06	Miyagi
07	Fukushima

8 - Hokkaido

01	Hokkaido

9 - C Honshu

28	Toyama
29	Fukui
30	Ishikawa

Ø - C Honshu

08	Niigata
09	Nagano

JAPANESE PREFECTURES (KENS)
PREFIX: JA-JS 7J-7N 8J-8N
HAM OUTLINE MAP No 2

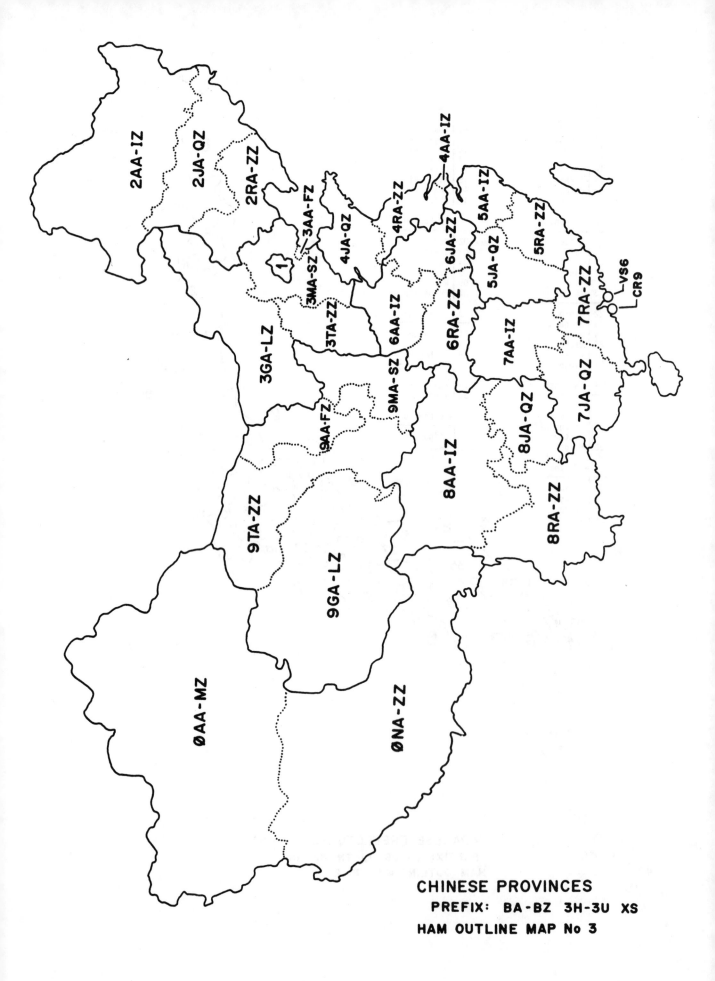

CHINESE PROVINCES
PREFIX: BA-BZ 3H-3U XS
HAM OUTLINE MAP No 3

RETURN POSTAGE CHART FOR WAS QSL CARDS

(Please include sufficient funds, as follows, for the safe return of your cards)

Foreign Small-Packet Air Mail

	Canada	Mexico	Central America, South America, Caribbean Islands	Europe European Russia	Pacific Rim, Asia, Africa, Asiatic Russia
WAS	$1.75	$2.00	$2.25	$3.25	$4.00
5BWAS	$7.00	$13.00	$10.00	$16.75	$23.25

United States

	Reg.	Cert.	First Class
WAS	$6.75	$2.75	$1.50
5BWAS	$9.00	$5.50	$4.50

All of the above are US funds
For Registered Small-Packet Air Mail add $4.95 to the above fees.
IRCs are valued at $0.50 US
These rates are subject to change at any time.

For the latest rates, see http://www.usps.gov

RETURN POSTAGE CHART FOR DXCC QSL CARDS

(Please include sufficient funds, as follows, for the safe return of your cards)

Award	Registered	Certified	First Class
DXCC (100 card new application)	$8.50 ($14.25[1])	$5.00	$4.00 ($9.25[1])
DXCC (10/25 endorsement)	$6.75 ($8.25[1])	$3.25	$2.25 ($3.75[1])
5BDXCC	$10.25 ($23.75[2])	$6.75	$5.75 ($19.25[2])

[1]Foreign small-packet Air Mail [2]Foreign surface Mail

All of the above are US funds.   **Registered or Certified Mail is recommended.**
IRCs are valued at $0.50 US.   These rates are subject to change at any time.

DXCC Fee Schedule

ALL amateurs initial first-time registration—$10. DXCC participants before Oct. 1, 1990 are exempt from this fee.
ALL League members 1st submission in a calendar year—FREE.
Non-League members outside of USA/Canada 1st submission in a calendar year—$10.
ALL League members 2nd submission in a calendar year—$10.
Non-League members outside of USA/Canada 2nd submission in a calendar year—$20.
MCS-16

ARRL **WAS** APPLICATION FORM

Please print or type **CLEARLY**:

NAME, CALLSIGN _____
(Print exactly as you want it on certificate)

List any ex-calls used on any cards _____

ADDRESS _____

_____
(City)   (State)   (Zip)

Membership ID number (from **QST** label) _____

<table>
<tr><td>

FUNDS ENCLOSED
For QSL return postage
Money/SASE
   $ _____

Return QSL's VIA:
❑ Registered
❑ Certified (US Only)
❑ First Class mail

For Certificate $_____
For Endorsement $ _____

</td><td>

❑ Initial Application (check one)
❑ Endorsement ❑ Original WAS Number held is _____
I am applying for ONE of the following WAS Awards:
(each is separately numbered)
❑ Basic Award (Mixed band and/or Modes)
❑ 50 MHz   ❑ 144 MHz   ❑ 432 MHz
❑ 160 Meters   ❑ 222 MHz   ❑ RTTY
❑ Satellite   ❑ SSTV
I am applying for the following ENDORSEMENTS: (Check all that app
❑ SSB ❑ CW ❑ Novice ❑ QRP ❑ Packet ❑ EME
❑ Single Band (circle one) **10 12 15 17 20 40 80** (Meter

</td></tr>
</table>

I have read, understood and followed all the rules of the WAS (MSD-264)

_____
Applicant's Signature   Date

* * * * * * * * * * * * * * * * * * **HF AWARDS MANAGER VERIFICATION** * * * * * * * * * * * * * * * * *

I have personally inspected the confirmations with all 50 states and verify that this application is correct and tru
This application is for the following SPECIALTY awards or ENDORSEMENTS: _____ (Write **NONE** if non

_____   _____   H-_____   _____
Signature   Callsign   (Mgr code#)   Date

* * * * * * * * * * * * * * * * * * * Directions to Applicant * * * * * * * * * * * * * * * * * * *

1. Read WAS Rules (MSD-264) carefully.
2. Fill out both sides of this application.
3. Sort cards by state as listed on the back of this application.
4. Present application and cards to your ARRL HF Awards Manager for verification. Applications from DX stations may be
certified by the Awards Manager of your IARU member-society.
5. Send application to ARRL HQ with the appropriate fee(s) (see MSD-264 Rule 10(d)). Send cards ONLY if there is no
local HF Awards Manager to verify your application.
6. If mailing cards, enclose sufficient postage for return of your cards. Mail to: ARRL WAS Award, 225 Main Street,
Newington, CT 06111, USA.

MSD-217 (10/96

WAS RECORD SHEET

Applicant's callsign _____   List any ex-calls used on any cards submitted: _____

STATE	CALL	DATE	BAND	MODE	
Alabama					
Alaska					
Arizona					
Arkansas					
California					
Colorado					
Connecticut					
Delaware					
Florida					
Georgia					
Hawaii					
Idaho					
Illinois					
Indiana					
Iowa					
Kansas					
Kentucky					
Louisiana					
Maine					
Maryland (D.C.)					
Massachusetts					
Michigan					
Minnesota					
Mississippi					
Missouri					
Montana					
Nebraska					
Nevada					
New Hampshire					
New Jersey					
New Mexico					
New York					
North Carolina					
North Dakota					
Ohio					
Oklahoma					
Oregon					
Pennsylvania					
Rhode Island					
South Carolina					
South Dakota					
Tennessee					
Texas					
Utah					
Vermont					
Virginia					
Washington					
West Virginia					
Wisconsin					
Wyoming					

ARRL 5BWAS AWARD APPLICATION

Please PRINT.

Call: _____   Name _____
                                        (Print exactly as you want it to appear on certificate)

List any ex-calls used on   Address _____
any cards submitted: (number and street)

_____   _____
                                        (City. State/Country, ZIP/PC)

┌─────────────────────────────┐   W/VEs: Membership control number
│   **FOR RETURN POSTAGE**   │   (from your **QST** wrapper)   : _____
│   Cash/Check/Stamps/IRCs   │
│   enclosed.   │   Is this your first WAS application? ☐ YES   ☐ NO
│   │
│ $ _____   │   If NO, my original WAS number held is _____
│   │
│   **Return my cards via:**   │   I was first licensed in (year): _____
│   ☐ Registered   │
│   ☐ Certified ● U.S. only   │   ☐ **$25 is enclosed for a plaque**
│   ☐ First-class mail   │
│   │   The undersigned affirms that he/she has read, understood, and abided by the
└─────────────────────────────┘   rules of the 5BWAS award printed below and the WAS Rules (MCS-264).

                                   Signature _____ Date _____ 19 _____

- -

HF AWARDS MANAGER VERIFICATION

I hereby verify that I have personally inspected the confirmations with all 50 states on five different bands and verify that
this application is correct and true.

_____   _____ H- _____ _____ 19 _____
        (signature)   (call)   (Mgr. 4-digit code #)   (date)

- -

5BWAS RULES

1. The 5BWAS certificate and plaque (see below) will be issued for having submitted confirmations with each of the 50
   United States for contacts **dated January 1, 1970,** or after, on five amateur bands (10, 18 and 24 MHz excluded).
   Phone and CW segments of a band do not count as separate bands.
2. WAS Rules (MCS-264), that do not conflict with these 5BWAS rules, also apply to the 5BWAS award.
3. There are no specialty 5Band awards or endorsements.
4. The 5BWAS certificate is offered free of charge. A handsome 9 × 12 ″ personalized walnut plaque is available for a fee of
   $25 US (check or money order).

- -

DIRECTIONS TO APPLICANT

1. Read 5BWAS rules above and WAS rules (MCS-264) carefully.

2. Fill out BOTH sides of this application.

3. Sort cards alphabetically by state and band.

4. Present application and cards to your ARRL Special Service Club HF Awards Manager for verification. Applications from DX stations may be certified by the Awards Manager of your IARU member-society.

5. Send application to ARRL HQ. Send cards also ONLY if there is no local HF Awards Manager.

6. If mailing cards, enclose postage sufficient for return of cards (see attached CD-16). MAIL TO:

   ARRL 5BWAS Award
   225 Main Street
   Newington, CT 06111 USA

7. Enclose $25 (check) if you want the plaque.

MCS-225(589)

5B-WAS RECORD SHEET

DIRECTIONS: Enter callsigns in the appropriate boxes by state and band.

QSLs must be for contacts **on or after January 1, 1970.**

STATE	80	40	20	15	10
Alabama					
Alaska					
Arizona					
Arkansas					
California					
Colorado					
Connecticut					
Delaware					
Florida					
Georgia					
Hawaii					
Idaho					
Illinois					
Indiana					
Iowa					
Kansas					
Kentucky					
Louisiana					
Maine					
Maryland (D.C.)					
Massachusetts					
Michigan					
Minnesota					
Mississippi					
Missouri					
Montana					
Nebraska					
Nevada					
New Hampshire					
New Jersey					
New Mexico					
New York					
North Carolina					
North Dakota					
Ohio					
Oklahoma					
Oregon					
Pennsylvania					
Rhode Island					
South Carolina					
South Dakota					
Tennessee					
Texas					
Utah					
Vermont					
Virginia					
Washington					
West Virginia					
Wisconsin					
Wyoming					

MCS-225(589)

ARRL VUCC AWARD APPLICATION FORM

Please print or type **clearly** and use a separate application for **each** award.

Callsign:_____ Ex Callsigns:_____

Name: _____
    Print exactly as you want to appear on certificate

Address:_____

City:_____ State:_____ Zip Code:_____

Membership ID Number:_____ Expiration Date:_____

Initial Application   Endorsement

Band: (please check only one)

50MHz	144MHz	222MHz	432MHz
902MHz	1296MHz	2.3GHz	3.4GHz
5.7GHz	10GHz	24GHz	47GHz
75GHz	119GHz	142GHz	241GHZ
Laser	Satellite		

Initial Applicants_____Number of Grid Squares

Endorsements

_____ + _____ = _____
Previous Total   Number added with   New Total
                this endorsement

√ "I affirm that I have observed all the VUCC rules as well as all pertinent government regulations established for Amateur Radio in my country. I agree to be bound by the ARRL Awards Committee (Decisions of the ARRL Awards Committee are final)."

√ Signature:_____ Callsign:_____ Date:_____

VHF Awards Manager Verification

"I have verified these contacts as set forth by the rules of the VUCC Program."

_____   _____ _____
Signature   Callsign   Verification Number
    Date

*** **Application Directions** ***

1) Complete all fields above and read MSD-261 detailing program rules
2) Enclose award fees ($10.00 for each new award)
3) Enclose $2.00 for return postage of paperwork, field sheets, and a 9X12 SASE (Endorsements only) ($4.00 foreign)
4) Sort cards alphabetically by field and numerically within the field from 00-99
5) Contact your VHF Awards Manager before sending cards to assure they are available to make arrangements for checking
6) Give Awards Manager all fees and mailing costs. They are responsible for sending paperwork to ARRL
7) **DO NOT SEND CARDS TO ARRL HEADQUARTERS**

MSD-260(1/97)

Field:
(first 2 letters
of locator)

Field:
(first 2 letters
of locator)

SQ.	Callsign	SQ.	Callsign
00		50	
01		51	
02		52	
03		53	
04		54	
05		55	
06		56	
07		57	
08		58	
09		59	
10		60	
11		61	
12		62	
13		63	
14		64	
15		65	
16		66	
17		67	
18		68	
19		69	
20		70	
21		71	
22		72	
23		73	
24		74	
25		75	
26		76	
27		77	
28		78	
29		79	
30		80	
31		81	
32		82	
33		83	
34		84	
35		85	
36		86	
37		87	
38		88	
39		89	
40		90	
41		91	
42		92	
43		93	
44		94	
45		95	
46		96	
47		97	
48		98	
49		99	

ARRL VUCC AWARD   CALLSIGN:

☐ 50 ☐ 144 ☐ 220 ☐ 432 ☐ 902 ☐ 1296 ☐ 2.3 ☐ 3.4

☐ 5.7 ☐ 10 ☐ 24 ☐ 47 ☐ Laser ☐ Satellite

DIRECTIONS: This side of tally sheet good for two fields. 1. Enter your callsign and check band. 2. Enter two-letter field. 3. Enter callsign of stations worked in appropriate grid square. 4. Total grid squares for each field.

SQ.	Callsign	SQ.	Callsign
00		50	
01		51	
02		52	
03		53	
04		54	
05		55	
06		56	
07		57	
08		58	
09		59	
10		60	
11		61	
12		62	
13		63	
14		64	
15		65	
16		66	
17		67	
18		68	
19		69	
20		70	
21		71	
22		72	
23		73	
24		74	
25		75	
26		76	
27		77	
28		78	
29		79	
30		80	
31		81	
32		82	
33		83	
34		84	
35		85	
36		86	
37		87	
38		88	
39		89	
40		90	
41		91	
42		92	
43		93	
44		94	
45		95	
46		96	
47		97	
48		98	
49		99	

TOTAL number of grid squares
worked in this field =

TOTAL number of grid squares
worked in this field =

MSD-259(392)

WAC AWARD APPLICATION

Name_____Callsign_____

Mailing Address_____

City/Town_____State_____ZIP Code_____

This application is for:

☐ Basic certificate (mixed mode)
☐ CW certificate
☐ Phone certificate
☐ SSTV certificate
☐ RTTY certificate
☐ FAX certificate
☐ Satellite certificate
☐ 5-band certificate

☐ QRP endorsement
   (5 watts output or less)
☐ 1.8-MHz endorsement
☐ 3.5-MHz endorsement
☐ 50-MHz endorsement
☐ 144-MHz endorsement
☐ 432-MHz endorsement
☐ 6-band endorsement

☐ My 5-band certificate is dated_____

Enclosed are QSL cards from the following stations:

(Enter band(s) and callsigns)

	MHz	MHz	MHz	MHz	MHz
N. America					
S. America					
Oceania					
Asia					
Europe					
Africa					

The undersigned has abided by all rules set forth for this award.
My ARRL membership does not expire until_____

Signature_____Callsign_____

Date_____   11/94

United States Counties

(information courtesy of CQ's Counties Award Record Book)

Alabama (67 counties)

Autauga
Balwin
Barbour
Bibb
Blount
Bullock
Butler
Calhoun
Chambers
Cherokee
Chilton
Choctaw
Clarke
Clay
Cleburne
Coffee
Colbert
Conecuh
Coosa
Covington
Crenshaw
Cullman
Dale
Dallas
Dekalb
Elmore
Escambia
Etowah
Fayette
Franklin
Geneva
Greene
Hale
Henry
Houston
Jackson
Jefferson
Lamar
Lauderdale
Lawrence
Lee
Limestone
Lowndes
Macon
Madison
Marengo
Marion
Marshall
Mobile
Monroe
Montgomery
Morgan
Perry
Pickens
Pike
Randolph
Russel
Saint Clair
Shelby
Sumter
Talladega
Tallapoosa
Tuscaloosa
Walker
Washington
Wilcox
Winston

Alaska (4 counties)

Southeastern
Northwestern
South Central
Central

Arizona (15 counties)

Apache
Cochise
Coconino
Gila
Graham
Greenlee
Lez Paz
Maricopa
Mohave
Navajo
Pima
Pinal
Santa Cruz
Yavapai
Yuma

Arkansas (75 counties)

Arkansas
Ashley
Baxter
Benton
Boone
Bradley
Calhoun
Carroll
Chicot
Clark
Clay
Cleburne
Cleveland
Columbia
Conway
Craighead
Crawford
Crittenden
Cross
Dallas
Desha
Drew
Faulkner
Franklin
Fulton
Garland
Grant
Greene
Hempstead
Hot Spring
Howard
Independence
Izard
Jackson
Jefferson
Johnson
Lafayette
Lawrence
Lee
Lincoln
Little River
Logan
Lonoke
Madison
Marion
Miller
Mississippi
Monroe
Montgomery
Nevada
Newton
Ouachita
Perry
Phillips
Pike
Poinsett

Polk
Pope
Prairie
Pulaski
Randolph
St. Francis
Saline
Scott
Searcy
Sebastian
Sevier
Sharp
Stone
Union
Van Buren
Washington
White Woodruff
Yell

California (58 counties)

Alameda
Alpine
Amador
Butte
Calaveras
Colusa
Contra Costa
Del Norte
El Dorado
Fresno
Glenn
Humboldt
Imperial
Inyo
Kern
Kings
Lake
Lassen
Los Angeles
Madera
Marin
Mariposa
Mendocino
Merced
Modoc
Mono
Monterey
Napa
Nevada
Orange
Placer
Plumas
Riverside
Sacramento
San Benito
San Bernardino
San Diego
San Francisco
San Joaquin
San Luis Obispo
San Mateo
Santa Barbara
Santa Clara
Santa Cruz
Shasta
Sierra
Siskiyou
Solano
Sonoma
Stanislaus
Sutter
Tehama
Trinity
Tulare

Tuolumne
Ventura
Yolo
Yuba

Colorado (63 counties)

Adams
Alamosa
Arapahoe
Archuleta
Baca
Bent
Boulder
Chaffee
Cheyenne
Clear Creek
Conejos
Costilla
Crowley
Custer
Delta
Denver
Dolores
Douglas
Eagle
Elbert
El Paso
Fremont
Garfield
Gilpin
Grand
Gunnison
Hinsdale
Huerfano
Jackson
Jefferson
Kiowa
Kit Carson
Lake
La Plata
Larimer
Las Animas
Lincoln
Logan
Mesa
Mineral
Moffat
Montezuma
Montrose
Morgan
Otero
Ouray
Park
Phillips
Pitkin
Prowers
Pueblo
Rio Blanco
Rio Grande
Routt
Saguache
San Juan
San Miguel
Sedgwick
Summit
Teller
Washington
Weld
Yuma

Connecticut (8 counties)

Fairfield
Hartford
Litchfield

Middlesex
New Haven
New London
Tolland
Windham

Delaware (3 counties)

Kent
New Castle
Sussex

Florida (67 counties)

Alachua
Baker
Bay
Bradford
Brevard
Broward
Calhoun
Charlotte
Citrus
Clay
Collier
Columbia
Dade
De Soto
Dixie
Duval
Escambia
Flagler
Franklin
Gadsden
Gilchrist
Glades
Gulf
Hamilton
Hardee
Hendry
Hernando
Highlands
Hillsborough
Holmes
Indian River
Jackson
Jefferson
Lafayette
Lake
Lee
Leon
Levy
Liberty
Madison
Manatee
Marion
Martin
Monroe
Nassau
Okaloosa
Okeechobee
Orange
Osceola
Palm Beach
Pasco
Pinellas
Polk
Putnam
St. Johns
St. Lucie
Santa Rosa
Sarasota
Seminole
Sumter
Suwannee
Taylor
Union
Volusia
Wakulla
Walton
Washington

Georgia (159 counties)

Appling
Atkinson
Bacon
Baker
Baldwin
Banks
Barrow
Bartow
Ben Hill
Berrien
Bibb
Bleckley
Brantley
Brooks
Bryan
Bulloch
Burke
Butts
Calhoun
Camden
Candler
Carroll
Catoosa
Charlton
Chatham
Chattahoochee
Chattooga
Cherokee
Clarke
Clay
Clayton
Clinch
Cobb
Coffee
Colquitt
Columbia
Cook
Coweta
Crawford
Crisp
Dade
Dawson
Decatur
De Kalb
Dodge
Dooly
Dougherty
Douglas
Early
Echols
Effingham
Elbert
Emanuel
Evans
Fannin
Fayette
Floyd
Forsyth
Franklin
Fulton
Gilmer
Glascock
Glynn
Gordon
Grady
Greene
Gwinnett
Habersham
Hall
Hancock
Haralson
Harris
Hart Heard
Henry
Houston
Irwin
Jackson

Jasper
Jeff Davis
Jefferson
Jenkins
Johnson
Jones
Lamar
Lanier
Laurens
Lee
Liberty
Lincoln
Long
Lowndes
Lumpkin
McDuffie
McIntosh
Macon
Madison
Marion
Meriwether
Miller
Mitchell
Monroe
Montgomery
Morgan
Murray
Muscogee
Newton
Oconee
Oglethorpe
Paulding
Peach
Pickens
Pierce
Pike
Polk
Pulaski
Putnam
Quitman
Rabun
Randolph
Richmond
Rockdale
Schley
Screven
Seminole
Spalding
Stephens
Stewart
Sumter
Talbot
Taliaferro
Tattnall
Taylor
Telfair
Terrell
Thomas
Tift
Toombs
Towns
Treutlen
Troup
Turner
Twiggs
Union
Upson
Walker
Walton
Ware
Warren
Washington
Wayne
Webster
Wheeler
White
Whitfield
Wilcox

Wilkes
Wilkinson
Worth

Hawaii (5 counties)

Hawaii
Honolulu
Kalawao
Kauai
Maui

Idaho (44 counties)

Ada
Adams
Bannock
Bear Lake
Benewah
Bingham
Blaine
Boise
Bonner
Bonneville
Boundary
Butte
Camas
Canyon
Caribou
Cassia
Clark
Clearwater
Custer
Elmore
Franklin
Fremont
Gem
Gooding
Idaho
Jefferson
Jerome
Kootenai
Latah
Lemhi
Lewis
Lincoln
Madison
Minidoka
Nez Perce
Oneida
Owyhee
Payette
Power
Shoshone
Teton
Twin Falls
Valley
Washington

Illinois (102 counties)

Adams
Alexander
Bond
Boone
Brown
Bureau
Calhoun
Carroll
Cass
Champaign
Christian
Clark
Clay
Clinton
Coles
Cook
Crawford
Cumberland
De Kalb
De Witt

Douglas
Du Page
Edgar
Edwards
Effingham
Fayette
Ford
Franklin
Fulton
Gallatin
Greene
Grundy
Hamilton
Hancock
Hardin Henderson
Henry
Iroquois
Jackson
Jasper
Jefferson
Jersey
Jo Daviess
Johnson
Kane
Kankakee
Kendall
Knox
Lake
La Salle
Lawrence
Lee
Livingston
Logan
McDonough
McHenry
McLean
Macon
Macoupin
Madison
Marion
Marshall
Mason
Massac
Menard
Mercer
Monroe
Montgomery
Morgan
Moultrie
Ogle
Peoria
Perry
Piatt
Pike
Pope
Pulaski
Putnam
Randolph
Richland
Rock Island
St. Clair
Saline
Sangamon
Schuyler
Scott
Shelby
Stark
Stephenson
Tazewell
Union
Vermilion
Wabash
Warren
Washington
Wayne
White
Whiteside
Will
Williamson

Winnebago
Woodford

Indiana (92 counties)
Adams
Allen
Bartholomew
Benton
Blackford
Boone
Brown
Carroll
Cass
Clark
Clay
Clinton
Crawford
Daviess
Dearborn
Decatur
De Kalb
Delaware
Dubois
Elkhart
Fayette
Floyd
Fountain
Franklin
Fulton
Gibson
Grant
Greene
Hamilton
Hancock
Harrison
Hendricks
Henry
Howard
Huntington
Jackson
Jasper
Jay
Jefferson
Jennings
Johnson
Knox
Kosciusko
Lagrange
Lake
La Porte
Lawrence
Madison
Marion
Marshall
Martin
Miami
Monroe
Montgomery
Morgan
Newton
Noble
Ohio
Orange
Owen
Parke
Perry
Pike
Porter
Posey
Pulaski
Putnam
Randolph
Ripley
Rush
St. Joseph
Scott
Shelby
Spencer
Starke

Steuben
Sullivan
Switzerland
Tippecanoe
Tipton
Union
Vanderburgh
Vermillion
Vigo
Wabash
Warren
Warrick
Washington
Wayne
Wells
White
Whitley

Iowa (99 counties)
Adair
Adams
Allamakee
Appanoose
Audubon
Benton
Black Hawk
Boone
Bremer
Buchanan
Buena Vista
Butler
Calhoun
Carroll
Cass
Cedar
Cerro Gordo
Cherokee
Chickasaw
Clarke
Clay
Clayton
Clinton
Crawford
Dallas
Davis
Decatur
Delaware
Des Moines
Dickinson
Dubuque
Emmet
Fayette
Floyd
Franklin
Fremont
Greene
Grundy
Guthrie
Hamilton
Hancock
Hardin
Harrison
Henry
Howard
Humboldt
Ida
Iowa
Jackson
Jasper
Jefferson
Johnson
Jones
Keokuk
Kossuth
Lee
Linn
Louisa
Lucas
Lyon

Madison
Mahaska
Marion
Marshall
Mills
Mitchell
Monona
Monroe
Montgomery
Muscatine
O'Brien
Osceola
Page
Palo Alto
Plymouth
Pocahontas
Polk
Pottawattamie
Poweshiek
Ringgold
Sac
Scott
Shelby
Sioux
Story
Tama
Taylor
Union
Van Buren
Wapello
Warren
Washington
Wayne
Webster
Winnebago
Winneshiek
Woodbury
Worth
Wright

Kansas (105 counties)
Allen
Anderson
Atchison
Barber
Barton
Bourbon
Brown
Butler
Chase
Chautauqua
Cherokee
Cheyenne
Clark
Clay
Cloud
Coffey
Comanche
Cowley
Crawford
Decatur
Dickinson
Doniphan
Douglas
Edwards
Elk
Ellis
Ellsworth
Finney
Ford
Franklin
Geary
Gove
Graham
Grant
Gray
Greeley
Greenwood
Hamilton

Harper
Harvey
Haskell
Hodgeman
Jackson
Jefferson
Jewell
Johnson
Kearny
Kingman
Kiowa
Labette
Lane
Leavenworth
Lincoln
Linn
Logan
Lyon
McPherson
Marion
Marshall
Meade
Miami
Mitchell
Montgomery
Morris
Morton
Nemaha
Neosho
Ness
Norton
Osage
Osborne
Ottawa
Pawnee
Phillips
Pottawatomie
Pratt
Rawlins
Reno
Republic
Rice
Riley
Rooks
Rush
Russell
Saline
Scott
Sedgwick
Seward
Shawnee
Sheridan
Sherman
Smith
Stafford
Stanton
Stevens
Sumner
Thomas
Trego
Wabaunsee
Wallace
Washington
Wichita
Wilson
Woodson
Wyandotte

Kentucky (120 counties)

Adair
Allen
Anderson
Ballard
Barren
Bath
Bell
Boone
Bourbon
Boyd

Boyle
Bracken
Breathitt
Breckenridge
Bullitt
Butler
Caldwell
Calloway
Campbell
Carlisle
Carroll
Carter
Casey
Christian
Clark
Clay
Clinton
Crittenden
Cumberland
Daviess
Edmonson
Elliott
Estill
Fayette
Fleming
Floyd
Franklin
Fulton
Gallatin
Garrard
Grant
Graves
Grayson
Green
Greenup
Hancock
Hardin
Harlan
Harrison
Hart
Henderson
Henry
Hickman
Hopkins
Jackson
Jefferson
Jessamine
Johnson
Kenton
Knott
Knox
Larue
Laurel
Lawrence
Lee
Leslie
Letcher
Lewis
Lincoln
Livingston
Logan
Lyon
McCracken
McCreary
McLean
Madison
Magoffin
Marion
Marshall
Martin
Mason
Meade
Menifee
Mercer
Metcalfe
Monroe
Montgomery
Morgan
Muhlenberg

Nelson
Nicholas
Ohio
Oldham
Owen
Owsley
Pendleton
Perry
Pike
Powell
Pulaski
Robertson
Rockcastle
Rowan
Russell
Scott
Shelby
Simpson
Spencer
Taylor
Todd
Trigg
Trimble
Union
Warren
Washington
Wayne
Webster
Whitley
Wolfe
Woodford

Louisiana (64 parishes)

Acadia
Allen
Ascension
Assumption
Avoyelles
Beauregard
Bienville
Bossier
Caddo
Calcasieu
Caldwell
Cameron
Catahoula
Claiborne
Concordia
De Soto
E. Baton Rouge
East Carroll
East Feliciana
Evangeline
Franklin
Grant
Iberia
Iberville
Jackson
Jefferson
Jefferson Davis
Lafayette
Lafourche
La Salle
Lincoln
Livingston
Madison
Morehouse
Natchitoches
Orleans
Ouachita
Plaquemines
Pointe Coupee
Rapides
Red River
Richland
Sabine
St. Bernard
St. Charles
St. Helena

St. James
St. John the Baptist
St. Landry
St. Martin
St. Mary
St. Tammany
Tangipahoa
Tensas
Terrebone
Union
Vermilion
Vernon
Washington
Webster
W. Baton Rouge
West Carroll
West Feliciana
Winn

Maine (16 counties)

Androscoggin
Aroostook
Cumberland
Franklin
Hancock
Kennebec
Knox
Lincoln
Oxford
Penobscot
Piscataquis
Sagadahoc
Somerset
Waldo
Washington
York

Maryland (24 counties)

Allegany
Anne Arundel
Baltimore
Baltimore City
Calvert
Caroline
Carroll
Cecil
Charles
Dorchester
Frederick
Garrett
Harford
Howard
Kent
Montgomery
Prince Georges
Queen Annes
St. Marys
Somerset
Talbot
Washington
Wicomico
Worcester

Massachusetts (14 counties)

Barnstable
Berkshire
Bristol
Dukes
Essex
Franklin
Hampden
Hampshire
Middlesex
Nantucket
Norfolk
Plymouth
Suffolk
Worcester

Michigan (83 counties)

Alcona
Alger
Allegan
Alpena
Antrim
Arenac
Baraga
Barry
Bay
Benzie
Berrien
Branch
Calhoun
Cass
Charlevoix
Cheboygan
Chippewa
Clare
Clinton
Crawford
Delta
Dickinson
Eaton
Emmet
Genesee
Gladwin
Gogebic
Grand Traverse
Gratiot
Hillsdale
Houghton
Huron
Ingham
Ionia
Iosco
Iron
Isabella
Jackson
Kalamazoo
Kalkaska
Kent
Keweenaw
Lake
Lapeer
Leelanau
Lenawee
Livingston
Luce
Mackinac
Macomb
Manistee
Marquette
Mason
Mecosta
Menominee
Midland
Missaukee
Monroe
Montcalm
Montmorency
Muskegon
Newaygo
Oakland
Oceana
Ogemaw
Ontonagon
Osceola
Oscoda
Otsego
Ottawa
Presque Isle
Roscommon
Saginaw
St. Clair
St. Joseph
Sanilac
Schoolcraft
Shiawassee
Tuscola
Van Buren
Washtenaw
Wayne
Wexford

Minnesota (87 counties)

Aitkin
Anoka
Becker
Beltrami
Benton
Big Stone
Blue Earth
Brown
Carlton
Carver
Cass
Chippewa
Chisago
Clay
Clearwater
Cook
Cottonwood
Crow Wing
Dakota
Dodge
Douglas
Faribault
Fillmore
Freeborn
Goodhue
Grant
Hennepin
Houston
Hubbard
Isanti
Itasa
Jackson
Kanabec
Kandiyohi
Kittson
Koochiching
Lac Qui Parle
Lake
L. of the Woods
Le Sueur
Lincoln
Lyon
McLeod
Mahnomen
Marshall
Martin
Meeker
Mille Lacs
Morrison
Mower
Murray
Nicollet
Nobles
Norman
Olmsted
Otter Trail
Pennington
Pine
Pipestone
Polk
Pope
Ramsey
Red Lake
Redwood
Renville
Rice
Rock
Roseau
St. Louis
Scott
Sherburne
Sibley
Stearns
Steele
Stevens
Swift
Todd
Traverse
Wabasha
Wadena
Waseca
Washington
Watonwan
Wilkin
Winona
Wright
Yellow Medicine

Mississippi (82 counties)

Adams
Alcorn
Amite
Attala
Benton
Bolivar
Calhoun
Carroll
Chickasaw
Choctaw
Claiborne
Clarke
Clay
Coahoma
Copiah
Covington
De Soto
Forrest
Franklin
George
Greene
Grenada
Hancock
Harrison
Hinds
Holmes
Humphreys
Issaquena
Itawamba
Jackson
Jasper
Jefferson
Jefferson Davis
Jones
Kemper
Lafayette
Lamar
Lauderdale
Lawrence
Leake
Lee
Leflore
Lincoln
Lowndes
Madison
Marion
Marshall
Monroe
Montgomery
Neshoba
Newton
Noxubee
Oktibbeha
Panola
Pearl River
Perry
Pike
Pontotoc
Prentiss
Quitman
Rankin
Scott
Sharkey
Simpson
Smith
Stone
Sunflower
Tallahatchie
Tate
Tippah
Tishomingo
Tunica
Union
Walthall
Warren
Washington
Wayne
Webster
Wilkinson
Winston
Yalobusha
Yazoo

Missouri (115 counties)

Adair
Andrew
Atchison
Audrain
Barry
Barton
Bates
Benton
Bollinger
Boone
Buchanan
Butler
Caldwell
Callaway
Camden
Cape Girardeau
Carroll
Carter
Cass
Cedar
Chariton
Christian
Clark
Clay
Clinton
Cole
Cooper
Crawford
Dade
Dallas
Daviess
De Kalb
Dent
Douglas
Dunklin
Franklin
Gasconade
Gentry
Greene
Grundy
Harrison
Henry
Hickory
Holt
Howard
Howell
Iron
Jackson
Jasper
Jefferson
Johnson
Knox
Laclede
Lafayette
Lawrence
Lewis
Lincoln

Linn
Livingston
McDonald
Macon
Madison
Maries
Marion
Mercer
Miller
Mississippi
Moniteau
Monroe
Montgomery
Morgan
New Madrid
Madrid
Newton
Nodaway
Oregon
Osage
Ozark
Pemiscot
Perry
Pettis
Phelps
Pike
Platte
Polk
Pulaski
Putnam
Ralls
Randolph
Ray
Reynolds
Ripley
St. Charles
St. Clair
St. Francois
St. Louis
St. Louis City
Ste. Genevieve
Saline
Schuyler
Scotland
Scott
Shannon
Shelby
Stoddard
Stone
Sullivan
Taney
Texas
Vernon
Warren
Washington
Wayne
Webster
Worth
Wright

Montana (56 counties)

Beaverhead
Big Horn
Blaine
Broadwater
Carbon
Carter
Cascade
Chouteau
Custer
Daniels
Dawson
Deer Lodge
Fallon
Fergus
Flathead
Gallatin
Garfield
Glacier

Golden Valley
Granite
Hill
Jefferson
Judith Basin
Lake
Lewis and Clark
Liberty
Lincoln
McCone
Madison
Meagher
Mineral
Missoula
Musselshell
Park
Petroleum
Phillips
Pondera
Powder River
Powell Prairie
Ravalli
Richland
Roosevelt
Rosebud
Sanders
Sheridan
Silver Bow
Stillwater
Sweet Grass
Teton
Toole
Treasure
Valley
Wheatland
Wibaux
Yellowstone

Nebraska (93 counties)
Adams
Antelope
Arthur
Banner
Blaine
Boone
Box Butte
Boyd
Brown
Buffalo
Burt
Butler
Cass
Cedar
Chase
Cherry
Cheyenne
Clay
Colfax
Cuming
Custer
Dakota
Dawes
Dawson
Deuel
Dixon
Dodge
Douglas
Dundy
Fillmore
Franklin
Frontier
Furnas
Gage
Garden
Garfield
Gosper
Grant
Greeley
Hall

Hamilton
Harlan
Hayes
Hitchcock
Holt
Hooker
Howard
Jefferson
Johnson
Kearney
Keith
Keya Paha
Kimball
Knox
Lancaster
Lincoln
Logan
Loup
McPherson
Madison
Merrick
Morrill
Nance
Nemaha
Nuckolls
Otoe
Pawnee
Perkins
Phelps
Pierce
Platte
Polk
Red Willow
Richardson
Rock
Saline
Sarpy
Saunders
Scotts Bluff
Seward
Sheridan
Sherman
Sioux
Stanton
Thayer
Thomas
Thurston
Valley
Washington
Wayne
Webster
Wheeler
York

Nevada (16 counties)
Churchill
Clark
Douglas
Elko
Esmeralda
Eureka
Humboldt
Lander
Lincoln
Lyon
Mineral
Nye
Pershing
Storey
Washoe
White Pine

New Hampshire (10 counties)
Belknap
Carroll
Cheshire
Coos
Grafton

Hillsboro
Merrimack
Rockingham
Strafford
Sullivan

New Jersey (21 counties)
Atlantic
Bergen
Burlington
Camden
Cape May
Cumberland
Essex
Gloucester
Hudson
Hunterdon
Mercer
Middlesex
Monmouth
Morris
Ocean
Passaic
Salem
Somerset
Sussex
Union
Warren

New Mexico (33 counties)
Bernalillo
Catron
Chaves
Cibola
Colfax
Curry
De Baca
Dona Ana
Eddy
Grant
Guadalupe
Harding
Hidalgo
Lea
Lincoln
Los Alamos
Luna
McKinley
Mora
Otero
Quay
Rio Arriba
Roosevelt
Sandoval
San Juan
San Miguel
Santa Fe
Sierra
Socorro
Taos
Torrance
Union
Valencia

New York (62 counties)
Albany
Allegany
Bronx
Broome
Cattaraugus
Cayuga
Chautauqua
Chemung
Chenango
Clinton
Columbia
Cortland
Delaware
Dutchess

Erie
Essex
Franklin
Fulton
Genesee
Greene
Hamilton
Herkimer
Jefferson
Kings
Lewis
Livingston
Madison
Monroe
Montgomery
Nassau
New York
Niagara
Oneida
Onondaga
Ontario
Orange
Orleans
Oswego
Otsego
Putnam
Queens
Rensselaer
Richmond
Rockland
St. Lawrence
Saratoga
Schenectady
Schoharie
Schuyler
Seneca
Steuben
Suffolk
Sullivan
Tioga
Tompkins
Ulster
Warren
Washington
Wayne
Westchester
Wyoming
Yates

North Carolina (100 counties)
Alamance
Alexander
Alleghany
Anson
Ashe
Avery
Beaufort
Bertie
Bladen
Brunswick
Buncombe
Burke
Cabarrus
Caldwell
Camden
Carteret
Caswell
Catawba
Chatham
Cherokee
Chowan
Clay
Cleveland
Columbus
Craven
Cumberland
Currituck
Dare

Davidson
Davie
Duplin
Durham
Edgecombe
Forsyth
Franklin
Gaston
Gates
Graham
Granville
Greene
Guilford
Halifax
Harnett
Haywood
Henderson
Hertford
Hoke
Hyde
Iredell
Jackson
Johnston
Jones
Lee
Lenoir
Lincoln
McDowell
Macon
Madison
Martin
Mecklenburg
Mitchell
Montgomery
Moore
Nash
New Hanover
Northampton
Onslow
Orange
Pamlico
Pasquotank
Pender
Perquimans
Person
Pitt
Polk
Randolph
Richmond
Robeson
Rockingham
Rowan
Rutherford
Sampson
Scotland
Stanly
Stokes
Surry
Swain
Transylvania
Tyrrell
Union
Vance
Wake
Warren
Washington
Watauga
Wayne
Wilkes
Wilson
Yadkin
Yancey

North Dakota (53 counties)
Adams
Barnes
Benson
Billings
Bottineau

Bowman
Burke
Burleigh
Cass
Cavalier
Dickey
Divide
Dunn
Eddy
Emmons
Foster
Golden Valley
Grand Forks
Grant
Griggs
Hettinger
Kidder
La Moure
Logan
McHenry
McIntosh
McKenzie
McLean
Mercer
Morton
Mountrail
Nelson
Oliver
Pembina
Pierce
Ramsey
Ranson
Renville
Richland
Rolette
Sargent
Sheridan
Sioux
Slope
Stark
Steele
Stutsman
Towner
Traill
Walsh
Ward
Wells
Williams

Ohio (88 counties)
Adams
Allen
Ashland
Ashtabula
Athens
Auglaize
Belmont
Brown
Butler
Carroll
Champaign
Clark
Clermont
Clinton
Columbiana
Coshocton
Crawford
Cuyahoga
Darke
Defiance
Delaware
Erie
Fairfield
Fayette
Franklin
Fulton
Gallia
Geauga
Greene

Guernsey
Hamilton
Hancock
Hardin
Harrison
Henry
Highland
Hocking
Holmes
Huron
Jackson
Jefferson
Knox
Lake
Lawrence
Licking
Logan
Lorain
Lucas
Madison
Mahoning
Marion
Medina
Meigs
Mercer
Miami
Monroe
Montgomery
Morgan
Morrow
Muskingum
Noble
Ottawa
Paulding
Perry
Pickaway
Pike
Portage
Preble
Putnam
Richland
Ross
Sandusky
Scioto
Seneca
Shelby
Stark
Summit
Trumbull
Tuscarawas
Union
Van Wert
Vinton
Warren
Washington
Wayne
Williams
Wood
Wyandot

Oklahoma (77 counties)
Adair
Alfalfa
Atoka
Beaver
Beckham
Blaine
Bryan
Caddo
Canadian
Carter
Cherokee
Choctaw
Cimarron
Cleveland
Coal
Comanche
Cotton
Craig

Creek
Custer
Delaware
Dewey
Ellis
Garfield
Garvin
Grady
Grant
Greer
Harmon
Harper
Haskell
Hughes
Jackson
Jefferson
Johnston
Kay
Kingfisher
Kiowa
Latimer
Le Flore
Lincoln
Logan
Love
McClain
McCurtain
McIntosh
Major
Marshall
Mayes
Murray
Muskogee
Noble
Nowata
Okfuskee
Oklahoma
Okmulgee
Osage
Ottawa
Pawnee
Payne
Pittsburg
Pontotoc
Pottawatomie
Pushmataha
Roger Mills
Rogers
Seminole
Sequoyah
Stephens
Texas
Tillman
Tulsa
Wagoner
Washington
Washita
Woods
Woodward

Oregon (36 counties)

Baker
Benton
Clackamas
Clatsop
Columbia
Coos
Crook
Curry
Deschutes
Douglas
Gilliam
Grant
Harney
Hood River
Jackson
Jefferson
Josephine
Klamath

Lake
Lane
Lincoln
Linn
Malheur
Marion
Morrow
Multnomah
Polk
Sherman
Tillamook
Umatilla
Union
Wallowa
Wasco
Washington
Wheeler
Yamhill

Pennsylvania (67 counties)

Adams
Allegheny
Armstrong
Beaver
Bedford
Berks
Blair
Bradford
Bucks
Butler
Cambria
Cameron
Carbon
Centre
Chester
Clarion
Clearfield
Clinton
Columbia
Crawford
Cumberland
Dauphin
Delaware
Elk
Erie
Fayette
Forest
Franklin
Fulton
Greene
Huntingdon
Indiana
Jefferson
Juniata
Lackawanna
Lancaster
Lawrence
Lebanon
Lehigh
Luzerne
Lycoming
McKean
Mercer
Mifflin
Monroe
Montgomery
Montour
Northampton
Northumberland
Perry
Philadelphia
Pike Potter
Schuylkill
Snyder
Somerset
Sullivan
Susquehanna
Tioga
Union

Venango
Warren
Washington
Wayne
Westmoreland
Wyoming
York

Rhode Island (5 counties)

Bristol
Kent
Newport
Providence
Washington

South Carolina (46 counties)

Abbeville
Aiken
Allendale
Anderson
Bamberg
Barnwell
Beaufort
Berkeley
Calhoun
Charleston
Cherokee
Chester
Chesterfield
Clarendon
Colleton
Darlington
Dillon
Dorchester
Edgefield
Fairfield
Florence
Georgetown
Greenville
Greenwood
Hampton
Horry
Jasper
Kershaw
Lancaster
Laurens
Lee
Lexington
McCormick
Marion
Marlboro
Newberry
Oconee
Orangeburg
Pickens
Richland
Saluda
Spartanburg
Sumter
Union
Williamsburg
York

South Dakota (66 counties)

Aurora
Beadle
Bennett
Bon Homme
Brookings
Brown
Brule
Buffalo
Butte
Campbell
Charles Mix
Clark
Clay
Codington

Corson
Custer
Davison
Day
Deuel
Dewey
Douglas
Edmunds
Fall River
Faulk
Grant
Gregory
Haakon
Hamlin
Hand
Hanson
Harding
Hughes
Hutchinson
Hyde
Jackson
Jerauld
Jones
Kingsbury
Lake
Lawrence
Lincoln
Lyman
McCook
McPherson
Marshall
Meade
Mellette
Miner
Minnehaha
Moody
Pennington
Perkins
Potter
Roberts
Sanborn
Shannon
Spink
Stanley
Sully
Todd
Tripp
Turner
Union
Walworth
Yankton
Ziebach

Tennessee (95 counties)

Anderson
Bedford
Benton
Bledsoe
Blount
Bradley
Campbell
Cannon
Carroll
Carter
Cheatham
Chester
Claiborne
Clay
Cocke
Coffee
Crockett
Cumberland
Davidson
Decatur
DeKalb
Dickson
Dyer
Fayette
Fentress

Franklin
Gibson
Giles
Grainger
Greene
Grundy
Hamblen
Hamilton
Hancock
Hardeman
Hardin
Hawkins
Haywood
Henderson
Henry
Hickman
Houston
Humphreys
Jackson
Jefferson
Johnson
Knox
Lake
Lauderdale
Lawrence
Lewis
Lincoln
Loudon
McMinn
McNairy
Macon
Madison
Marion
Marshall
Maury
Meigs
Monroe
Montgomery
Moore
Morgan
Obion
Overton
Perry
Pickett
Polk
Putnam
Rhea
Roane
Robertson
Rutherford
Scott
Sequatchie
Sevier
Shelby
Smith
Stewart
Sullivan
Sumner
Tipton
Trousdale
Unicoi
Union
Van Buren
Warren
Washington
Wayne
Weakley
White
Williamson
Wilson

Texas (254 counties)

Anderson
Andrews
Angelina
Aransas
Archer
Armstrong
Atascosa

Austin
Bailey
Bandera
Bastrop
Baylor
Bee
Bell
Bexar
Blanco
Borden
Bosque
Bowie
Brazoria
Brazos
Brewster
Briscoe
Brooks
Brown
Burleson
Burnet
Caldwell
Calhoun
Callahan
Cameron
Camp
Carson
Cass
Castro
Chambers
Cherokee
Childress
Clay
Cochran
Coke
Coleman
Collin
Collingsworth
Colorado
Comal
Comanche
Concho
Cooke
Coryell
Cottle
Crane
Crockett
Crosby
Culberson
Dallam
Dallas
Dawson
Deaf Smith
Delta
Denton
De Witt
Dickens
Dimmit
Donley
Duval
Eastland
Ector
Edwards
Ellis
El Paso
Erath
Falls
Fannin
Fayette
Fisher
Floyd
Foard
Fort Bend
Franklin
Freestone
Frio
Gaines
Galveston
Garza
Gillespie

Glasscock
Goliad
Gonzales
Gray
Grayson
Gregg
Grimes
Guadelupe
Hale
Hall
Hamilton
Hansford
Hardeman
Hardin
Harris
Harrison
Hartley
Haskell
Hays
Hemphill
Henderson
Hidalgo
Hill
Hockley
Hood
Hopkins
Houston
Howard
Hudspeth
Hunt
Hutchinson
Irion
Jack
Jackson
Jasper
Jeff Davis
Jefferson
Jim Hogg
Jim Wells
Johnson
Jones
Karnes
Kaufman
Kendall
Kenedy
Kent
Kerr
Kimble
King
Kinney
Kleberg
Knox
Lamar
Lamb
Lampasas
La Salle
Lavaca
Lee
Leon
Liberty
Limestone
Lipscomb
Live Oak
Llano
Loving
Lubbock
Lynn
McCulloch
McLennan
McMullen
Madison
Marion
Martin
Mason
Matagorda
Maverick
Medina
Menard
Midland

Milam
Mills
Mitchell
Montague
Montgomery
Moore
Morris
Motley
Nacogdoches
Navarro
Newton
Nolan
Nueces
Ochiltree
Oldham
Orange
Palo Pinto
Panola
Parker
Parmer
Pecos
Polk
Potter
Presidio
Rains
Randall
Reagan
Real
Red River
Reeves
Refugio
Roberts
Robertson
Rockwall
Runnels
Rusk
Sabine
San Augustine
San Jacinto
San Patricio
San Saba
Schleicher
Scurry
Shackelford
Shelby
Sherman
Smith
Somervell
Starr
Stephens
Sterling
Stonewall
Sutton
Swisher
Tarrant
Taylor
Terrell
Throckmorton
Titus
Tom Green
Travis
Trinity
Tyler
Upshur
Upton
Uvalde
Val Verde
Van Zandt
Victoria
Walker
Waller
Ward
Washington
Webb
Wharton
Wheeler
Wichita
Wilbarger
Willacy

Williamson
Wilson
Winkler
Wise
Wood
Yoakum
Young
Zapata
Zavala

Utah (29 counties)

Beaver
Box Elder
Cache
Carbon
Daggett
Davis
Duchesne
Emery
Garfield
Grand
Iron
Juab
Kane
Millard
Morgan
Piute
Rich
Salt Lake
San Juan
Sanpete
Sevier
Summit
Tooele
Uintah
Utah
Wasatch
Washington
Wayne
Weber

Vermont (14 counties)

Addison
Bennington
Caledonia
Chittendon
Essex
Franklin
Grand Isle
Lamoille
Orange
Orleans
Rutland
Washington
Windham
Windsor

Virginia (95 counties)

Accomack
Albemarle
Alleghany
Amelia
Amherst
Appomattox
Arlington
Augusta
Bath
Bedford
Bland
Botetourt
Brunswick
Buchanan
Buckingham
Campbell
Caroline
Carroll
Charles City
Charlotte
Chesterfield

Clarke
Craig
Culpeper
Cumberland
Dickenson
Dinwiddie
Essex
Fairfax
Fauquier
Floyd
Fluvanna
Franklin
Frederick
Giles
Gloucester
Goochland
Grayson
Greene
Greensville
Halifax
Hanover
Henrico
Henry
Highland
Isle of Wight
James City
King and Queen
King George
King William
Lancaster
Lee
Loudoun
Louisa
Lunenberg
Madison
Mathews
Mecklenburg
Middlesex
Montgomery
Nelson
New Kent
Northampton
Northumberland
Nottoway
Orange
Page
Patrick
Pittsylvania
Powhatan
Prince Edward
Prince George
Prince William
Pulaski
Rappahannock
Richmond
Roanoke
Rockbridge
Rockingham
Russell
Scott
Shenandoah
Smyth
Southampton
Spotsylvania
Stafford
Surry
Sussex
Tazewell
Warren
Washington
Westmoreland
Wise
Wythe
York

Washington (39 counties)

Adams
Asotin
Benton

Chelan
Clallam
Clark
Columbia
Cowlitz
Douglas
Ferry
Franklin
Garfield
Grant
Grays Harbor
Island
Jefferson
King
Kitsap
Kittitas
Klickitat
Lewis
Lincoln
Mason
Okanogan
Pacific
Pend Oreille
Pierce
San Juan
Skagit
Skamania
Snohomish
Spokane
Stevens
Thurston
Wahkiakum
Walla Walla
Whatcom
Whitman
Yakima

West Virginia (55 counties)

Barbour
Berkeley
Boone
Braxton
Brooke
Cabell
Calhoun
Clay
Doddridge
Fayette
Gilmer
Grant
Greenbrier
Hampshire
Hancock
Hardy
Harrison
Jackson
Jefferson
Kanawha
Lewis
Lincoln
Logan
McDowell
Marion
Marshall
Mason
Mercer
Mineral
Mingo
Monongalia
Monroe
Morgan
Nicholas
Ohio
Pendleton
Pleasants
Pocahontas
Preston
Putnam
Raleigh

Randolph
Ritchie
Roane
Summers
Taylor
Tucker
Tyler
Upshur
Wayne
Webster
Wetzel
Wirt
Wood
Wyoming

Wisconsin (72 counties)

Adams
Ashland
Barron
Bayfield
Brown
Buffalo
Burnett
Calumet
Chippewa
Clark
Columbia
Crawford
Dane
Dodge
Door
Douglas
Dunn
Eau Claire
Florence
Fond du Lac
Forest
Grant
Green
Green Lake
Iowa
Iron
Jackson
Jefferson
Juneau
Kenosha
Kewaunee
La Crosse
Lafayette
Langlade
Lincoln
Manitowoc
Marathon
Marinette
Marquette
Menominee
Milwaukee
Monroe
Oconto
Oneida
Outagamie
Ozaukee
Pepin
Pierce
Polk
Portage
Price
Racine
Richland
Rock
Rusk
St. Croix
Sauk
Sawyer
Shawano
Sheboygan
Taylor
Trempealeau
Vernon

Vilas	
Walworth	
Washburn	
Washington	
Waukesha	
Waupaca	
Waushara	
Winnebago	
Wood	

Wyoming (23 counties)

Albany	Hot Springs	Sublette	
Big Horn	Johnson	Sweetwater	
Campbell	Laramie	Teton	
Carbon	Lincoln	Uinta	
Converse	Natrona	Washakie	
Crook	Niobrara	Weston	
Fremont	Park		
Goshen	Platte		
	Sheridan	**Total US Counties = 3076**	

Notes: Since USA-CA started, counties that have been absorbed include Princess Ann, Norfolk and Nansemond in Virginia; Ormsby in Nevada; Washabaugh in South Dakota. Carson City, Nevada, is now considered an independent city. Richmond County in New York is now called Staten Island County. A new county has been added to New Mexico— Cibola (used to be part of Valencia County).

The following is a list of independent cities and the county for which each can be used.

Independent City   Counts as_____ County

[VIRGINIA]

Independent City	County
Alexandria	Arlington or Fairfax
Bedford	Bedford
Bristol	Washington
Buena Vista	Rockbridge
Charlottesville	Albermarle
Chesapeake	Isle of Wight
Clifton Forge	Alleghany
Colonial Heights	Chesterfield or Prince George
Covington	Alleghany
Danville	Pittsylvania
Emporia	Greensville
Fairfax	Fairfax
Falls Church	Fairfax
Fort Monroe	York
Franklin	Southampton
Fredericksburg	Spotsylvania
Galax	Carroll or Grayson
Hampton	York
Harrisonburg	Rockingham
Hopewell	Prince George
Lexington	Rockbridge
Lynchburg	Amherst or Bedford or Campbell
Manassas	Prince William
Manassas Park	Prince William
Martinsville	Henry
Newport News	York
Norfolk	Isle of Wight
Norton	Wise
Petersburg	Chesterfield or Dinwiddie or Prince George
Poquoson	York
Portsmouth	Isle of Wight
Radford	Montgomery
Richmond	Chesterfield or Henrico
Roanoke	Roanoke
Salem	Roanoke
South Boston	Halifax
Staunton	Augusta
Suffolk	Isle of Wight or Southampton
Virginia Beach	Isle of Wight
Waynesboro	Augusta
Williamsburg	James City
Winchester	Frederick

Carson City, Nevada	Douglas or Lyon or Story or Washoe
Washington, DC	Montgomery or Prince Georges, Maryland

ARRL Field Organization

The United States is divided into 15 ARRL Divisions. Every two years the ARRL full members in each of these divisions elect a director and a vice director to represent them on the League's Board of Directors. The Board determines the policies of the League, which are carried out by the Headquarters staff. A director's function is principally policymaking at the highest level, but the Board of Directors is all-powerful in the conduct of League affairs.

The 15 divisions are further broken down into 69 sections, and the ARRL full members in each section elect a Section Manager (SM). The SM is the senior elected ARRL official in the section, and in cooperation with the director, fosters and encourages all ARRL activities within the section. A breakdown of sections within each division (and counties within each split-state section) follows:

ATLANTIC DIVISION: *Delaware, Eastern Pennsylvania* (Adams, Berks, Bradford, Bucks, Carbon, Chester, Columbia, Cumberland, Dauphin, Delaware, Juniata, Lackawanna, Lancaster, Lebanon, Lehigh, Luzerne, Lycoming, Monroe, Montgomery, Montour, Northhampton, Northumberland, Perry, Philadelphia, Pike, Schuylkill, Snyder, Sullivan, Susquehanna, Tioga, Union, Wayne, Wyoming, York); *Northern New York* (Clinton, Essex, Franklin, Fulton, Hamilton, Jefferson, Lewis, Montgomery, Schoharie, St. Lawrence); *Maryland -D.C.; Southern New Jersey* (Atlantic, Burlington, Camden, Cape May, Cumberland, Gloucester, Mercer Ocean, Salem); *Western New York* (Allegany, Broome, Cattaraugus, Cayuga, Chautauqua, Chemung, Chenango, Cortland, Delaware, Erie, Genesee, Herkimer, Livingston, Madison, Monroe, Niagara, Oneida, Onondaga, Ontario, Orleans, Oswego, Otsego, Schuyler, Seneca, Steuben, Tioga, Tompkins, Wayne, Wyoming, Yates); *Western Pennsylvania* (those counties not listed under Eastern Pennsylvania).

CENTRAL DIVISION: *Illinois; Indiana; Wisconsin.*

DAKOTA DIVISION: *Minnesota; North Dakota; South Dakota.*

DELTA DIVISION: *Arkansas; Louisiana; Mississippi; Tennessee*

GREAT LAKES DIVISION: *Kentucky; Michigan; Ohio.*

HUDSON DIVISION: *Eastern New York* (Albany, Columbia, Dutchess, Greene, Orange, Putnam, Rensselaer, Rockland, Saratoga, Schenectady, Sullivan, Ulster, Warren, Washington, Westchester); *N.Y.C.-L.I.* (Bronx, Kings, Nassau, New York, Queens, Staten Island, Suffolk); *Northern New Jersey* Bergen, Essex, Hudson, Hunterdon, Middlesex, Monmouth, Morris, Passaic, Somerset, Sussex, Union, Warren).

MIDWEST DIVISION: *Iowa; Kansas; Missouri; Nebraska.*

NEW ENGLAND DIVISION: *Connecticut; Maine, Eastern Massachusetts* (Barnstable, Bristol, Dukes, Essex, Middlesex, Nantucket, Norfolk, Plymouth, Suffolk); *New Hampshire; Rhode Island; Vermont; Western Massachusetts* (those counties not listed under Eastern Massachusetts).

NORTHWESTERN DIVISION: *Alaska; Idaho; Montana; Oregon; Eastern Washington* (Adams, Asotin, Benton, Chelan, Columbia, Douglas, Ferry, Franklin, Garfield, Grant, Kittitas, Klickitat, Lincoln, Okangogan, Pend Oreille, Spokane, Stevens, Walla, Walla, Whitman, Yakima); *Western Washington* (Challam, Clark Cowlitz, Grays Harbor Island, Jefferson, King, Kitsap, Lewis, Mason, Pacific, Pierce, San Juan, Skagit, Skamania, Snohomish, Thurston, Wahkiakum, Whatcom).

PACIFIC DIVISION: *East Bay* (Alameda, Contra Costa, Napa, Solano); *Nevada; Pacific* (Hawaii and U.S. possessions in the Pacific); *Sacramento Valley* (Alpine, Amador, Butte, Colusa, El Dorado, Glenn, Lassen, Modoc, Nevada, Placer, Plumas, Sacramento, Shasta, Sierra, Siskiyou, Sutter, Tehama, Trinity, Yolo, Yuba); *San Franciso,* (Del Norte, Humboldt, Lake, Marin, Mendocino, San Franciso, Sonoma); *San Joaquin Valley* (Calaveras, Fresno, Kern, Kings, Madera, Mariposa, Merced, Mono, San Joaquin, Stanislaus, Tulare, Tuolumne); *Santa Clara Valley* (Monterey, San Benito, San Mateo, Santa Clara, Santa Cruz).

ROANOKE DIVISION: *North Carolina; South Carolina; Virginia; West Virginia.*

ROCKY MOUNTAIN DIVISION: *Colorado; Utah; New Mexico; Wyoming.*

SOUTHEASTERN DIVISION: *Alabama; Georgia; Northern Florida* (Alachua, Baker, Bay, Bradford, Calhoun, Citrus, Clay, Columbia, Dixie, Duval, Escambia, Flagler, Franklin, Gadsden, Gilchrist, Gulf, Hamilton, Hernando, Holmes, Jackson, Jefferson, Lafayette, Lake, Leon, Levy, Liberty, Madison, Marion, Nassau, Okaloosa, Orange, Pasco, Putnam, Santa Rosa, Seminole, St. Johns, Sumter, Suwanee, Taylor, Union, Volusia, Wakulla, Walton, Washington); *Southern Florida* (those counties not listed under Northern Florida); *Puerto Rico; U.S. Virgin Islands.*

SOUTHWESTERN DIVISION: *Arizona; Los Angeles; Orange* (Inyo, Orange, Riverside, San Bernardino); *San Diego* (Imperial, San Diego); *Santa Barbara* (San Luis Obispo, Santa Barbara, Ventura).

WEST GULF DIVISION: *North Texas* (Anderson, Archer, Baylor, Bell, Bosque, Bowie, Brown, Camp, Cass, Cherokee, Clay, Collin, Comanche, Cooke, Coryell, Dallas, Delta, Denton, Eastland, Ellis, Erath, Falls, Fannin, Franklin, Freestone, Grayson, Gregg, Hamilton, Harrison, Henderson, Hill, Hood, Hopkins, Hunt,

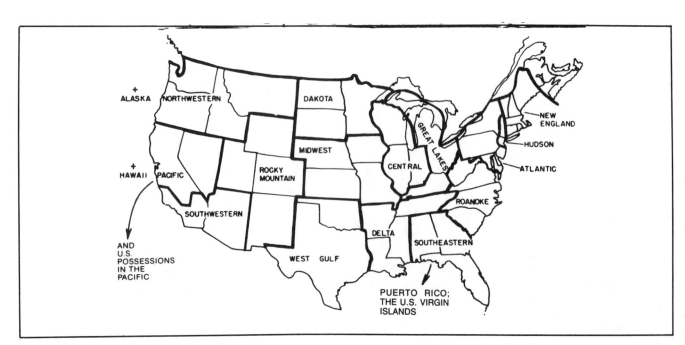

Jack, Johnson, Kaufman, Lamar, Lampasas, Limestone, McLennan, Marion, Mills, Montague, Morris, Nacogdoches, Navarro, Palo Pinto, Panola, Parker, Rains, Red River, Rockwall, Rusk, Shelby, Smith, Somervell, Stephens, Tarrant, Throckmorton, Titus, Upshur, Van Zandt, Wichita, Wilbarger, Wise, Wood, Young); *Oklahoma; South Texas* (Angelina, Aransas, Atacosa, Austin, Bandera, Bastrop, Bee, Bexar, Blanco, Brazoria, Brazos, Brooks, Burleson, Burnet, Caldwell, Calhoun, Cameron, Chambers, Colorado, Comal, Concho, DeWitt, Dimmitt, Duval, Edwards, Fayette, Fort Bend, Frio, Galveston, Gillespie, Goliad, Gonzales, Grimes, Guadalupe, Hardin, Harris, Hays, Hidalgo, Houston, Jackson, Jasper, Jefferson, Jim Hogg, Jim Wells, Karnes, Kendall, Kenedy, Kerr, Kimble, Kinney, Kleberg, LaSalle, Lavaca, Lee, Leon, Liberty, Live Oak, Llano, Madison, Mason, Matagorda, Maverick, McCulloch, McMullen, Medina, Menard, Milam, Montgomery, Newton, Nueces, Orange, Polk, Real, Refugio, Robertson, Sabine, San Augustine, San Jacinto, San Patricio, San Saba, Starr, Travis, Trinity, Tyle, Uvalde, Val Verde, Victoria, Walker, Waller, Washington, Webb, Wharton, Willacy, Williamson, Wilson, Zapata, Zavala); *West Texas* (Andrews, Armstrong, Bailey, Bordon, Brewster, Briscoe, Callahan, Carson, Castro, Childress, Cochran, Coke, Coleman, Collingsworth, Cottle, Crane, Crockett, Crosby, Culberson, Dallam, Dawson, Deaf Smith, Dickens, Donley, Ector, El Paso, Fischer, Floyd, Foard, Gaines, Garza, Glasscock, Gray, Hale, Hall, Hansford, Hardeman, Hartley, Haskell, Hemphill, Hockley, Howard Hudspeth, Hutchinson, Irion, Jeff Davis, Jones, Kent, King, Knox, Lamb, Lipscomb, Loving, Lubbock, Lynn, Martin, Midland, Mitchell, Moore, Motley, Nolan, Ochiltree, Oldham, Parmer, Pecos, Potter, Presidio, Randall, Reagan, Reeves, Roberts, Runnels, Schleicher, Scurry, Shackelford, Sherman, Sterling, Stonewall, Sutton, Swisher, Taylor, Terrell, Terry, Tom Green, Upton, Ward, Wheeler, Winkler, Yoakum).

RAC RADIO AMATEURS OF CANADA-CANADA: *Alberta; British Columbia; Manitoba; Maritime* (Nova Scotia, New Brunswick, Prince Edward Island, Labrador, Newfoundland); *Ontario; Quebec; Saskatchewan.*

SECTION MANAGER

In each ARRL section there is an elected Section Manager (SM) who will have authority over the Field Organization in his or her section, and in cooperation with his Director, foster and encourage ARRL activities and programs within the section. Details regarding the election procedures for SMs are contained in "Rules and Regulations of the ARRL Field Organization," available on request to any ARRL member. Election notices are posted regularly in the Happenings section of *QST*.

Any candidate for the office of Section Manager must be a resident of the section, a licensed amateur of the Technician class or higher, and a full member of the League for a continuous term of at least two years immediately preceding receipt of a petition for nomination at Hq. If elected, he or she must maintain membership throughout the term of office.

The following is a detailed resume of the duties of the Section Manager. In discharging his responsibilities, he:

a) Recruits and appoints nine section-level assistants to serve under his general supervision and to administer the following ARRL programs in the section: emergency communications, message traffic, official observers, affiliated clubs, public information, state government liaison, technical activities and on-the-air bulletins.

b) Supervises the activities of these assistants to ensure continuing progress in accordance with overall ARRL policies and objectives.

c) Appoints qualified ARRL members in the section to volunteer positions of responsibility in support of section programs, or authorizes the respective section-level assistants to make such appointments.

d) Maintains liaison with the Division Director and makes periodic reports to him regarding the status of section activities; receives from him information and guidance pertaining to matters of mutual concern and interest; serves on the Division Cabinet and renders advice as requested by the Division Director; keeps informed on matters of policy that affect section-level programs.

e) Conducts correspondence or other communications, including personal visits to clubs, hamfests and conventions, with ARRL members and affiliated clubs in the section; either responds

to their questions or concerns or refers them to the appropriate person or office in the League organization; maintains liaison with, and provides support to, representative repeater frequency-coordinating bodies having jurisdiction in the section.

f) Writes, or supervises preparation of, the monthly Section News column in *QST* to encourage member participation in the ARRL program in the section.

g) Recruits new amateurs and ARRL members to foster growth of Field organization programs and the amateur service's capabilities in support of public service.

[Note: Move of permanent residence outside the section from which elected will be grounds for declaring the office vacant.]

ARRL LEADERSHIP APPOINTMENTS

Field Organization leadership appointments from the Section Manager are available to qualified ARRL full members in each section. These appointments are as follows: Assistant Section Manager, Section Emergency Coordinator, Section Traffic Manager, Official Observer Coordinator, State Government Liaison, Technical Coordinator, Affiliated Club Coordinator, Public Information Officer, Bulletin Manager, District Emergency Co-ordinator, Emergency Coordinator and Net Manager. Holders of such appointments may wear the League emblem pin with the distinctive deep-green background. Functions of these leadership officials are described below.

Assistant Section Manager

The ASM is an ARRL section-level official appointed by the Section Manager, in addition to the Section Manager's eight section-level assistants. An ASM may be appointed if the Section Manager believes such an appointment is desirable to meet the goals of the ARRL Field Organization in that section. Thus, the ASM is appointed at the complete discretion of the Section Manager, and serves at the pleasure of the Section Manager.

1) The ASM may serve as a general or as a specialized assistant to the Section Manager. That is, the ASM may assist the Section Manager with general leadership matters as the Section Manager's general understudy, or the ASM may be assigned to handle a specific important function not within the scope of the duties of the Section Manager's eight assistants.

2) At the Section Manager's discretion, the ASM may be designated as the recommended successor to the incumbent Section Manager, in case the Section Manager resigns or is otherwise unable to finish the term of office.

3) The ASM should be familiar with "Guidelines for the ARRL Section Manager," which contains the fundamentals of general section management.

4) The ASM must be an ARRL full member, holding at least a Novice class license.

Section Emergency Coordinator

The SEC must hold a Technician class license or higher and is appointed by the Section Manager to take care of all matters pertaining to emergency communications and the Amateur Radio Emergency Service (ARES) on a sectionwide basis. The duties of the SEC include the following:

1) The encouragement of all groups of community amateurs to establish a local emergency organization.

2) Recommendations to the SM on all section emegency policy and planning, including the development of a section emergency communications plan.

3) Cooperation and Coordination with the Section Traffic Manager so that emergency nets and traffic nets in the section present a unified public sevice front. Cooperation and coordiantion should also be maintained with other section leadership officials as appropriate, particularly the State Governmnent Liaison and the Public Information Coordinator.

4) Recommendations of candidates for Emergency Coordinator and District Emergency Coordinator appointments (and cancellations) to the Section Manager and determinations of areas of jurisdiction of each amateur so appointed. At the SM's discretion, the SEC may be directly in charge of making (and cancelling) such appointments. In the same way, the SEC can handle the Official

Emergency Station program.

5) Promotion of ARES membership drives, meetings, activities, tests, procedures, etc., at the section level.

6) Collection and consolidation of Emergency Coordinator (or District Emergency Coordinator) monthly reports and submission of monthly progress summaries to ARRL Hq.

7) Maintenance of contact with other communication services and liaison at the section level with all agencies served in the public interest, particularly in connection with state and local government, civil preparedness, Red Cross, Salvation Army and the National Weather Service. Such contact is maintained in cooperation with the State Government Liaison.

Section Traffic Manager

The STM is appointed by the Section Manager to supervise traffic handling organization at the section level—that is, of coordinating the activities of all traffic nets, both National Traffic System-affiliated and independents, so that routings within the section and connections with other nets to effect orderly and efficient traffic flow are maintained. The STM should be a person at home and familiar with traffic handling on all modes, must have at least a Technician class license, and should possess the willingness and ability to devote equal consideration and time to all section traffic matters. The duties of the STM include the following:

1) Establish, administer, and promote a traffic handling program at the section level, based on, but not restricted to, National Traffic System networks.

2) Develop and implement one or more effective training programs within the section that addresses the needs of both traditional and digital modes of traffic handling. Ensure that Net Managers place particular emphasis on the needs of amateurs new to formal network traffic handling, as well as those who receive, send, and deliver formal traffic on a "casual" basis, via RTTY, Amtor, and Packet based message storage and bulletin board systems.

3) Cooperate and coordinate with the Section Emergency Coordinator so that traffic nets and emergency nets in the section present a unified public service front.

4) Recommend candidates for Net Managers and Official Relay Station appointments to the SM. Issue FSD-211 appointment/cancellation cards and appropriate certificates. At the SM's discretion, the STM may directly make or cancel NM and ORS appointments.

5) Ensure that all traffic nets within the section are properly and adequately staffed, with appropriate direction to Net Managers, as required, which results in coverage of all Net Control liaison functions. Assign liaison coverage adequate to ensure that all digital bulletin boards and message storage systems within the section are polled on a daily basis, to prevent misaddressed, lingering, or duplicated radiogram-formatted message traffic.

6) Maintain familiarity with proper traffic handling and directed net procedures applicable to all normally used modes within the section.

7) Collect and prepare accurate monthly net reports and submit them to ARRL Headquarters, either directly or via the Section Manager, but in any case on or prior to the established deadlines.

Affiliated Club Coordinator

The ACC is the primary contact and resource person for each Amateur Radio club in the section, specializing in providing assistance to clubs. The ACC is appointed by, and reports to, the Section Manager. Duties and qualifications of the ACC include:

1) Volunteer a great deal of time in getting to know the Amateur Radio clubs' members and officers person to person in his or her section. Learn their needs, strengths and interests and work with them to make clubs effective resources in their communities and more enjoyable for their members.

2) Encourage affiliated clubs in the section to become more active and, if the club is already healthy and effective, to apply as a Special Service Club (SSC).

3) Supply interested clubs with SSC application forms.

4) Assist clubs in completing SSC application forms, if requested.

5) Help clubs establish workable programs to use as SSCs.

6) Approve SSC application forms and pass them to the SM.

7) Work with other section leadership officials (Section Emergency Coordinator, Public Information Coordinator, Technical Coordinator, State Government Liason, etc.) to ensure that clubs are involved in the mainstream of ARRL Field Organization activities.

8) Encourage new clubs to become ARRL affiliated.

9) Ensure that annual progress reports (updated officers, liaison mailing addresses, etc.) are forthcoming from all affiliated clubs.

10) Novice Class license; ARRL membership required.

Bulletin Manager

Rapid dissemination of information is the lifeblood of an active, progressive organization. The ARRL Official Bulletin Station network provides a vital communications link for informing the amateur community of the latest developments in Amateur Radio and the ARRL. The ARRL Bulletin Manager is responsible for recruiting and supervising a team of Official Bulletin Stations to disseminate such news and information of interest to amateurs in the section and to provide a means of getting the news and information to all OBS appointees. The bulletins should include the content of ARRL bulletins (transmitted by W1AW), but should also include items of local, section and regional interest from other sources, such as ARRL section leadership officials, as well as information provided by the Division Director.

A special effort should be made to recruit an OBS for each major repeater and packet bulletin board in the section. This is where the greatest "audience" is to be found, many of whom are not sufficiently informed about the latest news of Amateur Radio and the League. Such bulletins should be transmitted regularly, perhaps in conjunction with a repeater net or on a repeater "bulletin board" (tone-accessed recorded announcements for repeater club members).

Although the primary mission of OBS appointees is to copy ARRL bulletins directly from W1AW, in some sections the Bullein Manager may take on the responsibility of retransmitting ARRL bulletins (as well as other information) for the benefit of OBS appointees, on a regularly scheduled day, time and frequency. An agreed-upon schedule should be worked out in advance. Time is of the essence when conveying news; therefore a successful Bulletin Manager will develop ways of communicating with the OBS appointees quickly and efficiently.

Bulletin Managers should be familiar with the position description of the Official Bulletin Station, which appears later. The duties of the Bulletin Manager include the following:

1) The Bulletin Manager must have a Technician class license or higher, and maintain League membership.

2) The Bulletin Manager is appointed by the SM and is required to report regularly to the SM concerning the section's bulletin program.

3) The Bulletin Manager is responsible for recruiting (and, at the discretion of the SM, appointing) and supervising a team of Official Bulletin Stations in the Section. A special effort should be made to recruit OBSs for each major repeater and PBBS in the section.

4) The Bulletin Manager must be capable of copying ARRL bulletins directly from W1AW on the mode(s) necessary. The Bulletin Manager may, in some cases, be required to retransmit ARRL bulletins for OBS appointees who might be unable to copy them directly from W1AW.

5) The Bulletin Manager is also responsible for funneling news and information of a local, section and regional nature to OBS appointees. In so doing, the Bulletin Manager must maintain close contact with other section-level officials, and the Division Director, to maintain an organized and unified information flow within the section.

Official Observer Coordinator

The Official Observer Coordinator is an ARRL section-level leadership official appointed by the Section Manager to supervise the Official Observer program in the section. The OO Coordinator must hold a Technician class (or higher) amateur license and be licensed as a Technician or higher for at least four years.

The Official Observer program has operated for more than half a century, and in that time, OO appointees have assisted thousands of amateurs whose signals, or operating procedures, were not in compliance with the regulations. The function of the OO is to *listen* for amateurs who might otherwise come to the attention of the FCC and to advise them by mail of the irregularity observed. The OO program is, in essence, for the benefit of amateurs who *want* to be helped. Official Observers must meet high standards of expertise and experience. It is the job of the OO Coordinator to recruit, supervise and direct the efforts of OOs in the section, and to report their activity monthly to the Section Manager and to ARRL Hq.

The OO Coordinator is a key figure in the Amateur Auxiliary to the FCC's Field Operations Bureau, the foundation of which is an enhanced OO program. Jointly created by the FCC and ARRL in response to federal enabling legislation (Public Law 97-259), the Auxiliary permits a close relationship between FCC and ARRL Field Organization volunteers in monitoring the amateur airwaves for potential rules discrepancies/violations. (Contact your Section Manager for further details on the Amateur Auxiliary program.)

Public Information Coordinator

The ARRL Public Information Coordinator is a section-level official appointed by the Section Manager to be the section's expert on public information and public relations matters. The Public Information Coordinator is also responsible for organizing, guiding and coordinating the activities of the Public Information Officers within the section.

The Public Information Coordinator must be a full member of the ARRL and, preferably, have professional public relations or journalism experience or a significantly related background in dealing with the public media.

The purpose of public relations goes beyond column inches and minutes of air time. Those are means to an end—generally, telling a specific story about hams, ham radio or ham-related activities for a specific purpose. Goals may range from recruiting potential hams for a licensing course to improving public awareness of amateurs' service to the community. Likewise, success is measured not in column inches or air time, but in how well that story gets across and how effectively it generates the desired results.

For this reason, public relations are not conducted in a vacuum. Even the best PR is wasted without effective follow-up. To do this best, PR activities must be well-timed and well-coordinated within the amateur community, so that clubs, Elmers, instructors and so on are prepared to deal with the interest the PR generates. Effective PICs will convey this goal-oriented perspective and attitude to their PIOs and help them coordinate public relations efforts with others in their sections.

Recruitment of new hams and League members is an integral part of the job of every League appointee. Appointees should take advantage of every opportunity to recruit a new ham or a member to foster growth of Field Organization programs, and our abilities to serve the public.

Specific Duties of the Public Information Coordinator

1) Advises the Section Manager on building and maintaining a positive public image for Amateur Radio in the section; keeps the SM informed of all significant events which would benefit from the SM's personal involvement and reports regularly to the SM on activities.

2) Counsels the SM in dealing with the media and with government officials, particularly when representing the ARRL and/or Amateur Radio in a public forum.

3) Maintains contact with other section level League officials, particularly the Section Manager and others such as the State Government Liaison, Section Emergency Coordinator and Bulletin Manager on matters appropriate for their attention and to otherwise help to assure and promote a coordinated and cohesive ARRL Field Organization.

4) Works closely with the section Affiliated Club Coordinator and ARRL-affiliated clubs in the section to recruit and train a team of Public Information Officers (PIOs). With the approval of the Section Manager, makes PIO appointments within the section.

5) Works with the SM and other PICs in the division to:

a) develop regional training programs for PIOs and club publicity chairpersons;

b) coordinate public relations efforts for events and activities which may involve more than one section, and

c) provide input on matters before the League's Public Relations Committee for discussion or action.

6) Establishes and coordinates a section-wide Speakers Bureau to provide knowledgeable and effective speakers who are available to address community groups about Amateur Radio, and works with PIOs to promote interest among those groups.

7) Helps local PIOs to recognize and publicize newsworthy stories in their areas. Monitors news releases sent out by the PIOs for stories of broader interest and offers constructive comments for possible improvement. Helps local PIOs in learning to deal with, and attempting to minimize, any negative publicity about Amateur Radio or to correct negative stories incorrectly ascribed to Amateur Radio operators.

8) Working with the PIOs, develops and maintains a comprehensive list of media outlets and contacts in the section for use in section-wide or nationwide mailings.

9) Helps local PIOs prepare emergency response PR kits containing general information on Amateur Radio and on local clubs, which may be distributed in advance to local Emergency Coordinators and District Emergency Coordinators for use in dealing with the media during emergencies.

10) Works with PIOs, SM and ARRL staff to identify and publicize League-related stories of local or regional interest, including election or appointment of ARRL leadership officials, scholarship winners/award winners, *QST* articles by local authors or local achievements noted or featured in *QST*.

11) Familiarize self with ARRL Public Service Announcements (PSAs), brochures and audio-visual materials; assists PIOs in arranging air time for PSAs; helps PIOs and speakers choose and secure appropriate brochures and audio-visual materials for events or presentations.

12) At the request of the Section Manager or Division Director, may assist in preparation of a section or division newsletter.

13) Encourages, organizes and conducts public information/public relations sessions at ARRL hamfests and conventions.

14) Works with PIOs to encourage activities that place Amateur Radio in the public eye, including demonstrations, Field Day activities, etc. and assures that sponsoring organizations are prepared to follow-up on interest generated by these activities.

Most public relations activities are conducted on a local level by affiliated clubs, which generally are established community organizations. PICs should encourage clubs to make public relations a permanent part of their activities.

With the Section Manager's approval, the PIC may appoint club publicity chairpersons or other individuals recommended by affiliated clubs as PIOs. Where the responsibility cannot or will not be assumed by the club, the PIC is encouraged to seek qualified League members who are willing to accept the responsibility of PIO appointments.

Appointees should take advantage of every opportunity to recruit a new ham or member to foster growth of Field Organization programs, and our abilities to serve the public.

State Government Liaison

The State Government Liaison (SGL) shall be an amateur who is aware (at a minimum) of state legislative proposals in the normal course of events and who can watch for those proposals having the potential to affect Amateur Radio without creating a conflict of interest.

The SGL shall collect and promulgate information on state ordinances affecting Amateur Radio and work (with the assistance of other ARRL members) toward assuring that they work to the mutual benefit of society and the Amateur Radio Service.

The SGL shall guide, encourage and support ARRL members in representing the interests of the Amateur Radio Service at all levels. Accordingly, the SGL shall cooperate closely with other section-level League officials, particularly the Section Emergency Coordinator and the Public Information Coordinator.

When monitoring state legislative dockets, SGL's should watch for key words that could lead to potential items affecting Amateur Radio. Antennas (dish, microwave, towers, structures, satellite, television, lighting), mobile radio, radio receivers, radio interference, television interference, scanners, license plates, cable television, ham radio, headphones in automobiles, lightning protection, antenna radiation and biological effects of radio signals are a few of the examples of what to look for.

In those states where there is more than one section, the Section Managers whose territory does not encompass the state capital may simply defer to the SGL appointed by their counterpart in the section where the state capital is located. In this case, the SGL is expected to communicate equally with all Section Managers (and Section Emergency Coordinators and other section-level League officials). In sections where there is more than one government entity, i.e., Maryland, DC, Pacific, there may be a Liaison appointed for each entity.

Technical Coordinator

The ARRL Technical Coordinator (TC) is a section-level official appointed by the Section Manager to coordinate all technical activites within the section. The technical Coordinator must hold a Novice class or higher amateur license. The Technical Coordinator reports to the Section Manager and is expected to maintain contact with other section-level appointees as appropriate to ensure a unified ARRL Field Organization within the section. The duties of the Technical Coordinator are as follows:

1) Supervise and coordinate the work of the section's Technical Specialists (TSs).

2) Encourage amateurs in the section to share their technical achievements with others through the pages of QST, and at club meetings, hamfests and conventions.

3) Promote technical advances and experimentation at VHF/UHF and with specialized modes, and work closely with enthusiasts in these fields within the section.

4) Serve as an advisor to radio clubs that sponsor training programs for obtaining amateur licenses or upgraded licenses in cooperation with the ARRL Affiliated Club Coordinator.

5) In times of emergency or disaster, function as the coordinator for establishing an array of equipment for communications use and be available to supply technical expertise to government and relief agencies to set up emergency communication networks, in cooperation with the ARRL Section Emergency Coordinator.

6) Refer amateurs in the section who need technical advice to local TSs.

7) Encourage TSs to serve on RFI and TVI committees in the section for the purpose of rendering technical assistance as needed, in cooperation with the ARRL OO/Coordinator.

8) Be available to assist local technical program committees in arranging suitable programs for ARRL hamfests and conventions.

9) Convey the views of section amateurs and TSs about the technical contents of QST and ARRL books to ARRL HQ. Suggestions for improvements should also be called to the attention of the ARRL HQ technical staff.

10) Work with the appointed ARRL TAs (technical advisors) when called upon.

11) Be available to give technical talks at club meetings, hamfests and conventions in the section.

District Emergency Coordinator

The DEC is an ARRL full member of at least Technician class experienced in emergency communications who can assist the SEC by taking charge in the area of jurisdiction especially during an emergency.
The DEC shall:

1) Coordinate the training, organization and emergency participation of Emergency Coordinators in the area of jurisdiction.

2) Make local decisions in the absence of the SEC or through coordination with the SEC concerning the allotment of available amateurs and equipment during an emergency.

3) Coordinate the interrelationship between local emergency plans and between communications networks within the area of jurisdiction.

4) Act as backup for local areas without an Emergency Coordinator and assist in maintaining contact with governmental and other agencies in the area of jurisdiction.

5) Provide direction in the routing and handling of emergency communications of either a formal or tactical nature.

6) Recommend EC appointments to the SEC and advise on OES appointments.

7) Coordinate the reporting and documentation of ARES activities in the area of jurisdiction.

8) Act as a model emergency communicator as evidenced by dedication to purpose, reliability and understanding of emergency communications.

EMERGENCY COORDINATOR

The ARRL Emergency Coordinator is a key team player in ARES on the local emergency scene. Working with the Section Emergency Coordinator, the DEC and Official Emergency Stations, the EC prepares for, and engages in management of communications needs in disasters. EC duties include:

1) Promote and enhance the activities of the Amateur Radio Emergency Service (ARES) for the benefit of the public as a voluntary, non-commercial communications service.

2) Manage and coordinate the training, organization and emergency participation of interested amateurs working in support of the communities, agencies or functions designated by the Section Emergency Coordinator/Section Manager.

3) Establish viable working relationships with federal, state, county, city governmental and private agencies in the ARES jurisdictional area which need the services of ARES in emergencies. Determine what agencies are active in your area, evaluate each of their needs, and which ones you are capable of meeting, and then prioritize these agencies and needs. Discuss your planning with your Section Emergency Coordinator and then with your counterparts in each of the agencies. Ensure they are all aware of your ARES group's capabilities, and perhaps more importantly, your limitations.

4) Develop detailed local operational plans with "served" agency officials in your jurisdiction that set forth precisely what each of your expectations are during a disaster operation. Work jointly to establish protocols for mutual trust and respect. All matters involving recruitment and utilization of ARES volunteers are directed by you, in response to the needs assessed by the agency officials. Technical issues involving message format, security of message transmission, Disaster Welfare Inquiry policies, and others, should be reviewed and expounded upon in your detailed local operations plans.

5) Establish local communications networks run on a regular basis and periodically test those networks by conducting realistic drills.

6) Establish an emergency traffic plan, with Welfare Traffic inclusive, utilizing the National Traffic System as one active component for traffic handling. Establish an operational liaison with local and section nets, particularly for handling Welfare traffic in an emergency situation.

7) In times of disaster, evaluate the communications meeds of the jurisdiction and respond quickly to those needs. The EC will assume authority and responsibility for emergency response and performance by ARES personnel under his jurisdiction.

8) Work with other non-ARES amateur provider-groups to establish mutual respect and understanding, and a coordination mechanism for the good of the public and Amateur Radio. The goal is to foster an efficient and effective Amateur Radio response overall.

9) Work for growth in your ARES program, making it a stronger, more valuable resource and hence able to meet more of the agencies' local needs. There are thousands of new Technicians coming into the amateur service that would make ideal additions to your ARES roster. A stronger ARES means a better ability to serve your communities in times of need and a greater sense of pride for Amateur Radio by both amateurs and the public.

10) Report regularly to the SEC, as required.

Recruitment of new hams and League members is an integral part of the job of every League appointee. Appointees should take advantage of every opportunity to recruit a new ham or member to foster growth of Field Organization programs, and our abilities to serve the public.

Net Manager

For coordinating and supervising traffic-handling activities in the section, the SM may appoint one or more Net Managers, usually on recommendation of the Section Traffic Manager. The number of NMs appointed may depend on a section's geographical size, the number of nets operating in the section, or other factors having to do with the way the section is organized. In some cases, there may be only one Net Manager in charge of the one section net, or one NM for the phone net, one for the CW net. In larger or more traffic-active sections there may be several, including NMs for the VHF net or nets, for the RTTY net, or NTS local nets not controlled by ECs. All ARRL NMs should work under the STM in a coordinated section traffic plan.

Some nets cover more than one section but operate in NTS at the section level. In this case, the Net Manager is selected by agreement among the STMs concerned and the NM appointment conferred on him by his resident SM.

NMs may conduct any testing of candidates for ORS appointment (see below) that they consider necessary before making appointment recommendations to the STM. Net Managers also have the function of requiring that all traffic handling in ARRL recognized nets is conducted in proper ARRL form.

Remember: All appointees or appointee candidates must be ARRL full members.

ARRL STATION APPOINTMENTS

Field Organization station and individual appointments from the Section Manager are available to qualified ARRL full members in each section. These appointments are as follows: Official Relay Station, Official Emergency Station, Official Bulletin Station, Public Information Officer, Official Observer and Technical Specialist. All appointees receive handsome certificates from the SM and are entitled to wear ARRL membership pins with the distinctive blue background. All appointees are required to submit regular reports to maintain appointments and to remain active in their area of specialty.

The report is the criterion of activity. An appointee who misses three consecutive monthly reports is subject to cancellation by the SM of the appropriate section leadership official, who cannot know what or how much you are doing unless you report. An appointee whose appointment is cancelled for this or other reasons must earn reinstatement by demonstrating activity and adherence to the requirements. Reinstatement of cancelled appointments, and indeed judgment of whether or not a candidate meets the requirements, is at the discretion of the SM and the section leadership.

The detailed qualifications of the six individual "station" appointments are given below. If you are interested, your SM will be glad to receive your application. Use application form FSD-187, reproduced nearby.

Official Relay Station

This is a traffic-handling appointment that is open to all licenses. This appointment applies equally to all modes and all parts of the spectrum. It is for traffic handlers, regardless of how or in what part of the spectrum they do it.

The potential value of the operator who has traffic know-how to his country and community is enhanced by his ability and the readiness of his station to function in the community interest in case of emergency. Traffic awareness and experience are often the signs by which mature amateurs may be distinguished.

Traditionally, there have been considerable differences between procedures for traffic handling by CW, phone, RTTY, ASCII and other modes. Appointment requirements for ORS do not deal with these, but with factors equally applicable to all modes. The appointed ORS may confine activities to one mode or one part of the spectrum if he wishes although versatility does indeed make it possible to perform a more complete public service. The expectation is that the ORS will set the example in traffic handling, however it is done. To the degree that he is deficient in performing traffic functions by any mode, to that extent he does not meet the qualifications for the appointment. Here are the basic requirements:

1) Full ARRL membership and Novice class license or higher.
2) Code and/or voice transmission.
3) Transmission quality, by whatever mode, must be of the

highest quality, both technically and operationally. For example, CW signals must be pure, chirpless and clickless, and code sending must be well spaced and properly formed. Voice transmission must be of proper modulation percentage or deviation, precisely enunciated with minimum distortion. RTTY must be clickless, proper shift, etc.

4) All ORSs are expected to follow standard ARRL operating practices (message form, ending signals, abbreviations or prowords, courtecy, etc.).

5) Regular participation in traffic activities, either freelance or ARRL-sponsored. The latter is encouraged, but not required.

6) Handle all record communications speedily and reliably and set the example in efficient operating procedures. All traffic is relayed or delivered promptly after receipt.

7) Report monthly to the STM, including a breakdown of traffic handled during the past calendar month.

Official Emergency Station

Amateur operators may be appointed as an Official Emergency Station (OES) by their Section Emergency Coordinator (SEC) or Section Manager (SM) at the recommendation of the EC, or DEC (if no EC) holding jurisdiction. The OES appointee must be an ARRL member and set high standards of emergency preparedness and operating. The OES appointee makes a deeper commitment to the ARES program in terms of functionality than does the rank-and-file ARES registrant.

The requirements and qualifications for the position include the following: Full ARRL membership; experience as an ARES registrant; regular participation in the local ARES organization including drills and tests; participation in emergency nets and actual emergency situations; regular reporting of activities.

The OES appointee is appointed to carry out specific functions and assignments designated by the appropriate EC or DEC. The OES appointee and the presiding EC or DEC, at the time of the OES appointment, will mutually develop a detailed, operational function/assignment and commitment for the new appointee. Together, they will develop a responsibility plan for the individual OES appointee that makes the best use of the individual's skills and abilities. During drills and actual emergency situations, the OES appointee will be expected to implement his/her function with professionalism and minimal supervision.

Functions assigned may include, but are not limited to, the following four major areas of responsibility:

This form is available to apply for an ARRL station appointment.

OPERATIONS–Responsible for specific, pre-determined operational asignments during drills or actual emergency situations. Examples include: Net Control Station or Net Liaison for a specific ARES net; Manage operation of a specified ARES VHF or HF digital BBS or MBO, or point-to-point link; Operate station at a specified emergency management office, Red Cross shelter or other served agency operations point.

ADMINISTRATION–Responsible for specific, pre-determined administrative tasks as assigned in the initial appointment commitment by the presiding ARES official. Examples include: Recruitment of ARES members; liaison with Public Information Officer to coordinate public information for the media; ARES registration data base management; victim/refugee data base management; equipment inventory; training; reporting; and post-event analysis.

LIAISON–Responsible for specific, pre-determined liaison responsibilities as assigned by the presiding EC or DEC. Examples include: Maintaining contact with assigned served agencies; Maintaining liaison with specified NTS nets; Maintaining liaison with ARES officials in adjacent jurisdictions; Liaison with mutual assistance or "jump" teams.

LOGISTICS–Responsible for specific, pre-determined logistical functions as assigned. Examples include: transportation; Supplies management and procurement (food, fuel, water, etc.); Equipment maintenance and procurement–radios, computers, generators, batteries, antennas.

MANAGEMENT ASSISTANT–Responsible for serving as an assistant manager to the EC, DEC or SEC based on specific functional assignments or geographic areas of jurisdiction.

CONSULTING–Responsible for consulting to ARES officials in specific area of expertise.

OES appointees may be assigned to pre-disaster, post-disaster, and recovery functions. These functions must be specified in the OES's appointment commitment plan.

The OES appointee is expected to participate in planning meetings, and post-event evaluations. Following each drill or actual event, the EC/DEC and the OES appointee should review and update the OES assignment as required. The OES appointee must keep a detailed log of events during drills and actual events in his/her sphere of responsibility to facilitate this review.

Continuation of the appointment is at the discretion of the appointing official, based upon the OES appointee's fulfillment of the tasks he/she has agreed to perform.

Recruitment of new hams and League members is an integral part of the job of every League appointee. Appointees should take advantage of every opportunity to recruit a new ham or member to foster growth of Field Organization programs, and our abilities to serve the public.

Official Bulletin Station

Rapid dissemination of information is the lifeblood of an active, progressive organization. The ARRL Official Bulletin Station network provides a vital communications link for informing the amateur community of the latest developments in Amateur Radio and the League. ARRL bulletins, containing up-to-the minute news and information of Amateur Radio, are issued by League Hq as soon as such news breaks. These bulletins are transmitted on a regular schedule by ARRL Hq station W1AW.

The primary mission of OBS appointees is to copy these bulletins directly off the air from W1AW—on voice, CW or RTTY/ASCII—and retransmit them locally for the benefit of amateurs in the particular coverage area, many of whom may not be equipped to receive bulletins directly from W1AW.

ARRL bulletins of major importance or of wide-ranging scope are mailed from Hq to each Bulletin Manager and OBS appointee. However, some bulletins, such as the ARRL DX Bulletin (transmitted on Fridays UTC), are disseminated only by W1AW because of time value. Thus, it is advantageous for each OBS to copy W1AW directly. In some sections, the Bulletin Manager may assume the responsibility of copying the bulletins from W1AW; therefore, individual OBSs should be sure to meet the Bulletin Manager on a regular, agreed-upon schedule to receive the latest bulletins.

Inasmuch as W1AW operates on all bands (160-2 meters), the need for OBSs on HF has lessened somewhat in recent times. However, OBS appointments for HF operation can be conferred by the Section Manager (or the Bulletin Manager, depending on how the SM organizes the section) if the need is apparent. More important, to serve the greatest possible "audience," OBS appointees who can send ARRL bulletins over VHF repeaters are of maximum usefulness and are much in demand. If possible, an OBS who can copy bulletins directly from W1AW (or the Bulletin Manager) should be assigned to each major repeater in the section. Bulletins should be transmitted regularly, perhaps in conjunction with a VHF repeater net, on a repeater bulletin board (tone-accessed recorded announcements for repeater club members), or via a local RTTY (computer) mailbox. Duties and requirements of the OBS include the following:

1) OBS candidates must have Novice class license or higher.

2) Retransmission of ARRL bulletins must be made at least once per week to maintain appointment.

3) OBS candidates are appointed by the Section Manager (or by the Bulletin Manager, if the SM so desires) and must adhere to a schedule that is mutually agreeable, as indicated on appointment application form FSD-187.

4) OBS appointees should send a monthly activity report (such as FSD-210, under "Schedules and Net Affiliations") to the Bulletin Manager, indicating bulletin transmissions made and generally updating the Bulletin Manager to any OBS-related activities. This reporting arrangement may be modified by the Bulletin Manager as he/she sees fit.

5) As directed by the Bulletin Manager, OBSs will include in their bulletin transmissions news of local, section and regional interest.

Public Information Officer

Public Information Officers (PIOs) are appointed by and report to the ARRL section Public Information Coordinator (PIC) generally upon the recommendation of an affiliated club and with the approval of the Section Manager (SM). PIOs are usually club publicity chairpersons and must be full ARRL members. Training for PIOs should be provided regularly on a sectional or regional basis by the PIC and/or other qualified people.

Good "grass roots" public relations activities involve regular and frequent publicizing of amateur activities through local news media plus community activities; school programs; presentations to service clubs and community organizations; exhibits and demonstrations; and other efforts which create a positive public image for Amateur Radio.

Public relations are not conducted in a vacuum. Even the best PR is wasted without effective follow-up. To do this best, PR activities must be well-timed and well-coordinated within the amateur community, so that clubs, Elmers, instructors and so on are prepared to deal with the interest the PR generates.

Recruitment of new hams and League members is an integral part of the job of every League appointee. Appointees should take advantage of every opportunity to recruit a new ham or member to foster growth of Field Organization programs, and our abilities to serve the public.

Specific Duties of the Public Information Officer

1) Establishes and maintains a list of media contacts in the local area; strives to establish and maintain personal contacts with appropriate representatives of those media (editors, news directors, science reporters and so on).

2) Becomes a contact for the local media and assures that editors/reporters who need information about Amateur Radio know where to find it.

3) Works with Local Government Liaisons to establish personal contacts with local government officials where possible and explain to them, briefly and non-technically, about Amateur Radio and how it can help their communities.

4) Keeps informed of activities by local hams and identifies and publicizes those that are newsworthy or carry human interest appeal. (This is usually done through news releases or suggestions for interviews or feature stories.)

5) Attempts to deal with and minimize any negative publicity about Amateur Radio and to correct any negative stories which are incorrectly ascribed to Amateur Radio operators.

6) Generates advance publicity through the local media of scheduled activities of interest to the general public, including

licensing classes, hamfests, club meetings, Field Day operations, etc.

7) Works with the section PIC to identify and publicize League-related stories of local news interest, including election and appointment of hams to leadership positions, *QST* articles by local authors or local achievements noted or featured in *QST*.

8) Maintains contact with other League officials in the local area, particularly the Emergency Coordinator and/or District Emergency Coordinator. With the PIC, helps prepare an emergency response PR kit, including general brochures on Amateur Radio and specific information about local clubs. Distributes them to ECs and DECs before an emergency occurs. During emergencies, these kits should be made available to reporters at the scene or at a command post. The PIO should help summarize Amateur Radio activity in an ongoing situation, and follow up any significant emergency communications activities with prompt reporting to media of the extent and nature of Amateur Radio involvement.

9) Assists the section PIC in recruiting hams for the section's Speakers Bureau; promotes interest among community and service organizations in finding out more about Amateur Radio through the bureau and relays requests to the PIC.

10) Helps individual hams and radio clubs to develop and promote good ideas for community projects and special events to display Amateur Radio to the public in a positive light.

11) Attends regional training sessions sponsored by section PICs.

12) Becomes familiar with ARRL Public Service Announcements (PSAs), brochures and audio-visual materials; contacts local radio and TV stations to arrange airing of Amateur Radio PSAs; secures appropriate brochures and audio-visual materials for use in conjunction with planned activities.

13) Keeps the section PIC fully informed on activities and places PIC on news release mailing list.

Official Observer

The Official Observer (OO) program has been sponsored by the League for over 50 years to help amateurs help each other. Official observer appointees have aided thousands of amateurs to maintain their transmitting equipment and operating procedures in compliance with the regulations. The object of the OO program is to notify amateurs by mail of operating/technical irregularities before they come to the attention of the FCC.

The ARRL commitment to volunteer monitoring has been greatly enhanced by the creation of the Amateur Auxiliary to the FCC Field Operations Bureau, designed to enable amateurs to play a more active and direct role in upholding the traditional high standard of conduct on the amateur bands. The OO is the foundation of the Amateur Auxiliary, carrying out the all-important day-to-day maintenance monitoring of the amateur airwaves. Following recommendation by the Section Manager, potential members of the Amateur Auxiliary are provided with training materials, and all applicants must successfully complete a written examination to be enrolled as Official Observers. For further information, please contact your SM.

The OO performs his function by *listening* rather than transmitting, keeping a watchful ear out for such things as frequency instability, harmonics, hum, key clicks, broad signals, distorted audio, overdeviation, out-of-band operation, etc. The OO completes his task once the notification card is sent. Reimbursement for postage expenses are provided for through the SM. The OO:

1) Must be an ARRL full member and have been a licensee of Technician class or higher for at least four years.

2) Must undergo and complete successfully the Amateur Auxiliary training and certification procedure.

3) Must report to the OO Coordinator regularly on FSD-23.

4) Maintain regular activity in sending out notices as observed.

The OO program is one of the most important functions of the League. A sincere dedication to helping our brother and sister amateurs is required for appointment. Only the "very best" are sought.

Technical Specialist

Appointed by the SM, or TC under delegated authority from the SM, the TS supports the TC in two main areas of responsibility: Radio Frequency Interference, and Technical Information. The TS must hold full ARRL membership and at least a Novice class license. TSs can specialize in certain specific technical areas, or can be generalists. Here is a list of specific job duties:

1) Serve as a technical oracle to local hams and clubs. Correspond by telephone and letter on tech topics. Refer correspondents to other sources if specific topic is outside TS's knowledge.

2) Serve as advisor in radio frequency interference issues. RFI can drive a wedge in neighbor and city relations. It will be the TS with a cool head who will resolve problems. Local hams will come to you for guidance in dealing with interference problems.

3) Speak at local clubs on popular tech topics. Let local clubs know you're available and willing.

4) Represent ARRL at technical symposiums in industry; serve on CATV advisory committees; advise municipal governments on technical matters.

5) Work with other ARRL officials and appointees when called upon for technical advice especially in emergency communications situations where technical prowess can mean the difference in getting a communications system up and running, the difference between life and death.

6) Handle other miscellaneous technically related tasks assigned by the Technical Coordinator.

Local Government Liaison

The Local Government Liaison (LGL) is primarily responsible for monitoring proposals and actions by local government bodies and officials which may affect Amateur Radio; for working with the local PIO to alert section leadership officials and area amateurs to any such proposals or actions, and for coordinating local responses. In addition, the LGL serves as a primary contact for amateurs encountering problems dealing with local government agencies, for those who want to avoid problems and for local officials who wish to work with amateurs or simply learn more about Amateur Radio. The most effective LGL will be able to monitor local government dockets consistently, muster local, organized support quickly when necessary, and be well known in the local amateur community as the point person for local government problems. The LGL must be a Full Member of the ARRL. LGLs are appointed by and report to the Section Manager, or State Government Liaison (acting under delegated authority from the SM).

Specific Responsibilities:

1) Monitor proposals and actions of town/city councils zoning appeals boards, and any other legislative or regulatory agencies or officials below the state level whose actions can directly or indirectly affect Amateur Radio.

2) Attend meetings of those bodies when possible, to become familiar with their policies, procedures and members. Assist local amateurs in their dealings with local boards and agencies.

3) Be available to educate elected and appointed officials, formally and informally, about the value of Amateur Radio to their community.

4) Work with the PIO or PIC to inform local amateurs, the SGL and the SM of any proposals of actions which may affect Amateur Radio, and report regularly on the progress or lack thereof.

5) Work with the PIO to organize the necessary local response to any significant proposals or actions, either negative or positive, and coordinate that response.

6) Refer amateurs seeking ARRL Volunteer Counsels to HQ.

7) Register on mailing list for Planning Commission meeting agendas.

8) Work with the PIO and local clubs to build and/or maintain good relations between Amateur Radio and local officials. (For example, invite the mayor to a club dinner or council members to Field Day.)

N5KR AZIMUTHAL EQUIDISTANT MAPS

The following computer-generated azimuthal maps centered on 21 different world locations were produced especially for *The ARRL Operating Manual* by William D. Johnston, N5KR, PO Box 370, White Sands, NM, 88002. Similar maps centered on your QTH or other locations can be ordered from N5KR (please include an SASE with your request for information).

N5KR AZIMUTHAL EQUIDISTANT MAP CENTERED ON
W1AW

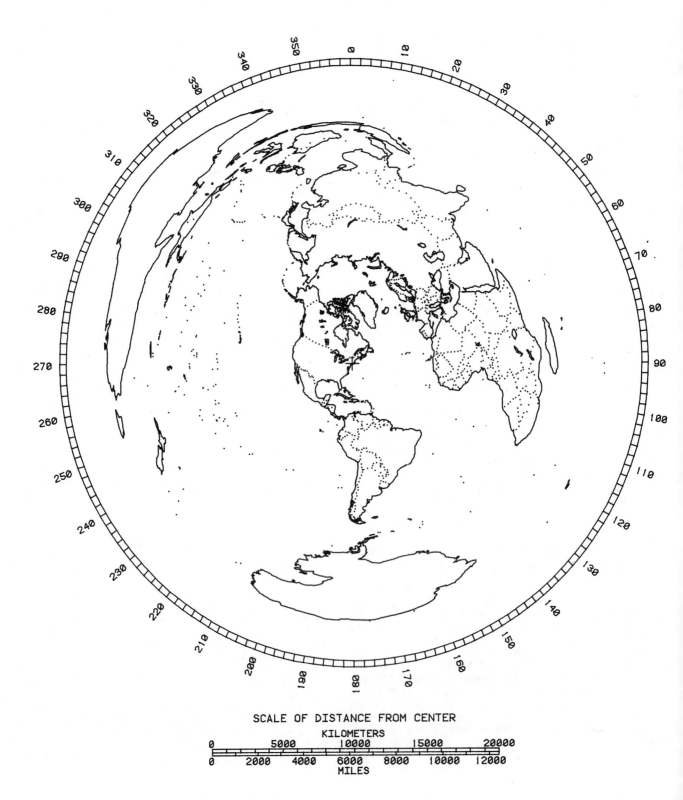

SCALE OF DISTANCE FROM CENTER

N5KR AZIMUTHAL EQUIDISTANT MAP CENTERED ON
EASTERN USA

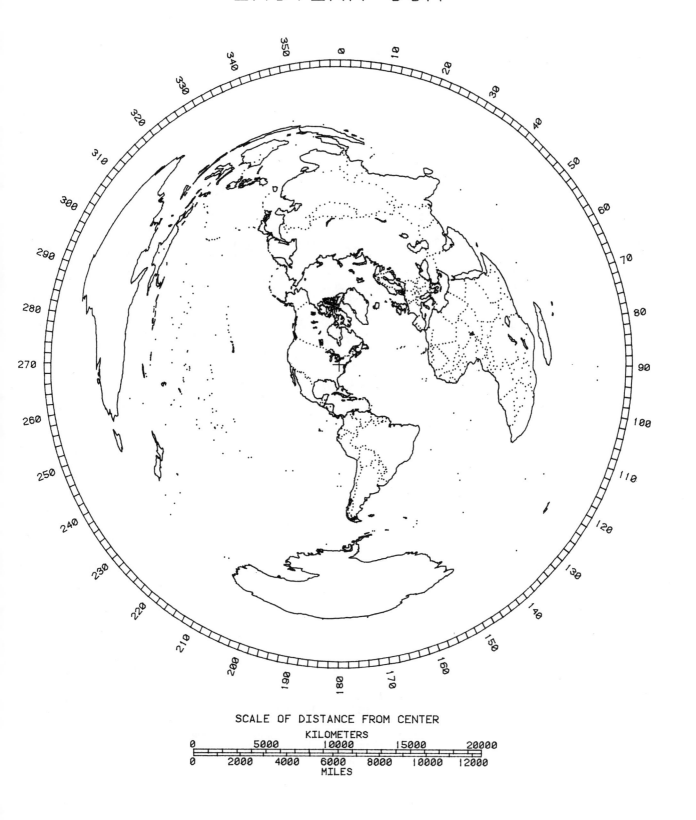

SCALE OF DISTANCE FROM CENTER

KILOMETERS

| 0 | 5000 | 10000 | 15000 | 20000 |

MILES

| 0 | 2000 | 4000 | 6000 | 8000 | 10000 | 12000 |

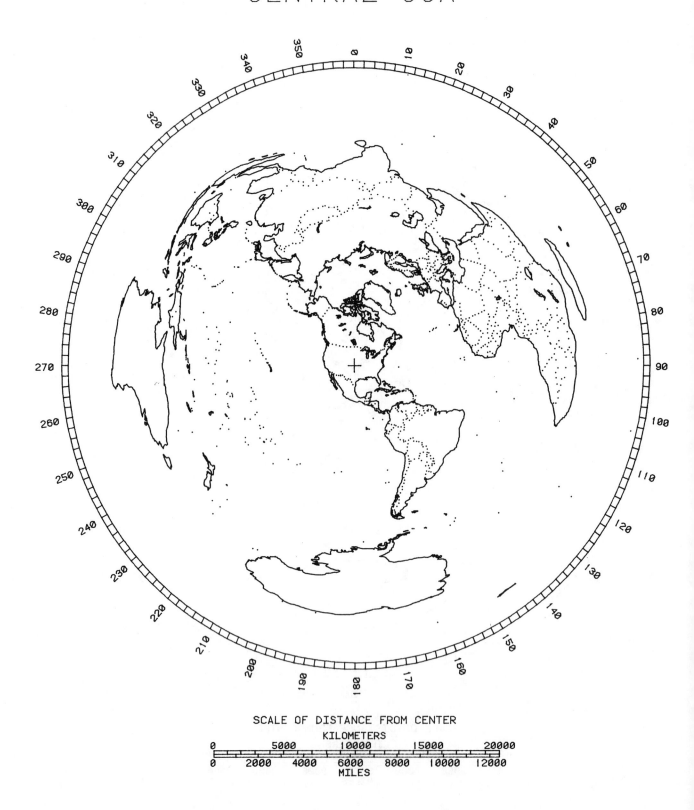

SCALE OF DISTANCE FROM CENTER

KILOMETERS

| 0 | 5000 | 10000 | 15000 | 20000 |

| 0 | 2000 | 4000 | 6000 | 8000 | 10000 | 12000 |

MILES

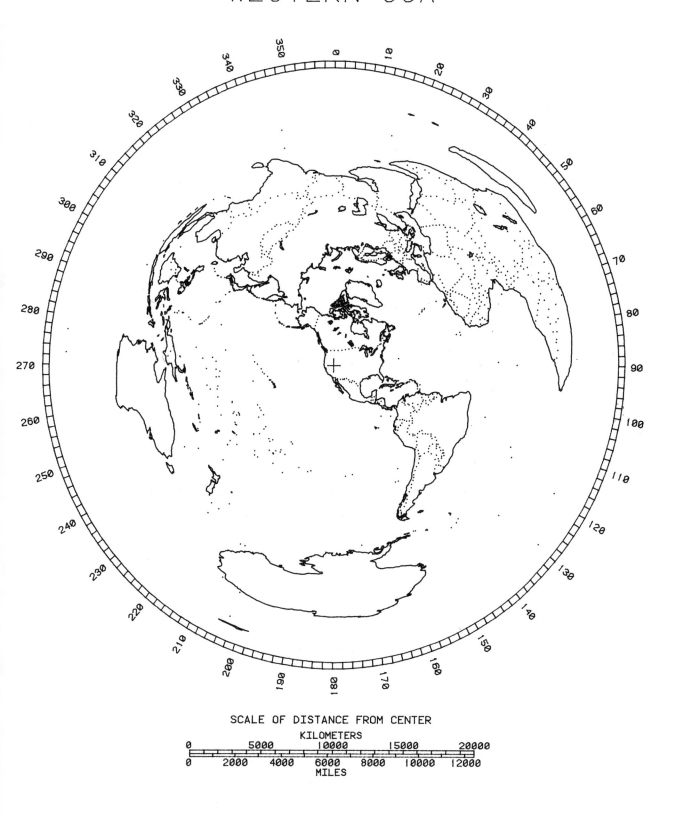

SCALE OF DISTANCE FROM CENTER

KILOMETERS

N5KR AZIMUTHAL EQUIDISTANT MAP CENTERED ON
ALASKA

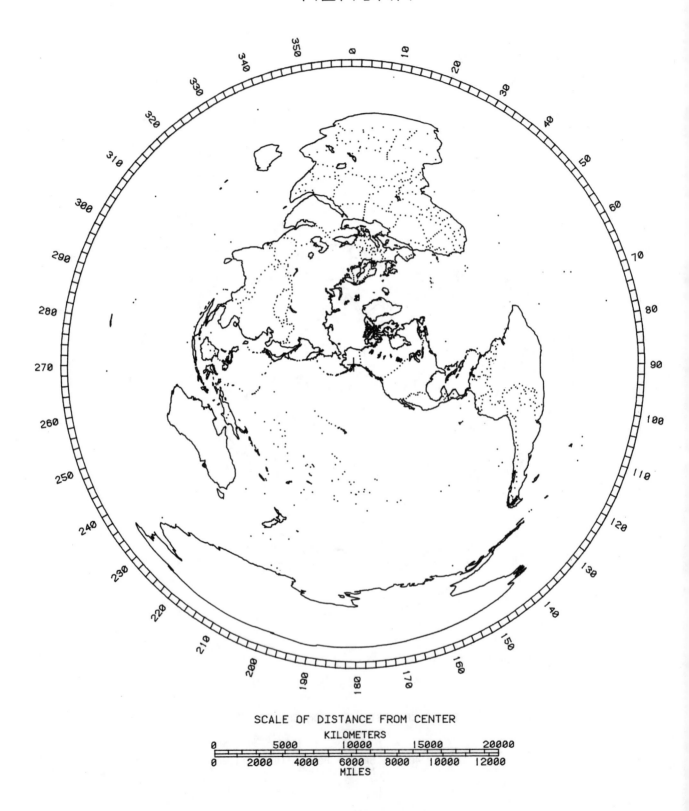

SCALE OF DISTANCE FROM CENTER

KILOMETERS

0	5000	10000	15000	20000

0	2000	4000	6000	8000	10000	12000

MILES

N5KR AZIMUTHAL EQUIDISTANT MAP CENTERED ON
HAWAII

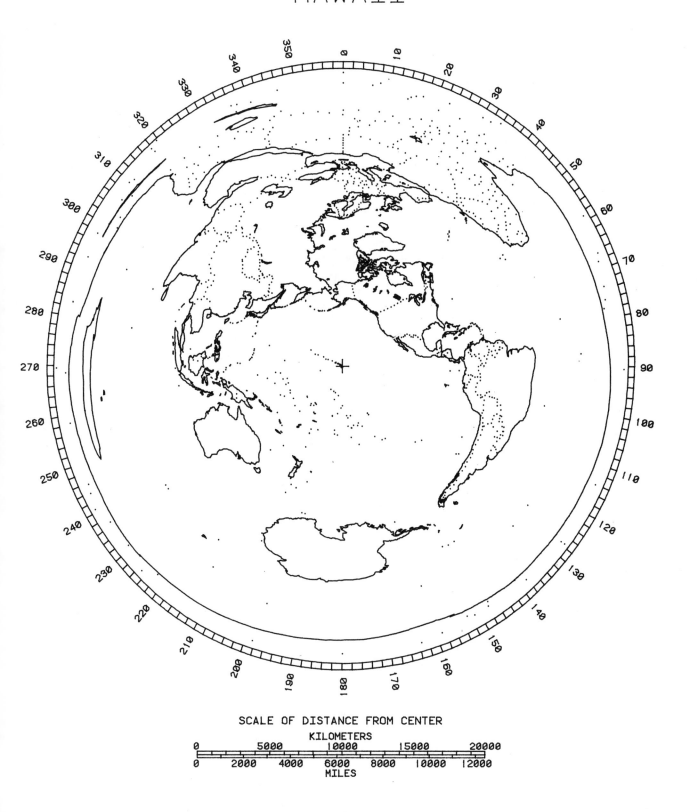

SCALE OF DISTANCE FROM CENTER
KILOMETERS

| 0 | 5000 | 10000 | 15000 | 20000 |

| 0 | 2000 | 4000 | 6000 | 8000 | 10000 | 12000 |
MILES

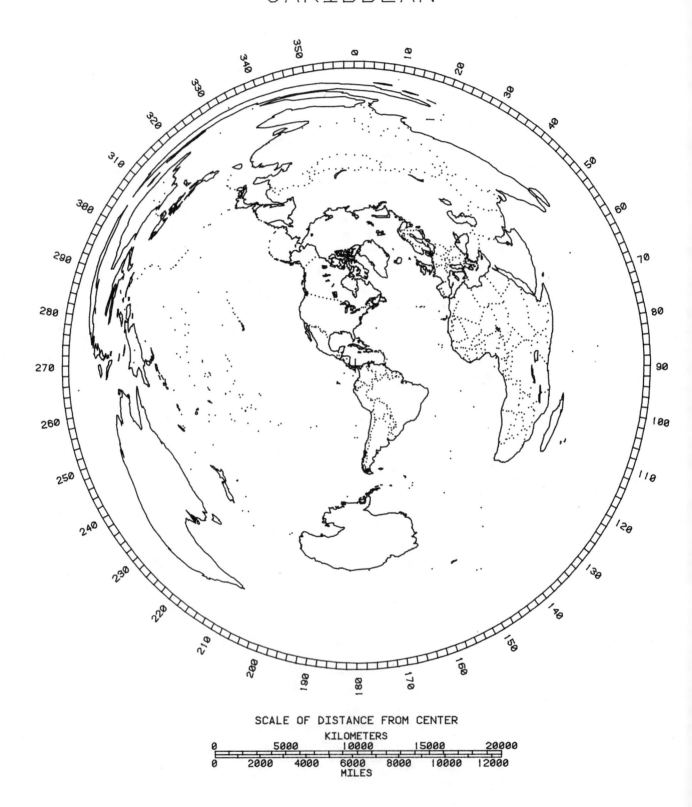

N5KR AZIMUTHAL EQUIDISTANT MAP CENTERED ON
CARIBBEAN

SCALE OF DISTANCE FROM CENTER
KILOMETERS

0		5000		10000		15000		20000

0	2000	4000	6000	8000	10000	12000

MILES

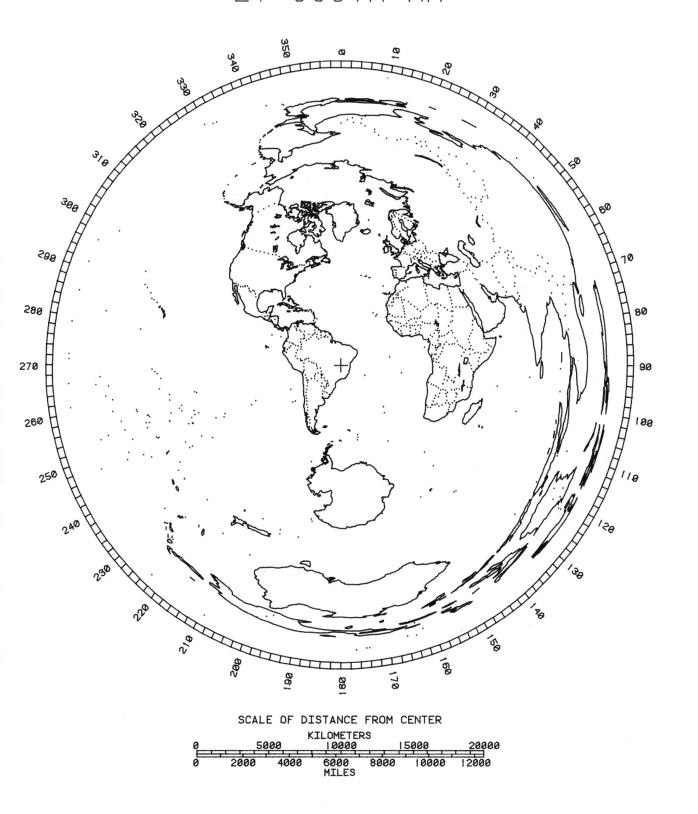

SCALE OF DISTANCE FROM CENTER

KILOMETERS

| 0 | 5000 | 10000 | 15000 | 20000 |

| 0 | 2000 | 4000 | 6000 | 8000 | 10000 | 12000 |

MILES

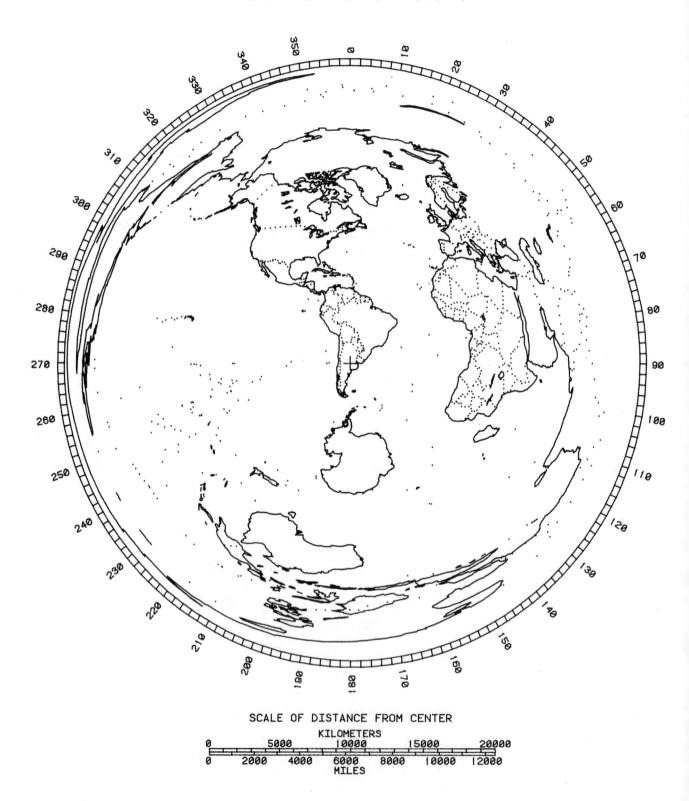

SCALE OF DISTANCE FROM CENTER
KILOMETERS

| 0 | 5000 | 10000 | 15000 | 20000 |

| 0 | 2000 | 4000 | 6000 | 8000 | 10000 | 12000 |
MILES

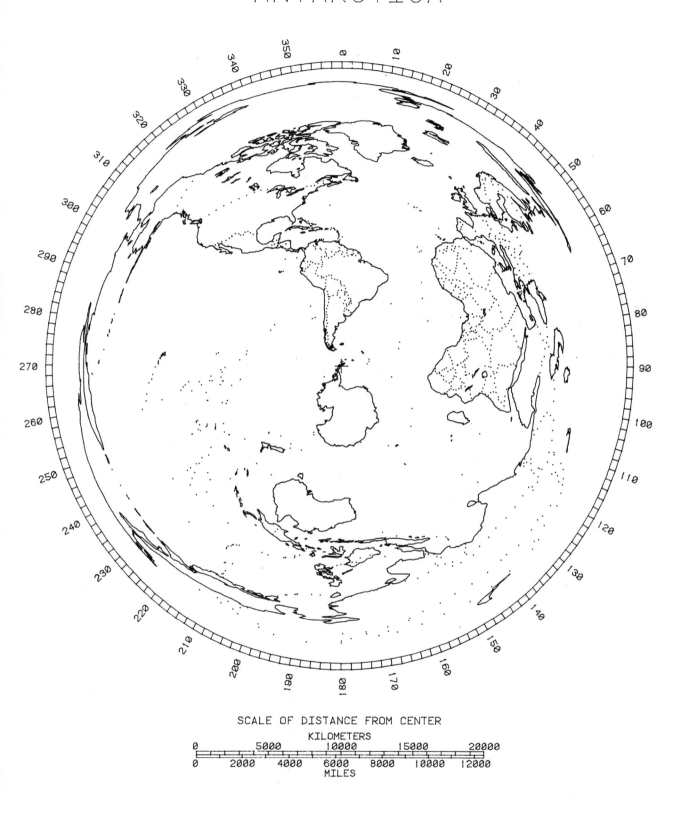

N5KR AZIMUTHAL EQUIDISTANT MAP CENTERED ON
ANTARCTICA

SCALE OF DISTANCE FROM CENTER

KILOMETERS

| 0 | 5000 | 10000 | 15000 | 20000 |

| 0 | 2000 | 4000 | 6000 | 8000 | 10000 | 12000 |

MILES

N5KR AZIMUTHAL EQUIDISTANT MAP CENTERED ON
W. EUROPE

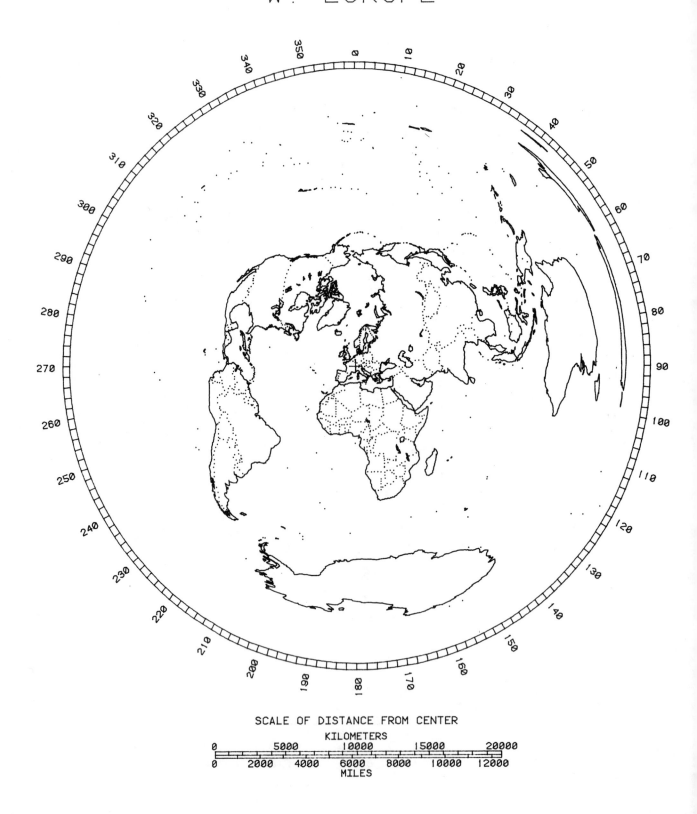

SCALE OF DISTANCE FROM CENTER
KILOMETERS
0   5000   10000   15000   20000
0   2000   4000   6000   8000   10000   12000
MILES

N5KR AZIMUTHAL EQUIDISTANT MAP CENTERED ON
E. EUROPE

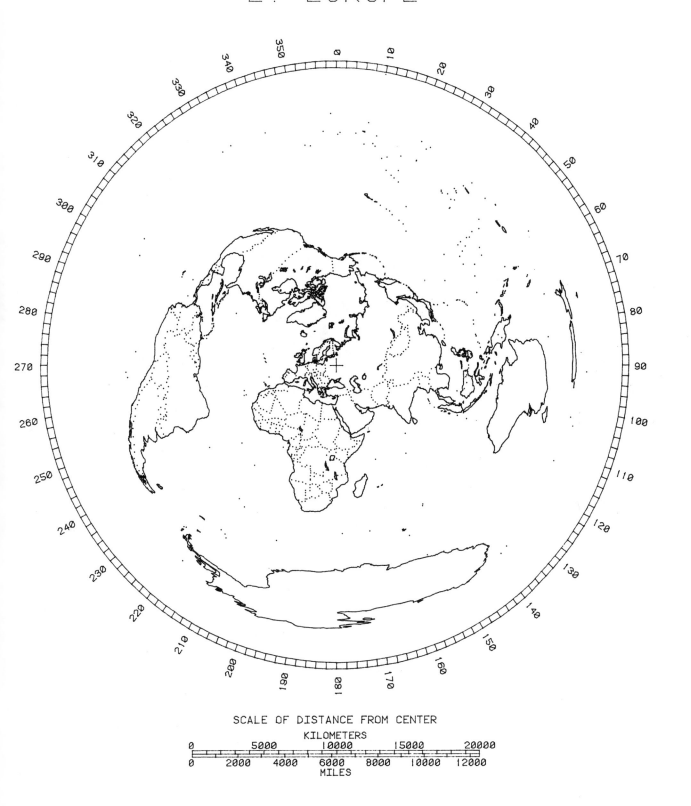

SCALE OF DISTANCE FROM CENTER
KILOMETERS

0   5000   10000   15000   20000

0   2000   4000   6000   8000   10000 12000
MILES

N5KR AZIMUTHAL EQUIDISTANT MAP CENTERED ON
WEST AFRICA

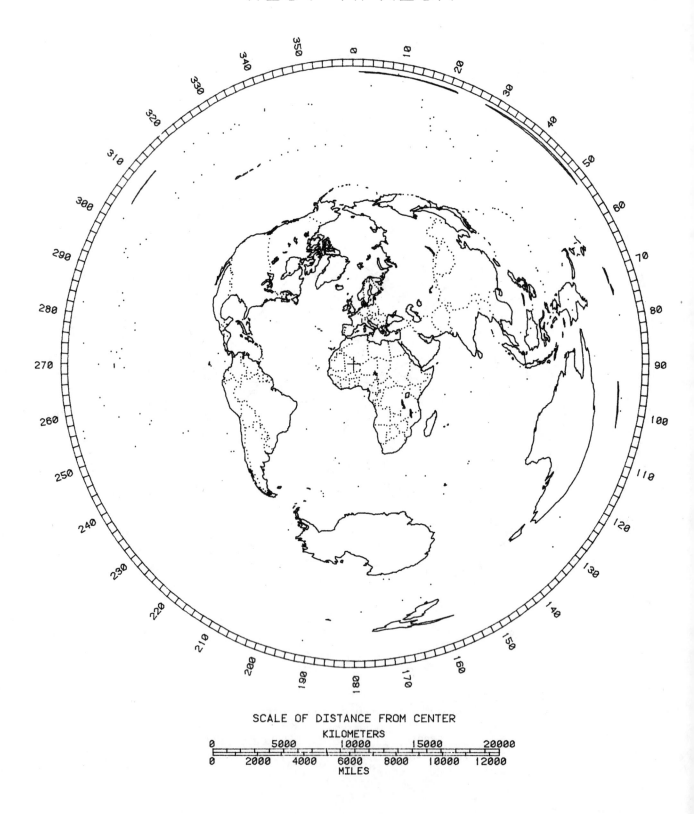

SCALE OF DISTANCE FROM CENTER

KILOMETERS

| 0 | 5000 | 10000 | 15000 | 20000 |

| 0 | 2000 | 4000 | 6000 | 8000 | 10000 | 12000 |

MILES

N5KR AZIMUTHAL EQUIDISTANT MAP CENTERED ON
EAST AFRICA

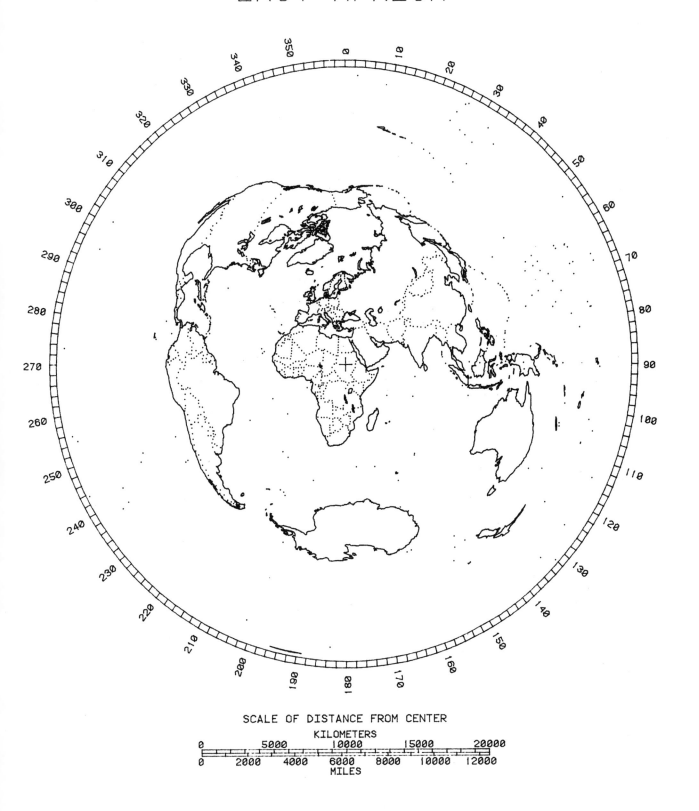

SCALE OF DISTANCE FROM CENTER

KILOMETERS

| 0 | 5000 | 10000 | 15000 | 20000 |

| 0 | 2000 | 4000 | 6000 | 8000 | 10000 | 12000 |

MILES

SOUTHRN AFRI

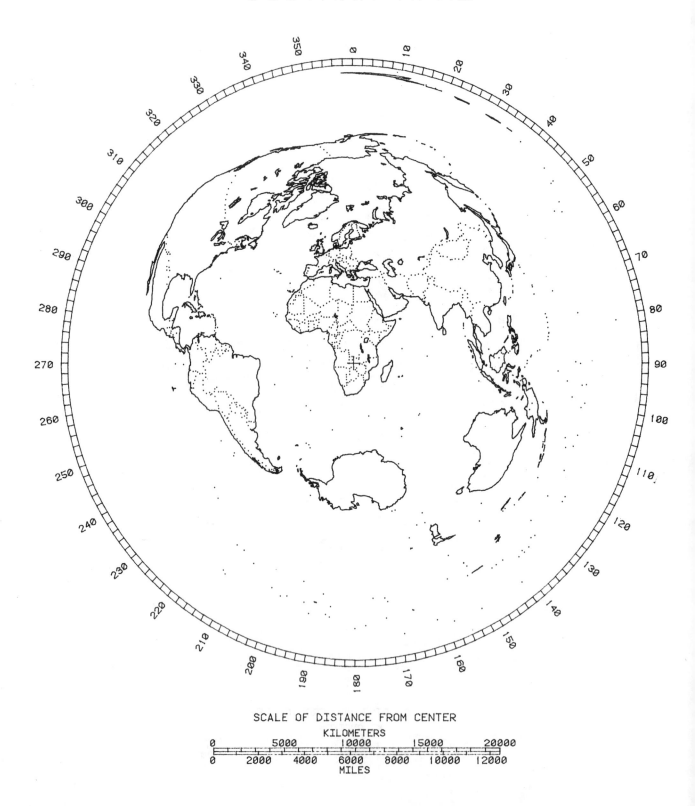

SCALE OF DISTANCE FROM CENTER
KILOMETERS

NEAR EAST

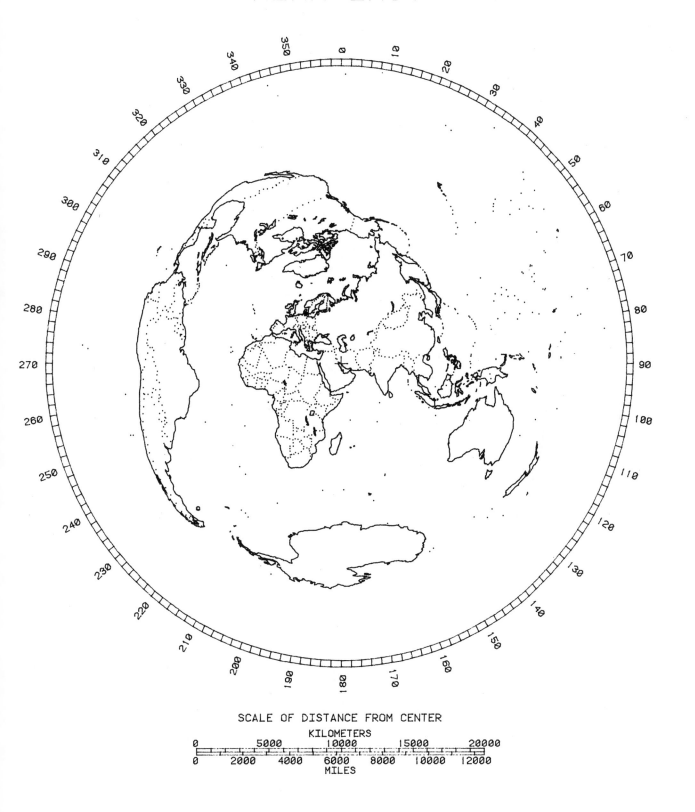

SCALE OF DISTANCE FROM CENTER
KILOMETERS

0	5000	10000	15000	20000

0	2000	4000	6000	8000	10000	12000

MILES

N5KR AZIMUTHAL EQUIDISTANT MAP CENTERED ON
SOUTH ASIA

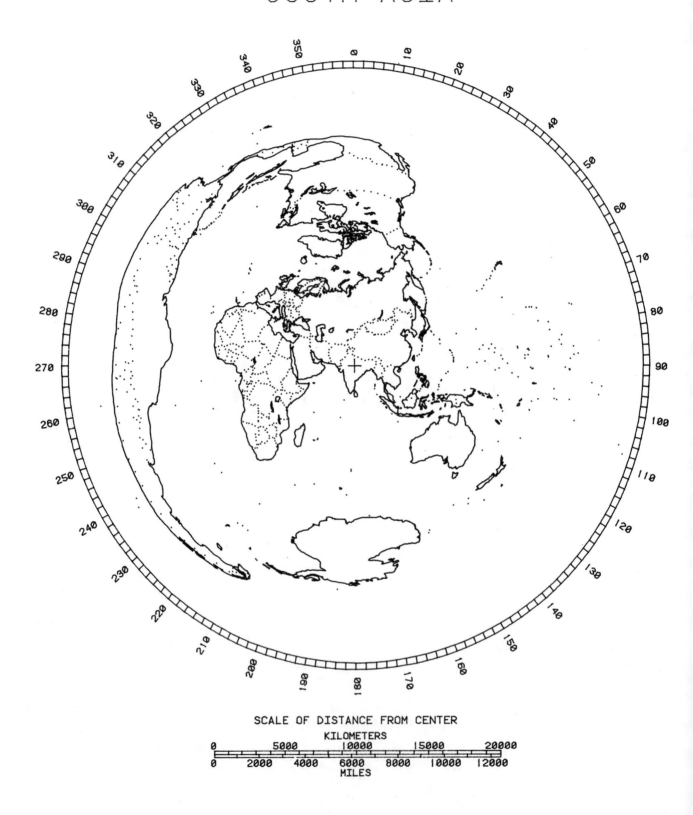

SCALE OF DISTANCE FROM CENTER
KILOMETERS

| 0 | 5000 | 10000 | 15000 | 20000 |

| 0 | 2000 | 4000 | 6000 | 8000 | 10000 | 12000 |

MILES

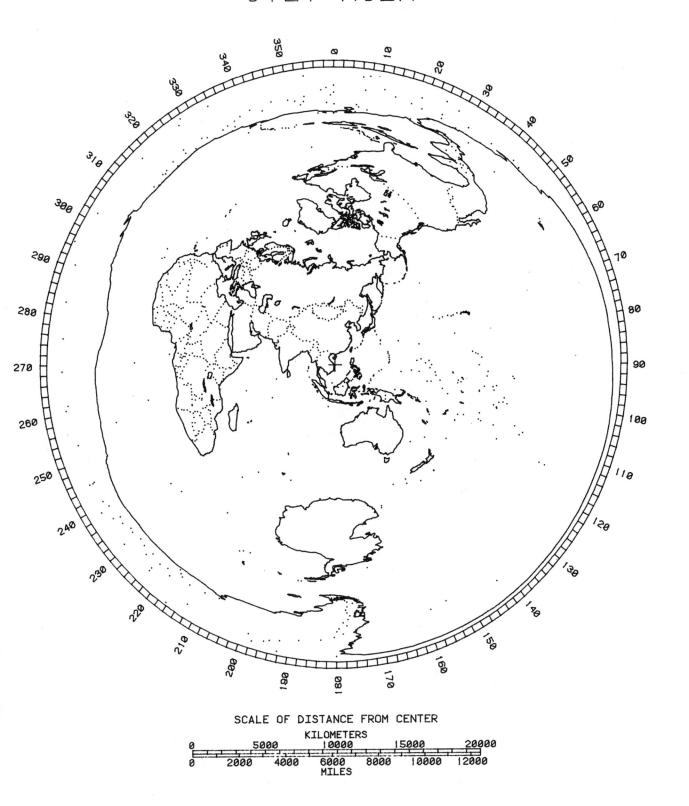

SCALE OF DISTANCE FROM CENTER
KILOMETERS

0	5000	10000	15000	20000

0	2000	4000	6000	8000	10000	12000
MILES

FAR EAST

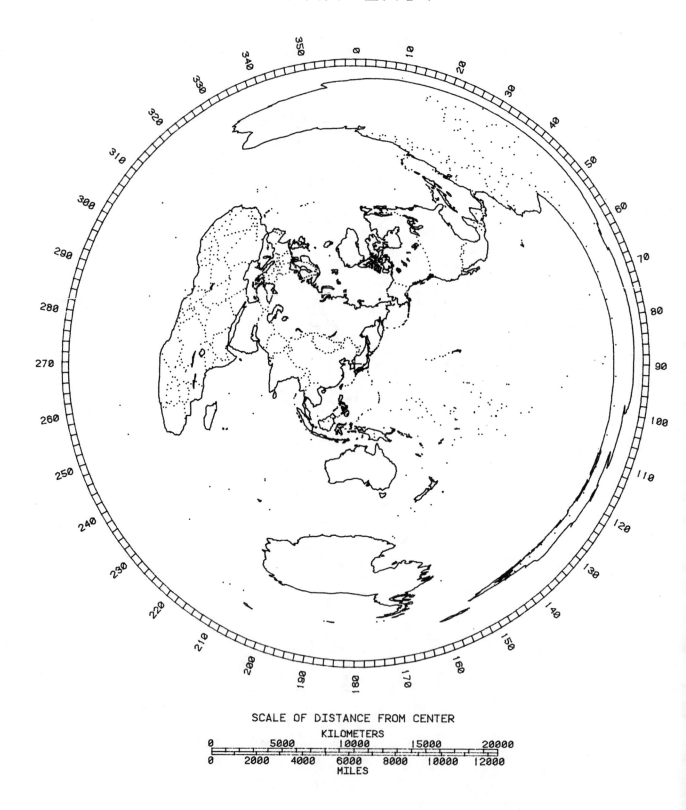

SCALE OF DISTANCE FROM CENTER

KILOMETERS

| 0 | 5000 | 10000 | 15000 | 20000 |

| 0 | 2000 | 4000 | 6000 | 8000 | 10000 | 12000 |

MILES

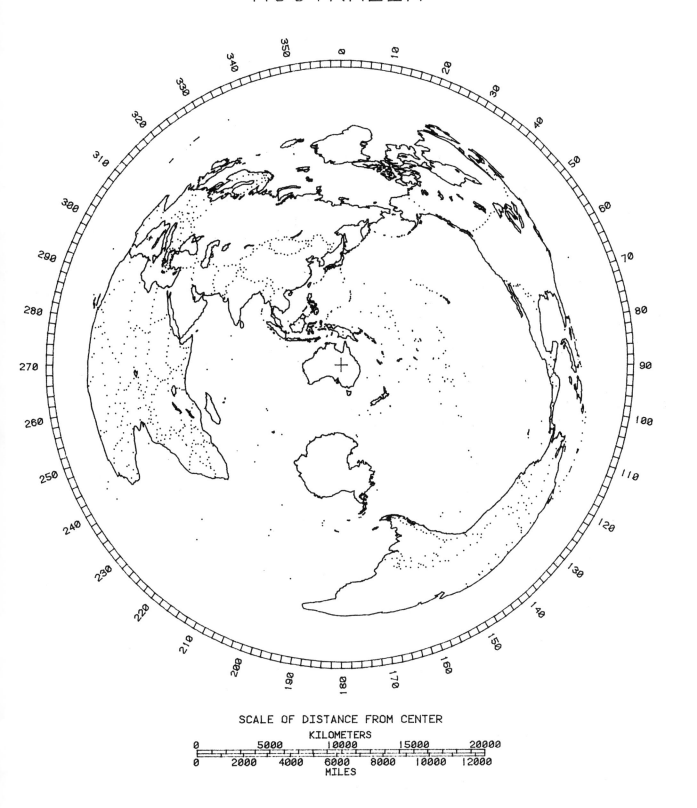

SCALE OF DISTANCE FROM CENTER

KILOMETERS

0	5000	10000	15000	20000

0	2000	4000	6000	8000	10000	12000

MILES

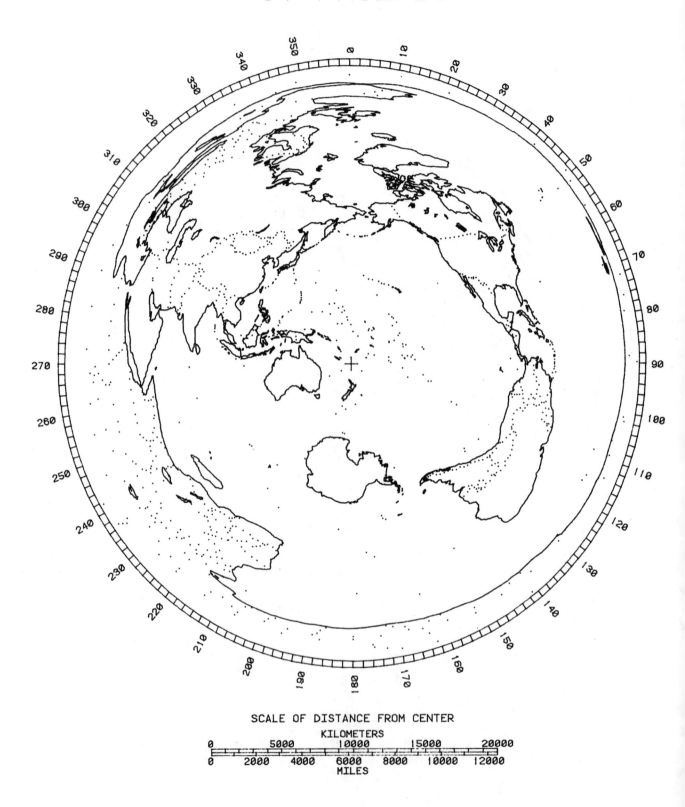

SCALE OF DISTANCE FROM CENTER

KILOMETERS

0	5000	10000	15000	20000

0	2000	4000	6000	8000	10000	12000

MILES

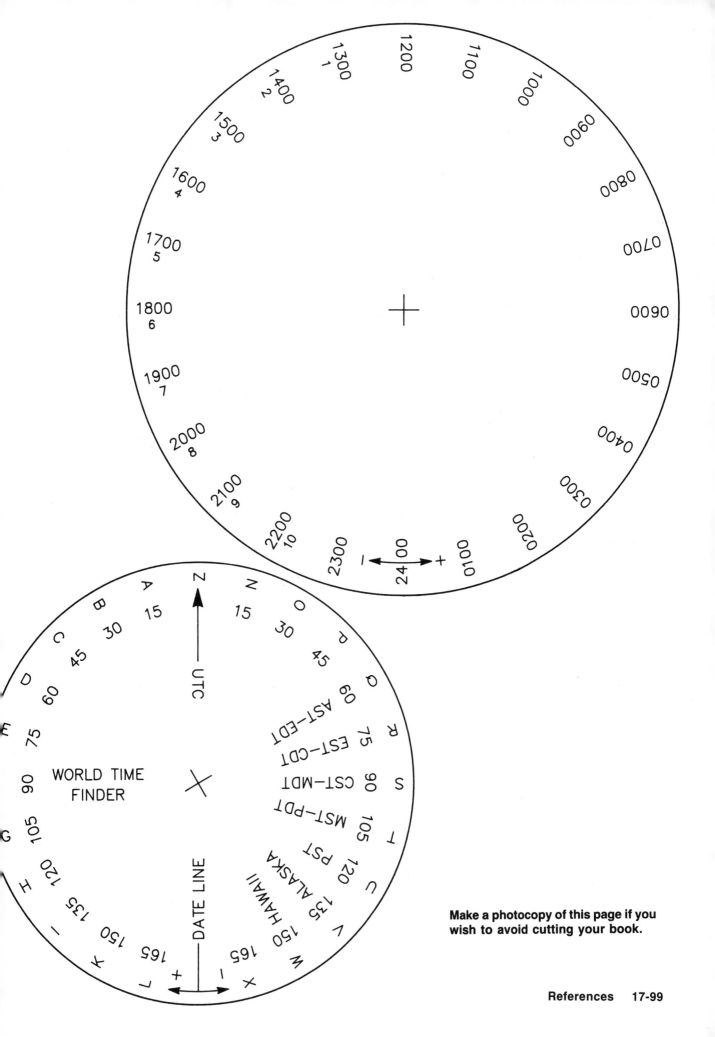

Make a photocopy of this page if you wish to avoid cutting your book.

Index

Note: HDR refers to the Ham Desktop Reference.

A

A-1 Operator Club Award, ARRL: 8-4
Abbreviations, CW: 3-15
Access: 2-2
Access code: 11-3
Accessories, equipment: 3-7
Accidents and hazards: 14-13
Activities, on the air: 3-20
Activity nights, VHF/UHF: 3-17, 12-2
Adjacent channel: 2-2
Advanced DX intelligence gathering: 5-9
Advanced DXing: 5-8
Advanced DXpeditioning: 5-11
Aeronautical communications: 1-13
AFSK: 9-3, 9-12
AIRS: 2-2
Alligator: 2-2
Allocation, frequency band: 2-2
AM: 1-7, 3-26
    Broadcast Band: 1-3
    International: 3-26
    Press/Exchange: 3-26
    Radio Network: 3-26
Amateur bands, US: 3-13
Amateur frequency allocations and band plans: 2-16
Amateur operator licenses: 3-27
Amateur Radio Emergency Service (ARES): 14-2, 14-6, 14-7
Amateur Radio Map of the World, ARRL: 4-3
Amateur Radio on the Internet: 6-9
    Businesses: 6-9
    Clubs: 6-9
    Information: 6-9
Amateur satellite earth station: 2-2
Amateur Satellite Service: 2-2
Amateur Satellite Space Station: 2-2
Amateur Service: 2-2
Amateur station: 2-2

Amateur television (ATV): 14-6, 16-1
    Activity: 16-6
Amateur Television Quarterly: 16-12
Amplifier: 5-6
Amplitude Compandored Single Sideband (ACSSB): 2-2
AMPRNET/Internet gateways: 6-10
AMSL: 2-2
AMSAT: 13-7, 13-9, 16-11
AMSAT Journal, The: 13-7, 13-12
AMTOR: 9-1, 9-2, 9-7, 9-12, 15-1
Analog satellites: 13-1
Analog transponder frequencies: 13-5, HDR-4
Annobon Is: 5-9
Antenna Anthology, The ARRL: 5-3
Antenna Book, The ARRL: 1-5, 1-21, 3-8, 3-9, 7-5
Antenna Compendium Vol 5, The ARRL: 5-3
Antenna
    Lengths, dipole: 3-9
    Parts: 3-9
    Polarization: 16-2
    Problems and cures: 3-9
    Tuner: 3-7
    Wire: 3-9
Antennas: 3-8
    Beam: 2-2
    DXpedition: 5-10
    First VHF/UHF: 3-8
    For contesting: 7-6
    Locating: 3-8
    Phased multiple: 7-7
    Shortwave: 1-8
    Simple ground plane: 3-8
    Simple HF: 3-8
Antique Wireless Association (AWA): 3-32
Applet: 6-10
Applications of packet radio: 10-16
APRS: 10-19

ARES: 14-2, 14-6, 14-7
ARES personal checklist: 14-13
ARL checks: 15-6
ARQ: 9-6
ARRL: 3-2, 3-32, 5-3, 8-1
   About the: .... HDR Cover 3
   Awards: 8-4
   BBS: 13-12
   Contest Branch: 7-14
   Field Organization: 17-70
   Home Page: 5-18, 6-10, 13-12
   How to contact the: HDR-23
   Overseas QSL Service: 3-21
ARRLWeb: 5-18, 6-10, 13-12
ASCII: 9-7
Assigned frequency: 2-2
Assigned frequency band: 2-2
Asignment, frequency: 2-2
Associated Public Safety Communications Officers: 14-10
Attenuation ratio: 2-2
Audio FSK (AFSK): 9-3, 9-12
Aurora: 12-6
Authorized bandwidth: 2-2
Authorized power: 3-30
Automatic control: 2-2
Automatic power control: 2-2
Automatic Transmitter Identification System (ATIS): 2-2
Autopatch: 11-3, 11-7, 14-15
Autopatch guidelines, ARRL: 11-8
Availability: 2-2
Awards, ARRL: 8-1, 8-4
   A-1 Operator Club: 8-4
   Code Proficiency: 8-2
   CQ US Counties: 8-11
   DXCC: 8-6
   Five Band DXCC: 8-7
   Five Band WAS: 8-3
   Friendship: 8-2
   Old Timer's Club: 8-4
   Operating: 3-21
   Rag Chewers Club (RCC): 8-1
   VHF/UHF Century Club (VUCC): 8-4
   Worked All Continents: 8-3
   Worked All States (WAS): 8-2
Awards, non-ARRL: 8-1
   ARCI: 8-12
   CQ: 8-8
   CQ DX: 8-10
   Hunting: 3-20
   Ten-Ten: 8-13
   WAZ: 8-8
   WNZ: 8-9
   World ITU Zone: 8-13
   WPX: 8-10
Awards, VHF: 12-17
   DXCC: 12-18
   VUCC: 12-18
   WAC: 12-17
   WAS: 12-17
Awards Manager, ARRL..8-3
AX.25: 10-4, 13-8
Azimuth: 13-5

Azimuthal maps [also see KØOST Azimuthal Equidistant
   Maps]: 4-2

B

Back scatter: 5-17
Balanced system: 2-2
Band: 2-2
Band openings: 12-7
Band openings and DX: 12-5
Band plan, 2 meter: 11-6
Band plans, amateur: 2-16
Bands, amateur
   US: 3-13
   10 meters: 3-23
   160 and 80 meters: 3-22
   17, 15 and 12 meters: 3-22
   20 meters: 3-22
   40 and 30 meters: 3-22
   50 MHz and above: 3-23
Bandwidth: 2-2
Baudot radioteletype: 9-1
BBS operation: 10-16
Beacon, engineering: 2-2
Beacon, packet radio: 2-2
Beacon, propagation: 2-15
Beam antenna: 2-2
Beamwidth: 2-2
Beat frequency oscillator (BFO): 1-6
Bicycle mobile: 3-31
Block diagram, FM repeater: 11-2
Book messages: 15-10
Break: 11-3
British Broadcasting Corporation (BBC): 1-3, 1-9
Broadcasting: 2-2
Buckmaster Callbook CD-ROM: 5-14
Bulletin board: 5-9
Bulletin board systems (BBSs): 1-11, 10-2
Bulletins, on-the-air: 3-2
Business communications: 11-8
Business rules: 3-30

C

Cable television interference (CATVI): 12-18
Call sign
   Directories: 10-21
   Lookup: 6-10
   Prefixes: 3-28
Call signs, allocation of international: HDR-10
Call signs, US amateur: 3-28
Calling frequencies, North American: 12-3
Calling frequency: 12-1, 3-17
Canadian licensing requirements: 3-28
Carrier: 2-2
Carrier operated relay (COR): 11-2, 11-3
Carrier power: 2-2
Carrier to noise ratio: 2-2
CCIR: 2-2
CCIR-625: 9-7
CCITT: 2-2
Cell: 2-2
Cellular: 2-3
Cellular phone frequencies: 1-18

Centrimetric waves: 2-3
Challenger: 16-9
Channel: 11-3
   Loading: 2-3, 2-13
   Spacing: 2-3
   Voice: 2-3
Characteristic frequency: 2-3
China: 5-11
CHU: 1-5
CISPR: 2-3
Class of emission: 2-3
Closed repeater: 2-3, 11-3, 11-5
CLOVER: 9-1, 9-2, 9-11
   Conversations: 9-12
   Handshaking: 9-11
   Station equipment: 9-11
Clubs, Amateur Radio on the Internet: 6-9
Coaxial cable: 3-9
Cochannel interference: 2-3
Code Division Multiple Access (CDMA): 2-3
Code Proficiency Award, ARRL: 8-2
Coded squelch: 2-3
Codeless Technician license: 11-1
Command set, Packet Cluster: 10-20
Command set, WØRLI mailbox: 10-16
Commercial fixed stations: 1-13
Commodore C64 computer: 10-3
Communications Act of 1934: 1-13, 1-18
Computer logging: 3-19
Computer resources for SWLs: 1-11
Computers: 3-7
Computers and contesting: 7-10
Connecting and disconnecting (packet radio): 10-7
Connectors: 3-9
   PL-259: 3-10
Considerate Operator's Frequency Guide: 3-12, HDR Cover 2
Contest Corral column in *QST*: 7-2
Contest operating: 7-14
Contest rules and entry forms: 7-16
Contesting: 3-20, 7-1, 11-5
   Antennas for: 7-6
   For beginners: 7-2
   Multiop: 7-12
   Preparing your station for: 7-6
   Station accessories for: 7-9
Contests
   ARRL DX: 7-1, 7-2, 7-4
   ARRL Field Day: 7-1, 7-2
   ARRL Sweepstakes: 7-1, 7-2, 7-4
   CQ VHF WPX: 7-4
   CQ World Wide: 7-2
   Major (table): 7-3
   Specialty: 7-4
   Worked All Europe: 7-2
Contests, VHF/UHF: 12-12
   EME Contest, ARRL: 12-13
   June and September VHF QSO Parties, ARRL: 12-13
   Spring Sprints: 12-13
   UHF Contest, ARRL: 12-13
   VHF Sweepstakes, ARRL: 12-13
   Worldwide VHF WPX Contest, *CQ*: 12-13

Control characters: 10-5
Control operator: 11-3
Coordinated station operation: 2-3
Coordinated Universal Time (UTC): 1-4
Coordination: 2-3
   Area: 2-3
   Contour: 2-3
   Distance: 2-3
Countries List Criteria: 8-7
Courtesy beeper: 11-3
Coverage: 11-3
CQ DX awards: 8-10
   DX Zones of the World: HDR-13
   Magazine awards: 8-8
   VHF WPX Contest: 7-4
   World Wide Contest: 7-2
Critical frequency: 2-3
Crossband: 2-3
Crosslinked repeaters: 11-1
CT contesting software: 7-11
CTCSS: 11-7
CTCSS (PL) tone frequencies: 11-11
CTCSS frequencies, 10 meter: 11-11
CW [also see Morse Code]
   Abbreviations: HDR-21
   Nets: 15-2
   Operating Procedure: 3-14

D

Decametric: 2-3
Decimetric waves: 2-3
Declination: 4-1
Delivering a message: 15-9
Deutsche Welle: 1-12
Deviation, frequency: 2-3
Digipeater: 10-9, 11-3
Digital Dimension column in *QST*: 6-12
Digital readouts: 1-7
Digital transponder frequencies and modes: 13-12
Digital transponders: 13-11
Dipole antenna
   Building a: 3-10
   Lengths: 3-9
Direct FSK: 9-3, 9-12
Direction and distance by trigonometry: 4-3
Direction finding: 11-10
Directories, computer: 6-8
Disaster communication: 2-3
Distress calling: 14-9
District Emergency Coordinator: 14-8
Diversity reception: 12-19
DO-17: 13-10
Domestic contests: 7-4
Doppler shift: 2-3, 13-4, 13-6
DOVE satellite: 13-10
Downlink signal, finding: 13-11
Drift, frequency: 1-7, 2-3
Drills and tests: 14-3
DTE: 10-2
DTE parameters: 10-4
Duplex operation: 2-3, 11-2, 11-3

Duty cycle: 2-3
DX
   Contest, ARRL: 7-1, 7-2, 7-4
   Contests: 7-4
   Information: 5-7
   Operating Code: 17-24
   Pileups on RTTY: 9-5
   Reflector, The: 5-7
   Window: 12-1
DX Century Club (DXCC), ARRL: 5-2, 5-13, 5-18, 8-6
   Application: 17-21
   Countries List, ARRL: 5-2, 8-11, 17-7
   Field Representatives: 17-16
   Rules: 17-2
DXing: 3-2, 3-20
   Advanced: 5-8
   Beginning: 5-2
   Intermediate: 5-5
   On shortwave bands: 1-3
   With image communications: 16-2
DXpeditions: 5-7, 5-10
   Operating permission and documentation: 5-12
   Transportation: 5-12
Dynamic frequency sharing: 2-3
Dynamic range: 2-3

E

E-mail: 1-11, 5-10, 6-5
E-mail lists: 6-6
Earth station: 2-3
Eastern Area Net (EAN): 15-12
Effective isotropic radiated power (EIRP): 2-3
Effective isotropically radiated power (EIRP): 2-3
Effective radiated power (ERP): 7-8
EIA-232: 10-2
EIA-422: 10-3
Electric Radio: 3-26
Electromagnetic compatibility: 2-3
Electronic Communications Privacy Act of 1986: . 1-13, 1-18
Electronic contest entries: 7-14
Elevation: 13-5
EME (earth-moon-earth): 5-17, 12-7, 12-11
EME Contest, ARRL: 12-13
Emergency communication: 2-3, 14-1
   HF network for: 14-6
Emergency Coordinator (EC), ARRL: 14-1, 14-8
Emergency Coordinator and Certification Course: 14-3
Emergency Coordinator's Manual, ARRL: 14-2, 14-9
Emergency
   Operations Center (EOC): 14-4
   Procedures: 14-2
   Training: 14-2
Emission: 2-3
Emission designators: 3-19
Equipment:
   Accessories: 3-7
   Advanced DXing: 5-8
   Beginning DXing: 5-2
   Buying: 3-3
   HF: 3-4
   HF transceivers: 3-6
   Homemade: 3-6
   Image communications: 16-2
   Intermediate DXing: 5-6

Low power (QRP): 3-4
Sources of information: 3-3
SSTV: 16-8
Ultimate DXpeditioning: 5-12
Used: 3-5
VHF packet: 3-4
VHF/UHF: 3-4
ERP: 7-8
European Council of Postal and Telecommunications
   Administrations: 2-3
Extremely high frequencies: 2-3

F

Facsimile (fax): 2-3, 16-1, 16-9
   Formats and standards: 16-10
   Hardware and software: 16-9
   Weather satellites: 16-10
Fading: 2-3
   Flat: 2-3
   Selective: 2-3
Fast-scan amateur TV (FSTV): 16-1
Fast-scan TV (ATV): 3-23
FCC-allocated prefixes for areas outside the continental
   US: HDR-12
FCC Rule Book, The: 3-26, 3-27, 16-2
FCC Rules [see Federal Communications Commission Rules]
FEC: 9-9, 9-12
Federal Communications Commission (FCC): 1-6, 1-16,
                                                  2-9, 11-5
   Rules: 3-11, 12-18, 14-10, 15-7
Federal Emergency Management Agency (FEMA): 14-9
Ferrell's Confidential Frequency List: 1-4, 1-19
Field Day, ARRL: 7-1, 7-2, 14-2
Field Organization, ARRL: 14-10
File Transfer Protocol (FTP): 6-2, 6-4, 6-8
Finding a repeater: 11-3, 11-4
First contact, making: 3-3
Five Band DXCC Award, ARRL: 8-7
Five Band WAS Award, ARRL: 8-3
Five Band WAZ award, *CQ*: 8-8
Fixed stations and prime time: 11-5
Fixed transceivers: 3-4
FM repeaters: 11-1
   Block diagrams: 11-2
   Operating: 11-2
Footswitch: 5-6
Form 610: 3-6
Formal message traffic: 14-6
Formats and standards, fax 16-10
Frequencies:
   Mystic Star: 1-12
   Packet: 10-10
   Shortwave: 1-2, 1-20
Frequency:
   Agility: 2-3
   Allocations and band plans, amateur: 2-11, 2-16
   Coordination: 2-14, 11-5
   Coordinator: 2-3, 11-3
   Diversity: 2-4
   Division Multiple Access (FDMA): 2-4
   Guide, Considerate Operator's: .... 3-12, HDR Cover 2
   Input: 2-4, 3-19, 11-1, 11-3
   Offset: 2-4, 11-11
   Optimum de Travail (FOT): 2-4

[Frequency, continued]

Output: 2-4, 3-19, 11-1, 11-3
Registration: 2-4
Reuse: 2-4
Sharing: 2-4
Shift: 9-3
Shift keying (FSK): 9-1, 13-8
Shift telegraphy: 2-4
Tolerance: 2-4
Friendship Award, ARRL: 8-2
FSK: 9-1, 13-8
FTP software: 6-4
Fuji-OSCAR satellites: 13-3
Full carrier single sideband emission: 2-4
Full duplex: 11-3, 13-1
Full quieting: 11-3
Fun of Short Wave Listening, The: 1-19
Fundamental overload: 12-18

G

G-TOR: 9-1, 9-2, 9-12
General class license: 11-2
General coverage receiver: 1-5
GeoClock: 4-4
Geostationary satellite: 2-4
    Orbit: 2-4
Geosynchronous satellite: 2-4
Geratol Net: 8-12
Global Positioning System (GPS): 1-4
Glorioso Is: 5-12
GOES satellite: 16-11
Graphical user interface (GUI): 6-1
Gray-line propagation: 5-15
Great circle bearing and distance: 4-4
Great circle map: 4-3, 5-11
Greenwich Mean Time (GMT): 1-9
Grid location, finding your: 12-4
Grid Locator for North America, ARRL: 8-4
Grid locator map: 12-5
Grid squares: 12-3
GRIDLOC: 12-3
Ground-plane antenna: 3-8
Ground reflection patterns: 7-7
Grounding: 3-7
Guard band: 2-4
Guard channel: 2-4
Guest operating: 3-29

H

Half duplex: 11-3
Ham Radio: 12-9, 12-11
Hand held: 11-3
Hand held transceivers: 3-4
Handbook, The ARRL: 1-5, 1-21, 3-4, 3-6, 3-9, 3-19, 3-31,
                                       7-5, 16-10, 16-12
Handi-Ham Courage Center: 3-6
Handling traffic by packet radio: 15-16
Harmful interference: 2-4
HCJB, Quito, Ecuador: 1-12
Headset: 5-6
Health and welfare traffic: 14-15
Hectometric waves: 2-4
High band: 2-4

Highest possible frequency: 2-4
High frequency: 2-4
High frequency (HF) bands: 1-1
    Antenna, simple: 3-8
    Ham bands: 1-5
    Network for emergency communications: 14-6
    Packet: 9-10
    Packet operation: 10-15
    Repeaters: 11-9
Hilltopping and portable operation: 12-11
History of amateur satellites: 13-2
Homemade equipment: 3-6
Hospital communications: 14-17
How to contact the ARRL: HDR-23
Hypertext link: 6-2
HyperText Transfer Protocol (http): 6-2

I

Identification requirements: 3-28
Image communications: 14-6
    Equipment: 16-2
    License requirements and operating frequencies: .. 16-2
Image frequency: 2-4
In-band: 2-4
Incident Command System (ICS): 14-4
Incoming QSL Bureau System, ARRL: 5-20
    Addresses: 5-21
Independent sideband: 2-4
Input: 11-1
Input frequency: 3-19, 11-3
InstantTrak: 13-9
Insulators: 3-9
Interdepartment Radio Advisory Committee (IRAC): 2-10
Interfacing packet equipment: 10-2
Interference: 2-4, 12-18
    Cable TV (CATVI): 12-18
    Fundamental overload: 12-18
    Radio frequency (RFI): 3-31, 12-18
    Television (TVI): 12-18
Internal signals: 1-8
International Amateur Radio Union (IARU): 2-4, 2-7
Inter-American Telecommunication (CITEL): 2-4, 2-9
International Frequency Registration Board (IFRB): 2-4
International Reply Coupons (IRCs): 5-14
International shortwave broadcasters: 1-8
International Telecommunication Union (ITU): . 1-2, 2-1, 2-7
    Regions: HDR-12
Internet: 6-1, 10-21
    Amateur Radio on the: 6-9
    Explorer: 6-4
    Fax page: 16-10
    Service Providers (ISPs): 6-2
Intersymbol Interference (ISI): 2-4
Inverse Square Law: 2-4
Ionosphere: 1-3, 7-6
Ionospheric scatter: 12-8
IRCs: 8-1
IRS: 9-7, 9-12
Iskra satellites: 13-3
Islands on the Air (IOTA): 5-2, 5-13
ISP: 6-2
ISS: 9-7, 9-13
ITU Zones: HDR-13

J - K

Jamming: 2-4
Java: 6-9
javAPRS: 6-9
June and September VHF QSO Parties, ARRL: 12-13
KØOST Outline Maps: 17-27
    African Countries: 17-47
    Antarctica and Offshore Islands: 17-32
    Argentine Provinces: 17-31
    ARRL Field Organization: 17-27
    Asian Countries: 17-48
    Belgium/Dutch Provinces: 17-43
    Brazilian States: 17-30
    British Counties: 17-41
    California Counties: 17-28
    Chinese Provinces: 17-50
    Eastern Russian Oblasts: 17-46
    Estonia: 17-39
    European Countries: 17-34
    French Departments: 17-35
    Italian Provinces: 17-38
    Japanese Prefectures: 17-49
    Mexican States: 17-33
    Polish Provinces: 17-44
    Slovakia: 17-40
    South American Countries: 17-29
    Spanish Provinces: 17-42
    Swedish Laens: 17-37
    Swiss Cantons: 17-36
    Western Russian Oblasts: 17-45
KA-Node: 10-11
Keplerian elements: 13-8
Key clicks: 2-4
Key up: 11-3
Keyer: 5-6, 7-10
Keys, keyers and paddles: 3-7
Kilometric waves: 2-4
Kingman Reef: 5-10
KISS mode: 11-11
KITSAT satellites: 13-3
Klingenfuss Guide to Utility Stations: 1-13, 1-19

L

Language lessons on shortwave bands: 1-3
Latitude and longitude of various cities: 4-5, HDR-6
Left-hand polarized wave: 2-4
Library of Congress: 3-6
License requirements for image communications: 16-2
License, amateur: 3-26
    Applying for a: 3-27
    Renewal/modification: 3-27
Lighting for image communications: 16-2
Line voltage: 5-10
Linear transponders: 13-1, 13-10
Links, Web: 6-5
Listening for DX stations: 5-3, 5-6, 5-8
Listserver: 6-6
Local clubs: 3-2
Log Book, ARRL: 3-18, 3-19
Logging: 3-30, 7-12
Logging, computer: 3-20
Logs: 7-14, 7-16
Long path: 5-7
Longwave Club of America: 1-2

Low band: 2-4
Low earth orbit (LEO): 2-4, 13-3
Lowest usable frequency (LUF): 2-4
Low frequencies (LF): 2-4
Low Profile Amateur Radio: 1-8

M

Machine: 2-4, 11-3
Macintosh computer: 6-1
Mag mount: 11-3
Magnetic north: 4-1
Magnetic variation: 4-1
Mail forwarding, HF packet: 9-10
Mailer: 6-4
Major Contests (table): 7-3
Manned space missions: 13-11
Maps [see also KØOST Outline Maps, N5KR Azimuthal
    Equidistant Maps]
    Amateur Radio Map of the World: 4-3
    Azimuthal: 4-2
    Great circle: 4-3, 5-11
    Grid locator: 12-5
Maritime communications: 1-13
Maximum usable frequency (MUF): 2-4
MAYDAY: 14-9
Mean power: 2-4
Medium frequencies (MF): 2-4
Medium wave: 1-1
Mellish Reef: 5-7
Memories: 1-7
Message centers: 14-10
    Form: 15-5
    Handling: 15-4
    Handling instructions: 15-5
Messages:
    Book: 15-10
    Checking: 15-6
    Delivering: 15-9
    Handling: 15-4
Meteor burst communications: 2-4
Meteor scatter: 2-4, 12-6, 12-8, 12-19
Meteor showers: 12-10
METEOSAT: 16-11
Metric waves: 2-4
Mexico-OSCAR satellite: 13-3
Michigan QRP Club: 3-25
Microcell: 2-4
Microwaves: 2-4
Military Affiliate Radio System (MARS): 15-6
Military transmissions: 1-13
Millimetric waves: 2-4
MiniProp Plus: 4-4
Mir space station: 13-11
Mobile transceivers: 3-4
Mode A (satellites): 13-4
Mode B (satellites): 13-9
Mode K (satellites): 13-4
Modem, computer: 6-2
Modulation: 2-4
Monitoring: 16-6
Monitoring mode: 10-8
Monitoring Times: 1-19
Moonbounce (EME): 12-11
    Operating procedures: 12-12

Morse code (CW): 1-1, 1-6, 3-28, 7-10
   Abbreviations: 3-14
   Exam exemption: 3-6
   Operating: 3-11
   Operating procedures: 3-14
   Procedural signals: 3-14
Morse Code for Other Languages: 17-24
Morse Code: The Essential Language: 3-14
Mountaintopping, VHF: 12-14
Multimedia: 6-4
Multimode communications processor (MCP): 9-1, 9-12
Multiop contesting: 7-11
Multipath: 2-4
Multipliers: 7-4, 7-15
Music on shortwave bands: 1-3
Myriametric waves: 2-4
Mystic Star Frequencies: 1-12

N

N5KR Azimuthal Equidistant Maps: 17-78
   Alaska: 17-82
   Antarctica: 17-87
   Australia: 17-97
   Caribbean: 17-84
   Central USA: 17-80
   Eastern Africa: 17-91
   Eastern Europe: 17-89
   Eastern South America: 17-85
   Eastern USA: 17-79
   Far East : 17-96
   Hawaii: 17-83
   Near East: 17-93
   South Pacific: 17-98
   Southeast Asia: 17-95
   Southern Africa: 17-92
   Southern Asia: 17-94
   Southern South America: 17-86
   W1AW: 17-78
   Western Africa: 17-90
   Western Europe: 17-88
   Western USA: 17-81
N7BH World Time Finder: 17-99
NA contesting software: 7-11
Narrow-band: 2-5
National
   Calling frequency: 11-5
   Communications System (NCS): 14-10
   Institute of Standards and Technology (NIST): 1-4
   Internet Service Providers: 6-3
   Radio Quiet Zone: 2-5
   Simplex frequency: 11-5
   Traffic System (NTS): 14-3, 14-7, 14-9, 15-1, 15-12, 15-14
   Volunteer Organizations Active in Disaster
     (NVOAD): 14-10
   Weather Service (NWS): 14-2, 14-10
Natural disasters and calamities: 14-12
Necessary bandwidth: 2-5
NET: 11-11
Net control sheet: 15-11
Net control station (NCS): 2-5, 15-1, 15-2, 15-10, 15-11
Net Directory, ARRL: 3-21, 11-7, 14-9, 15-1
Net manager: 2-5
NET/ROM and TheNet: 10-10

Nets: 3-21
   CW: 15-2
   SSB: 15-3
   VHF: 12-16
   VHF/UHF: 12-2
Netscape Navigator: 6-4
Network node: 10-10
Network relays on 219-220 MHz: 10-8
Networking, packet: 10-9
News reader: 6-4
News reports via SWLing: 1-2
Newsgroups: 1-11, 6-8
NiCd: 11-3
Node: 11-9
Node, network: 10-10
North American calling frequencies: 12-3
Northern California Contest Club: 7-5
NOS: 11-11
Now You're Talking!: 3-4, 3-9
NTIA: 2-5, 2-10
Numbers stations: 1-15

O

Occupied bandwidth: 2-5
Official Emergency Station: 14-8
Official Relay Station: 15-15
Offsets, repeater frequency: 11-11
Old Timer's Club, ARRL: 8-4
Older receivers: 1-7
Open repeater: 2-5, 11-3, 11-5
Operating
   Awards, ARRL: 3-21
   Beginning DXing: 5-4
   In a foreign country: 3-30
   Procedures, CW: 3-14
   Procedures, Phone: 3-15
   Procedures, VHF/UHF SSB: 3-17
   SSTV: 16-8
Optimum Working Frequency (OWF): 2-5
Organizations, packet radio: 10-13
OSCAR 13: 13-6
OSCAR 4: 13-2
OSCAR Satellite Report: 13-12
OSCAR satellites: 13-2
Out of band emission: 2-5
Outgoing QSL Service, ARRL: 5-13, 5-19
Outline map of US states, Canadian provinces and ARRL
   sections: HDR-16
Output: 11-1
Output frequency: 3-19, 11-3
Over: 11-3

P

Packet Cluster: 7-10, 7-12, 10-19
   Command set: 10-20
Packet radio: 3-21, 9-2, 10-1, 14-6, 15-1
   Amateur satellites: 10-16
   Applications: 10-16
   APRS: 10-19
   BBS operation: 10-16
   Connecting and disconnecting: 10-7
   Equipment: 1-15, 10-3
   Frequencies: 10-10, HDR-2

[Packet radio, continued]
Handling NTS traffic by: 15-16
HF operation: 10-15
Networking: 10-9
Organizations: 10-13
PBBS procedures: 10-18
Protocols and commands: 10-4
Public service communications: 10-18
Repeaters: 11-9
Sending mail: 10-16
Space communications: 10-16
Station equipment: 10-1
Suggested frequencies for VHF and HF: HDR-2, 3
VHF/UHF: 10-9
VHF/UHF operation: 10-15
Packet radio, HF: 9-10
Bulletin boards and mailboxes: 9-10
Mail Forwarding: 9-10
Packet terminal: 10-2
PacketCluster: 5-7, 5-18
PACSATs (packet satellites): 13-7, 13-11
PACTOR: 9-1, 9-2, 9-7, 9-8, 9-12
PACTOR BBS: 9-9
PACTOR II: 9-7, 9-8
Paddle: 7-10
Paging: 2-5
Pair, frequency: 2-5
Parades: 14-10
Passport to World Band Radio: 1-19
PBBS: 10-15
PBBS procedures: 10-18
Peak envelope power (PEP): 2-5
Period, satellite: 2-5
Permitted communications: 3-29
Personal Communications System (PCS): 1-20
Personal Computers in the Ham Shack: 4-5, 6-12, 7-10, 16-8
Phase 3 satellites: 13-4
Phase 3D frequencies: 13-8, HDR-5
Phase 3D satellite: 13-6
Phase 4 satellites: 13-7
Phased multiple antennas: 7-7
Phone operating procedures: 3-15
Phonetic alphabet, ITU: 3-29, 3-30, HDR-20
Phonetic alphabets: 17-24
Pilot: 2-5
Pirates and clandestine broadcasters: 1-16
Pirates, policemen, LIDs and jammers: 5-4
PL-259 connector: 3-10
Plenipotentiary conferences: 2-7
Polarization: 2-5
Discrimination: 2-5
Diversity: 2-5
Pole star: 4-2
Police, assisting the : 14-16
Popular Communications: 1-19
Portable operating, VHF: 12-14
Portable operation: 12-11
PoSAT satellite: 13-3
Postage for QSLs: 5-14
Power flux density: 2-5
Power on repeaters: 11-7
Practical Packet Radio: 3-23
Pratas Is: 5-11
Prefixes, call sign: 3-28

Preparing your station for contesting: 7-6
Print resources, shortwave: 1-19
Procedural signals, ARRL: 3-15, HDR-16
*Proceedings of the 9th Computer Networking
Conference*: 13-8
Product Review column in *QST*: 7-8
Propagation: 1-3, 2-5, 5-7, 5-15, 7-6, 7-15
Beacons: 2-15
Gray Line 5-15
Indicators: 12-10
Sporadic E: 12-9
Tropospheric: 12-8
VHF: 7-13, 12-19
VHF/UHF: 12-3
10 meters: 5-17
12 meters: 5-17
15 meters: 5-16
160 meters: 5-15
17 meters: 5-16
2 meters: 5-18
20 meters: 5-16
40 meters: 5-16
6 meters: 5-17
80 meters: 5-16
Protection ratio: 2-5
Public events, handling events at: 15-10
Public Law 97-259: 3-31
Public service: 16-5
Agencies: 14-16
Communications: 10-18
Events: 14-10
Public Service Communications Manual: 15-14
Push to talk (PTT): 3-17, 11-7

Q
Q signals: 3-12, 17-26, HDR-17
QN signals: 15-3
QRP
Amateur Radio Club International awards: 8-12
ARCI: 3-25
Awards: 3-25
Operating: 3-21, 3-24
QRP Power: 3-25
QRPp: 5-2
QRZ DX: 5-7, 5-9
QSL cards: 5-19, 8-1
QSL Service, ARRL Overseas: 3-21
QSLing: 1-16, 3-20, 5-13
QST: 3-1, 3-2, 3-3, 3-9, 3-26, 3-32, 4-5, 6-12, 7-5,
10-3, 11-5, 11-7, 14-9, 16-12
Contest Corral column 7-2
Digital Dimension column 6-12
Product Reviews: 5-5, 7-8
Public Service column: 15-15

R
Rabbit: 2-5
RACES: 14-9
Radiation: 2-5
Radiation inversion: 12-6
Radio Amateur Callbook: 7-4
Radio Buyer's Sourcebook, The ARRL: 3-4
Radiocommunication: 2-5
Service: 2-5

Radio control: 3-23
Radio direction finding (RDF): 11-3
Radio France Internationale: 1-12
Radio frequency interference (RFI): 3-31
Radio Frequency Interference—How to Find it and
  Fix It: 3-7, 3-31, 12-18
Radio Habana Cuba: 1-12
Radio Netherlands: 1-12
Radio parameters: 10-5
Radiograms, ARRL numbered: HDR-19
Radio Regulations Board (RRB): 2-5
Radioteletype (RTTY): 1-1, 1-13, 1-15, 3-21, 9-6
  Nets 15-4
  On repeaters: 11-7
Rag Chewers Club (RCC): 8-1
Rag chewing: 3-20
Raster: 2-5
Receiver performance: 5-5
Receivers: 7-8
  General coverage: 1-5
  Older: 1-7
Receiving traffic: 15-8
Reciprocal operating: 3-31
Reciprocity: 2-5
Reduced carrier single sideband emission: 2-5
Red Cross: 14-2, 14-9
Reference frequency: 2-5
Regulatory Information Branch, ARRL: 5-10
Religious broadcasters: 1-10
Repeater:
  Coordination: 2-14
  Input: 11-1
  Output: 11-1
Repeater Directory, ARRL: 3-23, 11-3, 11-5, 11-7, 11-9, 16-2
Repeaters: 3-17, 11-1
  Finding: 11-3
  HF: 11-9
  Identification: 11-4
  Operating 11-2
  Packet radio: 11-9
  10 meter: 11-9
Resources for DXers: 5-18
Return Postage Chart for DXCC and WAS QSL Cards: 17-51
Roaming: 2-5
ROSE: 10-11
Roving: 2-5
RS satellites: 13-2
RS-10 satellite: 13-4
RST system, The: 3-15, 7-25, HDR-16
RTTY nets: 15-4
RTTY Roundup Contest: 7-4
RTTY WAZ award, *CQ*: 8-8
Rules and regulations: 3-26
Russian Oblasts: 17-23

S

Safety: 3-7, 14-8
Salvation Army: 14-9
SAREX: 13-12
Satellite:
  Gateway: 11-11
  GOES: 16-11
  METEOSAT: 16-11
  WAZ award, *CQ*: 8-8

Satellite Anthology, The ARRL: 13-12
Satellite Experimenter's Handbook,
  The ARRL: 13-6, 13-8, 13-12
Satellite modes: 13-4
  Mode A: 13-4
  Mode B: 13-9
  Mode K: 13-4
Satellites: 13-1
  Analog: 13-1
  Analog transponder frequencies: 13-5, HDR-4
  DO-17: 13-10, 13-11
  Finding: 13-8
  Fuji-OSCAR 13-3
  History of amateur: 13-2
  KITSAT: 13-3
  Mode B: 13-9
  Operating through: 13-10
  OSCAR 13: 13-6
  Phase 3: 13-4
  Phase 3D: 13-6
  Phase 3D frequencies: 13-8, HDR-4
  Phase 4: 13-7
  RS-10: 13-4
  Setting up a station: 13-9
  Software: 13-9, 13-11
  WO-18: 13-10
Satellites vs shortwave: 1-17
Scanners: 1-6, 1-19, 2-5
  Equipment: 1-21
Scanning, VHF/UHF: 1-18
Scatter: 5-17
Scatter modes: 12-6
Search and rescue operations: 14-16
Search engines: 6-8
Secondary station identification (SSID): 10-7
Section Emergency Coordinator (SEC): 14-3, 14-8
Section Manager (SM), ARRL: 11-7, 14-7, 14-9
Section Traffic Manager (STM), ARRL: 11-7, 14-3, 15-15
SELCAL: 9-7
Selectivity: 2-5
Sending mail on packet radio: 10-16
Sending traffic: 15-8
Separate transmitter and receiver: 3-4
Separation: 11-3
Service area: 2-5
Setting up an SSTV studio: 16-8
Severe weather emergencies: 14-1
Severe weather spotting and reporting: 14-12
Shift, frequency: 2-5
Shortwave
  Antennas: 1-8
  Broadcasters, international: 1-8
  Computer resources: 1-11
  Frequencies: 1-20
  Frequency Guide: 1-2
  Print resources: 1-19
  Radios 1-5
  Shows for the hobbyist: 1-12
  Voice Utility Sampler: 1-14
Shortwave listening (SWLing): 1-1
  Programs: 1-3
  Logs: 1-16
Shortwave Magazine: 1-6
Shortwave Propagation Handbook, The New: 1-19

Shortwave radios 1-5
    Portable: 1-7
    Table-top: 1-6
Shuttle Amateur Radio Experiment (SAREX): 16-5
Sideband: 2-5
Signal to interference ratio: 2-5
Signal to noise ratio: 2-5
Signal reporting with image communications: 16-2
Silent period: 2-5
Simplex: 11-2, 11-3, 11-5
Simulated emergencies: 14-1
Simulated Emergency Test (SET): 14-3
Simulcast: 2-5
Single Band WAZ award, *CQ*: 8-8
Single sideband (SSB): 1-6
    Emission: 2-5
SINPO code: 1-17
Site notice: 2-5
SKYWARN: 14-10, 14-12
Slow scan TV (SSTV): 16-1, 16-6
    Calling frequencies: 16-7
    Formats: 16-7
    Identifying: 16-8
    License requirements and operating frequencies: 16-7
    Operating: 16-8
SM7PKK DXpedition Home Page: 5-18
SMIRK: 12-2
Soccoro Is: 5-13
Software:
    CT contesting: 7-11
    Facsimile: 15-9
    FTP: 6-4
    GRIDLOC: 12-3
    NA contesting: 7-11
    Satellite: 13-9, 13-11
    Telnet: 6-4
    Terminal emulation: 10-2
    UHF/Microwave Experimenter's: 12-3
Sound format for image communications: 16-2
Source Book for the Disabled, ARRL: 3-6
South Orkney Is: 5-15
Space:
    Communications, packet radio: 10-16
    Station: 2-5
    Telecommand: 2-5
    Telemetry: 2-5
Special Events Communications Manual, ARRL: 14-11
Special Interest Groups (SIGs): 1-11
Specialty contests: 7-4
Spectrum:
    Efficiency: 2-5
    Engineering: 2-5
    Management: 2-6
    Management glossary: 2-2
    Metric: 2-6
    Occupancy: 2-6
Splatter: 2-6
Split: 11-3
    Frequency: 5-12
    Working: 9-6
Sporadic E propagation: 12-6, 12-9
Sports events: 14-10
Spread spectrum: 2-6

Spring Sprints, ARRL: 12-13
Spurious emission: 2-6
SSB nets: 15-3
State QSO Parties: 7-2
Station: 2-6
    Accessories for contesting: 7-9
    Building a: 3-3
    Electronics: 7-8
    Setup: 3-7
Store and forward: 13-7
Stress management (for emergency workers): 14-5
Subband: 2-6
Sunspots: 1-3, 5-7, 5-17, 7-5
Super high frequencies (SHF): 2-6
Suppressed carrier single sideband emission: 2-6
Sweepstakes, ARRL: 7-1, 7-2, 7-4
SWL's Handbook, The: 1-19
SWOT: 12-2
SWR: 3-8
    High: 3-9
    Meter: 3-7, 3-9
Synchronous detection: 1-7

T

Tactical traffic: 14-5
Tailending: 5-9
TCP/IP: 10-12, 11-11
Technical Information Service (TIS), ARRL: 5-5
Technician license: 1-1, 11-2
Telecommand: 2-6
Telecommunication: 2-6
Telecomputing, getting started in: 1-11
Telegraphy: 2-6
Telemetry: 2-6
Telephony: 2-6
Television: 2-6
Television interference (TVI): 12-18
Telnet: 6-8
Telnet software: 6-4
Temporary designator: 3-29
Ten-Ten Award: 8-13
Terminal emulation software: 10-2
    Apple II: 10-2
    Apple Macintosh: 10-2
    Atari: 10-2
    IBM-PC: 10-2
    Tandy Color Computer: 10-2
Terminal node controller (TNC): 10-1, 10-3, 11-9, 13-9
Terminal, packet: 10-2
TexNet: 10-14
Third parties: 15-7
Third party
    Communications: 3-29
    Traffic outside the US: 3-30
    Traffic, international: HDR-15
Time: 3-18
    Bandwidth product: 2-6
    Conversion chart: HDR-20
    Division multiple access (TDMA): 2-6
    Out: 11-3, 11-4
Timer: 11-3
TNC parameters: 10-15
Tone frequencies, CTCSS (PL): 11-11

Tone pad: 11-3
Top Band Reflector: 5-7, 5-18
Toxic spills and hazardous materials: 14-17
Traffic:
    Handling: 3-16, 3-21, 11-7, 15-1, 15-10
    Receiving: 15-8
    Sending: 15-8
    Tactical: 14-5
Transceivers: 7-8
Transcontinental Corps (TCC): 15-13
Transequatorial propagation: 12-7
Transmission line loss: 13-9
Transmitter power: 2-6
Treaties and agreements, international: 2-1
Trigonometry, direction and distance by: 4-3
Tropo: 12-5
Tropo scatter: 12-6
Tropospheric propagation: 12-8
True north
    Determining: 4-1
    Determining by the Pole Star: 4-1
    Determining by the sun: 4-2
Trunked radio: 1-20
Trunking: 2-6
Tucson Amateur Packet Radio (TAPR): 6-10, 10-3, 13-10
Tuning: 2-6

U

UHF Contest, ARRL: 12-13
UHF/Microwave Experimenter's Software: 12-3
Ultimate DXpeditioning: 5-12
    Equipment: 5-12
    Packing: 5-13
Ultra high frequencies (UHF): 2-6
Universal Coordinated Time (UTC): 3-18
Unwanted emissions: 2-6
UoSAT satellites: 13-3
URL: 6-4
US
    Amateur bands: HDR-1
    Counties: 17-59
    Counties Award, *CQ*: 8-11
USENET: 1-11
UTC: 3-18, 7-9
Utility monitoring: 1-13

V

Vanity call system: 5-15
VE7TCP DX Reflector: 5-18
Very high frequencies (VHF): 2-6
    Contests: 7-4
    Mountaintopping: 12-14
    Nets: 12-16
    Packet radio equipment: 3-4
    Propagation: 7-13
    Sweepstakes, ARRL: 12-13
VHF-UHF Advisory Committee, ARRL: 11-6
VHF/UHF:
    Activity nights: 3-17, 12-2, HDR-3
    Antenna, first: 3-8
    Bands: 12-1
    Band openings and DX: 12-5
    Century Club (VUCC) Award, ARRL: 8-4

Nets: 12-2
Operating procedures: 3-17
Operation: 12-7
Packet operation: 10-9, 10-14
VHF/UHF/EHF calling frequencies: HDR-3
Very low frequencies (VLF): 2-6
Vestigial sideband: 2-6
Voice of America (VOA): 1-2, 1-10
Voice of Russia (Radio Moscow): 1-9
Volunteer examiners: 3-27
VOM: 3-7
VOX: 3-17, 7-14
Voyager spacecraft: 9-12, 16-5
VUCC application: 17-56

W

WØRLI mailbox command set: 10-16
W1AW: 1-1, 3-2, 3-14, 3-26, 4-2, 13-9, 14-2
    Schedule: HDR-22
W1FB's Antenna Notebook: 3-9
W1FB's QRP Notebook: 3-25
WAC Application: 17-58
WARC Bands WAZ award, *CQ*: 8-8
WAS Application: 17-52
WAZ Award, *CQ*: 8-8
Weak signal CW and SSB: 11-5
Weather nets: 14-12
Weather Satellite Handbook, The ARRL: 16-12
Weather satellites: 16-10
Weather warnings: 14-13
Web browser: 6-4
WF5E DX QSL Service: 5-18
White Pages: 10-2
Wide band: 2-6
Wire antenna: 3-9
WISP: 13-11
WNZ award, *CQ*: 8-8
WO-18: 13-10
WOM: 1-17
Worked All Continents (WAC) Award, ARRL: 8-3
Worked All Europe Contest: 7-2
Worked All States (WAS) Award, ARRL: 8-2
Worked All Zones (WAZ), *CQ*: 5-2
Working frequency: 2-6
Working split: 9-6
World Administrative Radio Conference (WARC): ... 2-6, 2-8
World Grid Locator Atlas: 12-3
World Grid Locator, ARRL: 8-4
World ITU Zone Award: 8-13
World map, ARRL: 5-2
World Radiocommunication Conference (WRC): 2-6, 2-8
World Radio TV Handbook: 1-13, 1-19
World Wide Web (WWW): 1-11, 4-1, 6-2
Worldwide VHF WPX Contest, *CQ*: 12-13
WPX award, *CQ*: 8-10
WWV: 1-4, 3-6, 12-10
    Web site: 1-4
WWVH: 1-4, 3-6

Y

Your Ham Antenna Companion: 1-8, 3-8, 5-3, 7-6
Your Packet Companion: 3-23, 3-26

Numbers

2 meter band plan: 11-6
2-meter repeater frequency pairs: 11-4
5BWAS Application: 17-54
6 meters: 12-1
10 meter CTCSS frequencies: 11-11
10 meter FM: 11-2
10 meter repeaters: 11-9
73, The origin of: 17-25
160-meter WAZ award, *CQ*: 8-8

Notes

Notes

Notes

Notes

Notes

Notes

FEEDBACK

Please use this form to give us your comments on this book and what you'd like to see in future editions, or e-mail us at **pubsfdbk@arrl.org** (publications feedback).

here did you purchase this book?
☐ From ARRL directly   ☐ From an ARRL dealer

there a dealer who carries ARRL publications within:
☐ 5 miles   ☐ 15 miles   ☐ 30 miles of your location?   ☐ Not sure.

cense class:
☐ Novice   ☐ Technician   ☐ Technician Plus   ☐ General   ☐ Advanced   ☐ Amateur Extra

ame _____   ARRL member? ☐ Yes ☐ No

_____ Call Sign _____

aytime Phone (   ) _____ Age _____

ddress _____

ty, State/Province, ZIP/Postal Code _____

licensed, how long? _____

her hobbies_____

_____

ccupation _____

From _____

_____

_____

EDITOR, OPERATING MANUAL
AMERICAN RADIO RELAY LEAGUE
225 MAIN STREET
NEWINGTON CT 06111-1494

— — — — — — — — — — — — — — — — — please fold and tape — — — — — — — — — — — — — — — — —